BIOMASS – DETECTION, PRODUCTION AND USAGE

Edited by **Darko Matovic**

Biomass – Detection, Production and Usage
Edited by Darko Matovic

Published by InTech
Janeza Trdine 9, 51000 Rijeka, Croatia

Publishing Process Manager Niksa Mandic

Technical Editor Teodora Smiljanic

Cover Designer InTech Design Team

Additional hard copies can be obtained from orders@intechopen.com

Biomass – Detection, Production and Usage, Edited by Darko Matovic
p. cm.
ISBN 978-953-307-492-4

For EU product safety concerns:
IN TECH d.o.o., Prolaz Marije Krucifikse Kozulić 3, 51000 Rijeka, Croatia,
info@intechopen.com or visit our website at intechopen.com.

Meet the editor

Darko Matovic, Ph.D., P. Eng. is a professor at Queen's University, Kingston, Canada where he teaches at Mechanical and Materials Engineering department. His research interests include sustainable energy from biomass, (via direct combustion, gasification and pyrolysis) and the production and use of biochar as a viable technique for carbon dioxide removal from the atmosphere and a valuable resource material in soil improvement and remediation of contaminated soils.

Contents

Preface

Biomass has been an intimate companion of humans from the dawn of civilization to the present. Its use as food, energy source, body cover and as construction material established the key areas of biomass usage that extend to this day. With the emergence of agriculture the soil productivity increased dramatically, especially with cultivation of new plant varieties and with emergence of intensive soil fertilization. In that context, the emergence and use of fossil fuels for energy and raw material in chemical industry is but a flick on the human history horizon. The amount of energy that humans used in the last two decades is roughly equal to the total amount of energy in the past. This enormous increase of energy use was made possible by extensive depletion of fossil reserves and is clearly unsustainable. Does it mean that once these reserves are depleted the amount of energy available to humans will be similar to the pre-fossil fuel era? Not necessarily. Currently, the total energy used by humanity amounts to 1/5500 fraction of the total solar energy incident on earth. In theory, significant percentage of that energy can be used for human needs, before it is let to complete the energy flow cycle (i.e. to be dissipated to space). Some of it can be harnessed and used as a direct solar energy, but other pathways uses natural photosynthesis to create biomass that can be seen as a form of chemically stored solar energy. Of course, biomass is also food and this brings about the key trade-off in biomass usage: the food vs. fuel controversy. Given these two primary uses of biomass the proper resolution of this tradeoff is essential for acceptable and beneficial biomass usage in the future. The glaring example of biomass for energy misuse is ethanol production from corn, a relatively inefficient conversion process that is also in a direct collision course with the corn as food pathway. Still, in 2009, about 15% of world corn production was converted into ethanol fuel. More subtle examples emerge when an inedible biomass is the energy source, but its production still competes with food supply chain. Recent world food price hikes, especially in 2008 have been blamed partly on diversion of food staples towards biomass fuel production. As humanity currently uses or appropriates (through deforestation and land use change) about 40% of land productive capacity, the accurate account of all existing and potential biomass usage pathways is critical for charting the way forward at the global scale, and in different regions.

Given the complexities of biomass as a source of multiple end products, food included, this volume sheds new light to the whole spectrum of biomass related topics by highlighting the new and reviewing the existing methods of its detection, production and usage. We hope that the readers will find valuable information and exciting new material in its chapters.

Since biomass means so many things to so many people, it is no wonder that the original book title, *Remote Sensing of Biomass* has attracted a wide range of papers, many of them very remote from the remote sensing theme. If there were few odd submissions that could not fit the theme at all, the choice would be simple. Check the quality of the paper and if it is good, suggest to the authors that it would be better to submit it elsewhere. InTech publishing is a wonderful open source publisher that published more than 180 volumes in 2010 alone, on such diverse topics as Virtual Reality, Biomedical Imaging or Globalization. Thus, an odd author who went astray could be stirred towards more suitable publication. And indeed, there were few that fell into that category. However, majority of submissions had a broad linkage to biomass, but not to its remote sensing. The wide range of themes, all related to biomass, prompted us to reconsider if the originally envisioned scope was perhaps understood by biologists and food scientists differently than by engineers? Is the simple act of examining biomass via a microscope a form of remote sensing? Is an indirect inference about details of physiological or genetic makeup of a subject biomass another form of remote sensing as well? Questions like these, and the desire to better reflect the scope and coverage of the book chapters led us to a new title, *Biomass - Detection, Production and Usage*. It reflects an even balance between these three areas of the biomass science and practice.

Dr. Darko Matovic
Queen's University, Kingston,
Canada

Part 1

Detection

1

Lidar for Biomass Estimation

Yashar Fallah Vazirabad and Mahmut Onur Karslioglu
Middle East Technical University
Turkey

1. Introduction

Great attention has been paid to biomass estimation in recent years because biomass can simply be converted to carbon storage which is very important to understand the carbon cycle in the environment. Biomass is typically defined as the oven-dry mass of the above ground portion of a group of trees in forestry (Brown, 1997, 2002; Bartolot and Wynne, 2005; Momba and Bux, 2010). However there are a few studies about below ground biomass estimation. Conventionally, it is estimated using measurements which are recorded on the ground. On the other hand, the large number of studies have confirmed that Lidar as a kind of active remote sensing system is able to estimate biomass properly (Popescu, 2007). Hence time-consuming field works can be avoided and unavailable regions become accessible using a relatively low cost and automated Lidar system. (Nelson et al., 2004; Drake et al., 2002, 2003; Popescu et al., 2003, 2004).
Traditional remote sensing systems detect vegetation cover using active and passive optical imaging sensors (Moorthy et al., 2011). Passive systems depend on the variability in vegetation spectral responses from the visible and near-infrared spectral regions. Widely accepted algorithms such as the Normalized Difference Vegetation Index (NDVI) have been empirically correlated to structural parameters (Jonckheere et al., 2006; Solberg et al., 2009; Morsdorf et al., 2004, 2006) such as Leaf Area Index (LAI) of canopy-level. On the contrary to passive optical imaging sensors, which are only capable of providing detailed measurements of horizontal distributions in vegetation canopies, Lidar systems can produce more accurate data in both the horizontal and vertical dimensions (Lim et al., 2003). Lidar-based instruments from space-borne, airborne, and terrestrial platforms provide a direct means of measuring forest characteristics which were unachievable previously by passive remote sensing imagery.
Developments in remote sensing technologies, in particular laser scanning techniques, have led to innovative methods and models in the estimation of forest inventories in terms of efficiency and scales (Hudak et al., 2008; Tomppo et al., 2002; Tomppo and Halme, 2004; Zhao et al., 2009; Koch, 2010; Yu et al., 2011). Lidar experiments and researches within the remote sensing community are now focusing to develop robust methodologies. These methods and models employ very precise 3D point cloud data (Omasa et al., 2007) to direct process and retrieve vegetation structural attributes which are validated by in situ measurements of vegetation biophysical parameters (Maas et al., 2008; Cote et al., 2011).
Laser scanning systems have been used to extract various kinds of parameters, such as tree height, crown size, diameter at breast height (dbh), canopy density, crown volume, and tree

species (Donoghue et al., 2007; Means et al., 1999, 2000; Magnussen et al. 1999). Most authors concentrate on the above-ground biomass while there are a few known studies focusing on the below-ground biomass (Kock, 2010; Nasset, 2004).

Bortlot and Wynne (2005) used Lidar data to generate canopy height models. Tree heights detected from image processing are entered as variables in a stepwise multiple linear regression to find an equation for biomass estimation. The method skips detecting small trees. They are not included in the process of estimation. A previous work by Lefsky et al. (1999) presented the prediction of two forest structure attributes, crown size and aboveground biomass from Lidar data. They analyzed the full waveform of the return pulses to define the beginning of canopy return. Linear regression was used to develop biomass estimation equation based on a defined canopy height index. Finally, they proposed stepwise multiple regression model to predict canopy volume and relatively biomass. They concluded that tree height is highly correlated with dbh in a square power function.

Van Aardt et al. (2008) evaluated the potential of an object oriented approach to forest classification as well as volume and biomass estimation using small footprint, multiple return Lidar data. A hierarchical segmentation method was applied to a canopy height model (CHM). An empirical model is employed to estimate the canopy volume and biomass. They performed stepwise discriminant analysis as a part of classification steps for variable reduction. Fallah Vazirabad and Karslioglu (2009) investigated the biomass estimation based on single tree detection method. This method is used to locate trees and detect the height of each tree top. Diameter at breast height is extracted from the close relation to the tree height which is defined by field measurements. A Log transformed model is applied for biomass estimation taking into account the dbh variable.

Airborne lidar is confirmed as the most ideal technology to obtain accurate CHM over large forested areas because of its high precision and its ability to receive ground returns over vegetated areas. Spaceborne geoscience laser altimeter system (GLAS) data on the other hand are intended to use mainly for scientific studies of sea ice elevation (Zwally et al., 2002; Kurtz et al., 2008; Xing et al., 2010), but it is also suitable for the estimation of the canopy height map (Lefsky et al., 2005; Simard et al., 2008; Chen, 2010; Duncanson et al., 2010).

The reason for the applications of GLAS data to canopy height mapping is to estimate the dynamic global carbon stock. Xing et al. (2010) analyzed the deforestation and forest degradation as a carbon source estimation model. They also investigated the forest growth model for afforestation and reforestation. Forest carbon stocks, fluxes, and biomass are directly related to each other (Garcia-Gonzalo et al., 2001; Widlowski et al., 2004). Therefore, accurate estimation of biomass of stocks and fluxes is essential for terrestrial carbon content and greenhouse gas inventories (Muukkonen and Heiskanan, 2007; Xing et al, 2010).

A general overview of forest applications is provided by recent studies (Hyyppä et al., 2009; Dees and Koch, 2008; Mallet and Bretar, 2009; Koch, 2010). They show that the information related to the height or structure of forests can be extracted with high quality.

Apart from the land cover classification Lidar intensity data can be used to differentiate materials such as asphalt, grass, roof, and trees (Hasegawa, 2006; Donoghue et al., 2007; Kim, 2009; Song et al., 2002). To identify the position and diameter of tree stems within a forest the intensity of Lidar returns has been successfully used (Lovell et al., 2011). Hopkinson and Chasmer (2009) compared four lidar-based models of canopy fractional cover and found that those models which included the intensity of the returns were less

affected by differences in canopy structure and sensor configuration. This is because the intensity measurements provide some quantification of the surface areas interacting with the laser beam. Reitberger et al. (2008) used a waveform decomposition method to extract intensity and concluded that detection of small trees below the main canopy was improved.
The ability to acquire laser pulse echoes from the bottom part of vegetation canopies is restricted in the spaceborne and airborne Lidar system. This is reffered to the system properties such as laser footprint size, recording frequency, as well as the natural placement of the crown elements, for example dense or open canopies. But to provide detailed specification of canopy and individual tree crowns characterization it is logical to introduce a terrestrial platform which has a much higher resolution laser pulse records than others. However, terrestrial data for tree 3D models have some problems such as overlapping crowns and under-story vegetation which cause shadowing effects.
Deriving forest data from Lidar data to model the canopy height distribution and its statistical analysis was proposed by (Holmgren and Persson, 2004; Lim et al., 2003, 2004; Næsset, 2002). The single tree detection, its location and characteristics on the basis of statistical analysis have been studied by (Hyyppä and Inkinen, 1999; Fallah Vazirabad and Karslioglu, 2010; Yu et al. 2011).

2. Lidar for biomass estimation

This section comprises two parts: systems and data acquisition. In the first part space-borne, airborne, and terrestrial systems and their sensors in relation to the biomass estimation are presented. The appropriate and useful laser band for vegetation detection is also discussed in the same part. In the second part, types of laser data acquisition such as first return, last return and multi-return are described and the applications of each type are discussed. Additionally, the new technology of light detection, namely full waveform and its utilization will be emphasized as the state of the art. The results of recent researches and studies related to the waveform for the feature extraction are highlighted.

2.1 Systems

Lidar systems make use of the time of flight principle or phase-based differences to measure the distances of objects. For this, the time interval is detected between sent and return laser pulses which are backscattered from an abject. Lidar point cloud of returns generate a 3D digital representation of the vegetation structure in which each point is characterized by XYZ coordinates (Maas et al., 2008; Cote et al, 2011).
Lidar System consists of a laser ranging unit, a scanning instrument like an oscillating mirror or rotating prism and a direct geo-referencing navigation unit (using global positioning system - GPS and inertial navigation system - INS). The choice of the platform depends mainly on the application. Space-borne systems map the globe for researches and experimental purposes. Airborne systems are collecting the data for national or regional investigations. Terrestrial platforms are frequently used to produce 3D models of man-made structures or natural resources like trees. Thus, the basic principle and technical specification for a sensor installed on a platform such as Earth orbiting satellite, airplane, helicopter, tripod, or vehicles change due to the variety of the applications (Shan and Toth, 2009). Some engineering and environmental studies require information about the shallow water basin. The Bathymetric Lidar systems are capable to provide this information in the coastal zones

or rivers deep to 50 meters in clear water (Bathymetric system is irrelevant to our discussions so, we will have no further dealings with it in this chapter).
Generally, commercial systems are designed to receive data from small-footprint (0.20-3.00m diameter, depending on flying height and beam divergence) with higher repetition frequency (Mallet and Bretar, 2009). These systems acquire a high point density and an accurate height determination. However, small-footprint systems often miss tree tops which cause under estimation in tree height. Therefore, it is hard to define whether the ground has been detected under dense vegetation or not. Consequently, ground and tree heights cannot be well estimated (Dubayah and Blair, 2000). Large-footprint systems (10-70 m diameter) increase the chance to both hit the ground and the tree top and eliminate the biases of small-footprint systems. Thus, the return waveform gives a record of vertical distribution of the captured surface within a wider area which provides important information for biomass estimation. First experimental full waveform topographic systems were large-footprint systems and mostly carried by satellite platforms. With a higher flying height, pulses must be fired at a lower frequency and with a higher energy to penetrate into the forest canopy as much as possible (Mallet and Bretar, 2009).

2.1.1 Space-borne systems

The geoscience laser altimeter system (GLAS) is the only Lidar operating space-borne system. GLAS is the important part of NASA earth science enterprise carried on the ice, cloud and land elevation satellite (ICESat) from 12 January 2003 (Afzal et al., 2007). This instrument has three lasers, each of which has a 1064 nm lidar channel for surface altimetry and dense cloud heights, and a 532 nm lidar channel for the vertical distribution of clouds and aerosols (NASA, 2007). The three lasers have been operated one at a time, sequentially throughout the mission. The mission mode involved 33 day to 56 day campaign, numerous times per year, to extend the operation life. The main objective of the GLAS instrument is to measure the ice sheet elevations and changes in elevation through time. Second objective is the cloud detections and measurements, atmospheric aerosol vertical profiles, terrain elevation, vegetation cover, and sea ice thickness. The figure 1 shows the world elevation maps for 2009 ICESat elevation data (national snow and ice data center, NSIDC, available online at: http://nsidc.org/data/icesat/world_track_laser2F.html)
Nevertheless, only a small number of studies have used airborne lidar data to evaluate the DTM which was derived from satellite laser altimetry GLAS data over forested areas. GLAS which is only operating on board ICESat, records the full waveform returns, and provides a high precision elevation data with nearly global spatial coverage at a low end user cost (Fricker et al., 2005; Martin et al., 2005; Schutz et al., 2005; Magruder et al., 2007; Neuenschwander et al., 2008). Space-borne data are mainly used to model the global canopy height for evaluating carbon budget (Xing et al., 2010).
Recently, Duong et al. (2007, 2009) compared terrain and feature heights derived from the satellite (GLAS) observations with a nationwide airborne lidar dataset (the Actual Height model of the Netherlands: AHN). They found that the average differences between GLAS- and AHN-derived terrain heights are below 25 cm over bare ground and urban areas. Over forests, the differences are even smaller but with a slightly larger standard deviation of about 60 cm (Chen, 2010). Harding et al. (2001) utilized GLAS full waveform data to generate the average forest CHM, and the results presented the variations of important canopy attributes, such as height, depth, and the over-story, mid-story, and under-story

forest layers. Sun et al. (2007,2008) applied GLAS waveforms to estimate the forest canopy height in the flat area in Northern China mountains, and found that the ICESat-derived forest height indices was well correlated with the field-measured maximum forest height $R^2 = 0.75$ where R^2 is the coefficient of determination .

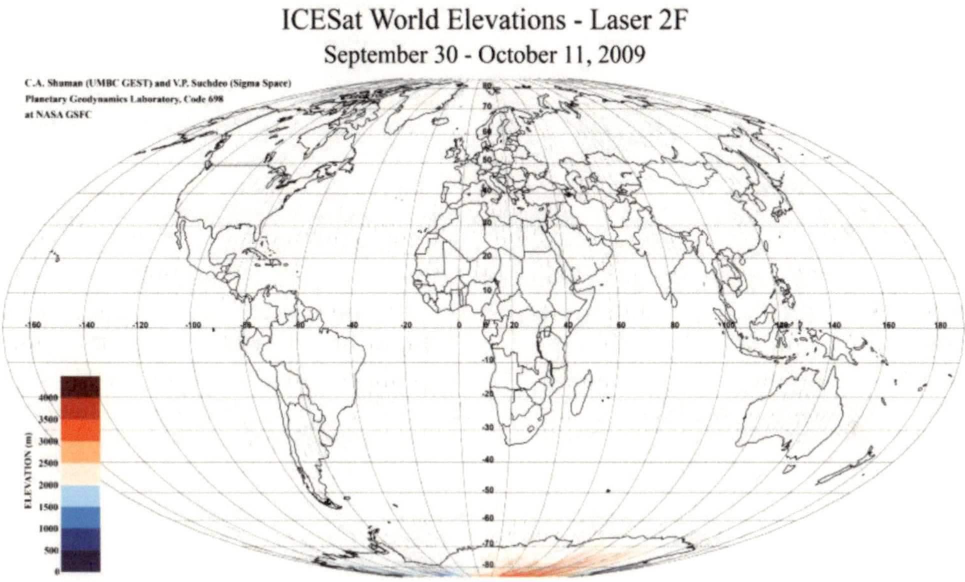

Fig. 1. Example of ICESat World Elevation Map

2.1.2 Airborne systems

An extensive test of laser profiler was performed at the Stuttgart University (1990) where Differential Global Positioning System (DGPS) and Inertial Measurement Unit (IMU) was integrated in the laser system for the first time to provide precise positioning and orientation (attitude) of the airborne platform. Soon after that, the scanning mechanism was designed by Optech company (Canada - ALTM system)

Laser profiler was developed in the forestry research by NASA's Goddard space flight center (GSFC) on the basis of Riegl laser rangefinder with 20 ns wide laser pulse and repetition rate of 2 kHz. There are three main commercial suppliers of airborne laser scanning systems, Optech International Inc., Leica Geosystem, and Riegl which are producing the data for the forest inventory and biomass estimation researches.

Generally, other companies completed their systems which utilize these three laser scanner instruments. Besides these commercial systems, a number of other systems built by US government research agencies are offered for scientific research purposes, like NASA, ATM, RASCAL, SLICER, Laser Vegetation Imaging Sensor (LVIS), and ScaLARS. LVIS has been developed by NASA for the topography mapping, elevation and the forest growing on it. A special design of scanning system such as the full waveform is required for the scanning of vegetation covered regions to capture the reflected pulse in different returns. This scanner has been used in USA (California, eastern states), Central America (Costa Rica and Panama). It was also applied in Amazonian forests of Brazil to generate direct measurements of canopy height and relatively aboveground biomass map. (Shan and Toth, 2009)

2.1.3 Terrestrial systems

The primary classification with respect to measuring principle is described by two techniques namely pulse ranging or time of flight (TOF) and phase measuring technique. Another classification is also available in accordance with the angular scanning technique and coverages of scanner which consist of Panorama, Hybrid, and Camera scanners (). Panorama scanners carry out distance and angular measurements providing 360° angular coverage within the horizontal plane. Types of laser scanners, which perform unrestricted scanning around the rotation axis, fall in the category of Hybrid scanners. The third category of scanners carrying out distance and angular measurements over a limited angular range and in a specific field of view is called Camera scanners (Shan and Toth, 2009). For the range measurements, it is necessary to obtain information about the exterior orientation elements (positions and orientation or attitude angles) of platforms of the terrestrial laser scanner. Precise exterior orientation elements can be detected during the calibration procedure.
Sensitivity of tree volume estimates which are related to different error sources in the spatial trajectory of the terrestrial Lidar has been analyzed by (Palleja et al. ,2010). Their tests have demonstrated that the tree volume is very sensitive to the errors in the determinations of distance and the orientation angle. Cote et al. (2011) proposed to estimate the tree structure attributes by means of terrestrial Lidar. They concluded that the main limitation of the use of terrestrial system was the effect of object shading and wind. In context with the precise biomass estimation terrestrial laser scanning can be considered as a support system for airborne and space borne Lidar.

2.2 Data acquisition

Measurement process of laser scanner can be represented by the frequency, intensity, phase and the travel time of the sent and returned signal. The transmitted and received energy are formulated similar to the Radar (radio detection and ranging) equation (Shan and Toth, 2009). This can be expressed as an integral (Mallet and Bretar, 2009) and the range is measured in pulsed systems as $R = c.t/2$, where c is the speed of light, t is two way laser light travel time, R is the distance to be measured (Shan and Toth, 2009). The equation of the continuous waveform is $R = 0.5\ (\phi/2\pi)\lambda$, where ϕ is the phase difference and λ is the wavelength which is operationally between 600 and 1000 nm (Electromagnetic infrared range). This interval is not eye-safe. Therefore, the optimum performance has to be balanced against safety considerations.
In addition to positional data, each Lidar observation must also contain the scan angle for each shot together with the measurement of reflectance from the target. Since the calculation of range for the detected pulse involves the elapsed time the precision of time measurement is of vital importance considering that 7 ns sensivitiy is needed to distinguish 1 m object. This plays in turn a decisive role in the scanning of vegetated areas. In some methods they use a fraction which is a constant in the sent and return pulse. But, in others, they take the centroid of the pulses as a time reference.
The characteristics of forest inventory from both discrete return (first, last, multi returns) and full waveform recordings are extensively studied by different Lidar approaches such as tree crown detection and biomass estimation (Harding et al., 2001; Coopes et al., 2004; Jang et al., 2008; Brantberg et al., 2003).

2.2.1 First return, last return

Lidar systems can be categorized by the way they process the waveform reflections for each pulse and also by the size of the footprint they record. Systems that record footprints up to

100 cm are often called small footprint systems typically at frequencies around 15 kHz (Heritage and Large, 2009). Early small footprint systems recorded the range only up to the first reflecting object or the first pulse in discrete returns. In principle, the map of all first pulses results in such a model showing only the height of all surface objects. This requires to record the last reflecting object in each return signal if there is more than one reflectance, which is often referred to the last pulse. Although the last pulse data has clearly the potential to penetrate vegetation canopies, it can never be guaranteed that the last pulse can reach the ground and is not reflected from the higher point of canopy. Furthermore, where low vegetation is involved, the first and last pulse may be too close together to generate a reliable range and leads consequently to over estimation of the terrain height.

Coopes et al. (2004) used airborne discrete returns to indicate canopy crown and height. Lim and Treitz (2004) collected the airborne discrete first and last returns for above ground biomass estimation. In Jang et al. (2008) the apple tree inventory are extracted from discrete return without explaining their effect on the results. First and last returns are used by Thomas et al. (2006) but the effects of which are not explained on the results of canopy height models.

Fallah Vazirabad and Karslioglu (2010) extracted the tree tops empirically from the first pulse data because it contains more canopy returns than the ground ones. In discrete return systems, the small diameter of footprints and the high repetition rates of these systems made possible to have high spatial resolution, which can yield dense distributions of sampled points. Thus, discrete return systems are preferred for detailed mapping of ground and canopy surface. Finally, these data are readily and widely available, with ongoing and rapid development in forestry.

2.2.2 Multi return

The capability of detecting different returns in the closely placed terrain surfaces depends on instrument parameters such as the laser pulse width (the shorter the better), detector sensitivity, response time, the system signal to noise performance, and others. In case of discrete returns more detectors are needed. With this technology the number of pulses between first pulse and last pulse can be recorded as many as the number of detectors. Thus, there are systems with second and third pulse beside first and last pulse record. In contrast to small footprint systems, large footprint systems (10-100 m) open up the possibility of recording the entire return pulse. Discrete return airborne laser systems (ALS) have the benefit of providing data over a large area, but are restricted by their laser pulse return density as $points/m^2$ ratio. Multiple return recording capabilities of system produce point cloud density between 1 and 20 $points/m^2$ optimistically. Often this level of point density is unsatisfactory to produce a comprehensive 3D model, especially in the vertical view (Moorthy et al. 2011).

2.2.3 Full waveform

The problems which are mentioned before in first and last pulse systems for vegetated regions can be solved with full waveform technology making an important contribution to biomass estimation (Shan and Toth, 2009). The waveform is usually digitized by recording the amplitude of the return signal at fixed time intervals (figure 2). To analyze the signal of emitted short duration laser pulse with only a few ns pulse-width, higher digitizer sampling rate is required. These devices have been primarily designed for measuring vegetation properties. Extensive researches (Harding et al, 2001; Lefsky et al., 2001, 2002; Reitberger et al., 2009) have shown that waveform shape is directly related to canopy biophysical parameters including canopy height, crown size, vertical distribution of canopy, biomass, and leaf area index.

Harding et al. (2001) discussed about canopy height profile detection from full waveform raw data provided by SLICER. They studied the laser energy from the full waveform Gaussian distribution. The advantages of full waveform recording include an enhanced ability to characterize canopy structure, the ability to concisely describe canopy information over increasingly large areas, and the availability of global data sets. The examples of these data are airborne like SLICER and LVIS, and satellite data like GLAS. The other advantage of full waveform systems is that they record the entire time varying power of the return signal from all illuminated surfaces on canopy structure. It should also be stated that Lidar data, which is collected from space globally, provides only full waveform recordings (Lefsky et al., 2002).

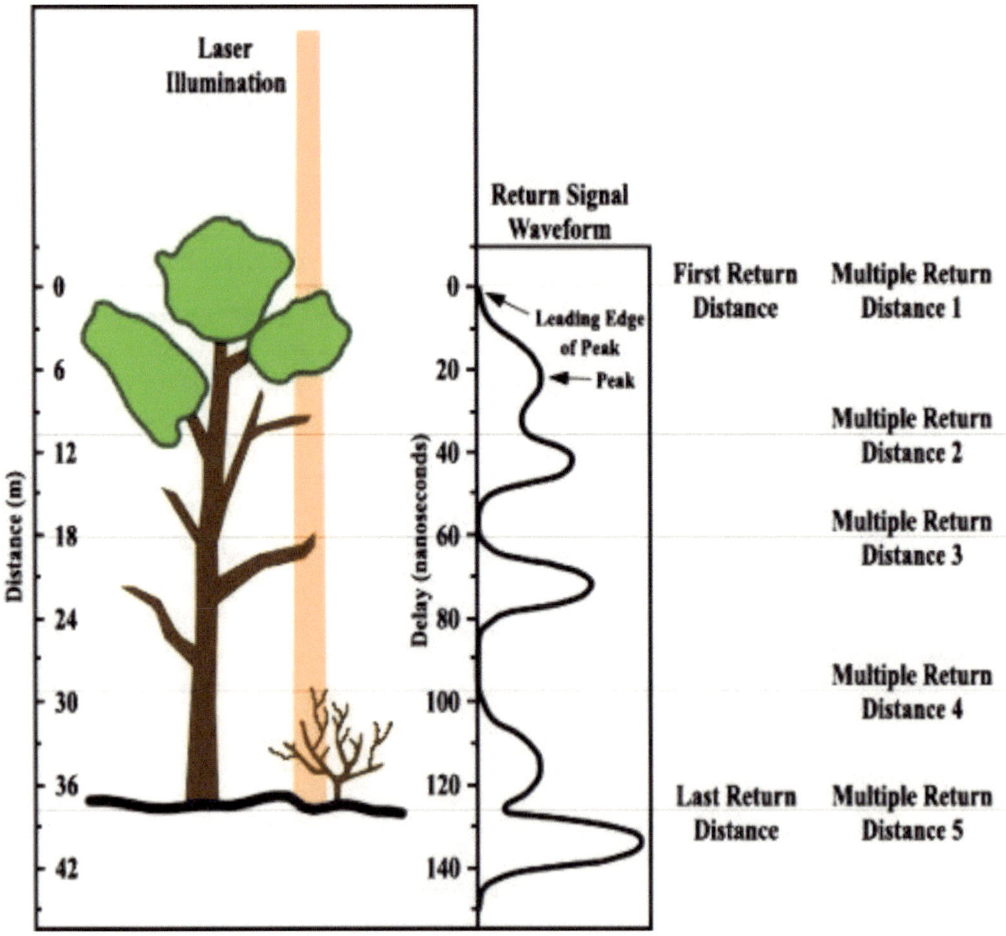

Fig. 2. Return pulse forms (Harding et al, 2001)

3. Methods and models for Biomass estimation

This section is organized in terms of three subsections containing data pre-processing, methods and models, and applications.
Data pre-processing methods in turn are divided into four parts. For the filtering methods some efficient algorithms are explained. Apart from different interpolation methods the generation of the digital terrain model (DTM), digital surface model (DSM), and canopy height model (CHM) is treated. Quality assessment of laser data is carried out within another subsection. Additionally, the quality of filtering methods, interpolation methods, DTMs, DSMs, CHMs results and their performances are also evaluated. The subsection ´´methods and models´´ consider the methods and models in biomass estimation, among others single tree and tree characteristics detection. The last subsection presents applications of Lidar using the models for biomass estimation to recognize the advantages of Lidar systems in the biomass estimation.

3.1 Data pre-processing

The critical step in using Lidar data is the data pre-processing. Choosing the proper filtering method plays an important role in the quality of results. Actually, it cannot be expected that the quality of the result should be better than the data accuracy itself. On the other side, all interpolation methods have no difficulties to generate precise 3D models since dense enough Lidar data is available.

3.1.1 Filtering

The purpose of filtering is to remove the vegetation points. Figure 3 shows all points before filtering (figure 3, left) and terrain points after filtering (figure 3, right).

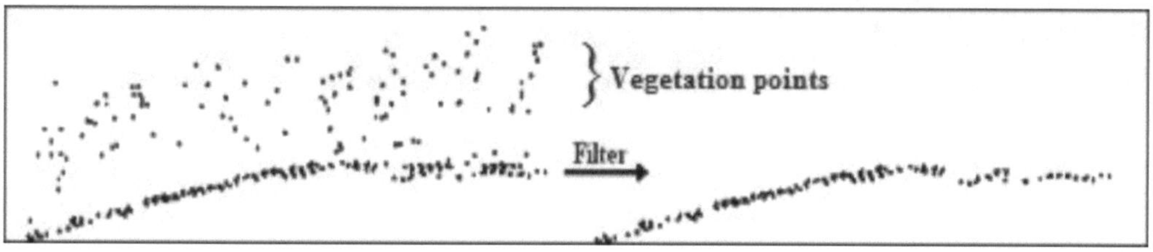

Fig. 3. Removing vegetation points

The terrain points extracted from the point cloud of Lidar data set are used as an input to generate a DTM. The first pulse data sets contain vegetation points and terrain points in the forest area. Numerous kinds of filtering methods are developed to classify the terrain and vegetation points in the point cloud (Pfeifer et. al., 2004; Tovari and Pfeifer, 2005). Different concepts for filtering, with different complexity and performance characteristics have been proposed in mainly four categories such as morphological, progressive densification, surface based, segmentation based filter. There are also developments, extensions, and variants for these filter methods.
The morphological filter was derived by Vosselman (2000) from the mathematical morphology definition. It works in such a way that the smaller are the distances between a ground point and its neighboring points, the lesser is the height difference. Based on this criterion the method can properly eliminate the outliers. The progressive densification filter is developed by Axelsson (Axelsson, 2000). This filter works progressively by classifying

points which belong to the ground. Surface based filters assume at the beginning that all points lying on the ground form a surface. Then a fitting procedure is applied to extract the points which do not belong to the ground. This method goes back to Pfeifer et al. (2001). Segmentation filters are developed as the fourth category. Segment is a group of points which are located within defined thresholds such as the distance and height difference between neighbor points. Sithole (2005) introduced a segment classification method by performing region growing techniques referring to Tovari and Pfeifer (2005). It works by classifying segments into as many classes as possible (Filin and Pfeifer, 2006).
The experimental comparison of filtering algorithms with manual methods for DTM extraction is introduced by Sithole and Vosselman (2004) to show the suitability of filters with the terrain shape. In comparison with other filtering methods, segment base filter is turned out to be a more reliable method in steep slope terrain extraction using a surface growing method (Sithole and Vosselman 2005).

Fig. 4. Segmentation method, point cloud from vertical view

The most important part in this method is the accuracy assessment and parameter tuning. These processes for the segmentation method are performed by Vazirabad and Karslioglu (2009) as shown in figure 4. Segmented terrain points are coloured as brown and green while white points are assumed to be the vegetation points in forest area.

3.1.2 Interpolation

Interpolation is necessary to produce digital models from Lidar point cloud. The simple idea of the interpolation is referred to the nearest neighbor interpolation method to estimate the elevation (Maune, 2007). It searches for the set of nearest points, thus the new elevation value is selected as the same value of the nearest point instead of taking the average of all points. An important problem here is the zigzag appearance of the surface. This is in fact due to the selecting of the nearest point method by defining Voroni diagrams or Theissen polygons. For this reason, some kinds of averaging methods should be applied to the set of known nearest elevation points. Therefore, a weighted average like inverse distance weighting (IDW) is introduced which is working with the distances between these points (Monnet et al, 2010; Bater and Coops, 2009).

In Lidar data especially in vegetated areas distances are not related to the elevations. In contrast, kriging or geostatistical approaches provide better results (Heritage and Large, 2009). However, they require more mathematically complex and computationally intensive algorithms. Since dense data is always available, rapid interpolation methods such as the nearest neighbor are prefered to use for the rough surfaces in the forest areas (Fallah Vazirabad and Karslioglu, 2010).
Riano et al. (2003) investigated the performances of spline and nearest neighbor interpolation methods to generate DTM. Spline interpolation is a special form of piecewise polynomial. The interpolation error in the DTM can be small even applying the low degree polynomial. They concluded that there were no large differences between the spline and nearest neighbor results while the spline computation was three times slower. Hollaus et al. (2010) described the derivation of DSM employing the least square fitting method to compare it with kriging interpolation. They introduced a moving least square fitting technique which selects the highest points in the search window as surface points. This technique finds the best fitting surface to the set of points by minimizing the sum of squares of the residuals of the points from surface. The results of this study showed that the least square fitting technique produced high precision DSM on rough surfaces while it needs more computational time.

3.1.3 DTM, DSM, CHM

The terrain model function $z = f(x, y)$ is computed from 3D points, $p_i = (x_i, y_i, z_i), i = 1, \dots, n$, where n is the number of points (Shan and Toth, 2009). Heights are stored at discrete, regularly aligned points, and the interpolated height as the height of the grid has to be given within a grid mesh. These grid heights are obtained by interpolation methods explained before in the subsection 3.1.2. These methods consist of nearest neighbor, IDW, kriging, spline, and least square fitting.
An alternative method to the interpolations is so called triangular irregular network (TIN) data structure. The original points are used for reconstructing the surface in the form of TIN. For large point sets, triangular networks are more effective than the time consuming methods which are mentioned before. Digital surface model (DSM) is generated from noise removed Lidar data and represents the canopy top model. Digital terrain model (DTM) is basically produced by the laser pulse returns which are assumed to be on the terrain. (van Aardt et al., 2008). By subtracting DTM from DSM, CHM can be obtained which is presented in figure 5. Hence, CHM is a digital description of the difference between tree canopy points and the corresponding terrain points.

3.1.4 Quality assessment

The quality assessment is necessary for each step of the pre-processing. Pfeifer et al. (2004) reported an RMSE of 57 cm for DTM in wooded areas using data point spacing about 3 m. Hyyppa et al. (1999) reported a random error of 22 cm for fluctuating forest terrain using data point density 10 pts/m^2. They analyzed the effects of the date, flight attitude, pulse mode, terrain slope, and forest cover within plot variation on the DTM accuracy in the boreal forest zone. Hyyppa and Inkinen (1999) reported the CHM with an RMSE of 0.98 m and a negative bias of 0.41 m (nominal point density about 10 pts/m^2). Yu et al. (2004) reported a systematic underestimation of CHM of 0.67 m for the data acquired in 2000 and 0.54 for another acquisition in 1998. The filtering methods mentioned before are likely to fail

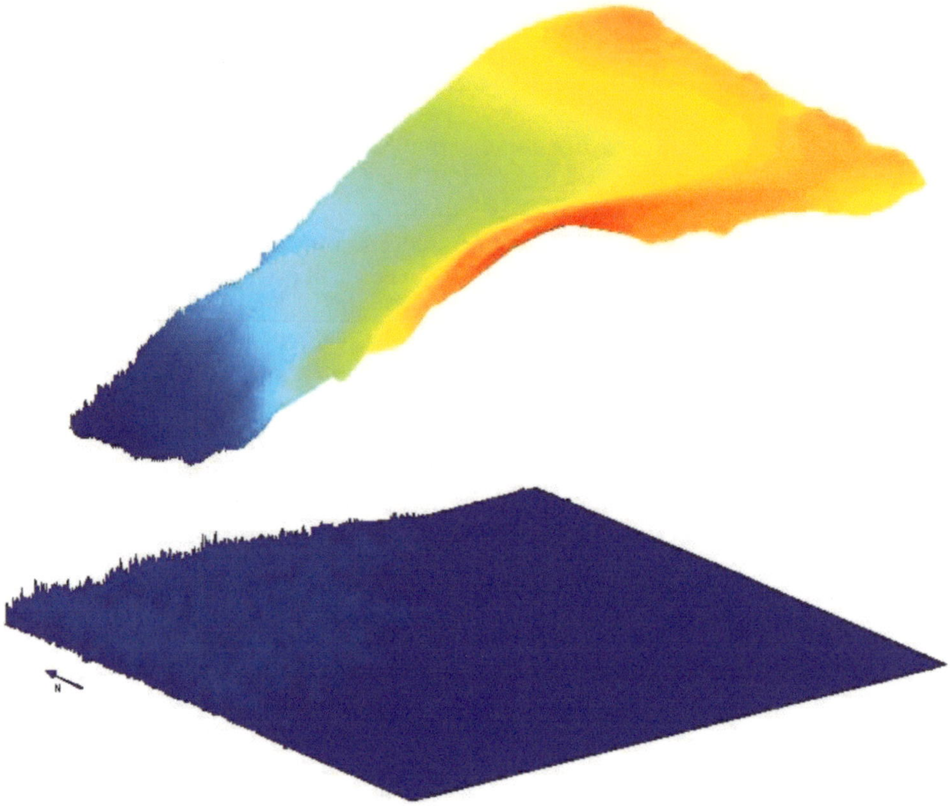

Fig. 5. DSM (up) and CHM (down)

Filter		Reduced		Sum
		Terrain	Off-terrain	
Original	Terrain	A	B	A+B
	Off-terrain	C	D	C+D
Sum		A+C	B+D	(Total) T=(A+B+C+D)
Type I = (B*100)/(A+B) & Type II = (C*100)/(C+D) Total Errors = (B+C)*100/T				

Table 1. Type I and Type II errors

facing with (i) outliers in the data, (ii) complexity of the terrain, (iii) small vegetation which is completely attached to the terrain like bushes. Most of filter algorithms start with the minimum height in data. Thus the most effective error is the negative outliers which are originated from multi path errors and errors in range finder. The vegetation on the slope also produces difficulties in filter algorithms because of the reflected pulses returning from the neighbor points. Therefore, filtering methods need some initial threshold values, which are usually defined by experience and a-priory information about the data and terrain characteristics.

Fallah Vazirabad and Karslioglu (2011) demonstrate that the quality of segmentation filter deteriorates with increasing point spacing of ALS point cloud looking at Type I and Type II errors (table 1). Large Type I error leads to a reduced DTM accuracy as a consequence, because many vegetation points will be included in DTM generation. The Type II error induces some effects resulting from the fact that measured elevation values in Lidar data are replaced by interpolated values for DTM, which cause a zig-zag pattern in the DTM modeling (figure 6).

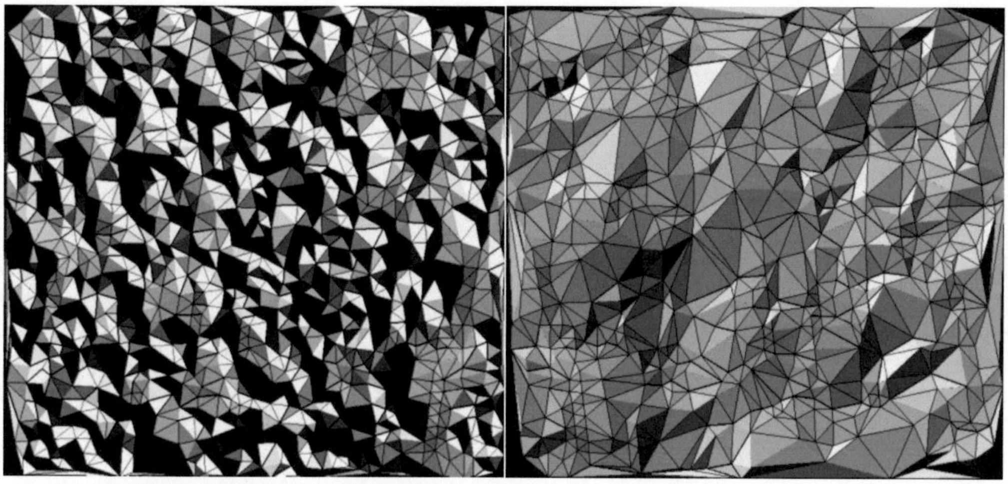

Fig. 6. Poorly filtered (left), good filtered (right).

3.2 Methods and models

Extracting the forest characteristics from Lidar data for biomass estimation is classified into two categories, height distribution with its statistical analysis, and single tree detection containing its location and characteristics.

3.2.1 Methods and models used in biomass estimation

A conventional model of biomass estimation is introduced by Thomas et al. (2006), which is given as: $b \times dbh^2 \times height$, where b is the coefficient. This equation was developed for the whole tree as well as the components of the stem wood, stem bark, branches, and foliage. As soon as the metrics (dbh and height) are measured for each plot, the equation can be established to estimate biomass and biomass components. The coefficient b is a variable which is related to the species of trees. Measurements for the deriving forest biomass are destructive sampling which is the input of regression modeling. For this, sample trees are measured and then cut and weighted (Popescu et al, 2004). The mass of components of each tree is regressed to one or more dimensions of the standing tree. As discussed in the introduction section, biomass has also been estimated by means of previously developed models using Lidar which relies on tree characteristics extraction like height, dbh, and crown size. Crown size is not used directly in the estimation procedure but it is useful for extracting the tree species. All developed models and their parameters for biomass estimation must be calibrated on the basis of tree characteristics. For this, four models were studied by Salmaca (2007). These are power function, Log transformed model, fractional power transformation, and explanatory function. The Power function is developed for North of USA, the Log transformed model is described by a linear function, the fractional power transformation is referred to linearized curvilinear model, and the explanatory function is constituted by a polynomial model. Under these models the Log transformed model is recommended which delivers the results with the unit of kilogram per every tree (Fallah Vazirabad, 2007). Consequently, tree characteristics extraction by Lidar data plays an important role in the biomass estimation model.

Bortlot et al. (2005) proposed to locate trees by image processing module assuming that the tree crown is circular, trees are taller than surroundings, and tree tops tend to be convex. They used the data of small footprint Lidar system. The algorithm starts by generating a CHM and works by shadow search method to find the crown boundaries which is related to tree tops. After defining a threshold and fitting the circles to the smoothed and generalized CHM, the circles should present the top of actual trees. The algorithm eliminates the small trees which are close to tall ones, because it searches for related high point neighboring. They conclude that tree heights are associated with canopy volume and therefore should be related to the biomass. They used the tree heights detected from image processing as variables for a stepwise multiple linear regression to find an equation for biomass prediction. They evaluated the results with highly significant (>95%) carrying out an efficient field measurement to calibrate the number of trees which are detected by an algorithm based on their height. Small trees are not included in this evaluation.

Lefsky et al. (1999) developed equations relating height indices to canopy area and biomass. They indicated that there are some differences in the predictive ability of the height indices; these differences are small, and statistically nonsignificant. However, the canopy structure information which is summarized in the median, mean, and quadratic mean canopy height indices, improved the stand canopy estimation related to the maximum canopy height. They defined the relation between tree height, H and dbh as: $dbh = (H/19.1)^{2.1}$. They concluded that the result of the model using stepwise multiple regressions causes a higher variance value than those from the simple linear regression referring to the CHM. But, the predictions of the stand attributes were less applicable to the CHM than the height indices. Stepwise multiple regressions of basal area and biomass using the canopy height profile vector as independent variables increase the importance of the field measured regression equations.

Fallah Vazirabad and Karslioglu (2009) investigated the biomass estimation with the method of single tree detection. Lidar data segmentation filtering method is applied to point clouds to distinguish canopy points from the terrain points which are used for the generation of a DTM. The CHM is obtained by subtracting the DSM (from original data) from DTM. A single tree detection method is employed to locate trees and detect the height of each tree top. Diameter at breast height (at 1.37 m from ground) is extracted from the close relation with the tree height which is defined by field measurements for the evaluation. A Log transformed model is applied for biomass estimation on the basis of the dbh variable.

3.2.2 Single tree detection, tree characteristics detection

The objective of many previous studies was to validate the tree detection, tree height estimation, crown size estimation for volume and biomass estimation of different forest types. Nelson et al (1988) used discrete Lidar data to collect forest canopy height data. Two logarithmic equations were tested to find the best model. They used a height distribution method and analyzed a statistical approach. Falkowski et al (2006) described and evaluated spatial wavelet analysis techniques to estimate the location, height, and crown diameter of individual trees from Lidar data. Two dimensional hat wavelets were convolved with a CHM to identify local maxima within the wavelet transformation image. Maltamo et al. (2004) examined the CHM local maxima search method for high dense forest regions to detect individual trees. Because of the dense understory tree layer in most area, about 40% of all trees were detected. However, the detected tree heights were obtained with an accuracy of ±50 cm.

Anderson et al. (2006) developed a methodology for acquiring accurate individual tree height field measurements within 2 cm accuracy using a total station instrument. They utilized these measurements to establish the expected accuracy of tree height derived from small and large footprint Lidar data. It turned out that the accuracy of small footprint Lidar data changes according to the tree species. The comparison has shown that tree heights which are retrieved from small footprint Lidar are more accurate than the result of large footprint data. Hopkinson (2007) investigated the influence of flight altitude, beam divergence, and pulse repetition frequency on laser pulse return intensities and vertical frequency distributions within a vegetated environment. The investigation showed that the reduction in the pulse power concentration by widening the beam, increasing the flight altitude, or increasing the pulse repetition frequency results in (i) slightly reduced penetration into short canopy foliage and (ii) increased penetration into tall canopy foliage, while reducing the maximum canopy return heights.

Yu et al. (2004) demonstrated the applicability of small footprint, multi return Lidar data for forest change detection like forest growth or harvested trees. An object oriented algorithm was used for tree detections referred to the tree to tree matching method and statistical analysis. The small trees could not be detected by the algorithm. The forest growth is estimated about 5 cm in canopy crown and 10-15 cm in tree height.

Fallah Vazirabad and Karslioglu (2010) used a technique based on the searching for the local maximum canopy height to detect individual tree with variable window size and shape. the method detects tree location, number of trees, and the height of each single tree. The variable window size and shape solved the problems of small tree detection and not detectable CHM margin regions. The importance of field measurements and reference information (like orthophoto) are emphasized for evaluation. Popescu and Zhao (2008) developed a method for assessing crown base height for individual tree using Lidar data in forest to detect single tree crown. They also investigated the Fourier and wavelet filtering, polynomial fit, and percentile analysis for characterizing the vertical structure of individual tree crowns. Fourier filtering used for smoothing the vertical crown profile. The investigation resulted in the detection of 80% of tree crown correctly.

Moorthy et al. (2011) utilized terrestrial laser scanning to investigate the individual tree crown. From the observed 3D laser pulse returns, quantitative retrievals of tree crown structure and foliage were obtained. Robust methodologies were developed to characterize diagnostic architectural parameters, such as tree height ($R^2 = 0.97, rmse = 0.21\ m$), crown width ($R^2 = 0.97,\ rmse = 0.13\ m$), crown height (($R^2 = 0.86,\ rmse = 0.14\ m$), crown volume ($R^2 = 0.99,\ rmse = 2.6\ m^3$). It seems that the first pulse return from the upside view of an individual tree in terrestrial laser scanning brought about the low performance in crown height while the other characteristics are detected well.

Riano et al. (2004) estimated leaf area index (LAI) and crown size using Lidar data. They concluded that LAI was better estimated using larger search windows while the crown size was better estimated using small window size. They generated the vegetation height above the ground for each laser pulse using interpolated values extracted from DTM. DTM was produced using the bisection principle. They also applied spline function interpolation in order to obtain the height above the ground. But in this work it is not obvious whether first or last return has been used to extract the canopy height, effecting the result significantly.

3.3 Applications

To provide reliable results on tree location, height, and number of detected trees the local maximum detection method is introduced by Vazirabad and Karslioglu (2009). This method determines the canopy height by applying a variable window size. The window size selection is related to the height and density of trees. High trees were easier to detect with large windows while short trees were easier to detect with small windows. The derivation of the appropriate window size to search for tree tops relies on the assumption that there is a relation between the height of trees and their crown size. In the 100*100 m test area, the correctness of single tree detection was calculated approximately 91%. The main reason for 9% error is referred to the not detected trees which are located in the corners and edges of the searched patch. To deal with this problem, the standard rectangle windows, variable size and variable shape are recommended (figure 6).

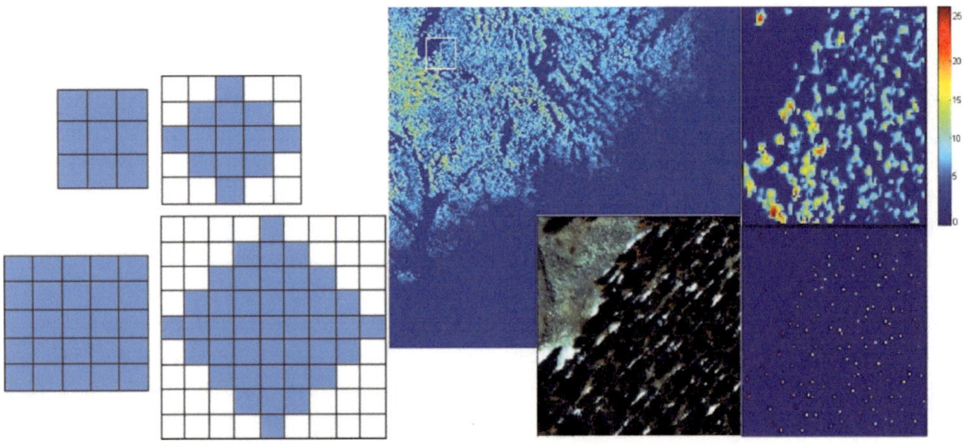

Fig. 6. Search windows (left); Single tree detection, CHM horizontal view (right-back), test patch 5 (right-top corner), respected orthophoto (center), and result (right-bottom)

Four window sizes such as standard 3*3 m, standard 5*5 m, rotated 3*3 m (5*5 m), and rotated 5*5 m (9*9 m) are employed (each pixel represents one meter). Tree heights from CHM show that they vary between 2 m to 25 m (figure 6, right). The single tree detection method works in several steps. First generation of a tree height model is required to obtain the tree height. In this model the algorithm looks for all nonzero values and then creates a sorted list depending on the point height above ground (reducing data makes searching procedure faster). In the second step a tree height specific filtering is accomplished, by moving the window pixel by pixel over the tree height model. By changing the window size and shape repeatedly the procedure is continuing up to the end. Six reference patches are provided for counting manually the number of trees by using orthophotos. Density and height of trees are variable inside the patches. The total 7479 trees are detected in whole 1*1 km^2. Tree height, dbh, and crown diameters are estimated in the whole area. All this information is adapted to the Log Transformed model for biomass estimation. Hence the total biomass which is given in kilograms for every tree in vegetation cover area is calculated as 1,966,123.3 kg.

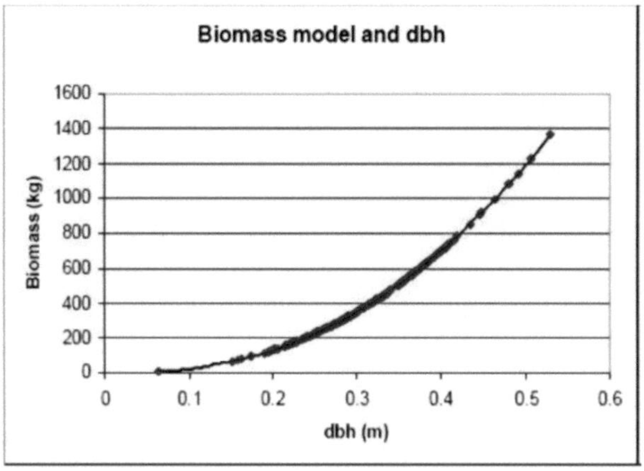

Fig. 8. Biomass model and dbh

4. Conclusion

A comprehensive review has been done within this chapter concerning the use of Lidar for biomass estimation. As a consequence it can be said that the reasons for the underestimation of biomass in relation to the tree height need further studies. The development of large footprint Lidar systems on the spaceborne platform GLAS will allow the biomass estimations on a global scale. Spaceborne systems are restricted to record regional and detailed forest data mainly due to the ground track resolution of the system. However, since they receive data continuously, biomass estimation and carbon storage studies are possible every time which can be regarded as a great benefit. Airborne Lidar has the advantages of variable height flying systems and hence collects more precise data with respect to the shape of the terrain. Taking advantages of intensity information from Lidar data provides more information about the interpretation of the ground surface. There are several full waveform airborne Lidar operational systems. But some substantial challenges still exist such as the huge data processing and the interpretation of waveform for complex objects like trees. The fast progresses in computer technologies will help overcome such problems. On the other hand, the high point density in terrestrial systems can help to evaluate the results of other systems. Besides, it allows to model vegetation canopy characteristics particularly concerning tree species estimations in detail. From the data acquisition point of view, it is obvious that models and methods need to exploit the whole potential of the full waveform data for biomass estimation in future. The investigation on the point density in Lidar data represents that having a sufficient number of points has a large impact on the filtering results. The result of the segmentation filtering shows a high capability of adaptation in different landscapes. But it requires choosing correct segmentation parameters by considering the point density. Point spacing plays also an important role for the selection of the interpolation method with respect to the DTM, DSM, and CHM resolution. The methods for individual tree detection which are described and evaluated in the application part are performing well, but they are still under development. Hence more empirical studies are required for improving the quality of the approaches.

5. References

Afzal R.S.; Yu A.W.; Dallas J.L.; Melak A.; Lukemire A.T.; Ramos-Izqueirdo L. & Mamakos W. (2007). The Geoscience Laser Altimeter System (GLAS) Laser Transmitter, *IEEE Journal of Selected Topics in Quantum Electronics*, Vol. 13, No. 3, p. 511

Axelsson P. (2000). DEM generation from laser scanner data - algorithms and application, *ISPRS Journal of Photogrammetry and Remote Sensing*, Vol. 54(2-3), pp. 138-147

Andersen, H.E.; Reutebuch, S.E. & McGaughey, R.J. (2006). A rigorous assessment of tree height measurements obtained using airborne lidar and conventional field methods, *Canadian Journal of Remote Sensing*, Vol. 32 (5), pp. 355–366

Bater and Coops, 2009, Evaluating error associated with lidar-derived DEM interpolation, *Computer and Geosciences*, Vol. 35, pp. 289-300

Bortolot, Z.J. & Wynne, R.H. (2005). Estimating forest biomass using small footprint LiDAR data: An individual tree-based approach that incorporates training data, *ISPRS Journal of Photogrammetry and Remote Sensing*, Vol. 59, pp. 342–360

Brandtberg, T.; Warner, T.A.; Landenberger, R.E. & McGraw, J.B. (2003). Detection and analysis of individual leaf-off tree crowns in small footprint, high sampling density lidar data from the eastern deciduous forest in North America, *Remote Sensing of Environment*, Vol. 85, pp. 290–303

Brown, S., (1997). Estimating biomass and biomass change in tropical forests: a primer. *FAO Forestry Paper, Food and Agriculture Organization of the United Nations*, Rome, Vol. 134

Brown, S. (2002). Measuring carbon in forests: current status and future challenges, *Environmental Pollution*, Vol. 116, pp. 363–372

Chen, Q. (2010). Assessment of terrain elevation derived from satellite laser altimetry over mountainous forest areas using airborne lidar data, *ISPRS Journal of Photogrammetry and Remote Sensing*, Vol. 65, pp. 111-122

Coops, N.; Wulder, M.; Culvenor, D. & St-Onge, B. (2004). Comparison of forest attributes extracted from fine spatial resolution multispectral and lidar data, *Canadian Journal of Remote Sensing*, Vol. 30 (6), pp. 855–866

Cote et al, 2011; An architectural model of trees to estimate forest structural attributes using terrestrial LiDAR, Environemtnal Modelling & Software XXX, pp. 1-17.

Dees, M. and Koch, B. (2008). Forestry applications, In: Li, Chen and Baltsavias (Eds.) Advances in Photogrammetry, Remote Sensing and Spatial Information Sciences, *2008 ISPRS Congress Book*, ISBN 978-0-415-47805-2, pp. 439-468

Donoghue, D.N.M.; Watt, P.J.; Cox, N.J. & Wilson, J. (2007). Remote sensing of species mixtures in conifer plantations using LiDAR height and intensity data, *Remote Sensing of Environment*, Vol. 110, pp. 509–522

Drake, J.B.; Dubayah, R.O.; Clark, D.B.; Knox, R.G.; Blair, J.B.; Hofton, M.A.; Chazdon, R.L.; Weishampel, J.F. & Prince, S. (2002). Estimation of tropical forest structural characteristics using large-footprint lidar, *Remote Sensing of Environment*, Vol. 79 (2–3), pp. 305– 319

Drake, J.B.; Knox, R.G.; Dubayah, R.O.; Clark, D.B.; Condit, R.; Blair, J.B. & Hofton, M. (2003). Above-ground biomass estimation in closed canopy Neotropical forests using LiDAR remote sensing: factors affecting the generality of relationships, *Global Ecology and Biogeography*, Vol. 12 (2), pp. 147–159

Dubayah, R. & Blair, J. (2000). Lidar remote sensing for forestry applications, *Journal of Forestry*, Vol. 98 (6), pp. 44-46

Duncanson L.I.; Niemann, K.O. & Wulder, M.A. (2010). Estimating forest canopy height and terrain relief from GLAS waveform metrics, *remote sensing of environment*, Vol. 114, pp. 138-154

Duong, H.; Lindenbergh, R.; Pfeifer, N. and Vosselman, G. (2009). ICESat full-waveform altimetry compared to airborne laser scanning altimetry over the Netherlands, IEEE Trnasaction on Geoscience and Remote Sensing, Vol. 47, No. 10, pp. 3365-3378

Duong, H.; Pfeifer, N. and Lindenbergh, R. (2007). Full waveform analysis: ICESat laser data for land cover classification, IAPRS Vol. XXXVI, part 7

Falkowski, M.J.; Smith, A.M.S.; Hudak, A.T.; Gessler, P.E.; Vierling, L.A. & Crookston, N.L. (2006). Automated estimation of individual conifer tree height and crown diameter via two-dimensional spatial wavelet analysis of LiDAR data, *Canadian Journal of Remote Sensing*, Vol. 32 (2), pp. 153–161

Fallah Vazirabad Y. (2007). Automatic snow depth calculation in Lidar data, M.Sc. thesis, University of Twente, Faculty of geo-information science and earth observation, Enschede, the Netherlands.

Fallah Vazirabad Y. & Karslioglu M.O. (2009). Airborne laser scanning data for snow covered biomass estimation, *Journal of Applied Remote Sensing*, Vol. 3, 033525; doi: 10.1117/1.3127447

Fallah Vazirabad Y. & Karslioglu M.O. (2010). Airborne laser scanning data for single tree characteristics detection, *ISPRS Journal of Photogrammetry and Remote Sensing, Istanbul Workshop, Modelling of Optical Airborne and Space Borne Sensors*, WG I/4, Oct. 11-13, IAPRS Vol. XXXVIII, part 1/W4 Istanbul, Turkey.

Fallah Vazirabad Y. & Karslioglu M.O. (2011). Toward automatic parameter tuning in segmentation method for airborne laser scanning data filtering, *5th International Conference on Recent Advances in Space Technologies, RAST2011*, Istanbul, Turkey.

Filin, S. & Pfeifer, N. (2006). Segmentation of airborne laser scanning data using a slope adaptive neighborhood, *ISPRS Journal of Photogrammetry and Remote Sensing*, Vol. 60, pp. 71-80

Fricker, H.A.; Borsa, A.; Minster, B.; Carabajal, C.; Quinn, K. & Bills, B. (2005). Assessment of ICESat performance at the salar de Uyuni, Boliva, *Geophysical Research Letters*, Vol. 32, doi:10.1029/2005GL023423

Garcia-Gonzalo, J.; Peltola, H.; Zubizarreta Gerendiain, A. & Kellomäki, S. (2007). Impacts of forest landscape structure and management on timber production and carbon stocks in the boreal forest ecosystem under changing climate, *Forest Ecology and Management*, Vol. 241, pp. 243-257

Harding, D.; Lefsky, M.; Parker, G. & Blair, J. (2001). Laser altimeter canopy height profiles—methods and validation for closed-canopy, broadleaf forests, *Remote Sensing of Environment*, Vol. 76 (3), pp. 283–297

Hasegawa, H. (2006). Evaluations of LIDAR reflectance amplitude sensitivity towards land cover conditions, *Bulletin of the Geographical Survey Institute*, Vol. 53

Heritage G.L. & Large A.R.G. (2009). Laser scanning for the environmental science, *Wiley-Blackwell, A John Wiley & Sonss, Ltd, Publication*. Chapter 4, pp. 49-66

Hollaus, M.; Mandlburger, G.; Pfeifer, N. and Mücke, W. (2010). Land cover dependent derivation of digital surface models from airborne laser scanning data, In: Paparoditis N., Pierrot-Deseilligny M., Mallet C., Tournaire O. (Eds), IAPRS, Vol. XXXVIII, Part 3A - Saint-Mandé, France, September 1-3

Holmgren, J. & Persson, A. (2004). Identifying species of individual trees using airborne laser scanner, *Remote Sensing of Environment*, Vol. 90 (4), pp. 415–423

Hopkinson, C. (2007). The influence of flying altitude, beam divergence, and pulse repetition frequency on laser pulse return intensity and canopy frequency distribution, *Canadian Journal of Remote Sensing*, Vol. 33 (4), pp. 312–324

Hopkinson, C. & Chasmer, L. (2009). Testing LiDAR models of fractional cover across multiple forest ecozones, *Remote Sensing of Environment*, Vol. 113 (1), pp. 275–288

Hudak, A.T.; Crookston, N.L.; Evans, J.S.; Hall, D.E. & Falkowski, M.J. (2008). Nearest neighbour imputation of species-level, plot-scale forest structure attributes from LiDAR data, *Remote Sensing of Environment*, Vol. 112 (5), pp. 2232–2245

Hyyppa, J.; Yu X.; Rannholm P.; Kaartinen H. & Hyyppa H. (1999). Dectecting and stimating attributes for single trees using laser scanner, *The Photogrammetric Journal of Finland*, Vol. 16, pp. 27-42

Hyyppa, J.; Hyyppa, H.; Yu, X.; Kaartinen, H.; Kukko, A. & Holopainen, M. (2009). In: Shan, J. & Toth, C.K. (Eds.), Forest Inventory Using Small Footprint Airborne Topographic Laser Ranging and Scanning Principles, *CRC Press, Boca Raton*, pp. 335–370.

Hyyppa, J. & Inkinen, M. (1999). Detecting and estimating attributes for single trees using laser scanner, *Photogrammetric Journal of Finland*, Vol. 16 (2), pp. 27- 42

Jang, J.D.; Payan, V.; Viau, A.A. & Devost, A. (2008). The use of airborne lidar for orchard tree inventory, *International Journal of Remote Sensing*, 29 (6), pp. 1767- 1780

Jonckheere, I.; Nackaerts, K.; Muys, B.; van Aardt, J. & Coppin, P. (2006). A fractal dimension-based modelling approach for studying the effect of leaf distribution on LAI retrieval in forest canopies, *Ecological Modelling*, Vol. 197, pp. 179-195

Kim, S.; McGaughey, R.J.; Anderson, H.E. & Schreuder, G. (2009). Tree species differentiation using intensity data derived from leaf-on and leaf-off airborne laser scanner data, *Remote Sensing of Environment*, Vol. 113, pp. 1575-1586, doi:10.1016/j.rse.2009.03.017

Koch, B. (2010). Status and future of laser scanning, synthetic aperture radar and hyperspectral remote sensing data for forest biomass assessment, *ISPRS Journal of Photogrammetry and Remote Sensing*, Vol. 65, pp. 581-590

Kurtz, N.T.; Markus, T.; Cavalieri, D.J.; Krabill, W.; Sonntag, J.G. & Miller, J. (2008). Comparison of ICESat data with airborne laser altimeter measurements over Arctic sea ice, *IEEE Transactions on Geoscience and Remote Sensing*, Vol. 46 (7), pp. 1913-1924

Lefsky, M.A.; Harding, D.; Cohen, W.B.; Parker, G. & Shugart, H.H. (1999). Surface Lidar remote sensing of basal area and biomass in deciduous forests of eastern Maryland, USA, *Remote Sensing of Environment*, Vol. 67 (1), pp. 83–98

Lefsky, M.A.; Cohen, W.B.; Harding, D.; Parker, G.; Acker, S.A. & Gower, S.T. (2001). Remote sensing of aboveground biomass in three biomes, *International Archives of the Photogrammetry Remote Sensing and Spatial Information Sciences*, Vol. 34, Part 3/W4, pp. 155–160

Lefsky, M. A.; Cohen, W. B.; Parker, G. G. & Harding, D. J. (2002). Lidar remote sensing for ecosystem studies, *Bioscience*, Vol. 52, pp. 19–30

Lefsky, M.A.; Harding, D.J.; Keller, M.; Cohen, W.B.; Carabajal, C.C.; Del Espirito- Santo, F.; Hunter, M.O.; de Oliveira Jr.R. & de Camargo, P. (2005). Estimates of forest canopy height and aboveground biomass using ICESat, *Geophysical Research Letters*, Vol. 32, doi:10.1029/2005GL023971

Lim, K.S. & Treitz, P.M. (2004). Estimation of above ground forest biomass from airborne discrete return laser scanner data using canopy-based quantile estimators, *Scandinavian Journal of Forest Research*, Vol. 19, pp. 558–570

Lim, K.; Treitz, P.; Baldwin, K.; Morrison, I. & Green, J. (2003). Lidar remote sensing of biophysical properties of tolerant northern hardwood forests, *Canadian Journal of Remote Sensing*, Vol. 29, pp. 658–678

Lovell, J.L.; Jupp, D.L.B.; Newnham, G.J. & Culvenor, D.S. (2011). Measuring tree stem diameters using intensity profiles from ground based scanning lidar from a fixed viewpoint, *ISPRS Journal of Photogrammetry and Remote Sensing*, Vol. 66, pp. 46-55, doi:10.1016/j.isprsjprs.2010.08.006

Maas, H.G.; Bienert, A.; Scheller, S. & Keane, E. (2008). Automatic forest inventory parameter determination from terrestrial laser scanner data, *International Journal of Remote Sensing*, Vol. 29 (5), pp. 1579–1593

Mallet, C. & Bretar, F. (2009). Full-waveform topographic lidar: State-of-the-art, *ISPRS Journal of Photogrammetry and Remote Sensing*, Vol. 64, pp. 1-16

Maltamo, M.; Mustonen, K.; Hyyppa, J.; Pitkänen, J. & Yu, X. (2004). The accuracy of estimating individual tree variables with airborne laser scanning in boreal nature reserve, *Canadian Journal of Forest Research*, Vol. 34 (9), pp. 1791–1801

Martin, C.F.; Thomas, R.H.; Krabill, W.B. & Manizade, S.S. (2005). ICESat range and mounting bias estimation over precisely-surveyed terrain, *Geophysical Research Letters*, Vol. 32, doi:10.1029/2005GL023800

Magruder, L.; Webb, C.; Urban, T.; Silverberg, E. & Schutz, B. (2007). ICESat altimetry data product verification at white sands space harbor, *IEEE Transactions on Geoscience and Remote Sensing*, Vol. 45 (1), pp. 147-155

Maune, D. (2007). Digital elevation model technologies and applications: the DEM user manual, 2nd edition, American society for photogrammetry and remote sensing, ISBN: 1-57083-082-7

McRae, B.H.; Schumaker, N.H.; McKane, R.B.; Busing, R.T.; Solomon, A.M. & Burdick, C.A. (2008). A multi-model framework for simulating wildlife population response to land-use and climate change, *Ecological Modelling*, Vol. 219, pp. 77-91

Means, J.; Acker, S.,; Harding, D.; Blair, J.; Lefsky, M.; Cohen, W.; Harmon, M. & McKee, W. (1999). Use of large-footprint scanning airborne lidar to estimate forest stand characteristics in the western cascades of Oregon, *Remote Sensing of Environment*, Vol. 67 (3), 298–308

Means, J.; Acker, S.; Fitt, B.; Renslow, M.; Emerson, L. & Hendrix, C. (2000). Predicting forest stand characteristics with airborne scanning lidar, *Photogrammetric Engineering and Remote Sensing*, Vol. 66 (11), 1367–1371

Momba, M. & Bux, F. (2010). Biomass, *Sciyo*, Croatia, ISBN 978-953-307-113-8, pp. 27-78

Monnet J.M.; Mermin, E.; Chanussot, J. and Berger, F. (2010). Using airborne laser scanning to assess forest protection function against rockfall, *Interpraevent 2010*, Taiwan, Province Of China

Moorthy, I.; Miller, J.R.; Berni, J.A.J.; Zarco-Tejada, P.; Hu, B. & Chen, J. (2011). Field characterization of olive (Olea europaea L.) tree crown architecture using terrestrial laser scanning data, *Agriculturea and Forest Meteorology*, Vol. 151, 204-214

Morsdorf, F.; Kotz, B.; Meier, E.; Itten, K. I. & Allgower, B. (2006). Estimation of LAI and fractional cover from small footprint airborne laser scanning data based on gap fraction, *Remote Sensing of Environment*, Vol. 104, 50–61

Morsdorf, F.; Meier, E.; Kotz, B.; Itten, K.; Dobbertin, M. & Allgower, B. (2004). Lidar based geometric reconstruction of boreal type forest stands at single tree level for forest and wildland fire management, *Remote Sensing of Environment*, Vol. 92 (3), 353–362

Muukkonen, P. & Heiskanen, J. (2007). Biomass estimation over a large area based on standwise forest inventory data and ASTER and MODIS satellite data: a possibility to verify carbon inventories, *Remote Sensing of Environment*, Vol. 107, 617–624

Næsset, E. (2002). Predicting forest stand characteristics with airborne scanning laser using a practical two-stage procedure and field data. *Remote Sensing of Environment*, Vol. 80 (1), 88–99

Næsset, E. (2004). Estimation of above- and below-ground biomass in boreal forest ecosystems, *International Archives of Photogrammetry, Remote Sensing and Spatial Information Sciences*, Vol. 36, Part 8/W2, 145–148

NASA, (2007). Report from the ICESat-II Workshop, 27–29 June, Linthicum, USA

Nelson, R.; Krabill, W. & Tonelli, J. (1988). Estimating forest biomass and volume using airborne laser data, *Remote Sensing of Environment*, Vol. 24 (2), 247–267

Nelson, R.; Short, A. & Valenti, M. (2004). Measuring biomass and carbon in Delaware using an airborne profiling LiDAR, *Scandinavian Journal of Forest Research*, Vol. 19 (6), 500–511

Neuenschwander, A.L.; Urban, T.J.; Gutierrez, R. & Schutz, B.E. (2008). Characterization of ICESat/GLAS waveforms over terrestrial ecosystems: Implications for vegetation mapping, *Journal of Geophysical Research*, Vol. 113, doi:10.1029/2007JG000557

Omasa, K.; Hosoi, F. & Konishi, A. (2007). 3D lidar imaging for detecting and understanding plant responses and canopy structure, *Journal of Experimental Botany*, 58 (4), 881–898

Palleja, T.; Tresanchez, M.; Teixido, M.; Sanz, R.; Rosell, J.R. and Palacin, J. (2010). Sensitivity of tree volume measurement to trajectory errors from a terrestrial LIDAR scanner, *Agricultural and Forest Meteorology*, Vol. 150, pp. 1420-1427

Patenaude, G.; Hill, R.; Milne, R.; Gaveau, D.; Briggs, B. & Dawson, T. (2004). Quantifying forest above ground carbon content using lidar remote sensing, *Remote Sensing of Environment*, Vol. 93 (3), 368–380

Pfeifer, N.; Gorte, B. & Oude Elberink, S. (2004). Influences of vegetation on laser altimetry analysis and correction approaches, *International Archives of Photogrammetry and Remote Sensing XXXVI*, 8/W2

Pfeifer N.; Stadler P. & Briese C. (2001). Derivation of digital terrain models in SCOP++ environment, *OEEPE Workshop on Airborne Lasescanning and Interferometric SAR for Detailed Digital Elevation Models*, Stockholm

Popescu, S.C.; Wynne, R.H. & Nelson, R.H. (2003). Measuring individual tree crown diameter with LiDAR and assessing its influence on estimating forest volume and biomass, *Canadian Journal of Remote Sensing*, Vol. 29 (5), 564– 577

Popescu, S.C.; Wynne, R.H. & Scrivani, J.A. (2004). Fusion of smallfootprint LiDAR and multispectral data to estimate plot-level volume and biomass in deciduous and pine forests in Virginia, USA, *Forest Science*, Vol. 50 (4), 551– 565

Popescu, S.C. (2007). Estimating biomass of individual pine trees using airborne LiDAR, *Biomass and Bioenergy*, Vol. 31 (9), 646–655

Popescu, S.C. & Zhao, K. (2008). A voxel-based lidar method for estimating crown base height for deciduous and pine trees, *Remote Sensing of Environment*, Vol. 112 (3), 767–781

Reitberger, J.; Krzystek, P. & Stilla, U. (2008). Analysis of full waveform lidar data for the classification of deciduous and coniferous trees, *International Journal of Remote Sensing*, Vol. 29 (5), 1407–1431

Reitberger, J.; Schnorr, Cl.; Krzystek, P. & Stilla, U. (2009). 3D segmentation of single trees exploiting full waveform lidar data, *ISPRS Journal of Photogrammetry and Remote Sensing*, Vol. 64, pp. 561-574, doi:10.1016/j.isprsjprs.2009.04.002

Riano, D.; Meier, E.; Allgower, B.; Chuvieco, E. & Ustin, S.L. (2003). Modeling airborne laser scanning data for the spatial generation of critical forest parameters in fire behaviour modelling. *Remote sensing of Environment*, Vol. 86, 177-186

Riano, D.; Valladares, F.; Conds, S. & Chuvieco, E. (2004). Estimation of leaf area index and covered ground from airborne laser scanner (lidar) in two contrasting forests. *Agricultural and Forest Meteorology*, Vol. 124 (3–4), pp. 269–275

Salas, C.; Ene, L.; Gregoire, T.G.; Næsset, E. & Gobakken, T. (2010). Modelling tree diameter from airborne laser scanning derived variables: A comparison of spatial statistical models, *Remote Sensing of Environment*, Vol. 114, pp. 1277-1285

Salmaca I.K. (2007). Estimation of forest biomass and its error: a case study in Kalimantan, Indonesia. M.Sc. thesis, University of Twente, Faculty of geo-information science and earth observation, Enschede, the Netherlands

Schutz, B. E.; Zwally, H. J.; Shuman, C. A.; Hancock, D. & DiMarzio, J. P. (2005). Overview of the ICESat Mission. *Geophysical Research Letters*, Vol. 32, L21S01

Shan J. & Toth C.K. (2009). Topographic laser ranging and scanning: principles and processing, *CRC Press, Taylor and Francis Group*, Chapter 2 and 3, pp. 29-127

Simard, M.; Rivera-Monroy, V.H.; Ernesto Mancera-Pineda, J.; Castañeda-Moya, E. & Twilley, R.R. (2008). A systematic method for 3D mapping of mangrove forests based on shuttle radar topography mission elevation data, ICEsat/GLAS waveforms and field data: Application to Ciénaga Grande de Santa Marta, Colombia, *Remote Sensing of Environment*, Vol. 112 (5), 2131_2144

Sithole G. (2005). Segmentation and classification of airborne laser scanner data, *Publication on Geodesy of the Netherlands Commission of Geodesy*, Vol. 59, Dissertation, TU DELFT, ISBN 90 6132 292 8

Sithole, G. & Vosselman, G. (2004). Experimental comparison of filter algorithms for bare earth extraction from airborne laser scanning point clouds. *International Society for Photogrammetry and Remote Sensing*, Vol. 59, (1-2), 85-101

Sithole, G. & Vosselman, G. (2005). Filtering of airborne laser scanner data based on segmented point clouds. *International Archives of Photogrammetry, Remote Sensing and Spatial Information Sciences XXXVI*, part 3/W19, pp. 66-71

Solberg, S.; Brunner, A.; Hanssen, K. H.; Lange, H.; Næsset, E. & Rautiainen, M. (2009). Mapping LAI in a Norway spruce forest using laser scanning. *Remote Sensing of Environment*, Vol. 113, 2317–2327

Song, J.H.; Han, S. H.; Yu, K. & Kim, Y.L. (2002). Assessing the possibility of land-cover classification using LIDAR intensity data. *ISPRS Commission III*, "Photogrammetric Computer Vision", Graz, Austria, Vol. 34(3B), pp. 259–262

Sun, G.; Ranson, K.J.; Kimes, D.S.; Blair, J.B. & Kovacs, K. (2008). Forest vertical structure from GLAS: an evaluation using LVIS and SRTM data, *Remote Sensing of Environment*, Vol. 112 (1), 107–117

Sun, G.; Ranson, K.J.; Masek, J.; Fu, A. & Wang, D. (2007). Predicting tree height and biomass from GLAS data, *Proceedings of the 10th International Symposium on Physical Measurements and Signatures in Remote Sensing*, Davos, Switzerland

Thomas, V.; Treitz, P.; McCaughey, J. & Morrison, I. (2006). Mapping stand-level forest biophysical variables for a mixedwood boreal forest using lidar: an examination of scanning density, *Canadian Journal of Forest Research*, Vol. 36 (1), pp. 34–47

Tomppo, E. & Halme, M. (2004). Using coarse scale forest variables as ancillary information and weighting of variables in k-NN estimation—a genetic algorithm approach, *Remote Sensing of Environment*, Vol. 92 (1), pp. 1–20

Tomppo, E.; Nilsson, M.; Rosengren, M.; Aalto, P. & Kennedy, P. (2002). Simultaneous use of Landsat-TM and IRS-1C WiFS data in estimating large area tree stem volume and aboveground biomass, *Remote Sensing of Environment*, Vol. 82 (1), pp. 156–171

Tovari, D. & Pfeifer, N. (2005). Segmentation based robust interpolation - A new approach to laser data filtering, *ISPRS International Society for Photogrammetry and Remote Sensing*, WG III/3, III/4, V/3 workshop, Enschede, the Netherlands

van Aardt, J.A.N.; Wynne, R.H. & Scrivani, J.A. (2008). LiDAR-based mapping of forest volume and biomass by taxonomic group using structurally homogenous segments. *Photogrammetric Engineering & Remote Sensing*, Vol. 74 (8), pp. 1033–1044

Vosselman, G. (2000). Slope based filtering of laser altimetry data, *IAPRS XXXIII*, B3/2, Amsterdam

Widlowski, J.L.; Pinty, B.; Gobron, N.; Verstraete, M.M.; Diner, D.J. & Davis, A.B. (2004). Canopy structure parameters derived from multi-angular remote sensing data for terrestrial carbon studies. *Climatic Change*, Vol. 67, pp. 403-415

Xing, Y.; de Gier, A.; Zhang, J. & Wang, L. (2010). An improved method for estimating forest canopy height using ICESat-GLAS full waveform data over sloping terrain A case study in Changbai mountains, China, *International Journal of Applied Earth Observation and Geoinformation*, Vol. 12, pp. 385-392, doi:10.1016/j.jag.2010.04.010

Yu, X.; Hyyppa, J.; Kaartinen, H.; & Maltamo, M. (2004). Automatic detection of harvested trees and determination of forest growth using airborne laser scanning, *Remote Sensing of Environment*, Vol. 90 (4), pp. 451–462

Yu, X.; Hyyppa, J.; Vastaranta, M.; Holopainen, M. & Viitala, R. (2011). Predicting individual tree attributes from airborne laser point clouds based on the random forests technique, *ISPRS Journal of Photogrammetry and remote sensing*, 66, 28-37

Zenner, E.K. & Hibbs, D.E. (2000). A new method for modeling the heterogeneity of forest structure, *Forest Ecology and Management*, Vol. 129, pp. 75-87

Zhao, K.; Popescu, S. & Nelson, R. (2009). LiDAR remote sensing of forest biomass: a scale-invariant estimation approach using airborne lasers, *Remote Sensing of Environment* Vol. 113 (1), pp. 182–196

Zwally, H.J.; Schutz, B.; Abdalati, W.; Abshire, J.; Bentley, C.; Brenner, A.; Bufton, J.; Dezio, J.; Hancock, D. and Harding, D. (2002). ICESat's laser measurements of polar ice, atmosphere, ocean, and land, *Journal of Geodynamics*, Vol. 34 (3–4), pp. 405-445

2

Field Measurements of Canopy Spectra for Biomass Assessment of Small-Grain Cereals

Conxita Royo and Dolors Villegas
IRTA (Institute for Food and Agricultural Research and Technology), Generalitat of Catalonia Centre, UdL-IRTA
Spain

1. Introduction

Small-grain cereals are the food crops that are most widely grown and consumed in the world. Wheat and rice jointly supply more than 55% of total calories for human nutrition, occupying about 59% of the total arable land in the world (225 and 156 million ha, respectively). Global production is around 682 million metric tons for wheat and 650 million metric tons for rice (FAOSTAT, 2008). Wheat is a very widely adapted crop, grown in a range of environmental conditions from temperate to warm, and from humid to dry and cold environments. Demand for wheat and rice will grow faster in the next few decades, and yield increases will be required to feed a growing world population. Because land is limited and environmental and economical concerns constrain the intensification of such crops, yield increases will have to come primarily from breeding efforts aimed at releasing new varieties that provide higher productivity per unit area.

The most integrative plant traits responsible for grain yield increases in small-grain cereals are the total biomass produced by the crop and the proportion of the biomass allocated to grains, the so-called harvest index (Van den Boogaard et al., 1996). The product of these traits provides a framework for expressing the grain yield in physiological terms and for contextualizing past yield gains in small-grain cereals, particularly wheat and barley. Retrospective studies conducted with wheat frequently associate increases in yield with increases in partitioning of biomass to the grain, with small or negligible increases (Austin et al., 1980, 1989; Royo et al., 2007; Sayre et al., 1997; Siddique et al; 1989; Waddington et al., 1986), or even significant decreases (Álvaro et al., 2008a) in total biomass production. Increases in biomass have been reported in spring wheat (Reynolds et al., 1999; 2001), winter bread wheat (Shearman et al., 2005), and durum wheat (Pfeiffer et al., 2000; Wadington et al., 1987).

Since harvest index has a theoretical maximum estimated to be 0.60 (Austin, 1980), increases in grain yield of more than 20 percent cannot be expected through increasing the harvest index above the maximum levels reached currently by some wheat genotypes (Reynolds et al., 1999; Richards, 2000; Shearman et al., 2005). It is therefore generally believed that future improvements in grain yield through breeding will have to be reached by selecting genotypes with higher biomass capacity, while maintaining the high partitioning rate of photosynthetic products (Austin et al., 1980; Hay, 1995).

Total dry matter is mainly determined by two processes: i) the interception of incident solar irradiance by the canopy, which depends on the photosynthetic area of the canopy; and ii) the conversion of the intercepted radiant energy to potential chemical energy, which relies on the overall photosynthetic efficiency of the crop (Hay & Walker, 1989). The relationship between above-ground biomass and yield has been demonstrated empirically in wheat. Positive associations (R^2=0.56, P<0.05) have been reported between biomass at maturity and yield in durum wheat (Waddington et al., 1987), and between biomass at anthesis and yield in bread wheat (Reynolds et al., 2005; Shearman et al., 2005; Singh et al., 1998; Tanno et al., 1985; Turner, 1997; Van der Boogaard et al., 1996), durum wheat (Royo et al., 2005), barley (Ramos et al., 1985) and rice (Turner, 1982). In a study conducted in Mediterranean conditions with 25 durum wheat cultivars, Villegas et al. (2001) found a strong association (R^2=0.75, P<0.001) of the biomass accumulated from the first node detectable stage with anthesis and yield. Vegetative growth before anthesis becomes particularly important when stresses during grain filling such as those caused by rising temperatures and falling moisture supply —usually occurring after anthesis in Mediterranean environments— limit the crop photosynthesis, forcing yield to depend greatly on the remobilization to the grain of pre-anthesis assimilates accumulated in leaves and stems (Álvaro et al., 2008b; Palta et al., 1994; Papakosta and Gagianas, 1991; Shepherd et al., 1987). The contribution of pre-anthesis assimilates to wheat grain yield and the efficiency of dry matter translocation to the filling grains seem to have increased in the last century as a consequence of breeding (Austin et al., 1980; Álvaro et al., 2008a,b).

Biomass assessment is thus essential not only for studies monitoring crop growth, but also in cereal breeding programs as a complementary selection tool (Araus et al., 2009). Tracking changes in biomass may also be a way to detect and quantify the effect of stresses on the crop, since stress may accelerate the senescence of leaves, affecting leaf expansion (Royo et al., 2004) and plant growth (Villegas et al., 2001).

Biomass assessment in breeding programs, in which hundreds of lines have to be screened for various agronomical traits in a short time every crop season, is not viable by destructive sampling because it is a time-and labor-intensive undertaking, it is subject to sampling errors, and samplings reduce the final area available for determining final grain yield on small research plots (Whan et al., 1991). Originally used in remote sensing of vegetation from aircraft and satellites, remote sensing techniques are becoming a very useful tool for assessing many agrophysiological traits (Araus et al., 2002). The measurement of the spectra reflected by crop canopies has been largely proposed as a quick, cheap, reliable and non-invasive method for estimating plant aboveground biomass production in small-grain cereals, at both crop level (Aparicio et al., 2000, 2002; Elliot & Regan, 1993; R.C.G. Smith et al., 1993) and individual plant level (Álvaro et al., 2007).

2. Growth patterns and biomass spectra

The growth cycle of small-grain cereals involves changes in size, form and number of plant organs. The external stages of cereal growth include germination, crop emergence, seedling growth, tillering, stem elongation, booting, inflorescence emergence, anthesis and maturity (Fig. 1). The classical monitoring of crop biomass requires destructive samplings of plants at different growth stages, counting of the number of plants contained in the sample and its weighing after oven-drying them. Crop biomass may be expressed as crop dry weight (CDW), which can be obtained from the plants sampled at a given stage as the product of average dry weight per plant (W, g) and the number of plants per unit area, and is frequently expressed as g m^{-2} (Villegas et al., 2001). The leaf area expansion of a cereal crop may be monitored through changes in its leaf area index (LAI, a dimensionless value), which is the ratio of leaf green area to the area of ground on which the crop is growing. LAI may be calculated as the product of the mean one-sided leaf area per plant (LAP, m^2 $plant^{-1}$) and the number of plants per unit area in the sample (plants m^{-2}). Changes in total green area of the crop may be described through the green area index (GAI, a dimensionless value), which is the ratio of total green area of the plants (leaves and stems, as well as spike peduncles and spikes when applicable) to the area of ground on which the crop is growing. It can be calculated as the product of total green area per plant (GAP, m^2 $plant^{-1}$) and the number of plants per unit area in the sample (plants m^{-2}) (Royo et al., 2004).

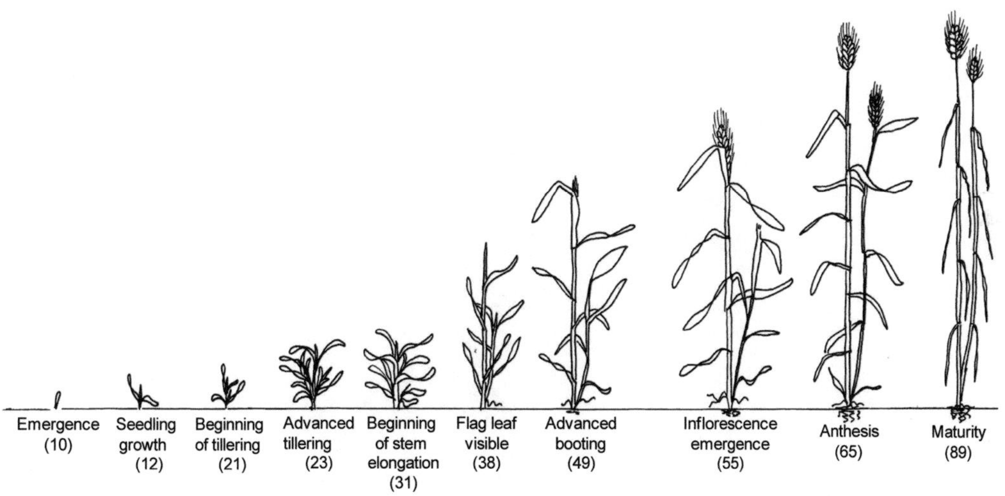

Fig. 1. Growth stages of small-grain cereals. Numbers correspond to the Zadoks scale (Zadoks et al., 1974)

Raw data from destructive sampling can be fitted to mathematical models, usually empirically based, to describe the growth pattern during the crop cycle. The logistic model of Richards (Richards, 1959), the expolinear equation of Goudriaan & Monteith (Goudriaan & Monteith, 1990), and the asymmetric logistic peak curve first used by Royo and Tribó (Royo & Tribó, 1997), have been used to describe the growth of crops. This last model has been useful for monitoring the biomass and leaf area expansion of triticale (Royo & Blanco, 1999) and durum wheat (Royo et al., 2004; Villegas et al., 2001). The mathematical models present the variation in dry matter production, leaf area or green area expansion over time, allowing variations between species (Fig. 2), genotypes, years and environmental conditions to be assessed (Fig. 3). Similarly to the case of grain yield, variability induced by the genetic background in the growth pattern of small-grain cereals has been found to be lower than the environmental variation caused by either year or site effects (Royo et al., 2004; Villegas et al., 2001).

Crop growth conditions can be monitored by measuring the spectra reflected by crop canopies in the visible (VIS, λ=400-700 nm) and near-infrared (NIR, λ =700-1300 nm) regions of the electromagnetic spectrum (Fig. 4). Given that the amount of green area of a canopy determines the absorption of photosynthetic active radiation by photosynthetic organs, spectral reflectance measurements can provide an instantaneous quantitative assessment of the crop's ability to intercept radiation and photosynthesize (Ma et al., 1996). Therefore, the absorption by the crop canopy of very specific wavelengths of electromagnetic radiation is associated with certain morphological and physiological crop attributes related to the development of the total photosynthetic area of the canopy.

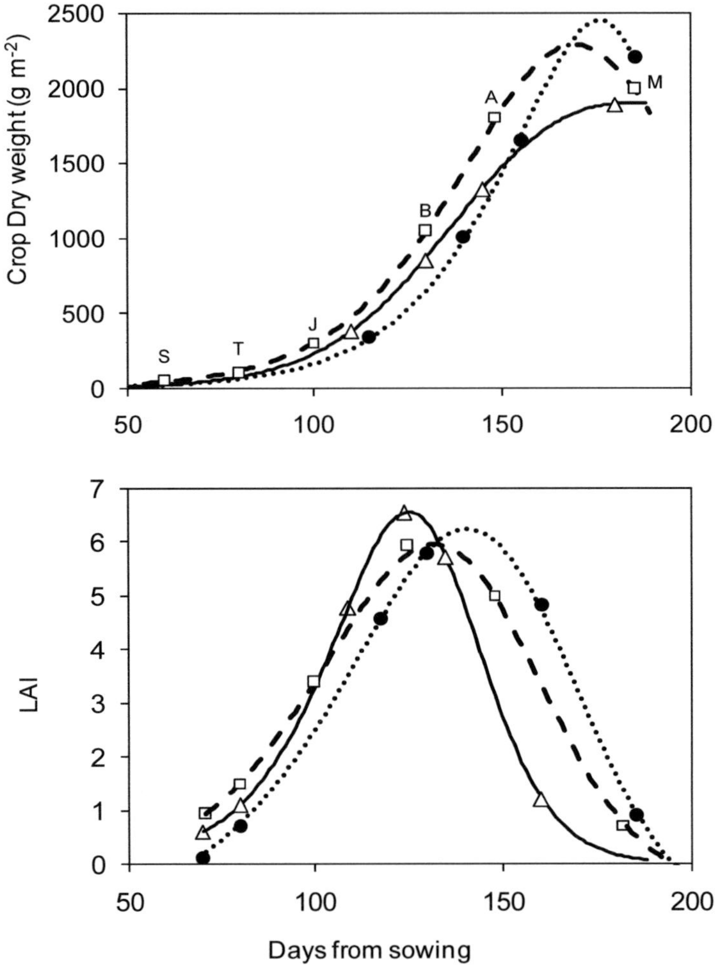

Fig. 2. Illustration of the differences between the patterns of biomass accumulation and leaf area expansion of barley (Δ), spring triticale (□), and winter triticale (•) from experiments conducted in 4 Mediterranean environments. Samples were taken at seedling (S), tillering (T), beginning of jointing (J), booting (B), anthesis (A), and physiological maturity (M). Biomass increased continually from anthesis to maturity in barley, but in triticale the peak of biomass took place between anthesis and maturity. The maximum LAI was reached at the booting stage in barley, but a little later in triticale. Adapted from Royo & Tribó (1997)

The reflectance spectra of a healthy crop-canopy shows a relative maximum around 550 nm, a relative minimum around 680 nm and an abrupt increase around 700 nm, remaining fairly constant beyond this point (Fig. 4). The spectral reflectance in the VIS wavelengths depends on the absorption of incident radiation by leaf chlorophyll and associated pigments such as carotenoid and anthocyanins. Crop reflectance is very low in the blue (400-500 nm) and red (600-700 nm) regions of the spectrum, because they contain the peaks of chlorophyll absorbance. Beyond 700 nm the reflectance of the NIR wavelengths is high since it is not absorbed by plant pigments and is scattered by plant tissues at different levels in the canopy (Knipling, 1970).

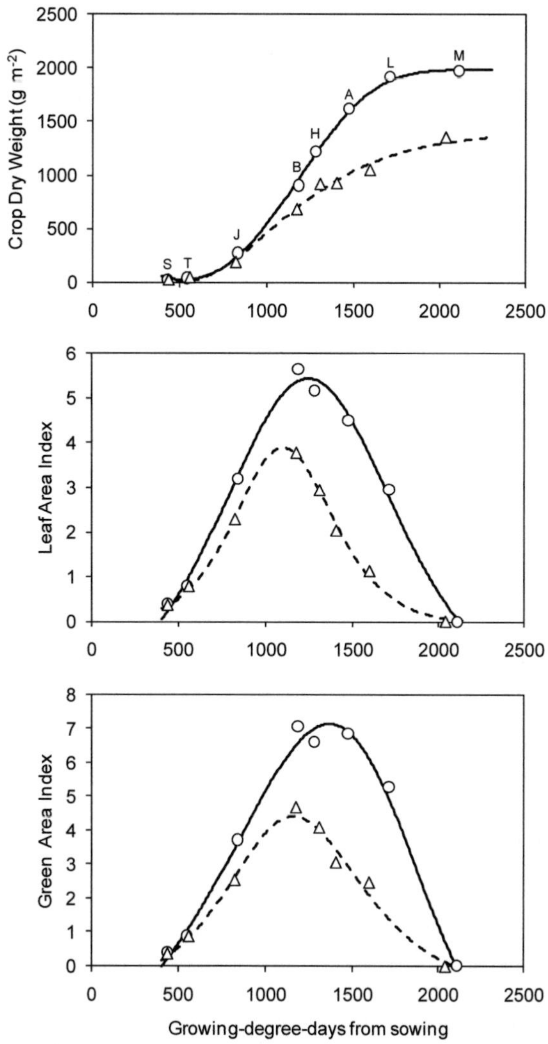

Fig. 3. Illustration of the effect of water input on the pattern of biomass accumulation (CDW), leaf area index (LAI), and green area index (GAI) of durum wheat grown under irrigated (o) and rainfed conditions (Δ). Data are means of 25 durum wheat cultivars grown in 1998 under Mediterranean conditions. The crop received 384 and 194 mm of water under irrigated and rainfed conditions, respectively. Samples were taken at seedling (S), tillering (T), beginning of jointing (J), booting (B), heading (H), anthesis (A), milk grain stage (L), and physiological maturity (M). Upper figure adapted from Villegas et al. (2001). LAI and GAI figures adapted from Royo et al. (2004)

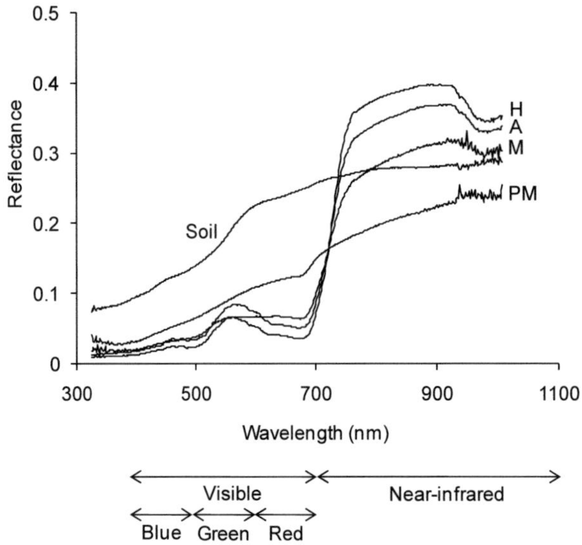

Fig. 4. Variation of the reflectance spectra of a healthy wheat canopy at different growth stages compared with the bare soil spectrum. H, heading; A, anthesis; M, milk-grain stage; PM, physiological maturity. The magnitude of the increase in reflectance at around 700 nm indicates differences in biomass

3. Methodology for capturing spectra

3.1 Field equipment

High spectral resolution devices have recently improved in sensitivity, decreased in cost, and increased in availability. The equipment for field measurements consists of a portable spectroradiometer, which measures the irradiance at different wavelengths with a band width of about 1-2 nm through the VIS and NIR regions of the spectrum. This unit is connected to a computer, which stores the individual scans, a fore-optics sensor for capturing the radiation, and some complements such as reference panels and supports (Fig. 5). The sensor appraises the radiation reflected by the crop canopy, delimiting the field of view to a given angle, generally between 10° and 25°, which limits the area of the crop scanned to 20-100 cm^2. The angle of incident light and the angle of observation of the sensor determine the proportion of elements in the observation field. The sensor is usually mounted on a fixed or hand-held tripod, which allows all measurements to be taken at the same angle and distance from the surface of the crop —usually from 0.5 m to around 1.0 m above the canopy facing the center of the plot. A fiber optic cable transmits the captured radiation to the spectrum analyzer. To convert captured spectra to reflectance units the spectra reflected by the crop canopy must be calibrated against light reflected from a commercially available white reference panel of $BaSO_4$ (Jackson et al., 1992). Each measurement takes around 1-2 s and between 5 and 10 scans are usually averaged per measurement.

The classical spectroradiometers measure about 250-500 bands, evenly spaced from a wavelength of 350 to 1110 nm, so a wide range of spectral reflectance indices can be calculated or the complete VIS/NIR reflectance spectra can be used. Cheaper units, such as Green Seeker^{TM}, which give only the basic spectroradiometric indices of green biomass, such as the normalized difference vegetation index (NDVI) and the simple ratio (SR, see section 4), have been designed more recently for diagnosing nitrogen status and biomass assessment (Li et al., 2010b). The methodology allows sampling at a rate of up to 1000 samples per day.

Fig. 5. Measurements of spectral reflectance on field plots and layout of the tube used by Álvaro et al. (2007) to capture the spectra of individual plants

3.2 Factors affecting the reflectivity of the canopy surface

Measurements of the reflectance spectra of crop canopies are affected by both sampling conditions and canopy features. The most important are detailed in the following sections.

3.2.1 Sensor position

The angles between sun, sensor and canopy surface may lead to the appearance of shadow or soil background in the field of view of the apparatus, causing disturbing effects in the spectra measured (Aparicio et al., 2004; Baret and Guyot, 1991; Eaton & Dirmhirn, 1979). The angle of the sun is more important in canopies with low LAI (Kollenkark et al., 1982; Ranson et al., 1985). Variability in reflectance due to variation in the sensor view angle has been reported to depend on the stage of development of the crop (J.A. Smith et al., 1975), the structure of the vegetative canopy (Colwell, 1974) and the leaf area index (Aparicio et al., 2004). Angles between the sensor azimuth and the sun azimuth of between 0° and 90° minimize the variability caused by changes in the elevation of the sensor or the sun (Wardley, 1984). However, when off-nadir view angles are used, the analysis of the remote sensing data could be complicated due to the non-Lambertian characteristics of vegetation (unequal reflection of incident light in all directions and reflection depending on the wavelength) (Ranson et al., 1985). The degree of canopy cover captured by the sensor is minimum at nadir position, and increases with the angle of observation. The effect of angle is particularly important in crops arranged in rows, which may have different orientations in relation to the solar angle and the observation angle (Ranson et al., 1985; Wanjura & Hatfield, 1987). The nadir position of the sensor (sensor looking vertically downward) is the most widely used, because it has a low interaction with sun position and row orientation and delays the time at which spectra become saturated by LAI (Araus et al., 2001).

3.2.2 Environmental conditions

Environmental factors can cause undesired variation in the captured spectra. Light intensity, sun position, winds or nebulosity may interfere with the way in which the interaction between solar irradiation and crop is captured (Baret & Guyot, 1991; Huete 1987; Jackson 1983; Kollenkark et al., 1982). Green biomass may be overestimated when measurements are taken on cloudy days because the increased diffuse radiation improves the penetration of light into the canopy. Brief changes in canopy structure caused by winds may also induce variations in the captured spectra (Lord et al., 1985). The presence of people or objects near to the target view area should be avoided, since they can cause alterations in the measured spectra by reflecting radiation. The instruments should be painted a dark color and people should preferable wear dark clothes (Kimes et al., 1983). As a means of minimizing the variability induced by sun position, it has also been recommended that measurements be taken at about noon on rows oriented east to west.

3.2.3 Canopy attributes

The reflectivity of a crop canopy may be affected by a number of internal and external factors. The crop species, its nutritional status, the phenological stage (Fig. 4), the glaucousness, the geometry of the canopy and the spatial arrangement of its constitutive elements greatly affect the optical properties of the canopy surface. Under severe nitrogen deficiencies, chlorosis in leaves causes plants to reflect more in the red spectral region (Steven et al., 1990). The presence of non-green vegetation or non-leaf photosynthetically active organs (such as spikes and leaf sheaths of cereals) and changes in leaf erectness can also affect the spectral signature of the canopy (Aparicio et al. 2002; Bartlett et al., 1990; Van Leeuwen & Huete, 1996); for high LAI values, the reflectivity decreases with greater leaf inclination in both the VIS and the NIR wavelengths (Verhoef & Bunnik, 1981). Radiation reflected perpendicularly from plant canopies has been reported to be greater for planophile than for erectophile canopies (Jackson & Pinter, 1986; Zhao et al., 2010).

3.2.4 Soil interferences

When the crop canopy does not cover the entire soil surface, the target view area may include measurements of soil background, which may disturb the spectra measurements. Soil reflectances in the red and NIR wavelengths are usually linearly related (Hallik et al., 2009). As shown in Fig. 4, reflectance of bare soil differs from that of the crop canopy, because green vegetation reduces the values of red reflectance and increases the values of NIR reflectance when compared with those of the soil background. A number of studies on the effect of the soil reflectivity on the crop reflectance (Colwell, 1974; Huete et al., 1985), concluded that the most important factors are the chemical composition and water content of the soil. Greater discrimination power between wheat plots differing in biomass has been found on dark soils than on light soils (Bellairs et al., 1996).

In an attempt to minimize the variability induced by external factors, reflectance values recorded by the spectroradiometer are seldom taken directly but rather used to calculate different indices –usually formulas based on simple operations between reflectances at given wavelengths.

4. Traditional and new spectral reflectance indices for biomass appraisal

Spectral reflectance indices were developed using formulations based on simple mathematical operations, such as ratios or differences, between the reflectance at given wavelengths. Most spectral indices use specific wavebands in the range 400 to 900 nm and their most widespread application is in the assessment of plant traits related to the photosynthetic size of the canopy, such as LAI and biomass.

The most widespread vegetation indices (VI), for measurements not only at ground level but also at aircraft and satellite level (Wiegand & Richardson, 1990) are the normalized difference vegetation index (NDVI = R_{NIR}-R_{RED} /R_{NIR} +R_{RED}) and the simple ratio (SR= R_{NIR}/R_{RED}) (see Table 1 for their definition). The ratio between the reflectances in the near-infrared (NIR) and red (RED) wavelengths is high for dense green vegetation, but low for the soil, thus giving a contrast between the two surfaces. For wheat and barley a wavelength (λ) of around 680 nm is the most commonly used for R_{RED}, and one of 900 nm for R_{NIR} (Peñuelas et al., 1997a). These indices have been positively correlated with the absorbed photosynthetically active radiation (PAR), the photosynthetic capacity of the canopy and net primary productivity (Sellers, 1987). According to Wiegand & Richardson (1984, as cited in Wiegand et al., 1991), the fraction of the incident radiation used by the crops for photosynthesis (FPAR) may be derived from vegetation indices through their direct relationship with LAI, according to Equation (1):

$$FPAR(VI) = FPAR(LAI) \times LAI(VI) \qquad (1)$$

For this reason, vegetation indices have proven to be useful for estimating the early vigor of wheat genotypes (Bellairs et al., 1996; Elliot & Regan, 1993), monitoring wheat tiller density (J.H. Wu et al., 2011), and assessing green biomass, LAI and the fraction of radiation intercepted in cereal crops (Ahlrichs & Bauer, 1983; Aparicio et al., 2000, 2002; Baret & Guyot, 1991; Elliott & Regan, 1993; Gamon et al., 1995; Peñuelas et al., 1993, 1997a; Price & Bausch, 1995; Tucker 1979; Vaesen et al., 2001). They tend to minimize spectral noise caused by the soil background and atmospheric effects (Baret et al., 1992; Collins, 1978; Demetriades-Shah et al., 1990; Filella & Peñuelas, 1994; Mauser & Bach, 1995).

Positive and significant correlations of SR and NDVI with LAI (Fig. 6), GAI and biomass (either on a linear or a logarithmic basis) have been reported in bread wheat and barley (Bellairs et al., 1996; Darvishzadeh et al., 2009; Fernández et al., 1994; Field et al., 1994; Peñuelas et al., 1997a). In a study conducted with 25 bread wheat genotypes, NDVI explained around 40% of the variability found in biomass (Reynolds et al., 1999). Studies involving 20-25 durum wheat genotypes have demonstrated a strong association between SR and NDVI and biomass under both rainfed and irrigated field conditions (Aparicio et al., 2000, 2002; Royo et al., 2003). Spectral reflectance measurements are also being used increasingly as a tool to detect the canopy nitrogen status and allow locally adjusted nitrogen fertilizer applications during the growing season (Mistele & Schmidhalter, 2010). Since grain yield is closely associated with crop growth and the vegetation indices are sensitive to canopy variables such as LAI and biomass that largely determine this growth, spectral data have also been proposed as suitable estimators in yield-predicting models (Aparicio et al., 2000; Das et al., 1993; Ma et al., 2001; Royo et al., 2003).

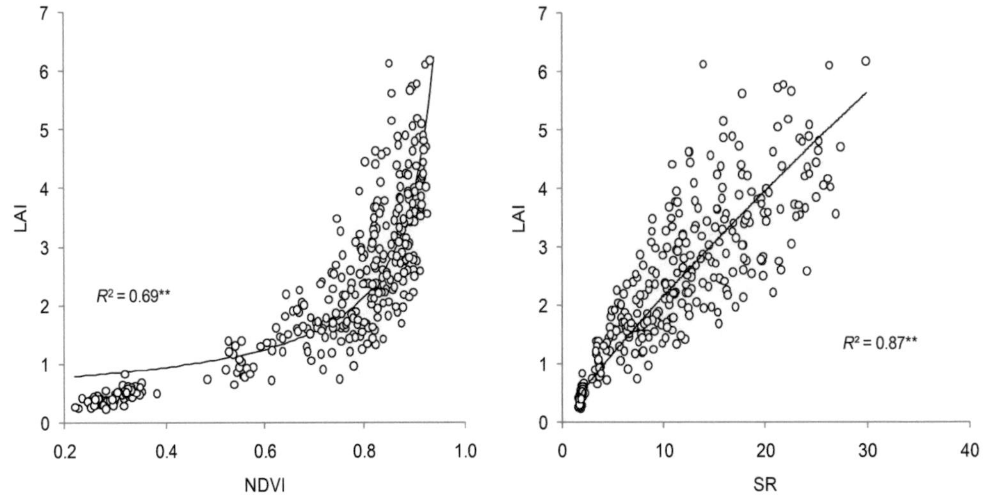

Fig. 6. Patterns of the relationships of leaf area index (LAI) with the normalized difference vegetation index (NDVI) and the simple ratio (SR). Data correspond to 7 field experiments involving 20-25 durum wheat genotypes and conducted under contrasting Mediterranean conditions for 2 years, with spectral reflectance measurements done at anthesis and milk-grain stage. Each point corresponds to the mean value of a genotype, experiment and growth stage. Adapted from Aparicio et al. (2002)

Another way to formulate the relationship between biomass and VI is to use the light use efficiency (ε) model (Kumar & Monteith, 1981) based on the fact that the growth rate of a crop canopy is almost proportional to the rate of interception of radiant energy. Thus, the crop dry weight of a crop canopy at a given moment (t) may be expressed as a function of the incident radiation (Io), the fraction of the radiation intercepted by the crop canopy (FPAR), and the radiation use efficiency (ε), as follows:

$$CDW = \int_0^t Io \times FPAR(LAI) \times \varepsilon \ dt \tag{2}$$

Small increases in biomass in a small period (expressed as days or thermal units) may then be calculated as a function of LAI from the derivative of Equation (2)

$$\frac{\delta CDW}{\delta t} = Io \times FPAR(LAI) \times \varepsilon \tag{3}$$

The incident radiation (Io) may be obtained from meteorological stations or, alternatively, it can be estimated from air temperatures (Allen et al., 1998). FPAR(LAI) may be calculated from vegetation indices on the basis of the linear relationship existing between vegetation indices and the FPAR of green canopies (Daughtry et al., 1992), and particularly between NDVI and FPAR (Bastiaansen & Ali, 2003). Radiation use efficiency (ε) is assumed to be constant during the crop growing season (Casanova et al., 1998). Values of radiation use efficiency have been summarized by Russell et al. (1989) for different crops and environmental conditions; moreover, ε-values can also be derived for a particular species and environment from the slope of the relationship between total aboveground biomass and absorbed PAR energy (Liu et al., 2004; Serrano et al., 2000).

An example of use of Kumar & Monteith's model to assess the pattern of changes in biomass from the LAI estimated from spectral reflectance measurements is shown in Fig. 7. In the example, LAI and CDW values were calculated from destructive samplings, and a comparison is made between the pattern of changes in CDW derived from the mathematical model and that assessed by destructive samplings (Fig. 7b). The model requires frequent reflectance measurements to accurately assess the pattern of changes in LAI over time (Christensen & Goudriaan, 1993), and proper estimations of the incident radiation.

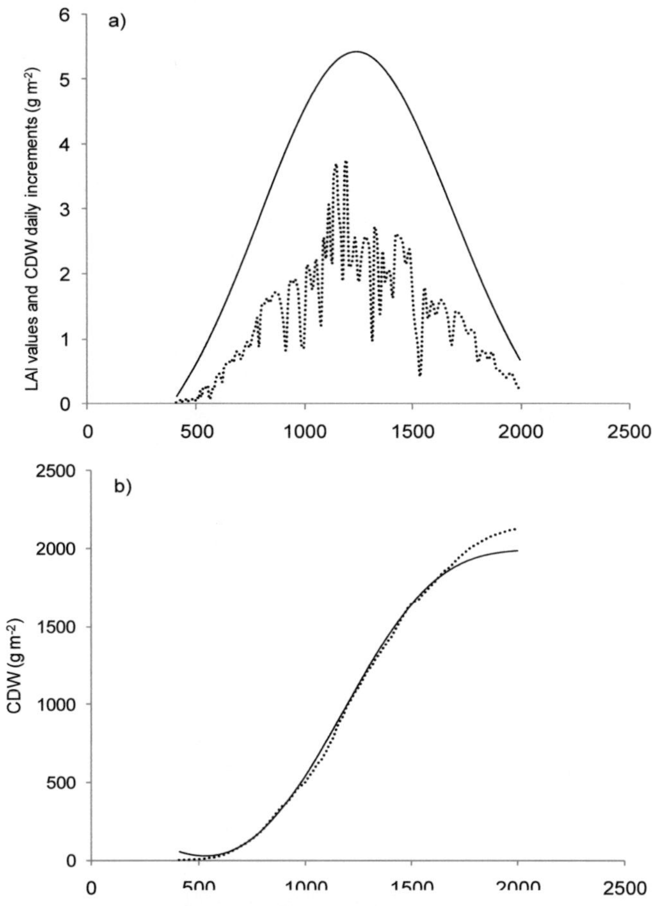

Fig. 7. Estimation of CDW from LAI data through the light use efficiency model (Kumar & Monteith, 1981). Fig. 7a. The solid line represents the mean pattern of changes in LAI of 25 durum wheat cultivars grown in 1998 under irrigated conditions, assessed through destructive biomass sampling (see Fig. 3). The discontinuous line shows daily increments in CDW, calculated from Eq. (3). Fig. 7b. The solid line shows the pattern of changes in CDW calculated from destructive sampling (see Fig.3), while the discontinuous line represents the CDW values calculated from the integration of the daily CDW increments represented in Fig. 7a

Studies conducted in bread wheat (Asrar et al., 1984; Serrano et al., 2000; Wiegand et al., 1992) and durum wheat (Aparicio et al., 2002) have demonstrated that SR increases linearly with increases in LAI, while NDVI shows a curvilinear response (Fig. 6). When the LAI of wheat canopies exceeds a certain level, the addition of more leaf layers to the canopy does not entail great changes in NDVI (Aparicio et al., 2000; Sellers, 1987), because the reflectance of solar radiation from the underlying soil surface or lower leaf layers is largely attenuated when the ground surface is completely obscured by the leaves (Carlson & Ripley, 1997). The consequence is that for LAI values higher than 3, NDVI becomes relatively insensitive to changes in canopy structure (Aparicio et al., 2002; Curran, 1983; Gamon et al., 1995; Serrano et al., 2000; Wiegand et al., 1992), which constitutes an important limitation for the use of NDVI to estimate LAI. In this context the linearity of the relationship between SR and LAI is not advantageous, because SR may be directly derived from NDVI as SR=(1+NDVI)/(1-NDVI), thus leading to similar statistical significances of both indices when LAI values are predicted (J.M. Chen & Cihlar, 1996). Because of the sensitivity of NDVI and SR to external factors —particularly the soil background at low LAI values—and the developments in the field of imaging spectrometry, a set of new vegetation indices have been developed in order to minimize the effect of disturbing elements in the capturing of the spectra (Baret & Guyot, 1991; Broge & Mortensen, 2002; Gilabert et al., 2002; Meza Diaz & Blackburn, 2003; Rondeaux et al., 1996).

In order to compare the suitability of the classical vegetation indices and the new ones mentioned in the literature as being appropriate for estimating growth traits in wheat and other cereals (P. Chen et al., 2009; Haboudane et al., 2004; Li et al., 2010a; Prasad et al., 2007), 83 hyperspectral vegetation indices were tested using durum wheat data from our own research. The indices were calculated from spectral reflectance measurements taken at different growth stages in 7 field experiments each involving 20-25 durum wheat genotypes, conducted under contrasting Mediterranean conditions for 2 years. Principal component analysis performed with the complete set of vegetation indices and LAI, GAI and CDW revealed that the vegetation indices most closely correlated with durum wheat growth indices were the 29 shown in Table 1. The correlation coefficients between growth traits and the selected indices are shown in Fig. 8. The results show that the majority of indices explained more than 50% of variation in LAI, GAI and CDW when determined at anthesis and milk grain stages, most correlation coefficients being statistically significant at $P<0.001$. However, the correlation coefficients were significant only for a small number of indices when measurements were taken at physiological maturity. From these results we can conclude that despite the large number of vegetation indices described to improve the appraisal of growth indices given by NDVI and SR, this objective was attained in only a few cases.

Fig. 8 shows that some indices changed from positive values determined at milk-grain to negative ones determined at physiological maturity, confirming that the utility of vegetation indices to assess growth traits decreases drastically when the crop starts to senesce (Aparicio et al., 2000). Young wheat plants normally absorb more photosynthetically active radiation and therefore reflect more NIR. As the plants progress in growth stage, new tissues are formed but older green tissues lose chlorophyll concentration, turning chlorotic and then necrotic. These senescent tissues increase reflectance at the visible wavelengths and decrease reflectance at the NIR wavelengths, causing a decrease in the values of the vegetation indices compared with that obtained at earlier growth stages. Aparicio et al. (2002) concluded that genotypic differences were maximized in durum wheat when growth traits were determined by spectral reflectance measurements taken at anthesis and milk-grain stage.

Identification	Definition	Equation	Reference
NDVI	Normalized difference vegetation index	$(R_{900}-R_{680})/(R_{900}+R_{680})$	Peñuelas et al. (1993)
SR	Simple ratio	R_{900}/R_{680}	Peñuelas & Filella (1998)
CI	Canopy index	R_{415}/R_{695}	Read et al. (2002)
CIG	Green chlorophyll index	$(R_{800}/R_{550})-1$	C.Y. Wu et al. (2010)
DD	Double difference index	$(R_{750}-R_{720})-(R_{700}-R_{670})$	Le Maire et al. (2004)
MCARI [705,750]	Modified chlorophyll absorption ratio index	$\left[(R_{750}-R_{705})-0.2\times(R_{750}-R_{550})\right]\times(\frac{R_{750}}{R_{705}})$	C.Y. Wu et al. (2008)
MCARI/OSAVI [705,750]	MCARI[705,750]/ OSAVI[705,750]	$\frac{\left[(R_{750}-R_{705})-0.2\times(R_{750}-R_{550})\right]\times(\frac{R_{750}}{R_{705}})}{(1+0.16)\times(R_{750}-R_{705})/(R_{750}+R_{705}+0.16)}$	C.Y. Wu et al. (2008)
MCARI2	Modified chlorophyll absorption ratio index 2	$\frac{1.5\ [2.5(R_{800}-R_{670})-1.3(R_{800}-R_{550})]}{\sqrt{(2R_{800}+1)2-(6R_{800}-5\sqrt{R_{670}})-0.5}}$	Haboudane et al. (2004)
mSR705	Modified simple ratio 705	$(R_{750}-R_{445})/(R_{705}-R_{445})$	Sims and Gamon (2002)
MTVI	Modified transformed vegetation index	$1.2\times[1.2\times(R_{800}-R_{550})-2.5\times(R_{670}-R_{550})]$	Haboudane et al. (2004)
ND705	Normalized difference vegetation index 705	$(R_{750}-R_{705})/(R_{750}+R_{705})$	Sims & Gamon (2002)
NDI1	Normalized difference index 1	$(R_{780}-R_{710})/(R_{780}-R_{680})$	Datt (1999)
NDI2	Normalized difference index 2	$(R_{850}-R_{710})/(R_{850}-R_{680})$	Datt (1999)
NDVI2	Normalized difference vegetation index 2	$(R_{800}-R_{600})/(R_{800}+R_{600})$	Ma et al. (1996)
NWI-1	Normalized water index-1	$(R_{970}-R_{900})/(R_{970}+R_{900})$	Prasad et al. (2007)
NWI-2	Normalized water index -2	$(R_{970}-R_{850})/(R_{970}+R_{850})$	Prasad et al. (2007)
NWI-3	Normalized water index -3	$(R_{970}-R_{920})/(R_{970}+R_{920})$	Prasad et al. (2007)
NWI-4	Normalized water index -4	$(R_{970}-R_{880})/(R_{970}+R_{880})$	Prasad et al. (2007)
OSAVI	Optimal soil adjusted vegetation index	$(1+0.16)\times(R_{800}-R_{670})/(R_{800}+R_{670}+0.16)$	Rondeaux et al. (1996)
OSAVI [705, 750]	Optimal soil adjusted vegetation index [705, 750]	$(1+0.16)\times(R_{750}-R_{705})/(R_{750}+R_{705}+0.16)$	C.Y. Wu et al. (2008)
PSNDc	Pigment specific normalized difference c	$(R_{800}-R_{470})/(R_{800}+R_{470})$	Blackburn (1998)
R_{780}/R_{740}	R_{780}/R_{740}	R_{780}/R_{740}	Mistele and Schmidhalter (2010)
RI	Ratio index	R_{810}/R_{560}	Xue et al. (2004)

RM	Red-edge model index	$(R_{750}/R_{720})-1$	Gitelson et al. (2005)
RR	Reflectance ratio	R_{740}/R_{720}	Vogelmann et al. (1993)
RTVI	Red-edge triangular vegetation index	$(100(R_{750}-R_{730})-10(R_{750}-R_{550}))\times\sqrt{(\frac{R_{700}}{R_{670}})}$	P. Chen et al. (2009)
SRPI	Simple ratio pigment index	R_{430}/R_{680}	Peñuelas et al. (1994) as read in Li et al. (2010a)
TVI	Transformed vegetation index	$0.5\times[120\times/R_{750}-R_{550})-200\times(R_{670}-R_{550})]$	Broge & Le Blanc (2000)
VI	Vegetation index	R_{750}/R_{550}	Gitelson et al. (1996)
WI	Water index	R_{900}/R_{970}	Peñuelas et al. (1997b)

Table 1. Definition of some of the spectral reflectance indices most closely associated with growth traits of small-grain cereals. R_n = reflectance at the wavelength (in nm) indicated by the subscript

Though a large number of studies demonstrate the utility of vegetation indices for assessing growth traits in small-grain cereals when there is a wide range of variability involved in the experimental data, the results indicate that the value of the indices decreases drastically when the range of variation caused by the environment or the crop canopies is low (Aparicio et al., 2002; Royo et al., 2003). In such cases the success of the indices at tracking changes in growth traits becomes much more experiment-dependent (Babar et al., 2006; Christensen & Goudriaan, 1993). Nevertheless, as stressed above, one of the practical applications of spectral reflectance may be its use as a routine tool for screening germplasm in breeding programs, when measurements are taken on a genotype basis, usually in one or a reduced number of experiments. Moreover, vegetation indices are more appropriate for assessing LAI than for estimating biomass (Aparicio et al., 2000, 2002; Serrano et al., 2000), particularly when measurements are taken with low variability backgrounds.

5. Field measurements of growth traits in individual plants

Biomass assessment of individual plants by conventional methodologies involves destructive sampling, which is inappropriate for studies aiming to monitor the growth of specific individuals during their growth cycle, or when the grain produced by the plant has to be harvested at ripening, as in breeding programs. In such cases growth traits such as dry weight per plant (W), green area per plant (GAP) and leaf area per plant (LAP) may be properly estimated through vegetation indices.

Since the devices commercially available at present only allow measurements at canopy level, spectral reflectance measurements of individual plants require some adaptation of common equipment to avoid background effects. In studies conducted with wheat by Casadesus et al. (2000) and with four cereal species by Álvaro et al. (2007), the plants were covered by a tube of reflecting walls provided by an artificial source of light (Fig. 5). In order to provide a homogeneous background, aluminum foil was placed around the base of each plant, covering the entire tube base. The spectroradiometer was fitted to a receptor for diffuse spectral irradiance, centered at the top of the tube. The spectra obtained were standardized with the spectrum previously sampled in the empty tube with the soil covered

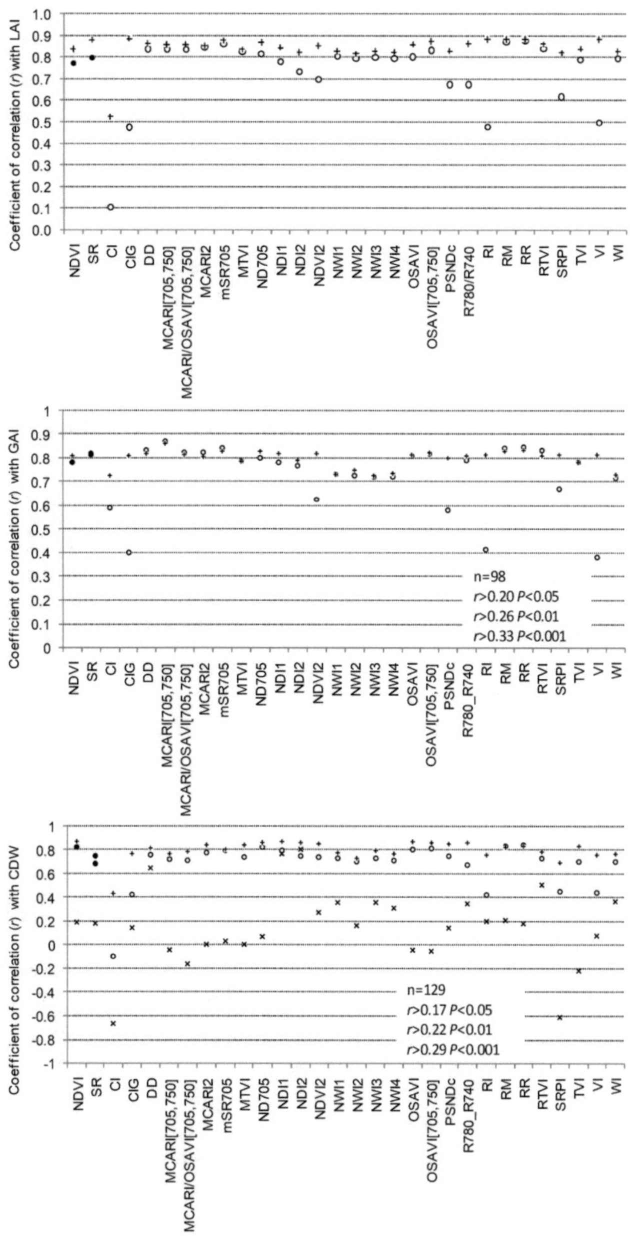

Fig. 8. Pearson correlation coefficients of some hyperspectral vegetation indices (see Table 1 for index definition)with the following durum wheat growth traits: a) leaf area index (LAI), b) green area index (GAI), and c) crop dry weight (CDW) considering pooled data of 7 field experiments involving 20-25 durum wheat genotypes, and conducted under contrasting Mediterranean conditions for 2 years. Destructive samples of biomass and reflectance measurements were taken at anthesis (○), milk-grain (+) and physiological maturity (x). Full symbols correspond to the classical vegetation indices, NDVI and SR. Unpublished data from Royo and Villegas

with a homogeneous white reflecting surface. This method allows measurements to be taken at any time of the day, regardless of the environmental conditions (sun light angle and intensity, weather conditions, etc.), while avoiding background disturbances such as soil color. In this case each spectral reflectance measurement takes 20-30 s and five scans per plant are sufficient to obtain reliable results.

Consistent associations of NDVI and SR with W (R^2=0.91, P<0.001), GAI (R^2=0.88-0.89, P<0.001) and LAP (R^2=0.66-0.69, P<0.001) measured on spaced plants (Álvaro et al., 2007) have been reported. The accuracy of reflectance measurements to detect differences between individual plants seems to be comparable to that obtained by destructive measurements of growth traits (Álvaro et al., 2007), so this methodology is a promising tool for assessing growth traits in spaced individual plants. However, the time needed to prepare the plants and to take measurements may constrain its extensive use.

6. Limitations and future challenges of using spectral reflectance field measurements for biomass assessment

Despite the possibilities that spectral reflectance measurements offer for monitoring growth traits in plots and individual plants (e.g. in breeding programs), their use until now has been very limited. One of the main reasons is that a wide range of variability must exist for the target growth traits within the experimental units to be detected by the apparatus (Royo et al., 2003). The strongest associations between growth traits and spectral reflectance indices have been found in studies in which a wide range of variability is induced by experimental treatments, such as rates of seed or nitrogen fertilizer, varying levels of water availability or soil salinity, or the combined analysis of data recorded at different plant stages. However, when the range of variation is low, particularly when the differences are only in the genetic background, and the predictive ability of vegetation indices is tested in specific environments and growth stages, the value of spectral reflectance measurements for estimating growth traits has proven to be much more limited (Aparicio et al., 2002; Royo et al., 2003). The fact that the pattern of changes in biomass is quite similar among modern wheat varieties (Villegas et al., 2001) may be an additional obstacle to the implementation of remote sensing techniques as a screening tool in breeding programs.

Another limitation to the extensive use of spectral reflectance measurements to track changes in biomass derives from the huge number of indices reported in the literature and their misleading use (Araus et al., 2009). In addition, the lack of equipment specially designed to take measurements at individual plant level restricts the use of spectral reflectance in breeding programs, where selection in early segregating generations involves the screening of thousands of individual plants or small plots, and only reliable, fast, and cheap screening tools may be helpful. Prediction models are not of general use and need to be developed for specific situations, such as in farmer's fields, where evidence indicates a decrease in the performance of classical and newly identified indices (Li et al., 2010b). Other great challenges are the development of functions to calculate sensor-specific spectral signal-to-noise ratios for a number of different conditions, which would allow the models to include the effects of sensor-related noise (Broge & Leblanc, 2000), and the development of new sensors more adapted to practical applications.

7. Conclusions

The use of spectral reflectance measurements for the assessment of growth traits in small-grain cereals offers several benefits. Their non-destructive nature allows repetitive measurements to be taken over time on the same plot or plant, so the grain produced on the measured plants is available at the end of their growth cycle. In addition, the method avoids the errors associated with destructive samplings of biomass, and is fairly quick. However, the use of canopy spectra for biomass assessment requires a thorough knowledge of the conditions of use and the constraints imposed by the measurement-related noise caused by the sensor system, the canopy structure, and the environment, which should be carefully taken into consideration in order to obtain reliable results.

8. Acknowledgements

This review was partially supported by Spanish projects CICYT AGL-2009-11187 and INIA RTA 2009-0085-00-00. Authors thank Dr. Nieves Aparicio and Dr. Fanny Álvaro for their valuable contribution to field experiments

9. References

Ahlrichs, J.S. & Bauer, M.E. (1983). Relation of agronomic and multispectral reflectance characteristics of spring wheat canopies. *Agronomy Journal*, Vol.75, No.6, (November-December 1983), pp. 987-993, ISSN 0002-1962

Allen, R.G.; Pereira, L.S.; Raes, D. & Smith, M. (1998). *Crop evapotranspiration. Guidelines for computing crop water requirements.* FAO Irrigation and drainage paper No. 56. FAO. ISBN 92-5-104219-5, Rome, Italy

Álvaro, F.; García del Moral, L.F. & Royo, C. (2007). Usefulness of remote sensing for the assessment of growth traits in individual cereal plants grown in the field. *International Journal of Remote Sensing*, Vol.28, No.11, (January 2007), pp. 2497-2512, ISSN 0143-1161

Álvaro, F.; Isidro, J.; Villegas, D.; García del Moral, L.F. & Royo, C. (2008a). Breeding effects on grain filling, biomass partitioning, and remobilization in Mediterranean durum wheat. *Agronomy Journal*, Vol.100, No.2 (March-April 2008), pp. 361-370, ISSN 0002-1962

Álvaro, F.; Royo, C.; García del Moral, L.F. & Villegas, D. (2008b). Grain filling and dry matter translocation responses to source-sink modifications in a historical series of durum wheat. *Crop Science*, Vol.48, No.4, (July-August 2008), pp. 1523-1531, ISSN 0011-183X

Aparicio, N.; Villegas, D.; Casadesús, J.; Araus, J.L. & Royo, C. (2000). Spectral vegetation indices as nondestructive tools for determining durum wheat yield. *Agronomy Journal*, Vol.92, No.1, (January-February 2000), pp. 83-91, ISSN 0002-1962

Aparicio, N.; Villegas, D.; Araus, J.L.; Casadesús, J. & Royo, C. (2002). Relationship between growth traits and spectral reflectance indices in durum wheat. *Crop Science*, Vol.42, No.5 (September-October 2002), pp. 1547-1555, ISSN 0011-183X

Aparicio, N.; Villegas, D.; Royo, C.; Casadesus, J. & Araus, J.L. (2004). Effect of sensor view angle on the assessment of agronomic traits by spectral reflectance measurements in durum wheat under contrasting Mediterranean conditions. *International Journal of Remote Sensing*, Vol.25, No.6, (March 2004), pp. 1131-1152, ISSN 0143-1161

Araus, J.L.; Casadesús, J. & Bort, J. (2001). Recent tools for the screening of physiological traits determining yield, In: *Application of physiology in wheat breeding*, M.P.

Reynolds, J.I. Ortiz-Monasterio & A. McNab (Eds.), pp. 59-77, CIMMYT, ISBN 970-648-077-3, Mexico D.F.

Araus, J.L.; Slafer, G.A.; Reynolds, M. & Royo, C. (2002). Plant Breeding and drought in C_3 cereals: What should we to breed for? *Annals of Botany*, Vol.89, Special Issue, (June 2002) pp. 925-940, ISSN 0305-7364.

Araus, J.L.; Slafer, G.A.; Reynolds, M.P. & Royo, C. (2009). Breeding for quantitative variables. Part 5: Breeding for yield potential. In: *Plant Breeding and Farmer Participation*, S. Ceccarelli, E.P. Guimaraes & E. Weltzien (Eds.), pp. 449-478, FAO, ISBN 978-92-5-106382-8, Rome, Italy

Asrar, G.; Fuchs, M.; Kanemasu, E.T. & Hatfield, J.L. (1984). Estimating absorbed photosynthetic radiation and leaf area index from spectral reflectance in wheat. *Agronomy Journal*, Vol.76, No.2, (March-April 1984), pp. 300-306, ISSN 0002-1962

Austin, R.B. (1980). Physiological limitations to cereals yields and ways of reducing them by breeding. In: *Opportunities for increasing crop yields*. R.G. Hurd, P.V. Biscoe & C. Dennis (Eds.), pp. 3-19, Association of Applied Biologists, Pitman Publishing, ISBN 0-273-08481-X, Boston, USA

Austin, R.B.; Bingham, J.; Blackwell, R.D.; Evans, L.T.; Ford, M.A.; Morgan, C.L. & Taylor, M. (1980). Genetic improvements in winter wheat yields since 1900 and associated physiological changes. *The journal of Agricultural Science*, Vol. 94, No.3, (June 1980), pp. 675-689, ISSN 0021-8596

Austin, R.B.; Ford, M.A. & Morgan, C.L. (1989). Genetic improvement in the yield of winter wheat: a further evaluation. *The Journal of Agricultural Science*, Vol.112, No.3 (June 1989), pp. 295-301, ISSN 0021-8596

Babar, M.A.; Reynolds, M.P.; van Ginkel, M.; Klatt, A.R.; Raun, W.R. & Stone, M.L. (2006). Spectral reflectance to estimate genetic variation for in-season biomass, leaf chlorophyll, and canopy temperature in wheat. *Crop Science*, Vol. 46, No. 3, (May-June 2006), pp. 1046-4057, ISSN 0011-183X

Baret, F. & Guyot, G. (1991). Potentials and limits of vegetation indices for LAI and APAR assessment. *Remote Sensing of Environment*, Vol.35, No.2-3, (February-March 1991), pp. 161-173, ISSN 0034-4257

Baret, F.; Jacquemoud, S.; Guyot, G. & Leprieur, C. (1992). Modeled analysis of the biophysical nature of spectral shifts and comparison with information content of broad bands. *Remote Sensing of Environment*, Vol.41, No.2-3, (August-September 1992), pp. 133-142, ISSN 0034-4257

Bartlett, D.S.; Whiting, G.J. & Hartman, J.M. (1990). Use of vegetation indices to estimate intercepted solar radiation and net carbon dioxide exchange of a grass canopy. *Remote Sensing of Environment*, Vol.30, No.2, (November 1989), pp. 115-128, ISSN 0034-4257

Bastiaansen, W.G. & Ali, S. (2003). A new crop yield forecasting model based on satellite measurements applied across the Indus Basin, Pakistan. *Agriculture, Ecosystems and Environment*, Vol.94, No.3, (March 2003), pp. 321-340, ISSN 0167-8809

Bellairs, S.M.; Turner, N.C.; Hick, P.T. & Smith, R.C.G. (1996). Plant and soil influences on estimating biomass of wheat in plant breeding plots using field spectral radiometers. *Australian Journal of Agricultural Research*, Vol. 47, No.7, pp. 1017-1034, ISSN 0004-9409

Blackburn, G.A. (1998). Quantifying chlorophylls and carotenoids at leaf and canopy scales: An evaluation of some hyper-spectral approaches. *Remote Sensing of Environment*, Vol.66, No.3, (December 1998), pp.273-285, ISSN 0034-4257

Broge, N.H. & Mortensen, J.V. (2002). Deriving green crop area index and canopy chlorophyll density of winter wheat from spectral reflectance data. Remote Sensing of Environment Vol.81, No.1, (July 2002), pp. 45-57, ISSN 0034-4257

Broge, N.H. & Leblanc, E. (2000). Comparing prediction power and stability of broadband and hyperspectral vegetation indices for estimation of green leaf area index and canopy chlorophyll density. *Remote Sensing of Environment*, Vol.76, No.2, (May 2001), pp. 156-172, ISSN 0034-4257

Carlson, T.N. & Ripley, D.A. (1997). On the relation between NDVI, fractional vegetation cover, and leaf area index. *Remote Sensing of Environment*, Vol.62, No. 3, (December 1997), pp. 241–252, ISSN 0034-4257

Casadesus, J.; Tambussi, E.; Royo, C. & Araus, J.L. (2000). Growth assessment of individual plants by an adapted remote sensing technique. In: *Durum wheat improvement in the Mediterranean region: New challenges*, C. Royo; M.M. Nachit; N. Di Fonzo, and J.L. Araus (Eds.), Vol.40, pp. 129-132. Options Méditerranéennes, Series A, Zaragoza, Spain

Casanova, D.; Epema, G.F. & Goudriaan, J. (1998). Monitoring rice reflectance at field level for estimating biomass and LAI. *Field Crops Research*, Vol.55, No.1-2, (January 1998), pp. 83-92, ISSN 0378-4290

Chen, J.M. & Cihlar, J. (1996). Retrieving Leaf Area Index of boreal conifer forests using Landsat TM images. *Remote Sensing of Environment*, Vol. 55, No. 2, (February 1996), pp. 153-162, ISSN 0034-4257

Chen, P.; Tremblay, N.; Wang, J. & Vigneault, P. (2009). New spectral index for corn green biomass estimation. In: *Proceedings of the 2nd International Conference on Earth Observation for Global Changes (EOGC)*, Q. Tong & D. Li (Eds.), pp. 507-514, Sichuan, China, May 25-29 2009

Christensen, S. & Goudriaan, J. (1993). Deriving light interception and biomass from spectral reflectance ratio. *Remote Sensing of Environment*, Vol. 43, No.1, (January 1993), pp. 87-95, ISSN 0034-4257

Collins, W. (1978). Remote sensing of crop type and maturity. *Photogrammetric Engineering and Remote Sensing*, Vol.44, No.1 (January 1978), pp 43-55, ISSN 0099-1112

Colwell, J.E. (1974). Vegetation canopy reflectance. *Remote Sensing of Environment*, Vol.3, No.3, pp. 175-183, ISSN 0034-4257

Curran, P. J. (1983). Multispectral remote sensing for the estimation of green leaf area index. *Philosophical Transactions of the Royal Society of London Series A*, Vol.309, No.1508, pp. 257–270, IDS RA253

Darvishzadeh, R.; Atzberger, C.; Skidmore, A.K. & Abkar, A.A. (2009). Leaf Area Index derivation from hyperspectral vegetation indices and the red edge position. *International Journal of Remote Sensing*, Vol.30, No.23, pp. 6199-6218, ISSN 0143-1161

Das, D.K.; Mishra, K.K. & Kalra, N. (1993). Assessing growth and yield of wheat using remotely-sensed canopy temperature and spectral indices. *International Journal of Remote Sensing*, Vol.14, No.17, (November 1993), pp. 3081-3092, ISSN 0143-1161

Datt, B. (1999). A new reflectance index for remote sensing of chlorophyll content in higher plants: Tests using eucaliptus leaves. *Journal of Plant Physiology*, Vol.154, No.1, (January 1999), pp. 30-36, ISSN 0176-1617

Daughtry, C.S.T.; Gallo, K.P.; Goward, S.N.; Prince, S.D. & Kustas, W.P. (1992). Spectral estimates of absorbed radiation and phytomass production in corn and soybean canopies. *Remote Sensing of Environment*, Vol.39, No.2, (February 1992), pp.141-152, ISSN 0034-4257

Demetriades-Shah, T.H.; Steven, M.D. & Clark, J.A. (1990). High resolution derivative spectra in remote sensing. *Remote Sensing of Environment*, Vol.33, No.1, (July 1990), pp. 55-64, ISSN 0034-4257

Eaton, F.D. & Dirmhirn, I. (1979). Reflected irradiance indicatrices of natural surfaces and their effect on albedo. *Applied Optics*, Vol.18, No.7, pp. 994-1008, ISSN 0003-6935

Elliott, G.A. & Regan, K.L. (1993). Use of reflectance measurements to estimate early cereal biomass production on sand plain soils. *Australian Journal of Experimental Agriculture*, Vol. 33, No.2, pp. 179-183, ISSN 0816-1089

FAOSTAT (2008). FAOSTAT © FAO Statisics Division.

Fernández, S.; Vidal, D.; Simón, E. & Solé-Sugranes, L. (1994). Radiometric characteristics of *Triticum aestivum* cv. Astral under water and nitrogen stress. *International Journal of Remote Sensing*, Vol.15, No.9 (June 1994), pp. 1867-1884, ISSN 0143-1161

Field, C.B.; Gamon, J.A. & Peñuelas, J. (1994). Remote sensing of terrestrial photosynthesis, In: *Ecophysiology of photosynthesis*, E.D. Schulze & M.M. Caldwell (Eds.), pp.511-528, Springer-Verlag, ISBN 0387585710, Berlin

Filella, I. & Peñuelas, J. (1994). The red edge position and shape as indicators of plant chlorophyll content, biomass and hydric status. *International Journal of Remote Sensing*, Vol.15, No.7, (May 1994), pp. 1459-1470, ISSN 0143-1161

Gamon, J.A.; Field, C.B.; Goulden, M.L.; Griffin, K.L.; Hartley, A.E.; Joel, G., Peñuelas, J. & Valentini, R. (1995). Relationships between NDVI, canopy structure and photosynthesis in three Californian vegetation types. *Ecological Applications, Vol.*5, No.1 (February 1995), pp. 28-41, ISSN 1051-0761

Gilabert, M.A.; González-Piqueras, J.; García-Haro, F.J. & Meliá, J. (2002). A generalized soil-adjusted vegetation index. *Remote Sensing of Environment*, Vol.82, No.2 (October 2002), pp. 303-310, ISSN 0034-4257

Gitelson, A.; Kaufman, Y. & Merzlyak, M. (1996). Use of a green channel in remote sensing of global vegetation from EOS-MODIS. *Remote Sensing of Environment*, Vol.58, No.3, (December 1996), pp. 289-298, ISSN 0034-4257

Gitelson, A.A.; Viña, A.; Ciganda, V.; Rundquist, D.C. & Arkebauer, T.J. (2005). Remote estimation of canopy chlorophyll content in crops. *Geophysical Research Letters*, Vol.32, No.8, (April 2005), Art.No. L08403, ISSN 0094-8276

Goudriaan J. & Monteith J.L. (1990). A mathematical function for crop growth based on light interception and leaf area expansion. *Annals of Botany*, Vol.66, No.6 (December 1990), pp. 695-701, ISSN 0305-7364

Haboudane, D.; Miller, J.R.; Pattey, E.; Zarco-Tejada, P.J. & Strachan, I.B. (2004). Hyperspectral vegetation indices and novel algorithms for predicting green LAI of crop canopies: Modeling and validation in the context of precision agriculture. *Remote Sensing of Environment*, Vol.90, No.3, (April 2004), pp. 337-352, ISSN 0034-4257

Hallik, L.; Kull, O.; Nilson, T. & Peñuelas, J. (2009). Spectral reflectance of multispecies herbaceous and moss canopies in the boreal forest understory and open field. *Canadian Journal of Remote Sensing*, Vol.35, No.5, (October 2009), pp. 474-485, ISSN 1712-7971

Hay, R. & Walker, K.M. (1989). *An introduction to the physiology of crop yield*, Addison Wesley Longman, ISBN 0-582-40808-3, Harlow, UK

Hay, R.K.M. (1995). Harvest index: a review of its use in plant breeding and crop physiology. *Annals of Applied Biology*, Vol.126, No.1, (February 1995), pp. 197-216, ISSN 0003-4746

Huete, A.R.; Jackson, R.D. & Post, D.F. (1985). Spectral response of a plant canopy with different soil backgrounds. *Remote Sensing of Environment*, Vol. 17, No. 1, (February 1985), pp. 37-53, ISSN 0034-4257

Huete, A. R. (1987). Soil-dependent spectral response in a developing plant canopy. *Agronomy Journal*, Vol.11, No.1 (January-February 1987), pp. 61-68, ISSN 0002-1962

Jackson, R.D. (1983). Spectral indices in n-space. *Remote Sensing of Environment*, Vol.13, No.5 (1983), pp. 409–421, ISSN 0034-4257

Jackson, R.D. & Pinter, P.J.,Jr. (1986). Spectral response of architecturally different wheat canopies. *Remote Sensing of Environment*, Vol.20, No.1, (August 1986), pp. 43-56, ISSN 0034-4257

Jackson, R.D.; Clarke, T.R. & Moran, M.S. (1992). Bidirectional calibration results for 11 spectralon and 16 $BaSO_4$ reference reflectance panels. *Remote Sensing of Environment*, Vol.40, No.3, (June 1992), pp. 231-239, ISSN 0034-4257

Kimes, D.S.; Kirchner, J.A. & Newcomb, W.W. (1983). Spectral radiance errors in remote-sensing ground studies due to nearby objects. *Applied Optics*, Vol.22, No.1, (January 1983), pp.8-10, ISSN 0003-6935

Knipling, E.B. (1970). Physical and physiological basis for the reflectance of visible and near-infrared radiation from vegetation. *Remote Sensing of Environment*, Vol.1, No.3, (Summer 1970), pp. 155-159, ISSN 0034-4257

Kollenkark, J.C.; Vanderbilt, V.C; Daughtry, C.S.T. & Bauer, M.E. (1982). Influence of solar illumination angle on soybean canopy reflectance. *Applied Optics*, Vol.21, No.7, (April 1982), pp. 1179-1184, ISSN 0003-6935

Kumar, M. & Monteith, J.L. (1981). Remote sensing of crop growth, In: *Plants and the Daylight Spectrum*, H.G. Smith (Ed.), pp. 133-144, Academic Press, ISBN-10: 0126509808, London

Le Maire, G.; François, C. & Dufrêne, E. (2004). Towards universal broad leaf chlorophyll indices using PROSPECT simulated database and hyperspectral reflectance measurements. *Remote Sensing of Environment*, Vol. 89, No.1, (January 2004), pp. 1-28, ISSN 0034-4257

Li, F.; Miao, Y.X.; Chen, X.P.; Zhang, H.L.; Jia, L.L. & Bareth, G. (2010a). Evaluating hyperspectral vegetation indices for estimating nitrogen concentration of winter wheat at different growth stages. *Precision Agriculture*, Vol.11, No.4, (August 2010), pp. 335-357 ISSN 1385-2256

Li, F.; Miao, Y.X.; Chen, X.P.; Zhang, H.L.; Jia, L.L. & Bareth, G. (2010b). Estimating winter wheat biomass and nitrogen status using an active crop sensor. *Intelligent automation and soft computing*, Vol.16, No.6 (Special Issue), pp. 1221-1230, ISSN 1079-8587

Liu, J.; Miller, J.R.; Pattey, E.; Haboudane, D.; Strachan, I.B. & Hinther, M. (2004). Monitoring crop biomass accumulation using multi-temporal hyperspectral remote sensing data. *IGARSS 2004: IEEE International Geoscience and Remote Sensing Symposium Proceedings*, ISBN 0-7803-8742-2, Anchorage, Alaska, September 2004

Lord, D.; Desjardins, R.L. & Dube, P.A. (1985). Influence of wind on crop canopy reflectance measurements. *Remote Sensing of Environment*, Vol.18, No.2, (October 1985), pp. 113-123, ISSN 0034-4257

Ma, B.L.; Dwyer, L.M.; Costa, C.; Cober, E.R. & Morrison, M.J. (2001). Early prediction of soybean yield from canopy reflectance measurements. *Agronomy Journal*, Vol.93, No.6, (November-December 2001), pp. 1227-1234, ISSN 0002-1962

Ma, B.L.; Morrison, M.J. & Dwyer, M.L. (1996). Canopy light reflectance and field greenness to assess nitrogen fertilization and yield of maize. *Agronomy Journal,* Vol.88, No.6, (November-December 1996), pp.915-920, ISSN 0002-1962

Mauser, W. & Bach, H. (1995). Imaging spectroscopy in hydrology and agriculture - determination of model parameters, In: *Imaging spectrometry - a tool for environmental observations*, J. Hill & J. Mégier (Eds.), pp. 261-283, Kluwer Academic Publishing, ISBN 0-7923-2965-1, Dordrecht, The Netherlands

Meza Díaz, B. & Blackburn, G.A. (2003). Remote sensing of mangrove biophysical properties: evidence from a laboratory simulation of the possible effects of background variation on spectral vegetation indices. *International Journal of Remote Sensing*, Vol.24, No.1, (January 2003), pp. 53-73, ISSN 0143-1161

Mistele, B. & Schmidhalter, U. (2010). Tractor-based quadrilateral spectral reflectance measurements to detect biomass and total nitrogen in winter wheat. *Agronomy Journal*, Vol.102 No.2, (March-April 2010), pp. 499-506, ISSN 0002-1962

Palta, J.A.; Kobata, T.; Fillery, I.R. & Turner, N.C. (1994). Remobilization of carbon and nitrogen in wheat as influenced by postanthesis water deficits. *Crop Science*, Vol. 34, No.1 (January-February 1994), pp.118–124, ISSN 0011-183X

Papakosta, D.K. & Gagianas, A.A. (1991). Nitrogen and dry matter accumulation, remobilization, and losses for Mediterranean wheat during grain filling. *Agronomy Journal* , Vol.83 No.5 (September-October 1991, pp. 864-870, ISSN 0002-1962

Peñuelas, J.; Gamon, J.A.; Griffin, K.L. & Field, C.B. (1993). Assessing community type, plant biomass, pigment composition, and photosynthetic efficiency of aquatic vegetation from spectral reflectance. *Remote Sensing of Environment*, Vol.46, No.2, (November 1993), pp.110-118 ISSN 0034-4257

Peñuelas, J.; Isla, R.; Filella, I. & Araus, J.L. (1997a). Visible and near-infrared reflectance assessment of salinity effects on barley. *Crop Science,* Vol.37, No.1 (January-February 1997), pp. 198-202, ISSN 0011-183X

Peñuelas, J.; Piñol, J.; Ogaya, R. & Filella, I. (1997b). Estimation of plant water concentration by the reflectance water index WI (R900/R970). *International Journal of Remote Sensing*, Vol.18, No.13, (September 1997), pp. 2869-2875, ISSN 0143-1161

Peñuelas, J. & Filella, I. (1998). Visible and near-infrared reflectance techniques for diagnosing plant physiological status. *Trends in Plant Science*, Vol.3, No.4, (April 1998), pp. 151-156, ISSN 1360-1385

Pfeiffer, W.H.; Sayre, K.D. & Reynolds, M.P. (2000). Enhancing genetic grain yield potential and yield stability in durum wheat. In: *Durum Wheat Improvement in the Mediterranean Region: New Challenges*. Royo, C., Nachit, M.M., Di Fonzo, N. &

Araus, J.L. (Eds.) *Options Mediterranéennes* Vol.40, pp. 83-93, ISBN 2-85352-212-1, Zaragoza, Spain, 12-14 April 2000

Prasad, B.; Carver, B.F.; Stone, M.L.; Babar, M.A.; Raun, W.R. & Klatt, A.R. (2007). Potential use of spectral reflectance indices as a selection tool for grain yield in winter wheat under Great Plains conditions. *Crop Science,* Vol.47 No.4 (July-August 2007), pp. 1426-1440, ISSN 0011-183X

Price, J.C. & Bausch, W.C. (1995). Leaf-area index estimation from visible and near-infrared reflectance data. *Remote Sensing of Environment,* Vol.52 No.1, (April 1995), pp. 55-65, ISSN 0034-4257

Ramos, J.M.; García del Moral, L.F. & Recalde, L. (1985). Vegetative growth of winter barley in relation to environmental conditions and grain yield. *The Journal of Agricultural Science* Vol.104 No.2 (April 1985), pp. 413-419, ISSN 0021-8596

Ranson, K.J.; Daughtry, C.S.T.; Biehl, L.L. & Bauer, M.E. (1985). Sun-view angle effects on reflectance factors of corn canopies. *Remote Sensing of Environment,* Vol.18, No.2, (October 1985), pp. 147–161, ISSN 0034-4257

Read, J.J.; Tarpley, J.M.M. & Reddy, K.R. (2002). Narrow waveband reflectance ratios for remote estimation of nitrogen status in cotton. *Journal of Environmental Quality,* Vol.31, No.5, (September-October 2002), pp. 1442-1452, ISSN 0047-2425

Reynolds, M.P.; Pellegrineschi, A. & Skovmand, B. (2005). Sink-limitation to yield and biomass: a summary of some investigations in spring wheat. *Annals of Applied Biology* Vol.146, No.1, (January 2005), pp. 39-49, ISSN 0003-4746

Reynolds, M.P.; Ortiz-Monasterio, J.L. & McNab, A. (Eds.) (2001). *Application of physiology in wheat breeding,* CIMMYT, ISBN 970-648-077-3, México D.F.

Reynolds, M.P.; Rajaram, S. & Sayre, K.D. (1999). Physiological and genetic changes of irrigated wheat in the post-green revolution period and approaches for meeting projected global demand. *Crop Science,* Vol.39, No.6, (November-December 1999), pp. 1611-1621, ISSN 0011-183X

Richards, F.J. (1959). A flexible growth function for empirical use. *Journal of Experimental Botany,* Vol.10, No.2 (June 1959), pp. 290-301, ISSN 0022-0957

Richards, R.A. (2000). Selectable traits to increase crop photosynthesis and yield of grain crops. *Journal of Experimental Botany,* Vol.51, Suppl. 1, (February 2000), pp. 447-458, ISSN 0022-0957

Rondeaux, G.; Steven, M. & Baret, F. (1996). Optimization of soil-adjusted vegetation indices. *Remote Sensing of Environment,* Vol.55, No.2, (February 1996), pp. 95-107, ISSN 0034-4257.

Royo, C. & Tribó, F. (1997). Triticale and barley for grain and for dual-purpose (forage + grain) in a Mediterranean-type environment. I. Growth analyses. *Australian Journal of Agricultural Research,* Vol.48, No.4, (1997) pp. 411-421, ISSN 0004-9409

Royo, C. & Blanco, R. (1999). Growth analysis of five spring and five winter triticale genotypes. *Agronomy Journal,* Vol.91, No.2, (March-April 1999), pp. 305-311, ISSN 0002-1962

Royo, C.; Aparicio, N.; Villegas, D.; Casadesus, J.; Monneveux, P. & Araus, J.L. (2003). Usefulness of spectral reflectance indices as durum wheat yield predictors under contrasting Mediterranean environments. *International Journal of Remote Sensing,* Vol.24, No.22 (November 2003), pp. 4403-4419, ISSN 0143-1161

Royo, C.; Aparicio, N.; Blanco, R. & Villegas, D. (2004). Leaf and green area development of durum wheat genotypes grown under Mediterranean conditions. *European Journal of Agronomy,* Vol.20, No.4, (April 2004), pp.419-430, ISSN 1161-0301

Royo, C.; García del Moral, L.F.; Slafer, G.; Nachit, M.M. & Araus, J.L. (2005). Selection tools for improving yield-associated physiological traits, In: *Durum Wheat Breeding: Current Approaches and Future Strategies*. Royo, C.; Nachit, M.N.; Di Fonzo, N.; Araus, J.L.; Pfeiffer, W.H. & Slafer, G.A. (Eds), pp. 563-598, Food Products Press, ISBN 1-56022-333-2, New York

Royo, C.; Álvaro, F.; Martos, V.; Ramdani, A.; Isidro, J.; Villegas, D. & García del Moral, L.F. (2007). Genetic changes in durum wheat yield components and associated traits in Italian and Spanish varieties during the 20th century. *Euphytica,* Vol.155, No.1-2 (May 2007), pp. 259-270, ISSN 0014-2336

Russell, G.; Jarvis, P.G. & Monteith, J.L. (1989). Absorption of radiation by canopies and stand growth. In: *Plant canopies: their growth, form and function,* G. Russell; J. Marshall & P.G. Davis (Eds.), pp. 21-40, Cambridge University Press, ISBN 0521328381, Cambridge, UK

Sayre, K.D., Rajaram, S. & Fischer, R.A. (1997). Yield potential progress in short bread wheats in Northwest Mexico. *Crop Science,* Vol.37, No.1, (January-February 1997), pp. 36-42, ISSN 0011-183X

Sellers, P.J. (1987). Canopy reflectance, photosynthesis, and transpiration. II. The role of biophysics in the linearity of their interdependence. *Remote Sensing of Environment,* Vol.21, No.2 (March 1987), pp. 143-183, ISSN 0034-4257

Serrano, L.; Filella, I. & Peñuelas, J. (2000). Remote sensing of biomass and yield of winter wheat under different nitrogen supplies. *Crop Science,* Vol.40, No.3 (May-June 2000), pp. 723-731, ISSN 0011-183X

Shearman, V.J.; Sylvester-Bradley, R.; Scott, R. K. & Foulkes, M.J. (2005). Physiological processes associated with wheat yield progress in the UK. *Crop Science,* Vol.45, No.1 (January-February 2005), pp.175-185, ISSN 0011-183X

Shepherd, K. D.; Cooper, P.M.J.; Allan, A.Y.; Drennan, D.S.H. & Keatinge, J.D.H. (1987). Growth, water use and yield of barley in Mediterranean-type environments. *The Journal of Agricultural Science,* Vol.108, No.2 (April 1987), pp. 365-378, ISSN 0021-8596

Siddique, K.H.; Kirby, E.J.M. & Perry, M.W. (1989). Ear:stem ratio in old and modern wheat varieties: relationship with improvement in number of grains per ear and yield. *Field Crops Research,* Vol.21, No.1 (June 1989), pp. 59-78, ISSN 0378-4290

Sims, D.A. & Gamon, J.A. (2002). Relationships between leaf pigment content and spectral reflectance across a wide range of species, leaf structures and developmental stages. *Remote Sensing of Environment,* Vol.81, No.2-3, (August 2002), pp. 337-354, ISSN 0034-4257

Singh, R.P.; Huerta-Espino, J.; Rajaram, S. & Crossa, J. (1998). Agronomic effects from chromosome translocations 7DL.7Ag and 1BL.1RS in spring wheat. *Crop Science,* Vol.38, No.1 (January-February 1998), pp. 27-33, ISSN 0011-183X

Smith, J.A.; Oliver, R.E. & Berry, J.K. (1975). *Signature extension for sun angle.* Final report NAS-9-14467, Department of Earth Resources, Fort Collins, CO

Smith, R.C.G.; Wallace, J.F.; Hick, P.T.; Gilmour, R.F.; Belford, R.K.; Portmann, P.A.; Regan, K.L. & Turner, N.C. (1993). Potential of using field spectroscopy during early growth for ranking biomass in cereal breeding trials. *Australian Journal of Agricultural Research*, Vol.44, No.8, pp. 1713-1730, ISSN 0004-9409

Steven, M.D.; Malthus, T.J.; Demetriades-Shah, T.H.; Daanson, F.M. & Clark, J.A. (1990). High-spectral resolution indices for crop stress, pp. 209-227. In: *Applications of remote sensing in agriculture*, M.D. Steven & and J.A. Clark (Eds.), Butterworths, ISBN 0-408-04767-4, Sevenoaks, Kent, UK

Tanno, H.; Komaki, Y. & Gotoh, K. (1985). The effectiveness of selection based on harvest index in spring wheat. *Memoirs of the Faculty of Agriculture*, Hokkaido University, Japan, Vol.14, pp. 352-356, ISSN 0367-5726

Tucker, C.J. (1979). Red and photographic infrared linear combinations for monitoring vegetation. *Remote Sensing of Environment* Vol.8, No.2, pp. 127-150, ISSN 0034-4257

Turner, N. C. (1997). Further progress in crop water relations. *Advances in Agronomy* Vol.58, pp. 293-338, ISSN 0065-2113

Turner, N.C. (1982). The role of shoot characteristics in drought resistance in crop plants, In: *Drought resistance in crops with emphasis on rice*, pp. 115-134 IRRI, ISBN 971-104-078-6, Los Baños, The Philippines

Vaesen, K.; Gilliams, S.; Nackaerts, K. & Coppin, P. (2001). Ground-measured spectral signatures as indicators of ground cover and leaf area index: the case of paddy rice. *Field Crops Research*, Vol.69, No.1, (January 2001), pp. 13-25, ISSN 0378-4290

Van den Boogaard, R., Veneklaas, E. J., & Lambers, H. (1996). The association of biomass allocation with growth and water use efficiency of two *Triticum aestivum* cultivars. *Australian Journal of Plant Physiology*, Vol.23, No.6, pp. 751-761, ISSN 0310-7841

Van Leeuwen, W.J.D. & Huete, A.R., (1996). Effects of standing litter on the biophysical interpretation of plant canopies with spectral indices. *Remote Sensing of Environment*, Vol.55, No.2 (February 1996), pp. 123-138, ISSN 0034-4257

Verhoef, W. & Bunnik, N.J.J. (1981). Influence of crop geometry on multispectral reflectance determined by the use of canopy reflectance models. In: *Photon-vegetation interactions: Applications in optical remote sensing and plant ecology*, Ross, J. & Myneni, R.B. (Eds.), pp.191-228, Springer-Verlag, ISBN 3540521089, Berlin

Villegas, D.; Aparicio, N.; Blanco, R. & Royo, C. (2001). Biomass accumulation and main stem elongation of durum wheat grown under Mediterranean conditions. *Annals of Botany*, Vol.88, No.4 (October 2001), pp. 617-627, ISSN 0305-7364

Vogelmann, J.E.; Rock, B.N. & Moss, D.M. (1993). Red edge spectral measurements from sugar maple leaves. *International Journal of Remote Sensing*, Vol.14, No.8, (May 1993), pp.1563-1575, ISSN 0143-1161

Waddington, S. R.; Ransom J. K., Osmazai, M. & Saunders, D.A. (1986). Improvement in the yield potential of bread wheat adapted to northwest Mexico. *Crop Science*, Vol.26, No.4, (July-August 1986), pp.698-703, ISSN 0011-183X

Waddington, S. R.; Osmanzai, M.; Yoshida, S. and J.K. Ranson. (1987). The yield of durum wheats released in Mexico between 1960 and 1984. *The Journal of Agricultural Science*, Vol.108, No.2, (April, 1987), pp. 469-477, ISSN 0021-8596

Wanjura, D.F. & Hatfield, J.L. (1987). Sensitivity of spectral vegetative indices to crop biomass. *Transactions of the ASAE*, Vol.30, No.3, (May-June, 1987), pp. 810-816, ISSN 0001-2351

Wardley, N.W. (1984). Vegetation index variability as a function of viewing geometry. *International Journal of Remote Sensing*, Vol.5, No.5, pp. 861-870, ISSN 0143-1161

Whan, B.R., Carlton, G.P., & Anderson, W.K. (1991). Potential for increasing early vigour and total biomass in spring wheat. I. Identification of genetic improvements. *Australian Journal of Agricultural Research*, Vol.42, No.3, pp. 347-361, ISSN 0004-9409

Wiegand, C.L. & Richardson, A.J. (1990). Use of spectral vegetation indices to infer leaf area, evapotranspiration and yield. II. Results. *Agronomy Journal,* Vol.82, No.3 (May-June, 1990), pp. 630-636, ISSN 0002-1962

Wiegand, C.L; Richardson, A.J.; Escobar, D.E. & Gerbermann, A.H. (1991). Vegetation indices in crop assessments. *Remote Sensing of Environment,* Vol.35, No.2-3, (February-March, 1991), pp. 105-119, ISSN 0034-4257

Wiegand, C.L.; Maas, S.J.; Aase, J.K.; Hatfield, J.L.; Pinter, P.J. Jr.; Jackson, R.D.; Kanemasu, E.T. & Lapitan, R.L. (1992). Multisite analyses of spectral-biophysical data for wheat. *Remote Sensing of Environment,* Vol.42, No.1, (October 1992), pp.1-21, ISSN 0034-4257

Wu, C.Y.; Niu, Z.; Tang, Q. & Huang, W.J. (2008). Estimating chlorophyll content from hyperspectral vegetation indices: Modeling and validation. *Agricultural and Forest Meteorology*, Vol.148, No.8-9, (July 2008), pp. 1230-1241, ISSN 0168-1923

Wu, C.Y.; Han, X.Z.; Ni, J.S.; Niu, Z. & Huang, W.J. (2010). Estimation of gross primary production in wheat from in situ measurements. *International Journal of Applied Earth Observation and Geoinformation,* Vol.12, No.3, (June 2010), pp.183-189, ISSN 0303-2434

Wu, J.H.; Yue, S.C.; Hou, P.; Meng, Q.F.; Cui, Z.L.; Li, F. & Chen, X.P. (2011). Monitoring winter wheat population dynamics using an active crop sensor. *Spectroscopy and Spectral Analysis*, Vol.31, No.2 (February 2011), pp. 535-538, ISSN 1000-0593

Xue, L.H.; Cao, W.X.; Luo, W.H.; Dai, T.B. & Zhu, Y. (2004). Monitoring leaf nitrogen status in rice with canopy spectral reflectance. *Agronomy Journal*, Vol.96, No.1, (January-February 2004), pp. 135-142, ISSN 0002-1962

Zadoks, J.C.; Chang, T.T. & Konzak, C.F. (1974). A decimal code for the growth stage of cereals. *Weed Research,* Vol.14, No. pp. 415-421, ISSN 0043-1737

Zhao, C.J.; Wang, J.H.; Huang, W.J. & Zhou, Q.F. (2010). Spectral indices sensitively discriminating wheat genotypes of different canopy architectures. *Precision Agriculture,* Vol.11, No.5, (October 2010), pp. 557-567, ISSN 1385-2256

3

SAR and Optical Images for Forest Biomass Estimation

Jalal Amini[1] and Josaphat Tetuko Sri Sumantyo[2]
[1]University of Tehran, Tehran,
[2]Chiba University, Chiba,
[1]Iran
[2]Japan

1. Introduction

Biomass, in general, includes the above-ground and below-ground living mass, such as trees, shrubs, vines, roots, and the dead mass of fine and coarse litter associated with the soil. Due to the difficulty in collecting field data of below-ground biomass, most previous researches on biomass estimation have been focused on the above-ground biomass (AGB). Different approaches have been applied for above ground biomass (AGB) estimation, where traditional techniques based on field measurement are the most accurate ways for collecting biomass data. A sufficient number of field measurements are a prerequisite for developing AGB estimation models and for evaluating its results. However, these approaches are often time consuming, labour intensive, and difficult to implement, especially in remote areas; also, they cannot provide the spatial distribution of biomass in large areas.

The advantages of remotely sensed data, such as in repetitively of data collection, a synoptic view, a digital format that allows fast processing of large quantities of data, and the high correlations between spectral bands and vegetation parameters, make it the primary source for large area AGB estimation, especially in areas of difficult access. Therefore, remote sensing-based AGB estimation has increasingly attracted scientific interest (Nelson et al., 1988; Sader et al., 1989; Franklin & Hiernaux, 1991; Steininger, 2000; Foody et al., 2003; Zheng et al., 2004; Lu, 2005). There are also other papers including (Dobson et al., 1992; Rignot et al., 1995; Rignot et al., 1994; Quinones & Hoekman, 2004) with SAR-based methods in above ground biomass estimation.

One strategy that can be used for AGB estimation is to combine synthetic aperture radar (SAR) image texture with optical images based on the classification analysis. Limitation on the used only optical data is the insensitivity of reflectance to the change in biomass and different stands. The use of the SAR data has the potential to overcome this limitation. But presence of the speckle in SAR data is also a barrier to the exploitation of image texture. Reducing the speckle would improve the discrimination among different land use types, and would make the textual classifiers more efficient in radar images. Ideally, the filters will reduce speckle without loss of information.

Many adaptive filters that preserve the radiometric and texture information have been developed for speckle reduction. Adaptive filters based upon the spatial domain are more widely used than frequency domain filters. The most frequently used adaptive filters include Lee, Frost, Lee-Sigma and Gamma-Map. The Lee filter is based on the multiplicative speckle model, and it can use local statistics to effectively preserve edges and features (Lee, 1980). The Frost filter is also based on the multiplicative speckle model and the local statistics, and it has similar performance to the Lee filter (Frost, 1982). The Lee-Sigma filter is a conceptually simple but effective alternative to the Lee filter, and Lee-Sigma is based on the sigma probability of the Gaussian distribution of image noise (Lee, 1980). Lopes (Lopes et al., 1990) developed the Gamma-Map filter, which is adapted from the Maximum a Posterior (MAP) filter (Kuan, 1987). Lee, Frost and Lee-Sigma filters assume a Gaussian distribution for the speckle noise, whereas Gamma-Map filter assumes a Gamma distribution of speckle (Lopes et al., 1990a; Lopes et al, 1990b). Modified versions of Gamma-Map have also been proposed (Nezry et al., 1991; Baraldi & Parmiggiani, 1995). Nezry (Nezry et al., 1991) combined the ratio edge detector and the Gamma-Map filter into the refined Gamma-Map algorithm. Baraldi and Parmiggiani (1995) proposed a refined Gamma-Map filter with improved geometrical adaptively. Walessa and Datcu combined the edge detection and region growing to segment the SAR image and then applied speckle filtering within each segment under stationary conditions. Dong et al. (2001) proposed an algorithm for synthetic aperture radar speckle reduction and edge sharpening. The proposed algorithm was functions of an adaptive-mean filter. Achim et al. (2006) proposed a novel adaptive de-speckling filter using the introduced heavy-tailed Rayleigh density function and derived a maximum a posterior (MAP) estimator for the radar cross section (RCS). The authors (Sumantyo & Amini, 2008) proposed a filter based on the least square method for speckle reduction in SAR images.

In this chapter, we develop a method for the forest biomass estimation based on (Amini & Sumantyo, 2009). Both SAR and optical images are used in a multilayer perceptron neural network (MLPNN) that relates them to the forest measurements on the ground. We use a speckle noise model that proposed by the authors in 2008 (Sumantyo & Amini, 2008) for reducing the speckle noise in the SAR image. Reducing the speckle would improve the discrimination among different land use types, and would make the textual classifiers more efficient in SAR images. We investigate both quantitative and qualitative criteria in speckle reduction and texture preservation to evaluate the performance of the proposed filter on the forest biomass estimation.

In summary, the objectives of this chapter are:

1. The efficiency of the de-speckling filter on forest biomass estimation and,
2. Improved the accuracy of forest biomass estimation when using both SAR images texture and optical images in a non-linear classifier method (MLPNN).

In the rest of the chapter, we will have a survey on de-speckling filters and then we will describe a method for the forest biomass estimation and we finally give the experimental results for the study area.

2. De-speckling filters on SAR images

Both the radiometric and texture aspects are less efficient for area discrimination in the presence of speckle. Reducing the speckle would improve the discrimination among different land use types, and would make the usual per-pixel or textual classifiers more efficient in radar images. Ideally, this supports that the filters reduce speckle without loss of information.

In the case of homogeneous areas (e.g. agricultural areas), the filters should preserve the backscattering coefficient values (the radiometric information) and edges between the different areas. In addition for texture areas (e.g. forest), the filter should preserve the spatial variability (textual information).

Many adaptive filters that preserve the radiometric and texture information have been developed for speckle reduction. Filtering techniques generally can be grouped into multi-look processing and posterior speckle filtering techniques. Multi-look processing is applied during image formation, and this procedure averages several statistically independent looks of the same scene to reduce speckle (Porcello et al. 1976). A major disadvantage of this technique is that the resulting images suffer from a reduction of the ground resolution that is proportional to the number of looks N (Martin and Turner 1993). To overcome this disadvantage, or to further reduce speckle, many posterior speckle-filtering techniques have been developed. These techniques are based on either the spatial or the frequency domain.

The Wiener filter (Walkup and Choens, 1974) and other filters with criteria of minimum mean-square error (MMSE) are examples of filtering algorithms that are based upon the frequency domain (Li 1988). The Wavelet approaches have been used to reduce speckle in SAR images, following Mallat's (1989a, b) theoretical basis for multi-resolution analysis. Gagnon and Jouan (1997), Fukuda and Hirosawa (1998), and Simard et al. (1998) have successfully applied wavelet transformation to reduce speckle in SAR images. Gagnon and Jouan (1997) presented a Wavelet Coefficient Shrinkage (WCS) filter, which performs as well as the standard filters for low-level noise and slightly outperforms them for higher-level noise. The wavelet filter proposed by Fukuda and Hirosawa (1998) has satisfactory performance in both smoothing and edge preservation.

There are also other filters less frequently used, such as the mean filter, the median filter, the Kalman filter (Woods and Radewan 1977), the Geometric filter (Crimmins 1985), the adaptive vector linear minimum mean-squared error (LMMSE) filter (Lin and Allebach 1990), the Weighting filter (Martin and Turner 1993), the EPOS filter (Hagg and Sties 1994), the Modified K-average filter (Rao et al. 1995) and a texture-preserving filter (Aiazzi et al. 1997).

2.1 Fundamentals of the speckle model

An electromagnetic wave scatters from two dimensional position (x, y) on the earth surface, the physical properties of the terrain cause changes in both the phase, $\phi(x,y)$, and amplitude, $A(x,y)$, of the wave. The SAR, in fact measures the number pair $(A\cos\phi, A\sin\phi)$ in the in-phase and quadrature channels of the receiver, weighted by the SAR PSF (point sprit function). The estimates of the local reflectivity at each pixel can also be represented by the complex number $Ae^{i\phi}$; in this form, the SAR data are known as the complex image.

From the complex image, a variety of other products can be formed. For example, images of the real part $A\cos\phi$ (the in-phase component), the imaginary part $A\sin\phi$ (the quadrature component), the amplitude A, the phase ϕ, the intensity $I = A^2$, or the log intensity $\log I$. The use of the word 'intensity' is by analogy with measurements at optical wavelengths and is synonymous with power or energy.

The real and imaginary images show some structure but appear extremely noisy, the phase image is noise-like and shows no structure, while the amplitude, intensity, and log images, though noisy, are clearly easier to interpret. The noise-like quantity characteristic of these types of images is known as *speckle*. It must be stressed that speckle is noise-like, but it is not noise; it is a real electromagnetic measurement, which is exploited, for example, in SAR interferometry (Oliver, and Quegan, 2004). Given that the SAR in making true measurements of the earth's scattering properties, why do such effect arise?

As the wave interacts with the target, each scatterer contributes a backscattered wave with a phase and amplitude change, so the total returned modulation or the observed value at each pixel of the incident wave is

$$Ae^{i\phi} = \sum_{k=1}^{N} A_k e^{i\phi k} \tag{1}$$

This summation is over the number of scatters illustrated by the beam. The individual scattering amplitudes A_k and phases ϕ_k are unobservable because the individual scatterers are on much smaller scales than the resolution of the SAR, and there are normally many such scatterers per resolution cell.

The observed intensity or power $I = A^2$ has a negative exponential distribution (Oliver, 1991).

$$P_I(I) = \frac{1}{\sigma}\exp\left(-\frac{I}{\sigma}\right) \quad I \geq 0 \tag{2}$$

with mean value and standard deviation both equal to σ, so that in this case the coefficient of variation (CV) defined as the standard deviation divided by the mean is equal CV=1. From (2) we can see that σ corresponds to the *average intensity*.

We need to distinguish the measured value at a pixel and the parameter value σ (σ is the Radar Cross Section (RCS) or backscattering coefficient). Equation (1) indicates that the observed value at each pixel is the resultant of interfering Huygens wavelets unless a single scatter completely dominates the return. Hence the value of σ is specific to each pixel; the measured value is just a sample from the distribution parameterized by σ.

A SAR image comprise of some variable, corresponding to local RCS, that is combined with speckle to yield the observed intensity at each pixel. The intensity is given by $I = \sigma n$ where n is the speckle contribution.

All the reconstruction methods for σ that are described require estimates of the sample mean and normalized variance over the window comprising N_W pixels, defined by:

$$\bar{x} = \frac{1}{N_W}\sum_{j=1}^{NW} x_j \qquad and \qquad \bar{V}_x = \frac{\operatorname{var} x}{\bar{x}^2} = \frac{\sum_{j=1}^{NW}(x_j - \bar{x})^2}{N_W \bar{x}^2} \tag{3}$$

Where x_j denotes the pixel value. In single-stage filters, x corresponds to intensity I. The size of window depends on the application (e.g. $3\times 3, 5\times 5, \ldots$).

The ideal filter should eliminate the speckle so that the original signal σ is retrieved. In practice, its behaviour depends on the heterogeneity of the considered area.

First, two classes can be considered: 1) the homogeneous class corresponding to the area where σ is constant; 2) the heterogeneous class corresponding to the area where σ varies and includes textured areas, edges, and point targets. The filter should have the following behaviour.

1. *Within the Homogeneous class*: The filter should restore σ. As the minimum variance unbiased estimator is the mean pixel value, the filter should assign to each pixel C the average of the pixels in a moving window centred at C for the image.
2. *Within the Heterogeneous class*: the filter should smooth the speckle and, at the same time, preserve edges and texture information (σ variations). This supposes that: i) The filter is based on good discriminators which allow a perfect separation between speckle and textural information; and ii) the conditions assumed for the filter establishment are satisfied.

In practice, these two conditions are not always satisfied. A third class is then pointed out where the filter is no longer reliable, and original pixel values are then preserved. In the case of an isolated point target, the filter should conserve the observed value *I*. This is also the case when there are a few scatterers within the resolution cells.

According to above consideration, the following classes are pointed out as a function of the coefficient of variation value.

1. Class to be averaged: if $C_I \le C_u$ then $\hat{\sigma} = \bar{I}$.
2. Class to be filters: if $C_u \prec C_I \prec C_{\max}$, than the filter should operate so that the more heterogeneous area [the larger C_I], the less it has to be smoothed.
3. Class to be preserved: If $C_I \ge C_{\max} \Rightarrow \hat{\sigma} = I$

Where $C_I = sqrt(\bar{V}_x)$

The threshold determination is given by the following consideration (Lopes, et al., 1990a). For an L-look image $C_u = 1/\sqrt{L}$ (an area is considered homogeneous). The threshold $C_{\max}$ is more difficult to determine. A theoretical and experimental study should be developed to determine exactly the $C_{\max}$ value as a function of the image patterns. One of the upper thresholds equal to $\sqrt{1+2/L}$ for an intensity image has been obtained for likelihood ratio edge detection (Touzi, et al., 1988).

2.2 The de-speckling model

The approach of this chapter for reconstruction of backscattering coefficient ($\hat{\sigma}$) is based on Bayes criterion relating the observed intensity *I* to the σ such that

$$P_{AP}(\sigma \mid I) = P(I \mid \sigma) P_{\sigma}(\sigma) / P_I(I) \tag{4}$$

Where $P_{AP}(\sigma \mid I)$ is the a posterior conditional probability of σ, which has a particular value given *I*, and $P(I \mid \sigma)$ is the likelihood function, which describes the effect of speckle during imaging. This is given by (Oliver, and Quegan, 2004).

$$P(I \mid \sigma) = \left(\frac{L}{\sigma}\right)^{L} \frac{I^{L-1}}{\Gamma(L)} \exp\left[-\frac{LI}{\sigma}\right] \tag{5}$$

for L-look SAR. $P_{\sigma}(\sigma)$ is the a priori PDF that encapsulates prior knowledge about the RCS. $P_I(I) = \int P(I \mid \sigma) P_{\sigma}(\sigma) d\sigma$ Only serves to normalize the expression and need not be included specifically in most instances. Generally we wish to provide an estimate of σ that represents it's most likely value given an observed *I*. This is equivalent to minimizing the log likelihood $\lambda = \ln P_{AP}(\sigma | I)$ with respect to σ. Two types of maximum will be considered. If there is not prior knowledge of the form of $P_\sigma(\sigma)$ we can only optimize with respect to the likelihood function in (4) leading the Maximum Likelihood Estimate (MLE). However, if the form of the a priori PDF is known, the optimum is referred to as the maximum a posterior (MAP) estimate. The latter is more precisely determined since it is based on more specific prior knowledge about the properties of the complete process.

The simplest approach to de-speckling is to average the intensity over several pixels within a window centred on a specific pixel. This is tantamount to assuming that the RCS is constant over the filter window. If this assumption is incorrect, the method is fundamentally flawed. The joint probability that all *N* pixels have this mean value is given by

$$P(\sigma | I_1, I_2, \dots, I_N) \propto \prod_{j=1}^{N} P(I_j | \sigma) = \prod_{j=1}^{N} \left(\frac{L}{\sigma}\right)^L \frac{I_j^{L-1}}{\Gamma(L)} \exp\left[-\frac{LI_j}{\sigma}\right] \tag{6}$$

for L-look SAR, where pixels are assumed independent, The MLE for σ is then given by $\sigma_{ML} = \bar{I}$ which is the average intensity over all the pixels in the window, corresponding to the multi-looking. Note that if this is applied to a single pixel the MLE is equal to the intensity of that pixel. Different values for the MLE in the de-speckling filters depend on constraints introduced by the model.

Multi-look de-speckling fails where the assumption of constant RCS within the window breaks down. The filter should then adapt to model the excess fluctuations compared with speckle within the window.

In this chapter, the approach that we developed for de-speckling is based on the least square method. If the original intensity of the centre pixel in a window is *I*, then its corrected value can be obtained by performing a first-order expansion in Taylor saris about the local mean $\bar{I}$ such that

$$\sigma_{LS} = \bar{I} + k(I - \bar{I}) + e \tag{7}$$

Where

e: is the error that must be optimized; *k*: is selected to minimized *e;* σ_{LS} : is the backscattering coefficient and $\bar{I} = \frac{1}{N}\sum_{j=1}^{N} I_j$

But a better estimate for σ can be obtained, if we have a prior knowledge about the PDF of the RCS. The Bayes rule in (4) shows how this priori PDF can be used to provide a MAP reconstruction when combined with the likelihood function. The RCS of natural clutter can be well represented by a Gamma distribution of the form

$$P_\sigma(\sigma) = \left(\frac{v}{\mu}\right)^v \frac{\sigma^{v-1}}{\Gamma(v)} \exp\left[-\frac{v\sigma}{\mu}\right] \tag{8}$$

Where μ *and* v are the mean RCS and order parameter, respectively. These parameters cannot be measured directly and must be estimated from the data. Hence, estimates for μ *and* v are obtained by passing a window over the original image and setting

$$\hat{\mu} = \overline{I} \quad and \quad \hat{v} = 1/\overline{V}_\sigma = (1+1/L)/(\overline{V}_I - 1/L)$$

The PDF of σ given intensity I when both likelihood and a priori PDF are available is given by

$$P_{AP}(\sigma | I) \propto P(I|\sigma)P_\sigma(\sigma) = \left(\frac{L}{\sigma}\right)^L \frac{I^{L-1}}{\Gamma(L)} \exp\left[-\frac{LI}{\sigma}\right]\left(\frac{v}{\mu}\right)^v \frac{\sigma^{v-1}}{\Gamma(v)} \exp\left[-\frac{v\sigma}{\mu}\right] \tag{9}$$

Hence, the log likelihood is given by

$$\begin{aligned}\lambda = \ln P(I|\sigma) + \ln P_\sigma(\sigma) = L\ln L - L\ln\sigma + (L-1)\ln I - \ln\Gamma(L) - LI/\sigma \\ + v\ln v - v\ln\mu + (v-1)\ln\sigma - \ln\Gamma(v) - \frac{v\sigma}{\mu}\end{aligned} \tag{10}$$

and the corresponding Gamma MAP solution for RCS (Kuan, at al., 1987; Oliver, 1991) is given by the quadratic:

$$\frac{v\sigma_{MAP}^2}{\mu} + (L+1-v)\sigma_{MAP} - LI = 0 \tag{11}$$

In regions of pure speckle, we would expect $\overline{V}_I \approx 1/L$ so that, $\hat{v} = \infty$ *and* $\sigma_{MAP} \approx \overline{I}$. However, statistical fluctuations cause the estimate for $\overline{V}_I$ to be less than $1/L$, so $\hat{v}$ becomes negative. Again, the reconstruction can be improved when this occurs by setting $\hat{v} = \infty$ so that $\sigma_{MAP} = \overline{I}$. In the opposite limit of small $\hat{v}$, provided that $\mu/I \succ\succ 4vL/(L+1)^2$, the solution becomes $\sigma_{MAP} = I/(1+1/L)$.

In this chapter, we improve the Gamma-MAP filter by introducing an algorithm that detects and adapts to structural features, such as edges, lines, and points using lease square method. The Gamma-MAP filter appears to give limited de-specking performance. Large windows yield good speckle reduction over homogeneous regions but lead to artifacts over a distance equal to the filter dimension in the presence of strong features. This means that background clutter has excess variations in the precisely those areas where one would like to accurately defined. Small windows are largely free of these artifacts but give inadequate speckle reduction. In our algorithm, iteration leads to a considerable reduction in the speckle.

In principal, it should be possible to base the iteration process on updating the current pixel value, denoted by x, rather than the original intensity I. However, this demands knowledge of the conditional probability $P(x|\sigma)$ relating the current pixel value x to the RCS σ. For residual speckle, this PDF would be expected to be gamma-distributed. Also any degradation in reconstruction will be retained, and probably exacerbated, during subsequent iterations. Thus it seems preferable to insist that the estimated RCS during each iteration should be consistent with the original intensity image described by the speckle conditional probability $P(I|\sigma)$. Though convergence is slower, there is less chance of progressively increasing radiometric distortion. Thus, we hope that x converges to σ and PDF for x converge to equation (8).

The equation (11) is nonlinear with respect to σ_{MAP}, we linearize equation (11) by Taylor series about the initial value for σ_{MAP} (σ_{MAP}^o) as follows:

$$f(\sigma_{MAP}) = \frac{v\sigma_{MAP}^2}{\mu} + (L+1-v)\sigma_{MAP} - Lx = 0$$

$$f(\sigma_{MAP}) = f(\sigma_{MAP}^0) + (\frac{\partial f}{\partial \sigma_{MAP}})^0 d\sigma_{MAP} + e = 0 \qquad (12)$$

$$f(\sigma_{MAP}) = \left\{\frac{v(\sigma_{MAP}^0)^2}{\mu} + (L+1-v)\sigma_{MAP}^0 - Lx\right\} + \left\{\frac{2v\sigma_{MAP}^0}{\mu} + (L+1-v)\right\} d\sigma_{MAP} + e = 0$$

Where x is the current pixel value, μ and v are estimated from the current iteration, so that $\mu = \bar{x}$ and $v = 1/\bar{V}_x$.

Thus, we can write N observation equations for pixels with intensity x_i (i=1, 2... N) in the current iteration within the moving window with size of N=w×w (here w =3) that centred on a specific pixel as follows:

$$\left\{\frac{v_W(\sigma_{MAP}^I)^2}{\mu_W} + (L+1-v_W)\sigma_{MAP}^I - Lx_i\right\} + \left\{\frac{2v_W\sigma_{MAP}^I}{\mu_W} + (L+1-v_W)\right\} d\sigma_{MAP} + e = 0;\ i = 1,2,...,N \qquad (13)$$

Fig 1 shows the process of the de-speckling model.

According to the diagram of Fig 1, a moving window, W, is placed in the top left centre of the SAR image to be filtered (Fig 2) and the mean and the standard deviation values of the pixels within the moving window centred on a specific pixel are computed. Based on the pixels in the window, a linear observation equation system is performed for all pixels in the window using the observation equation (13). The system is solved by using the least square method (LSM) to determine the correction $d\sigma_{MAP}$. This correction is added to the value, σ_{MAP}^I, and the new value , σ_{MAP}^{II}, is replaced in the output image (filtered image) at the point that is corresponding to the location of the specific pixel(see Fig 2).

The proposed algorithm in Fig 1 proceeds as following steps:

Step 1: Initialization stage

1. Set the parameters and consider the lth pixel with intensity I_l

Step 2: Perform intensity update (Filtered image)

1. Compute the mean and the standard deviation values of the moving window W centred in the lth pixel
2. Perform the linear observation equation system based on the equation (13) for all elements in the window W
3. Using the least square method to determine the correction $d\sigma_{MAP}$
4. Compute the new value $\sigma_{MAP}^{II}(\sigma_{MAP}^{II} = \sigma_{MAP}^I + d\sigma_{MAP})$ for lth pixel
5. Increment l and go to step 2 until $l = M_{im} \times N_{im}$, ($M_{im} \times N_{im}$ is the size of the image)

Step 3: Acceptance/ Rejection stage

1. Evaluation of the ratio of the original intensity image, I, to the derived RCS image, x_2, (Ratio image)
2. Estimate the mean, $\bar{r}$, and standard deviation, $SD[r]$, for the Ratio image as follows

$$\bar{r} = \frac{1}{N_{im} \times M_{im}} \sum_{l=1}^{Nim \times Mim} r_l \quad and \quad SD[r] = \sqrt{\frac{1}{N_{im} \times M_{im}} \sum_{l=1}^{Nim \times Mim} (r_l - 1)^2} \tag{14}$$

Where $r_l = I_l/(x_2)_l$ is the ratio of the pixel intensity I_l to the derived RCS $(x_2)_l$ at pixel l.

3. IF { $\bar{r}$ and $SD[r]$ values are remained almost the same in the previous iteration} THEN {stop the algorithm}
   ELSE {continue and go to step 1}.

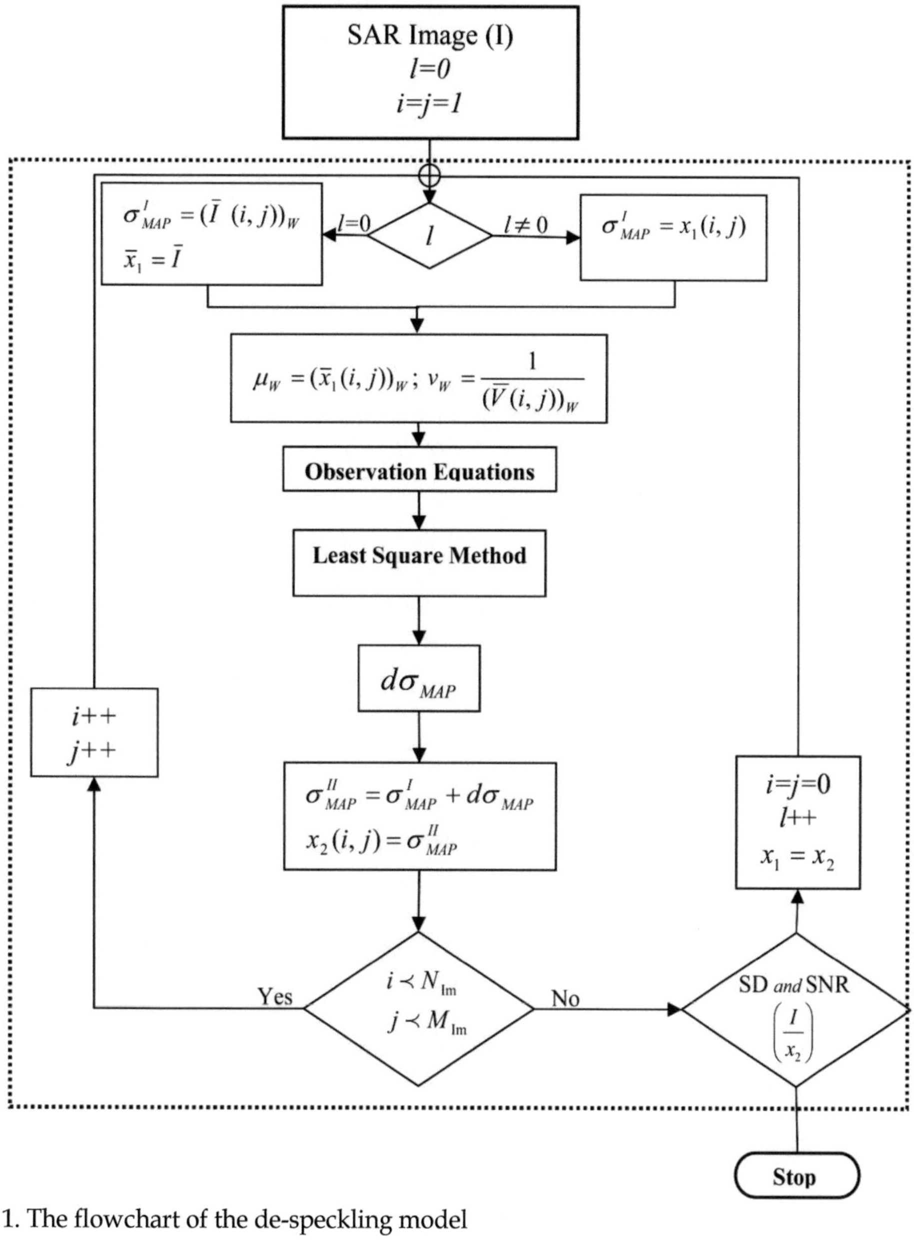

Fig. 1. The flowchart of the de-speckling model

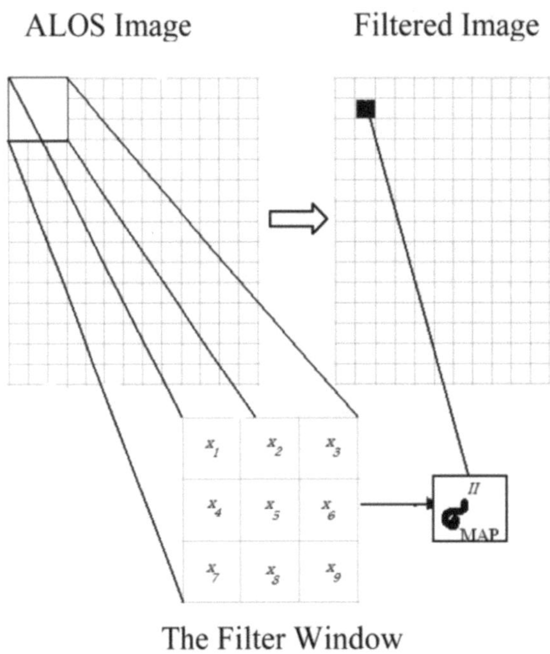

Fig. 2. Operation of the moving window with size of 3×3

3. Methodology and implementation

The methodology used for the forest inventory is distinct according to the vegetation type. In forest areas, different parameters are measured namely: diameter at breast height (DBH), total and commercial height, crown cover percent, and location of each plots. Total height is the height from the upper branches of a tree to the ground and the commercial height is the height of the main trunk of a tree. The crown cover percent is also percent of the number of trees in a hectare. We measured the total height during the field survey and used it in the allometric equation. In addition, the identification of botanical species is also conducted.

The field work consists of collecting some bio-physical and dendrometric parameters which allowed the biomass estimation of the plots and the physiognomic-structural characterization of the different vegetation types considered. The precise geographic coordinates of each plot are obtained using a high-precision Global Positioning System (GPS), which allows the localization of each plots, in the previously geo-referenced images.

The study area is located in the northern forests of Iran around the Rezvanshahr city (Fig. 3(a)). The dominant trees of these forests are: *Maple, Alder, Conifer, Beech, Hornbeam, Azedarach* and *Acorn*. Remote sensing data also consist of: AVNIR-2 and PRISM images from ALOS and a JERS-1 image. The JERS-1 image has a spatial resolution of approximately 13m and, AVNIR-2 and PRISM images have the spatial resolutions of 10m and 2.5m respectively. According to Fig. 3(b), the ground data is collected at five plots in the study area. Each plot

Fig. 3. (a) Study area of the north of Iran, (b) Plots in the study area indicated with circles.

was a square with size of 50m×50m with 25 subplots with size of 10m×10m approximately. The minimum DBH considered was of 37cm. The plots were mostly covered by two classes: Acorn and Azedarach. The distribution of the classes with numbers of stands where measured in each subplots are shown in Table 1. Table 1 summarizes some of the ground measurements and resulting calculations.

The biomass in Table 1 is modelled based on the direct DBH and the total height measurements performed during the field survey and included afterwards in the general allometric equation (15) (Brown et al., 1989).

$$biomass = 0.044 \times ((DBH)^2 \times height)^{0.9719} \tag{15}$$

Where: DBH is in cm, height is in m, and biomass is in kg/tree.

For speckle reduction in the SAR image, the de-speckling model apply on the JERS-1 image of the study area and then its result is compared with several of the most widely used adaptive filters including the Kuan, Gamma, Lee and Frost filters.

In order to investigate the performance of the model, we use some quantitative criteria including speckle smoothing measures and texture preservation to evaluate the performance of the model.

Plot	# of subplots for Acorn	Azedarach	Mean height (m)	Mean DBH (cm)	Mean Biomass (ton/tree)	# of stands for Acorn	Azedar	Total mean biomass (ton) for Acorn	Azedarach
1	20	5	28.5	40	1.484	15	05	26.712	07.420
2	07	18	34	55	3.275	08	13	25.960	42.575
3	19	06	26.5	35	1.066	24	10	25.584	10.660
4	15	10	29	45	1.897	14	09	26.558	17.073
5	04	21	27.5	38	2.373	06	24	14.238	56.952

Table 1. Field plots characteristics

The ratio of the original intensity image to the filtered image enable us to determine the extent to which the reconstruction filter introduces radiometric distortion so that the reconstruction departs from the expected speckle statistics. The mean and standard deviation (SD) can then be estimated over the ratio images. When the mean value differs significantly from one, it is an indication of radiometric distortion. If the reconstruction follows the original image too closely, the standard deviation would be expected to have a lower value than predicted. It would be larger than predicted if the reconstruction fails to follow genuine RCS variations. This provides a simple test that can be applied to any form of RCS reconstruction filters. Table 2, columns 2 and 3, shows the mean and standard deviation values of the ratio images for comparison of the filters.

Algorithm	*Ratio image*		*Filtered image*	
	Mean	S. D	ENL	VTO
The model	*0.991*	*0.037*	*26.78*	*643.12*
Kuan	0.968	0.195	4.96	90.12
Gamma	0.968	0.195	4.96	335.12
Enhanced Lee	0.968	0.195	4.96	234.26
Enhanced Frost	0.968	0.195	16.15	401.32

Table 2. Comparison of the mean and SD in the ratio images, ENL and variance texture operator of the filtered images

According to Gagnon and Jouan (1997), Equivalent number of Looks (ENL) is often used to estimate the speckle noise level in a SAR image and is equivalent to the number of independent intensity values that are used per pixel.
It is the mean-to-standard deviation ratio, which is a measure of the signal-to-noise ratio and is defined over a uniform area as follows:

$$ENL = \frac{(mean^2)_{UniformArea}}{(variance)_{UniformArea}} \tag{16}$$

ENL is used to measure the degree of speckle reduction in this study. The higher the ENL value concludes the stronger the speckle reduction.
Texture preservation is another measure that is important in a SAR image for interpretation and classification. Therefore, the texture preserving capability should play an important role in measuring the performance of a speckle filter. A second-order texture, variance (Iron & Petersen, 1981), is used to measure the retention of texture information in the original and the filtered images.
The ENL and the second-order texture values of the filtered images are shown in Table 2 columns 4 and 5 respectively. Of the four commonly used filters, Enhanced Frost filter has higher speckle-smoothing capabilities than Kuan, Gamma and Enhanced Lee filters. The ENL value of the model is 26.78 that it is comparable to Enhanced Frost filter. According column 5 ,Variance Texture Operator (VTO), in Table 2, the texture preservation of the proposed filter is better than, or comparable to, those of the commonly used speckle filters. We concluded the model is slightly better than the commonly used filters in terms of preserving details in forestry areas. Furthermore, the model also affects in smoothing speckles. This improvement in the accuracy of the speckle reduction can be played an important role in the forest biomass estimation.
After reduction the speckle noise, the texture of SAR image must be measured. Of the many describing texture methods, the grey-level co-occurrence matrix (GLCM) is the most common (Marceau et al., 1990; Smith et al., 2002; Zhang et al.,2003) in remote sensing.
Nine texture measures are calculated from the GLCM for a moving window with size of 5×5 pixels that centred in pixel *i, j* of the de-speckled JERS-1image. After the Gram-Schmidt process, just four texture measures: *contrast, correlation, maximum probability* and *standard-deviation* are selected as the optimum measures in this area.
The PRISM image is transformed in the universal transverse Mercator (UTM) projection with a WGS84 datum based on the GPS measurements and is used as the base map. Two GPSs measured the coordinates of points along the roads of the study area. To place all data sets in a unified coordinate system, the AVNIR and JERS-1 image are registered to this map. The co-registered and geo-referenced data sets contain PRISM, AVNIR and SAR images are used to extract intensity values and texture measures respectively.

4. Experimental results

Intensity value and texture measures from the co-registered and geo-referenced data sets are used in the algorithm to estimate the forest biomass. The data sets are related to the forest biomass through a classification analysis. The correspondence between the data sets and ground plots is made using PCI Geomatica software, where the ground plot GPS locations are superimposed on the data set. For each selected pixel (or point) from data set, a window with size of 5×5 pixels around the point is used and the average intensity values for the PRISM and three channels of the AVNIR images with four texture values of the JERS-1 image are calculated. Thus each selected point contains a vector with eight attributes where the first four elements are the average intensity values and the second four elements are the texture measures values. These vectors of data set construct the feature space. The vectors belong to the pixels of the ground plots and subplots are used as training patterns in the classification process.

The classification analysis is done with a MLPNN. A multi layers neural network is made up of sets of neurons assembled in a logical way and constituting several layers. Three distinct types of layers are present in the MLPNN. The input layer is not itself a processing layer but is simply a set of neurons acting as source nodes which supply input feature vector components to the second layer. Typically, the number of neurons in the input layer is equal to the dimensionality of the input feature vector. Then there is one or more hidden layers, each of these layers comprising a given number of neurons called hidden neurons. Finally, the output layer provides the response of neural network to the pattern vector submitted in the input layer. The number of neurons in this layer corresponds to the number of classes that the neural network should differentiate (Haykin, 1999; Miller et al., 1995; .

The network that is used in this study arrange in layers as following. The number of neurons in the output layer is taken to be equal to the number of classes desired for the classification. Here, the output layer of the network used to categorize the image in five classes should contain five neurons. The input layer contains eight neurons corresponding to the number of attributes in the input vectors. The input vector to the network for pixel i of the data sets is the form $v_{i_{os}} = \{v_{i1}, v_{i2}, ... v_{i8}\}$. Where the first four elements belong to the intensity values of PRISM and AVNIR images and the second four elements belong to the texture measures of JERS-1 image for a window with size of 5×5 around pixel i of the geo-referenced data sets. After the determination of the input layer, the number of hidden layers required as well as the number of neurons in these layers still needs to be decided upon. An important result, established by the Russian mathematician Kolmogorov in the 1950s, states that any discriminate function can be derived by a three-layer feed-forward neural network (Duda, 2001). Increasing the number of hidden layers can then improve the accuracy of the classification, pick up some special requirements of the recognition procedure during the training or enable a practical implementation of the network. However, a network with more than one hidden layer is more prone to be poorly trained than one with only one hidden layer.

Thus, a three-layer neural network with the structure 8-10-5 (eight input neurons, ten hidden neurons and five output neurons) is used to classify the data sets into five classes.

Training the neural network involves tuning all the synaptic weights so that the network learns to recognize given patterns or classes of samples sharing similar properties. The learning stage is critical for effective classification and the success of an approach by neural networks depends mainly on this phase. The network is trained by using back-propagation rule (Paola & Schowengerdt, 1995). After training the network, the parameters are selected as: Momentum value 0.9, Learning rate 0.1, and the number of iteration 2000. The numbers of training data are 200 patterns of the subplots that are selected randomly from the classes, in which each class is represented with at least 40 patterns. The set of training patterns is presented repeatedly to the neural network until it has learnt to recognize them. A training pattern is said to have been learnt when the absolute difference between the output of each output neuron and its desired value is less than a given threshold. Indeed, it is pointless to train the network to reach the target outputs 0 or 1 since the sigmoid function never attains its minimum and maximum (Masters, 1993). For classification of data sets into five classes, the threshold is set to 0.4. The network is trained when all training patterns have been learnt. Once the network is trained, the weights of the network are applied on the data sets to classify into five classes: class1 *Azedarach*, class2 *Acorn*, class3 *Beech*, class4 *Grassland* and class5 *None*. The result of the classified image is shown in Fig. 4.

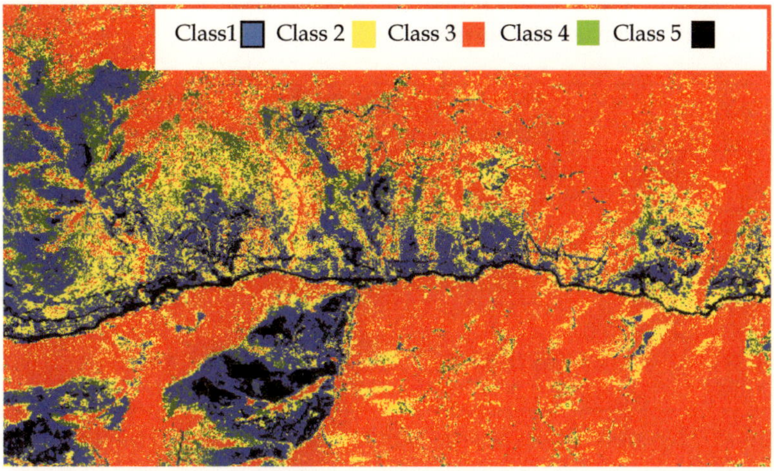

Fig. 4. The classified image with MLPNN.

After classification, it is needed to determine the degree of classification accuracy. The most commonly used method of representing the degree of accuracy of a classification is to build confusion matrix.

The confusion matrix is usually constructed by a test sample of patterns for each of the five classes. A set of test sample with 105 patterns based on the ground truth collection were randomly selected in the classified image for accuracy assessment. The values 70% and 65% are achieved for overall accuracy and kappa coefficient respectively. One reason for misclassification can be due to poor selection of training areas, so that some training patterns don't accurately reflect the characteristics of the classes used. Another reason can be due to poor selection of land cover categories, resulting in correct classification of areas from the point of view of the network, but not from that of the user. Thus the classification accuracy can be improved by redefining the training patterns and land cover categories.

In order to show the texture of SAR image and the neural network classifier improve the accuracy of the classification and then forest biomass estimation, we employ the Maximum Likelihood (ML) classifier method using only the intensity values of the PRISM and AVNIR images. The overall classification accuracy of 57% is achieved with ML classifier. The accuracy of 70% with the neural network is significantly better than the accuracy of 57% with ML.

In comparison between the MLPNN and ML classifiers, the advantages of MLPNN that is used in this study are:

i. It can accept all kind of numerical inputs whether or not these conform to statistical distribution or not.

ii. It can recognize inputs that are similar to those which have been used to train them.
Because the network consists of a number of layers of neurons, it is tolerant to noise present in the training patterns.
Thus, we can estimate the forest biomass of the classes in the classified image which has been classified based on the SAR image texture and the MLPNN classifier. We also evaluate the biomass for two classes based on the allometric equation (15) for the classic method based on the ML classifier and the proposed method. The results are shown in Table 3, where the classic method and the proposed method have been applied in the classified image to estimate the biomass for two classes.

	The classic method		**The proposed method**	
	Acorn	Azedarach	Acorn	Azedarach
Area (ha)	853.217	1129.552	937.312	1241.320
Mean height (m)	34	28.5	34	28.5
Mean DBH (cm)	55	45	55	45
# of tree (ha)	34	23	34	23
Mean biomass (kg/tree)	3272	1861.99	3272	1861.99
Total biomass (tons/ha)	94918.85	48374.08	104274.085	53160.484

Table 3. Estimated biomass for the classic method and the proposed method by both optical and sar data.

For the accuracy assessment of the proposed method, Table 4 shows how well the results agree with the ground measurements results from Table 1, when the classic method and the proposed method are used for biomass estimation. Table 4 shows the estimated biomass when both methods are used. The root mean square error (RMSE) of estimated biomass with both methods is indicated in the table. The RMSE values is decreased when the model is used (RMSE=2.175 ton) compared the classic method (RMSE=5.34 ton).

Plot	Measured biomass (ton) for		**The classic method** Estimated biomass (ton) for		**The proposed method** Estimated biomass (ton) for	
	Azedarach	Acorn	Azedarach	Acorn	Azedarach	Acorn
1	26.712	07.42	29.13	10.40	27.43	09.12
2	25.960	42.575	30.40	46.39	27.13	41.43
3	25.584	10.660	18.13	06.43	23.32	08.86
4	26.558	17.073	22.13	24.32	23.16	21.36
5	14.238	56.952	17.43	66.13	15.29	58.56
RMSE			***4.71***	***5.97***	***1.97***	***2.38***
Mean RMSE				*5.34*		*2.17*

Table 4. Accuracy assessment for the classic method and the proposed model using the ground measurements from Table 1.

From the above paragraphs, the accuracy of the proposed method is better than, or comparable to, the classic method used for biomass estimation. We conclude using both optical image and SAR image texture in a non-linear classifier method, neural network, significantly improve the accuracy of the forest biomass estimation.

5. Discussion

It is often difficult to transfer one model developed in a specific study area to other study areas because of the limitation of the model itself and the nature of remotely sensed data. Foody (Foody et al., 2003) discussed the problems encountered in model transfer. Many factors, such as uncertainties in the remotely sensed data (image preprocessing and different stages of processing), AGB calculation based on the field measurements, the disparity between remote sensing acquisition date and field data collection, and the size of sample plot compared with the spatial resolution of remotely sensed data, could affect the success of model transferability. Each model has its limitation and optimal scale for implementation. Models developed in one study area may be transferred to (1) across-scene data, which have similar environmental conditions and landscape complexity, to estimate AGB in a large area; and (2) multi-temporal data of the same study area for AGB dynamical analysis if the atmospheric calibration is accurately implemented. The spectral signatures, vegetation indices, and textures are often dependent on the image scale and environmental conditions. Caution must be taken to ensure that there is consistency between the images used in scale, atmospheric and environmental conditions. Calibration and validation of the estimated results may be necessary using reference data when using transferred models.

The data sources used for AGB estimation may include field-measured sample data, remotely sensed data, and ancillary data. A high-quality sample dataset is a prerequisite for developing AGB estimation models as well as for validation or assessment of the estimated results. Direct measurement of AGB in the field is very difficult. In general, AGB is calculated using the allometric equations based on measured DBH and/or height, or from the conversion of forest stocking volume. These methods generate many uncertainties and calibration or validation of the calculated AGB is necessary. Previous research has discussed the uncertainties of using the allometric equations (Brown & Gaston, 1995; Keller et al., 2001; Ketterings, 2001; Fearnside, 1992) and of conversion from stocking volume (Masters, 1993). It is important to ensure that the remote sensing data, ancillary data, and sample plots are accurately registered when ancillary data are used for AGB estimation. Understanding and identifying the sources of uncertainties and then devoting efforts to improving them are keys to a successful AGB estimation. More research is needed in the future for reducing the uncertainties from different sources in the AGB estimation procedure. Many remote sensing variables, including spectral signatures, vegetation indices, transformed images, and textures, may become potential variables for AGB estimation. However, not all variables are required because some are weakly related to AGB or they have high correlation with each other. Hence, selection of the most suitable variables is a critical step for developing an AGB estimation model. In general, vegetation indices can partially reduce the impacts on reflectance caused by environmental conditions and shadows, thus improving correlation between AGB and vegetation indices, especially in those sites with complex vegetation stand structures (LU, 2004). On the other hand, texture is an important variable for improving AGB estimation performance. One critical step is to identify suitable textures that are strongly related to AGB but are weakly related to each other. However, selection of suitable textures for AGB estimation is still a challenging task because textures vary with the characteristics of the landscape under investigation and images used. Identifying suitable textures involves the determination of appropriate texture measures, moving window sizes, image bands, and so on (Franklin & Hiernaux, 1991). Not all texture measures can effectively extract biomass information. Even for the same texture measure, selecting an appropriate window size and image band is crucial. A small window size, such as 3×3, often exaggerates the difference within the moving windows, increasing the noise content on the texture image. On the other hand, too large a window size, such as 11×11 or larger, cannot effectively extract texture information due to smoothing the textural variation too much. Also, a large window size implies more processing time. In practice, it is still difficult to identify which texture measures, window sizes, and image bands are best suited to a specific research topic and there is a lack of guidelines on how to select an appropriate texture. More research is needed to develop suitable techniques for identification of the most suitable textures for biomass estimation.

In addition to remotely sensed above ground biomass estimation in data, different soil conditions, terrain factors, and climatic conditions may influence AGB estimation because they affect AGB accumulation rates and development of forest stand structures. Incorporation of these ancillary data and remote sensing data may improve AGB estimation performance. Geographical Information System (GIS) techniques can be useful in developing advanced models through the combination of remote sensing and ancillary data.

6. Conclusion

In this chapter, we proposed a method for forest biomass estimation. One speckle noise model was used for reducing the speckle noise in SAR images. The speckle model was slightly better than the commonly used filters in terms of preserving details in forestry areas. A combination of spectral responses from optical images and textures from SAR images improved biomass estimation performance comparing pure spectral responses or textures. Intensity values of ALOS-AVNIR-2 and PRISM images and texture features of JERS-1 image were used in a multilayer perceptron neural network (MLPNN) that relates them to the forest variable measurements on the ground. We showed the biomass estimation accuracy was significantly improved when MLPNN was used in comparison to estimating the biomass by using classic method only. The RMSE values was decreased when the proposed method was used (RMSE=2.175 ton) compared the classic method (RMSE=5.34 ton).

7. References

Achim, A.; Kuruoglu, E.E. & Zerubia, J. (2006). SAR image filtering based on the heavy-tailed Rayleigh model, IEEE Trans. Image Processing, Vol. 15, no.9, pp. 2686- 2693.

Aiazzi, B.; Alparone, L.; Baronti, S. & Carla, R. (1997). Adaptive texture-preserving filtering of multitemporal ERS-1 images. Proceedings of IGARSS' 97, *IEEE International Geoscience and Remote Sensing Symposium*, 3–8 August 1997 (Singapore: IGARSS), pp. 2066–2068.

Amini, J. & Sumantyo, J.T.S. (2009). Employing a method on SAR and Optical Images for Forest Biomass Estimation, *IEEE Trans. on Geoscience and remote sensing*, Vol. 47, No. 12. pp. 4020-4026.

Baraldi, A. & Parmiggiani, F. (1995). A refined Gamma Map SAR speckle filter with improved geometrical adaptively, *IEEE Trans. Geosci. Remote Sensing*, 33, pp. 1245-1257.

Brown, S.; Gillespie, A.J.R. & Lugo, A.E. (1989). Biomass estimation methods for tropical forests with applications to forest inventory data, *Forest Science*, 35, pp. 881-902.

Brown, S. & Gaston, G. (1995). Use of forest inventories and geographic information systems to estimate biomass density of tropical forests: application to tropical Africa, *Environmental Monitoring and Assessment*, 38, pp. 157-168.

Crimmins, T. R. (1985), Geometric filter for speckle reduction. Applied Optics, 24, 1438-1443.

Dobson, M.C.; Ulaby, F.T.; LeToan, T.; Beaudoin, A; Kasischke, E.S. & Christensen, N. (1992). Dependence of radar backscatter on coniferous forest biomass, *IEEE Trans Geosc.Remote Sensing*, Vol. 30, no. 2, pp. 412 - 415.

Dong, Y.; Milne, A.K. & Forester, B.C. (2001). Toward edge sharpening: a SAR speckle filtering algorithm, *IEEE Trans. Geosci. Remote Sensing*, Vol. 39, no.4, pp. 851-863.

Duda, R.O.; Hart, P.E. & Stork, D.G. (2001). *Pattern Classification, 2nd edition,* New York: John Wiley and Sons.

Franklin, J. & Hiernaux, P. (1991). Estimating foliage and woody biomass in Sahelian and Sudanian woodlands using a remote sensing model, *Int. J. Remote Sensing*, 12, pp. 1387-1404.

Foody, G.M.; Boyd, D.S. & Cutler, M. (2003). Predictive relations of tropical forest biomass from Landsat TM data and their transferability between regions, *Remote Sensing of Environment*, 85, pp. 463-474.

Frost, V.S; Stiles, J.A; Schanmugan, K.S. & Holzman, J.C. (1982). A model for radar images and its application to adaptive digital filtering of multiplicative noise, *IEEE Trans. on Pattern Analysis and Machine Intelligence*, 4, pp. 157-166.

Fearnside, P.M. (1992). Forest biomass in Brazilian Amazoˆ nia: comments on the estimate by Brown and Lugo, *Interciencia*, 17, pp. 19-27.

Fukuda, S. & Hirosawa, H. (1998). Suppression of speckle in synthetic aperture radar images using wavelets. *International Journal of Remote Sensing*, 19, pp. 507-519.

Gagnon, L. & Jouan, A. (1997). Speckle filtering of SAR images—a comparative study between complex- wavelet-based and standard filters, *Proceedings of SPIE Wavelet Applications in Signal and Image Processing*, SPIE volume 3169, pp. 80-91.

Gagnon, L. & Jouan, A. (1997). Speckle filtering of SAR images—a comparative study between complex- wavelet-based and standard filters, *Proceedings of SPIE Wavelet Applications in Signal and Image Processing*, 1997, SPIE volume 3169, pp. 80-91.

Hagg, W. & Sties, M. (1994). Efficient speckle filtering of SAR images, *Proceedings of IGARSS' 94*, 1994 (California: IGARSS), pp. 2140-2142.

Haykin, S. (1999). Neural networks: *A Comprehensive Foundation, 2nd edition,* Upper Saddle River, New Jersey: Prentice Hall.

Iron, J.R. & Petersen, G.W. (1981). Texture transforms of remote sensing data, *Remote sensing of environment*, Vol. 11, pp. 359-370.

Keller, M.; Palace, M.; & Hurtt, G. (2001). Biomass estimation in the Tapajos National Forest, Brazil, Examination of sampling and allometric uncertainties, *Forest Ecology and Management*, 154, pp. 371-382.

Ketterings, Q.M.; Coe, C.; Van noordwijk, M; Ambagau, K. & Palm, C.A. (2001). Reducing uncertainty in the use of allometric biomass equations for predicting aboveground tree biomass in mixed secondary forests, *Forest Ecology and Management*, 146, pp. 199–209.

Kuan, D. T.; Sawchuk, A. A.; Strand, T. C. & Chavel, P. (1987). Adaptive restoration of images with speckle, *IEEE Transactions on Acoustics, Speech, and Signal Processing*, 35, pp. 373–383.

Kuan, D.T.; Sawchuk, A.A; Strand, T.C. & Chavel, P. (1987). Adaptive restoration of images with speckle, *IEEE Trans. on Acoustics, Speech, and Signal Processing*, 35, pp. 373–383.

Lu, D. (2005). Aboveground biomass estimation using Landsat TM data in the Brazilian Amazon Basin, *Int. J. Remote Sensing*, 26, pp. 2509–2525.

Lee, J.S. (1980). Digital image enhancement and noise filtering by use of local statistics, *IEEE Trans. on Pattern Analysis and Machine Intelligence*, 2, pp. 165–168.

Lopes, A.; Touzi, R; Nezry, E. (1990a). Adaptive speckle filters and scene heterogeneity, *IEEE Trans. Geosci. Remote Sensing*, Vol. 28, No. 6, pp. 1953–1962.

Lopes, A.; Nezry, E.; Touzi, R & Laur, E. (1990b). Maximum a posteriori speckle filtering and first order texture models in SAR images, *Proceedings of IGARSS' 90*, May 1990, vol. 3 (Maryland: IGARSS), pp. 2409-2412.

Lu, D.; Mausel, P.; Brondizio, E. & Moran, E. (2004). Relationships between forest stand parameters and Landsat Thematic Mapper spectral responses in the Brazilian Amazon Basin, *Forest Ecology and Management*, 198, pp. 149–167.

Li, C. (1988). Two adaptive filters for speckle reduction in SAR images by using the variance ratio. *International Journal of Remote Sensing*, 9, pp. 641–653.

Lin, Q. & Allebach, J. P. (1990). Combating speckle in SAR images: vector filtering and sequential classification based on a multiplicative noise model. *IEEE Transactions on Geoscience and Remote Sensing*, 28, pp. 647–653.

Lopes, A.; Touzi, R.; Nezry, E. (1990a). Adaptive speckle filters and scene heterogeneity. *IEEE Transactions on Geoscience and Remote Sensing*, Vol. 28, No. 6, pp. 1953–1962.

Marceau, D.J; Howarth, P.J.; Dubois, J.M. & Gratton, D.J. (1990). Evaluation of the grey-level co-occurrence matrix method for land-cover classification using SPOT imagery, *IEEE Trans. Geosci. Remote Sensing*, 28, pp. 513–519, 1990.

Miller, D.M.; Kaminsky, E.J. & Rana, S. (1995). "Neural network classification of remote sensing data", *Computers and Geosciences*, Vol. 21, pp: 377-386.

Masters, T. (1993). *Practical Neural Network Recipes in C++*, San Diego: Academic Press.

Mallat, S. G. (1989a). A theory for multiresolution signal decomposition: the wavelet epresentation. *IEEE Transactions on Pattern Analysis andMachine Intelligence*, 11, pp. 674–693.

Mallat, S. G. (1989b). Multifrequency channel decomposition of images and wavelet models. *IEEE Transactions on Acoustics, Speech, and Signal Processing*, 37, p. 2091–2110.

Martin, F. J.; & Turner, R. W. (1993). SAR speckle reduction by weighted filtering. *International Journal of Remote Sensing*, 14, pp. 1759–1774.

Nelson, R.; Krabill, W. & Tonelli, J. (1998). Estimating forest biomass and volume using airborne laser data, *Remote Sensing of Environment*, 24, pp. 247–267.

Nezry, E.; Lopes, A. & Touzi, R. (1991). Detection of structural and textural features for SAR images filtering, *Proceedings of IGARSS' 91*, June 1991, vol. 3 (Helsinki: IGARSS), pp. 2169–2172.

Oliver C. & Quegan S. (2004). Underestanding Synthatic Aperture Radar Images, Scitech Pub.,

Oliver, C. J. (1991). Review article-Information from SAR images, J. Phys. D: Appl. Phys., VOl. 24, pp. 1493-1514.

Paola, J. & Schowengerdt, R.A. (1995). "A review and analysis of back-propagation neural networks for classification of remotely-sensed multi-spectral imagery", *Int. J. Remote Sensing*, Vol. 16, pp: 3033-3058.

Porcello, L. J.; Massey, N. G.; Innes, R. B.; & Marks, J. M. (1976). Speckle reduction in synthetic radars. *Journal of the Optical Society of America*, 66, pp. 1305–1311.

Quinones, M.J. & Hoekman, D.H. (2004). Exploration of factors limiting biomass estimation by polarimetric radar in tropical forests, *IEEE Trans. Geosci. Remote Sensing*, Vol. 42, nu.1, pp. 86 - 104.

Rignot, E.J.; Zimmermann, R. & van Zyl, J.J. (1995). Spaceborne applications of P band imaging radars for measuring forest biomass, *IEEE Trans. Geosci. Remote Sensing*, Vol. 33, no. 5, pp. 1162 - 1169.

Rignot, E.; Way, J; Williams, C. & Viereck, L. (1994). Radar estimates of aboveground biomass in boreal forests of interior Alaska, *IEEE Trans. Geosci. Remote Sensing*, Vol. 32, no5, pp. 1117 - 1124.

Rao, P. V. N.; Vidyadhar, M. S. R. R.; Rao, T. C. M.; & Venkataratnam, L. (1995). An adaptive filter for speckle suppression in synthetic aperture radar images. *International Journal of Remote Sensing*, 16, pp. 877–889.

Sumantyo, J.T.S. & Amini, J. (2008). A model for removal of speckle noise in SAR images (ALOS PALSAR), *Canadian Journal of Remote Sensing*, Vol. 34, No. 6, pp. 503-515.

Simard, M.; Degrandi, G.; Thomson, P. B. & Benie, G. B. (1998). Analysis of speckle noise contribution on wavelet decomposition of SAR images. *IEEE Transactions on Geoscience and Remote Sensing*, 36, pp. 1953–1962.

Smith, A.M.S.; Wooster, W.J; Powell, A.K. & Usher, D. (2002). Texture based feature extraction: application to burn scar detection in Earth observation satellite sensor imagery, *Int. J. Remote Sensing*, 23, pp. 1733–1739.

Sader, S.A. ; Waide, R.B.; Lawrence, W.T. & Joyce, A.T. (1989). Tropical forest biomass and successional age class relationships to a vegetation index derived from Landsat TM data, *Remote Sensing of Environment*, 28, pp. 143–156.

Steininger, M.K. (2000). Satellite estimation of tropical secondary forest aboveground biomass data from Brazil and Bolivia, *Int. J. Remote Sensing*, 21, pp. 1139–1157.

Touzi, R.; Lopes, A.; & Bousquet, P. (1988). A statistical and geometrical edge detector for SAR images, *IEEE Trans. on Geoscience and Remote sensing*, Vol. 26, pp. 764-773.

Walkup, J. F. & Choens, R. C. (1974). Image processing in signal dependent noise. *Optical Engineering*, 13, pp. 250–266.

Woods, J. W. & Radewan, C. H. (1977). Kalman filtering in two dimensions. *IEEE Transactions on Information T heory*, 23, pp. 473–482.

Zheng, D.; Rademacher, J.; Chen, J.; Crow, T.; Bresee, M.; Moine, J.L. & Ryu, R. (2004). Estimating aboveground biomass using Landsat 7 ETM+ data across a managed landscape in northern Wisconsin, USA, *Remote Sensing of Environment*, 93, pp. 402–411.

4

Detection of Ammonia-oxidizing Bacteria (AOB) in the Biofilm and Suspended Growth Biomass of Fully- and Partially-packed Biological Aerated Filters

Fatihah Suja'
Universiti Kebangsaan Malaysia
Malaysia

1. Introduction

Nitrification is a two step process namely ammoniacal oxidation and nitrite oxidation. Oxidation of ammonium to nitrite is carried out by autotrophic bacterium mainly *Nitrosomonas* (e.g. *N. europaea, N.oligocarbogenes*) and Nitrosospira while conversion of nitrite to nitrate is performed by *Nitrobacter* (e.g. *N. agilis, N. winogradski*) and Nitrospira. However, ammoniacal oxidation is considered as the limiting or critical process in nitrification since the ammonia-oxidizing bacteria (AOB) has very low growth rate (Metcalf and Eddy 1991).
Various approaches, both culture dependent and independent have been applied to analyze and compare the microbial structure of biomass. However, culture dependent methods are biased by the selection of species which obviously do not represent the real dominant structure (Wagner et al 1995; Lipponen et al 2002). Recently, the development of culture independent molecular techniques, like fluorescence *in situ* hybridization (FISH), polymerase chain reaction (PCR) or denaturing gradient gel electrophoresis (DGGE) improved the analysis of environmental samples.
Whole cell fluorescene *in situ* hybridization (FISH) is a technique that uses fluorescently labelled phylogenetic oligonucleotide probes to detect specific whole cells/organisms in biological samples. It can be a valuable tool for the study of microbial dynamics in natural environments (Li et al 1999; Liu et al 2002, Eschenhagen et al 2003). These probes could be designed using the wealth of 16S and 23S rDNA sequence data available to target species, genera subdivisions or divisions in-situ and could be labelled with fluorescent groups, radioactive groups or antigens for immunological detection (Amann 1995).
A combination of the FISH approach with the application of scanning confocal laser microscopy (SCLM) allows non-destructive studies of the three dimensional arrangements of bacterial population identified and out-of-focus fluorescence (Wagner et al 1995).
Biological Aerated Filters (BAFs) also have a long history of successfully removing nitrogen in wastewater treatment plants (Chen et al 2000; Quyang et al 2000; Chui et al 2001). Biofilm in the reactors bears great potential for simultaneous and efficient removal of nitrogen (Fdz-Polanco et al 2000). Therefore, an assessment of nitrogen removal efficiency has been made to detect any deterioration to the performance. A possible adverse effect of reduced mass of biofilm in the partial-bed reactor was foreseen for the reason that the slow-growing nitrifiers

will be more easily washed out at lower mean solids retention times (SRT) (Gieseke et al 2002). The denitrification process may also be disrupted because the biofilm provides potential anaerobic conditions in which denitrification flourishes.

Fdz-Polanco et al (2000) pointed out the importance of understanding the spatial distribution of the microbial population, and its activity, for the optimisation of nitrogen removal performance in reactors treating wastewater. The performance of the full and partial-bed reactors for nitrogen removal has been examined (Fatihah 2004). It was verified that the full- and partial-bed reactors have the capacity to remove 79.3 ±7.7 % and 79.4 ±3.6 % nitrogen at carbon organic loadings of 5.71 ±0.16 kg COD/m^3.d, corresponding to nitrogen loadings of 0.24 ± 0.02 kg N/m^3.d. At this condition, the organic carbon removal efficiency was 5.34 kg COD/m^3.d for the full-bed and 5.22 kg COD/m^3.d for the partial-bed. The successful removal of nitrogen indicates the existence of ammonia-oxidizing bacteria (AOB) in both reactors.

From the perspective of engineering design, it is important to be able to predict the functional groups of bacteria that are most favoured by various applied reactor conditions. In this respect, knowledge of their activities is more important than that of the detailed microbial population (Beer and Muyzer 1995). The nitrogen removal process in such systems is typically initiated by chemoliautotrophic ammonia-oxidizing bacteria converting ammonia to nitrite and traces of oxidized nitrogen gases. Subsequently nitrite-oxidizing bacteria catalyse the oxidation of nitrite to nitrate, and the process is then completed by denitrification (Metcalf and Eddy 1991). Clearly the oxidation processes of nitrification are an essential prerequisite for the whole removal process. In addition, retaining a large amount of nitrifying bacteria within the reactor can be difficult to achieve, due to their relatively low rates of respiration, and their subsequent sensitivity to DO and temperature, thereby making nitrification the rate-determining microbial system in the entire nitrogen removal process (Tsuneda et al 2003).

Since the number and the physiological activity of the ammonia oxidizers are generally the rate-limiting parameters, the rapid and reliable identification of this autotrophy is an important task. The aerobic ammonia oxidizers belong to a very restricted group of autotrophs with Nitrosomonas and Nitrosospira being the best-known oxidizers (Sliekers et al 2002), dominated by β-Proteobacteria (Wagner et al 1995; Eschenhagen et al 2003). Rowan et al (2003) found that detection of ammonia-oxidizing bacteria using PCR amplified 16S rRNA gene in a laboratory-scale BAF reflects the dominant AOB within a full-scale plant.

If the partial-bed reactor exhibited comparable nitrogen removal performance, intriguing questions would arise: would the slow-growing nitrifying bacteria's preference for attachment on biofilm thereby enhancing sludge retention time (SRT), be challenged by bacterial growth in suspension: or would there be other factors related to reactor configuration that satisfied the need for nitrifying bacteria to grow in the partial-bed reactor. Since, for any high rate system, the AOBs need to reside within the biofilm that has a longer SRT than the suspended growth, it is interesting to locate the microorganisms along the height of both the full- and partial-bed reactors. The detailed aspects to be evaluated in this part include:

- to detect and enumerate the presence of AOBs in the biofilm and suspended growth biomass using fluorescence *in situ* hybridization (FISH) technique in combination with confocal laser scanning microscopy (CLSM)

- to correlate changes in the proportion of AOBs to all bacteria along the reactor heights in relation to the reactor configuration
- to associate factors that contribute to the changes in the AOB proportion

2. Experimental system

Two identical reactors were built; each reactor was 14 cm in diameter and 100 cm in height, providing an empty bed volume of 15 l. A small amount of freeboard or headspace (2.8 litres) was provided at the top of the reactor. The reactors were constructed from PVC, a non-transparent material that prevents the growth of phototrophic organisms. The columns were built with considerations for process air and influent supplies, backwashing air and water requirement and sampling outlets.

The control reactor was filled with 10.9 l cascade rings (Glitsch UK) whilst the second reactor was only partially packed with 5.5 l cascade rings. The media were stationary and held in place by a rigid polypropylene mesh with 15 mm diameter holes placed at the top and bottom of the packing. Three ports were placed along the height of the reactors for sample collection.

A synthetic waste prepared in the laboratory was used to provide a consistent organic substrate for all loadings. The basic make-up of the influent organic strength material used in the study was whey powder, glucose and meat extract (Lab Lemco powder) which contributed approximately 38%, 33% and 29% of the total soluble COD content of the substrate respectively. In order to guarantee that organic carbon was the limiting nutrient, a COD:N: P ratio of 25:5:1 was adopted. Nitrogen component of the feed came from whey powder (24.7%), meat extract (63.7%), and ammonium-dihydrogenphosphate (11.6%). 1 l of the prepared mixture produces a concentrated feed around 40000 mg/l COD.

2.1 Suspended biomass and biofilm sampling

The collection of samples for this study was carried out at the end of the steady-state condition of 0.24 ± 0.02 kg N/m^3.d nitrogen loadings. Samples of the biofilm and suspended growth biomass were taken at different depths of the reactors. The in-situ characterization followed a top-bottom approach. Fig. 1 illustrates the exact locations where the samples of suspended biomass and biofilm were obtained from the reactors.

Samples of suspended biomass were taken from port 1, port 2 and port 3 respectively. At each port, about 50 ml of reactor aliquot was wasted before sample collection to ensure that any debris or anaerobic bacteria residing in the pipeline was discarded. A 10 mL volume of aliquot was taken and immediately fixed with 1:1 absolute ethanol. Samples were then stored at -20° C.

For sampling the biofilm, the liquid was first drained from port 1 in order to allow access into the upper bed layer. Tongs were used carefully to remove the media from the upper layer. A random piece of media from the specified level was chosen. The biofilm was gently scraped off the plastic material using a sterile surgical knife before washing the media with 10 ml phosphate-buffered saline (PBS) solution. This procedure was repeated four times until all the biofilm attached to the media was completely removed. To homogenize the biofilm, the sample was sonicated for 2 minutes using an ultrasonic homogenizer (Bandelin Electronics D-1000, Germany). 10 ml of the aliquot was put in a universal bottle and fixed with 1:1 absolute ethanol before storing at -20° C. The sampling of biofilm at the second location was subsequently continued by draining the liquid from port 2. The same procedures were repeated until the media at the bottom were sampled. To detect the AOB in the samples, the FISH technique (Coskunur 2000) was applied in order to produce the fluorescent sites in the cells, and these were detected through the use of confocal scanning laser microscopy (CSLM).

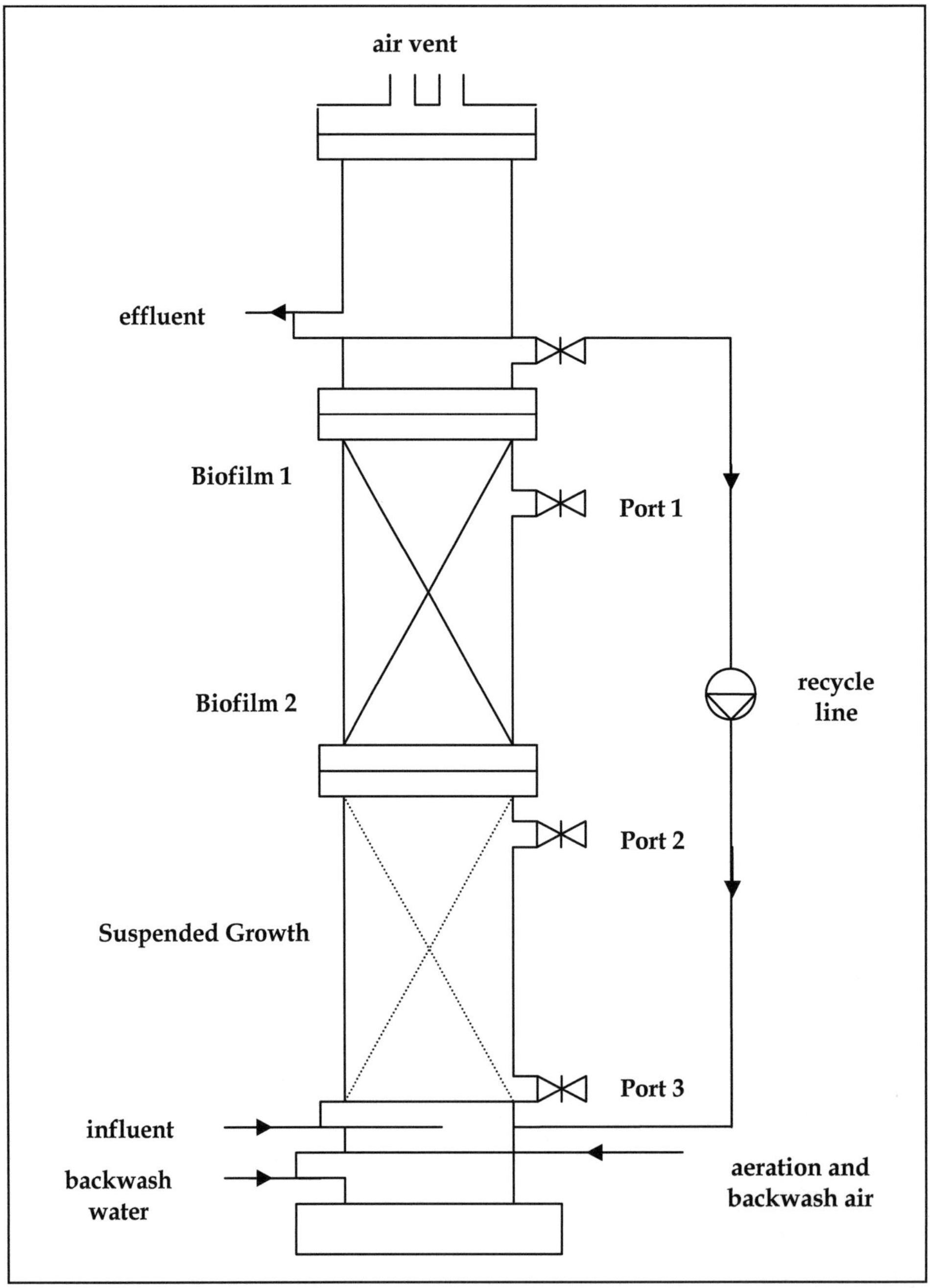

Fig. 1. Sampling locations for biofilm and suspended growth biomass along the reactor's height

2.2 Fluorescent *in situ* *h*ybridization (FISH) technique (coskunur 2000)

This method was applied to determine the presence of ammonia oxidizing bacteria (AOB) and to quantify them in the reactors. The steps involved fixation of the samples, permeabilization and hybridisation with probes, and finally detection with confocal laser scanning microscope (CLSM).

2.2.1Paraformaldehyde Fixation and Permeabilization

Generally, the samples used for this technique have undergone short term fixation where absolute ethanol was added in a volume ratio of 1 sample: 1 ethanol in sterile universal bottles and stored at -20° C.

A 1 ml volume of the stored sample was transferred to a 1.5 ml eppendorf tube and centrifuged at 13000 x g for 3 minutes. The supernatant was removed and the sample was washed with phosphate buffered saline (PBS) by adding 1 ml of the solution, mixing using vortex and centrifuging at 13000 x g for 3 minutes before removing the supernatant again. The resulting pellet was resuspended in 0.25 ml PBS and 0.75 ml PFA fixative and vortexed. A 4 % paraformaldehyde fixative solution was prepared fresh for every time of use, the procedure of which tabulated in Appendix 4.1. The suspension was incubated for at least 3 hours, or overnight, at 4°C.

After fixation, the cells were washed by centrifuging at 13000 x g for 3 minutes, removing the supernatant, adding 1 ml PBS and mixing. The samples were centrifuged again at 13000 x g for 3 minutes. The supernatant was removed and the sample was kept with PBS and absolute ethanol at 1:1 (v/v) and mixed. It was then stored at -20°C.

2.2.2 Hybridization

A volume of 250 µl of fixed sample was centrifuged at 13000 x g for 3 minutes and the supernatant was removed. The sample was washed once by adding 1 ml PBS and centrifuged again. The sample was then divided into four tubes: a negative control containing no probe to observe autofluorescence, a negative control to observe non-specific binding events, a positive control where a universal eubacterial probe was added (Bact 338) and a sample to be hybridised by a specific AOB detection probe. The samples were serially dehydrated in successively increasing concentrations of molecular grade ethanol (60%, 80%, 100% v/v). After adding 1 ml of the ethanol solution, the sample was vortexed and left for 3 minutes. The sample was then centrifuged at 13000 x g for 3 minutes and the supernatant was removed.

The following step is to hybridize the samples. Hybridisation buffer (HB) was prepared according to Amann et al (1990). HB was added so that the final volume including the probe will be 40 µl. Thus, for the negative control for autofluorescence, 40 µl HB is added. For a hybridisation containing only one probe (2ul), 38ul HB is added. For a hybridisation containing two probes (2+2 µl) 36 µl HB is added. The samples were prehybridized for 15 minutes at the hybridisation temperature. After prehybridisation, 2 µl of probe (50 ng/µl) was added to the samples that were then incubated at the optimal hybridisation temperature for the given probe (Table 1) for at least 4 hours (or overnight).

Following hybridisation, the samples were centrifuged at 13000 x g for 3 minutes and the supernatant was removed. A volume of 0.5 ml of wash buffer was added and the sample was mixed using a pipette before being incubated for 15 minutes at the same temperature as the hybridisation step. The washing step was again repeated.

Probe	Sequence	rRNA target	Target	Formamide ; Temperature	Reference
nonEUB	ACTCCTACGG GAGGCAGC		None (negative control)	0% ; 37°C	Amann et al (1990)
EUB338	5'GCTGCCTCCC GTAGGAGT-3'	16S	Eubacteria	20% ; 37°C	Amann et al (1990)
Nso1225	5'- CGCGATTGTAT TACGTGTGA-3'	16S	Ammonia oxidizing β-Proteobacteria	35% ; 51°C	Mobarry et al (1990)

Table 1. Features and conditions of probes during hybridisation

The samples were centrifuged again at 13000 x g for 3 minutes, the supernatant was removed and 1 ml of MilliQ water was added. Finally, the samples were centrifuged, the supernatant removed and the samples resuspended in 100 ul MilliQ water.

A 10 ul aliquot of the sample was added to a gelatine-coated slide with Teflon-coated wells of a known diameter (Appendix 4.1) and allowed to dry in a hybridization oven at 30°C. The sample spot on the slide was mounted in a small drop of the antifadent-Citifluor (AFI, Canterbury, UK). A cover glass was sealed carefully on the top of the slide by applying clear nail varnish to the edges to prevent movement during microscopy. The slide was then stored at -20°C in the dark and was prepared for viewing.

2.2.3 Scanning on a confocal laser microscope

The distribution of hybridized cells was subsequently visualised by means of a Leica TCS SP2 UV confocal laser scanning microscope (CLSM) equipped with Leica DMRXA microscope. Images were captured and processed using LCS V2.5.1040-1 software. For observation x 60 Na 1.32 lenses were applied.

The CLSM was run in the following mode: single channel for Fluorescene and double channel for Carbocyanine-5. Fluorescene was detected using excitation at 488 nm and a long pass emission filter in the range of 500-530 nm. Cy5 was detected using excitation at 633 nm and a long pass emission filter of 650-680 nm. The artificial colours green and red were assigned to the monochrome images acquired in the fluorescene and Cy5 channels respectively. The LCS software actively mixed colours so that a cell emitting red and green (the AOB) would appear yellow. For each sample, only 5 fields of view were randomly recorded in view of the time and budget available for the process.

2.2.4 Enumeration technique

An Excel spreadsheet constructed by Coskunur (2000) was used to carry out the calculation based on Equation 1 below:

$$K = \frac{(Nx2xA1)}{(A2x0.01x10xODF)} \tag{1}$$

where

K = average number of microcolonies in one ml of sample

A1 = area of sample spot (the area can be calculated from the diameter of the sample spot , [$\pi(D/2)^2$])

A2 = area of one field view
N = average number of ammonia oxidizer microcolonies/field of view
V = volume of sample applied
Vo = original volume of sample
ODF = other dilution factors not considered above may be required (e.g. volume of sample spun down). Where no ODF, default value = 1

The spreadsheet was designed for the quantification of AOB population in wastewater treatment plants following FISH and quantification typically using CLSM produced images. It requires that the user inputs data concerning the number of AOB microcolonies, the shortest and longest diameter of the microcolonies, area measurements of the fields of view and sample spots and dilution factors used in FISH. The spreadsheet returns the average number of microcolonies and geometric mean diameter. This data sheet can also be used to calculate the concentration of AOB in mg/l, the % AOB in terms of total bacterial population (measured by volatile suspended solids, VSS), following an empirically determined conversion factor, in terms of total cell numbers.

3. Comparison of AOB Cells in the biofilm and suspended growth samples

3.1 Cluster size

The relative frequencies of AOB cluster diameters for all the samples investigated are presented in Fig. 2.

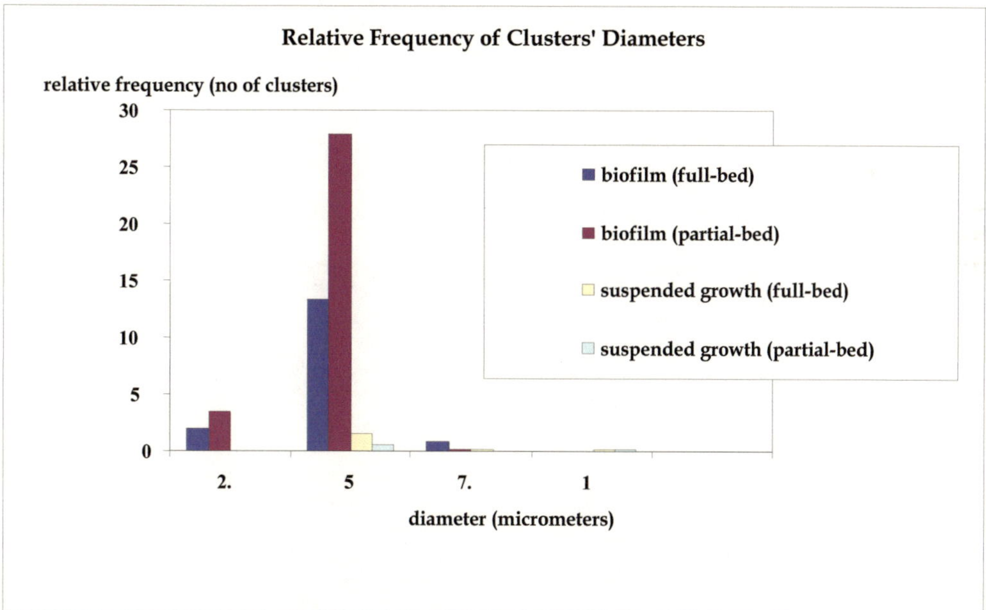

Fig. 2. Size distribution of cell clusters in the full- and partial-bed reactors

The results show that the majority of the clusters had diameters of 5 µm with the largest being 10 µm. These findings are quite consistent with the results obtained by Kloep et al (2000). Using probe Nsm 156, the majority of the hybridized clusters was found to be smaller than 10 µm and only a few were larger than 15 µm. Wagner et al (1995) also detected clusters hybridized with probe Neu 23 having diameters between 3 µm and 20 µm from samples of municipal sewage treatment plants. Nitrifier agglomerates are therefore small, for example well below those particle sizes (>100 µm) effectively removed by conventional primary sedimentation (Kiely 1998). Their retention in the system must therefore be mainly due to interactions with the biofilm attached to the media elements in the bed.

By visual observation, yellow clusters emerge on all biofilm samples as shown on Plates 1- 4. The AOB appear yellow due to double bindings of the fluorescene-labelled probe EUB 338 (emitted as green) and Cy5-labelled probe Nso 1225 (emitted as red). The formation of cluster growths is a feature of ammonia-oxidizing bacteria, in particular Nitrosomonas sp (Wagner et al 1995; Mobarry et al 1996). The clusters were spherical to oval shaped and appeared over diameters ranging from approximately 2.5 to 12.5 µm.

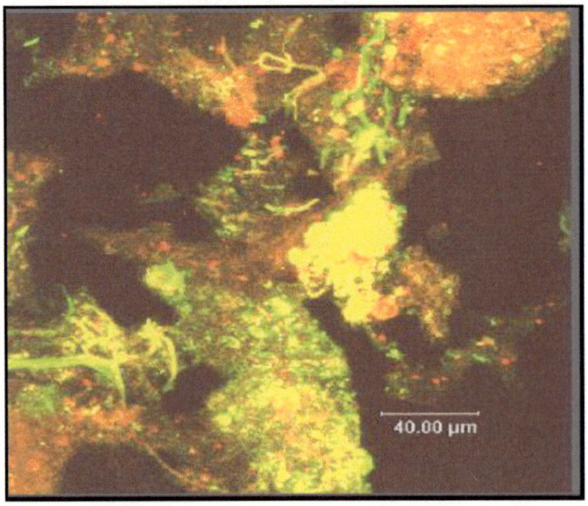

Plate 1. CLSM image of a biofilm sample from the top of the full-bed reactor

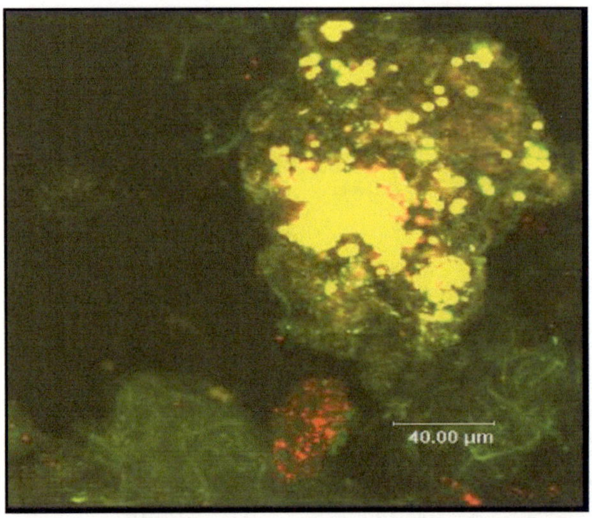

Plate 2. CLSM image of a biofilm sample from the middle of the full-bed reactor

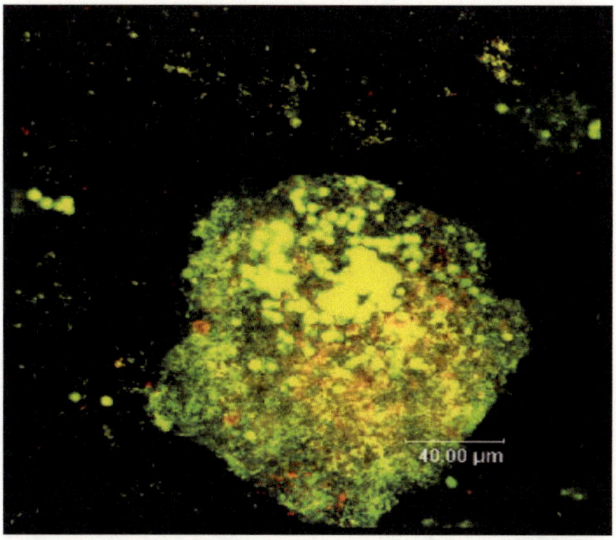

Plate 3. CLSM image of a biofilm from the top of the partial-bed reactor

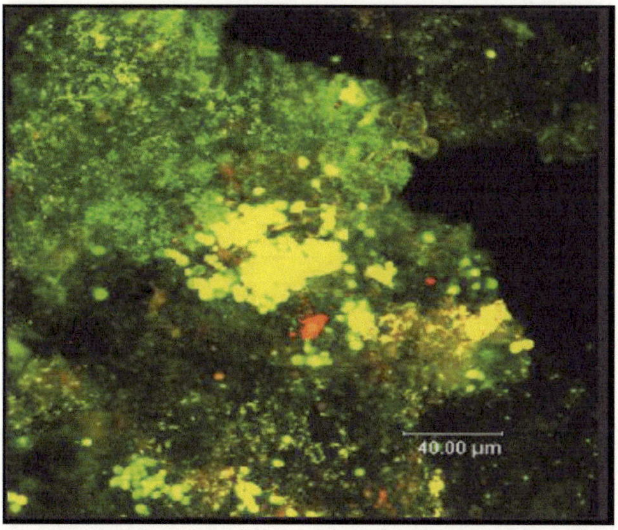

Plate 4. CLSM image of a biofilm from the middle of the partial-bed reactor

Plates 5 - 7 of suspended growth samples from the full-bed reactor show fewer AOB clusters than Plates 1 - 4. Layers of filamentous bacteria can be seen dominating, especially the suspended biomass samples from the top and middle parts of the reactors.

For the CLSM images of the suspended growth biomass samples from the partial-bed reactor, intense diffuse, green coloured fluorescence was often observed. This could have been due to debris, inorganic particles or the bacterial cells. A large number of coccoid structures was detected using the EUB 338 probe. They usually occurred in characteristic clumps and appeared ring shaped. MacDonald and Brozel (2000) observed the same phenomena in their study of bacterial biofilms in a simulated recirculating cooling-water reactor and suggested that this could result from dense chromosomal material at the cell center, leading to a concentration of ribosomes at the periphery of the cells.

Plate 5. CLSM image of suspended growth biomass from the top of the full-bed reactor

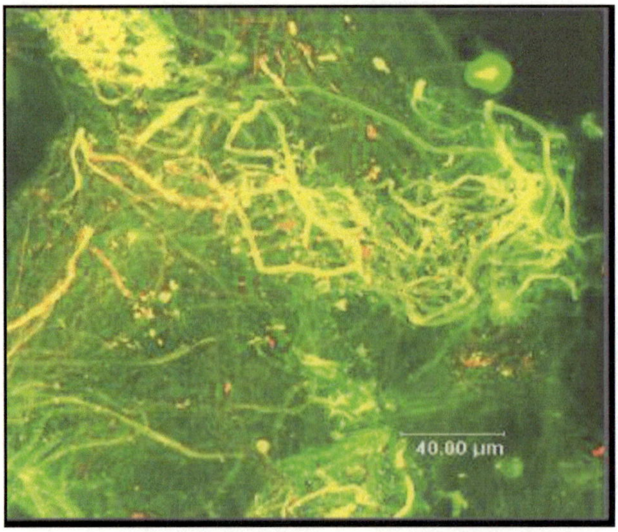

Plate 6. CLSM image of suspended growth biomass from the middle of the full-bed reactor

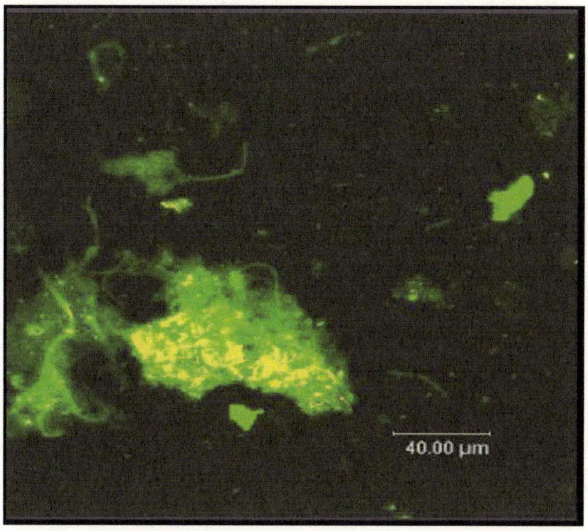

Plate 7. CLSM image of suspended growth biomass from the bottom of the full-bed reactor

3.2 Enumeration of ammonia-oxidizing bacteria

The number of AOB cells per ml of biomass was calculated from the counts based on cluster diameters using an Excel spreadsheet developed by Coskunur (2000). The numbers of AOB cells obtained are given in Table 2 below:

	Full-bed		Partial-bed	
	Biofilm	Suspended growth	Biofilm	Suspended growth
Top	1.720×10^5	2.149×10^4	5.589×10^5	1.075×10^4
Middle	2.204×10^5	1.344×10^4	2.929×10^5	ND
Bottom	6.451×10^4	1.345×10^4		8.075×10^3

Table 2. Number of AOB cells per ml of biomass in the biofilm and suspended growth samples

The higher number of AOB cells present in the biofilm samples than in the suspended growth samples could be due to the fact that AOB are slow-growing bacteria that need long mean solids' retention times to become established. Nitrifying bacteria, when compared with the heterotrophic organisms, are very much slower growing. Watson et al (1989) observed that the doubling times of these bacteria range from 8 hours to several days and that they have a tendency to attach to surfaces and to grow in cell aggregates referred to as zoogloeae or cysts (Lipponen et al 2002). In order to maintain an effective population of nitrifying bacteria within a biological reactor, a long retention time is required (Barber and Stuckey 2000). This is in accordance with the results obtained by Hidaka et al (2003), who discovered that in a biofiltration process for the advanced treatment of sewage, attached biomass contributed to most nitrification activity. Gerceker (2002) reported the loss of nitrification between SRTs of 0.9 and 2.4 days in a closely controlled jet-looped membrane bioreactor. Noguiera et al (2002) found that competition in biofilm results in a stratified biofilm structure, the fast-growing heterotrophic bacteria being drawn to the outer layers where both substrate concentration and detachment rate are high, whilst the slow-growing nitrifying bacteria stay deeper inside the biofilm. The heterotrophic layer has a positive effect on the nitrifiers by protecting them from detachment as long as the bulk oxygen concentration is high enough to preclude its depletion in the biofilm.

It is a fact that biofilm is significant in controlling long SRTs in a system. The full-bed reactor, which has a higher mass of biofilm than the partial-bed, as a result of the greater volume and surface area of the fully packed reactor, has SRTs of 21.2, 27.5 and 11.1 days at the three backwashing rates used in the study. The partial-bed reactor, on the other hand, had much shorter SRTs of 3.3, 3.9 and 2.7 days. Meanwhile, the biofilm in the partial-bed reactor was kept thin and stable, and therefore was not easily washed out during the backwash operation. Therefore, the retention time of biofilm in the partial-bed reactor is actually longer than the overall SRT of the system. Chuang et al (1997) pointed out that satisfactory nitrogen removal is achieved at SRT > 10 days.

The suspended growth biomass in the reactors, and especially that of the partial-bed reactor, was always subject to being washed out by the backwashing operation and lost in the effluent.

3.3 Significance of AOB Cells in the biofilm and suspended growth cultures

Tests carried out to compare the significance of AOB cells in both types of cultures were based on nonparametric methods of one-way ANOVA. Table 3 lists the results obtained.

	Full-bed		Partial-bed	
	Biofilm	Suspended growth	Biofilm	Suspended growth
Mean	$1.523 \times 10^5 \pm 7.979 \times 10^4$	$1.613 \times 10^4 \pm 4.645 \times 10^3$	$4.259 \times 10^5 \pm 1.881 \times 10^5$	$6.275 \times 10^3 \pm 5.596 \times 10^3$
Pooled s.d.	5.651×10^4		1.0867×10^5	
p-value	0.042		0.024	

Table 3. Results of variance analysis of AOB cells (no. AOB cells/ml sample) in the biofilm and suspended growth samples

Table 3 indicates that in both reactors there is a significant difference in the number of AOB cells in the biofilm and suspended growth samples. At 95% confidence levels, the p-value for the full-bed reactor is 0.042 whilst that of the partial-bed reactor is 0.024. Since the p-values obtained are smaller than 0.05, this means that in both reactors, specific cell concentrations of AOB were found to be significantly higher in the biofilm samples as compared to the suspended growth samples.

It was found that the AOB cells are more numerous in the biofilm samples than in the suspended growth samples of both the full- (p=0.042) and the partial-bed (p=0.024) reactors. It is therefore interesting to compare the significance of the overall AOB cells in the full- and partial-bed configurations, knowing that the mass of biofilm is lower in the partial-bed reactor due to the reduced media volume compared to the full-bed reactor.

Table 3 also indicates that there is no significant difference between the concentrations of AOB cells in the biofilm samples of the full- and partial-bed reactors (p=0.099), and also in the suspended growth samples (p=0.079). To put the overall abundance of AOB cells in the full and partial-bed reactors side-by-side, the AOB cells in the biofilm and suspended growth samples for each reactor were combined, giving total concentrations of AOB cells for that particular configuration. The p-value of specific AOB concentrations comparing the full- and partial-bed configuration is p=0.427. The value indicates an almost comparable AOB relative abundance in both the full- and partial-bed reactors. Higher mean AOB cells of the biofilm in the partial-bed reactor equate with the higher mean value of suspended growth samples in the full-bed reactor, resulting in almost equivalent mean AOB cells in both reactors.

Lazarova et al (1994) made a point that the balance between biofilm losses and growth processes on the outside of the media was dominated by shear forces, exerted by the liquid as it flowed past the media surfaces in the reactor. In a study to evaluate the essential role of hydrodynamic shear force in the formation of biofilm, Liu and Tay (2002) pointed out that biofilm density quasi-linearly increases with the increase of shear stress. Chang et al (1991) discovered that the medium concentration and the turbulence indicated by Reynolds numbers, significantly affected biofilm density and thickness of a fluidized bed biofilm reactor. In this type of reactor, increasing medium concentration can be associated with increasing attrition due to particle-to-particle contacts and increasing turbulence correlates flow fluctuations that could create forces normal to the biofilm, i.e. the shear stress. Table 4 illustrates the results obtained in their study.

Glass beads concentrations (g/l)	Reynolds number	Shear stress (dyne/cm^2)	Biofilm density (mg VS/cm^3)	Biofilm thickness (µm)
664.0	0.55	8.30	56.0	10.6
457.0	0.61	6.77	18.5	32.0
463.0	0.61	6.82	21.0	31.3
684.4	0.55	8.42	41.50	8.8
604.1	0.56	7.90	30.5	15.4
609.4	0.56	7.90	28.5	15.3
502.9	0.79	8.26	52.0	11.0
542.0	0.78	8.58	62.0	7.1
269.7	1.16	7.44	14.5	21.4
258.6	1.17	7.31	14.0	23.2
265.2	1.16	7.38	9.9	22.1

Table 4. Measured and calculated values for experimental runs with the fluidised bed biofilm reactor (Chang et al 1991)

In this study, since the medium is fixed, there is no attrition effect. Therefore turbulence effect could be the major factor that increases the detachment pressures, and caused the biofilm to become denser and thinner.

3.4 Relative concentration of AOB at different filter heights of the full- and partial-bed reactors

Fig. 3 illustrates the percentage values of AOB concentrations with respect to VSS concentrations in biofilm samples from the full-bed reactor.

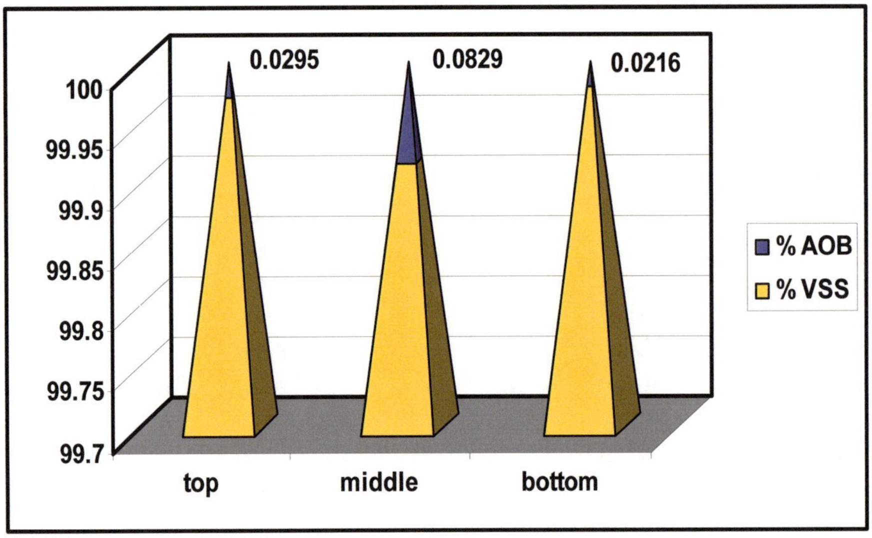

Fig. 3. Percentage values of AOB in the biofilm samples of the full-bed reactor

The highest percentage of AOB was found in a sample from the middle of the full-bed reactor (0.0829%), followed by the top part (0.0295%), whilst very little was found in the bottom part (0.0216%). A low percentage of AOB was obtained at the bottom despite the fact that the substrate and oxygen sources were supplied from here. This anomaly could best be explained by the fact that competition between heterotrophic and nitrifying bacteria for substrates (oxygen and ammonia) and space in the biofilms resulted in the fast-growing heterotrophic bacteria dominating the bottom part of the reactor. Plate 8 of biofilm sample from the bottom of the full-bed reactor show that AOB clusters are not dense as in Plates 1- 2 of the top and the middle positions.

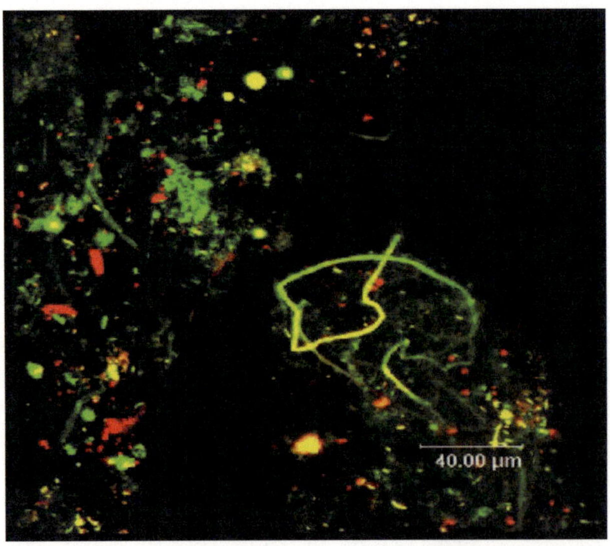

Plate 8. CSLM image of a biofilm sample from the bottom of the full-bed reactor

The trend of AOB growth in the biofilm samples of the full-bed reactor was followed through for the partial-bed reactor (Fig. 4):

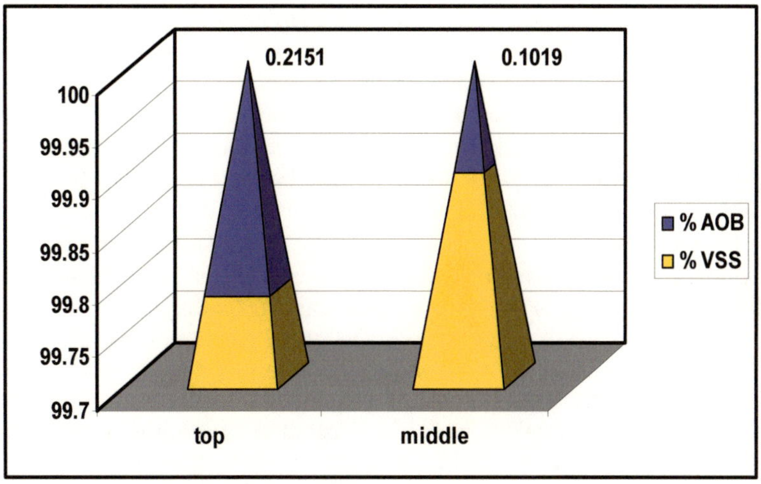

Fig. 4. Percentage values of AOB in the biofilm samples of the partial-bed reactor

The same argument of competition for substrates and space between heterotrophic bacteria and nitrifiers explained the lower percentage of AOB obtained in the middle (0.1019%) compared to the top part of the partial-bed reactor (0.2151%).
To validate the hypothesis made on AOB distribution in both the full and partial-bed reactors, a previous work by Wijeyekoon et al (2000) was used to investigate the effect of organic loading rates on nitrification activity. Table 5 summarizes the reactor conditions of their study.

Biofilters	A	B	C
Diameter (cm)	5	5	5
Height (cm)	50	50	50
Influent flow (l/h)	1.6	0.8	0.4
Influent conc. (mg/l TOC)	5	5	5
Influent nitrogen (mg/l NH_4^+-N)	5	5	5
OLR (kg COD/m^3.d)	0.19	0.098	0.097

Table 5. Unit dimensions and operating conditions of downflow biological filters (Wijeyekoon et al 2000)

The three reactors, packed with the same weights of anthracite, were equipped with sampling ports at depths of 6 cm (port 1), 18.5 cm (port 2) and 37.5 cm (port 3) from the top end of the filters. The specific rate of NH_4^+-N oxidation in the reactors was determined by the biomass extracted from those ports. It was discovered that the highest rates in filter A and B were obtained at the effluent ends of the reactors, but in filter C, the rates were comparably high from all ports. Also, among the three reactors, filter C produced the highest rates, with an average of 48.1 and 56.4 g N/(mg protein.hr) for ports 1 and 2 respectively. The conclusion derived from the study is that at high organic carbon loadings nitrifiers are non-uniformly distributed along the length of a filter, with excessive growth of heterotrophs near the feed end and nitrifiers at the effluent end under the influence of comparatively higher organic loading. Meanwhile, at low organic loadings, the heterotrophs and autotrophs can coexist. Filter C had the lowest organic carbon loading and consequently had the lowest biomass density. Therefore, the nitrifiers in filter C may have experienced less competitive pressure from the faster-growing heterotrophic organisms for oxygen and space. The displacement of the nitrifying population by the heterotrophs is caused by the varying ratio of carbon and nitrogen entering the reactor.

The carbon loading used in this part of study, 5.71 ±0.16 kg COD/m^3.d, was much higher than the loadings used by Wijeyekoon (Table 9.4), and therefore nitrifiers were not only displaced further away from the feed source, but also buried deeper into the biofilm (Ohashi et al 1995). Fdz-Polanco et al (2000) also observed that as the amount of organic carbon entering the filter increases, the nitrification activity is displaced to the upper part of the filter in an upflow process. Quyang et al (2000) also argued that the differences in biological activity at different filter heights were due to their varying loadings.

Rowan et al (personal communication) also investigated the percent value of AOB in a full-scale BAF plant treating municipal wastewater and obtained a value of 0.65%. This value is almost three times higher than the highest percentage obtained in this study (0.2151% from Figure 9.4). The difference in values could be attributed to a number of factors including carbon loading, nitrogen loading, pH, DO, media type and size, direction of flow, backwashing regime and thus mean SRT and biofilm attachment characteristics.

4. Conclusion

The extent of comparable nitrogen removal in the two reactor configurations needs further microbiological evidence, specifically that of the existence of AOB. The formation of a dense biofilm as a result of higher turbulence would account for the higher number of AOB cells enumerated in the biofilm samples from the partial-bed reactor ($4.259 \times 10^5 \pm 1.881 \times 10^5$ no of AOB cells/ml sample) as compared to those from the full-bed reactor ($1.523 \times 10^5 \pm 7.979 \times 10^4$ no of AOB cells/ml sample). Although biomass was washed out in the treated effluent and during backwash operation, the SRT at the high organic loading of 5.71±0.16 kg COD/m^3.d was still maintained at 4.2 days for the partial-bed reactor and 7.6 days for the full-bed reactor. These SRTs were still longer than the limit noted by Sastry et al (1999), who claimed that a mean cell residence time > 3 days is desirable for nitrifiers to reach a stable population for effective nitrification, and Gerçeker (2002) who recorded a loss of nitrification below 2.5-2.7 days at an OLR of 5 kg COD/m^3.d and a temperature of 25°C.

5. Acknowledgement

This chapter of the book could not have been written without the help of my PhD supervisor Prof Tom Donnelly who not only served as my supervisor but also encouraged and challenged me throughout my academic program. He and the other faculty members, Dr. Davenport and Dr Joana of University of Newcastle upon Tyne guided me through the process, never accepting less than my best efforts. I thank them all. And last but not least the Government of Malaysia for the sponsorship of my study.

6. References

Amann, R.I. (1995). Fluorescently labelled, rRNA-targeted Oligonucleotide Probes in the Study of Microbial Ecology. *Molecular Ecology*, Vol.4, pp.543-554.

Amann R.I., Krumholz L. and Stahl D.A. (1990). Fluorescent-oligonucleotide Probing of Whole cells for Determinative, Phylogenetic and Environmental Studies in Microbiology, *Journal of Bacteriology*, Vol.172, pp.762-770

Barber W.P. and Stuckey D.C. (2000). Nitrogen Removal in a Modified Anaerobic Baffled Reactor (ABR): 1, Denitrification, *Water Research*, Vol 34, No.9, pp.2413-2422.

Beer D. and Muyzer G. (1995). Multispecies Biofilms : Report from the Discussion Session. *Water Science and Technology*, Vol.32, No.8, pp. 269-270.

Chang H.T., Rittmann B.E., Amar D., Heim R., Ehlinger O. and Lesty Y. (1991). Biofilm Detachment Mechanisms in a Liquid-fluidized Bed. *Biotechnology and Bioengineering*, Vol.38, pp. 499-506.

Chen J.J., McCarty D., Slack D. and Rundle H. (2000). Full Scale Case Studies of a Simplified Aerated Filter (BAF) for Organics and Nitrogen Removal. *Water Science and Technology*, Vol.41, No.4-5, pp. 1-4.

Chuang S.H., Ouyang C.F., Yuang H.C. and You S.J. (1997). Effects of SRT and DO on Nutrient Removal in a Combined AS-biofilm Process, *Water Science and Technology*, Vol.36, No.12, pp. 19-27.

Chui P.C., Terashima Y., Tay J.H. and Ozaki H. (2001). Wastewater Treatment and Nitrogen Removal Using Submerged Filter Systems. *Water Science and Technology*, Vol.43, No.1, pp. 225-232.

Christensen B.E. and Characklis W.G. (1990). Physical and Chemical Properties of Biofilms. In: *Biofilms*, W.G. Characklis and K.C. Marshall, (Eds), John Willey, New York.

Coskunur G. (2000). *The Use of Fluorescent in-situ Hybridisation for the Study of Nitrification in Activated Sludge*. PhD Thesis, University of Newcastle upon Tyne, Civil Engineering Department, Newcastle UK.

Eschenhagen M., Schuppler M. and Roske I. (2003). Molecular Characterization of the Microbial Community Structure in Two Activated Sludge Systems for the Advanced Treatment of Domestic Effluents. *Water Reserach*, Vol.37, pp.3224-3232.

Fatihah S. (2004). *Effect of Full and Partial Bed Configuration on the Performance Characteristics of Biological Aerated Filters*. PhD Thesis. University of Newcastle Upon Tyne.

Fdz-Polanco F., Mendez E., Uruena M.A., Villaverde S. and Garcia P.A. (2000). Spatial Distribution of Heterotrophs and Nitrifiers in a Submerged Biofilter for Nitrification. *Water Research*, Vol.34, No.16, pp.4081-4089.

Gerçeker M. (2002). *Effect of Low Solids Retention Time on Macro-Nutrient Removal and Microbial Diversity in a Jet-Loop Membrane Bioreactor*. PhD thesis. University of Newcastle Upon Tyne.

Gieseke A., Arnz P., Amann R., Schramm A. (2002). Simultaneous P and N Removal in a Sequencing Batch Biofilm Reactor: Insights from Reactor and Microscale Investigations. *Water Research*, Vol.36, pp.501-509.

Hidaka T., Tsuno H. and Kishimoto N. (2003) Advanced Treatment of Sewage by Pre-coagulation and Biological Filtration Process. *Water Research*, Vol.37, pp.4259 - 4269.

Kiely G. (1998). *Environmental Engineering*. McGraw Hill. Singapore.

Kloep F., Roske I. and Neu T.R. (2000). Performance and Microbial Structure of a Nitrifying Fluidized Bed Reactor. *Water Research*, Vol.34, No.1, pp. 311-319.

Lazarova V., Capdeville B. and Nikolov L. (1994). Influence of Seeding Conditions on Nitrite Accumulation in a Denitrifying Fluidized Bed Biofilm Reactor. *Water Research*, Vol.28, No.5, pp.1189-1197.

Li J-H, Purdy K.J., Takii S. and Hayashi H. (1999). Seasonal Changes in Ribosomal RNA of Sulphate-reducing Bacteria and Sulphate Reducing Activity in a Freshwater Lake Sediment. *FEMS Microbiology Ecology*, Vol.28, pp.31-39.

Lipponen M.T.T., Suutari M.H. and Martikainen P.J. (2002). Occurrence of Nitrifying Bacteria and Nitrification in Finnish Drinking Water Distribution System. *Water Research,* Vol.36, pp.4319-4329.

Liu J. and Tay J.H. (2002). The Essential Role of Hydrodynamic Shear Force in the Formation of Biofilm and Granular Sludge. *Water Research,* Vol.36, pp.1653-1665.

Liu W-T., Chan O-C., and Fang H.H.P. (2002). Characterisation of microbial community in granular sludge treating brewery wastewater. *Water Research,* Vol.36, pp.1767-1775.

MacDonald R. and Brozel V.S. (2000). Community Analysis of Bacterial Biofilms in a Simulated Recirculating Cooling-water System by Fluorescent in Situ Hybridization with rRNS-Targeted Oligonucleotide Probes. *Water Research,* Vol.34, No.9, pp. 2439-2446.

Metcalf and Eddy (1991). *Wastewater Engineering: Treatment, Disposal and Reuse* (3rd edition), Mc Graw Hill Inc., New York.

Mobarry B.K., Wagner M., Urbain V., Rittmann B.E. and Stahl D.A. (1996). Pylogenetic Probes for Analyzing Abundance and Spatial Organization of Nitrifying Bacteria. *Applied Environmental Microbiology,* Vol. 62, pp. 2156-2162.

Nogueiro R., Mido L.F., Parkhold U., Wuertz S., Wagner M. (2002). Population Dynamics in Biofilm Reactors: Effects of Hydraulic Retention Time and the Presence of Organic Carbon. *Water Research,* Vol.36, pp.69-481.

Ohashi A., de Silva D.G.V., Mobarry B., Manem J.A., Stahl D.A. and Rittmann B.E. (1995). Influence of Substrate C/N Ratio on the Structure of Multi-species Biofilms Consisting of Nitrifiers and Heterotrophs. *Water Science and Technology,* Vol.32, No.8, pp.75-84.

Quyang C.F., Chiou R.J. and Lin C.T. (2000). The Characteristics of Nitrogen Removal by the Biofilm System. *Water Science and Technology,* Vol.42, No.12, pp. 137-147.

Rowan, A.K., Snape J.R. Fearnside, D., Curtis,,T.P., Barer, M.R., and Head, I.M. (2003). Composition and Diversity of Ammonia-oxidising Bacterial Communities in Wastewater Treatment Reactors of Different Design Treating Identical Wastewater. *FEMS Microbial Ecology,* Vol.43, pp.195-206.

Sastry, B., DeLosReyes A. Jr., Rusch K., and Malone R. (1999). Nitrification Performance of a Bubble-washed Bead Filter for Combined Solids Removal and Biological Filtration in a Recirculating Aquaculture System. *Journal of Aquacultural Engineering,* Vol.19, pp. 105-117.

Sliekers O., Derwort N., Gomez J.L.C., Strous M., Kuenen J.G. and Jetten M.S.M. (2002). Completely Autotrophic Nitrogen Removal Over Nitrate in One Single Reactor. *Water Research,* Vol.36, pp. 2475-2485.

Tsuneda S., Nagano T., Hoshino T., Ejiri Y., Noda N. and Hirata A. (2003). Characterisation of Nitrifying Granules Produced in an Aerobic Upflow Fluidised Bed Reactor. *Water Research,* Vol.37, pp. 4965 – 4973.

Wagner M., Rath G., Amann R., Koops H and Schleifer K. (1995). In situ Identification of Ammonia-oxidising Bacteria. *System Applied Microbiology,* Vol.18, pp. 251-264.

Wang Shi He and Zhou Ping (1994). Organic Matter Degradation Kinetics in a Fluidized Bed Bioreactor, *Water Research,* Vol.28, No.9, pp.2021-2028.

Wijeyekoon S., Mino T., Satoh H. and Matsuo T. (2000). Fixed Bed Biological Aerated Filtration for Secondary Effluent Polishing-effect of Filtration Rate on Nitrifying Biological Activity Distribution. *Water Science and Technology,* Vol. 41, No. 1, pp. 187-195.

5

A Combination of Phenotype MicroArray™ Technology with the ATP Assay Determines the Nutritional Dependence of *Escherichia coli* Biofilm Biomass

Preeti Sule, Shelley M. Horne and Birgit M. Prüß
North Dakota State University
USA

1. Introduction

Biofilms are defined as sessile communities of bacteria that form on surfaces and are entrapped in a matrix that they themselves produce. Biofilms cause severe problems in many natural (Ferris et al., 1989; Nyholm et al., 2002), clinical (Nicolle, 2005; Rice, 2006), and industrial settings (Brink et al., 1994; McLean et al., 2001; Wood et al., 2006), while being beneficial for waste water treatment and biofuel production (Wang and Chen, 2009). In addition, the bioremediation of crude oil spills involves a biofilm of oil degrading microbes, potentially supplemented by marine flagellates and ciliates (Gertler et al., 2010). Identifying the environmental conditions that prevent or support biofilm formation, as well as understanding the regulatory pathways that signal these conditions, is a pre-requisite to both, the solving of biofilm-associated problems and the use for beneficial purposes. In a previous study by our laboratory (Prüβ et al., 2010), it was determined that nutrition ranked among the more important environmental factors affecting biofilm-associated biomass in *Escherichia coli* K-12. The key to this study was a high-throughput experiment, where biofilm biomass was determined in a collection of cell surface organelle and global regulator mutants under a variety of combinations of environmental conditions. The cell surface organelles each represented a distinct phase of biofilm formation (Sauer et al., 2002). Flagella are required for reversible attachment (phase I), curli or type I fimbriae are characteristic of irreversible attachment (phase II), and a polymeric capsule forms the matrix that permits the maturation of the biofilm (phase III). Eventually, flagellated bacteria are released from the biofilm (phase IV). Phases III and IV are particularly problematic for the disease progression. Bacteria that are located deep within the mature biofilm are particularly resistant to antibiotics and dispersed bacteria tend to serve as a reservoir that continuously feed the infection. Please, see Figure 1 for the distinction of biofilm phases.

The global regulators included in our previous study (Prüβ et al., 2010) are involved in the co-ordinate expression and synthesis of biofilm-associated cell surface organelles. Many of them are components of two-component systems (2CSTS), each consisting of a histidine kinase and a response regulator (for reviews on 2CSTS signaling, please, see Galperin, 2004; Parkinson, 1993; West & Stock, 2001). In response to an environmental stimulus, the sensor kinase uses ATP as a phosphodonor to auto-phosphorylate at a conserved histidine, then transferring the phosphate to the response regulator at a conserved aspartate residue. In addition, many response regulators can be phosphorylated in a kinase independent manner by the activated acetate intermediate acetyl phosphate (for a review on acetyl phosphate as a signaling molecule, please, see Wolfe, 2005). One 2CSTS that is involved in the formation of biofilms is EnvZ/OmpR, regulating the synthesis of flagella (Shin and Park, 1995), type I fimbriae (Oshima et al., 2002), and curli (Jubelin et al., 2005). RcsCDB is involved in the formation of biofilms, serving as an activator of colanic acid production (Gottesman et al., 1985). RcsCDB constitutes a rare phosphorelay, consisting of three proteins and four signaling domains (Appleby et al., 1996). Much of the effect of EnvZ/OmpR, and RcsCDB upon biofilm formation involves FlhD/FlhC (Prüβ et al., 2006), which was initially described as a flagella master regulator (Bartlett et al., 1988) and later recognized as a global regulator of bacterial gene expression (Prüβ & Matsumura, 1996; Prüβ et al., 2001, 2003).

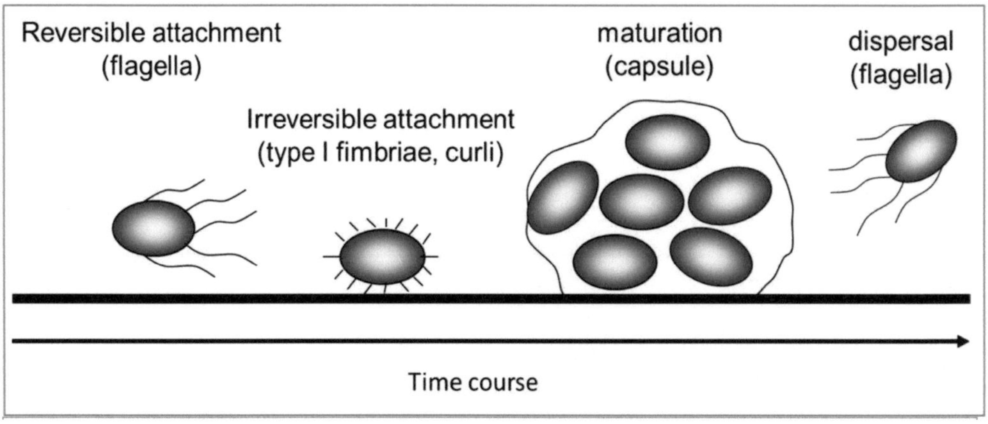

Fig. 1. Time course of biofilm formation

An early review article (Prüβ et al., 2006) summarized the portion of the transcriptional network of regulation that centered around FlhD/FlhC. This partial network contained 16 global regulators, among them many 2CSTSs, such as EnvZ/OmpR, RcsCDB, and CpxR. The regulation of approximately 800 genes was affected by the network. Since many of these encoded components of the biofilm-associated cell surface organelles, it was hypothesized that the network may affect biofilm formation. This hypothesis was confirmed by the high-throughput study that led to the identification of nutrition as one of the more instrumental factors in determining biofilm biomass (Prüβ et al., 2010). The global regulators that were part of the network led to the mutant collection for the experiment. Among the tested environmental conditions were temperature, nutrition, inoculation density, and incubation time. Temperature and nutrition were more important in determining biofilm biomass than were inoculation density and incubation time. The mutant screen was consistent with the idea that acetate metabolism may act as a nutritional sensor, relaying information about the environment to the development of biofilms. This hypothesis was confirmed by scanning electron microscopy. A new 2CSTS, DcuS/DcuR, was identified as important in determining the amount of biofilm-associated biomass (Prüβ et al., 2010).

The high-throughput experiment merely determined that nutrient rich bacterial growth media are more supportive of biofilm formation than are nutrient poor media. Specific nutrients that are supportive or inhibitory to biofilm formation were not determined and are the next logical step. This will be dependent on an assay system that quantifies biofilm biomass in the presence of an array of single nutrients. With this study, we will introduce such a system that quantifies biofilm biomass formed by *Escherichia coli* mutants in the presence of single nutrients by combining the Phenotype MicroArray™ technology from BioLog (Hayward, CA) with the ATP quantitative biofilm assay that was previously developed by our own lab (Sule et al., 2009), followed up by statistical analysis of the data, and metabolic modeling.

The BioLog Phenotype MicroArray (PM) technology has been developed for the determination of bacterial growth phenotypes (Bochner, 2009; Bochner et al., 2001, 2008). The PM technology consists of 96 well plates with 95 single nutrients dried to the base of each of 95 wells (the additional well constitutes the negative control). When used with the tetrazolium dye that is provided by the manufacturer and indicative of respiration, the PM system is used to determine growth of bacterial strains on single nutrients. Since the total system consists of 20 of such plates, the user is enabled to screen growth under close to 2,000 conditions. The plates are designated PM1 through PM20, with PM1 and PM2 containing carbon sources, PM3 containing nitrogen sources, and PM4 containing sulfur and phosphorous sources. The remaining plates can be used to determine the pH range of growth or resistance to antibiotics or other harsh conditions. Liquid growth media are supplied together with the respective plates.

With respect to bacterial growth, PMs have been used in numerous previous studies (Baba et al., 2008; Edwards et al., 2009; Mascher et al., 2007; Mukherjee et al., 2008; Zhou et al., 2003). However, use of this technology for the investigation of biofilms has been limited (Boehm et al., 2009). In *E. coli*, the use of PM technology for the quantification of biofilm biomass has not been reported. In addition, the previous use of PM technology in biofilm studies has been based on the use of the crystal violet assay for the quantification of biomass. There are, however, many more assays that have been developed for the quantification of biofilm-associated biomass, each of which serves a different purpose. The different quantitative biofilm assays are compared in Table 1.

Crystal violet is a non-specific protein dye that stains the bacterial cells and their exopolysaccharide matrix for dead and live bacteria alike. Biofilms are cultivated on 96 well plates and stained with 0.1% crystal violet in H_2O. In a second step, crystal violet is solubilized with a mix of ethanol and acetone (80:20) and measured spectrophotometrically (O'Toole et al., 1999; Pratt & Kolter, 1998). The assay was developed as a high-throughput assay that is suitable for robotic instrumentation (Kugel et al., 2009; Stafslien et al., 2006, 2007). ATP (adenosine triphosphate) (Sule et al., 2008, 2009) and XTT (4-nitro-5-sulfophenyl-5-[(phenylamino) carbonyl]-2H-tetrazolium hydroxide) (Cerca et al., 2005) are both assays that quantify the energy metabolism of the bacteria. Therefore, only biomass of live bacteria is considered. ATP is converted by the enzyme luciferase into a bioluminescence signal, XTT is reduced by NADH to an orange colored water-soluble formazan derivative. Similar to crystal violet, fluoro-conjugated lectins quantify the biomass of live and dead bacteria alike (Burton et al., 2006). Lectins are highly-specific carbohydrate binding proteins that have been utilized to quantify different cell wall components, as well as extracellular matrix (Stoitsova et al., 2004). Specifically, wheat germ agglutinin (WGA) and soybean agglutinin (SBA) selectively complex lipooligosaccharides and colanic acid, respectively. For our experiments, we needed an assay that quantifies biofilm biomass in live bacteria that is also suitable for high-throughput experimentation, cost effective, and rapid. The ATP assay appeared as the most suitable assay among the five compared assays (Table 1).

Assay	Live/dead cells	Detected material	High-throughput suitability	Reference
Crystal violet	Live and dead cells	Exopolysaccharide	Yes	(Kugel et al., 2009; Stafslien et al., 2006, 2007)
ATP	Live cells	Energy (ATP)	Yes	(Sule et al., 2008, 2009)
XTT	Live cells	Energy (NADH)	Yes	(Cerca et al., 2005)
WGA	Live and dead cells	Lipooligosaccharide	Not tested	(Burton et al., 2006; Stoitsova et al., 2004)
SBA	Live and dead cells	Colanic acid	Not tested	(Burton et al., 2006; Stoitsova et al., 2004)

Table 1. Comparison of different quantitative biofilm assays

In the past, ATP has been used as a measure of biomass (Monzón et al., 2001; Romanova et al., 2007; Takahashi et al., 2007) because its concentration is relatively constant across many growth conditions (Schneider & Gourse, 2004). For the quantification of biofilms, the BacTiter Glo™ assay from Promega (Madison WI) has been used for biomass determination in *Pseudomonas aeruginosa* (Junker & Clardy, 2007) and *E. coli* (Sule et al., 2008, 2009). In *E. coli*, we established that a two fold increase in bioluminescence did indeed relate to a two fold increase in the ATP concentration and a 2 fold increase in the number of bacteria (Sule et al., 2008). Across eight isogenic *E. coli* strains (one parent strain and seven mutants), differences in biofilm biomass that were determined with the ATP assay were paralleled by observations made with scanning electron microscopy (Sule et al., 2009).

The protocol involves the formation of the biofilms on 96 well micro titer plates, incubation at the desired temperature, and washing of the biofilms with phosphate buffered saline (PBS). Special attention is needed to distinguish the pellicle that forms at the air-liquid interface from the biofilm that forms at the bottom of the wells. In particular, the AJW678 derivatives that we are working with form a solid pellicle that covers the entire surface of the culture (Wolfe et al., 2003). For users who like to include the pellicle into their study, the growth medium and the PBS will be pipetted off carefully from each well. Users who wish to discard of the pellicle can flip the entire 96 well plate over and remove the liquid this way. Eventually, 100 µl of BacTiter Glo reagent are added to each well. After 5 min of incubation, bioluminescence is measured.

For this study, we will use the ATP assay to quantify biofilm biomass that forms on the PM1 plate of BioLog's PM system. The PM1 plate contains 95 single carbon sources in addition to the negative control. Besides the fact that the use of PM technology for the determination of the nutritional requirements of biofilm has not been reported in *E. coli* yet, the combination of PM technology with the ATP assay is novel. The combination of both, PM technology and ATP assay, together with the statistical analysis and metabolic modeling, enables the rapid screening of thousands of nutrients for their ability to support or inhibit growth and biofilm formation in one experimental setup. The described technique is not only cost-efficient and easy to perform, but also high-throughput in nature, providing valuable insight into the nutritional requirements during biofilm formation.

2. Materials and methods

2.1 Bacterial strains and growth conditions

The bacterial strains used in this study were the *E. coli* parental strain AJW678, which was characterized as an efficient biofilm former (Kumari et al., 2000) and its isogenic *flhD, fliA, fimA,* and *fimH* mutants. The *flhD* mutant was constructed by P1 transduction, using MC1000 *flhD*:kan (Malakooti, 1989) as a donor and AJW678 as a recipient. This resulted in strain BP1094. AJW2145 contained a *fliA*::*Tn5* insertion, AJW2063 a *fimA*::*Kn* mutation, and AJW2061 a *fimH*::*kn* mutation, all in AJW678 (Wolfe et al., 2003). The mutations abolish expression of FlhD/FlhC, FliA, FimA, and FimH, respectively. As a consequence, mutants in *flhD* and *fliA* are non-motile, whereas mutants in *fimA* are lacking the major structural subunit and mutants in *fimH* the mannose specific adhesive tip of the type I fimbrium. Bacterial strains were stored at -80°C in 8% dimethylsulfoxide, plated onto Luria Bertani plates (LB; 1% tryptone, 0.5% yeast extract, 0.5% NaCl, 1.5% agar) prior to use, and incubated overnight at 37°C. Bacterial strains are summarized in Table 2.

Strain	Relevant genotype	Reference
AJW678	*thi-1 thr-1*(am) *leuB6 metF159*(am) *rpsL136* Δ*lacX74*	(Kumari et al., 2000)
BP1094	AJW678 *flhD*::*kn*	(Prüß et al., 2010)
AJW2145	AJW678 *fliA*::*Tn5*	(Wolfe et al., 2003)
AJW2063	AJW678 Δ*fimA*::*kn*	(Wolfe et al., 2003)
AJW2061	AJW678 *fimH*::*kn*	(Wolfe et al., 2003)

Table 2. Bacterial strains used for this study

2.2 Strain selection for the biofilm experiment

For this study, a mutation was needed that would abolish one of the early cell surface organelles that contribute to the biofilm, while still permitting the formation of biofilms. We performed scanning electron microscopy (SEM) to determine the ability of the five bacterial strains (parental strain, *flhD* mutant, *fliA* mutant, *fimA* mutant, *fimH* mutant) to form biofilms. Biofilms were grown for 38 h at 37°C on glass cover slips with tryptone broth (TB; 1% tryptone, 0.5% NaCl) as a growth medium. Biofilms were fixed in 2.5% glutaraldehyde and prepared for SEM as described (Sule et al., 2009). Images were obtained with a JEOL JSM-6490 LV scanning electron microscopy (SEOL Ltd., Tokyo, Japan) at 3,000 fold magnification. 10 to 15 images were obtained per bacterial strain from at least three independent biological samples. One representative image is shown per bacterial strain.

2.3 Biofilm quantification with PM technology and the ATP assay

We used the PM1 plate of the BioLog PM system that contains 95 single carbon sources. When used with the tetrazolium dye that is provided by the manufacturer and indicative of respiration (Bochner et al., 2001), the PM system can be used for measuring growth of bacterial strains on single nutrients. We here describe a protocol for the determination of biofilm amounts (Figure 2).

As recommended by the manufacturer for the determination of growth phenotypes, the bacterial cultures were streaked from LB plates onto R2A plates (to deplete nutrient stores) and incubated at 37°C for 48 hours. Bacteria were removed from the plates with a flocked

swab (Copan, Murrieta CA), resuspended and then further diluted with IF-0a GN/GP Base (BioLog, Hayward CA) inoculation fluid to an optical density (OD_{600}) of 0.1. Leucine, methionine, threonine and thiamine were added at a final concentration of 20 µg/ml, the redox dye that is used for the determination of growth phenotypes was omitted for biofilm quantification. 100 µl of the inoculum was then dispensed into each of the 96 wells of the PM1 plates. The inoculated plates were wrapped with parafilm to minimize evaporation and incubated at 37°C for 48 hours. Biofilm amounts were quantified using the previously described ATP based technique (Sule et al., 2008, 2009). Briefly, the growth medium was carefully aspirated out of each well, minimizing loss of biofilm at the air liquid interface. The biofilms were then washed twice with phosphate buffered saline (PBS) in order to remove any residual media components. The biofilms were air dried and quantified using 100 µl BacTiter Glo™ reagent (Promega, Wisconsin, WI). The biofilms were incubated with the reagent for 10 min at room temperature and the bioluminescence was recorded using a TD 20/20 luminometer from Turner Design (Sunnyvale, CA). The bioluminescence was reported as relative lux units (RLU).
The determination of biofilm amounts in the presence of single nutrients was performed four times for each strain. In addition, growth on these carbon sources was determined in three independent replicate experiments, following the protocol that is described for the determination of growth phenotypes and including the redox dye (Bochner et al., 2001). Carbon sources on which both strains grew to an average OD_{600} of 0.5 or more were selected for the *t*-test analysis and carbon sources on which each strain grew to an average OD_{600} of 0.5 or more were selected for the ANOVA/Duncan analysis of biofilm amounts (see below).

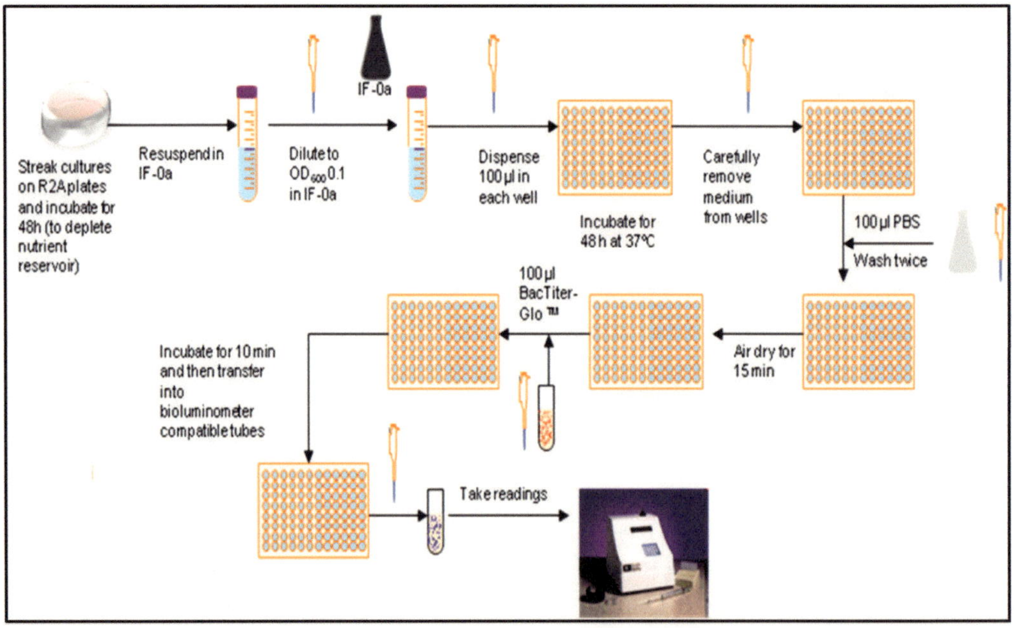

Fig. 2. Work flow for the determination of biofilm amounts on PM plates with the ATP assay

2.4 Data analysis

Prior to the statistical analysis, the biofilm amounts from each strain were normalized for experiment specific variation; total bioluminescence across each experiment was summed

up and the fold variation was calculated, using the lowest experiment as a norm (1 fold). Data points in each experiment were divided by the respective fold variation. The normalized experimental data sets were subjected to two independent types of statistical analysis, all done using SAS software (SAS Institute Inc., 2009). First, we performed Student's *t*-test on all those carbon sources on which both strains grew to an average OD_{600} of 0.5 or more to determine statistically significant differences between the amounts of biofilm that were formed on a given carbon source between the two strains. Since this analysis yielded more carbon sources than we could comprehend on a physiological level, we then analyzed each strain individually and then compared biofilm amounts on individual carbon sources for specific nutrient categories of structurally related carbon sources. For this analysis, the normalized biofilm data from each strain were subjected to separate one way ANOVAs, followed up with Duncan's multiple range tests. The tests compared the means of the amount of biofilm formed in the presence of each carbon source to all the other carbon sources within each strain. Carbon sources whose mean was different from the means of all the other carbon sources with statistical significance formed their own group in the Duncan's test. Carbon sources whose mean difference from the other carbon sources was not statistically significant formed overlapping groups.
Performing Duncan's test on the parent strain, two carbon sources formed groups A and B. Among the remaining carbon sources, we determined those that were structurally related to group A and B carbon sources. This was done after a determination of the respective chemical structures with the Kyoto Encyclopedia of Genes and Genomes (KEGG; Kanehisa & Goto, 2000; KEGG, 2006). Biofilm amounts formed by the *flhD* mutant were compared to the parent strain for all these carbon sources. In a second analysis, one carbon source formed group A in the Duncan's test for the *flhD* mutant. Among the remaining carbon sources, we identified two carbon sources that were structurally related. Biofilm amounts for these three carbon sources were compared between the two strains. For both analyses, data were summarized in a Table (3 and 4).

2.5 Metabolic modeling

Metabolic pathways that lead to the degradation of all the carbon sources that are discussed in this study were determined with KEGG. Metabolic intermediates that were common between different pathways were used to construct metabolic maps. Pathways for both strains were combined in Figures 5 and 6.

3. Results

3.1 Strain selection using electron micrographs

To determine the ability to form biofilm, electron microscopy was performed with the five strains that were listed in Materials and Methods. Figure 3 depicts one representative illustration of the 10 to 15 images that were obtained per bacterial strain. Most of these strains formed biofilm despite mutations affecting cell surface organelles of either reversible (flagella) or irreversible (type I fimbriae) attachment. The sole exception was the *fimH* mutant which only showed a small number of scattered bacteria attached across the slide. The *fimA* mutant exhibited a large number of filamentous appendages. We are currently unable to explain these appendages.

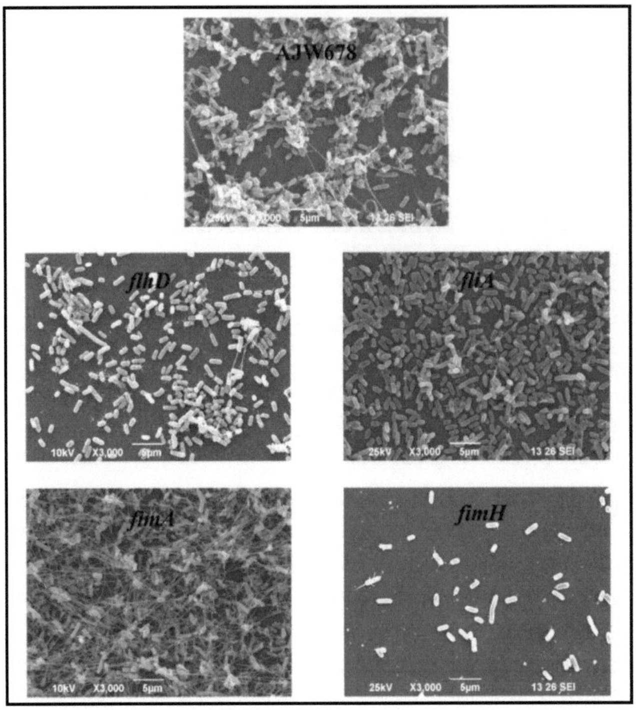

Fig. 3. Electron micrographs at 3,000 fold magnification for the AJW678 parent strain, and its isogenic mutants in *flhD*, *fliA*, *fimA*, and *fimH*

We wanted a strain for the phenotype microarray experiment that was able to form biofilm on complex media, while lacking one of the cell surface organelles. Since the amount of biofilm formed by the *flhD* mutant was similar to that of the parental strain in the electron micrographs, the *flhD* mutant was selected for further testing using the PM1 plates. The *flhD* mutant has as an additional advantage that much of the regulation by FlhD/FlhC has been previously described. This vast amount of information will help us to analyze the complex metabolic data.

3.2 Biofilm quantification with PM technology and statistical analysis

Biofilms that formed on the PM1 plates were quantified with the ATP assay and compared between the two strains with the *t*-test. The analysis did not yield any carbon sources that supported more biofilm in the parent strain than in the mutant. The 25 carbon sources that yielded significantly higher amounts of biofilm in the *flhD* mutant are demonstrated in Figure 4. Since the carbon sources that supported biofilm formation by the mutant more so than by the parent are numerous, we decided to analyze each strain statistically first and focus the comparison between the strains to specific structural categories of carbon sources. These are designated 'nutrient categories' throughout this manuscript.

3.2.1 Carbon sources that formed their own duncan's group for the parent strain

The normalized data set from the parent strain was subjected to Duncan's multiple range test. According to this test, the two carbon sources that were the best biofilm supporters for the parent *E. coli* strain, maltotriose and maltose, formed exclusive groups A and B. Without

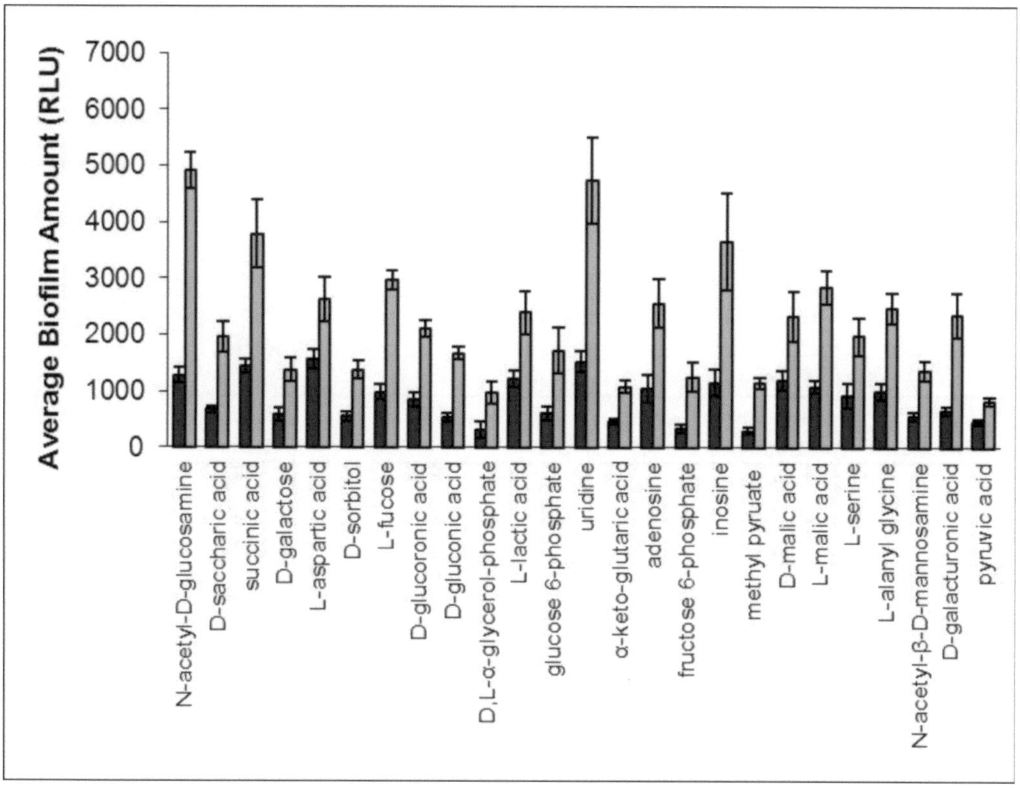

Fig. 4. Biofilm formation in the parent strain and the *flhD* mutant were compared using a *t*-test. The dark shaded bars resemble the parent strain, the lighter bars the mutant. The error bars in the graph indicate the standard deviation. Note that only carbon sources were included in this analysis that supported growth to at least 0.5 OD_{600} in both strains.

forming its own Duncan group, ribose was the carbon source that supported the smallest amount of biofilm among all carbon sources tested, while still supporting growth. The parent strain also formed good amounts of biofilm on the remaining C6-sugars. Interestingly, the amount of biofilm that formed on maltotriose (trisaccharide of glucose) was roughly three times the amount of biofilm that formed on glucose. The amount of biofilm that formed on maltose (disaccharide of glucose) was about twice the amount that formed on glucose. The C5-sugars xylose and lyxose did not support growth of the parental strain to the cutoff of 0.5 OD_{600}. For all these carbon sources, biofilm amounts formed by the *flhD* mutant were compared to the parent strain (Table 3). In contrast to the parental strain, the *flhD* mutant did not grow well on C6-sugars and their oligosaccharides. Unlike the parental strain, the mutant did not grow well on ribose, but grew to the cut off of 0.5 OD_{600} on lyxose and xylose. Still, the amount of biofilm formed by this strain on C5-sugars was low (<1,000 RLU). An interesting phenomenon was observed for sugar phosphates and sugar acids. Sugar phosphates supported biofilm production by the mutant more so (>1,200 RLU) than for the parent strain (<600 RLU). Likewise, sugar acids were found to be good supporters of biofilm for the *flhD* mutant strain (1,500 to 2,500 RLU), but not for the parent (500 to 800 RLU). This was even more remarkable, considering the fact that the parental strain ($OD_{600} \sim 1.0$) grew better on sugar acids than the *flhD* mutant (OD_{600} of 0.2 to 0.8).

Nutrient category	Nutrients	AJW678	*flhD* mutant
		Biofilm Amount (RLU)	**Biofilm Amount (RLU)**
Trisaccharide	Maltotriose	4,935	NA*
Disaccharide	Maltose	2,928	NA*
C6-sugars	Glucose	1,615	NA*
	Fructose	1,500	NA*
	Mannose	1,745	NA*
	Rhamnose	873	NA*
C5-sugars	Ribose	147	NA*
	Lyxose	NA	650
	Xylose	NA	544
Sugar phosphates	Glucose 6-P	614	1,722
	Fructose 6-P	338	1,258
Sugar acids	D-galacturonic acid	668	2,358
	D-gluconic acid	532	1,679
	D-glucuronic acid	852	2,110

Table 3. Biofilm amounts on carbon sources which formed their own Duncan's grouping for the parent strain and structurally related carbon sources. Columns 1 and 2 indicate the nutrient categories and single carbon sources for which data are included. Columns 3 and 4 represent biofilm amounts for the parent strain and the mutant on carbon sources that permitted growth to more than 0.5 OD_{600}. NA denotes 'not applicable', where the strain grew to an OD_{600} below 0.5.

3.2.2 Carbon source that formed its own duncan's group for the *flhD* mutant

The amount of biofilm formed on each carbon source by the *flhD* mutant was quantified and subjected to Duncan's multiple range test. According to the Duncan's grouping, the sole carbon source that formed its own group A for the *flhD* mutant was N-acetyl-D-glucosamine. Structurally related carbon sources that were included in the PM1 plate are D-glucosaminic acid and N-acetyl-β-D-mannosamine. Biofilm amounts formed on these three carbon sources were compared between the two strains (Table 4).

Nutrient category	Nutrients	*flhD* mutant	AJW678
		Biofilm Amount (RLU)	**Biofilm Amount (RLU)**
Sugar amines	N-acetyl-D-glucosamine	4,911	1,285
	D-glucosaminic acid	660	NA
	N-acetyl-β-D-mannosamine	1,368	559

Table 4. Biofilm amounts on carbon sources which formed their own Duncan's grouping for the *flhD* strain and structurally related carbon sources. Columns 1 and 2 indicate the nutrient category and single carbon sources for which data are included. Columns 3 and 4 represent biofilm amounts for the *flhD* mutant and its parent strain on carbon sources that permitted growth to more than 0.5 OD_{600}. NA denotes 'not applicable', where the strain grew to an OD_{600} below 0.5.

On N-acetyl-D-glucosamine, the *flhD* mutant (4,900 RLU) formed a significantly larger amount of biofilm than the parent strain (1,300 RLU), while both strains grew to approximately 1 OD_{600}. On D-glucosaminic acid, the parent strain did not grow to the cutoff OD of 0.5. The *flhD* mutant grew well, but the amount of biofilm biomass was poor (~600 RLU). For N-acetyl-β-D-mannosamine, both strains grew well, the *flhD* mutant expressed more than twice the ability to form biofilm than its isogenic parent.

3.3 Metabolic modeling

Metabolic pathways were drawn for the degradation of all those carbon sources that supported amounts of biofilm larger than 1,000 RLU for one of the tested strains. These are carbon sources of the nutrient categories C6-sugars, sugar phosphates, sugar acids, and sugar amines. C6-sugars all have pathways that feed into the Embden-Meyerhof pathway, sugar phosphates are intermediates of this pathway. As shown in Figure 5, mannose, fructose, and N-acetyl D-glucosamine feed into fructose 6-phosphate. Gluconate, glucuronate, galacturonate, and rhamnose feed into glyceraldehyde 3-phosphate. This leads to the production of acetyl-CoA, acetyl phosphate and acetate (Figure 6).

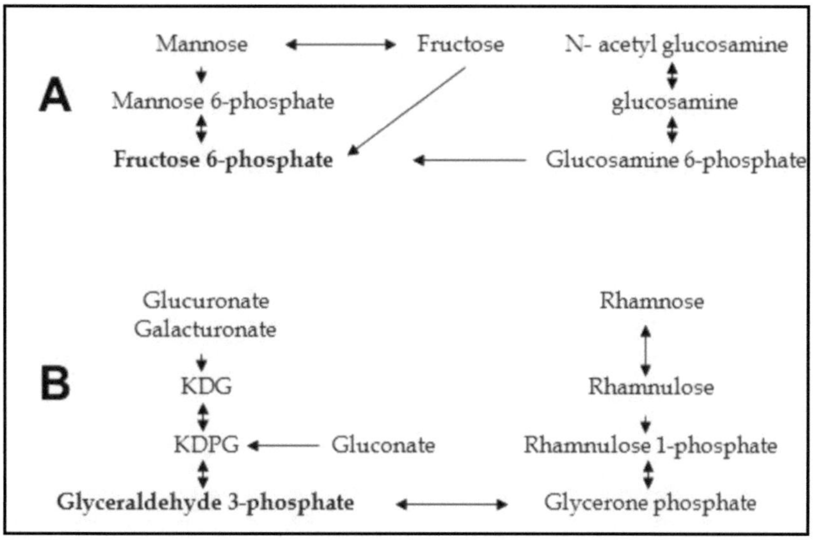

Fig. 5. Metabolic pathways from the top biofilm producing carbon sources for both *E. coli* strains, feeding into the Embden-Meyerhof pathway.

4. Discussion

4.1 Development of the combination assay

Altogether, we present an assay that builds upon two previous assays, the PM technology and the ATP assay. Both assays have been used in much different contexts previously. PM plates have been commonly used to discover various bacterial characteristics based on phenotypic changes (Bochner et al., 2008). Studies involving PM plates include the evaluation of the alkaline stress response induced changes in the metabolism of *Desulfovibrio vulgaris* (Stolyar et al., 2007). PMs have also been used for the identification of bacterial species (Al-Khaldi & Mossoba, 2004). The use of PM technology in biofilm research is

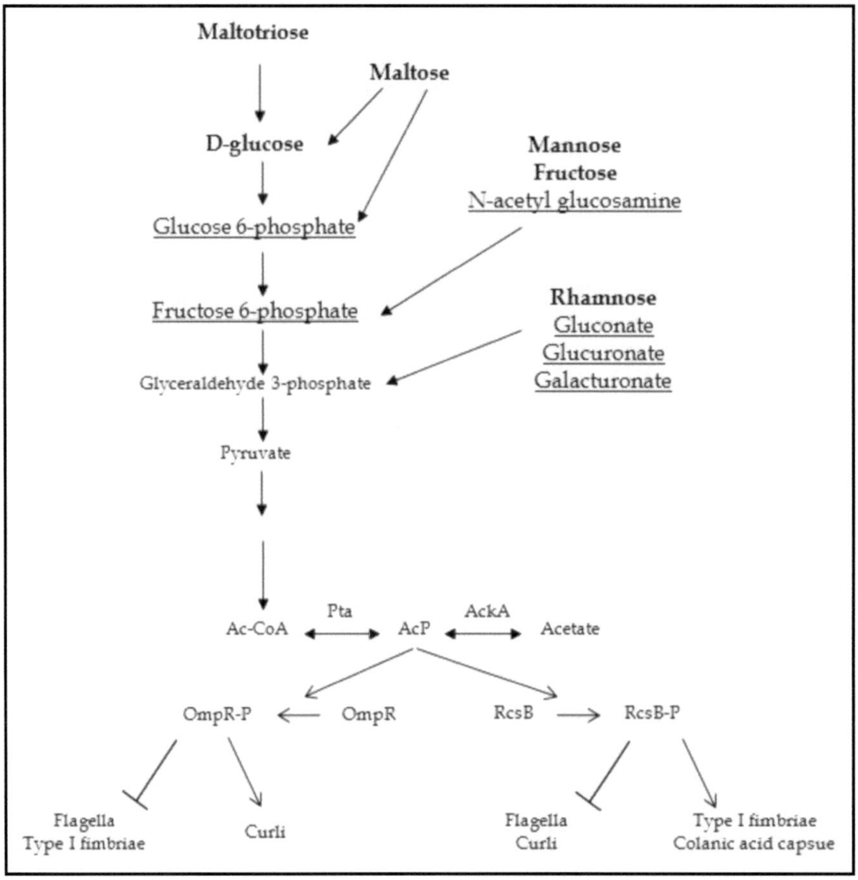

Fig. 6. Metabolic pathways from the top biofilm producing carbon sources for both strains to the production of acetate. Carbon sources that are printed in bold were top biofilm supporters for the parent strain. Carbon sources that are underlined were top biofilm supporters for the *flhD* mutant. The effect of acetyl phosphate on RcsB and OmpR on the synthesis of flagella, curli, fimbriae, and capsule is indicated.

limited to a study of the ability of *E. coli* to form biofilm upon ribosomal stress (Boehm et al., 2009). That study used the crystal violet assay as a detection tool for the amount of biofilm. Here we report for the first time a combination of the established ATP assay along with the PM technology to assess nutritional dependence of *E. coli* during biofilm formation. Since the statistics approach alone (*t*-test) yielded no more than a list of data that were difficult to interpret, we decided for a combined statistics/metabolism approach to analyze the complex data. The combination of the two experimental parts of the assay together with the two analysis parts enables the user to rapidly screen hundreds and thousands of single nutrients for their ability to inhibit growth and biofilm formation in one experimental setup. Integrating different mutants into the study will yield valuable insight into the regulatory mechanisms that are involved in the signaling of these nutrients. The described technique is not only cost-efficient and easy to perform, but also high-throughput in nature. It is ideally suited to provide valuable insight into the nutritional requirements that determine biofilm biomass, as well as the respective signaling pathways.

4.2 Biological analysis of the data

In the described study, we observed that the FlhD mutants made quantitatively higher amounts of biofilms on numerous carbon sources. Interestingly, the parental strain did not form higher quantities of biofilm than the mutant on any of the tested carbon sources. These observations shed light into the ongoing controversial debate, elucidating the role of motility in biofilm formation. In certain bacterial species including *Yersinia enterocolitica,* the presence of motility has been shown to be beneficial for biofilm formation (Wang et al., 2007). Several previous studies from our lab demonstrate that the absence of motility enhances the ability of *E. coli* to form substantial amounts of biofilm. As one example, strains transformed with the FlhD expressing plasmid pXL27 showed diminished biofilm forming capabilities (Prüß et al., 2010). Additionally, ongoing studies carried out in the lab with *E. coli* O157:H7 and the *E. coli* K-12 strains MC1000 and AJW678 point in the same direction, exemplifying our belief that FlhD and motility are detrimental to biofilm formation for our bacterial strains and under the conditions of our experiments (Sule et al., unpublished data).

As a second observation, carbon sources that supported maximal biofilm formation by either strain all fed into glycolysis eventually, and produced actetate. Although the carbon sources that promoted the highest biofilm amounts were different for the two strains, they still were in the same pathway. The previous high-throughput experiment that had pointed towards nutriition as instrumental in determining biofilm associated biomass had also postulated acetate metabolism as one of the key players in biofilm formation (Prüß et al., 2010). Phosphorylation of OmpR and RcsB by the activated acetate intermediate acetyl phosphate (Kenney et al., 1995) and acetylation of RcsB by acetyl-CoA (Thao et al., 2010) have been described in the past. These activated 2CSTS response regulators then affect the expression level of biofilm associated cell surface organelles, such as flagella, type I fimbriae, curli, and capsule (Ferrieres & Clarke, 2003; Francez-Charlot et al., 2003; Oshima et al., 2002; Prüß, 1998; Shin & Park, 1995) (Figure 6). The positive effect on biofilm amounts of carbon sources that lead to the production of acetate can be explained with the combined inhibitory effect of acetyl phosphate and acetyl-CoA on flagella through OmpR and RcsB and the above described disadvantage of flagella and motility during biofilm formation. We however do not state that acetate is the sole controlling mechanism as the complexity of the bacterial system cannot be explained based on a small number of signaling molecules.

The most striking observation obtained from our studies pertains to the pattern of growth and biofilm formation on sugar acids. It was observed that the FlhD mutants grew to lower optical densities on sugar acids, but formed much higher amounts of biofilm as compared to the parental strain. Previous work from the Prüß lab had shown similar defects in growth of *flhD* mutants on sugar acids (Prüß et al., 2003), biofilm formation was not tested in that study. The inverse effect of sugar acids on growth and biofilm amounts may have implications in the intestine. Mutants in *flhD* have an early disadvantage in colonization, but recover after prolonged incubation (Horne et al., 2009). They even take over the population after more than two weeks (Leatham et al., 2005). The initial lack of colonization could be explained by the inability of the *flhD* mutant to degrade the numerous sugar acids present in the intestine (Peekhaus & Conway, 1998). On the other hand, the ability to take over the bacterial population at a later stage may have to do with the lack of the flagellin, which is a potent cytokine inducer (McDermott et al., 2000). The here discovered ability to make an increased amount of biofilm may add to the long term survival of *flhD* mutants in the intestine. Bacteria deep within the biofilm will be protected from the immune system, while metabolizing very slowly and not needing much nutrition.

Among the carbon sources that were the least supportive of biofilm formation, the inability of the C5-sugars to support growth and/or biofilm formation was the most striking. Ribose supported growth by the parent strain, but yielded the lowest biofilm amount of all tested carbon sources. The *flhD* mutant did not even grow on ribose. According to Fabich and coworkers (Fabich et al., 2008), ribose is not among the carbon sources that the *E. coli* K-12 strain MG1655 utilizes when bacteria colonize the intestine. Our data are consistent with this observation. Since *E. coli* O157:H7 EDL933 does actually utilize ribose in the intestine, ribose utilization may constitute a mechanism by which pathogenic *E. coli* can find a niche in the intestine to co-exist with the commensal *E. coli* strains.
The inability to grow on lyxose is also consistent with previous observations, where only a mutation in the *rha* locus enabled the bacteria to grow on lyxose via the rhamnose pathway (Badia et al., 1991). Normally, *E. coli* are unable to grow on lyxose. Most interesting is the behavior of the two strains on xylose. The parent *E. coli* strain was unable to grow on xylose. The *flhD* mutant did grow, while producing moderately low amounts of biofilm. Co-utilization of glucose and xylose by *E. coli* strains is of upmost importance during the production of biofuels, since the fermented plant material contains both, cellulose (polymer of glucose) and hemicellulose (polymer of glucose and xylose), in addition to lignin. Much research is currently dedicated to the genetic modification of *E. coli* that enables the bacteria to utilize xylose more efficiently (Balderas-Hernandez et al., 2010; Hanly & Henson, 2010). It would be interesting to see whether a mixture of our parent strain and its isogenic *flhD* mutant would be able to co-utilize glucose and xylose, particularly since the mutant produced a moderate amount of biofilm which can also be beneficial to the production of biofuels.

5. Conclusion

In summary, we developed an assay system that quantifies biofilm biomass in the presence of distinct nutrients. The assay enables the user to screen a large number of such nutrients for their effect on biofilm amounts. Examples of metabolic analysis relate back to previous literature, as well as giving raise to new hypotheses. Yielding further evidence for the previous hypothesis that acetate metabolism was important in determining biofilm amounts can serve as a positive control that the assay actually yields data of biological significance. Particularly with respect to life in the intestine and the production of biofuels, the data open new avenues of research by providing testable hypotheses. Overall, there is no limit to extensions of the assay into different bacterial species or serving the development of high-throughput data mining algorithms that will computerize the statistic/metabolic analysis that we started in this study.

6. Acknowledgement

The authors like to thank Dr. Alan J. Wolfe (Loyola University Chicago, Maywood IL) for providing the bacterial strains that were used for this study, Dr. Jayma Moore (Electron Microscopy Lab, NDSU) for help with the scanning electron microscopy, Dr. Barry Bochner (BioLog, Hayward CA) for helpful discussions during the development of the combination assay, and Curt Doetkott (Department of Statistics, NDSU) for performing the statistical analyses of our data and helping us with their interpretation. The work was funded by an earmark grant on Agrosecurity: Disease Surveillance and Public Health through

USDA/APHIS and the North Dakota State Board of Agricultural Research and Education. Figure 2 was created using Motifolio (Motifolio Inc., Ellicott MD).

7. Reference

Al-Khaldi, S. & Mossoba, M. (2004). Gene and bacterial identification using high-throughput technologies: genomics, proteomics, and phenomics. *Nutrition*, Vol.20, No.1, (January 2004), pp. 32-38, ISSN 0899-9007

Appleby, J.; Parkinson, J. & Bourret, R. (1996). Signal transduction via the multi-step phosphorelay: not necessarily a road less traveled. *Cell*, Vol.86, No.6, (September 20,1996), pp. 845-848, ISSN 0092-8674

Baba, T.; Huan, H.; Datsenko, K.; Wanner, B. & Mori, H. (2008). The applications of systematic in-frame, single-gene knockout mutant collection of *Escherhichia coli* K-12, In: *Microbial Gene Essentiality: Protocols and Bioinformatics*, (*Methods in Molecular Biology Vol.416*), A. Osterman and S. Gerdes, (Eds.), pp. 183-194, Humana Press, ISBN 978-1-58829-378-7, New York, USA

Badia, J.; Gimenez, R.; Baldomá, L.; Fessner, W. & Aguilar J. (1991). L-lyxose metabolism employs the L-rhamnose pathway in mutant cells of *Escherichia coli* adapted to grow on L-lyxose. *Journal of Bacteriology*, Vol.173, No.16, (August 1991), pp. 5144-5150, ISSN 0021-9193

Balderas-Hernández, V,; Hernández-Montalvo, V.; Bolivar, F.; Gosset, G. & Martinez, A. (2010). Adaptive evolution of *Escherichia coli* inactivated in the phosphotransferase system operon improves co-utilization of xylose and glucose under anaerobic conditions. *Applied Biochemistry and Biotechnology*, PMID: 20740380 (August 26, 2010), ISSN 1559-0291

Bartlett, D.; Frantz, B. & Matsumura, P. (1988). Flagellar transcriptional activators FlbB and FlaI: gene sequences and 5' consensus sequences of operons under FlbB and FlaI control. *Journal of Bacteriology*, Vol.170, No.4, (April 1988), pp. 1575-1581, ISSN 0021-9193

Bochner, B.; Gadzinski, P. & Panomitros, E. (2001). Phenotype microarrays for high-throughput phenotypic testing and assay of gene function. *Genome Research*, Vol.11, No.7, (July 1, 2001), pp. 1246-1255, ISSN 1088-9051

Bochner, B.; Giovannetti, L. & Viti, C. (2008). Important discoveries from analysing bacterial phenotypes. *Molecular Microbiology*, Vol.70, No.2, (October 2008), pp. 274-280, ISSN 0950-382X

Bochner, B. (2009). Global phenotypic characterization of bacteria. *FEMS Microbiology Reviews*, Vol.33, No.1, (January 2009), pp. 191-205, ISSN 0168-6445

Boehm, A.; Steiner, S.; Zaehringer, F.; Casanova, A.; Hamburger, F.; Ritz, D.; Keck, W.; Ackermann, M.; Schirmer, T. & Jenal, U. (2009). Second messenger signalling governs *Escherichia coli* biofilm induction upon ribosomal stress. *Molecular Microbiology*, Vol.72, No.6, (June 2009) pp. 1500-1516, ISSN 0950-382X

Brink, D.; Vance, I. & White, D. (1994). Detection of *Desulfobacter* in oil field environments by non-radioactive DNA probes. *Applied Microbiology and Biotechnology*, Vol.42, No.2 - 3, (November 1994), pp. 469-475 ISSN 0175-7598

Burton, E.; Yakandawala, N.; LoVetri, K. & Madhyastha, M. (2006). A microplate spectrofluorometric assay for bacterial biofilms. *Journal of Industrial Microbiology & Biotechnology*, Vol.34, No.1, (January 2007), pp. 1-4, ISSN 1367-5435

Cerca, N.; Martins, S.; Cerca, F.; Pier, G.; Oliveira, R. & Azeredo, J. (2005). Comparative assessment of antibiotic susceptibility of coagulase-negative staphylococci in biofilm versus planktonic culture as assessed by bacterial enumeration or rapid XTT colorimetry. *Journal of Antimicrobial Chemotherapy*, Vol.56, No.2, (August 2005), pp. 331-336, ISSN 0305-7453

Edwards, R.; Dalebroux, Z. & Swanson, M. (2009). *Legionella pneumophila* couples fatty acid flux to microbial differentiation and virulence. *Molecular Microbiology*, Vol.71, No.5, (March 2009), pp. 1190-1204, ISSN 0950-382X

Fabich, A.; Jones, S.; Chowdhury, F.; Cernosek, A.; Anderson, A.; Smalley, D.; McHargue, W.; Hightower, G.; Smith, J.; Autieri, S.; Leatham, M.; Lins, J.; Allen, R.; Laux, D.; Cohen, P. & Conway, T. Comparison of carbon nutrients for pathogenic and commensal *Escherichia coli* strains in the mouse intestine. *Infection and Immunity*, Vol.76, No.3, (March 2008), pp. 1143-1152, ISSN 0019-9567

Ferrieres, L. & Clarke, D. (2003). The RcsC sensor kinase is required for normal biofilm formation in *Escherichia coli* K-12 and controls the expression of a regulon in response to growth on a solid surface. *Molecular Microbiology*, Vol.50, No.5, (December 2003), pp. 1665-1682, ISSN 0950-382X

Ferris, F.; Schultz, S.; Witten, T.; Fyfe, W. & Beveridge, T. (1989). Metal interactions with microbial biofilms in acidic and neutral pH environments. *Applied and Environmental Microbiology*, Vol.55, No.5, (May 1989), pp. 1249-1257, ISSN 0099-2240

Francez-Charlot, A.; Laugel, B.; Van Gemert, A.; Dubarry, N.; Wiorowski, F.; Castanié-Cornet, M.; Gutierrez, C. & Cam, K. (2003). RcsCDB His-Asp phosphorelay system negatively regulates the *flhDC* operon in *Escherichia coli*. *Molecular Microbiology*, Vol. 49, No.3, (August 2003), pp. 823-832, ISSN 0950-382X

Galperin, M. (2004). Bacterial signal transduction network in a genomic perspective. *Environmental Microbiology*, Vol.6, No.6, (June 2004), pp. 552-567, ISSN 1462-2920

Gertler, C.; Näther, D.; Gerdts, G.; Malpass, M. & Golyshin, P. (2010). A mesocosm study of the changes in marine flagellate and ciliate communities in a crude oil bioremediation trial. *Microbial Ecology*, Vol.60, No.1, (July 2010), pp. 180-191, ISSN 0095-3628

Gottesman, S.; Trisler, P. & Torres-Cabassa, A. (1985). Regulation of capsular polysaccharide synthesis in *Escherichia coli* K-12: characterization of three regulatory genes. *Journal of Bacteriology*, Vol.162, No.3, (March 1985), pp. 1111-1119, ISSN 0021-9193

Hanly, T. & Henson, M. (2011). Dynamic flux balance modeling of microbial co-cultures for efficient batch fermentation of glucose and xylose mixtures. *Biotechnology and Bioengineering*, Vol.108, No.2, (February 2011), pp. 376-385, ISSN 0006-3592

Horne, S.; Mattson, K. & Prüß, B. (2009). An *Escherichia coli aer* mutant exhibits a reduced ability to colonize the streptomycin-treated mouse large intestine. *Antonie van Leeuwenhoek*, Vol.95, No.2 (February 2009), pp. 149-158, ISSN 0003-6072

Jubelin, G.; Vianny, A.; Beloin, C.; Ghigo, J.; Lazzaroni, J.; Lejeune, P. & Dorel, C. (2005). CpxR/OmpR interplay regulates curli gene expression in response to osmolarity in *Escherichia coli*. *Journal of Bacteriology*, Vol.187, No.6, (March 2005), pp. 2038-2049, ISSN 0021-9193

Junker, L. & Clardy, J. (2007). High-throughput screens for small-molecule inhibitors of *Pseudomonas aeruginosa* biofilm development. *Antimicrobial Agents and Chemotherapy*, Vol.51, No.10, (October 2007), pp. 3582-3590, ISSN 0066-4804

Kanehisa, M. & Goto, S. (2000). KEGG: kyoto encyclopedia of genes and genomes. *Nucleic Acids Research.*, Vol.28, No.1, (January 1, 2000), pp. 27-30, ISSN 0305-1048

KEGG (2006). Kyoto Encyclopedia of Genes and Genomes, KEGG, In: *http://www.genome.jp/kegg,* Accessed October 21, 2010, Available from: http://www.genome.jp/kegg

Kenney, L.; Bauer, M. & Silhavy, T. (1995). Phosphorylation-dependent conformational changes in OmpR, an osmoregulatory DNA-binding protein of *Escherichia coli. Proceedings of the National Academy of Sciences*, Vol.92, No.19, (September 12, 1995), pp. 8866-8870, ISSN 0027-8424

Kugel, A.; Jarabek, L.; Daniels, J.; Vander Wal, L.; Ebert, S.; Jepperson, M.; Stafslien, S.; Pieper, R.; Webster, D.; Bahr, J. & Chisholm, B. (2009). Combinatorial materials research applied to the development of new surface coatings XII: Novel, environmentally friendly antimicrobial coatings derived from biocide-functional acrylic polyols and isocyanates. *Journal of Coatings Technology and Research*, Vol.6, No.1, (March 2009), pp. 107-121, ISSN 1547-0091

Kumari, S.; Beatty, C.; Browning, D.; Busby, S.; Simel, E.; Hovel-Miner, G. & Wolfe, A. (2000). Regulation of acetyl coenzyme A synthetase in *Escherichia coli. Journal of Bacteriology*, Vol.182, No.15, (August 2000), pp. 4173-4179 ISSN 0021-9193

Leatham, M.; Stevenson, S.; Gauger, E.; Krogfelt, K.; Lins, J.; Haddock, T.; Autieri, S.; Conway, T. & Cohen, P. (2005). Mouse intestine selects non-motile *flhDC* mutants of *Escherichia coli* MG1655 with increased colonizing ability and better utilization of carbon sources. *Infection and Immunity*, Vol.73, No.2, (December 2005), pp. 8039-8049, ISSN 0019-9567

Malakooti, J. (1989). Molecular analysis of the *Escherichia coli* motor components and regulation of internal control elements of the *flaA* operon. *Ph.D. thesis, University of Illinois at Chicago.*

Mascher, T.; Hachmann, A. & Helmann, J. (2007). Regulatory overlap and functional redundancy among *Bacillus subtilis* extracytoplasmic function sigma factors. *Journal of Bacteriology*, Vol.189, No.19, (October 2007), pp. 6919-6927, ISSN 0021-9193

McDermott, P; Ciacci-Woolwine, F.; Snipes, J. & Mizel, S. (2000). High-affinity interaction between gram-negative flagellin and a cell surface polypeptide results in human monocyte activation. *Infection and Immunity*, Vol.68, No.10, (October), pp. 5525-5529, ISSN 0019-9567

McLean, R.; Cassanto, J.; Barnes, M. & Koo, J. (2001). Bacterial biofilm formation under microgravity conditions. *FEMS Microbiology Letters*, Vol.195, No.2, (February 2001), pp. 115-119, ISSN 0378-1097

Monzón, M.; Oteiza, C.; Leiva, J. & Amorena, B. (2001). Synergy of different antibiotic combinations in biofilms of *Staphylococcus epidermidis. Journal of Antimicrobial Chemotherapy*, Vol.48, No.6, (December 2001), pp. 793-801, ISSN 0305-7453

Mukherjee, A.; Mammel, M.; LeClerc, J. & Cebula, T. (2008). Altered utilization of N-acetyl-D-galactosamine by *Escherichia coli* O157:H7 from the 2006 spinach outbreak. *Journal of Bacteriology*, Vol.190, No.5, (March 2008), pp. 1710-1717 ISSN 0021-9193

Nicolle, L. (2005). Catheter-related urinary tract infection. *Drugs & Aging*, Vol.22, No.8, (August 2005), pp. 627-639, ISSN 1170-229X

Nyholm, S.; Deplancke, B.; Gaskins, H.; Apicella, M. & McFall-Ngai, M. (2002). Roles of *Vibrio fischeri* and nonsymbiotic bacteria in the dynamics of mucus secretion during symbiont colonization of the *Euprymna scolopes* light organ. *Applied and Environmental Microbiology*, Vol.68, No.10, (October 2002), pp. 5113-5122, ISSN 0099-2240

Oshima, T.; Aiba, H.; Masuda, Y.; Kanaya, S.; Sugiura, M.; Wanner, B.; Mori, H. & Mizuno, T. (2002). Transcriptome analysis of all two-component regulatory system mutants of *Escherichia coli* K-12. *Molecular Microbiology*, Vol.46, No,1, (October 2002), pp. 281-291, ISSN 0950-382X

O'Toole, G.; Pratt, L.; Watnick, P.; Newman, D.; Weaver, V. & Kolter, R. (1999). Genetic approaches to study of biofilms, In: *Methods in Enzymology, Vol.310*, R. Dole, J. Abelson, & M. Simon, (Eds.), pp. 91-109, Academic Press, ISBN 978-0-12-182211-8, San Diego, CA, USA

Parkinson, J. (1993). Signal transduction schemes of bacteria. *Cell*, Vol.73, No.5, (June 4, 1993), pp. 857-871, ISSN 0092-8674

Peekhaus, N. & Conway, T. (1998). What's for dinner?: entner-doudoroff metabolism in *Escherichia coli*. *Journal of Bacteriology*, Vol.180, No.14, (July 1998), pp. 3495-3502, ISSN 0021-9193

Pratt, L. & Kolter, R. (1998). Genetic analysis of *Escherichia coli* biofilm formation: roles of flagella, motility, chemotaxis and type I pili. *Molecular Microbiology*, Vol.30, No.2, (October 1998), pp. 285-293, ISSN 0950-382X

Prüß, B. & Matsumura, P. (1996). A regulator of the flagellar regulon of *Escherichia coli, flhD*, also affects cell division. *Journal of Bacteriology*, Vol.178, No.3, (February 1996), pp. 668-674, ISSN 0021-9193

Prüß, B. (1998). Acetyl phosphate and the phosphorylation of OmpR are involved in the regulation of the cell division rate in *Escherichia coli*. *Archives of Microbiology*, Vol.170, No.3, (August 1998), pp. 141-146, ISSN 0302-8933

Prüß, B.; Liu, X.; Hendrickson, W. & Matsumura, P. (2001). FlhD/FlhC-regulated promoters analyzed by gene array and *lacZ* gene fusions. *FEMS Microbiology Letters*, Vol.197, No.1, (April 2001), pp. 91-97, ISSN 0378-1097

Prüß, B.; Campbell, J.; Van Dyk, T.; Zhu, C.; Kogan, Y. & Matsumura, P. (2003). FlhD/FlhC is a regulator of anaerobic respiration and the Entner-Doudoroff pathway through induction of the methyl-accepting chemotaxis protein Aer. *Journal of Bacteriology*, Vol.185, No.2, (January 2003), pp. 534-543, ISSN 0021-9193

Prüß, B.; Basemann, C.; Denton, A. & Wolfe, A. (2006). A complex transcription network controls the early stages of biofilm development by *Escherichia coli*. *Journal of Bacteriology*, Vol.188, No.11, (June 2006), pp. 3731-3739, ISSN 0021-9193

Prüß, B.; Verma, K.; Samanta, P.; Sule, P.; Kumar, S.; Wu, J.; Christianson, D.; Horne, S.; Stafslien, S.; Wolfe, A. & Denton, A. (2010). Environmental and genetic factors that contribute to *Escherichia coli* K-12 biofilm formation. *Archives of Microbiology*, Vol.192, (June 18, 2010), pp. 715-728, ISSN 0302-8933

Rice, L. (2006). Unmet medical needs in antibacterial therapy. *Biochemical Pharmacology*, Vol.71, No.7, (March 30, 2006), pp. 991-995, ISSN 0006-2952

Romanova, N.; Gawande, P.; Brovko, L. & Griffiths, M. (2007). Rapid methods to assess sanitizing efficacy of benzalkonium chloride to *Listeria monocytogenes* biofilms. *Journal of Microbiological Methods*, Vol.71, No.3, (December 2007), pp. 231-237, ISSN 0167-7012

Sauer, K.; Camper, A.; Ehrlich, G.; Costerton, J. & Davies, D. (2002). *Pseudomonas aeruginosa* displays multiple phenotypes during development as a biofilm. *Journal of Bacteriology*, Vol.184, No.4, (February 2002), pp. 1140-1154, ISSN 0021-9193

Schneider, D. & Gourse, R. (2004). Relationship between growth rate and ATP concentration in *Escherichia coli*: a bioassay for available cellular ATP. *Journal of Biological Chemistry*, Vol.279, No.9, (February 27, 2004), pp. 8262-8268, ISSN 0021-9258

Shin, S. & Park, C. (1995). Modulation of flagellar expression in *Escherichia coli* by acetyl phosphate and the osmoregulator OmpR. *Journal of Bacteriology*, Vol.177, No.16, (August 1995), pp. 4696-4702, ISSN 0021-9193

Stafslien, S.; Bahr, J.; Feser, J.; Weisz, J.; Chisholm, B.; Ready, T. & Boudjouk, P. (2006). Combinatorial materials research applied to the development of new surface coatings I: a multiwell plate screening method for the high-throughput assessment of bacterial biofilm retention on surfaces. *Journal of Combinatorial Chemistry*, Vol.8, No.2, (March 1, 2006), pp. 156-162, ISSN 1520-4766

Stafslien, S.; Daniels, J.; Chisholm, B. & Christianson, D. (2007). Combinatorial materials research applied to the development of new surface coatings III. Utilisation of a high-throughput multiwell plate screening method to rapidly assess bacterial biofilm retention on antifouling surfaces. *Biofouling*, Vol.23, No.1-2, pp. 37-44, ISSN 0892-7014

Stoitsova, S.; Ivanova, R. & Dimova, I. (2004). Lectin-binding epitopes at the surface of *Escherichia coli* K-12: examination by electron microscopy, with special reference to the presence of a colanic acid-like polymer. *Journal of Basic Microbiology*, Vol.44, No.4, (July 2004), pp. 296-304, ISSN 0233-111X

Stolyar, S.; He, Q.; Joachimiak, M.; He, Z.; Yang, Z.; Borglin, S.; Joyner, D.; Huang, K.; Alm, E.; Hazen, T.; Zhou, J.; Wall, J.; Arkin, A. & Stahl, D. (2007). Response of *Desulfovibrio vulgaris* to alkaline stress. *Journal of Bacteriology*, Vol.189, No.24, (December 2007), pp. 8944-8952, ISSN 0021-9193

Sule, P.; Wadhawan, T.; Wolfe, A. & Prüß, B. (2008). Use of the BacTiter-Glo™ microbial cell viability assay to study bacterial attachment in biofilm formation. *Promega Notes*, 99, (May 2008), pp. 19-21

Sule, P.; Wadhawan, T.; Carr, N.; Horne, S.; Wolfe, A. & Prüß, B. (2009). A combination of assays reveals biomass differences in biofilms formed by *Escherichia coli* mutants. *Letters in Applied Microbiology*, Vol.49, No.3, (September 2009), pp. 299-304, ISSN 0266-8254

Takahashi, N.; Ishihara, K.; Kato, T. & Okuda, K. (2007). Susceptibility of *Actinobacillus actinomycetemcomitans* to six antibiotics decreases as biofilm matures. *Journal of Antimicrobial Chemotherapy*, Vol.59, No.1, (January 2007), pp. 59-65, ISSN 0305-7453

Thao, S; Chen, C.; Zhu, H. & Escalante-Semerena, J. (2010). N^{ε}-lysine acetylation of a bacterial transcription factor inhibits its DNA-binding activity. *PLoS One*, e15123, Vol.5, No.12, (December 31, 2010), pp. 1-9, ISSN 1932-6203

Wang, Y.; Ding, L.; Hu, Y.; Zhang, Y.; Yang, B. & Chen, S. (2007). The *flhDC* gene affects motility and biofilm formation in *Yersinia pseudotuberculosis*. Science in China Series C: Life Sciences, Vol.50, No.6, (December 2007), pp. 814-821, ISSN 1006-9305

Wang, Z. & Chen, S. (2009). Potential of biofilm-based biofuel production. *Applied Microbiology and Biotechnology*, Vol.83, No.1, (May 2009), pp. 1-18, ISSN 0175-7598

West, A. & Stock, A. (2001). Histidine kinases and response regulator proteins in two-component signaling systems. *Trends in Biochemical Sciences*, Vol.26, No.6, (June 1, 2001), pp. 369-376, ISSN 0968-0004

Wolfe, A. (2005). The acetate switch. *Microbiology and Molecular Biology Reviews*, Vol.69, No.1, (March 2005), pp. 12-50, ISSN 1092-2172

Wolfe, A.; Chang, D.; Walker, J.; Seitz-Partridge, J.; Vidaurri, M.; Lange, C.; Prüß, B.; Henk, M.; Larkin, J. & Conway, T. (2003). Evidence that acetyl phosphate functions as a global signal during biofilm development. *Molecular Microbiology*, Vol.48, No.4, (May 2003), pp. 977-988, ISSN 0950-382X

Wood, T.; Gonzalez Barrios, A.; Herzberg, M. & Lee, J. (2006). Motility influences biofilm architecture in *Escherichia coli*. *Applied Microbiology and Biotechnology*, Vol.72, No.2, (September 2006), pp. 361-367, ISSN 0175-7598

Zhou, L.; Lei, X.; Bochner, B. & Wanner, B. (2003). Phenotype microarray analysis of *Escherichia coli* K-12 mutants with deletions of all two-component systems. *Journal of Bacteriology*, Vol.185, No.16, (August 2003), pp. 4956-4972, ISSN 0021-9193

6

Changes in Fungal and Bacterial Diversity During Vermicomposting of Industrial Sludge and Poultry Manure Mixture: Detecting the Mechanism of Plant Growth Promotion by Vermicompost

Prabhat Pramanik[1], Sang Yoon Kim[1] and Pil Joo Kim[1,2]*
[1]*Division of Applied Life Science (BK21 Program), Gyeongsang National University, Jinju, 660-701*
[2]*Division of Applied Life Sciences, Gyeongsang National University, 900 Gazwa, Jinju 660-701,*
[1]*South Korea*
[2]*Republic of Korea*

1. Introduction

Agriculture is facing a challenge to develop strategies for sustainability that can conserve non-renewable natural resources, such as soil and enhance the use of renewable resources such as organic wastes. It has been estimated that more than 18 metric tons of organic sludge was generated every day in Korea in 2003 (Anonymous, 2004)while it was 105 metric tonnes per year in India (Chitdeshwari and Savithri, 2004). Among different options for recycling this sludge, application to agricultural land is probably the most reliable and cost-effective technique to supply organic matter to field crops (Coker et al., 1987). But direct application of this sludge to agricultural land might cause heavy metal contamination (McGrath, 1994). Under this perspective, industrial sludge (IS) was recycled after bioremediation involving earthworms.

Unlike several chemical methods, removal of heavy metals by biological means is more specific, eco-friendly and economical. Begum and Krishna (2010) revealed that heavy metal content in organic wastes reduced after passage through earthworm guts. Therefore, industrial sludge could be recycled through vermicomposting to produce nutrient rich plant amendment. Vermicomposting is the stabilization of organic substrates by microorganisms in presence of earthworms. Though earthworms consume fungi with organic substrates to fulfil their nitrogen requirement, the viable fungal count in earthworm casts was generally higher than that of initial waste substrates during vermicomposting (Edwards and Bohlem, 1996). Ergosterol, marker molecule of fungal cell membrane, is frequently used in microbiology to quantify fungal biomass in infected media. Madan et al. (2002) estimated fungal biomass in soil by FAME assay. Hill et al. (2000) also quantified fungal specific FAME (18:19ωc) to estimate fungal biomass in compost. Yasir et al. (2009) revealed that bacterial biomass also played important role during organic matter decomposition. Muramic acid could be used as a marker molecule for bacterial biomass determination (King and White, 1977). The objectives of this study were to (i) standardize recycling technique of IS through vermicomposting, (ii) evaluate fungal and bacterial diversity during vermicomposting and (iii) determine plant growth promoting mechanism of vermicompost.

2. Materials and methods

2.1 Substrates used and experiment design

The vermicompost experiment was conducted in polythenelined earthen pots (5 L capacity). Poultry manure was used as the initial energy source for earthworms. Poultry manure (PM) was procured from the nearby poultry farm and industrial sludge (IS) was procured from the industrial region, Tangra, Kolkata, India. Initial chemical and microbiological properties of poultry manure and industrial sludge were presented in Table 1.

Parameters studied	Industrial sludge	Poultry manure
Total organic carbon (mg g^{-1})	305.26	371.53
Total Kjeldahl nitrogen (mg g^{-1})	3.74	4.97
C/ N ratio	81.62	74.75
Total phosphorus (mg g^{-1})	3.51	4.18
Total potassium (mg g^{-1})	3.84	4.22
Total chromium (µg g^{-1})	859.97	108.49
Total copper (µg g^{-1})	471.08	241.92
Total lead (µg g^{-1})	64.83	9.07

Table 1. Some chemical properties of poultry manure (PM) and industrial sludge (IS)

Fresh PM was air-dried and autoclaved at 15 lb/in^2 pressure for 30 min. Industrial sludge was concentration by air-drying and concentrated IS and PM mixture was used for vermicomposting. In this experiment, PM was mixed with IS in three different proportions i.e., 5% PM (T_1), 10% PM (T_2) and 20% PM (T_3) along with control (T_0) and the waste mixtures were allowed to pass through earthworm guts for vermicomposting. One and half kilogram of those waste mixtures were taken in each pot and 25 almost equal maturity (mean weight 0.48 ± 0.06 g) earthworms (*Eisenia fetida*) were introduced in each treatment pot. The moisture content of the organic substrates in each pot was maintained between 60% and 65% throughout the study period by sprinkling water after every 10-12 hours. The experiment was conducted following complete randomized design with three replications. Total organic carbon (TOC), total Kjeldahl nitrogen (TKN), total phosphorus (TP), total potassium (TK) and total concentration of some heavy metals (Cr, Cu and Pb) were measured initially and after completion of vermicomposting process. During vermicomposting, the feed materials from each treatment were analyzed after 15, 30, 45, 60 days after initiation of the process and on stabilization of the process (73 days) for estimating microbial biomass C, ergosterol, total fatty acid methyl esters (FAMEs) and muramic acid content.

2.2 Chemical analysis

Total organic carbon (TOC) of the vermicompost was estimated using the standard dichromate oxidation method of Nelson and Sommers (1982). Total Kjeldahl nitrogen (TKN) was estimated after digesting the sample with concentrated H2SO4 (1:20, w/v) followed by distillation (Bremner and Mulvaney, 1982). Total phosphorus (TP) and total potassium (TK) were analyzed from the wet digest [tri-acid (HNO_3-H_2SO_4-$HClO_4$) mixture was used for digestion] of vermicompost (Jackson, 1973). Total phosphorus (TP) was estimated by the colorimetric method using ammonium molybdate in hydrochloric acid and total potassium (TK) was determined by flame photometer (Bansal and Kapoor, 2000).

2.3 Microbial analysis

Microbial biomass was determined by the chloroform fumigation-extraction (FE) method (Vance et al., 1987). For fumigation, organic substrates were incubated with ethanol-free chloroform in desiccators. The TOC analyzer was used to determine total organic C (C_{org}) and total N in 0.5 M K_2SO_4 extracts of non-fumigated and fumigated soils. The microbial biomass carbon (MBC) was calculated as MBC = (C_{org} in fumigated soil - C_{org} in non-fumigated soil)/k_c; where, k_c = 0.33, the factor used to convert the extracted organic C to MBC (Sparling and West, 1988).

An analysis for ergosterol estimation was performed with 50 mg of lyophilized organic waste or vermicompost sample. Ergosterol was extracted from leaf litter by 30 min refluxing in alcoholic base (Gesser et al., 1991) and purified by solid-phase extraction. Final purification and quantification of ergosterol was achieved by high-performance liquid chromatography (HPLC). The system was run with HPLC grade methanol at a flow rate of 1.5 ml min^{-1}. Ergosterol eluted after 7:11 min and detected at 282 nm; peak identity was checked on the basis of retention times of commercial ergosterol (98% purity).

The FAME analysis was performed using the modified procedure of Schutter and Dick (2000). Before analysis, fresh samples were lyophilized and three grams of lyophilized sample was treated with 10 mL of 0.2 M KOH in methanol and incubated at 37°C for 1 hr. After incubation, the pH of the system was adjusted to 7.0 with 1.0 M acetic acid, 10 mL of n-hexane was mixed and then it was vortexed. After centrifugation at 1600 rpm for 20 min., 5 mL of n-hexane layer was evaporated by N_2 gas. The residue was dissolved in 170 µL of 1:1 mixture of n-hexane and methyl t-buthyl ether with 30 µL of 0.01M methyl nonadecanoate ($C_{19:0}$) as internal standard for FAME and analyzed with a Hewlett-Packard 5890 Series II (Palo Alto, CA) equipped with an HP Ultra 2 capillary column (5% diphenyl-95% dimethylpolysiloxane, 25 m by 0.2m) and a flame ionization detector. For FAME analysis, the oven temperature was raised from 170°C to 270°C at 5°C min^{-1} and kept at 270°C for 2 minutes.

Amino sugars in biomass suspensions, chloroform-fumigation-extraction (CFE) extracts and in incubated organic wastes were determined following standard method of Zhang and Amelung (1996). Sample aliquots corresponding to a about 50 mg microbial biomass, with 100 µg myo-inositol added as internal standard, were hydrolyzed with 10 ml of 6M HCl at 105 °C for 8 h. The CFE extracts were freeze-dried prior to hydrolysis. The released amino sugars were separated from impurities by neutralization with 0.4M KOH. Prior to derivatization, 100µg of methylglucamine was added as recovery standard. Derivatization was carried out according to (Guerrant and Moss, 1984). In brief, aldononitrile derivatives of the amino sugars were prepared by heating the samples in 0.3 ml of a derivatization reagent (32 mg hydroxylamine hydrochloride ml^{-1} and 40 mg 4-(dimethylamino) pyridine ml^{-1} in pyridine–methanol 4/1) at 75 °C for 30 min. After acetylation with 1 ml of acetic anhydride at 75–80 °C for 20 min, dichloromethane was added, and excess derivatization reagents were removed by washing with 1 ml of 1 M HCl and 1 ml of water two times each. The remaining organic phase was dried under an air stream at 45 °C and dissolved in 0.3 ml ethyl acetate-hexane (1/1). The amino sugar derivatives were separated on a HP 6890 GC equipped with a HP-5 fused silica column (30 m×0.25 mm ID with 0.33 µm film thickness) and a flame ionization detector. Amino sugars were quantified using inositol as the internal standard and methylglucamine as recovery standard.

2.4 Plant growth promotion study

Vermicompost was extracted with ethyl acetate (vermicompost: ethyl acetate = 1: 5, w/v) and the extract the centrifuged at 7000 rpm for 15 minutes. The supernatant was used for radish bioassay. Five radish seeds were taken on 2mm x 2mm sterile Whatman filter paper and 750 µl of that extract applied on radish seeds under aseptic condition and incubated at 25±1 ^{0}C for 5 days. After 5 days incubation, root and shoot length of extract applied seedlings were compared with that of control treatment.

After finding the presence of plant growth promoting compound, the ethyl acetate extract was fractionated by column chromatography using different proportions of hexane, dichloromethane and methanol to obtain 24 fractions, each of 50 ml. The fractions were then concentrated to 2-3 ml by rotary evaporator at a temperature below 40 ^{0}C. All the fractions were then tested by radish bioassay. The active three fractions (please follow the result below) were then analysed by HPLC and methanol water mixture (60: 40, v/v) was used as mobile phase for this analysis.

Vermicompost was then extracted with sterile water (vermicompost: water = 1: 100, w/v) under aseptic condition. The extract was then serially diluted 10^3 fold and incubated in broth medium with different amount of tryptophan at 30^{0}C for 7 days. After incubation, cell pellets were removed by centrifugation at 6000 rpm for 10 minutes. The supernatant was treated with Salkosky reagent and pink colour intensity was measured at 420 nm.

3. Results

3.1 Chemical properties

Chemical analysis revealed that total concentrations of nitrogen, phosphorus and potassium of all the treatments were increased due to vermicomposting. Addition of poultry manure (PM) significantly ($P < 0.05$) increased nitrogen content in final vermicompost as compared to control treatment (Table 2). Data revealed that total nitrogen and phosphorus content of final vermicompost was increased with increasing PM proportion in initial waste mixtures. Addition of PM with IS significantly ($P < 0.05$) increased total potassium content after vermicomposting, however, its values in T_2 and T_3 treatments were statistically at per.

Parameters studied	T_0	T_1	T_2	T_3
Total organic carbon (mg g^{-1})	201.0±5.4	177.6±3.3	168.9±4.7	158.4±6.1
Total Kjeldahl nitrogen (mg g^{-1})	7.62±0.40	8.35±0.43	9.47±0.23	9.91±0.49
Total phosphorus (mg g^{-1})	7.05±0.41	8.75±0.56	9.23±0.44	9.89±0.39
Total potassium (mg g^{-1})	6.89±0.49	8.16±0.33	8.94±0.40	9.23±0.57
Total chromium (µg g^{-1})	618.2±21.7	573.4±14.9	559.41±17.5	548.7±15.4
Total copper (µg g^{-1})	325.1±9.4	293.9±10.6	291.7±13.4	286.4±11.8
Total lead (µg g^{-1})	41.6±1.08	34.44±0.97	32.06±1.83	30.69±1.58

Table 2. Changes in nutrient content and heavy metal concentrations due to vermicomposting of different proportions of IS and PM proportions

Total heavy metal content of the organic substrates decreased due to vermicomposting (Table 2). The extent of decrease in heavy metal content was proportionately increased with the amount of PM added to IS. Among different heavy metals, zinc recorded the maximum decrease in total concentration after vermicomposting followed by Cr, Cu and Pb. Though vermicomposting significantly ($P < 0.05$) reduced total content of different heavy metals, the values were not significantly affected by different PM proportions.

3.2 Microbial biomass

Total microbial biomass of the organic wastes was significantly ($P < 0.05$) increased due to vermicomposting (Fig. 1). Periodical analysis indicated an exponential nature of biomass dynamics in organic substrates during vermicomposting. Addition of PM significantly ($P < 0.05$) increased microbial biomass in final vermicompost. The highest MBC content was registered within 15-30 days of vermicomposting. MBC of vermicomposts, prepared from T_1 and T_2 were statistically at par. Vermicompost of T_3, however, recorded significantly ($P < 0.05$) higher MBC as compared to other treatments.

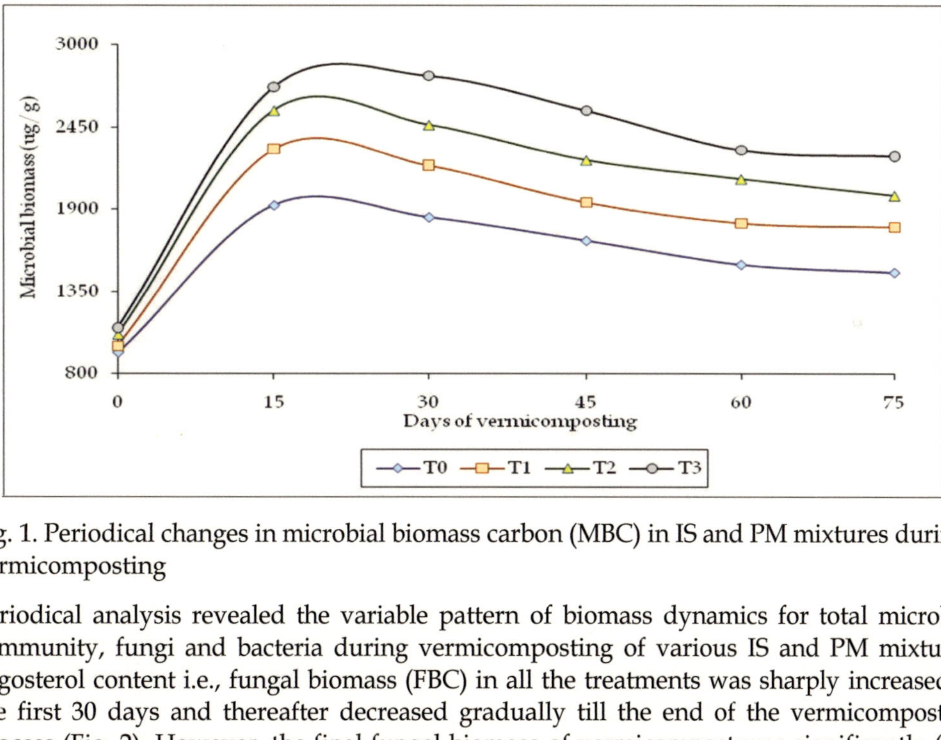

Fig. 1. Periodical changes in microbial biomass carbon (MBC) in IS and PM mixtures during vermicomposting

Periodical analysis revealed the variable pattern of biomass dynamics for total microbial community, fungi and bacteria during vermicomposting of various IS and PM mixtures. Ergosterol content i.e., fungal biomass (FBC) in all the treatments was sharply increased in the first 30 days and thereafter decreased gradually till the end of the vermicomposting process (Fig. 2). However, the final fungal biomass of vermicompost was significantly ($P < 0.05$) higher than that of initial organic substrates. Addition of PM with IS, significantly ($P < 0.05$) increased fungal biomass of final vermicompost. Vermicompost prepared from T_3 recorded significantly ($P < 0.05$) higher FBC as compared to other treatments and FBC values of vermicomposts, prepared from T_1 and T_2, were statistically at par.

Periodical analysis results revealed that total FAME content in vermicompost followed almost same of ergosterol content (Fig. 3). The highest FAME was recorded in T_3 treatment and it was significantly higher than other treatments.

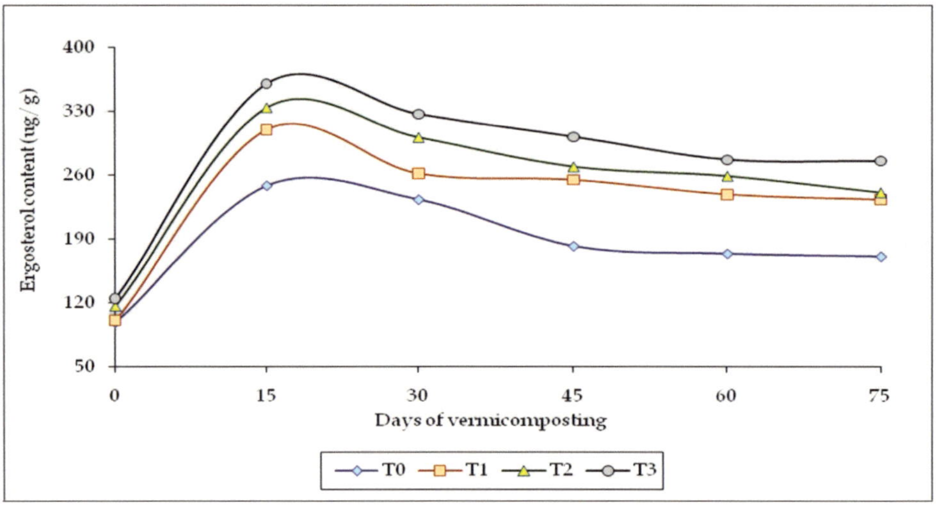

Fig. 2. Periodical changes in ergosterol content in IS and PM mixtures during vermicomposting

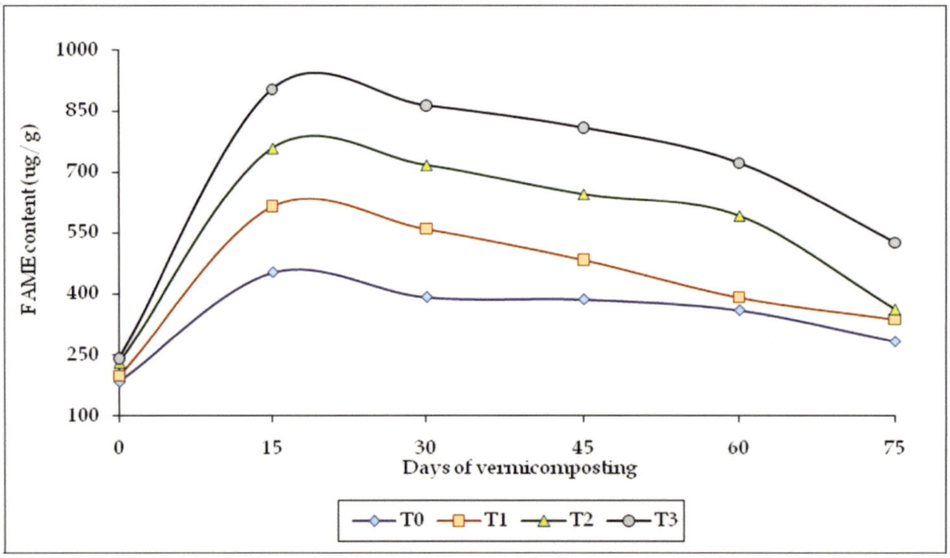

Fig. 3. Periodical changes in total fatty acid methyl esters (FAMEs) content in IS and PM mixtures during vermicomposting

Muramic acid was estimated as an indicator of bacterial biomass. Periodical estimation of muramic acid in the waste mixture revealed a steady increase in the muramic acid content up to 45 days of the process and thereafter it decreased till the end of the process. The final muramic acid contents of vermicomposts, prepared from T_2 and T_3, were significantly ($P < 0.05$) higher than that of their initial waste mixtures. In case of T_0 and T_1 treatments, muramic acid contents of vermicomposts were statistically at par with that of initial wastes. Analysis revealed that muramic acid contents of vermicomposts, prepared from T_0 and T_1 treatments, did not differ statistically among them.

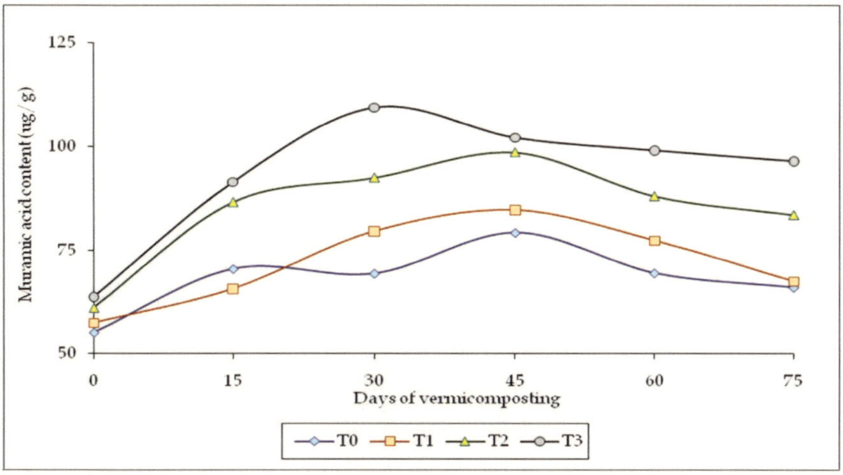

Fig. 4. Periodical changes in muramic acid content in IS and PM mixtures during vermicomposting

3.3 Plant growth promotion

Incubation of radish seeds with ethyl acetate extract of vermicomposts for 5 days significantly ($P < 0.05$) increased root and shoot length of radish as compared to control. Column chromatography of concentrated ethyl acetate extract of vermicomposts yielded 24 fractions. Radish bioassay with all these fractions revealed that 3 fractions (5th, 7th and 8th) out of 24 fractions were able to increase radish root and shoot length as compared to control as well as other fractions (Fig. 5). The root and shoot length of all fractions were presented in Fig. 6. Vigor index, summation of root length and shoot length, is a good indicator for plant-growth promotion and its highest value was recorded in fraction 5.

Fig. 5. Radish bioassay test results of different fractions of vermicompost extract

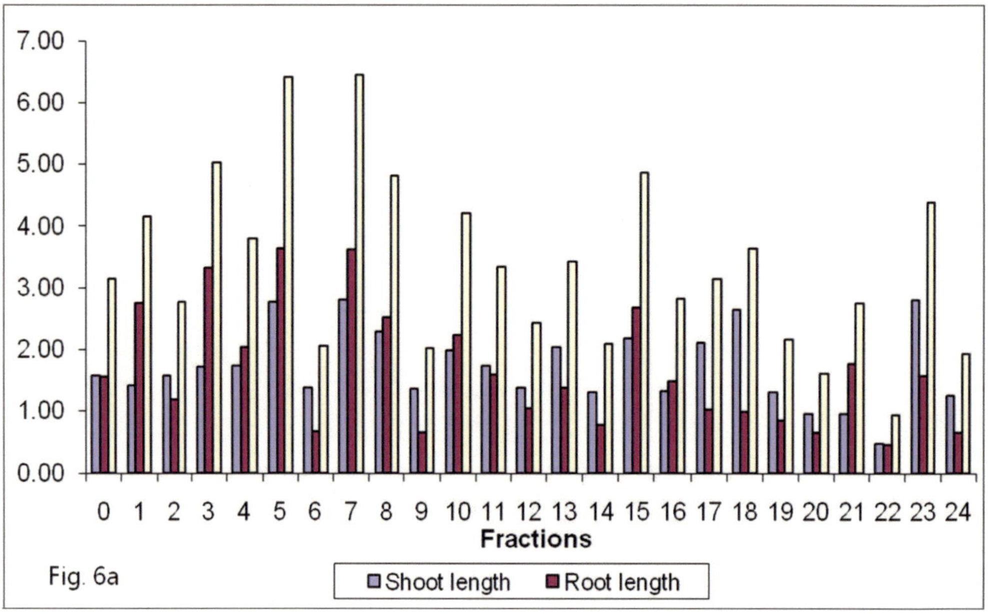

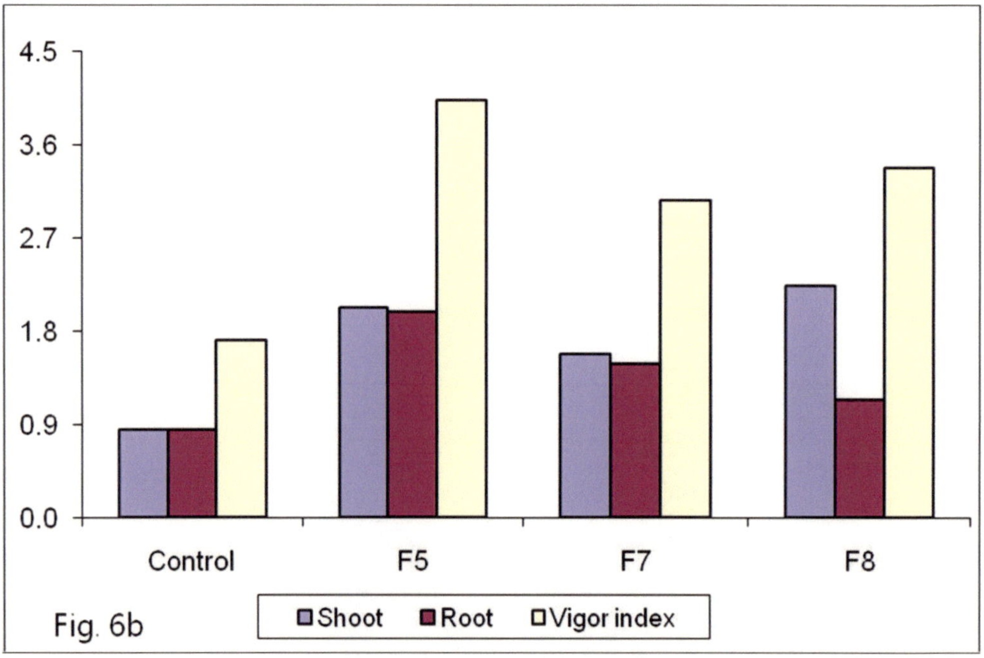

Fig. 6. Root, shoot lengths (cm) vigor indexes of radish seedlings as affected by different fractions obtained after column chromatography

HPLC analysis of these three fractions confirmed the presence of indole acetic acid (IAA) in 5th fraction (Fig. 7). Incubation of serially diluted vermicomposts extract in tryptophan-amended broth medium revealed pink colouration after 7 days incubation. Colorimetric analysis indicated the presence of 137 µg IAA L^{-1} medium after 7 days.

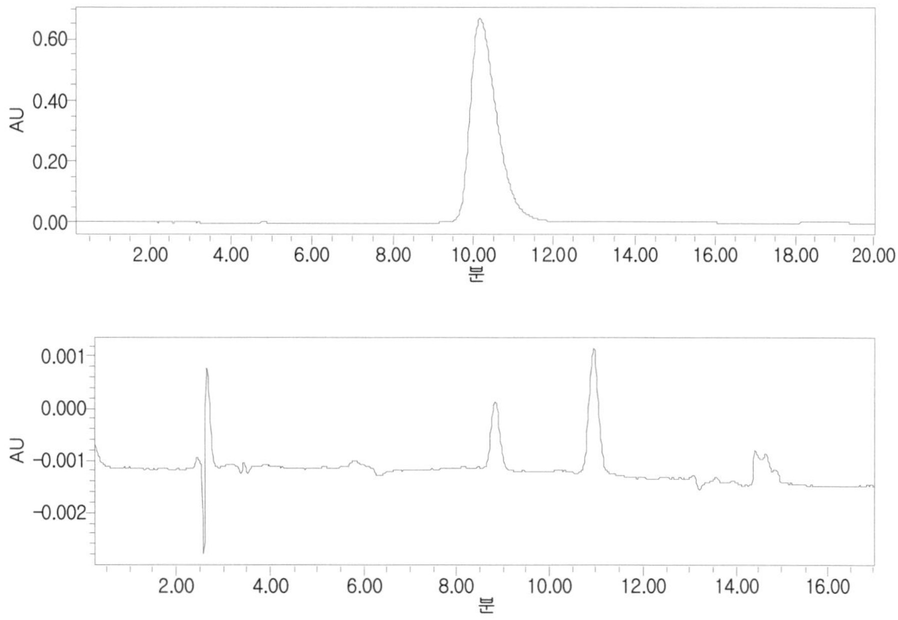

Fig. 7. HPLC chromatogram of standard and fraction 5 for IAA analysis

4. Discussion

Vermicomposting is the controlled oxidative decomposition of organic substrates by mutual interaction between earthworm and microorganisms. Cow manure was generally mixed with initial organic wastes to provide easily available energy source and favourable environment to the earthworms. In this experiment, cow manure was replaced by poultry manure (PM) to recycle industrial sludge (IS). Data indicated that proportion of PM determined the quality of final vermicompost. Addition of 10% and 20% PM with IS yielded vermicomposts which have significantly ($P < 0.05$) higher NPK content and lower heavy metals content as compared to other treatments, however, values of these two treatments were statistically at per. PM mixing enhanced the earthworm activity which in turn increased the rate of organic substrate decomposition. During mineralization, dry mass of organic substrates was lost as CO_2 by oxidative decomposition (Viel et al., 1987). Addition of PM lowered the C/N ratio of initial waste mixture. Tripathi and Bhardwaj (2004) proposed that narrower C/N ratio facilitates earthworm feeding, which in turn enhanced the rate of organic matter decomposition.

Organic substrates were stabilized by action of microorganisms in the presence of earthworms during vermicomposting (Edwards and Fletcher, 1988). Epigeic earthworms are generally used for organic waste decomposition and they consume microorganisms specially fungi to satisfy their nitrogen requirement. Pramanik and Chung (2011) also found similar results during vermicomposting of fly ash and vinasse mixture. This increase in microbial biomass indicated that vermicomposting facilitates microbial proliferation in final stabilized product. Ergosterol content of organic substrates was multiplied by conversion factor 5.4 (Klamer and Baath, 2004) to calculate fungal biomass (FB) in it. Though

earthworms selectively consume fungi as their food, increased fungal biomass during vermicomposting suggested that not all the fungi were killed during passage through earthworm guts, in fact the rate of germination of fungal spores was probably enhanced under favourable condition of earthworm guts (Hendrikson, 1990). Comparisons of fungal biomass, calculated from ergosterol content, with total FAME content of decomposing substrates gave a significantly positive correlation value (r = 0.921*). The ratio of these two parameters could be arranged following a linear regression with a mean value 2.71 (standard deviation = 0.48). Since FAME analysis is more precise method to estimate FBC, this conversion factor (2.71) could be used to calculate FBC of vermicompost from its FAME values.
Muramic acid occurs naturally as N-acetyl derivatives in peptidoglycan, the characteristic polysaccharide composing bacterial cell wall. In this experiment, muramic acid was estimated as a marker molecule for bacterial biomass in decomposing waste mixture. Data of periodical muramic acid content indicated a steady increase in bacterial biomass during vermicomposting. Muramic acid content was proportionately increased with increasing PM ratio in initial waste mixture and 20% PM addition recorded significantly ($P < 0.05$) higher muramic acid content in final vermicomposts. Though addition of 10% and 20% PM with IS produced vermicomposts having significantly higher NPK content, but based on microbial status of vermicomposts, it could be concluded that 20% PM mixing with IS was probably the optimum combination to obtain the best quality vermicomposts.
In this study, conversion factor of muramic acid to bacterial biomass was biomass was estimated by assuming that fungi and bacteria are the major microbial community present in vermicompost and bacterial biomass was calculated by subtracting fungal biomass from total microbial biomass. This bacterial biomass was compared with muramic acid content and it had shown significant correlation (r = 0.918*) between these two parameters. Analysing indicated that ratios of calculated bacterial biomass and muramic acid had the mean value 8.22 with standard deviation 0.88. Therefore, this value (8.22) could be used as a conversion factor for calculating bacterial biomass from muramic acid of vermicompost.
Several researchers found that application of vermicompost had hormone-like effect on plants (Arancon et al., 2004). The results of this experiment confirmed that vermicompost possessed IAA-producing microorganisms which in turn facilitated plant growth through IAA production.

5. Conclusion

Vermicomposting is a rapid and safe process to recycle IS and PM mixture into nutrient-rich soil amendment. Passage of organic substrates through earthworm guts also reduced total heavy metal content in it. Microbiological diversity of organic substrates was also modified during vermicomposting. Both fungal and bacterial biomass was increased during vermicomposting of IS and PM mixture. Results indicated the presence of IAA-producing bacteria in vermicomposts, which enabled it to promote plant growth. Mixing of 10% PM with IS was probably the optimum condition to obtain the best quality vermicomposts.

6. Acknowledgement

This work was supported by the Institute of Agriculture and Life Sciences, Gyeongsang National University, South Korea and also by scholarships from the BK21 program, Ministry of Education and Human Resources Development, South Korea.

7. References

Anonymous (2004). *Statistics on Production and Outlets of Sewage Sludge.* Ministry of Environment, Republic of Korea, pp. 11–12.

Arancon, N.Q., Edwards, C.A., Atiyeh, R., Metzger, J.D. (2004). Effect of vermicomposts produced from food waste on the growth and yields of greenhouse peppers. *Bioresource Technology*, volume 93, issue 2 (June 2004), pp. 139-144, ISSN: 0960-8524.

Bansal, S., Kapoor, K.K. (2000). Vermicomposting of crop residues and cattle dung with *Eisenia foetida. Bioresouorce Technology*, volume 73, issue 1 (May 2000), pp. 95-98, ISSN: 0960-8524.

Begum, A., Krishna, H. (2010). Management of municipal sewage sludge by vermicomposting technique. *International Journal of ChemTech Research*, volume 2, issue 3 (July-September 2010), pp. 1521-1525, ISSN: 0974-4290.

Bremner, J.M., Mulvaney, R.G. (1982). Total Nitrogen. In: *Methods of Soil Analysis*, Page, A.L., Miller, R.H., Keeney, D.R. (Eds.), Part 2, American Society of Agronomy, Madison, pp. 575 - 624.

Chitdeshwari, T., Savithri, P. (2004). Quality of sewage biosolid composts and their impact on the yield of maize. *Indian Journal of Agricultural Research*, volume 38, issue 2 (June 2004), pp. 147-150, ISSN: 0367-8245.

Coker, E.G., Hall, J.E., Carlton-Smith, C.H., Davis, R.D. (1987). Field investigations into the manurial value of lagoon-matured digested sewage sludge. *The Journal of Agricultural Science*, volume 109, issue 3 (December 1987), pp. 467–478.

Edwards, C.A., Fletcher, K.E. (1988). Interaction between earthworms and microorganisms in organic matter breakdown. In: *Biological Interactions in Soil*, Edwards, C.A. (Ed.), Elsevier, New York, pp. 235–247.

Gesser, M.O., Bauchrowitz, M.A., Escautier, M. (1991). Extraction and quantification of ergosterol as a measure of fungal biomass in leaf litter. Microbial Ecology, volume 22, issue 1 (December 1991), pp. 285–291, ISSN: 1432-184X.

Guerrant, G.O., Moss, C.W. (1984). Determination of monosaccharides as aldononitrile, O-methyloxime, alditol and cyclitol acetates derivatives by gas chromatography. *Analytical Chemistry*, volume 56, issue 4 (April 1984), pp. 633–638, ISSN: 0003-2700.

Hendrikson, N.B. (1990). Leaf litter selection by detritivore and geophagous earthworms. *Biology and Fertility of Soils*, volume 10, issue 1 (October 1990), pp. 17–21, ISSN: 1432-0789.

Hill, G.T., Mitkowski, N.A., Aldrich-Wolfe, L., Emele, L.R., Jurkonie, D.D., Ficke, A., Malodonado-Ramirez, S., Lynch, S.T., Nelson, E.B. (2000). Methods for assessing the composition and diversity of soil microbial communities. *Applied Soil Ecology*, volume 15, issue 1 (August 2000), pp. 25-36, ISSN: 0929-1393.

Jackson, M.L. (1973). *Soil Chemical Analysis*. Prentice Hall Private Ltd., New Delhi, India.

King, J.D., White, D.C. (1977). Muramic acid as a measure of microbial biomass in estuarine and marine samples. *Applied and Environmental Microbiology*, volume 33, issue 4 (April 1977), pp. 777-783, ISSN: 0099-2240.

Klamer, M., Baath, E., 2004. Estimation of conversion factors for fungal biomass determination in compost using ergosterol and PLFA 18:2ω6,9. *Soil Biology and Biochemistry*, volume 36, issue 1 (January 2004), pp. 57–65, ISSN: 0038-0717.

McGrath S.P. (1994). Effects of heavy metals from sewage sludge on soil microbes in agricultural ecosystems. In: *Toxic Metals in Soil-Plant Systems*, Ross S. M., (ed.), Chichester, John Wiley, pp. 242-274.

Nelson, D.W. and Sommers, L.E., 1982. Total carbon and organic carbon. In: In: *Methods of Soil Analysis*, Page, A.L., Miller, R.H., Keeney, D.R. (Eds.), Part 2, American Society of Agronomy, Madison, pp. 539–579.

Pramanik, P., Chung, Y.R. (2011). Changes in fungal population of fly ash and vinasse mixture during vermicomposting by *Eudrilus eugeniae* and *Eisenia fetida*: Documentation of cellulose isozymes in vermicompost. *Waste Management*, in Press (DOI: 10.1016/j.wasman.2010.12.017).

Schutter, M.E., Dick, R.P. (2000). Comparison of fatty acid methyl ester (FAME) methods for characterizing microbial communities. *Soil Science Society of America Journal*, volume 64, issue 5 (September - October 2000), pp. 1659-1668, ISSN: 1435-0661.

Sparling, G.P., West, A.W. (1988). Modifications to the fumigation extraction technique to permit simultaneous extraction and estimation of soil microbial C and N. *Communications in Soil Science and Plant Analysis*, volume 19, issue 3 (February 1988), pp. 327–344, ISSN: 1532-2416.

Triphathi, G., Bhardwaj, P. (2004). Decomposition of kitchen waste amended with cow dung using an epigeic species (*Eisenia fetida*) and an anecic species (*Lampito mauritii*). *Bioresource Technology*, volume 92, issue 2 (April 2004), pp. 215–218, ISSN: 0960-8524.

Vance, E.D., Brookes, P.C., Jenkinson, D.S. (1987). An extraction method for measuring soil microbial biomass C. *Soil Biology and Biochemistry*, volume 19, issue 6 (March 1987), pp. 689-696, ISSN: 0038-0717.

Viel, M., Sayag, D., Andre, L. (1987). Optimization of agricultural, industrial waste management through in-vessel composting. In: *Compost: Production, Quality and Use*, deBertodi, M. (Ed.), Elsevier Applied Sciences, Essex, pp. 230–237.

Yasir, M., Aslam, Z., Kim, S.W., Lee, S.-W., Joen, C.O., Chung, Y.R. (2009). Bacteria community composition and chitinase gene diversity of vermicompost with antifungal activity. *Bioresource Technology*, volume 100, issue 19 (October 2009), pp. 4396–4403, ISSN: 0960-8524.

Zhang, X., Amelung, W. (1996). Gas chromatographic determination of muramic acid, glucosamine, mannosamine, and galactosamine in soils. *Soil Biology & Biochemistry*, volume 28, issue 9 (September 1996), pp. 1201–1206, ISSN: 0038-0717.

7

Genetic and Functional Diversities of Microbial Communities in Amazonian Soils Under Different Land Uses and Cultivation

Karina Cenciani[1], Andre Mancebo Mazzetto[2], Daniel Renato Lammel[1],
Felipe Jose Fracetto[2], Giselle Gomes Monteiro Fracetto[2],
Leidivan Frazao[2], Carlos Cerri[1] and Brigitte Feigl[1]
[1]*Centro de Energia Nuclear na Agricultura/Laboratorio de Biogeoquimica Ambiental*
[2]*Escola Superior de Agricultura "Luiz de Queiroz"*
Brazil

1. Introduction

Amazonia is a natural region formed by the Amazon River Basin and covered by the largest equatorial forest in the world, covering an area of 6,915,000 km^2, of which 4,787,000 km^2 are in Brazil. Due to the large size and low population density, it is considered to be the best-well preserved Brazilian biome. Amazonian tropical forest soils are supposed to hold high microbial biodiversity, however the human impact has been extensive in the last decades, coupled with uncontrolled wood removal and the concomitant advancement of agricultural frontier (Fearnside, 2005).

Under the current scenario it is notorious the importance of Amazonia to the Brazilian ecosystem and even worldwide. Precisely because of this the images of slash-and-burn of the forest produce a strong impact on the public opinion. More than 60 million hectares were deforested. Of this total an estimated 35 million hectares were replaced by pastures for beef production, one million hectares were occupied with perennial crops, three million hectares with annual crops, and more than 20 million hectares support secondary vegetation called "capoeira" or fallow (Fig. 1).

Fig. 1. Conversion of forest (A) to well-managed pasture (B) and the fallow site (C) in the Amazon Forest.

What's occurring in the pastures at the Amazonia, as well in other tropical regions is the loss of the productive capacity after 4 to 10 years of use due to overgrazing, invasion of unpalatable weed species, loss of soil fertility and cultivation of inadequate grass species (Fernandes et al., 2002). It is estimated that 30 to 50% of pastures in the Brazilian Amazon are in advanced stage of degradation, giving rise to the fallow sites. In general, the establishment of pastures is done with simple technology and no use of fertilizers. Its maintenance depends almost exclusively on the nutrients contained in the ashes produced during burning of the original vegetation. Fallows also play an essential role for recovery of native species, as it reassimilates part of carbon and nitrogen that were released when slash-and-burn of native vegetation was used (Fernandes et al., 2002; Schroth et al., 2002).

The quality and soil fertility are defined from the point of view of some essential attributes that maintain the agricultural productivity, namely as: soil ability to promote plant growth, water supply and nutrient processing, efficient gases exchange in the atmosphere-soil interface and the activity of micro and macro organisms (Dilly & Nannipieri, 2001). In this context it is highlighted the role of soil microbial biomass (SMB), defined as the living portion of soil organic matter, excluding roots and larger organisms than, approximately, 5000 μm^3 (Cenciani et al., 2009).

In recent years many technological advances and the development of new and independent cultivation techniques led microbiologists to explore more precisely the "black box" of soil microbial diversity. This new knowledge is contributing to our better understanding of the distribution and abundance of soil microorganisms, the effect of community structure on ecosystem functioning, the effects of land use changes on microbial communities and hence in the ecosystem.

Traditional methods were usually based on specific cultivation media in laboratory conditions, in which only 1-3% of the soil microbes present conditions for growth. For this reason much research have been developed using generic properties, such as the microorganisms basal respiration, enzymatic activity, mineralization of soil organic matter, among others, that under controlled laboratory conditions represent rough estimates of the metabolic functions of microbial biomass, reflecting its physiology as whole soil community (Ananyeva et al., 2008).

Considered one of the most important "hot spots" in the world, Amazonia has an important role in the discovery of new species of plants, animals and microorganisms, which may be important for the functionality of different ecosystems. However there are limited studies addressing the impacts of land use changes under the Amazonia microbial communities and their functions in the soil. Within this context bacterial and fungal communities, considered the most abundant groups of microorganisms in the soil, can act as important indicators of environmental stresses induced by the use of Amazonian soils.

Soil microbial diversity is usually assessed as species and genetic diversity rather than as structural and functional diversity. However, in terms of soil quality, these two last forms of diversity may be equally important due to the microorganism's functional redundancy. The importance of functional and catabolic diversity lies in the fact that only based on changes in the genetic diversity; it is not possible to infer whether some functions of soils were lost or not (Mazzetto et al., 2008).

A soil with high redundancy of functions is probably able to maintain well-balanced its ecological processes, even under a disturbance. This approach, defined as resilience, refers to the buffering effects of external disturbances to the ecosystem. In a soil system the

reduction of microbial diversity can be an important indicator of the loss of resilience and, consequently, the soil quality. The abundance of some species of microorganisms seems not to be as important as the maintenance of their genetic and functional diversities. This is because the abundance reflects more immediately the short-term microbial fluctuation, while the diversity reveals the balance between the number of microorganisms and the functional domains in the soil (Kennedy, 1999; Lavelle, 2000).
The main objective of our chapter is to describe the relationship between the genetic and functional diversity approaches to study the microbial ecology and the impact of different land uses under soil microorganisms in Amazonia.

2. Microbial biomass in amazonian soils

The Amazon Basin covers almost 25% of South America. With about 7.5 million km^2, it extends into the territory of nine countries and accounts for 70% of tropical forests around the globe. Only in Brazil the total area is 5.1 million km^2 (Fearnside, 2005). Despite its great beauty and exuberance, the Amazon rainforest is found in soils of low fertility, while its maintenance depends on the cycling of nutrients from vegetation covering (Cenciani et al., 2009).
The quality and soil fertility are defined from the point of view of some essential attributes that maintain the agricultural productivity, namely as: soil ability to promote plant growth, water supply and nutrient processing, efficient gases exchange in the atmosphere-soil interface and the activity of micro and macro organisms (Dilly & Nannipieri, 2001). In this context it is highlighted the role of soil microbial biomass (SMB), defined as the living portion of soil organic matter, excluding roots and larger organisms than, approximately, 5000 μm^3. The microbial biomass comprises the dormant and the metabolically active organisms in the soil; performing a primary role for maintenance and the products of microbial recycling are then absorbed by plant roots (Cenciani et al., 2009).
Soil quality or even "soil health" can be analyzed by the activity of microbial biomass, one of few active fractions of organic matter, sensitive to tillage and that can be quantified. Overall SMB comprises about 2-3% of total organic carbon in the soil, thus indicating it to be a sensible parameter to evaluate the quality of soils submitted to different management strategies, or to pollution impacts. The development of indirect methods for measurement of SMB such as the incubation-fumigation (IF) (Jenkinson & Powlson, 1976), the substrate induced respiration (SIR) (Anderson & Domsch, 1978), the content of ATP in microbial cells (Jenkinson & Ladd, 1981) and the extraction-fumigation (EF) method (Vance et al., 1987) facilitated the assessment of the SMB compartment.
Some studies previously carried out in chronosequences forest to pasture in Amazonia have shown that SMB is reduced after 3 years of establishing pastures, but their levels are raised in older pastures, and reach similar contents in the native forest. Several studies quantified the main elements (C, N, P, S) immobilized into microbial cells at different soil depths (Feigl et al., 1995 a,b; Fernandes et al., 2002; Cenciani et al., 2009).
Overall SMB reflects the contents of total organic matter, representing an efficient and sensitive parameter in assessing the quality of soils under different management or impacts of pollution. In Brazil, some studies realized in chronosequences forest to pastures in Amazonia have shown that microbial biomass is reduced in the early years (about three to five years), but increases in older pastures reaching levels similar to those of the native forest (Feigl et al., 1995 a,b; Fernandes, 1999). The ability of SMB to increase again in older

pastures, reaching values closer to the native forest suggests that the microorganisms of such soils have high resilience, or the capacity for growth and physiological activity, even after the impact of slash-and-burn of the native forest.
The stability of a system determines its ability to continue working under stress conditions, for both natural and those induced by human action (Orwin & Wardle, 2004). Since the microorganisms are the key players of the conversion of soil organic matter and the availability of nutrients, its resilience directly affects plant productivity and the stability of forest and agricultural ecosystems (Orwin & Wardle, 2005). For this reason it is essential to understand how microorganisms respond to environmental disturbances, as well as the factors involved in this response.

3. Diversity approach applied to soil microorganisms

Amazonian tropical forest soils are supposed to hold high microbial biodiversity, since they support by litter recycling one of the most luxuriant ecosystems. However anthropogenic practices of slash-and-burn, mainly for pasture establishment, induce deep changes in the biogeochemical cycles, and possibly in the composition and function of microbial species (Cenciani et al., 2009).
While the diversity of microorganisms in the soil is immense, only a very low percentage is cultivable (around 1%) under laboratory conditions. The limited range between the bacteria species, for example, hampers the detection by microscopy techniques. Additionally the methods of obtaining bacteria in culture medium are not very effective for its quantification, due to difficulties in reproducing the conditions that every species or groups require in their natural habitats (Felski & Akkermans, 1998). Estimates of the global diversity of fungi indicate that a small percentage is described in the literature, especially due to limitations found in techniques of cultivation to assess the diversity of fungi. Apart from this the lack of taxonomic knowledge hinders the identification of bacterial and fungal species found in the soil (Kirk et al., 2004).
The study of prokaryote diversity is extremely complex because the definition of species for these organisms is a question still open. Currently a prokaryotic species is regarded as a group of strains including the standard strain, characterized by some degree of phenotypic consistency showing 70% or more DNA-DNA homology and more than 95% similarity between the 16S rRNA gene sequences. In this context we highlight the importance of polyphasic taxonomy, which aims to integrate different datasets and phenotypic, genetic and phylogenetic information about the microorganisms (Gevers et al., 2005).
With the advance of molecular biology it became possible to identify bacteria, fungi and other microorganisms in the soil and plants without need to isolate them. One of cultivation-independent molecular tool that has often been used to analyze the diversity and dynamics of microbial populations in the environment is the polyacrylamide gel electrophoresis in denaturing gradient (DGGE). The DNA is extracted and purified and only a fragment of the rRNA gene is amplified by the polymerase chain reaction (PCR). The amplification products are analyzed by gel electrophoresis, which allows the separation of small PCR products, commonly up to 400 bp according to their contents of guanine plus cytosine (G+C) Consequently, the fingerprinting pattern is distributed along a linear denaturing gradient (Muyzer & Ramsing, 1995; Courtois et al., 2001; Cenciani et al., 2009).

3.1 *Fungi* diversity assessed by PCR-DGGE

The role of fungi in the soil is complex and fundamental to maintain the functionality of the biome. Fungi play an active role in nutrient cycling and develop pathogenic or symbiotic associations with plants and animals, besides interacting with other microorganisms (Anderson & Cairney, 2004).

Working with soils in the Amazonia, Monteiro et al. (2007) described the changes in the genetic profiling of soil fungal communities caused by different land use systems (LUS): primary forest, secondary forest, agroforestry, agriculture and pasture. The author conducted her study in the following sequence: DNA extraction - total DNA was extracted using the Fast DNA kit (Qbiogene, Irvine, CA, USA), according to the manufacturer's instructions; PCR - a fragment of the 18S rRNA gene (1700 bp) of fungi was amplified by PCR according to Oros-Sichler et al. 2006; DGGE - amplicons were separated on an acrylamide gel containing bisacrilamide and a linear gradient of urea and formamide (Fig. 2).

Diversity Database program (BioRad) was used to determine the richness of amplicons. The non-metric multidimensional scaling (NMDS) tool was used to determine the effect of land use changes under the fungi communities through the PRIMER 5 program (PRIMER-E Ltd., 2001).

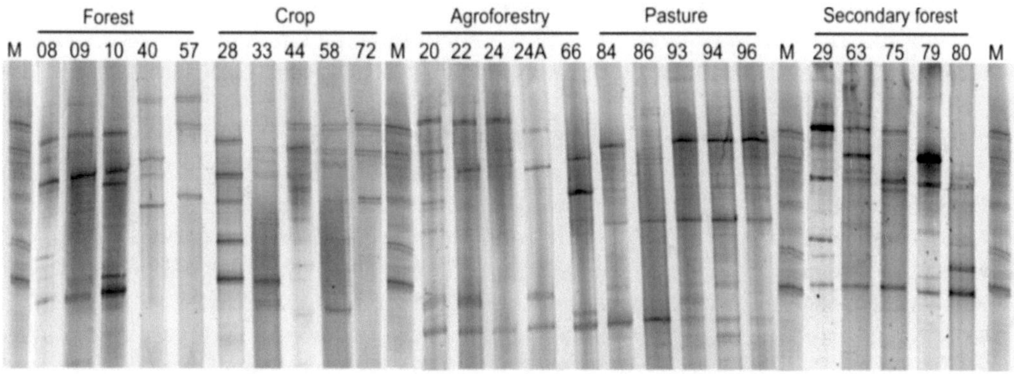

Fig. 2. DGGE gel of 25-38% urea and formamide, generated by separation of 18S rRNA gene fragments amplified from samples of natural soils under different LUS. M – molecular marker.

The DGGE of the 18S rRNA gene combined with NMDS statistic tool showed the presence of distinct communities in each of the areas analyzed, with the presence of single bands. Results indicated the dominance of specific fungal groups in every treatment, especially in the area converted to pasture, distant from the other systems of land use (Fig. 2).

Following this pattern the authors asserted that the banding profile generated by DGGE represent fungi communities from different soils, and were shown to be more similar among samples from the same system of land use than among samples of different systems of land use. However the clustering of samples through NMDS showed that there is a tendency for samples from pasture be different of the other sites, which are closest relatives among them (Fig. 3). Finally the results obtained by the authors show that changes in the land use affected the community structure of soil fungi; as well it is also possible that the type of vegetation covering has a key role in such changes (Monteiro et al., 2007).

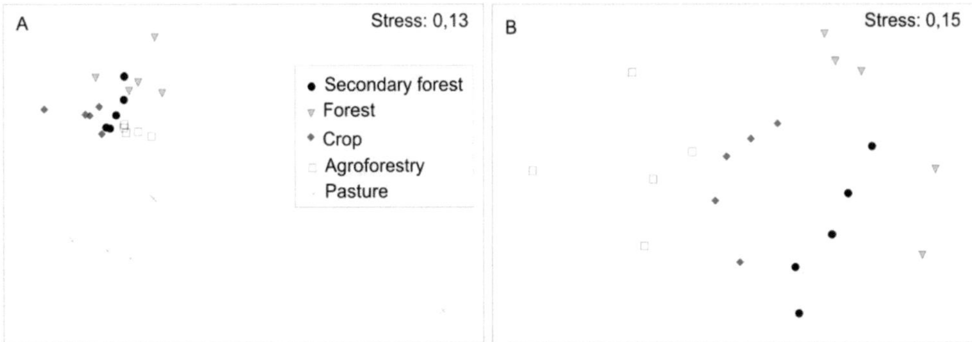

Fig. 3. Non-metric multidimensional scaling (NMDS) of 18S rRNA gene amplicons from soils of the Amazon forest under different systems of land use: secondary forest, forest, crop, agroforestry and pasture - stress 0.13 (A); and secondary forest, forest, crop, agroforestry and pasture - stress 0.15 (B).

Although molecular fingerprinting approaches such as cloning and sequencing are being used increasingly for evaluation of fungal communities, there are scarce studies reaching the diversity of fungi in soils of native forests, and in the same soils but impacted by agricultural management. Within this context changes in the genetic profile of fungi according to each system of land use, and the environmental stress can provide valuable information for the sustainable management of forest soils (Monteiro et al., 2007).

3.2 *Bacteria* diversity assessed by PCR-DGGE

Advances in molecular approach such as the DNA profiling through PCR-DGGE can also provide information regarding the composition of bacterial populations in soils. Cenciani et al. 2009 examined how the clearing of Amazonian rainforest for pasture and the seasonality affected the diversity of *Bacteria* domain. The aim of this study was to assess the extension that land use changes in Amazonia had on the structure of *Bacteria* domain.

According to Cenciani et al. (2009) field works were developed at Nova Vida Ranch (62º49`27``W; 10º10`5``S), in the central region of Rondonia state (Fig. 4). The predominant soil is classified as Argissolos in the Brazilian classification system (Empresa Brasileira de Pesquisa Agropecuaria - EMBRAPA, 2006) and as Ultisols (Kandiuldults) in the US soil taxonomy. It is a representative soil of Amazonian basin covering almost 22% of the Brazilian Amazonian basin. The Nova Vida Ranch covers an area of approximately 22.000 ha, consisting of a mixture of native forest and pastures of different ages. Pastures were established with no mechanical machinery nor chemical fertilization and soil acidity correction. Wood weeds were controlled by cutting the aerial part, removing the residues and burning them to reduce volume and incorporate the ashes into the soil (Feigl et al., 2006).

A sequence was chosen at Nova Vida: (1) a 3-ha plot of native forest, (2) a well-established pasture of 20 years (*Brachiaria brizantha* and *Pannicum maximum*), and (3) a fallow site (Fig. 1). The botanical composition of fallow includes 15-18% of woody species (*Tabebuia* spp., *Erisma uncinatum* and *Vismia guianensis*), 12% of Babaçu palm (*Orbignya phalerata* Mart), herbaceous weeds 4-11%, and 63.5% of a mixture of *Brachiaria brizantha* and *Pannicum maximum* (Feigl et al., 2006). Soil samples were taken at surface layer (0-10 cm) in the rainy season and 6 months later, in the dry season.

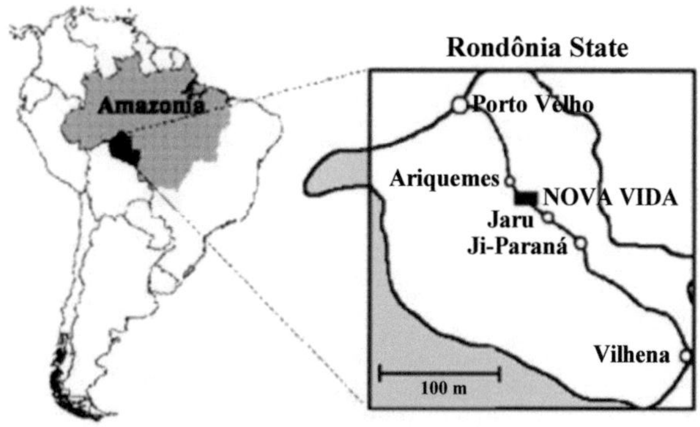

Fig. 4. Map of the study site located in Rondonia State.

Total soil DNA extraction and PCR products were generated according to conditions described by Øvreas et al., 1997. PCR products (300 ng) were resolved using DGGE to provide the molecular profiles of bacterial communities. The structure of similarity for *Bacteria* was generated from binary data. Dendrograms representing hierarchical linkage levels were constructed based on the Euclidean distance coefficient using Systat 8.0 software.

As expected PCR with specific primer sets including the forward primer coupled with a GC clamp resulted in a single 180-bp fragment. PCR products were separated by DGGE to assess the qualitative bacterial composition. Some groups of bands, exemplified as I to VI, were chosen to better compare similar and/or different band profiles (Figs. 5 and 6).

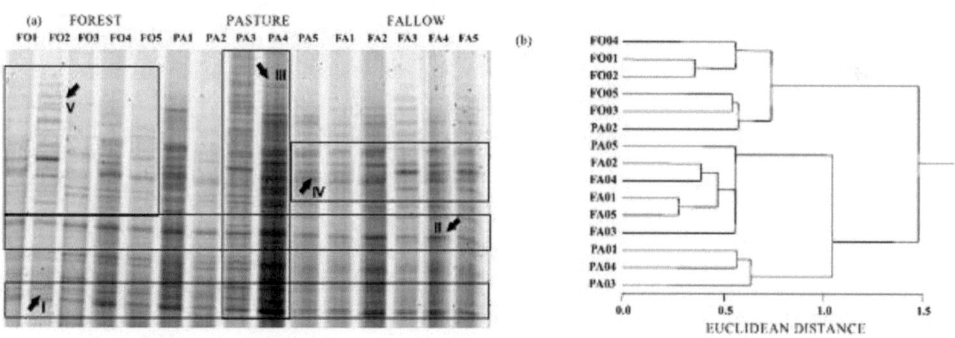

Fig. 5. DGGE (a) and cluster analysis (b) of the 16S rRNA gene in Amazonian soil samples, collected in the wet season.

In the Figure 5a (wet season), some bands were found in all soil replicates (I, II). It means that they were present in the DNA extracted from each sample and it indicated the presence of the same bacterial community in the three sites. Pasture was characterized by the presence of band patterns concentrated in PA3 and PA4 (III), and IV is a band profile found in the fallow and in the PA5 replicate of pasture. Forest contained replicates with high variability of band patterns; therefore FO2 contained more bands than the others (V).

DGGE profiling in the dry season (Figure 6a) revealed more visible differences in the bacterial structure among the sites than in the wet season. Band patterns I and II were presented in almost all samples, except FA1 to FA4. Group III represented bands common to pasture and fallow, while IV and V were bands specific to replicates FO1 to FO4 and FO1, respectively. VI was a particular banding pattern from pasture. It was not found a band profile presented specifically in the fallow site. Independently of sampling period, similar bands were found among the sites; as well each site had its own particular bands along DGGE profile.

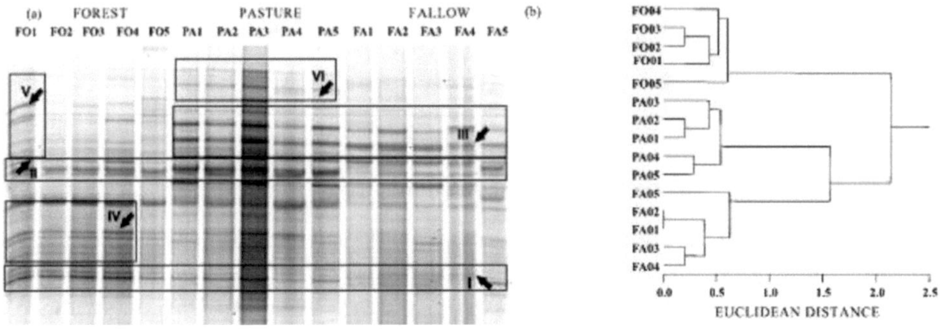

Fig. 6. DGGE (a) and cluster analysis (b) of the 16S rRNA gene in Amazonian soil samples collected in the dry season.

In the cluster analysis of PCR-DGGE products, the three sites clustered at 65% level of similarity for both wet and dry seasons. Data presented in Figure 5B shows that, in the wet season, bacterial communities were separated in three clusters, except PA2 replicate that tended to group together forest cluster; whereas PA5 replicate fell into the fallow cluster. In the Figure 6B, the effect of low water content plus history of soil use contributed to separate completely the bacterial populations from each site during the dry season. The variation in the composition of microbial community DNA between replicate soil samples was found to be as great as the variation between treatments in field based studies. The reasons for such variability are not clear, however it is likely that are attributable to the effect of soil chemical attributes plus the contents and composition of organic matter (Clayton et al., 2005; Ritz et al., 2004).

According to the authors the DGGE profiling revealed lower number of bands per area in the dry season, but differences in the genetic diversity of bacterial communities along the sequence forest to pasture was better defined than for wet season. The few research works using molecular approaches to investigate the diversity of microorganisms in Amazonia have shown that, in fact, a tiny fraction of their microbial diversity is known (Cenciani et al., 2009).

3.3 Other molecular tools applied to microbial diversity in amazonian soils

Soil microbial diversity is still a difficult field to study, especially due to the several limitations of techniques. Since 95-99% of organisms cannot be cultivated by culture based-methodologies, the microbial diversity of soils shall be assessed by molecular biology techniques (Elsas & Boersama, 2011).

New DNA and RNA sequencing techniques provide high resolution information, especially using depth sequencing of metagenomic samples. Most of times a high amount of the obtained sequences are related with unknown genes or unknown organisms, involving a high cost per sample. Since soils imply in most of times in high spatial variability, which means high number of samples and replicates, fingerprinting techniques are recommended prior to sequencing in order to reduce costs for the high resolution techniques.

The first study of microbial diversity in Amazon soils using molecular techniques, by means of clone library, showed a high prokaryotic diversity (Borneman & Tripplett, 1997). Analyzing 100 sequences, differences between mature forest and pasture were detected, and about 18% of sequences were related to unknown Bacteria. A decade after, analyzing 654 clones similar results were detected in other study site, in which 7% of sequences could not be classified in any bacterial phyla (Jesus et al., 2009). In both studies land use changes was an important factor, and the unknown species were surveyed showing that depth sequencing should be used to better characterize the Amazon soils.

The most popular techniques for soil microbial communities fingerprinting are DGGE and the terminal restriction fragments length polymorphism (T-RFLP), which should be complemented by sequencing information to provide an overview of the study sites. Such techniques consist in extraction of nucleic acids from the soil samples; followed by amplification by PCR, aiming to target specific microbial groups according to the primers chosen (i.e. a universal primer for 16S rRNA gene will give a general prokaryotic overview of the samples). After PCR the amplicons should be analyzed by denaturizing gel separation (DGGE) or digestion with restriction enzymes and analysis of the dye labeled fragments (T-RFLP), or DNA sequencing. In turn metagenomics techniques allow sequencing without preview amplification by PCR and other techniques to be considered (Elsas & Boersama, 2011).

T-RFLP consists in a PCR using dye labeled primers followed by a digestion with restriction enzymes, purification and reading in a DNA sequencer. The PCR amplifies a specific gene (mainly the 16S rRNA gene for prokaryotic diversity), and the restriction enzymes fragment the PCR products according to its polymorphism. The sequencer separates the fragments by length reading them in an electrophoresis run. So the presence of distinct fragment sizes found in different soil samples allows the diversity separation among them (Jesus et al., 2009). Clone libraries consist in cloning the PCR amplicons into bacterial vectors, followed by DNA sequencing. Since the PCR from environmental samples amplify different DNA sequences of different organisms at the same time, cloning technique allows the separation of amplicons and the sequencing of individual sequences (Borneman & Tripplett, 1997). Different studies using other molecular approaches to access the diversity of Amazon soils (Table 1) are described below.

In Western Amazon a T-RFLP analysis of the bacterial communities showed how it was influenced by soil attributes correlated to land use (Jesus et al., 2009). Community structure changed with pH and nutrient concentration. By DNA sequencing, bacterial communities presented clear differences among the different sites. Pasture and one of crops presented the highest diversity. Secondary forest presented similar diversity with the community structure of the primary forest, showing that bacterial community can be restored after agricultural use of the soils. Using the automated ribosomal intergenic spacer amplification (ARISA) technique distinct microbial structures were also observed between agricultural and forest soils (Navarrete et al., 2010). Seasonal changes in the two different years of sampling and distinct band patterns were observed for fungal, bacterial and archaeal richness.

Different patterns between Terra Preta soil (Dark Earth or Anthrosols) and an adjacent soil were observed in the Southwestern Amazon using 16S rRNA gene sequencing (Kim et al., 2007). *Acidobacteria* were predominant in both sites but 25% greater species richness was observed in the Antrosol. In other study in three Dark Earth sites near Manaus, "Lago Grande", "Hatahara" and "Açutuba", a cultivable bacteria survey showed a higher richness in Antrosols than in the adjacent soils (O'Neill, 2009). Several bacteria were isolated using rich media or soil-extract media and genetic groups were separated by RFLP. By sequencing, *Bacillus* was the most abundant genera.

Main Aim of the Study	Technique(s)	Localization (States of Brazil)	Reference
Compare *Bacteria* diversity in forest and pasture soils	Clone Library	Paragominas, Para (2°599S; 47°319W)	Borneman & Tripplett., 1997
Investigate Dark Earth bacterial diversity	Clone Library	Jamari, Rondonia (8°45'0S; 63°27'0W)	Kim et al., 2007
Compare Bacterial communities in Anthrosols and adjacent soils	Bacteria isolation + RFLP + Sequencing	Manaus, Amazonas (3°08'S; 59°52'W)	O'Neill, 2009
Investigate land use impact on soil *Bacteria* structure	T-RFLP + Clone Library	Benjamin Constant, Amazonas (4°21S,69°36W; 4°26S,70°1W)	Jesus et al., 2009
Compare Anthrosols with adjacent soils	DGGE followed bands Sequencing + T-RFLP	Manaus, Amazonas (3°08`S; 59`52'W)	Grossman et al., 2010
Investigate microbial communities in agricultural systems	ARISA + T-RFLP + Pyrosequencing	Benjamin Constant, Amazonas (4°21S, 69°36W; 4°26S,70°1W) + Iranduba, Amazonas (03°16'28.45"S; 60°12'17.14"W)	Navarrete et. al., 2010
Land use in *Archaeal* and *amo*A structures in Dark Earths	T-RFLP + Qpcr + Clone Library	Manaus, Amazonas (from 02°01'52.50"S, 26'28.30"W; to 03°18'05.01"S, 60°32'07.38"W)	Taketani, 2010
Investigate Archaeal structure in a wetland soil	Clone Library + methanogenic bacteria isolation	Santarem, Para (02°23'20"S; 54°19'39.5"W)	Pazinato et al., 2010
Investigate the influence of different land uses on the bacterial structure of Cerrado and Forest Soils	T-RFLP	Sinop (Tropical Forest - S120553.3W; 552846.0) and Campo Verde (Cerrado - S 151588.8; W 550700.0), Mato Grosso	Lammel et al., 2010

Table 1. Diversity studies using other molecular biology techniques in Amazon soils

Grossman et al. (2010) studying the three same Dark Earths sites, including one additional site, "Dona Stella", and using different molecular techniques also found difference among the samples.. T-RFLP of the 16S rRNA genes provided clear distinction between the two types of soils, and the same result was observed using DGGE and 16S rRNA sequencing. While T-RFLP provided a good fingerprinting between Anthrosols and Adjacent soils, 16S rRNA sequencing provided better resolution of the changes, indicating *Verrucomicrobia* as an important group to the Anthrosols, *Proteobacteria* and *Cyanobacteria* for Adjacent soils; while *Pseudomonas, Acidobacteria* and *Flexibacter* were found in both sites.

Studying the "Hatahara" site, differences in bacterial communities were also observed among Amazonian Dark Earth, black carbon and an adjacent oxisol by T-RFLP (Navarrete et al., 2010). By pyrosequencing it was shown that the most predominant phyla were *Proteobacteria, Acidobacteria, Actinobacteria* and *Verrucomicrobia*. About one-third of the sequences corresponded to unclassified *Bacteria*. For archaeal structure comparison by T-RFLP the soil attributes were more important than the type of soil, if it was Terra Preta or adjacent soils (Taketani, 2010). DNA sequencing showed that *Candidatus spp.* was the most abundant genera in both types of soils. An *amo*A clone library showed differences among the sampled sites, but also did not show differences between Terra Preta and the adjacent soil.

Using T-RFLP of bacterial 16S rRNA, distinct patterns were observed among biomes and land uses in the Southwestern Amazon (Lammel et al., 2010). Southwestern Amazon is divided in two mainly biomes, Tropical Forest and Cerrado (Brazilian Savanna). Over the last three decades these natural vegetations have been converted to pasture and agriculture. Land use was the most important factor to distinguish the bacterial communities, and it was correlated with the soil chemical changes: pH - due to liming and chemical fertility - due to fertilizers application. Pristine Tropical Forest and Cerrado formed distinct clusters, but they were more similar to each other than in relation to pasture or soybean field (Fig. 7).

Fig. 7. Different land uses (native forest, native cerrado, soybean field and pasture) studied by Lammel et al. (2010).

In Eastern Amazon wetland soils Archaeal community was characterized by 16S rRNA gene libraries and by isolation of methanogenic Archaea (Pazinato et al., 2010). Archaeal diversity decreased with depth and the most of sequences belonging to *Crenarchaeota, Methanosarcina* and *Metahnobacteriam* genera were isolated from the sites.

These different techniques showed a high microbial diversity on Amazon soils. Fingerprinting techniques, such as T-RFLP and ARISA, were sensitive tools to detect difference in the microbial structure among the different sites and land uses. However only DNA sequencing provided a better resolution of the diversity, i.e. identify taxonomic groups and report unknown *Bacteria* that probably belong to new taxonomic groups. These pioneer studies showed, in general, that diversity does not decrease from pristine vegetation to agricultural uses, but the structure of microbial community as a whole is affected by land use changes. They can be restored after stopping the soil cultivation followed by secondary forest growth. The Amazon region is a "hot spot" regarding the soil microbial diversity.

3.4 Arbuscular mycorrhizal fungi

Arbuscular mycorrhizal fungi (AMF) are also an important microbial group in soil, since they can form symbiosis with most of the plants, contributing to plant health and nutrition. AMF is beneficial to tropical plants and presents potential influence on soil processes and plant diversity, increasing the interest For studying this group this group, especially in Amazon where little is known about them (Stürmer & Siqueira, 2010).

Most of AMF studies consist on identification of its spores from soil samples. Since AMF produce spores significantly bigger than the other fungi species, it is possible to separate them from soil samples by sieve and centrifugation in a sucrose gradient. Up to now, the studies in Brazilian Amazon were made using this approach (Leal et al., 2009; Mescolotti et al. 2010; Stürmer and Siqueira, 2010).

In Southwestern Amazon an AMF study compared three land uses: native vegetation, soybean fields and pastures, in two regions: Sinop (Forest) and Campo Verde (Cerrado), both in Mato Grosso State, Brazil (Mescolotti et al., 2010). Comparing Forest with Cerrado different patterns were observed. The largest amount of spores was found in soybean fields in the Forest region, and the number of spores was the same for the three land uses in the Cerrado region. *Glomus* spp. was the most common specie found (Fig. 8.).

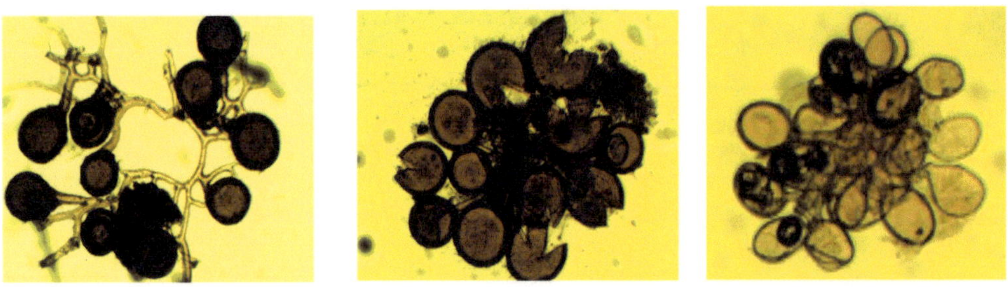

Glomus sp1 *Glomus* sp2 *Glomus* sp3

Fig. 8. AMF surveyed in Southwestern Amazon. *Glomus* spp was the most common (Mescolotti et al., 2010).

In Western Amazon different AMF patterns were observed in different land uses (Stürmer & Siqueira, 2010). A total of 61 AMF morphotypes were recovered and 30% could not be classified as known species. *Acaulospora* and *Glomus* were the most common genera identified in the sites and higher AMF richness values were found in agriculture and pasture sites, than in the pristine areas. AMF patterns were also influenced by land use in a survey using different trap cultures in the same region (Leal et al., 2009). Among all trap plants and land uses, a higher number of spores were found in pasture and young secondary forest. In total 24 AMF species were recovered. *Acaulospora* spp. (10 species) was the most common genera followed by *Glomus* spp (5 species). Both studies showed that in Amazon soils the land use change from pristine vegetation to pasture and crops did not reduce the AMF diversity and probably new AMF species were found.

3.5 Catabolic diversity profile

Catabolic diversity profile (CDP) is a method aiming to measure the similarity of the catabolic functions of microbial communities in different soils or changes in the same soil under different treatments or land uses, or yet the intensity of respiratory responses to a range of substrates tested (Table 2). The richness (variety) of catabolic diversity is given by the total number of substrates that could potentially be used by the microbial community. The higher is the index of similarity, the greater is the diversity of microbial population; as it is maintained the ability of soil microorganisms to give an intense respiratory response to all substances (substrates) tested. With a reduction of microbial diversity, it is lost some species able to metabolize certain functional groups, and with it, the ability of the system to react (resilience) in the form of CO_2 emission decreases. The lower is the index of similarity; the lower is the diversity of microbial population (Van Heerden et al., 2002).

Substrates	Amine	Carbohydrate	Aminoacid	Carboxilic Acid
Glutamine	X			
Glucosamine	X			
Glucose		X		
Manose		X		
Arginine			X	
Asparagine			X	
Glutamic Acid.			X	
Histidine			X	
Lisine			X	
Serine			X	
Citric Acid				X
Ascorbic Acid				X
Glucomic Acid				X
Fumaric Acid				X
Malonic Acid				X
Malic Acid				X
Ketoglutaric Acid				X
Ketobutiric Acid				X
Pantotenic Acid				X
Quinic Acid				X
Succinic Acid				X
Tartaric Acid				X

Table 2. Substrates used in the catabolic diversity profile of soil microorganisms.

The two most common methods to measure the utilization of substrates by microorganisms are Biolog (Garland & Mills, 1991; Zak et al., 1994) and the respiratory response to addition of substrates, known as substrate induced respiration (SIR) (Degens & Harris, 1997; Degens et al., 2001). The authors claim that these techniques are sensitive enough to distinguish changes in the catabolic diversity that occur over short periods of time, as well as large differences that occur in the soil after a few years (Graham & Haynes, 2005). The main substrates used for SIR analysis are shown in Table 2. The diverse substrates are dissolved in 2 ml of solution for each equivalent of 1g dry soil and incubated in sealed bottles. The flow of CO_2 for each sample is usually measured in an Infra-Red Gas Analyser (IRGA), after incubation of bottles for 4 hours at 25°C.

Few studies have been carried out in the Amazon region. Among these is the work of Mazzetto et al. 2008. This research evaluated the possibility to check whether there are catabolic patterns in the Amazon soils under agricultural cultivation, native forest and pasture. A total of 60 areas were chosen distributed as: 20 native forest, 20 agricultural lands and 20 pasture sites in the regions of Mato Grosso and Rondonia, which are part of the Brazilian Amazon.

At first analyses were performed only in the native areas, which could be separated in Amazon rainforest, Cerrado and Cerradão. The low catabolic response obtained in the Cerrado soils may be linked to the frequent firing process that this biome suffers (Fig. 9). According to Arocena & Opio (2003), fire has a major impact on the physical (aggregate stability, clay content) and chemical (pH) soil properties, with significant influence on the microbial biomass. According to Hart (2005) fire alters the structure of microbial biomass, this being a selection factor in areas exposed to periodic events. Campbell et al. (2008) demonstrated in their studies that the use of carbonated substrates decreases with burning of area, suggesting a lower resistance/resilience of the microbial community. Among the substrates that can be influenced by burning of vegetation is arginine, which has a low response in Cerrado and Cerradão soils. The use of arginine in the microbial metabolism requires the presence of deaminase arginine enzyme, which is inhibited by fire.

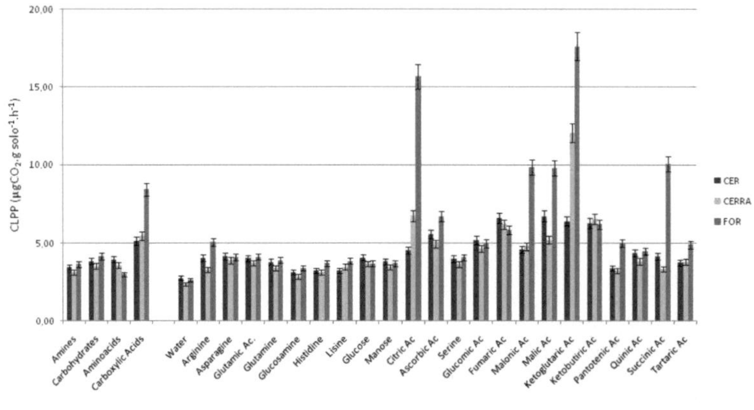

Fig. 9. Catabolic profile of soil microbial biomass in native areas: Cerrado (CER), Cerradão (CERRA) and Forest (FOR).

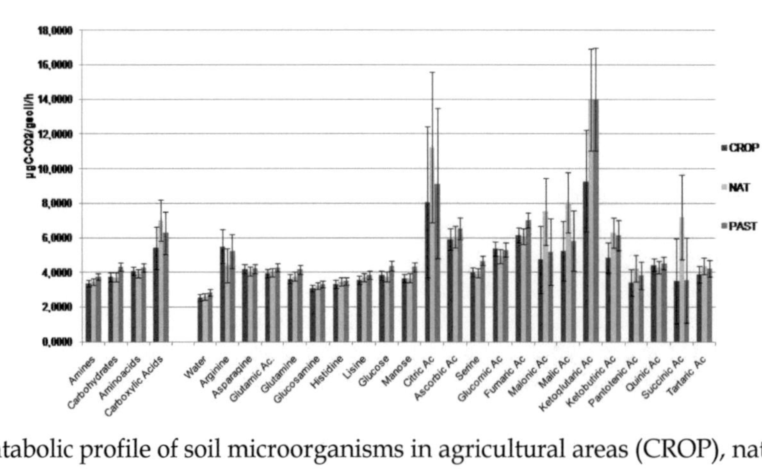

Fig. 10. Catabolic profile of soil microorganisms in agricultural areas (CROP), native areas (NAT) and pasture areas (PAST).

Regarding the disturbed areas analysis were realized aiming to characterize the diversity of soil microbial biomass at these sites (Fig. 10), and to check the possible separation of the areas through multivariate statistical analysis (Fig. 11).

Soils under pasture had significant catabolic responses to amine and carbohydrate, and individually to the substrates glutamic acid, glutamine, glucose, mannose, serine and fumaric acid. In contrast soils under native vegetation had significant responses to malonic acid, malic acid and succinic acid. Soils under agriculture use did not show significant responses to any substrate examined, however they showed expressive responses to the aminoacids group, but not statistically different from the pasture soil (Fig. 10).

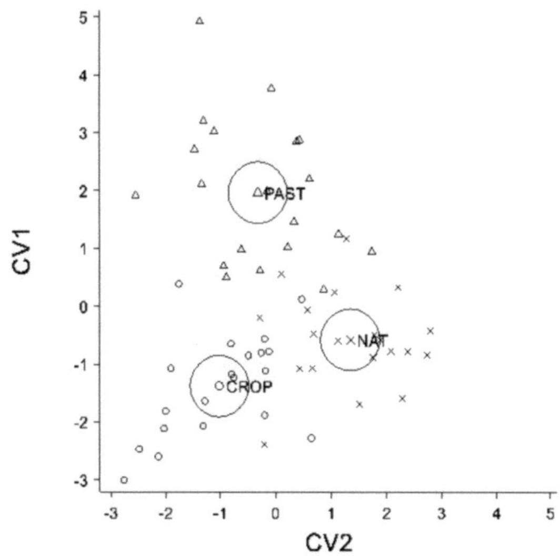

Fig. 11. Canonical analysis of the catabolic profile of microorganisms. Coefficient variation 1 (CV1) explained 67.50% of variability, while CV2 explained 32.50%. (Δ) Pasture, (○) Agricultural Areas, (x) Native Areas.

The canonical analysis showed that datasets related to CDP had great success in distinguishing the three land uses analyzed (Fig. 11). CV1 explained 67.5% of the variability observed, separating pastures from native areas and agriculture. Averages of native and agriculture areas were negative (-1.38 and -0.58, respectively) for CV1, while the average of pasture was positive (1.96). Asparagine, histidine and quinic acid with highly negative values were closely tied to native areas and agriculture, while glutamic acid and glucosamine had great representation in relation to pasture. CV2 explained 32.5% of the variability observed, separating native areas from agriculture and pastures. The average of native areas for the second axis was positive (1.34), while those of agriculture and pastures were negative (-1.02 and -0.32, respectively). The main substrates that provided this separation were serine and quinic acid, which showed negative values (linked to pasture and agriculture), and the tartaric acid, considered the more representative substrate related to native areas.

Among the major substrates involved, serine is documented as present in root exsudates (Bolton et al., 1992), quinic acid is a component of plant tissues (Gebre & Tchaplinski, 2002), and tartaric acid is one of main intermediary compounds of the Krebs cycle, in the basic metabolism of aerobic microorganisms (Tortora et al., 2005).

When only one ecoregion (Alto Xingu) was selected for analysis results of the CDP approach was even more significant (Fig. 12). CV1 explained 66.5% of the variability, separating native areas (-7.87 - negative score) of areas under agriculture and pasture (4.33 and 0.49 - positive scores, respectively). The main substrates involved in such axis were: succinic acid and malonic acid, with negative values. With positive values quinic acid and glucose also contributed to the separation observed. CV2 explained the remaining 33.5% of the variability, separating areas under pasture (4.84 - positive score) of native and agricultural areas (-2.04 and -2.65 - negative scores, respectively). Among the major substrates in this axis are highlighted asparagine and tartaric acid showing negative values, while lysine and pantothenic acid had positive values (Fig. 12).

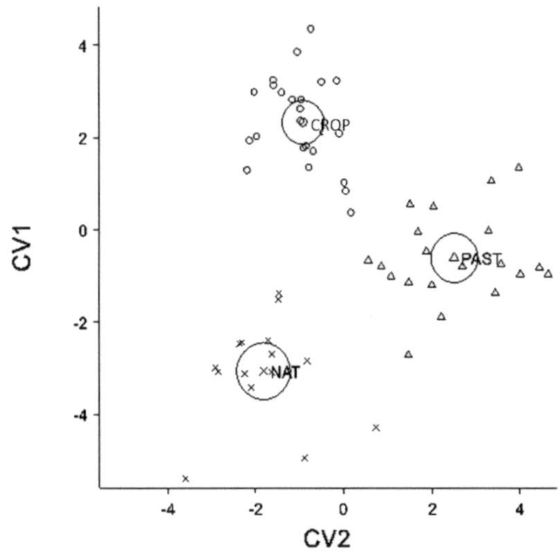

Fig. 12. Canonical analysis of the catabolic profile of microorganisms in the Alto Xingu ecoregion. CV1 explained 66.5% of variability, while CV2 explained 33.50%. (Δ) Pasture, (○) Agricultural Areas, (x) Native Areas.

Taking into account only data corresponding to the agricultural areas present in the database, we could distinguish areas under perennial crops, tillage and conventional tillage. By means of discriminant analysis the reallocation of data was performed in order to observe if datasets was homogeneous among the land uses analyzed. Data from areas under conventional tillage were relocated with 70% success, while data from conventional tillage and perennial cultivation showed higher percentage (98% and 100%, respectively). The same analysis was performed for pasture data that could be reallocated according to the following classification: typical pasture (100% success), improved pasture (95% success) and degraded pasture (91% success). This high percentage of reallocation of data shows that the microbial communities analyzed by CDP have high correlation with the use of land deployed. According to Mazzetto et al. 2008 the application of substrate induced respiration was efficient in distinguishing the land uses. The composition of microbial community revealed, through CDP approach, a close relationship with vegetation cover, regardless of climatic factors or the soil type.

As highlighted by Tótola & Chaer (2002) and San Miguel et al. (2007), the importance of functional and catabolic diversity lies in the fact that only based on changes in the genetic diversity it is not possible infer whether some functions of soil were lost or not. The physiological profile of microbial community allows accessing the metabolic capacity of the microbial biomass as a whole, through tests realized with specific carbon sources defined in the laboratory.

4. Conclusion

Soil microbial diversity is still a difficult field to study, since 95-99% of organisms cannot be cultivated by culturing methodologies. The most popular techniques for soil microbial communities fingerprinting are DGGE and T-RFLP, which should be complemented by sequencing information to provide an overview of the study areas, especially those with high spatial variability that requires the collection of a high number of samples and replicates. New DNA and RNA sequencing provide high resolution information especially using depth sequencing of metagenomic samples.

Using DGGE, T-RFLP and other approaches, it has been clear that land use changes influenced significantly the diversity and structure of microbial communities in the Amazonian soils. Data available of DNA sequencing provided a high resolution view pointing changes of specific microbial groups and also the high quantities of unknown microorganisms. Catabolic diversity profile was efficient in distinguishing the land uses. The composition of microbial community revealed, through CDP approach, a close relationship with vegetation cover, regardless of climatic factors or the soil type.

Land use changes modify the genetic structure of microbial communities in the Amazonian soils, but they do not reduce the diversity in the areas affected by deforestation and conversion for pasture and crops, in comparison with the native areas. Also many new species are to be discovered in such areas.

5. Acknowledgments

The authors are indebted to Coordenacao de Aperfeicoamento de Pessoal de Nivel Superior (CAPES), to Fundacao de Amparo a Pesquisa do Estado de Sao Paulo (FAPESP) and to Fundacao de Amparo a Pesquisa do Estado de Minas Gerais (FAPEMIG) for concession of scholarships and financial resources.

6. Appendix

Acronyms and Abbreviations

AMF - Arbuscular Mycorrhizal Fungi
ARISA – Automated Ribosomal Intergenic Spacer Amplification
CAPES - Coordenacao de Aperfeicoamento de Pessoal de Nivel Superior
CDP – Catabolic Diversity Profile
DGGE - Gel Electrophoresis in Denaturing Gradient
EF – Extraction-Fumigation
EMBRAPA – Empresa Brasileira de Pesquisa Agropecuaria
FAPEMIG - Fundacao de Amparo a Pesquisa do Estado de Minas Gerais
FAPESP – Fundacao de Amparo a Pesquisa do Estado de Sao Paulo
IF – Incubation-Fumigation

IRGA - Infra-Red Gas Analyser
LUS - Land Use Systems
NMDS - Non-Metric Multidimensional Scaling
PCR - Polymerase Chain Reaction
RFLP - Restriction Fragments Length Polymorphism
SIR - Substrate Induced Respiration
SMB - Soil Microbial Biomass
T-RFLP - Terminal Restriction Fragments Length Polymorphism

7. References

Ananyeva, N. D.; Susyan, E. A.; Chernova, O. V. & Wirth, S. (2008). Microbial respiration activities of soil from different climatic regions of European Russia. *European Journal of Soil Biology*, Vol. 44, pp. 147-157, ISSN 11645563.

Anderson, J. P. & Domsch, K. H. (1978). A physiological method for the quantitative measurement of microbial biomass in soils. *Soil Biology and Biochemistry*, Vol.10, pp. 215-221, ISSN 00380717.

Anderson, I. C. & Cairney, J. W. G. (2004). Diversity and ecology of soil fungal communities: increased understanding through the application of molecular techniques. *Environmental Microbiology*, Vol. 6, pp. 769-779, ISSN 0006-3568.

Arocena, J. M. & Opio, C. (2003). Prescribed fire-induced changes in properties of sub-boreal forest soils. *Geoderma*, Vol. 113, pp. 1–16, ISSN 0016-7061.

Bolton, H. Jr.; Fredrickson, J. K. & Elliott, L. F. (1992). Microbial ecology of the rhizosphere, In: *Soil microbial ecology: applications in agricultural and environmental management*, F.B. Metting, (Ed.), 27–63, New York Dekker, ISBN 0824787374 , New York, USA.

Borneman, J. & Triplett, E. W. (1998). Molecular microbial diversity in soils from Eastern Amazonia: evidence for unusual microorganisms and microbial population shifts associated with deforestation. *Applied and Environmental Microbiology*, Vol.63, pp. 2647-2653, ISSN 1098-5336.

Campbell, C. D.; Cameron, C. M.; Bastias, B. A.; Chen, C. & Cairney, J. W. G. (2008). Long term repeated burning in a wet sclerophly forest reduces fungal and bacterial biomass and responses to carbon substrates. *Soil Biology and Biochemistry*, Vol.40, pp. 2246-2252, ISSN 00380717.

Cenciani, K.; Lambais, M. R.; Cerri, C. C.; Azevedo, L. C. B. & Feigl, B. J. (2009). Bacteria diversity and microbial biomass in forest, pasture and fallow soils in the southwestern Amazon basin. *Revista Brasileira de Ciência do Solo*, Vol.33, pp. 907-916, ISSN 01000683.

Clayton, S. J.; Clegg, C. D.; Murray, P. J. & Gregory, P. J. (2005). Determination of the impact of continuous defoliation of *Lolium perenne* and *Trifolium repens* on bacterial and fungal community structure in rhizosphere soil. *Biology and Fertility of Soils*, Vol. 41, pp. 109-115, ISSN 0178-2762.

Courtois, S.; Frostegard, A.; Goransson, P.; Depret, G.; Jeannin, P. & Simonet, P. (2001). Quantification of bacterial subgroups in soil: comparison of DNA extracted directly from soil or from cells previously released by density gradient centrifugation. *Environmental Microbiology*, Vol.3, pp. 431-439, ISSN 1462-2920.

Degens, B. P. & Harris J. A. (1997). Development of a physiological approach to measuring the catabolic diversity of soil microbial communities. *Soil Biology and Biochemistry*, Vol.29, pp. 1309-1320, ISSN 00380717.

Degens, B. P.; Schipper, L. A.; Sparling, G. P. & Duncan, L. C. (2001). Is the microbial community in a soil with reduced catabolic diversity less resistant to stress or disturbance? *Soil Biology and Biochemistry*, Vol.33, pp. 1143-1153, ISSN 00380717.

Dilly, O. & Nannipieri, P. (2001). Response of ATP content, respiration rate and enzyme activities in an arable and a forest soil to nutrient additions. *Biology and Fertility of Soils*, Vol.34, pp. 34-64, ISSN 0178-2762.

Elsas, J.D. & Boersma, F. G. H. (2011). A review of molecular methods to study the microbiota of soil and the mycosphere. *European Journal of Soil Biology*, Vol.47, pp. 77-87, ISSN 1164-5563.

EMBRAPA SOLOS. (2006). *Sistema Brasileiro de Classificação de Solos*, Embrapa Informação Tecnológica, ISBN 85-85864-19-2. Brasília, Brazil.

Fearnside, P.M. 2005. Deforestation in Brazilian Amazonia: history, rates and consequences. *Conservation Biology*, Vol.19, pp. 680-688, ISSN 0888-8892.

Feigl, B. J.; Melillo, J. M. & Cerri, C. C. (1995a) Changes in the origin and the quality of soil organic matter after pasture introduction in Rondonia (Brazil). *Plant and Soil*, Vol.175, pp. 21-29, ISSN 0032-079X.

Feigl, B. J.; Sparling, G.; Ross, D. & Cerri, C. C. (1995b). Soil microbial biomass in Amazonian soils: evaluation of methods and estimates of pools sizes. *Soil Biology and Biochemistry*, Vol.27, pp. 1467-1472, ISSN 00380717.

Feigl, B.; Cerri, C.; Piccolo, M.; Noronha, N.; Augusti, K.; Melillo, J.; Eschenbrenner, V. & Melo, L. (2006). Biological survey of a low-productivity pasture in Rondonia state, Brazil. *Outlook on Agriculture*, Vol. 35, pp. 199-208, ISSN 0030-7270.

Felske, A. & Akkermans, D. L. (1998). Spatial homogeneity of abundant bacterial 16S rRNA molecules in grassland soils. *Microbiology Ecology*, Vol.36, pp. 31-36 ISSN 1432-184X.

Fernandes, S. A. P.; Bernoux, M.; Cerri, C. C.; Feigl, B. J. & Piccolo, M. C. (2002). Seasonal variation of soil chemical properties and CO_2 and CH_4 fluxes in unfertilized and P-fertilized pastures in an Ultisol of the Brazilian Amazon. *Geoderma*, Vol.107, pp. 227 - 241, *ISSN* 0016-7061.

Garland, J. L. & Mills, A. L. (1991). Classification and characterization of heterotrophic microbial communities on the basis of patterns on community-level, sole-carbon-source utilization. *Applied and Environmental Microbiology*, Vol. 57, pp. 2351–2359, ISSN 1098-5336.

Gebre, G. M. & Tschaplinski, T. J. (2002). Solute accumulation of chestnut oak and dogwood leaves in response to throughfall manipulation of an upland oak forest. *Tree Physiology*, Vol. 22, pp. 251–260, ISSN 1758-4469.

Gevers, D.; Cohan, F. M.; Lawrence, J. G.; Spratt, B. G.; Coenye, T.; Feil, E. J.; Stackebrandt, E.; Van de Peer, Y.; Vandamme, P.; Thompson, F. & Swings, J. (2005). Re-evaluating prokaryotic species. *Nature Reviews Microbiology*, Vol. 3, pp. 733-739, ISSN 1740-1526.

Graham, M. H. & Haynes, R. J. (2005). Catabolic diversity of soil microbial communities under sugarcane and other land uses estimated by Biolog and substrate-induced respiration methods. *Applied Soil Ecology*. Vol. 29, pp. 155-164, ISSN 0929-1393.

Grossman, J. M.; O'Neill, B.; Tsai, S. M.; Liang, B.; Neves, E.; Lehmann, J. & Thies, E. J. (2010). Amazonian Anthrosols support similar microbial communities that differ distinctly from those extant in adjacent, unmodified soils of the same mineralogy. *Environmental Microbiology*, Vol.60, pp. 192-205, ISSN 1462-2912.

Hart, S. C.; De Luca, T. H.; Newman, G. S.; Mackenzie, M. D. & Boyle, S. I. (2005). Post-fire vegetative dynamics as drives of microbial communities structure and function in Forest soils. *Forest Ecology and Management*, Vol. 220, pp. 166-184, ISSN 0378-1127.

Jenkinson, D. S. & Powlson, D. S. (1976). The effects of biocidal treatment on metabolism in soil-V. A method for measuring soil biomass. *Soil Biology and Biochemistry*, Vol. 8, pp. 209-213, ISSN 0038-0717.

Jenkinson, D. S. & Ladd, J. N. (1981). Microbial biomass in soil. Measurement and turnover, In: *Soil Biochemistry*, E.A. Paul. & J.M. Ladd, (Ed.), Vol.5, pp.415-471, Dekker, ISBN 0-8247-11131-9, New York, USA.

Jesus, E. C.; Marsch, T. L.; Tiedje, J. M. & Moreira, F. M. S. (2009). Changes in lan use alter the structure of bacterial communities in Western Amazon soils. *The ISME Journal*, Vol.3, pp. 1004-1011, ISSN 1751-7362.

Kennedy, A. C. (1999). Bacterial diversity in agroecosystems. *Agriculture, Ecosystems and Environment*, Vol.74, pp. 65-76, ISSN 0167-8809.

Kim, J.; Sparovek, G.; Longo, R. M.; Melo, W. J. & Crowley, D. (2007). Bacterial diversity of terra preta and pristine forest soil from the Western Amazon. *Soil Biology & Biochemistry*, Vol.39, pp. 684-690, ISSN 0038-0717.

Kirk, J. L.; Beaudette, L. A.; Hart, M.; Moutoglis, P.; Klironomos, J. N.; Lee, H.; Trevors, J. T. (2004). Methods of studying soil microbial diversity. *Journal of Microbiological Methods*, Vol. 58, pp. 169-188, ISSN 0167-7012.

Lammel, D. R.; Cerri, C. C. & Tsai, S. M. (2010). Microbial structure in different land uses in Southwest Amazon by T-RFLP. *XIII International Symposium on Microbial Ecology* (ISME), Seattle, USA, August 22-27.

Lavelle, P. (2000). Ecological challenges for soil science. *Soil Science*, Vol.165, pp. 73-86, ISSN 1538-9243.

Leal, P. L.; Stürmer, S. L. & Siqueira, J. O. (2009). Occurrence and diversity of arbuscular mycorrhizal fungi in trap cultures from soils under different land use systems in the Amazon, Brazil. *Brazilian Journal of Microbiology*, Vol. 40, pp. 111-121, ISSN 1517-8382.

Mazzetto, A. M.; Feigl, B. J. & Cerri, C. C. (2009) Diversidade catabólica da biomassa microbiana do solo alterada pelo uso da terra. *XXV Congresso Brasileiro de Microbiologia*, Porto de Galinhas, Brazil, November 8-12.

Mescolotti, D.; Lammel, D. R.; Castro, T.; Cardoso, E. J. B. & Cerri, C. C. (2010). Ocorrência de FMAs em diferentes usos da terra no sudoeste da Amazônia e colonização micorrízica de *Glycine max* e *Brachiaria brizantha. Poster Presentation. XII Encontro Nacional de Microbiologia Ambiental - ENAMA.* Manaus, Brazil, December 5-8.

Monteiro, G. G.; Azevedo, L. C. B.; Armas R.; Lambais, M. R. & Pfenning, L.H. (2007). Estrutura da comunidade de fungos em solos da Amazônia sob diferentes sistemas de uso da terra., *XXXI Congresso Brasileiro de Ciência do Solo*, Gramado, Rio Grande do Sul, Brazil, August 05-10.

Muyzer, G. & Ramsing, N. B. (1995). Molecular methods to study the organization of microbial communities. *Water Science and Technology,* Vol. 8, pp. 1-9, ISSN 0273-1223.

Navarrete, A. A.; Cannavan, F. S.; Taketani, R. G. & Tsai, S. M. (2010). A molecular survey of the diversity of microbial communities in different Amazonian agricultural model systems. *Diversity,* Vol. 2, pp. 787-809, ISSN 1424-2818.

O'Neill, B.; Grossman, J.; Tsai, S. M.; Lehmann, J.; Peterson, J.; Neve E. & Thies, J. E. (2009). Bacterial community composition in Brazilian Anthrosols and adjacent soils characterized using culturing and molecular identification. *Microbial Ecology,* Vol. 58, pp. 23-35 ISSN 0095-3628.

Orwin, K. H. & Wardle, D. A. (2004). New indices for quantifying the resistance and resilience of soil biota to exogenous disturbances. *Soil Biology and Biochemistry,* Vol. 36, pp. 1907-1912, ISSN 0038-0717.

Orwin, K. H. & Wardle, D. A. (2005). Plant species composition effects on belowground properties and the resistance and resilience of the soil microflora to a drying disturbance. *Plant and Soil,* Vol. 278, pp. 205-221, ISSN 1573-5036.

Øvreås, L.; Forney, L.; Daae, F. L. & Torsvik, V. (1997). Distribution of bacterioplankton in meromictic lake saelevannet, as determined by denaturing gradient gel electrophoresis of PCR-amplified gene fragments coding for 16S rRNA. *Applied and Environmental Microbiology,* Vol. 63, pp. 3367-3373, ISSN 1098-5336.

Pazinato, J. M.; Paulo, E. N.; Mendes, L. W.; Vazoller, R. F. & Tsai, S. M. (2010). Molecular characterization of the archaeal community in an Amazonian wetland soil and culture-dependent isolation of methanogenic archaea. *Diversity,* Vol. 2, pp. 1026-1047, ISSN 1424-2818.

Ritz, K.; Mcnicol, J. M.; Nunan, N.; Grayston, S.; Millard, P.; Atkinson, D.; Gollotte, A.; Habeshaw, D.; Boag, B.; Clegg, C. D.; Griffiths, B. S.; Wheatley, R. E.; Glover, L. A.; Ccaig, A. E. & Prosser, J. I. (2004). Spatial structure in soil chemical and microbiological properties in an upland grassland. FEMS Microbiology Ecology, Vol. 49, pp.191-205, ISSN 1574-6941.

San Miguél, C. S.; Dulinski, M. & Tate, R. L. (2007). Direct comparison of individual substrate utilization from a CLPP study: a new analysis for metabolic diversity data. *Soil Biology and Biochemistry,* Vol. 39, pp. 1870-1877, ISSN 00380717.

Schroth, G.; D´ Angelo, S. A.; Teixeira, W. G.; Haag, D. & Lieberei, R. (2002). Conversion of secondary forest into agroforestry and monoculture plantations in Amazônia: consequences for biomass, litter and soil carbon stocks after 7 years. *Forest Ecology and Management,* Vol.163, pp. 131-150, ISSN 0378-1127.

Stürmer, S. L. & Siqueira, J. O. (2010). Species richness and spore abundance of arbuscular mycorrhizal fungi across distinct land uses in Western Brazilian Amazon. *Mycorrhiza,* Online, (July 2010), ISSN 1432-1890.

Taketani, R. G. & Tsai, S. M. (2010). The influence of different land uses on the structure of archaeal communities in Amazonian Anthrosols based on 16S rRNA and *amo*A Genes. *Soil Microbiology,* Vol. 59, pp. 737-747, ISSN 0095-3628.

Tortora, G. J.; Funke, B. R. & Case, C. L. (2005). *Microbiologia,* Art-Med, ISBN 853630488X, Porto Alegre, Brasil.

Tótola, M. R. & Chaer, G. M. (2002). Microrganismos e processos microbiológicos como indicadores da qualidade dos solos. In: *Tópicos em Ciência do Solo,* UFV (Ed.), 195-276, ISBN 1519-3934, Viçosa, Brasil.

Van Heerden, J.; Korf, C.; Ehlers, M. M. & Cloete, T. E. (2002). Biolog for the determination of diversity in microbial communities. *Water SA,* Vol. 28, pp. 29-35, ISSN 0378-4738.

Vance, E. D.; Brookes, P. C. & Jenkinson, D. S. (1987). An extraction method for measuring soil microbial biomass C. *Soil Biology and Biochemistry,* Vol. 19, pp. 703-707, ISSN 00380717.

Zak, J. C.; Willig, M. R.; Moorhead, D. L. & Wildman, H. G. (1994). Functional diversity of microbial communities: a quantitative approach. *Soil Biology and Biochemistry,* Vol. 26, pp. 1101-1108, ISSN 00380717.

8

Temporal Changes in the Harvest of the Brown Algae *Macrocystis pyrifera* (Giant Kelp) along the Mexican Pacific Coast

Margarita Casas-Valdez[1], Elisa Serviere-Zaragoza[2] and Daniel Lluch-Belda[1]
[1]*Centro Interdisciplinario de Ciencias Marinas-IPN (CICIMAR-IPN)*
[2]*Centro de Investigaciones Biológicas del Noroeste (CIBNOR, S. C.)*
México

1. Introduction

Macrocystis pyrifera (L.) C. Agardh "Sargazo gigante" is distributed along the west coast of Baja California Peninsula, from the border with the USA to Punta Prieta, Baja California Sur. This kelp forms dense submarine prairies that emerge from the sea covering areas of several hectares or square kilometers. *Macrocystis* has been harvested from Islas Coronado (32° 15´ N) to Bahía del Rosario (30° 30´ N) in 15 beds for 49 years, from 1956 to 2004. It was exported raw for alginate production. Recently, it has been harvested in smaller quantities to obtain extracts to be used as fertilizer (Casas-Valdez et al., 2003).

The *Macrocystis* seaweed was harvested by specially designed ships that cut the algae at a depth about of 1.2 m and then transported it. The ships "El Capitán" harvested from 1956 to 1966 (storage capacity of 300 t) and "El Sargacero" from 1967 to 2004 (storage capacity of 400 t). The ship operations were the same at all beds and did not change over the study period.

The biomass and standing crop of *Macrocystis* was evaluated in summer 1982 and in an annual cycle in 1985-1986 in their natural distribution (Casas-Valdez et al., 1985; Hernández-Carmona et al., 1989a, 1989b, 1991). The recruitment and effect of nutrient availability during the ENSO event of 1997-1998 at the southern limit of distribution of *Macrocystis* were studied by Lada et al. (1999), Hernández-Carmona et al. (2001) and Edwards & Hernández (2005). The relationship between environmental variables as temperature, upwelling, sea level and wind speed and the catch per unit effort (CPUE) of *Macrocystis* were analyzed by Casas-Valdez et al. (2003). They found an inverse correlation between temperature and harvest and concluded that temperature is the variable that best explained the variations in the *Macrocystis* harvest.

This is the first time that the temporal variability of harvest, effort, and harvest per unit effort (CPUE) as an indicator of the abundance in each of 15 harvested beds of *Macrocystis* has been analyzed.

2. Data and methodology

Daily records from 1956 to 1999 were provided by Productos del Pacifico, S. A. de C. V. These contained the information of harvest date, name of the bed, number of trips, and

harvest size (wet weight). Total data were extracted from 3 230 daily records. For the period 1993 to 1999, additional information was obtained of the time of harvest from 638 records. With these data we estimated the monthly, seasonal, and annual values of harvest, effort and harvest per unit of effort for each bed and for the full region. The difference in the storage capacity of the two ships was weighted according with Casas-Valdez et al. (2003).
We selected as units of effort: a) the number of trips and b) the time of harvest. The harvest per unit effort (CPUE) (volume of harvest per trip made by each ship or volume of harvest per hour of harvest), was calculated with the equation-

$$CPUE = C/f \quad (1)$$

Where: C = volume of *Macrocystis* harvested; f = effort
Seasonal harvest, and harvest and effort per category were compared using an ANOVA analysis with the software Statistic 7.0. The significant difference among treatments was determined using the Tuckey test. The relationship between the harvest of *Macrocystis* and the effort was determined through correlation analysis (Anderson, 1972).

3. Results

3.1 Harvest, effort, and CPUE in *Macrocystis* beds

Macrocystis was harvested from Islas Coronado (32° 15´ N) to Bahía del Rosario (30° 30´ N) from 1956 to 2004 at 15 beds: Islas Coronados (01), Playas de Tijuana (02), Punta Mezquite (03), Salsipuedes (04), Isla Todos Santos (05), San Miguel y Sauzal (06), Punta Banda (07), Bahía de La Soledad (08), Santo Tomás (09), Punta China (10), Punta San José (11), Punta San Isidro (12), Punta San Telmo (13), Punta San Martín (14) and Bahía del Rosario (15) (Fig. 1). These beds are located at a distance of 1-5 km of the coast.
The harvest of *Macrocystis* increased from 9,900 t in 1956 to 41,500 t in 1976-1977. The average harvest from 1978 to 1982 was 30,000 t, from 1984 to 1997 it was 32,000 t and from 1999 to 2004 was 28,000 t (Fig. 2). In the years 1958, 1983, and 1998 the harvest underwent drastic reductions due the high temperatures presented due to ENSO phenomena. The historical series of CPUE accordingly shows considerable decreases during 1958, 1983, and 1998. In all the other years it was almost a constant level at an average of 342 t/trip.
The historical series of harvest and effort of the 15 beds of *Macroystis* (Figs. 3, 4, and 5) show that there is ample variability among them. For example, the Punta Mezquite (03) bed was harvested for 40 years, with an effort of 741 trips (Fig. 6) and a total harvest of 257,000 t. The Punta Banda (07) bed, however, was only harvested for 5 years, with an effort of 5 trips (Fig. 6) and a total harvest of 1,800 t.
Considering the average harvest and the effort applied during 49 years the *Macrocystis* beds were grouped into three categories; I) with an average harvest of 2,160 t (1,800 - 76,450 t) and an effort of 6 trips/year (01, 02, 05, 06, 07, 11, 12, 13, 14 and 15); II) with an average harvest of 3,600 t (90,800 - 176,150 t) and an effort of 12 trips/year (04, 08, 09 and 10); III) with an average harvest of 6,400 t (257,000 t) and an effort of 19 trips/year (03). There are significant differences ($P < 0.05$) among categories. The variation of the CPUE of *Macrocystis* beds is shown in figures 7, 8 and 9. The CPUE was more stable in the beds where more effort was used (03, 04, 08, 09 and 10).

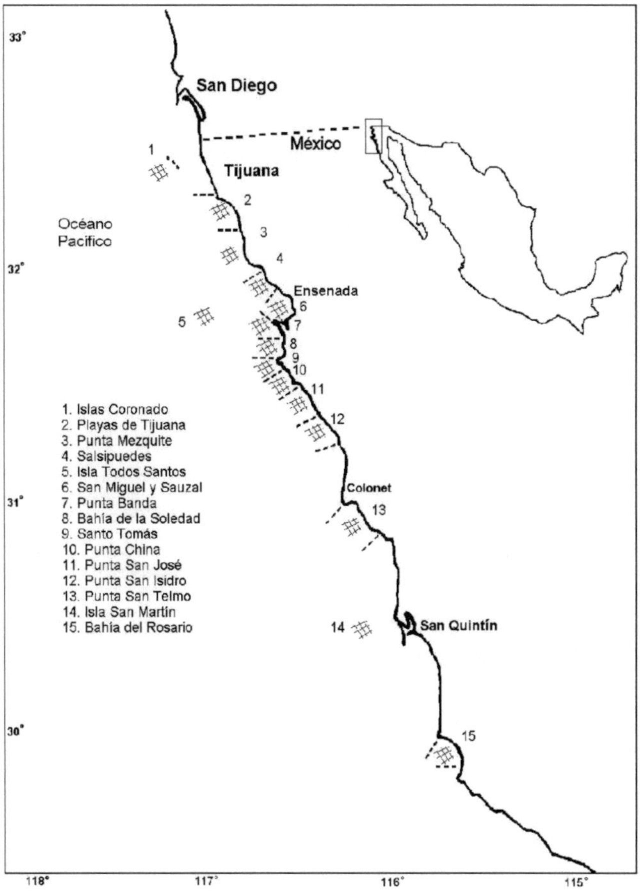

Fig. 1. Distribution of *Macrocystis pyrifera* beds harvested off the Baja California Peninsula (Taken from Casas-Valdez et al., 2003).

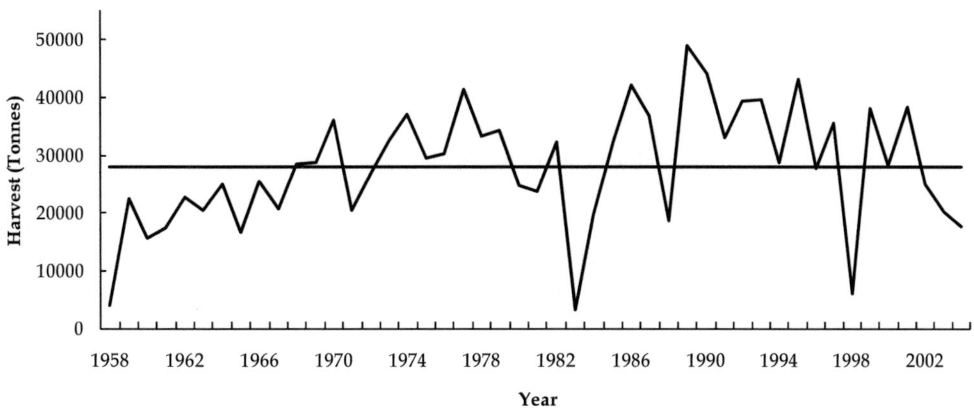

Fig. 2. Data series of harvest volume of *Macrocystis pyrifera*.

3.2 Seasonal variation

The harvest of *Macrocystis* off the Baja California Peninsula shows a seasonal pattern with minimum values in winter, and the maximum during spring and summer, then decreasing in autumn (Fig. 10). The spring and summer harvests were greater ($P < 0.05$) than winter and autumn, and the harvest of winter was the lowest ($P < 0.05$). In general the harvest of all beds had the same pattern. In the beds at Punta Mezquite (03), Salsipuedes (04), and Bahía de la Soledad (08), which were more frequently exploited, this pattern is evident, and less so the beds less harvested, such as Playas de Tijuana (02), Isla Todos Santos (05), and San Isidro (12). A similar behavior was found when the harvest obtained per hour of ship harvest (CPUE) for 1993 - 1999 was analyzed. The highest harvest/hour was during May to August (75 t/hour). These values were significantly different ($P < 0.05$) to both periods: February to April (62 t/hour) and September to December (54 t/hour), which were lower (Fig. 11).

3.3 Relation harvest-effort

During 1956 to 1999, the harvest of *Macrocystis* increased as a function of the level of effort (number of trips) ($r = 0.98$, Fig. 12) and similarly when the effort was measured as number of hours of ship harvest ($r = 0.85$) for 1993 –to 1999 (Fig. 13).

4. Discussion

From 1958 to 2004, the average harvest of *Macrocystis* was 26,000 t, which was about 50% of the standing crop estimated by Casas-Valdez et al. (1985) and Hernández et al. (1989a, 1989b, 1991), who evaluated the biomass and standing crop of *Macrocystis* using aerial photography and field work along the area of the distribution of this kelp. From Islas Coronado to Bahía del Rosario they estimated a standing crop of 40,000 t in summer 1985 and 63,000 t in summer 1986. This species of seaweed has a high growth rate (13 - 21 cm/day) (Hernández, 1996) and its regeneration rate is high.

The lowest harvest and effort recorded in category I can be related to: a) the harvest being suspended in beds 11 (1978), 06 (1985), 07 (1984), 02 (1991), and 01 (1993), b) the long distance from the beds to the base port, bed 12 (12 h 20 min), 13 (13 h), 14 (16,5 h), and 15 (20 h). The highest harvest and effort recorded in category III can be related to a) a high productivity of the bed and, b) the short distance from the bed to the base port (5 h). In relation to the previous information, Roberto Marcos (com. pers.) noted that the quantity of effort used at each bed depended on the productivity of the bed and its cost of operation, which are related principally to the distance that the ship most run from the base port to the bed. Guzmán et al. (1971) and Corona (1985) mention that the more productive beds for 1956 - 1968 and 1974 - 1985 were the beds 03, 04, 08, 09, and 10 that are in categories II and III of this study. The largest harvest of *Macrocystis* was in spring and summer and the lowest in winter.

Along the northwest coast of the Baja California Peninsula the greatest upwellings are during spring and summer (Casas-Valdez, 2001) and have high nutrient concentrations and lower temperatures (Lynn & Sympson, 1987; Parés & O'Brien, 1989) that favor the development of *Macrocystis* fronds (Tegner & Dayton, 1987; Tegner et al., 1996; Lada et al., 1999). Growth studies *in situ* showed that the lower temperatures of spring enhance the growth rate of *Macrocystis* (González et al., 1991) and also the increase of nutrients (Zimmerman & Kremer, 1986). Casas-Valdez et al. (1985) and Hernández-Carmona et al. (1989a, 1989b, 1991) evaluated the biomass and standing crop of *Macrocystis* along their natural distribution and found the largest surface and biomass of the beds in spring (45,000 t) and summer (63,000 t). They noted that these values were three times greater than those in winter (14,000).

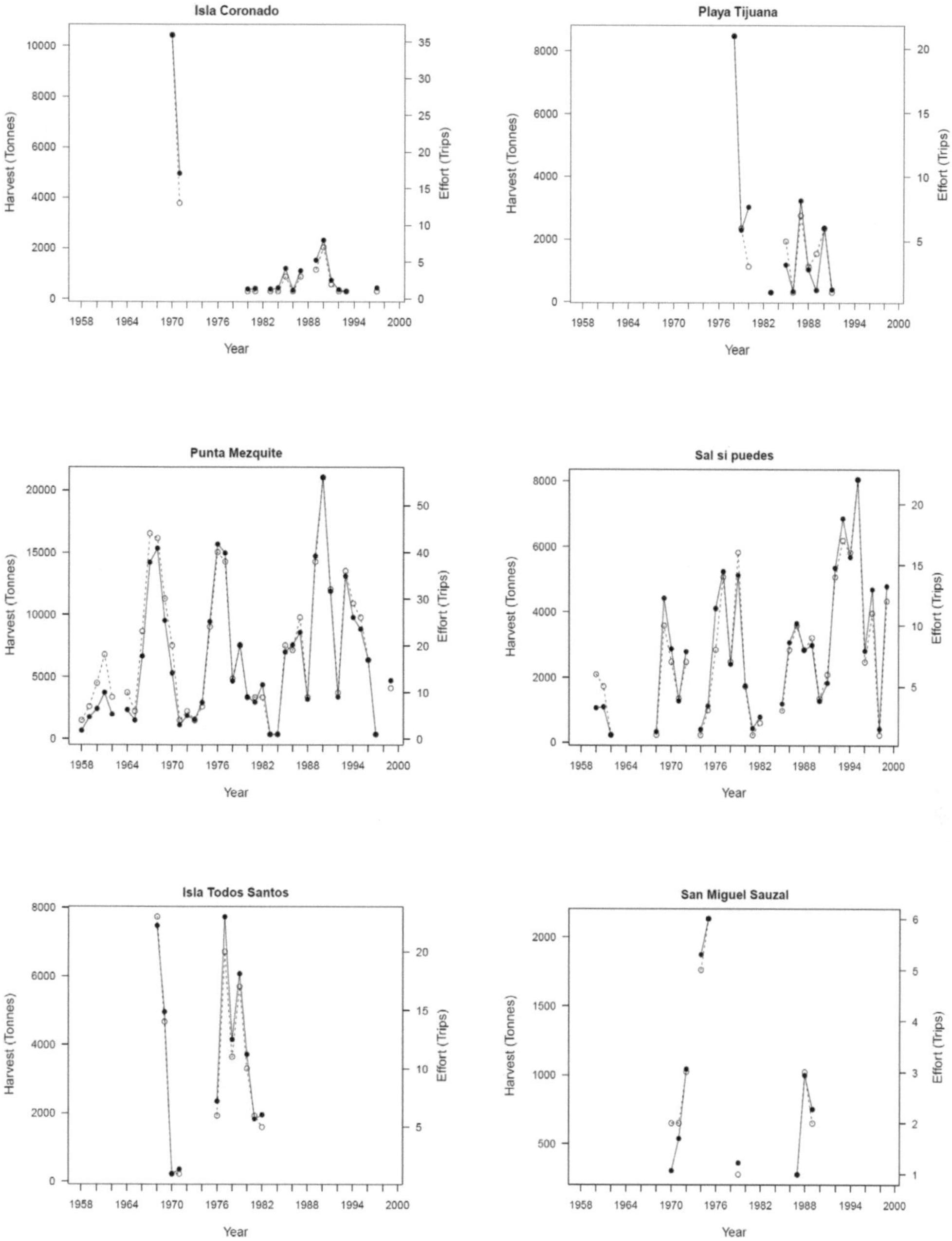

Fig. 3. Data series of harvest and effort of the *Macrocystis pyrifera* beds: Islas Coronados, Playas de Tijuana, Punta Mezquite, Salsipuedes, Isla Todos Santos and San Miguel y El Sauzal. Harvest —●—, effort - -○- -.

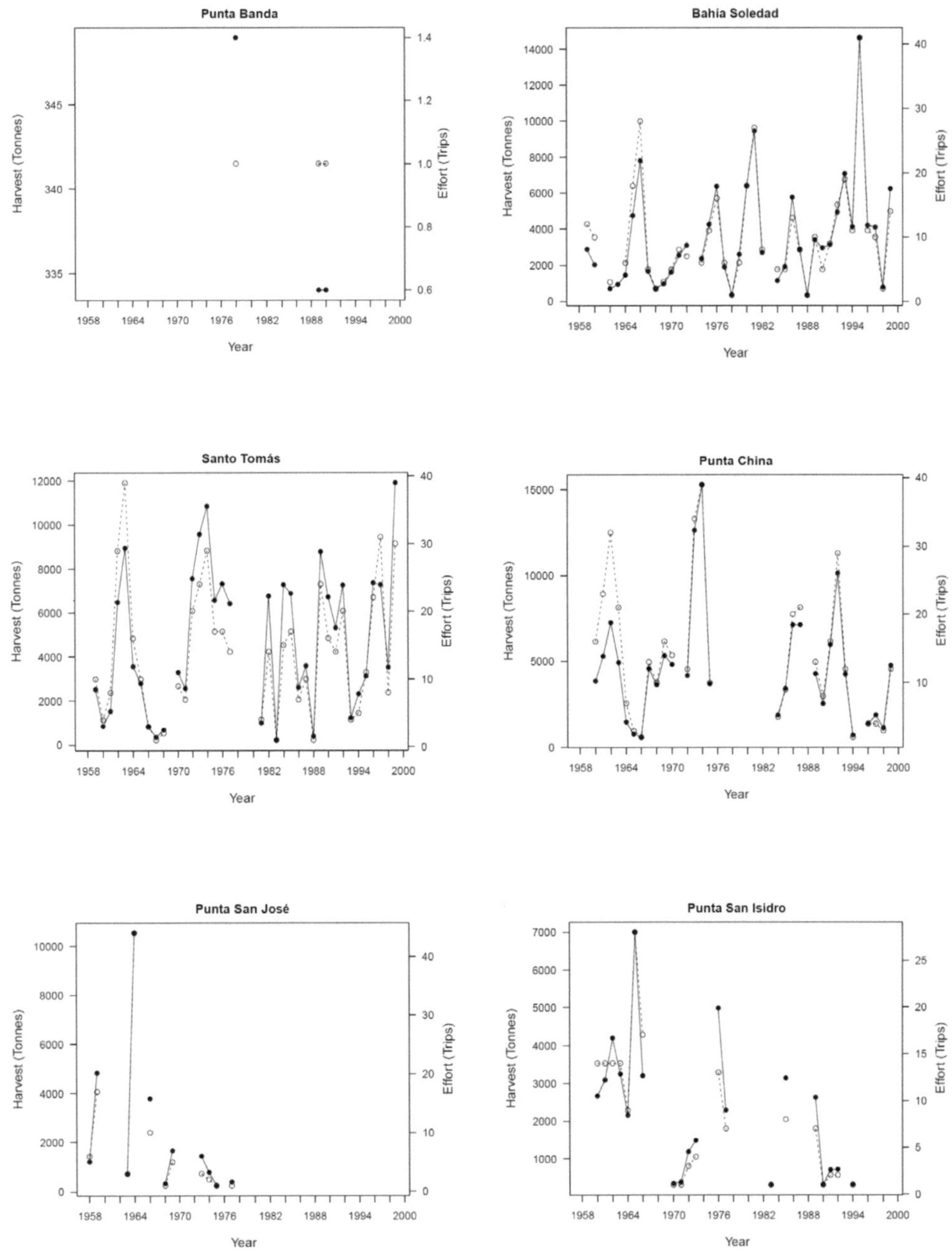

Fig. 4. Data series of harvest and effort of the *Macrocystis pyrifera* beds: Punta Banda, Bahía de la Soledad, Santo Tomás, Punta China, Punta San José and Punta San Isidro.
Harvest —●— , effort - -○- - .

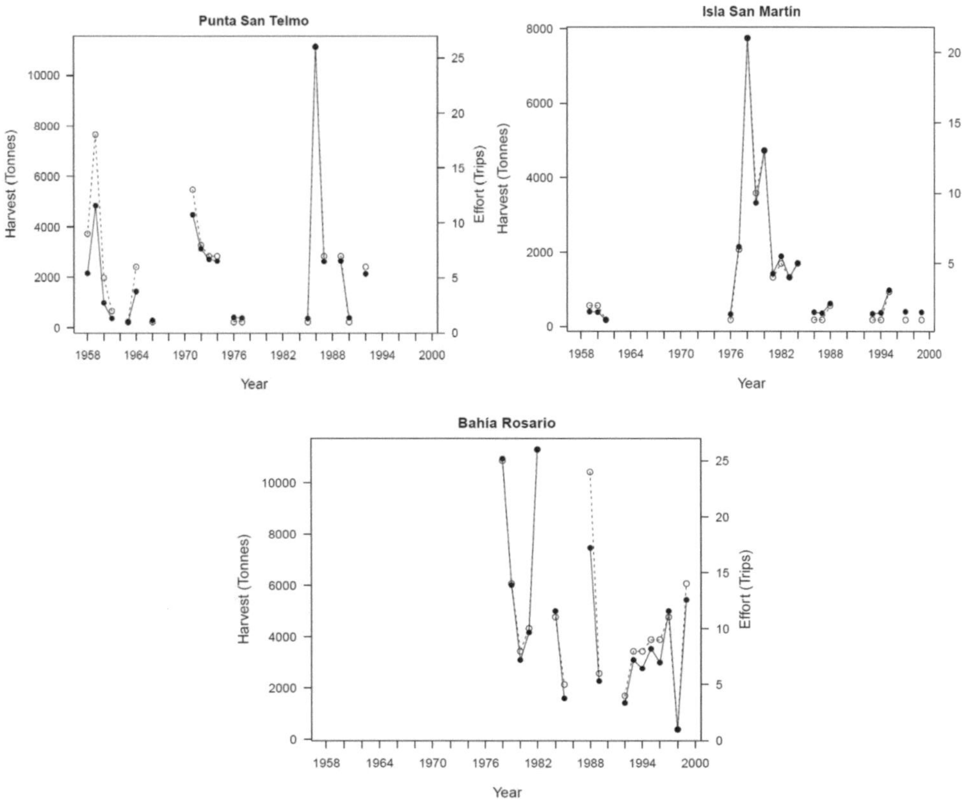

Fig. 5. Data series of harvest and effort of the *Macrocystis pyrifera* beds: Punta San Telmo, Isla San Martín and Bahía del Rosario. Harvest —●—, effort - -O- -.

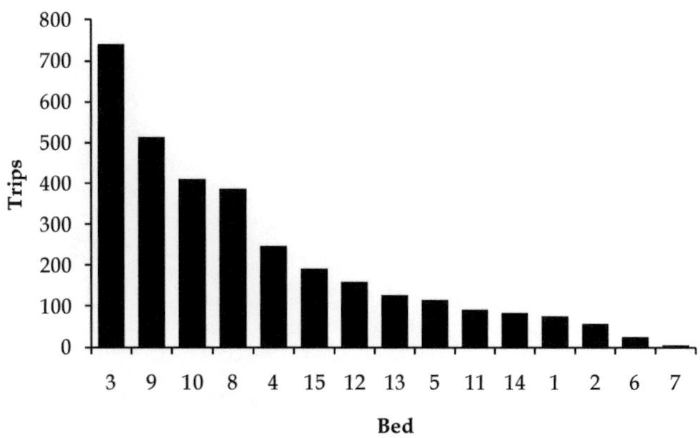

Fig. 6. Number total of trips in the beds: (01) Islas Coronados, (02) Playas de Tijuana, (03) Punta Mezquite, (04) Salsipuedes, (05) Isla Todos Santos, (06) San Miguel and El Sauzal, (07) Punta Banda, (08) Bahía de la Soledad,(09) Santo Tomás, (10) Punta China, (11) Punta San José, (12) Punta San Isidro, (13) Punta San Telmo, (14) Isla San Martín and (15) Bahía del Rosario.

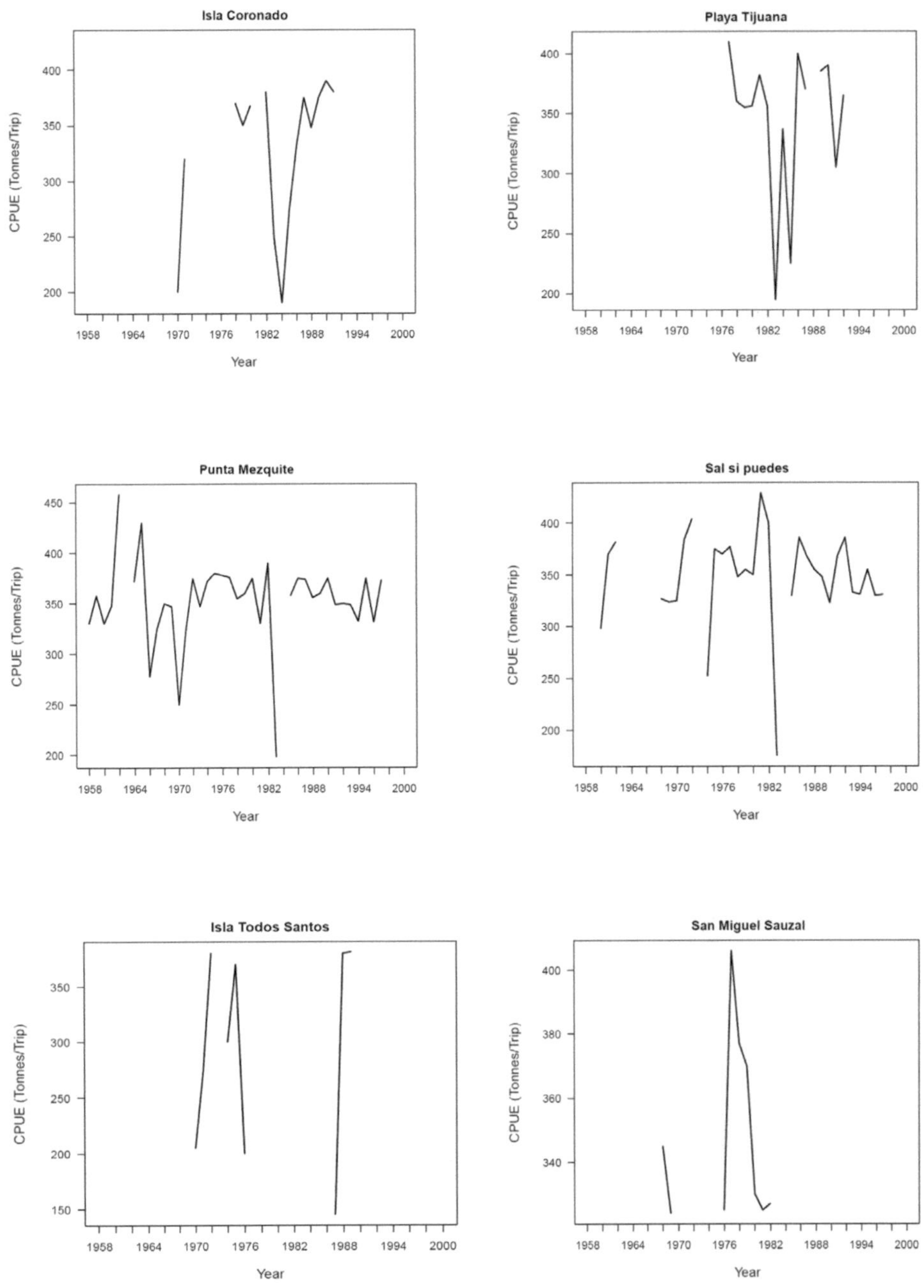

Fig. 7. Data series of harvest per unit effort (CPUE) of the *Macrocystis pyrifera* beds: Islas Coronados, Playas de Tijuana, Punta Mezquite, Salsipuedes, Isla Todos Santos and San Miguel y El Sauzal.

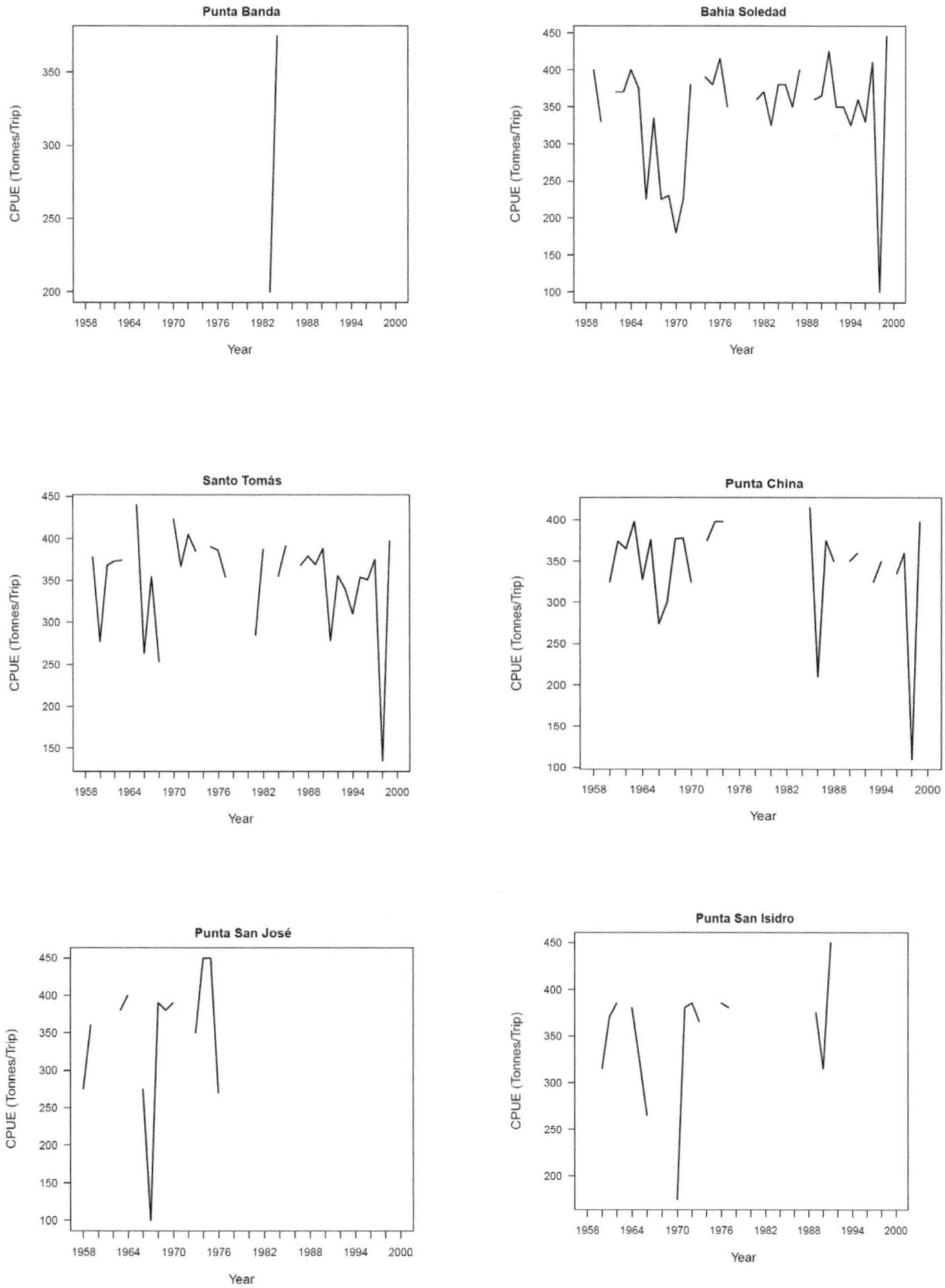

Fig. 8. Data series of harvest per unit effort (CPUE) of the *Macrocystis pyrifera* beds: Punta Banda, Bahía de la Soledad, Santo Tomás, Punta China, Punta San José and Punta San Isidro.

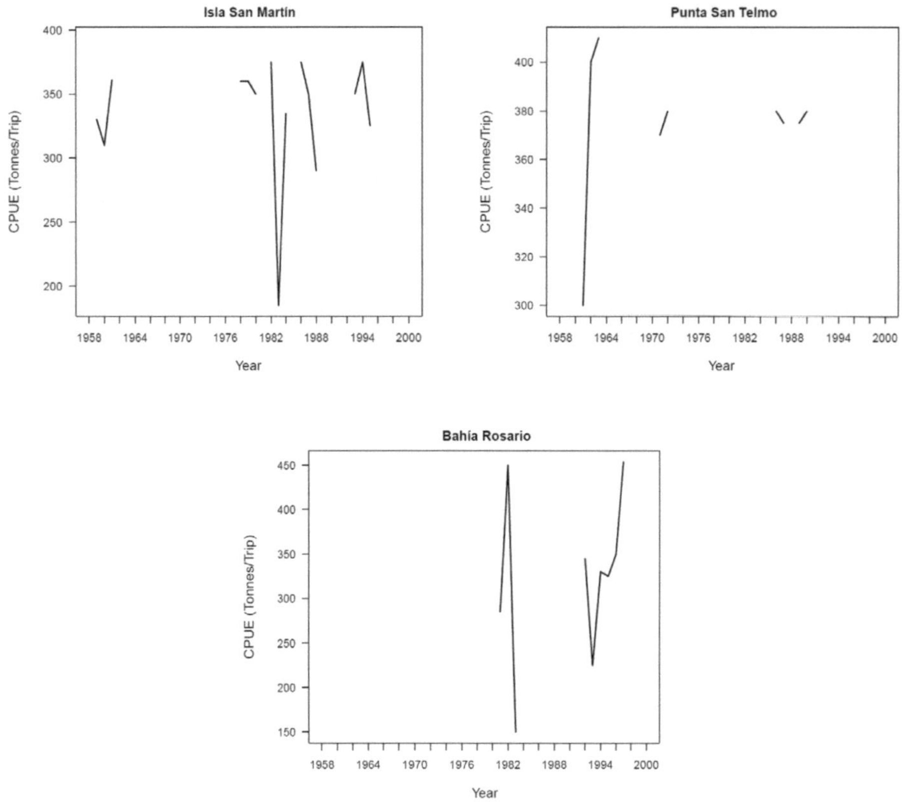

Fig. 9. Data series of harvest and effort of the *Macrocystis pyrifera* beds: Punta San Telmo, Isla San Martín and Bahía del Rosario.

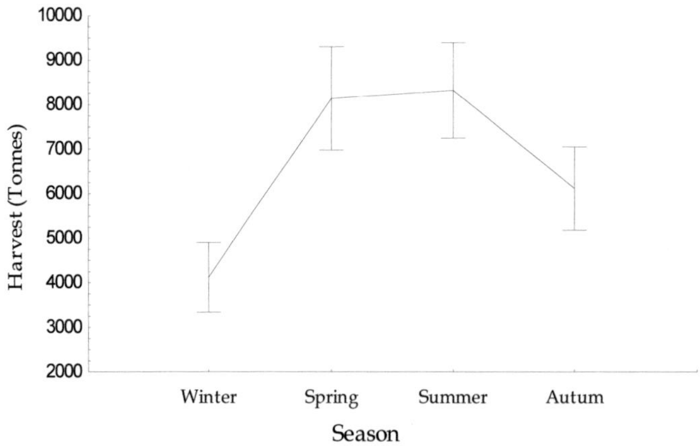

Fig. 10. Seasonal variation of the harvest of *Macrocystis pyrifera* in Baja California Peninsula. ± 2 SD.

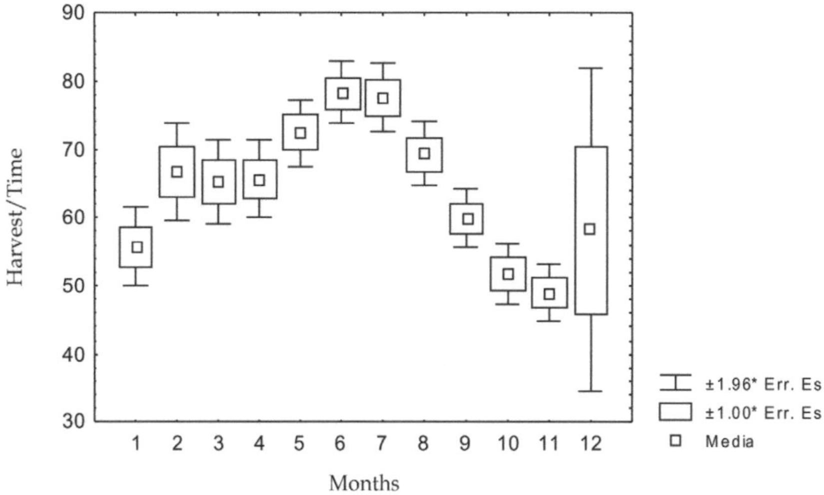

Fig. 11. Monthly average harvest per hour of *Macrocystis pyrifera* in the Baja California Peninsula for the period of 1993-1999.

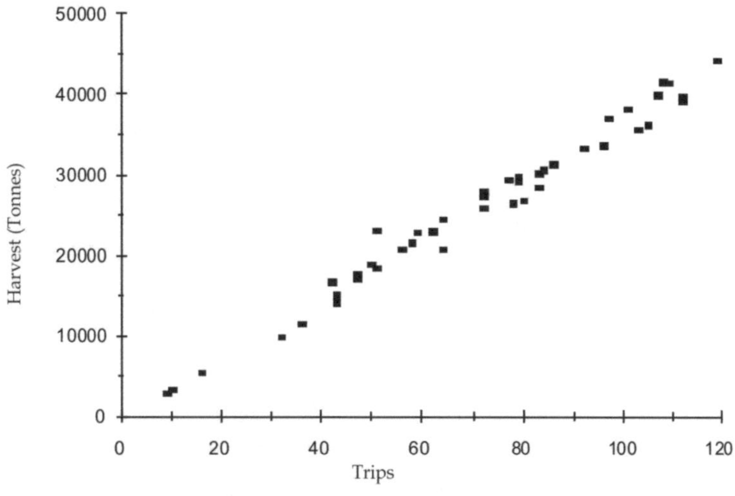

Fig. 12. Relationship of the harvest and effort (number of trips) of *Macrocystis pyrifera* for the period of 1956-1999.

The CPUE was used as indicator of abundance for *Gelidium robustum* a red seaweed that is harvested along in the west coast of the Baja California Peninsula from 1956 to the present. The unit of effort selected for this fishery was the fishing equipment (a boat with three fishermen) and the CPUE was expressed as harvest/boat (Casas-Valdez et al., 2001). They used the CPUE to determine the relationship of the abundance of *Gelidium* with both temperature and upwelling. As an indicator of the abundance of *Macrocystis*, Tegner et al. (1996) compared data on the maximum canopy of the kelp forest and size of the annual harvest of *Macrocystis* for California, and they chose harvest size as the most useful data to relate to environmental variables. They pointed out that harvest size was a reflection of changes in consumer demand, harvest productivity, and natural disturbances. They also noted that this variable has the advantage of integrating growth over a long period and has less subjectivity in its measurement.

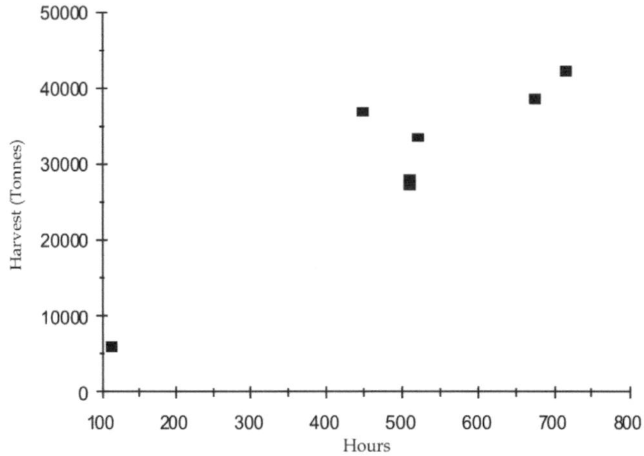

Fig. 13. Relationship of the harvest and effort (number of hours) of *Macrocystis pyrifera* for the period of 1956-1999.

In our study, we considered that the CPUE shows the changes in the abundance of *Macrocystis* better than only the harvest, because the size of the harvest varies according to the amount of effort used and not only as a function of the abundance. Furthermore, the use of the CPUE is cheaper than the use of aerial photography and field work to determine the variations in the abundance of this resource. Casas-Valdez et al. (2003) mentioned that the harvest/trip is a reasonable indicator of the *Macrocystis* abundance, because about 60% of the alga biomass is present in the surface canopy (North, 1968), and almost 95% of its production takes place in the first meter of the top of the water column, and the kelp is harvested at a maximum depth of 1.2 m. Furthermore the ship operations were the same at all beds and did not change over the study period. We considered that the harvest/hour is a better indicator.

The surplus production models of Schaefer and Fox were used to assess the fishery condition of *Gelidium* off the Baja California Peninsula from 1985 to 1997. The results have shown that the resource is not overexploited (Casas-Valdez et al., 2005). In this study we tried to use these surplus models for the data of *Macrocystis*, but the fit was not satisfactory. This occurred because an increased effort produced increased harvest. To fit these models, it is necessary to count, along with the catch, effort, and CPUE data, an ample range of fishing effort levels, preferably including those that correspond to the level of overexplotation in the curve (IATTC, 1999). The linear relation (correlation) found between the harvest and the effort used for the *Macrocystis* fishery means that the fishery was in the eumetric growth segment of the curve of the Schaefer model and therefore it is possible to conclude that there have not been negative effects of the harvest on the resource. It is considered that the effort has not been increased, due to the fact that the demand for *Macrocystis* has not been increased either. In fact, the harvest drastically decreased in 2005, when the principal company that was buying this kelp as raw material for the alginate production ceased buying it (Roberto Marcos com. pers.).

5. Conclusions

The *Macrocystis* fishery along the Mexican Pacific coast did not show signals of over exploitation due to increases in the effort corresponding to increases in the harvest, and the CPUE has been maintained almost constant since the begging of the harvesting of this resource until now (2004), with the exception of the years when "El Niño" event was present.

Along the northwest coast of the Baja California Peninsula, the highest harvest of *Macrocystis* was found in spring and summer, when the greatest upwellings ocurre in agreement with high nutrient concentrations and lower temperatures.

The harvest per unit of effort (CPUE) was more stable in the beds where more effort was used, as in the beds at Punta Mezquite, Salsipuedes, Bahía de La Soledad, Santo Tomás and Punta China, whereas in the beds where less effort was used the CPUE was more variable.

6. Acknowledgment

Thanks to Productos del Pacifico, S. A. de C. V. for providing the data of harvest of *Macrocystis*. We really appreciate the adviser of Roberto Marcos Ramírez. Thanks to Dr. Ellis Glazier for editing this English-language text. Margarita Casas Valdez and Daniel Lluch Belda are fellows of COFAA-IPN and EDI-IPN.

7. References

Anderson, T. (1972). *The Statistical Analysis of Time Series*. John Wiley & Sons, Inc. U.S.A.

Casas-Valdez, M., Hernández-Carmona G., Torres-Villegas R. & Sánchez-Rodríguez, I. (1985). Evaluación de los mantos de *Macrocystis pyrifera* (sargazo gigante) en la Península de Baja California (verano de 1982), *Investigaciones Marinas CICIMAR*, Vol.2, No.1, (December 1985), pp. 1-17. ISSN 0186-5102.

Casas-Valdez, M. (2001). Effect of the climatic variability on the abundance of *Macrocystis pyrifera* and *Gelidium robustum* in Mexico. Ph. D. Thesis. CICIMAR-IPN, 133 p. (August 2001)

Casas-Valdez, M., Serviere-Zaragoza, E., Ortega-García, S., Lora-Sánchez, D. & Hernández-Guerrero, C. (2001). The harvest per unit effort (cpue) of *Gelidium robustum* along Baja California Peninsula and its relationship with temperature and upwelling. *Anales de la Escuela Nacional de Ciencias Biológicas*, Vol. 47, No.1, (January 2001), pp. 73-83. ISSN 0365-0946.

Casas-Valdez, M., Serviere-Zaragoza, E., Lluch-Belda, D., Marcos-Ramírez, R. & Aguila-Ramírez, N. (2003). Effects of climatic change on the harvest of the kelp *Macrocystis pyrifera* at the Mexican Pacific coast. *Bulletin of Marine Science*, Vol.73, No.3, (September 2003), pp. 445-456. ISSN 007-4977.

Casas-Valdez, M., Lluch-Belda, D., Ortega-García, S., Hernández-Vazquez, S., Serviere-Zaragoza, E. & Lora-Sánchez, D. (2005). Estimation of maximum sustainable yield of *Gelidum robustum* seaweed fishery in Mexico. *Journal of the Marine Biological Association of the United Kingdom*, Vol.85, (June 2005), pp. 775-778. ISSN 0025-3154.

Corona, R. (1985). Estudio de la producción de *Macrocystís pyrifera* en la costa noroccidental de Baja California. Tesis de Licenciatura. Universidad Autónoma de Baja California, Ensenada, B. C. 57 p. (September 2005)

Edwards, M. & Hernández-Carmona, G. (2005). Delayed recovery of giant kelp near its southern range limit in the North Pacific following El Niño. *Marine Biology*, Vol.147, pp. 273-279. (n.d) ISSN 0025-3162.

González, J., Ibarra, S., & North, J. (1991). Frond elongation rates of shallow water *Macrocystís pyrífera* (L.) Ag. In northern Baja California, Mexico. *Journal of Applied Phycology*, Vol.3, (June 1991). pp. 311-318. ISSN 0021-9010.

Guzmán del Próo, S., De la Campa, S. & Granados, L. (1971). El sargazo gigante *Macrocystís pyrífera* y su explotación en Baja California. *Revista de la Sociedad Mexicana de Historia Natural*, Vol.32, (June 1971), pp. 15-49. ISSN 0370-7415.

Hernández-Carmona, G., Rodríguez, E., Torres, R., Sánchez, I. & Vilchis, M. (1989a). Evaluación de los mantos de *Macrocystís pyrifera* (Phaeophyta, Laminariales) en Baja California, México. I. Invierno 1985-1986. *Ciencias Marinas*, Vol. 15, No.2, (June 1989), pp. 1-27. ISSN 0185-3880.

Hernández-Carmona, G., Rodríguez, E., Torres, R., Sánchez, I., Vilchis, M. & García, O. (1989b). Evaluación de los mantos de *Macrocystís pyrifera* (Phaeophyta, Laminariales) en Baja California, México. II. Primavera1986. *Ciencias Marinas*, Vol.15, No.4, (December 2989), pp. 117-140. ISSN 0185-3880

Hernández-Carmona, G., Rodríguez, Casas-Valdez, M., R., Vilchis, M. & Sánchez, I. (1991). Evaluación de los mantos de *Macrocystís pyrifera* (Phaeophyta, Laminariales) en Baja California, México. III. Verano 1986 y variación estacional. *Ciencias Marinas*, Vol.17, No.4, (December 1991), 121-145. ISSN 0185-3880

Hernández-Carmona, G. (1996). Tasas de la elongación de frondas de *Macrocystís pyrífera* (L.) Ag. en Bahía Tortugas, Baja California Sur, México. *Ciencias Marinas*, Vol.22, No.1, (March 1996), pp. 57-72. ISSN 0185-3880

Hernández-Carmona, G., Robledo, D. & Serviere, E. (2001). Effect of nutrient availability on *Macrocystís pyrifera* recruitment survival near its southern limit of Baja California. *Botanica Marina*, Vol.44, (May 2001), pp. 221-229. ISSN 0006-8055.

IATTC. (1999). Annual Report of the Inter-American Tropical Tuna Commision, 1997. Annual Report IATTC, 310 p. (n.d)

Ladah, B., Zertuche, J. & Hernández-Carmona, G. (1999). Giant kelp *(Macrocystís pyrífera,* Phaeophyeeae) recruitment near its southern limit in Baja California after mass disappearance during ENSO 1997-1998, *Journal of Phycology*, Vol.35, (December 1999), pp. 1106-1112. ISSN 0303-3910.

Lynn, J. & Simpson, J. (1987). The California Current System: The Seasonal Variability of its Physical Characteristics. *Journal of Geophysical Research*, Vol.92, No.12, (December 1987), pp. 947-966. ISSN 0148-0227.

North, J. (1968). Concluding discussion. In North, J. & Hubbs, L. (ed). *Utilization of kelp-bed resources in Southern California*. California Department Fish Game, Fish Bulletin, Vol.139, pp. 255-259. (n.d)

Parés, A. & O'Brien, J. (1989). The seasonal and interannual variability of the California Current system: A numerical model, *Journal of Geophysical Research*, Vol.94, (December 1989), pp. 3159-3180. ISSN 0148-0227.

Tegner, M., & Dayton, P. (1987). El Niño effects on Southern California kelp forest communities. *Advances in Ecological Research*, Vol.17, pp. 243-279. (n.d) ISSN 065-2504.

Tegner, M., Dayton, P., Edwards, B. & Riser, L. (1996). Is there evidence for long-term climatic change in Souther California kelp forest?, *CalCOFI Report*, Vol.37, pp. 111-126. (n.d) ISSN 0575-3317.

Zimmerman, R. & Kremer, J. (1986). *In situ* growth and chemical composition of the giant kelp *Macrocystis pyrifera:* response to temporal change in ambient nutrient availability. *Marine Ecology Progress Series*, Vol.27, (August 1986), pp. 277-285. ISSN 0171-8630.

Part 2

Production

9

Supplying Biomass for Small Scale Energy Production

Tord Johansson
Swedish University of Agricultural Sciences,
Department of Energy and Technology,
Sweden

1. Introduction

Our sources of energy are constantly changing. In Sweden the focus is on nuclear and hydro power for producing electricity and total Swedish energy production amounts to about 612 TWh (Anon, 2010). Since Sweden has a cold climate, there is a high demand for energy to heat homes and energy sources other than oil and coal are required. Currently, fuel systems are based on oil and electrical power but there has been an increase in the use of biomass during recent decades. The support of biomass for heating provides 19% of the total Swedish energy output, (Fig. 1).

For centuries trees have been used in a domestic context for firewood and charcoal production. In Sweden, conventional forest management combined with bioenergy production has been practiced for the last 40-50 years. Currently, for economic reasons, bioenergy harvesting is mainly based on large areas of forest land. Tops and branches are harvested from clear cut areas and this biomass contributes greatly to the production of bioenergy. Special equipment is used to harvest biomass, which is used for energy production in direct heating plants. The infrastructure is well established. Most of the harvested material goes to heating plants close to cities, although some is used by individual households.

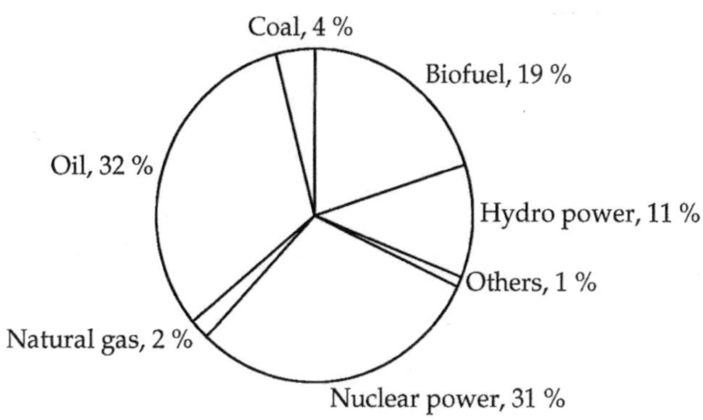

Fig. 1. Total energy use in Sweden in 2007 (Anon, 2010)

The management of forests is mainly directed towards producing pulpwood and timber. The remaining parts of the tree - branches and tops - represent raw material for bioenergy production. Over the last twenty years there has been an increased willingness to make use of these parts of the tree.

Biomass production on former farmland, using willows, poplar and hybrid aspens, is another option for energy production. In general, the Swedish people look favorably on such land use, as well as forest biomass production. There is strict regulation of the management of forest land to minimize the risks of nutrient loss, but no such regulations exist for farmland. Farmers and some sections of the public wish to maintain farmland as an open landscape and to continue with agricultural cultivation.

The Swedish government has twice proposed a reduction in farmland available for the production of cereals, in 1969 and 1986. The plan was to reduce the area by about one million hectares, out of the total of three million hectares. Both attempts failed, although since 1968 350,000 ha have been taken out of production. Some areas of this former farmland have been planted, mostly with Norway spruce and birches, but more than 200,000 hectares which were taken out of production in the period 1970-1980 have received no subsequent management. Today these areas are covered by broadleaved trees with a range of numbers of stems per hectare (Johansson, 1999a), but they are not managed to generate forest products.

2. Small-scale production of biomass

Currently, there are standard practices for the management and harvesting of biomass from large forest stands, used in state forests and by forestry companies. It is much more challenging, however, for small-scale forest owners to utilize forest biomass for bioenergy. The amount of biomass that can be harvested from forest land or farmland depends on various factors including site condition, species and management intensity. Few practical recommendations for small-scale owners have been published, and land owners may be unaware of appropriate practice. More information would enhance the use of resources available for bioenergy production.

Herein I present examples of activities and the management of farmland and forest land demonstrating how an owner can undertake small scale biomass production for their own consumption or to supply a local market (neighbors etc.).

The examples presented are:

- ingrowth, i.e. natural establishment of broadleaved trees on former farmland via seeds, sprouts or suckers;
- direct seeding on farmland;
- management of existing mixed stands;
- harvesting tops and branches after clear cutting; and
- establishing and using fast-growing species.

Finally, some recommendations for small scale bioenergy production are presented.

3. Ingrowth

The most important factors affecting the colonization of open areas by plants are: the year and season of abandonment; the physical state of the site; climate; soil; the existing flora and fauna; proximity and position of source material; opportunities for vegetative regeneration;

and the presence, within a range possible for seed dispersal, of an efficient generative reproduction and a rapid, rich and long-distance dispersal of seeds (Falinski, 1980; Harmer et al., 2001). Reviews by Osbornova et al. (1990) and Myster (1993) report many studies of tree generation on abandoned farmland. Natural colonization by trees and other species have been recorded since 1882 at the Broadbalk Wilderness, UK, which has established on former farmland (Harmer et al., 2001). The first tree plants were recorded 30 years after abandonment, i.e. in 1913. The main species regenerating in the area were: common ash (*Fraxinus excelsior* L.); sycamore (*Acer pseudoplatanus* L.); field maple (*Acer campestre* L.); suckers of wild cherry (*Prunus avium* L.); blackthorn (*Prunus spinosa* L.); pedunculate oak (*Quercus robur* L.) and hazel (*Corylus avellana* L.). In 1998 the dominant and most frequent tree species were pedunculate oak, common ash, wild cherry and sycamore.

Fig. 2. Naturally seeded birch (left), sucker from aspen (right) and naturally seeded grey alder (below)

The area of farmland no longer in agricultural production increases as land owners cease activities or direct their energies towards other forms of management. When farmland is abandoned it is invaded by herbs and broadleaved tree species (alder, aspen and birch). In general, one species dominates in the new stand. Most such farmland areas are owned by private individuals. In Sweden, Johansson (1999a) found up to 10,000 broadleaved tree stems ha^{-1} on about 100,000 hectares of former farmland.

Natural tree establishment in an open area is a slow process, and it may be 5-10 years before trees 2-5 old are seen (Werner and Harbeck, 1982). Most such areas in Northern Europe are small, amounting to 0.5-2.0 ha. In the initial phase, the areas are not noticeable from the surroundings, but later a dense stand is established and the landscape is changed. In general, these areas continue to develop unnoticed by the owner or the public. Eventually, former open areas become covered by forest. Such ingrowth can be the result of natural seeding, sprouting or suckering (Fig. 2).

3.1 Natural seeding

To produce conditions that will encourage establishment of a wide range of seedlings through natural seeding, and avoid revegetation failing, an understanding of certain abiotic and biotic factors is required. The main factors that affect establishment through natural seeding are: species present, soil type, moisture, competition by grasses and herbs, available seed trees, and weather conditions (heat, dryness etc). It is important to know the timing and periodicity of seed production and dispersal. Basic knowledge about the period for the high rates of seed dispersal is necessary when practicing natural regeneration. In order to encourage natural seeding, ground preparation must be undertaken prior to seed dispersal.
Specific characteristics of a species, such as number of seeds per tree, seed weight and frost resistance, greatly influence the establishment of seedlings. Seeds from some species are wind dispersed (e.g. birch and sallow (*Salix caprea* L.)) and others water dispersed (e.g. alder); a combination of methods may be used. Studies of wind-mediated seed dispersal for different species indicate the following order of decreasing dispersal:
birch>elm=maple>alder>hornbeam>beech>oak (Augspurger and Franson, 1987; Okubo and Levin, 1989; Willson, 1990; Karlsson, 2001). Table 1 contains data on birch and alder seed dispersal.

Distance from forest stand, m				Country	Reference
<50	50-100	100-150	>150		
			Birch		
>400			>100	Sweden[1]	Fries (1982)
>200	<100			Sweden[2]	Björkroth (1973)
58 % of total	10 % of total			USA[3]	Björkbom (1971)
10,450	4,200	400		USA[4]	Hughes and Fahey (1988)
			Alder		
78-94 % of total				Sweden[5]	Johansson and Lundh (2006)
90 % of total				Sweden[5]	Karlsson (2001)

1) *Betula pendula* Roth 2) *Betula pubescens* Ehrh. 3) *Betula papyrifera* March. 4) *Betula alleghaniensis* Brit. 5) *Alnus glutinosa* (L.) Gaertner

Table 1. Dispersal of birch and alder seeds into open areas, number of seeds m^{-2} $year^{-1}$

Both downy (*Betula pubescens* Ehrh.) and silver (*Betula pendula* Roth) birch produce many seeds. In Estonia, Uri et al. (2007) recorded 3060-36,200 8-year-old birches ha^{-1} that had been produced by natural seeding on farmland. Seeds from a birch growing at the edge of a clear cut area have been found to spread at a rate of about 100 seeds m^{-2} up to 200 m from the tree (Fries, 1984). Most of these birch seeds were dispersed during September, although the process continued until December. In a study of sweet birch (*Betula lenta* L.), Matlack (1989) reported seed were dispersed 3.3 times further than the distance measured by Fries (1984). In a study of silver birch in Estonia, 21 % of the seeds were dispersed in July, 77 % in August and 2 % in September (Kohh, 1936). Heikinheimo (1932, 1937), who reported the same dispersal periods, commented that the weather during summer and autumn is the main factor affecting the period of seed dispersal. Graber and Leak (1992) presented a study on seed fall for broadleaved species in New Hampshire. The mean seed fall (million ha^{-1}) in a study lasting 11 years was: 6.58 for yellow birch (*Betula alleghaniensis* Britton); 6.38 for paper birch (*Betula papyrifera* Marsh.); 4.11 for sugar maple (*Acer saccharum* Marsh.); and 0.17 for American beech (*Fagus grandifolia* Ehrh.). The seed viability was 30-50 %, depending on species.

Besides wind dispersal, there are some reports of secondary dispersal of seeds (Hesselman, 1934; Matlack, 1989; Greene and Johansson, 1997). The most common is by movement on snow, but for this to occur, seed fall must happen during winter months when snow is on the ground. The seeds can be damaged by friction on frozen snow, thus reducing viability.

The level of seed production by alder depends on the number of hours of sunshine in the period April-September in the year before fruiting, the number of hours of sunshine in the seeding year and the level of seed production in the preceding year (MacVean, 1955). According to MacVean (1955), common alder (*Alnus glutinosa* (L.) Gaertner) seeds are generally dispersed within a radius of 30-60 m of the mother tree. Karlsson (2001) found that 50 % of the total number of alder seeds produced fell within 5 m and 90 % within 20 m of the stand. In a study by Johansson and Lundh (2006), 50 % of the common alder seeds were found to have fallen before December and 75 % before February. Alder seeds can also be transported by water in spring at the time of snow melt.

Seeds from European aspen (*Populus tremula* L.) are extremely small (low weight) with a limited growing capacity (Blumenthal, 1942, Latva-Karjanmaa et al., 2006). A large aspen growing close to Tartu city, Estonia, produced 49 kg or 54 million seeds (Reim, 1930). Only a small proportion of the aspen seeds produced will grow; success depends on site conditions, seed size and the level of competition. Aspen seeds can grow on poor sandy sites, burned areas and small patches without vegetation (Blumenthal, 1942). Seeds of sallow are also small and have a plume to aid dispersal (Grime et al., 1988). Seeds of both species can be dispersed over long distances.

The most favorable soil types for rapid establishment of seedlings are fine sand, silt and light clay, sandy-silty till and light clay till. Even peat soils can provide an ideal site, providing there is sufficient water. A mixture of mineral soil and humus is common on farmland, where the area has been cultivated for many years.

Birch seeds establish well on undisturbed sites with a high level of moisture (Mork, 1948; Fries, 1982). During the first part of the growing season in Nordic countries (April-May) soil moisture tends to be low. The lack of rain combined with the sunshine during this period results in a dry soil. Therefore any soil treatment (plowing, harrowing or screefing) should be undertaken in autumn or very early in spring. Studies to determine the best soil treatment to ensure limited cover of competitive vegetation indicate that removal of topsoil is preferable (Karlsson, 1996).

3.2 Sprouting and suckering

The main difference between sprouting and suckering is that sprouts emerge from a stump whilst suckers originate from roots, (Fig. 3). Both types of regeneration result in fast-growing individual stems. In studies of dormant buds on birch, most have been found close to the ground: 0-10 cm above or 0-5 cm below ground level (Kauppi, 1989; Kauppi et al., 1987; 1988 Johansson, 1992a). The number of sprouts per living birch stump has been found to vary between 1 and 52, mean 10±8, decreasing to 3-8 sprouts per stump after five years (Johansson, 1992 b, c). Rydberg (2000) found the number of birch sprouts had decreased by >40 % of the initial number two years after stump creation nine years after cutting, Johansson (2008) found that the initial number of sprouting birch stumps had decreased to 61 and 55 % respectively for downy and silver birch stumps. In a study of downy birch growing in central Finland, the number of sprouts decreased from an average of 9.5 one year after cutting to 5 after three years and 3 after seven years. The sprouting abilities of red oak (*Quercus rubra* L.), white oak (*Quercus alba* L.), black cherry (*Prunus serotina* Ehrh.), sugar maple and yellow poplar (*Liriodendron tulipifera* L.) growing in West Virginia were studied by Wendel (1974). After ten years the number of sprouts per living stump was 15-20 % of the initial number produced. In another study of yellow poplar, the average number of sprouts recorded six years after cutting was 7.0 per stump (Beck, 1977). Sprouting capacity is highest when a tree is young (Johansson, 1992c). Kauppi et al. (1988) reported the poorest sprouting results from old (40 year) downy birch stumps. Older trees have thicker stem bark, so the buds cannot penetrate the bark and develop into sprouts (Mikola, 1942). Sprouting capacity may depend on carbohydrates in the roots. However, Johansson (1993) found no pronounced peaks in the carbohydrate content in birch roots during the year. Sprouting capacity may also depend on the cutting date. Johansson (1992b) found the highest number of living birch stumps producing sprouts cut in all months but June-October. Etholén (1974) found no effect of cutting time on the sprouting ability of young downy birch stumps.

Fig. 3. Sprouts of birch (left) and suckers of aspen (right)

In southeastern New York, Kays and Canham (1991) studied the sprouting ability of four hardwood species: red maple (*Acer rubrum* L.), gray birch (*Betula populifolia* Marsh.), white ash (*Fraxinus Americana* L.) and black cherry (*Prunus serotina* Ehrh.). They reported that gray birch had the highest mortality (87 %) of stumps after cutting in May but the other species only had mortalities of 10-20 % depending on cutting date. In a study of the suckering capacity of parent trees of American beech, a mean of 41,365 (3,924-89,765) suckers ha^{-1} was found (Jones and Raynal, 1986).

European aspen and trembling aspen (*Populus tremuloides* Michx.) are two *Populus* species with a high capacity for sucker production. The number of suckers after cutting the mother tree differs depending on the cutting date (Johansson, 1993) and on site, stand and management factors (Frey et al., 2003). The age of the mother tree also influences the suckering ability (Brinkman and Roe, 1975). A trembling aspen stand was found to produce 8000 suckers ha^{-1} after cutting (Tew, 1970). In a study by Alban et al. (1994) of trembling aspen growing in Minnesota, the number of suckers the first year after disturbance was >250,000 per hectare. The number had decreased to 40,000 after five years (Stone and Elioff, 1998). Trembling aspen stands growing on similar soils in Minnesota and British Columbia produced 50,000 suckers ha^{-1} after five years and the mean sucker height was 2.1 m (Stone and Kabzems, 2002). The root system of an individual aspen is widely spread, with root lengths up to 20 m (Reim, 1930). In a Swedish study, about 70 % of the suckers occurred within 10 m of the parent aspen tree (Bärring, 1988). In a study by Johansson (1993) the content of starch in roots of European aspens fluctuated during the year with the lowest levels in May-July. The same pattern has been reported for trembling aspen by Baker (1925), Zehngraff (1946), Tew (1970) and Brinkman and Roe (1975).The lowest content has been recorded in late May and early June. When aspen is cut in the winter the highest numbers of suckers are produced (Stoeckler and Macon, 1956; Steneker, 1976; Peterson and Peterson, 1992). In other studies (Shier and Zasada, 1973; Fraser et al., 2002) on trembling aspen, no relationships have been identified between carbohydrate content in roots and the number of suckers initiated.

Alder regenerate vegetatively by sprouts or suckers depending on species. In a study of red alder (*Alnus rubra* Bong.), the number of sprouts per living stump ranged between 5 and 9 (Harrington, 1989). In another study of the same species, the number of sprouts was in the range 9-13 (DeBell and Turpin, 1989). According to Rytter (1996), young grey alders (*Alnus incana* (L.) Moench) produce sprouts after cutting, but the old trees produce suckers. In a Finnish study, grey alder stumps sprouted within three weeks of cutting (Paukkkonen and Kauppi, 1992). Sucker production by grey alder is the main means of vegetative regeneration when the trees are more than 25-30 years old (Schrötter, 1983). In a study of seasonal variation of carbohydrates in the roots of common and grey alders, levels were found to be highest during September-November (Johansson, 1998). In a study of the influence of felling time on sprout and sucker production by common and grey alder, the carbohydrate content in the roots was found to influence biomass production (Johansson, 2009). The highest number of sprouts from common alder stumps was produced after cutting in August-October (23-24 sprouts $stump^{-1}$). Ten years later, the number of sprouts had decreased to 1.3-2.3 sprouts $stump^{-1}$. The average number of sprouts on living grey alder stumps was highest after cutting in March (3.0), August (3.4) and September (3.4), with a reduction to an average of 2.0 after five years. The number of grey alder suckers per m^2 was highest, 21.0, after cutting in September with a reduction to 1.5 after five years. The recommendation, therefore, is to cut grey alder in August and September ad common alder in August-October when the largest number of sprouts and suckers will result. In a study on the initial sprouting of 4-year-old red alders, the percentage of sprouting stumps was highest when the alders were cut in January (Harrington, 1984).

In a study of the spouting ability of *Eucalyptus* in plantations, the number of sprouts per living stump varied, but the highest number was 5-6 sprouts $stump^{-1}$ (Sims, 1999). The stumps have the capacity to resprout several times, depending on their vigor.

4. Direct seeding

When practicing direct seeding on forest land there are practical recommendations considering among others Norway spruce (*Picea abies* (L.) Karst.) , Scots pine (*Pinus sylvestris* L.), birch, beech (*Fagus sylvatica* L.) and oak in relation to the target species. There are, however, few recommendations available for seeding on farmland, although the factors associated with successful establishment are the same as for natural seeding (species, mineral soil, moisture, competition by grasses and herbs, and weather conditions).

The success of establishment of seedlings after direct seeding depends on the nature of the soil treatment and the date of seeding. The critical phase is the emergence of seedlings during the first days or weeks after seeding and the moisture conditions in the treated spots. Generally, precipitation is low in late spring and therefore seeding must be undertaken early in spring.

High quality seeds are expensive and therefore a natural seed source close to the planting site can allow collection from mature seed trees of the appropriate species. Birch and alder are suitable species for producing stands for bioenergy harvest, with subsequent vigorous sprouting or suckering. Depending on seeding method the amount of seeds is 0.5-1.0 kg ha^{-1}.

5. Management of mixed stands on farmland

Using a mixture of species in forest management has been common in Europe for the last three centuries. Hegre and Langhammer (1967) and Stewart et al. (2000) have presented overviews of the importance of mixed stands and their management in different countries worldwide.

Fig. 4. Mixed stand of alder and Norway spruce (left), aspen and Norway spruce (middle) and birch and Norway spruce (right)

In Finland and Norway, a forest stand is defined as being mixed if 20 % of its basal area is made up of broadleaved species, with conifers comprising the dominant species (Frivold, 1982). In Sweden, the proportion is 30 % and in Italy 10 % of the basal area. The Swedish definition of a mixed broadleaved and coniferous stand is "a type of stand in which the total percentage of broadleaved species is 30-70 % of the growing stock" (Anon., 2010). In Nordic countries mixed stands are the most frequent type of stand.

Mixed stands mostly establish spontaneously i.e. a planted or naturally regenerated conifer stand is mixed with naturally regenerated broadleaves. Areas of clear felling that are moist are readily colonized by broadleaves, which can establish from seeds, sprouts or suckers. The number of stems can amount to 5000 to 50,000 per hectare. However there is a conflict between broadleaf cover preventing frost damage to young spruce trees and the strong competition between broadleaves and conifer seedlings. In older stands, both species become established, competition is stabilized and the risk of frost damage declines (Johansson, 2003).

Mostly, Nordic forestry is focused on the management of stands for the production of softwood. A large number of young broadleaves are likely to compete with the conifer seedlings in such stands. In the past, the broadleaves were cut or treated with herbicides. Nowadays, with increasing interest in the supply of biomass for bioenergy production, other management systems have been introduced.

When managing mixed forest stands, a stratified mixture of shade-tolerant, late-successional species in the lower stratum and early successional species in the upper stratum is recommended (Assmann, 1970; Kelty, 1992). Mixed stands may contain alder, aspen or birch and Norway spruce (Johansson, 2003), (Fig. 4). The management of mixed stands is often based on stands which have not been cleaned at the correct time. The spontaneous establishment of broadleaved trees takes up to10 years.

5.1 Mixed forest management

A number of methods are practiced in the Nordic countries, most commonly the shelter method (Tham, 1988; Johansson and Lundh, 1991) and the "Kronoberg" method (Anon., 1985). The descriptions in the sections below are based on a mixed stand of birch and Norway spruce, since this is the most common situation, but the same techniques can be used for other broadleaved species with Norway spruce.

When managing this type of stand it is important that the density of the broadleaved stems is not too high once the spruces have been established. According to Braathe (1988), the competition is too strong for spruces if there are more than 1200 birches ha^{-1} and they are >3 m tall. In that case, he postulated a 30 % decrease in the height increment of the spruce.

5.1.1 The shelter method

This method is common in Finland, Norway and Sweden. It was introduced in Sweden by Tham (1988) with some modifications by Johansson and Lundh (1991). Currently, the same technique is used for birch and Norway spruce in Finland, Norway and Sweden. The principal aim is to create an initial mixed stand with an optimal density of birch.

The method involves two or three steps:

1. When the spruces are 1.5-2 m tall, the density of birch is reduced by cleaning to 800-1000 stems ha^{-1}.
2. The "birch shelter" is cut when the birches are 30-35 years old with a diameter at breast height (dbh) of 15-20 cm.
3. An alternative is to cut all 30-35-year-old birches except 50-100 stems ha^{-1}. The remaining stems should be evenly spread through the stand. These birches will produce high-quality timber during the following 20 years.

5.1.2 The "Kronoberg" method

This method was first introduced in southern Sweden (Anon., 1985). The aims are to avoid frost damage to Norway spruce plants and to control the number of sprouts that are able to establish after the removal of birch in each step.

The method involves three steps:

1. When the birches are 3-4 m tall the stand is cleaned. A total of 3000-4000 birch stems ha^{-1} should be retained. The Norway spruce is not cleaned.
2. When the birches are 6-9 m tall the stand is cleaned again. A total of 1000-1500 birch stems ha^{-1} should be retained; the dbh of the birches should be about 5 cm.
3. When the birch stand is 20-25 years old the birches are felled. They will be 8-12 m tall with a dbh of 8 cm. The mean height of the Norway spruce will be 3-4 m. The spruce stand should be thinned to 2000-2500 stems ha^{-1}.

Alternatively, instead of felling all the birches, 600-800 birches ha^{-1} could be left for 10-15 years. When the birches are finally cut, their mean dbh will be 15-20 cm.

5.1.3 Mixed stands of birch and Norway spruce

The most common type of young stands in Nordic countries is mixed birch and Norway spruce, Fig. 5. Many reports describe how to manage birch and Norway spruce. In Finland, Norway and Sweden the management of mixed stands is common (Mielikänen, 1985; Braathe, 1988; Tham, 1988; Mård, 1997; Klang and Ekö, 1999). Frivold and Groven (1996) discussed the importance of managing mixed stands for future high timber quality. The competition between the taller birches and Norway spruce may adversely affect spruce growth. Therefore the birches must be carefully managed with respect to both numbers of stems removed and controlling competition. A common recommendation is to leave 500-1000 stems ha^{-1} when the birches are 10-15 years old. A Finnish study of a mixed stand of birches (downy and silver) and Norway spruce examined the influence of competition (Valkonen and Valsta, 2001). A reduction of 7-15 % by volume production was reduced by 7-15 % in mixed stands with 1000 birches ha^{-1} compared to pure spruce stands.

Fig. 5. Managed mixed stand of birch and Norway spruce.

Below an experiment in mixed stands of birch and Norway spruce is described (Johansson, 2000b). The experiment was started in 1983 and was based on trials established at eight localities in central and southern Sweden. The experimental stands were 20-30 years old. They were dense, 1520-20,280 stems ha^{-1}, and self regenerated.

The experiment included three thinning regimes:

- Thinning of the birch overstory to create a shelter of 500 stems ha^{-1}.
- Total removal of the birch trees
- Only Norway spruces

At the first cutting, to create the shelter and the pure Norway stands, 1520 to 20,280 birch stems ha^{-1} with a mean diameter of 5.2 cm were removed. After 5 years, 373 to 507 birch stems ha^{-1} with a mean diameter of 15.7 cm were recorded.

Data collected five years after the experiment started are presented in Table 2. The competition by the birch shelter did not influence the growth of Norway spruce. As shown in the table, the mean diameter of the Norway spruce trees was almost the same in the shelter as in the pure stands, 7.6 and 7.0 cm respectively.

		dbh, cm	Height, m	Stocking level, stems ha^{-1}
		Shelter		
Birch				
	Mean ± SE	13.3±0.4	14.2±0.5	499±5
	Range	8.1-19.9	8.2-20.0	480-574
Norway spruce				
	Mean ± SE	7.6±0.3	9.7±0.5	2811±110
	Range	4.6-9.9	5.3-13.5	1693-3373
		No shelter		
Norway spruce				
Mean ± SE	Mean ± SE	7.0±0.1	8.5±1.0	2517±154
Range	Range	3.3-9.2	4.2-11.2	1293-3453

Table 2. Stand characteristics of the trees remaining five years after cutting

Fig. 6. Managed mixed stand of European aspen and Norway spruce

5.1.4 Mixed stands of aspen and Norway spruce

Mixed stands of European aspen and Norway spruce are usually established on rich soils, (Fig. 6). Hegre and Langhammer (1967) and Langhammer (1982) presented results from a

Norwegian experiment on farmland that involved planted European aspen and Norway spruce. Aspens and Norway spruces were planted each at a density of 2000 stems ha^{-1}. The aspens were thinned 30 years later and 580 stems ha^{-1} were retained. Recommendations based on the study stated that planting densities of 2000 Norway spruce and 1000 aspen ha^{-1} would avoid strong competition by the aspens.

5.1.5 Mixed stands of alder and Norway spruce

Naturally established mixed stands of alder are common on wet or moist sites, (Fig. 7). Few studies have examined mixed stands of alder and Norway spruce; those which do exist are based on stands that were not managed correctly during the first ten years after establishment (Lines, 1982; Johansson, 1999d).

Fig. 7. Managed mixed stand of grey alder and Norway spruce

6. Harvesting tops and branches after clear cutting

After clear cutting, tops and branches from felled trees are traditionally left on site together with small trees (Fig. 8). On nutrient-limited sites this slash should not be removed because that would reduce the nutrients present on site. The amount of biomass present in tops and branches is estimated to amount to 20-30 % of the total harvest. The supply of biomass from tops and branches is the main source of bioenergy production in Sweden.

Fig. 8. Clear cut area with branches and tops (left) and stacks of branches and tops (right)

7. Fast-growing species

Besides conventional forestry management, there is increasing interest in management of so-called fast-growing species. Depending on geographical location, different species can be considered fast-growing. There are at least three types of tree suitable and frequently used for management in Europe, the USA and Canada: *Salix* clones, poplar and hybrid aspen. In areas with higher temperatures than northern Europe, species of *Eucalyptus* are also planted.

7.1 Salix

In Sweden research on short rotations using *Salix* began in the end of 1900. Today 10,000-15,000 hectares of short rotation *Salix* stands have been established and are actively managed using advanced technology. The management is based on small-scale plots, where the farmer owns the stand and manages it. Harvesting is undertaken using machinery owned by entrepreneurs and the harvested material is sold to be used for district heating. Common rotation periods are 4-5 years with 5-6 repeated rotations; a plantation lasts a total of 20-30 years before a new one must be established. The plantations must be fertilized and in some cases treated with herbicides. Pathogens (fungi and insects) damaging the leaves and shoots will cause a reduction in growth. As the seedlings represent attractive wildlife habitat, the plantations must be fenced.

Fig. 9. Harvested area of *Salix* (left) and a stack of harvested coppice (right)

7.2 Poplar

Worldwide, and for a long time, poplars have been used for, *inter alia*, pulpwood and timber production. Currently, short rotation plantations intended for biomass production are being established. In Sweden poplars have been planted in experiments or plots for practical survey for the last 20 years. Poplar plantations covering small areas of 0.5-2 ha on former farmland can produce 80-100 tonnes ha^{-1} of wood in ten years (Mean annual increment (MAI): 8-10 tonnes ha^{-1} $years^{-1}$). If rotations are longer than 10 years, some of the material harvested will be suitable for use as pulpwood. Nowadays short rotation plantations aiming biomass production has been established. In Sweden poplars have been planted in experiments or plots for practical survey the last 20 years. After harvesting, regeneration of older trees by suckers or sprouts is limited. Certain clones and species produce no or only a few sprouts or suckers. This may be because poplars must be young when they are cut for sprouts to be initiated. The bark on the poplar stems is thick already when the alders are 15 years old, preventing any buds from growing into sprouts.

Fig. 10. Hybrid poplar stand

7.3 Hybrid aspen

Hybrid aspen is a hybrid between European aspen and trembling aspen (Wettstein, 1933). The hybrid was introduced into Sweden in 1939. Today plantations of hybrid aspen are a potential source of bioenergy, pulpwood and timber. The MAI for hybrid aspen is the same as for poplar, 10 tonnes ha^{-1} $year^{-1}$. A German study compared the biomass production in repeated five-year rotations of European, trembling and hybrid aspen (Liesebach, et al., 1999). After harvest of the 5-year-old plantation the biomass was: 7 tonnes ha^{-1} $year^{-1}$ from European aspen, 18 from trembling aspen and 16-34 from the four clones of hybrid aspen that were examined. The plants were then allowed to produce suckers, resulting in 165,000 suckers ha^{-1} during the first year and 45,000 suckers ha^{-1} five years later. During the second rotation, the production was 18 and 20 tonnes ha^{-1} for European and trembling aspen and 27-41 for the hybrid aspen clones. The amount of biomass after 5 and 10 years could amount to 50 and 100 tonnes ha^{-1} respectively. If longer rotations are preferred, the focus should be

Fig. 11. Hybrid aspen stand

on pulpwood and timber production, with bioenergy derived from tops and branches. After harvesting the trees, the stumps produce 50,000-100,000 suckers ha^{-1}. During the subsequent 5-10 year period the sucker biomass will amount to 50-100 tonnes ha^{-1}. However biomass production during a 10-year-old rotation was found to amount to 47, 51 and 87-124 tonnes ha^{-1} respectively for the aspen stands.

8. Biomass characteristics

The biomass fractions of a tree are the stump (including roots), stem, branches and foliage (needles and leaves). Broadleaved trees and conifers have different fractions of these aboveground components (Johansson 1999a, b). For birches, the mean aboveground fractions are: stem, 75 %; branches, 18 %; and leaves, 7 %. For conifers, the mean values are 63 %, 23 % and 14 % respectively (Johansson, 1999b, c). The percentage represented by needles is higher in young than old conifers, Fig. 12.

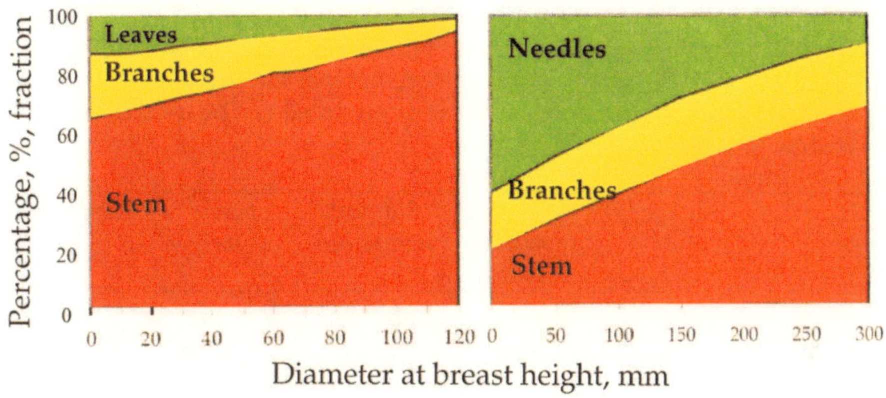

Fig. 12. Percentage biomass fractions by total d. w. %, of a tree at different diameters (DBH), mm

The effect of repeated harvesting on biomass production and sprouting of downy birches growing in central and northern Finland has been studied by Hytönen and Issakainen (2001). Different harvesting cycles of 1, 2, 4, 8, 12 and 16 years were examined. The main results were that downy birch is not suitable for biomass production using short rotations. Most of the stumps, 87 %, did not sprout in the one year rotations, but 8-year rotations produced the same number of sprouting stumps as the longer rotations.

Reim (1929) reported that European aspen growing along the borders of farmland may produce large numbers of suckers when cultivation ceases. In a study of repeated short rotations of aspen, the number of suckers per hectare decreased with every additional rotation (Perala, 1979). The study included rotations of four or eight years and, in both cases, the number of suckers decreased over the three rotations studied.

9. Conclusions

There are several establishment and management techniques available that can be applied to small-scale plots for biomass production on farmland and forest land.

The management methods presented here rely on the land owner having extensive and detailed knowledge of biological processes. The changes in growth of individual species and mixed stands must be known. Some of the methods are based on optimal rotation periods and adequate management of the stand, including cleaning and thinning at the correct time. Severe competition could drastically decrease tree growth. Besides the need for the site to be suitable for tree cultivation, the skill of the owners is important. The most important factor, however, is the enthusiasm and curiosity of the owner; without this, most of the methods will not produce the yields suggested in the present study.
Table 3 lists possible future management models for trees established on farmland and forest land. When operating on a small-scale, there are many alternatives and the owner can be more flexible than is possible in large-scale operations. As the possible rotation periods range from 5 to 40 years it is important to have stands of different ages to ensure a continuous supply. Efficient management of such small areas would make it possible to produce a certain amount of biomass for personal use or to sell to neighbors or local heating plants..
Figures for potential energy supply from different stand types and management options allow us to make comparisons and select appropriate ways to use available land.
Most of the methods are cheap, need a short time to establish and involve relatively straightforward management. The raw materials produced can be used to generate energy for the landowner or can be sold.

Activity	Rotation period, years	Biomass, tonnes ha^{-1}	MWh[1] ha^{-1}	Next generation
Ingrowth				
Natural seeding	10-20	50-110	115-255	Sprouts or suckers
Sprouting, suckering	5-15	50-120	115-275	Sprouts or suckers
Direct seeding	10-15	40-80	90-185	Sprouts or suckers
Mixed stands	35-40	100-150	230-345	
Harvesting tops and branches	-	50	135	
Fast-growing species	5-25	30-300	70-690	Sprouts or suckers

1) Conversion factor MWH/tonnes: 2.3

Table 3. Small-scale management of tree stands on farmland and forest land and possible biomass production

10. References

Agestam, E. (1985). En produktionsmodell för blandskog av tall, gran och björk i Sverige. Summary: A growth simulator for mixed stands for pine, spruce and birch in Sweden. Swedish University of Agricultural Sciences. Department of Forest Yield Research. Report 15, 150 pp.

Alban, D.H., Host, G.E., Elliof, J.D. and Shadis, D. (1994) Soil and vegetation response to soil compaction and forest floor removal after aspen harvesting. U.S.D.A. Forest Service Research Paper NC-315, 8 pp.

Anon. (1985). Lövröjning med skärmmetoden - Skötsel av granföryngringar med tätt lövsly. (Cleaning of broadleaves - Management of Norway spruce regenerations with a dense broadleaved stands). Skogsvårdsstyrelsen i Kronobergs län. Information sheet, 4 pp. (In Swedish).

Anon. (2010). Swedish statistical yearbook of forestry 2010. Swedish Forest Agency. Jönköping. 337 pp.

Assmann, E. (1970). The principles of forest yield study. Oxford. Pergamon Press, p. 506.

Augspurger, C.K. and Franson, S.E. (1987). Wind dispersal of artificial fruits varying in mass, area and morphology. Ecology 68(1), pp. 27-42.

Bärring, U. (1988). On reproduction of aspen (*Populus tremula* L.) with emphasis on its suckering ability. Scandinavian Journal of Forest Research 3, pp. 229-240.

Beck, D.E. (1977). Growth and development of thinned versus unthinned yellow-poplar sprout clumps. U.S.D.A. Forest Service Research Paper SE-173, 16 pp.

Björkbom, J.C. (1971). Production and Germination of Paper birch Seed and its Dispersal into a Forest opening. U.S.D.A. Forest Service. Research Paper NE-209, 14 pp.

Björkroth, G. (1973). Sådd och självsådd med björk på inägomark - resultat från ett 1-årigt försök i övre Norrland. (Seeding and natural seeding with birch on farmland - results from a 1-year-old trial). Royal College of Forestry. Department of Forest Regeneration. Note 1973-03-12, 4 pp. (In Swedish).

Blumenthal, B-E. (1942). Studier angående aspens förekomst och egenskaper i Finland. Referat: Untersuchungen über des Vorkommen und die Eigenschaften der Espe in Finnland. Silva Fennica 56, pp. 1-63.

Braathe, P. (1988). Utviklingen av gjenvekst och ulike blaningsforhold mellom barrtræer of løvtræer - II: Summary: Development of regeneration with different mixtures of conifers and broadleaves II. Norwegian Forest Research Institute. Report 8, 50 pp. (In Norwegian).

Brinkman, K.A. and Roe, E.I. (1975). Quaking aspen silvics and management in Lake States. U.S.D.A. Forest Service. Agriculture Handbook No. 486, 52 pp.

Børset, O. (1956). Rotskudd hos osp. Summary: Suckers i aspen. Tidskrift for Skogbruk IV, pp. 219-240. (In Norwegian).

Etholén, K. (1974). Kaatoajankohdan vaikutus koivun ja haavan vesomiseen taimistonhoitoaloilla Pohjois-suomessa. Summary: The effect of felling time of *Betula pubescens* and *Populus tremula* in the seedling stands in Northern Finland. Folia Forestalia 213, pp. 1-16. (In Finnish).

DeBell, D.S. and Turpin, T.C. (1989). Control of red alders by cutting. U.S.D.A. Forest Service. PNW-RP-414, 10 pp.

Falinski, J.B. (1980). Vegetative dynamics and sex structure of the populations of pioneer dioecious woody plants. Vegetatio 43, pp. 23-38.

Fraser, E.C., Lieffers, V.J., Landhäusser, S.M. and Frey, B.R. (2002). Soil nutrition and temperature as drivers of root suckering in trembling aspen. Canadian Journal of Forest Research 32, pp. 1685-1691.

Frey, B.R., Lieffers, V.J., Landhäusser, S.M., Comeau, P.G. and Greenway, K.J. (2003). An analysis of sucker regeneration of trembling aspen. Canadian Journal of Forestry Research 33, pp. 1169-1179.

Fries, C. (1984). Den frösådda björkens invandring på hygget. (The establishment of naturally seeded birch on clear cut areas). Sveriges Skogsvårdsförenings Tidskrift 82 (3-4), pp. 35-49. (In Swedish).

Frivold. L-H. (1982). Blandingsskogens status i europeisk skogbruk. Summary: Status of mixed forest in European forestry. Tidskrift for Skogbruk 90, pp. 250-261. (In Norwegian).

Frivold, L-H. and Groven, R. (1996). Yield and management of mixed stands of spruce, birch and aspen. In. J. Dietrichson (Ed.). *Silviculture for fuelwood.* Norwegian Journal of Agricultural Science. Supplement 24, pp. 21-28.

Graber, R.E. and Leak, W.B. (1992). Seed fall in an old-grown northern hardwood forest. U.S.D.A. Forest Service. Northeastern Forest Experiment Station. Research Paper NE-663, 11 pp.

Greene, D.F. and Johnsson, E .A. (1997). Secondary dispersal of tree seeds on snow. Journal of Ecology 85, pp. 329-340.

Grime, Hodgson, J.G. & Hunt, R. (1988). Comparative Plant Ecology. - A functional approach to common British species. Oxford University. Press, UK.

Harmer, R., Peterken, G., Kerr, G. and Poulton, P. (2001). Vegetation changes during 100 years of development of two secondary woodlands on abandoned arable land. Biological Conservation 101, 291-304.

Harrington, C.A. (1984). Factors influencing initial sprouting of red alder. Canadian Journal of Forest Research 14, pp. 357-361.

Hegre, A. and Langhammer, Aa. (1967). Et bidrag til diskusjonen om blandingskog. Zusammenfassung: Beitrag zur Diskussion über Mischwald. Scientific Report from the Agricultural College Norway 46 (9), 30 pp. (In Norwegian).

Heikinheimo, O. (1932). Skogarnas naturliga föryngring. (Natural regeneration in forest stands). Centralsällskapets förening för skogskultur. Skriftserie nr 3. Helsinki, 90 pp. (In Swedish).

Heikinheimo, O. (1937). Über die Besamoungsfähigkeit der Waldbäume II. Communicationes Instituti Forestalis Fenniæ, 24 (4), 67 pp. (In German).

Hesselman, H. (1934). Några studier over fröspridningen hos gran och tall och kalhyggets besåning. (Some aspects about seed dispersal by Norway spruce and Scots pine and the seeding of the clear cut area)Meddelande från Statens Skogsforsöksanstalt , 27. (In Swedish).

Hughes, J.W. and Fahey, T.J. (1989). Seed dispersal and colonization in a disturbed northern hardwood forest. Bulletin of the Torrey Botanical Club 115 (2), pp. 89-99.

Hytönen, J. (1994). Effect of cutting season, stump height and harvest damage on coppicing and biomass production of willow and birch. Biomass and Bioenergy 6 (5), pp. 349-357.

Hytönen, J. and Issakainen, J. (2001). Effect of repeated harvesting on biomass production and sprouting of *Betula pubescens*. Biomass and Bioenergy 20, pp. 237-245.

Johansson, T. (1992a). Dormant buds on *Betula pubescens* and *Betula pendula* stumps under different field conditions. Forest Ecology and Management 47, pp. 245-259.

Johansson, T.(1992b). Sprouting of 2- and 5-year-old birches (*Betula pubescens* Ehrh. and *Betula pendula* Roth) in relation to stump height and felling time. Forest Ecology and Management 53, pp. 263-281.

Johansson, T. (1992c). Sprouting of 10- and 50-year-old *Betula pubescens* in relation to felling time. Forest Ecology and Management 53, pp. 283-296.

Johansson, T. (1993). Seasonal changes in contents of root starch and soluble carbohydrate in 4-6-year old *Betula pubescens* and *Populus tremula*. Scandinavian Journal of Forest Research 8, pp. 94-106.

Johansson, T. (1998). Seasonal changes in contents of root starch and soluble carbohydrates in young *Alnus incana* and *Alnus glutinosa*. Swedish University of Agricultural Sciences. Department of Forest Yield Research. Report 44, 20 pp.

Johansson, T. (1999a). Förekomst av självföryngrade lövträd på nedlagd åkermark. Summary: Presence of self-regenerated broad-leaved trees growing on abandoned farmland. The Swedish University of Agricultural Sciences. Department of Forest Management and Products. Report 2, 83 pp. (In Swedish).

Johansson, T. (1999b). Biomass production of Norway spruce (*Picea abies* (L.) Karst.). Silva Fennica 33 (4), 261-280.

Johansson, T. (1999c). Biomass equations for determining fractions of pendula and pubescent birches growing on abandoned farmland and some practical implications. Biomass and Bioenergy 16, pp. 223-238.

Johansson, T. (1999d). Dry matter amounts and increment in 21-91-year-old common alder and grey alder and some practical implications. Canadian Journal of Forest Research 29, pp. 1679-1690.

Johansson, T. (2000a). Biomass equations for determining fractions of common and grey alders growing on abandoned farmland and some practical implications. Biomass and Bioenergy 18, pp. 471-480.

Johansson, T. (2000b). Regeneration Norway spruce under the shelter of birch on good sites might increase the bioenergy supply I Sweden. New Zealand Journal of Forestry Science 30 (1/2), pp. 16-28.

Johansson, T. (2003). Mixed stands in Nordic countries - a challenge for the future. Biomass and Bioenergy 24, 365-372.

Johansson, T. (2008). Sprouting ability and biomass production of downy and silver birch stumps of different diameters. Biomass and Bioenergy 32, pp. 944-951.

Johansson, T. (2009). Influence of felling time on the vegetative reproduction of 15-year-old *Alnus glutinosa* and 8-year-old *Alnus incana*: Implications for biomass production. Swedish University of Agricultural Sciences. Department of Energy and Technology. Report 8, 27 pp.

Johansson, T. and Lundh, J-E. (1991).Anläggning av blandskog. (Cultivation of management of mixed forests), Skog och Forskning 2, pp. 11-18. (In Swedish).

Johansson, T. and Lundh, J-E. (2006). Seed dispersal from a common alder (*Alnus glutinosa* (L.) Gaertner) stand. Swedish University of Agricultural Sciences. Department of Bioenergy. Report 10, 49 pp.

Jones, R.H. and Raynal, D.J. (1986). Spatial distribution and development of root sprouts in *Fagus grandifolia* (Fagaceæ). American Journal of Botany 73 (12), 1723-1731.

Karlsson, A. (1996). Site preparation of abandoned fields and early establishment of naturally and direct-seeded birch in Sweden. Studia Forestalia Suecica 199, 25 pp.

Karlsson, M. (2001). Natural regeneration of broadleaved tree species in southern Sweden: Effects of silvicultural treatment and seed dispersal form surrounding stands. Acta Universitalis Agriculturœ Suecicœ. Silvestria 196, 44 pp. Doctoral dissertation.

Kauppi, A. (1989). Sprouting in birches. Acta Universitas Ouly. A209, 32 pp.

Kauppi, A., Rinne, P. and Ferm, A. (1987). Initiation, structure and sprouting of dormant buds in *Betula pubescens*. Flora 179, pp. 55-83.

Kauppi, A., Rinne, P. and Ferm, A. (1988). Sprouting ability and significance for coppicing of dormant buds on *Betula pubescens* Ehrh. stumps, Scandinavian Journal of Forest Research 3, pp. 343-354.

Kays, J.S. and Canhem, C.D. (1991). Effects of time and frequency of cutting on h and wood root reserves and sprout growth. Forest Science 37 (2), pp. 524-539.

Kelty, J.M. (1992). Comparative productivity of monocultures and mixed-species stands. In. M.J. Kelty and C.D. Oliver (Eds). *Dordreht*. Kluwer Academic Publishers, pp. 125-141.

Klang, F. and Ekö, P-M. (1999). Tree properties and yield of *Picea abies* planted in shelterwoods. Scandinavian Journal of Forest Research 14, pp. 262-269.

Kohh, E. (1936). Beobachtungen über Reifen und Fallzeit der Samen im Lehr- und Versuchs-Forsamt v.j. 1930-1035. Mitteilungen der Forstwissenshaftlichen Abteilung der Universität. Tartu, 124 pp.

La Bastide, J.G.A. and van Vredenburch, C.L.H. (1970). The influence of weather conditions on the seed production of some forest trees in the Netherlands. Mede. Bosbouwproefstation, nr 102, 12 pp.

Langhammer, Aa. (1982). Refleksjoner omkring et plantefelt med osp (*Populus tremula*) and spruce (*Picea abies*) in Norway. Tidskrift for Skogbruk 90, pp. 102-110. (In Norwegian).

Latva-Karjanmaa, T., Suvanto, L., Leinonen, K. and Hannu, R. (2006). Sexual reproduction of European aspen (*Populus tremula* L.) at prescribed burned site: the effects of moisture conditions. New Forests 31, pp. 545-548.

Lines, R. (1982). Mixture experiments. Forestry Commission. Report on Forest Research. HMSO. London, pp. 13-14.

Lundh, J-E. (2003). Direct seeding of alder on farmland. –Effects of seed stratification on seedling emergence and height development. The Swedish of Agricultural Sciences. Department of Bioenergy. Licentiate thesis, 32 pp.

MacVean, D.N. (1955). Ecology of *Alnus glutinosa* (L.) Gaertn. Seed distribution and germination. Journal of Ecology 43, pp. 61-71.

MacVean, D.N. (1956). Ecology of *Alnus glutinosa* (L.) Gaertn. III. Seedling establishment. Journal of Ecology 44, pp. 195-218.

Matlack, G.R. (1988). Secondary dispersal of seed across snow in *Betula lenta*, a gap-colonizing tree species. Journal of Ecology 77, pp. 853-869.

Mård, H. (1997). Stratified mixtures of young Norway spruce and birch as an alternative to pure stands of Norway spruce. Acta Universitatis Sueciæ. Silvestria 35, 29. Doctorial thesis.

Mielikäinen, K. (1985). Koivuskoituksen vaikutus kuusikon rakenttseen ja kehitykseen. Summary: Effect of an admixture of birch on the structure and development of Norway spruce stands. Communicationes Institutii Forestalis Fenniæ 99 (3), 99 pp.

Mikola, P. (1942). Koivun vesomisesta ja sen metsänhoidollista merkityksestä. Referat: Über die Ausschlagbildung bei der Birke und ihre forstlige Bedeutung. Acta Forestaia Fenniœ 50, p0p. 92-101. (In Finnish)

Mork,E. (1944). Om björkfruktens bygning, modning, og spiring. (The anatomy and germination of birch seeds). Meddelser fra det Norske Skogsfosøgsvesen 30, pp. 423-471. (In Norwegian).

Myster, R.W. (1993). Tree invasion and establishment in old fields at the Hutcheson Memorial Forest. Botanical Review 59, 251-278.

Okubo. A. and Levin, S.A. (1989). A theoretical framework for data analysis of wind dispersal of seeds and pollen. Ecology, 70(2), pp. 329-338.

Osbornova, J., Kovarova, M., Leps, J. and Prach, K. (1990). Succession in abandoned fields. Kluwer Academic Publishers, Dordrecht.

Paukkonen, K. and Kauppic, A. (1992). Root and stump buds as structural faculties for reinvigoration in *Alnus incana* (L.) Moench. Flora 187, pp. 353-367.

Perala, D. (1979). Regeneration and productivity of aspen grown on repeated short rotations. U.S.D.A. Forest Service Research Paper NC-176, 7 pp.

Peterson, E.B. and Peterson, N.M. (1996). Ecology and silviculture of trembling aspen. In. P.G. Comeay, G.J. Harper, M.E. Blache, J.O. Boateng and K.D. Thomas (Eds.). *Ecology and management of B.C. hardwoods.* F.R.D.A. British Columbia. Ministry of Forests. Research Branch. Victoria, B.C. Report 255, pp. 31-52.

Reim, P. (1929). Die Vermehrungsbiologie der Aspe auf Grundlage des in Estland und Finnland gesammelten Untersuchungsmaterials. Ph. D. Thesis. The University of Tartu, Estonia, 60 pp. (In German).

Reim, P. (1930). Haava paljunemis-bioloogia. Zusammenfassung: Die Vermehrungsbiologie der Aspe auf Grundlage des in Estland und Finnland gesammelten Untersuchungsmaterials. Mitteilungen der Forstwissenschaftlichen Abteilung der Universität. Tartu 16, 188 p.

Rydberg, D. (2000). Initial sprouting, growth and mortality of European aspen and birch after selective coppicing in central Sweden. Forest Ecology and Management 130, pp. 27-35.

Rytter, L. (1996). Grey alder in forestry: a review. Norwegian Journal of Agricultural Sciences. Supplement no. 24, 65-84.

Schrötter, H. (1983). Waldbaulichertragskundliche Untersuchungen an Weisserle (*Alnus incana* (L.) Moench) im Jungpeistozän der DDR. Beiträge für die Fortwisschaft 17, pp. 89-102.

Shier, G.A. and Zasada, J.C. (1973). Role of carbohydrate reserves in the development of root suckers in *Populus tremuloides*. Canadian Journal of Forest Research 3, pp. 243-250.

Sims, R.E.H., Senetwa, K., Maiava, T. and Bullock, B.T. (1999). *Eucalyptus* species for biomass energy in New Zealand. - Part II: Coppice performance. Biomass and Bioenergy 17, pp. 333-343.

Steneker, G.A. (1976). Guide to the silvicultural management of trembling aspen in the prairie provinces. Canadian Information Report. NOR-X-164, 6 pp.

Stewart, J.D., Landhäusser, S.M., Stadt, K.J. and Lieffers, V.J. (2000). Regeneration of white spruce under aspen canopies: seeding, planting and site preparation. Western Journal of Applied Forestry 15 (4), pp. 177-182.

Stone, D.M. and Elioff, J.D. (1998). Soil properties and aspen development five years after compaction and forest floor removal. Canadian Journal of Soil Science 78, 51-58.

Stone, D.M. and Kabzems, R. (2002). Aspen development on similar soils in Minnesota and British Columbia after compaction and forest floor removal. The Forest Chronicle 78 (6), pp. 886-891.

Tew, R.K. (1970). Root carbohydrate reserves in vegetative reproduction of aspen. Forest Science 16, 318-320.

Tham, Å. (1988). Yield prediction after heavy thinning of birch in mixed stands of Norway spruce (*Picea abies* (L.). Karst.) and birch (*Betula pendula* Roth and *Betula pubescens* Ehrh.). University of Agricultural Sciences. Department of Forest Yield Researh. Report 33, 36 pp. Doctoral Thesis.

Uri, V., Vares, A., Tullus, H. and Kanal, A. (2007). Above-ground biomass production and nutrient accumulation in young stands of silver birch on abandoned agricultural land. Biomass and Bioenergy 31, 195-204.

Valkonen, S. and Valsta, L. (2001). Productivity and economics of mixed two-storied spruce and birch stands in Southern Finland simulated with empirical models. Forest Ecology and Management 140, 133-149.

Wendel, G.W. 1975. Stump sprout growth and quality of several Appalachian hardwood species after clearcutting. U.S.D.A. Forest Service. Research Paper NE-329, 9 pp.

Werner, P.A. and Harbeck, A.L. (1982). The pattern of tree seedling establishment relative to staghorn sumac cover in Michigan old fields. The American Midland Naturalist 108 (1), pp. 124-132.

Willson, M.F. and Traverset, A. (2000). The ecology of seed dispersal. In. M. Fenner (ed.), *Seeds –The Ecology of Regeneration in Plant Communities. 2nd edition.* CABI Publishing. UK, pp. 85-110.

Zehngraff, P.J. (1946). Season of cutting affects aspen sprouting. Lake States Forest Experimental Station. Technical Note No. 250, 4 pp.

10

Production of Unique Naturally Immobilized Starter: A Fractional Factorial Design Approach Towards the Bioprocess Parameters Evaluation

Andreja Gorsek and Marko Tramsek
University of Maribor, Faculty of Chemistry and Chemical Engineering
Slovenia

1. Introduction

Pure and/or mixed isolated microbial cultures, in the dairy sector known as starters, are widely used in the manufacture of numerous fermented (cultured) milk products as well as in butter and cheese making (Bylund, 1995). The starter is added to the sterilized milk-based fermentation media and allowed to grow under controlled and, if necessary, on-line regulated process conditions. During the fermentation, the pure or diversified microbial community produces organic substances which give the cultured milk products their characteristic organoleptic properties such as acidity (pH), flavour, aroma, colour and odour as well as consistency.

According to the basic definition known from the literature, the probiotics are food products and nutritional supplements containing live microorganisms and other components of microbial cells that have an extremely beneficial impact on the citizen's live and well-being of the host (Lahteenmaki & Ledeboer, 2006; Salminen et al., 1999). Therefore, it is not surprising that during the last few years, there has been a significantly increase in the worldwide sales of cultured products containing probiotic bacteria (Ostlie et al., 2005).

One of the dairy cultured products is also kefir (known also as kephir, kiaphur, kefer knapon, kipi and kippi), i.e. unique self-carbonated viscous dairy beverage with small quantities of alcohol and can be made with any kind of animal milk, such as those of cows, goats, sheep, camels and buffalos as well as coconut, rice and soy milk (Abraham & De Antoni, 1999; Farnworth, 1999; Koroleva, 1988; Kwak et al., 1996; Loretan et al., 2003; Otles & Cagandi, 2003). Original kefir contains among others also numerous bioactive ingredients that give its unique health benefits, such as, for instance, strengthening immune system (Vinderola et al., 2005), antitumor activity (Liu et al., 2002), improving intestinal immunity (Thoreux & Schmucker, 2001), antimicrobial activity (Garrote et al., 2000; Rodriguez et al., 2005), regulation of cholesterol metabolism (Liu et al., 2006a), improving anti-allergic resistance (Liu et al., 2006b), improving sugars digestion (Hetzler & Clancy, 2003) and antioxidant activity (Liu et al., 2005). Those kefir's health properties indicate that kefir may be an important, high quality and price-competitive targeted probiotic product.

Several methods for kefir production, which use pure and isolated starters, can be found in the literature (Assadi et al., 2000; Beshkova et al., 2003; Fontan et al., 2006). Nevertheless, the real and original kefir can only be produced using traditional methods of adding kefir grains to a quantity of milk (Otles & Cagandi, 2003; Tamine et al., 1999). Kefir grains are complex natural microbial community entrapped into matrix of protein and polysaccharide (kefiran) and is believed to have its origin in the Caucasian mountains (Bosch et al., 2006; Farnworth, 2005). They are white to light yellowish globular particles (masses) with a diameter (5–35) mm (Bosch et al., 2006; Garrote et al., 1997; Marshall, 1993). The shape of the grains is irregular. Plainly, they are similar to a piece of cauliflower. On the other side, their microflora is much more diverse and complex and therefore difficult to understand and scientifically prove.

During the last two decades, many studies have been focused on thorough analysis of kefir grains microbial composition (Angulo et al., 1993; Garrote et al., 2001; Irigoyen et al., 2005; Kwak et al., Loretan et al., 2003; 1996; Mainville et al., 2006; Marshall, 1993; Simova et al., 2002; Takizawa et al., 1998; Vancanneyt et al., 2004; Witthuhn et al., 2005; Witthuhn et al., 2004). Summarily, kefir grains contain gram-positive homo-fermentative and hetero-fermentative lactic and acetic acid bacteria (*Lactobacillus caucasicus, Lactobacillus brevis, Lactobacillus bulgaricum, Lactobacillus casei, Lactobacilus kefir, Lactobacillus acidophilus, Lactobacillus plantarum, Lactobacillus kefiranofaciens, Lactobacilus kefigranu, Lactobacillus helveticus* ssp. *jogurti, Lactubacillus lactis* ssp. *lactis, Lactobacillus fermentum, Lactobacillus cellobiosuss, Lactococci lactis* ssp. *lactis 1, Lactococci lactis* ssp. *lactis 2, Lactococcus lactis* ssp. *lactis* var. *diacetylactis, Lactococcus lactis* ssp. *cremoris, Streptococcus thermophilus, Lactococcus filant, Streptococcus durans, Leuconostoc dextranicum, Leuconostoc kefir, Leuconostoc lactis, Leuconostoc mesenteroides* ssp. *mesenteroides* and *Leuconostoc mesenteroides* ssp. *Cremoris*) gram-negative acetic acid bacteria (*Acetobacter* spp.) and both lactose fermenting and non-fermenting yeasts (*Kluyveromyces lactis, Kluyveromyces marxianus, Torula kefir, Saccharomyces cerevisiae, Saccharomyces unisporus, Candida keyfr, Saccharomyces rouxii, Torulaspora delbrueckii, Debaryomyces hansenii, Candida holmii, Zygosaccharomyces* sp., *Candida lipolytica* and *Cryptococcus humicolus*). Mentionable, the variegated natural microbial population found in kefir grains represent a pattern of symbiotic community (Lopitz-Otsoa et al., 2006).

The unique variegated microbial composition of kefir grains enables their application not only in large-scale kefir production but potentially also in another novel industrial food manufacturing bioprocesses or even in some specific innovative and visionary eco-efficient bioprocesses in sustainable production of safe, efficient as well as high quality fine biochemicals with the highest added value. For instance, different studies indicate that kefir grains can be used in bread production as a substitute for baker's yeast (Plessas et al. 2005) polysaccharide production as a natural source of exopolysacharide (kefiran) (Rimada and Abraham, 2001; Rimada and Abraham, 2003) and bioalcohol production as a natural immobilized kefir yeast cells (Athanasiadis et al., 1999). Moreover, they can also be used as natural variegated microbial starter in production of fermented soy milk powder (Kubow, S. & Sheppard, WO/2007/087722 A1) as well as in production of novel fermented low-alcoholics drink from mixture of whey and raisin extract (Athanasiadis et al., 2004; Koutinas et al., 2007).

Considering abovementioned scientifically proven potential industrial applications as well as other emerging innovative visionary applications which are currently under thorough screening, evaluation and assessment, it is realistic to expect that in the near future the global demand for grains will extremely increase. Therefore, the classical batch production of kefir grains using traditional propagation in milk with relatively low daily kefir grain increase mass fraction, $w_{KG,di}$ = (5–7) %/d, (Libudzisz & Piatkiewicz, 1990) has to be optimized and improved. When grains are produced commercially, it is critically important for optimization, as well as for monitoring and control of their batch production to know the impact of different bioprocess parameters on daily kefir grains increase mass and mass fraction.

Traditionally, the impact of various significant bioprocess parameters on batch bioprocess performance has been determined experimentally using through planning and time consuming as well as cost ineffective implementing experiments on large industrial scale. With the technological development and growth of the society, however, the bioprocess parameters assessment has been progressively transferred to laboratory scale, which resulted in increased effectiveness and reduced planning cost. Consequently, today almost all bioprocess development activities, which among others include also determination of the relative impact of various significant bioprocess parameters, are practically carried out in laboratory or pilot scale and afterwards, only scale up and tech-transfer into production line is performed.

The technique for the determination and investigation of the influential experiment (bioprocess) parameters at different levels is called the 'design of experiments' (DoE) (Ranjit, 1990). The selection of relevant DoE technique depends especially on the number of parameters influencing the product quality, and the type of the investigated problem. However, conventional full factorial DoE techniques involve altering of one parameter at a time keeping all other parameters constant. When we want to study any given system with a set of independent variables (bioprocess parameters) over a specific region of interest (levels region) and intend to improve the process planning strategy and quality optimization of the bioprocess parameters at the same time, we use the so-called 'Taguchi's approach' (Ranjit, 1990). The use of its algorithm is observed in various optimization problems, starting with optimization of diesel engine parameters (Nataraj et al., 2005), the leaching of non-sulphide zinc ore in the ammonium-sulphate solution (Moghaddam et al., 2005), to the production of clavulanic (Saudagar & Singhal, 2007) and citric acid (Shojaosadati & Babaeipour, 2002) as well as laccase by *Pleurotus ostreatus* 1804 (Prasad et al., 2005), etc. In contrast to the traditional DoE, the standardized Taguchi's experiment design methodology for two independent problem solution plans usually brings the same results, which enables determination of individual bioprocess parameters' relative impact on the final result. This methodology envisages implementation of a minimum number of experiments, which are defined by specific standard orthogonal arrays (OA). Selection of relevant OA is conditioned by the number of parameters and levels.

This chapter examines the traditional batch propagation of kefir grains in fresh high temperature pasteurized (HTP) whole fat cow's milk with some additions (glucose and baker's yeast) under different bioprocess conditions. The main objective of the contribution is to present and describe an experimental determination of the relative impacts of various significant bioprocess parameters that influence traditional batch propagation of kefir grains and daily kefir grain increase mass using the Taguchi's experiment design methodology.

2. Materials and methods

2.1 Equipment

Determination of the relative impact of various significant bioprocess parameters that influence traditional batch propagation of kefir grains and daily kefir grain increase mass using the Taguchi design methodology requires the performance of a series of experiments. In order to ensure the highest quality as well as repeatability of raw experimental data, it is desired to perform those experiments (batch propagations of kefir grains in enriched milk under different bioprocess conditions) in computer controlled state-of-the-art laboratory reactor or fermentor.

Perhaps one of the most user-friendly and at the same time the most efficient high quality aforementioned equipment is heat flow reaction calorimeter RC1 (Mettler Toledo, Greifensee, Switzerland). Basically, the RC1 system is actually both - state-of-the-art computer controlled, electronically safe-guarded bench-scale 'model' of a batch/semi-batch reactor or fermentor from pilot and/or industrial plant (automated lab reactor (ALR)) and at the same time a heat-flow reaction calorimeter. The RC1 system allows real time measurement, monitor and control of all important bioprocess parameters such as rotational frequency of the stirrer, temperature of reaction or fermentation media, reactor jacket temperature, pH value of reaction or fermentation media, mass concentration of dissolved oxygen, amount of added (dosed) material, etc. Primarily, it is designed for determination of the complete mass and heat balance over the course of the entire chemical reaction or physical transformation (e.g. crystallization, dissolution, etc.). In addition, using specific modifications, it can be employed for investigating thermal effects during bioprocess (Marison et al., 1998). This means that by using RC1 system it is possible to gain and/or determine wide range of process thermal data and constants such as specific heat capacity of reaction mixture, heat flow profile of the reaction or physical transformation, reaction enthalpy, maximum heat flow due to reaction or physical transformation, potential adiabatic temperature increase in case of cooling failure, heat accumulation, etc.. All obtained time-depended calorimetric data (heat flow data) can be further used for kinetic studies, etc. The RC1 system enables performance of chemical and also bio(chemical) reactions or physical transformation under different modes such as isothermal conditions, adiabatic conditions, etc. Using RC1 it is possible to perform distillations and reactions (transformations) under reflux with heat balancing. Last but not least, the RC1 system is a recipe driven (managed) which means that all process operations can be programmed or written by recipe beforehand and thus its maximum flexibility is assured. Finally, it is worldwide recognized as an industrial standard to gain safety data for a later scale-up to pilot or production plant.

2.2 Chemicals, kefir culture and culture medium

Daily kefir grain increase mass was studied using fresh HTP whole fat cow's milk (Ljubljanske mlekarne d.d.) as a culture medium. Its chemical composition is 3.2 % proteins, 4.6 % carbohydrates, 3.5 % fat and 0.13 % calcium. 3D-(+) Glucose anhydrous (Fluka) was obtained from commercial sources. Kefir grains, used as inocolum in this study, originate from Caucasian Mountain and were acquired from an internationally recognized local dairy (Kele & Kele d.o.o.). Their detailed microbial composition was not analyzed. Importantly, the microbial population (bacteria and yeasts) of kefir grains depends on many different factors (age, storage conditions and fermentation medium) and varies with the season. It is almost impossible to assure equal microbial composition during long term period, therefore for sets of experiments within one research, kefir grains with the same viability should be used.

2.3 Kefir grain biomass activation

Kefir grain biomass activation was performed in a glass lab beaker. The collected inactive kefir grains (γ_{KG} = 40 g/L) were inoculated in 1 L of fresh HTP whole fat cow's milk. After incubation at room temperature ($\vartheta = (22 \pm 2)$ °C) for 24 h, the grains were separated from the kefir beverages using a household sieve. After washing, they were reinoculated into the fresh milk. The same procedure was repeated over six subsequent days. After this procedure the kefir grains were considered active.

2.4 Analytical determination of kefir grain mass

For the determination of kefir grain mass, the gravimetric method was used. Therefore, kefir grains were separated first from the fermentation medium with plastic household sieve. Then the grains were washed with cold water and dried on filter paper to remove of bulk of adhered water. Finally, kefir grain mass was determined by weighting on Mettler-Toledo analytical balance (PG5002–S).

2.5 Taguchi's experiment design methodology

Dr. Genichi Taguchi has defined the optimization criterion quality as a consistency in achieving the desired or targeted value and minimization of the deviation (Ranjit, 1990). This goal is connected with the performance of a series of experiments with different bioprocess parameters at different levels. The bioprocess parameter is a factor affecting the optimization criterion quality, and its value is called the 'level'. The number of experiments and their sequence are determined by standard OA. When planning the experiments using four bioprocess parameters at four levels, we use the OA L_{16}. Such a plan envisages the performance of 16 experiments, which is significantly less when compared to the full factorial DoE with $4^4 = 128$ experiments.

Due to performing only a part of the envisaged experiments using the traditional full factorial DoE methodology, it is necessary to include an analysis of the results confidence. The standard statistical technique is used for this purpose, the so-called 'analysis of variance' (ANOVA), which recognizes the relative impact of the bioprocess parameters for the optimization criterion (in our case daily kefir grain increase mass) value.

The mathematical algorithm of the ANOVA statistical technique is based on calculation of the variance, which is an indicator of the optimization criterion quality. The ratio between the variance of the bioprocess parameter and the error variance shows whether the parameter affect on the product's quality. The equations required for calculating the relative impact of various significant bioprocess parameters affecting the optimization criterion are presented bellow. The meanings of symbols are described in the sub-chapter "Nomenclature".

$$S_{\mathrm{T}} = \sum_{i=1}^{N} Y_i^2 - \left(\sum_{i=1}^{N} Y_i\right)^2 \Big/ N \tag{1}$$

$$S_j = \sum_{k=1}^{L} \left(\left(\sum_{i=1}^{N_k} Y_i \right)^2 \Big/ N_k \right) - \left(\sum_{i=1}^{N} Y_i \right)^2 \Big/ N \tag{2}$$

$$S_{\mathrm{e}} = S_{\mathrm{T}} - \sum_{j=1}^{M} S_j \tag{3}$$

$$V_j = S_j / f_j \tag{4}$$

$$f_j = L - 1 \tag{5}$$

$$V_e = S_e / f_e \tag{6}$$

$$f_e = f_T - \sum_{j=1}^{M} f_j \tag{7}$$

$$f_T = M - 1 \tag{8}$$

$$F_j = V_j / V_e \tag{9}$$

$$X_j = \left(S_j - f_j V_e\right) 100 / S_T \tag{10}$$

$$X_e = \left(S_e + \sum_{j=1}^{M} f_j V_e\right) 100 \Big/ S_T \tag{11}$$

We compare variance ratio of bioprocess parameter j, F_j, to the standardized value at defined level of significance, $F_{m,n}$, which is obtained from the standard F tables (Ranjit, 1990), whereby m stands for the degree of freedom of bioprocess parameter j and n means the degree of freedom of error variance, and thus determine the bioprocess parameter impact accordingly. In the case where the variance ratio of bioprocess parameter j falls below $F_{m,n}$, the bioprocess parameter has no impact on the optimization criterion, therefore, it is pooled and ignored in the calculations. Consequently, the variance error changes, as the sum of squares and degree of freedom of the pooled bioprocess parameter are added to the error sum of squares and degree of freedom of error variance, respectively. By using the adjusted variance error, we determine new variance ratio of bioprocess parameter j and compare them again by the $F_{m,n}$. The process of pooling is sequential, which means that the parameter having the smallest impact on the optimization criterion should be pooled first, then we re-calculate the variance ratio of bioprocess parameter j and continue pooling until each bioprocess parameter meets the condition $F_j > F_{m,n}$. If the pooling process begins to perform, Taguchi recommends pooling bioprocess parameters until the degree of freedom of error variance is approximately half the total degree of freedom irrespective of significant test criterion validity $F_j > F_{m,n}$ for all remaining bioprocess parameters (Taguchi, 1987). When the pooling procedure is completed, the relative impact of bioprocess parameter j and error on optimization criterion can be calculated using Eqs. (10) and (11).

3. Experimental work

Experimentally determining the relative impact of various significant bioprocess parameters on the daily kefir grain increase mass, during 24 h incubation in cow's milk, based on Taguchi's fractional factorial design approach, requires the performance of a series experiments. It was established (Harta et al., 2004; Schoevers and Britt, 2003) that culture medium temperature, ϑ, glucose mass concentration, γ_G, baker's yeast mass concentration,

γ_Y, and the rotational frequency of the stirrer, f_m are the main influences bioprocess parameters. The bioprocess parameter in our case is a factor affecting daily kefir grain increase mass and its value is called the 'level'. We examined the relative impact of the selected bioprocess parameters at four different levels, as shown in Table 1.

Bioprocess parameter			Level			
			1	2	3	4
A:	Culture medium temperature	ϑ (°C)	20	22	24	26
B:	Baker's yeast mass concentration	γ_Y (g/L)	0	5	10	15
C:	Glucose mass concentration	γ_G (g/L)	0	10	20	30
D:	Rotational frequency of the stirrer	f_m (1/min)	0	50	70	90

Table 1. Proposed bioprocess parameters and their levels

Experiment	Bioprocess parameter[1]				
	A	B	C	D	E[2]
1	1	1	1	1	1
2	2	1	2	3	4
3	1	2	2	2	2
4	4	1	4	2	3
5	1	4	4	4	4
6	2	2	1	4	3
7	4	2	3	1	4
8	4	4	1	3	2
9	4	3	2	4	1
10	3	1	3	4	2
11	2	3	4	1	2
12	3	4	2	1	3
13	1	3	3	3	3
14	2	4	3	2	1
15	3	3	1	2	4
16	3	2	4	3	1

Table 2. Design of experiments – orthogonal array L_{16}

During the first stage of the experimental work, it is necessary to prepare the design of experiments. The DoE envisages determining the number of experiments, their performance conditions, and their sequence. Based on the assumption that the daily kefir grain increase mass would be affected by four bioprocess parameters being considered at four levels, we chose the L_{16} array as the most adequate OA requiring the performance of 16 experiments (Ranjit, 1990). The OA L_{16} is usually intended for the investigation of five bioprocess

[1] In our case bioprocess parameter E was not considered.

[2] Bioprocess parameters and values of their levels are indicated in Table 1.

parameters at four levels; however, it may also be used in our case (four parameters at four levels) by ignoring the bioprocess parameter E. The DoE is presented in Table 2. The first column presents the experimental serial number. Each experiment was defined by the bioprocess parameters (A, B, C, D and E) marked at specific levels by numbers from 1 to 4. During the second stage of the experimental work, we implemented the proposed DoE by performing the 24 h kefir grain biomass incubations in the RC1 system. The incubation procedure was the same for all experiments. Individual experiments were implemented by means of first charging the reactor by 1 L of fresh HTP whole fat cow's milk and adding the mass of glucose previously defined by the DoE. This fermentation medium was heated up to working temperature under the defined rotational frequency of the stirrer. After establishing the temperature steady state and dissolved glucose, we inoculated the fermentation medium with the mass of the baker's yeast also defined by DoE and with 40 g of active kefir grains, which corresponds to initial kefir grain mass concentration, γ_{KG} = 40 g/L. After the 24 h incubation was completed, the kefir grain increase mass was determined using the gravimetric method.

4. Results and discussion

The final kefir grain mass concentration in the culture medium, $\gamma_{KG,f}$, daily kefir grain increase mass, $m_{KG,di}$, and daily kefir grain increase mass fraction, $w_{KG,di}$, experimentally determined under different conditions proposed by the DoE (Table 2), are presented in Table 3. Daily kefir grain increase mass fraction, $w_{KG,i}$ is the quotient between the kefir grain increase mass concentration ($\gamma_{KG,f}$ - 40 g/L) and the initial kefir grain mass concentration (γ_{KG} = 40 g/L).

Experiment	$\gamma_{KG,f}$ (g/L)	$m_{KG,di}$ (g)	$w_{KG,di}$ (%)
1	40.40	0.40	1.00
2	45.83	5.83	14.58
3	46.51	6.51	16.28
4	45.44	5.44	13.60
5	43.39	3.39	8.48
6	45.55	5.55	13.88
7	42.06	2.06	5.15
8	53.10	13.10	32.75
9	50.14	10.14	25.35
10	60.62	20.62	51.55
11	41.70	1.70	4.25
12	41.90	1.90	4.75
13	52.60	12.60	31.50
14	58.06	18.06	45.15
15	55.93	15.93	39.83
16	52.56	12.56	31.40

Table 3. Experimental results - orthogonal array L_{16}

Table 3 shows that the highest daily kefir grain increase mass fraction ($w_{KG,i}$ = 51.5 %) was found at the rotational frequency of the stirrer, f_m = 90 (1/min), at culture medium temperature, ϑ = 24 °C, with a glucose mass concentration, γ_G = 20 g/L, and without baker's yeast (γ_Y = 0 g/L).

Moreover, the average impacts of the bioprocess parameters along with interactions at the assigned levels on the daily kefir grain increase mass are shown on Fig. 1. The difference between levels of each bioprocess parameters indicates their relative impact (Prasad et al., 2005). The larger the difference, the stronger is the influence.

It can be observed from Fig.1 that among bioprocess parameters studied rotational frequency of stirrer showed the strongest influence and followed by glucose mass concentration, culture medium temperature and baker's yeast mass concentration. However, the relative impact of the proposed influencing bioprocess parameters on daily kefir grain increase mass were estimated by ANOVA. The sum of squares or deviation, S_j, and the variance of individual bioprocess parameters, V_j, were calculated by equations (2) and (4), and the error value by equations (3) and (6), respectively. The variance ratio, F_j, is the ratio of variance due to the effect of an individual bioprocess parameter and variance due to the error term. It was calculated by equation (9). The results of ANOVA are shown in Table 4.

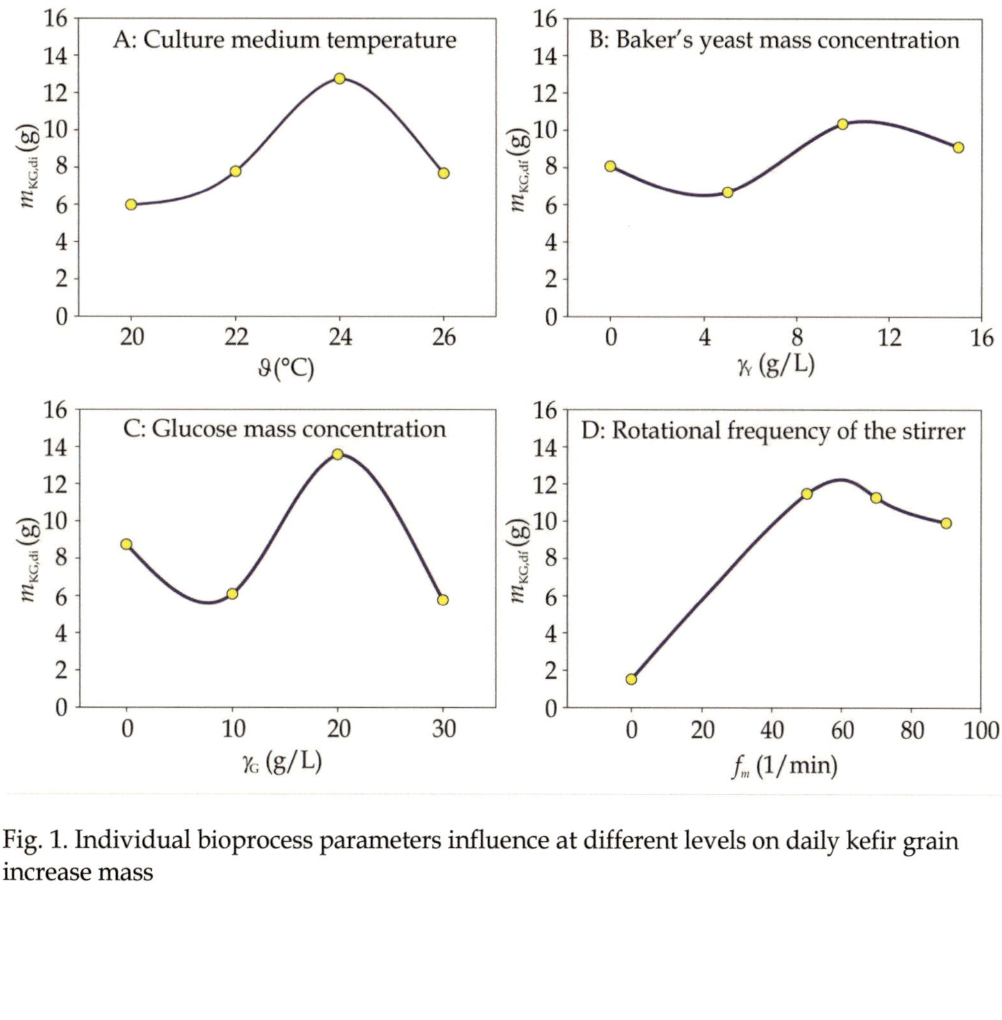

Fig. 1. Individual bioprocess parameters influence at different levels on daily kefir grain increase mass

The degrees of freedom of bioprocess parameter *j* and error variance equaled ($f_j = f_e = 3$) in all cases. At 90 % confidence (level of importance 0.1), the value $F_{3,3} = 5.3908$ was determined through standardized tables of F-statistics. Table 5 shows that the variance ratio of all bioprocess parameters fell below $F_{3,3}$. In accordance with the Taguchi's method algorithm, we pooled baker's yeast mass concentration from further statistical consideration as the least important bioprocess parameter, i.e., with the lowest variance ratio compared to $F_{3,3}$.

Bioprocess parameter	S_j	f_j	V_j	F_j
A: ϑ (°C)	102.52	3	34.17	1.893
B: γ_Y (g/L)	29.18	3	9.73	0.539
C: γ_G (g/L)	156.58	3	52.19	2.891
D: f_m (1/min)	269.57	3	89.86	4.978
Error	54.16	3	18.05	1.000
Total	612.01	15	–	–

Table 4. Analysis of variance - orthogonal array L_{16}

Pooling of the baker's yeast as an insignificant bioprocess parameter requires a repeated variance analysis, whereby the sum of squares and the degree of freedom of the pooled bioprocess parameter are added to the error sum of squares and the degree of freedom of error variance, respectively. The results in Table 5 show that, consequently, the variance ratios of the remaining bioprocess parameters increase. In spite of this, a repeated comparison of variance ratio of each bioprocess parameter indicated in Table 5 with the F-statistics value, $F_{3,6} = 3.2888$, shows that culture media temperature does not meets the $F_j > F_{3,8}$ condition. Nevertheless, regarding significant test criterion ($F_j > F_{m,n}$) and especially Taguchi's recommendation, we pooled only baker's yeast mass concentration as insignificant bioprocess parameter on daily kefir grain increase mass. The final results of ANOVA terms, which were modified after pooling baker's yeast mass concentration, are shown in Table 5. The relative influences of the bioprocess parameter j and error on the daily kefir grain increase mass were calculated using equations (10) and (11), respectively.

Bioprocess parameter	S_j	f_j	V_j	F_j	X_j
A: ϑ (°C)	102.52	3	34.17	2.460	9.9
B: γ_Y (g/L)	pooled				
C: γ_G (g/L)	156.58	3	52.19	3.758	18.8
D: f_m (1/min)	269.57	3	89.86	6.469	37.3
Error	83.34	6	13.89	1.000	34.0
Total	612.01	15	–	–	100.0

Table 5. Final results of variance analysis - orthogonal array L_{16}

The results, shown in Table 5, assign the highest relative influence on the daily kefir grain increase mass (37.3 %) during 24 h incubation to the rotational frequency of the stirrer. The impact of glucose mass contraction and culture medium temperature within the observed ranges (γ_G = (0–30) g/L and ϑ = (20–26) °C) show the lower ones, 18.8 % and 9.9 %, respectively. The remaining fraction represents error influence.
It is well known that kefir grains are bulky and awkward to handle (Bylund, 1994). Despite extensive and careful kefir grain biomass activation, their variegated symbiotic microbial community makes it impossible to retain the constant viability over a long time period. This fact, together with neglecting of possible secondary interactions between bioprocess parameters, mainly explains the relatively high error influence on daily kefir grain increase mass (34.0 %).

5. Conclusion

Using the Taguchi's fractional factorial design approach we analyzed the bioprocess parameters impacts on daily kefir grain increase mass during 24 h incubation in fresh high temperature pasteurized whole fat cow milk. Experiments proposed by the design of experiments (OA L_{16}) were performed in an RC1 reactor system. We determined those conditions which assure the highest kefir grain increase mass fraction and, using analysis of variance, estimated the relative impact of the proposed bioprocess parameters on daily kefir grain increase mass. In the observed bioprocess parameters ranges, we established that the yeast mass concentration was insignificant compared to the other bioprocess parameters. The most influential bioprocess parameter is found to be the rotational frequency of the stirrer (37.3 %), followed by the glucose mass concentration (18.8 %), and the medium temperature (9.9 %), while the remaining share represents an error.
Summarily, this chapter deals with the experimental determination of the relative impacts of various significant bioprocess parameters, that influence one of the most difficult bioprocesses in the dairy industry. The presented results confirm and, even more importantly, upgrade well-known findings about influence of various bioprocess parameters on kefir grain increase mass. On the other side, the presented results also confirm the tremendous importance of optimal kefir grain biomass managements. In addition, the results also clearly verify the fact, that inadequate combination of different significant critical bioprocess parameters has a strong negative influence on daily kefir grain increase mass. For instance, in the worst case the kefir grains growth is almost totally stopped. Last but not least, the presented chapter presents important cutting-edge and, in scientific and commercial society, shortfall basic knowledge needed either for kefir grains mass growth kinetic studies or designing, optimization and commercialization of modern batch or continuous industrial kefir grains production processes.

6. Nomenclature

ALR Automatic Lab Reactor
ANOVA ANalysis Of VAriance
DoE Design of Experiments
f_e degree of freedom of error variance (1)
F_j variance ratio of bioprocess parameter j (1)
f_j degree of freedom of bioprocess parameter j (1)

f_m	rotational frequency of the stirrer (1/min)
$F_{m,n}$	standardized value from the F tables at defined level of significance (1)
f_T	total degree of freedom of result (1)
HTP	High Temperature Pasteurized
L	number of levels (1)
M	number of bioprocess parameters (1)
$m_{KG,di}$	daily kefir grain increase mass (g)
N	total number of experiments (1)
N_k	number of experiments on k level (1)
OA	Orthogonal Array
S_e	error sum of squares (/)
S_j	sum of squares of bioprocess parameter j (/)
S_T	total sum of squares (/)
V_e	variance error (/)
V_j	mean square (variance) of bioprocess parameter j (/)
$w_{KG,di}$	daily kefir grain increase mass fraction (%/d)
X_e	relative impact of error on optimization criterion (%)
X_j	relative impact of bioprocess parameter j on optimization criterion (%)
Y_i	i value of optimization criterion (/)
γ_G	glucose mass concentration (g/L)
γ_{KG}	kefir grain mass concentration (g/L)
$\gamma_{KG,f}$	final kefir grain mass concentration in culture medium (g/L)
γ_Y	baker's yeast mass concentration (g/L)
ϑ	temperature (°C)

7. References

Abraham, A. G. & De Antoni, G. L. (1999). Characterization of Kefir Grains Grown in Cow's Milk and in Soya milk. *Journal of Dairy Research*, Vol.66, No.2, pp. 327-333, ISSN 0022-0299

Angulo, L.; Lopez, E. & Lema, C. (1993). Microflora Present in Kefir Grains of the Galician Region (North West of Spain). *Journal of Dairy Research*, Vol.60, No.2, pp. 263-267, ISSN 0022-0299

Assadi, M. M.; Pourahmad, R. & Moazami, N. (2000). Use of Isolated Kefir Starter Cultures in Kefir Production. *World Journal of Microbiology and Biotechnology*, Vol.16, No.6, pp. 541-543, ISSN 0959-3993

Athanasiadis, I.; Boskou, D.; Kanellaki, M. & Koutinas, A. A. (1999). Low-Temperature Alcoholic Fermentation by Delignified Cellulosic Material Supported Cells of Kefir Yeast. *Journal of Agricultural and Food Chemistry*, Vol.47, No.10, pp. 4474-4477, ISSN 0021-8561

Athanasiadis, I.; Paraskevopoulou, A.; Blekas, G. & Kiosseoglou, V. (2004). Development of a Novel Whey Beverage by Fermentation with Kefir Granules: Effect of Various Treatments. *Biotechnology Progress*, Vol.20, No.4, pp. 1091-1095, ISSN 8756-7938

Beshkova, D. M.; Simova, E. D.; frengova, G. I.; Simov, Z. I. & Dimitrov, Z. P. (2003). Production of Volatile Aroma Compounds by Kefir Starter Cultures. *International Dairy Journal*, Vol.13, No.7, pp. 529-535, ISSN 0958-6946

Bosch, A.; Golowczyc, M. A.; Abraham, A. G.; Garrote, G. L.; De Antoni, G. L. & Yantorno, O. (2006). Rapid Discrimination of *Lactobacili* isolated from Kefir grains by FT-IR spectroscopy. *International Journal of Food Microbiology*, Vol.111, No.3, pp. 280-287, ISSN 0168-1605

Bylund, G. (1995). *Dairy Processing Handbook*, Tetra Pak Processing systems AG, Lund, Sweden

Farnworth, E. R. (1999). Kefir: From Folklore to Regulatory Approval. *Journal of Nutraceuticals, Functional & Medical Foods*, Vol.1, No.4, pp. 57-68, ISSN 1089-4179

Farnworth, E. R. (2005). Kefir - A Complex Probiotic. *Food Science & Technology Bulletin: Functional Foods*, Vol.2, No.1, pp. 1-17, ISSN 1476-2137

Fontan, M. C. G.; Martinez, S.; Franco, I. & Carballo, J. (2006). Microbiological and Chemical Changes during the Manufacture of Kefir Made from Cows' Milk, Using a Commercial Starter Culture, *International Dairy Journal*, Vol.16, No.7, pp. 762-767, ISSN 0958-6946

Garrote, G. L.; Abraham, A. G. & De Antoni, G. L. (2000). Inhibitory Power of Kefir: The Role of Organic Acids. *Journal of Food Protection*, Vol.63, No.3, pp. 364-369, ISSN 0362-028X

Garrote, G. L; Abraham, A. G. & De Antoni, G. L. (2001). Chemical and Microbiological Characterisation of Kefir Grains. *Journal of Dairy Research*, Vol.68, No.4, pp. 639-652, ISSN 0022-0299

Garrote, G. L.; Abraham, A. G. & De Antoni, G. L. (1997). Preservation of Kefir Grains, a Comperative Study. *LWT - Food Science and Technology*, Vol.30, No.1, pp. 77-84, ISSN 0023-6438

Harta, O.; Iconomopoulou, M.; Bekatorou, A.; Nigam, P.; Kontominas, M. & Koutinas, A. A. (2004). Effect of Various Carbohydrate Substrates on the Production of Kefir Grains for Use as a Novel Baking Starter. *Food Chemistry*, Vol.88, No.2, pp. 237-242, ISSN 0308-8146

Hetzler, S. R. & Clancy, S. M. (2003). Kefir Improves Lactose Digestion and Tolerance in Adults. *Journal of the American Dietetic Association*, Vol.103, No.5, pp. 582-587, ISSN 0002-8223

Irigoyen, A.; Arana, I.; Castiella, M.; Torre, P. & Ibanez, F. C. (2005). Microbiological, Physicochemical, and Sensory Characteristics of Kefir during Storage. *Food Chemistry*, Vol.90, No.4, pp. 613-620, ISSN 0308-8146

Koroleva, N. S. (1988). Technology of kefir and Kumys. *Bulletin of the International Dairy Federation*, Vol.227, pp. 96-100, ISSN 0250-5118

Koutinas, A. A.; Athanasiadis, I.; Bekatorou, A.; Psarianos, C.; Kanellaki, M.; Agouridis, N. & Blekas, G. (2007). Kefir-Yeast Technology: Industrial Scale-up of Alcoholic Fermentation of Whey, Promoted by Raisin Extracts, Using kefir-Yeast Granular Biomass. *Enzyme and Microbiological Technology*, Vol.41, No.5, pp. 576-582, ISSN 0141-0229

Kubow, S. & Sheppard, J., *Use of Soy Kefir Powder for Reducing Pain, Blood pressure and Inflammation*, KLCM Research in Nutrition Inc. Canada, WO/2007/087722, World Intellectual Property Organization, 2007-08-09

Kwak, H. S.; Park, S. K. & Kim, D. S. (1996). Biostabilization of Kefir with a Nonlactose-Fermenting Yeast. *Journal of Diary Science*, Vol.79, No.6, pp. 937-942, ISSN 0022-0302

Lahteenmaki, L. & Ledeboer A.M. (2006). Probiotics - The Consumer Perspective. *Food Science & Technology Bulletin: Functional Foods*, Vol.3, No.5, pp. 47-50, ISSN 1476-2137

Libudzisz, Z. & Piatkiewicz, A. (1990). Kefir Production in Poland. *Dairy Industries International*, Vol.55, No.7, pp. 31-33, ISSN 0308-8197

Liu, J. R.; Chen, M. J. & Lin, C. W. (2005) Antimutagenic and Antioxidant properties of Milk-Kefir and Soymilk-Kefir. *Journal of Agricultural and Food Chemistry*, Vol.53, No.7, pp. 2467-2474, ISSN 0021-8561

Liu, J. R.; Wang, S. Y.; Chen, M. J.; Chen, H. L.; Yueh, P. Y. & Lin, C. W. (2006a). Hypocholesterolaemic Effects of Milk-Kefir and Soyamilk-Kefir in Cholesterol-Fedhamsters. *British Journal of Nutrition*, Vol.95, No.5, pp. 939-946, ISSN 0007-1145

Liu, J. R.; Wang, S. Y.; Chen, M. J.; Yueh, P. Y. & Lin, C. W. (2006b). The Anti-Allergenic Properties of Milk Kefir and Soymilk Kefir and Their Beneficial Effects on the Intestinal Microflora. *Journal of Science and Food and Agriculture*, Vol.86, No.15, pp. 2527-2533, ISSN 0022-5142

Liu, J. R.; Wang, S. Y.; Lin, Y. Y. & Lin, C. W. (2002). Antitumor Activity of Milk Kefir and Soy Milk. *Nutrition and Cancer*, Vol.44, No.2, pp. 182-187, ISSN 1532-7914

Lopitz-Otsoa, F.; Rementeria, A. Elguezabal, N. & Garaizar, I. (2006). Kefir: A Symbiotic Yeast-Bacteria Community with Alleged Healthy Capabilities. *Revista Iberoamericana Micologia*, Vol.23, No.2, pp. 67-74, ISSN 1130-1406

Loretan, T.; Mostert, J. F. & Viljoen, B. C. (2003). Microbial Flora Associated with South African Household Kefir. *South African Journal of Science*, Vol.99, pp. 92-94, ISSN 0038-2353

Mainville, I.; Robert, N.; Lee, B. & Farnworth, E. R. (2006). Polyphasic Characterization of the Lactic Acid bacteria in Kefir. *Systematic and Applied Microbiology*, Vol.29, No.1, pp. 59-68, ISSN 0723-2020

Marison, I.; Liu, J. S.; Ampuero, S.; Von Stockar, U. & Schenker, B. (1998). Biological Reaction Calorimetry: Development of High Sensitivity Bio-Calorimeters. *Thermochimica Acta*, Vol.309, No.1-2, pp. 157-173, ISSN 0040-6031

Marshall, V. M. (1993). Starter Cultures for Milk Fermentation and Their Characteristics. *Journal of Society of Dairy Technology*, Vol.46, No.2, pp. 49-56, ISSN 0037-9840

Moghaddam, J.; Sarraf-Mamoory, R.; Yamini, Y. & Abdollahy, M. (2005). Determination of the Optimum Conditions for the Leaching of Nonsulfide zinc ores (High-SiO_2) in Ammonium Carbonate Media. *Industrial & Engineering Chemistry Research*, Vol.44, No.24, pp. 8952-8958, ISSN 0888-5885

Nataraj, M.; Arunachalam, V. P. & Dhandapani, N. (2005). Optimizing Diesel Engine Parameters for Emission Reduction Using Taguchi Method: Variation Risk Analysis Approach - Part II. *Indian Journal of Engineering and Materials Sciences*, Vol.12, No.6, pp. 505-514, ISSN 0971-4588

Ostlie, H. M.; Treimo, J. & Narvhus, J.A. (2005). Effect of Temperature on Growth and Metabolism of Probiotic Bacteria in Milk. *International Dairy Journal*, Vol.15, No.10, pp. 989-997, ISSN 0958-6946

Otles, S. & Cagandi, O. (2003) Kefir: A Probiotic Dairy-Composition, Nutritional and Therapeutic Aspects. *Pakistan Journal of Nutrition*, Vol.2, No.2, pp. 54-59, ISSN 1680-5194

Plessas, S. Pherson, L.; Bekatorou, A.; Nigam P. & koutinas, A. A. (2005). Bread Making Using Kefir Grains as Baker's Yeast. *Food Chemistry*, Vol.93, No.4, pp. 585-589, ISSN 0308-8146

Prasad, K. K.; Mohan, S. V.; Rao, R. S.; Pati, B. R. & Sarma, P. N. (2005). Laccase Production by *Pleurotus Ostreatus 1804*: Optimization of Submerged Culture Conditions by Taguchi DoE Methodology. *Biochemical Engineeing Journal*, Vol.24, No.1, pp. 17-26, ISSN 1369-703X

Ranjit, K. R. (1990). *A Primer on the Taguchi Method*, Van Nostrand Reinhold, ISBN: 0-87263-468-X, New York, Unites States of America

Rimada, P. S. & Abraham, A. G. (2003). Comparative Study of Different Methodologies to Determine the Exopolysaccharide Produced by Kefir Grains in Milk and Whey, *Le Lait*, Vol.83, No.1, pp. 79-87, ISSN 0023-7302

Rimada, P. S. & Abraham, A. G. (2001). Polysaccharide Production by Kefir Grains during Whey Fermentation, *Journal of Dairy Research*, Vol.68, No.4, pp. 653-661, ISSN 0022-0299

Rodriguez, K. L.; Caputo, L. R. G.; Carvalho, J. C. T.; Evangelista, J. & Schneeforf, J. M. (2005). Antimicrobial and Healing Activity of Kefir and Kefiran Extract. *International Journal of Antimicrobial Agents*, Vol.25, No.5, pp. 404-408, ISSN 0924-8579

Salminen, S.; Ouwehand, A.; Benno, Y. & Lee, Y. K. (1999). Probiotics: How Should be They Defined. *Trends in Food Science & Technology*, Vol.10, No.3, pp. 107-110, ISSN 0924-2244

Saudagar, P. S. & Singhal, R. S. (2007). Optimization of Nutritional Requirements and Feeding Strategies for Clavulanic Acid Production by Streptomyces Clavuligerus. *Bioresource Technology*, Vol.98, No.10, pp. 2010-2017, ISSN 0960-8524

Shojaosadati, S. A. & Babaeipour, V. (2002). Citric Acid Production from Apple Pomace in Multi-Layer Packed Bed Solid-State Bioreactor. *Process Biochemistry*, Vol.37, No.8, pp. 909-914, ISSN 1359-5113

Schoevers, A. & Britz, T. J. (2003). Influence of Different Culturing Conditions on Kefir Grain Increase, *International Journal of Dairy Technology*, Vol.56, No.3, pp. 183-187, ISSN 1364-727X

Simova, E.; Beshkova, D.; Angelov, A.; Hristozova, T.; Frengova, G. & Spasov, Z. (2002). Lactic Acid Bacteria and Yeasts in Kefir Grains and Kefir Made from Them.

Journal of Industrial Microbiology & Biotechnology, Vol.28, No.1, pp. 1-6, ISSN 1367-5435

Taguchi, G. (1987). *System of experimental design*, UNIPUB/Kraus International Publications, ISBN: 0-527-91621-8, New York, Unites States of America

Takizawa, S.; Kojima, S.; Tamura, S.; Fujinaga, S.; Benno, Y. & Nakase, T. (1998). The Composition of the *Lactobacillus* Flora in Kefir Grains, *Systematic and Applied Microbiology*, Vol.21, No.1, pp. 121-127, ISSN 0723-2020

Tamine, A. Y.; Muir, D. D. & Wszolek, M. (1999). Kefir, Koumiss and Kishk. *Dairy Industries International*, Vol.64, No.5, pp. 32-3, ISSN 0308-8197

Thoreux, K. & Schmucker, D. L. (2001). Kefir Milk Enhances Interstinal Immunity in Young but not Old Rats. *The Journal of Nutrition*, Vol.131, No.3, pp. 807-812, ISSN 0022-3166

Vancanneyt, M.; Mengaud, J.; Cleenwerck, I.; Vanhonacker, K.; Hoste, B.; Dawyndt, P.; Degivry, M. C.; Ringuet, D.; Janssens, D. & Swings, J. (2004). Reclassification of *Lactobacillus Kefirgranum* Takizawa et al. 1994 as *Lactobacillus Kefiranofaciens* Subsp. *Kefirgranum* Subsp. Nov. and Emended Description of *L. Kefiranofaciens* Fujisawa et al.. *International Journal of Systematic and Evolutionary Microbiology*, Vol.72, pp. 551-556, ISSN 1466-5026

Vinderola, C. G.; Duarte, J.; Thangavel, D.; Perdigon, G.; Farnworth, E. & Matar, C. (2005). Immunomodulating Capacity of Kefir. *Journal of Dairy Research*, Vol.72, No.2, pp. 195-202, ISSN 0022-0299

Witthuhn, R. C.; Schoeman, T. & Britz, T. J. (2005) Characterisation of the Microbial Population at Different Stages of Kefir Production and Kefir Grain Mass Cultivation. *International dairy Journal*, Vol.15, No.4, pp. 383-389, ISSN 0958-6946

Witthuhn, R. C.; Schoeman, T. & Britz, T. J. (2004). Isolation and Characterization of the Microbial Population of Different South African Kefir Grains. *International Journal of Dairy Technology*, Vol.57, No.1, pp. 33-37, ISSN 1364-727X

11

Recent Advances in Yeast Biomass Production

Rocío Gómez-Pastor[1,2], Roberto Pérez-Torrado[2],
Elena Garre[1] and Emilia Matallana[1,2]
[1]*Departamento de Bioquímica y Biología Molecular, Universitat de València.*
[2]*Departamento de Biotecnología, Instituto de Agroquímica y Tecnología de Alimentos, Spain*

1. Introduction

Yeasts have been used by humans to produce foods for thousands of years. Bread, wine, sake and beer are made with the essential contribution of yeasts, especially from the species *Saccharomyces cerevisiae*. The first references to humans using yeasts were found in Caucasian and Mesopotamian regions and date back to approximately 7000 BC. However, it was not until 1845 when Louis Pasteur discovered that yeasts were microorganisms capable of fermenting sugar to produce CO_2 and ethanol. Ancient practices were based on the natural presence of this unicellular eukaryote, which spontaneously starts the fermentation of sugars. As industrialisation increased the manufacture of fermented products, the demand of yeast grew exponentially. At the end of the 19th century, addition of exogenous yeast biomass to produce bread and beer started to become a common practice. Wineries were more reluctant to alter traditional practices, and started using exogenous yeast inocula in the 1950's, especially in countries with less wine tradition (USA, South Africa, Australia and New Zealand). In the 1960's, yeast biomass-producing plants contributed to the technology of producing large amounts of active dry yeast (ADY), and its use rapidly spread to European countries (Reed and Nagodawithana, 1988).

Nowadays, modern industries require very large amounts of selected yeasts to obtain high quality reproducible products and to ensure fast, complete fermentations. Around 0.4 million metric tonnes of yeast biomass, including 0.2 million tonnes baker's yeast alone, are produced each year worldwide. Efficient and profitable factory-scale processes have been developed to produce yeast biomass. The standard process was empirically optimised to obtain the highest yield by increasing biomass production and decreasing costs. However in recent years, several molecular and physiological studies have revealed that yeast undergoes diverse stressful situations along the biomass production process which can seriously affect its fermentative capacity and technological performance.

In this chapter, we review the yeast biomass production process, including substrates, growth configuration, yield optimisation and the particularities of brewing, baker- or wine-yeasts production. We summarise the new studies that describe the process from a molecular viewpoint to reveal yeast responses to different stressful situations. Finally, we highlight the key points to be optimised in order to obtain not only high yields, but also the best biomass fermentative efficiency, and we provide future directions in the field.

2. Molasses: A suitable substrate

Beet or cane molasses are the main substrate used in yeast production plants. These materials were selected for two main reasons: first, yeasts grow very well using the sugars present in the molasses and second, they are economically interesting since they are a waste product coming from sugar refineries without any other application. Usually, molasses contain between 65% and 75% of sugars, mainly sucrose (Hongisto and Laakso, 1978); but the composition is highly variable depending on the sucrose-refining procedure and on the weather conditions of that particular year. Sucrose is extracellularly hydrolysed by yeasts in two monosaccharides, glucose and fructose, which are transported to and incorporated into the yeast metabolism as carbon sources. However, molasses are deficient in other essential elements for yeast growth. One of them is nitrogen since its molasses content is very poor (less than 3%). Yeasts can use some of the amino acids present in molasses, but addition of nitrogen sources is needed, generally in the form of ammonium salts or urea. Magnesium and phosphate elements are also supplemented in salt forms. Finally, three vitamins (biotin, thiamine and pantothenic acid), required for fast growth, must be supplemented since their content in molasses is also very low (Oura, 1974; Woehrer and Roehr, 1981). Another negative aspect of molasses being used as a substrate to produce yeasts is the presence of different toxics that can affect yeast growth. Variable amounts of herbicides, insecticides, fungicides, fertilizers and heavy metals applied to beet or cane crops can be found in molasses and in different stocks. Moreover bactericides, which are added during sugar production in refinery plants, can be found (Reed and Nagodawithana, 1988). All these toxics can decrease yeast performance by inhibiting growth (Pérez-Torrado, 2004). In fact, a common practice in yeast plants is to mix different stocks to dilute potential toxics.

The effects of molasses composition on yeast growth have been recently analysed at molecular level by determining the transcriptional profile of yeast growing in beet molasses and by comparing it to complete synthetic media (Shima *et al.*, 2005). The results revealed that yeast displays clear gene expression responses when grown in industrial media because of the induction of *FDH1* and *FDH2* genes to detoxify formate and the *SUL1* expression as a response to low sulphate levels. Thus it can be concluded that molasses are far from being an optimal substrate for yeast growth. Another interesting conclusion drawn is that molecular approaches can be especially suited to gain insight into the yeast biomass production process.

In the last years, the price of molasses has increased because of their use in other industrial applications such as animal feeding or bioethanol production (Arshad *et al.*, 2008; Kopsahelis *et al.* 2009; Xandé *et al.*, 2010), thus rendering the evaluation of new substrates for yeast biomass propagation a trend topic for biomass producers' research. New assayed substrates include molasses mixtures with corn steep liquor (20:80), different agricultural waste products (Vu and Kim, 2009) and other possibilities as date juice (Beiroti and Hosseini, 2007) or agricultural waste sources, also called wood molasses, that can be substrate only for yeast species capable of using xylose as a carbon source.

3. Scaling up: Bach and fed-bach

Nowadays, yeast biomass propagation of wine, distiller's and brewer's yeasts are usually produced in baker's yeast plants. The procedure is designed as a multistage-based fermentation, previously defined for the production of baker´s yeast (Chen and Chiger, 1985; Reed and Nagodawithana, 1991) using supplemented molasses as growth media. The first stage (F1) is initiated with a flask culture containing molasses, which is inoculated with the selected yeast strain. Production cultures may be periodically renewed from the stock cultures maintained under more stringent control procedures in a central quality control laboratory. Then, the initial culture is used to inoculate the first fermentor, and cells grow in various transient stages during the batch (F2-F4) and fed-batch (F5-F6) phases of the process. In a sequence of consecutive fermentations, the yeast biomass grown in small fermentors is used to inoculate larger tanks (Reed, 1982; Chen and Chiger, 1985; Reed and Nagodawithana, 1991; Degre, 1993).

In the initial batch phase (F2), cells are exposed to the high sugars concentration present in molasses. All the other nutrients are also present in the fermentor, and pH must be adjusted to 4.5-5.0 after sterilisation to be then monitored during batch fermentation. Once the batch phase has started, the only controllable parameters are temperature and aeration. Yeast propagation typically involves continuous aeration or oxygenation, but a relatively short aeration period has been suggested to suffice (Maemura *et al.*, 1998). However the presence of O_2 from the beginning of the process allows yeast cells to synthesise lipids, thereby revitalising the sterol-deficient cell population and ensuring that fermentation can proceed efficiently. Besides, those propagation experiments carried out in non-oxygenated media considerably reduce yeast growth and increase internal oxidative stress (Boulton, 2000; Pérez-Torrado *et al.*, 2009).

During batch fermentation (F2-F4), a growth lag phase takes place in which cells synthesise the enzymes involved in gluconeogenesis and the glyoxylate cycle (Haarasilta and Oura, 1975). During the subsequent exponential phase, a very small amount of glucose is oxidised in the mitochondria, but when the sugar concentration drops below a strain-specific level or the specific growth rate in aerobic cultures exceeds a critical value (μ_{crit}), a mixed respiro-fermentative metabolism occurs. This phenomenon has been described as the "Crabtree effect" (De Deken, 1966; Pronk *et al.*, 1996) and was originally considered a consequence of the catabolite repression and limited respiratory capacity of *S. cerevisiae* (Postma *et al.*, 1989; Alexander and Jeffries, 1990).It has also been suggested that there is no limitation in the respiratory capacity, as can be deduced from the increased respiratory capacity displayed by a PGK-overproducing mutant, indicating that the activity of respiration itself is not saturated and suggesting that it is not the main cause triggering ethanol production and inducing the long-term Crabtree effect (Van der Aar *et al.*, 1990). However, more recent works have showed that Crabtree effect is derived from the limited mitochondrial capacity to absorb the NADH produced in the glycolysis (Vemuri et al., 2007).

Alcoholic fermentation leads to a suboptimal biomass concentration because the ATP yield is much lower than the yield obtained during respiratory carbohydrate degradation (Verduyn, 1991; Rizzi *et al.*, 1997). However, pre-adaptation to large amounts of glucose during the batch phase is necessary to ensure the produced biomass' optimal fermentative capacity by accumulating several necessary reserve metabolites to be used in the fed-batch phase (Dombek and Ingram, 1987; Rizzi *et al.*, 1997; Pérez-Torrado *et al.*, 2009). In addition, prolonged growth in aerobic, glucose-limited chemostat cultures of *S. cerevisiae,* avoiding the batch phase, causes a partial loss of glycolytic capacity (Jansen *et al.*, 2005). The presence of O_2 during the process also allows yeast to oxidise alcoholic fermentation-produced ethanol when sucrose is exhausted, which triggers the metabolism to change from fermentation to respiration, and eliminates ethanol from the media. When ethanol is exhausted, the fed-batch phase starts (F5-F6). In the transition to the respiratory phase, an increase in the cAMP levels triggers the breakdown of storage carbohydrates and an increased influx of glucose into the glycolytic pathway. The resulting increase in the $NAD^+/NADH$ ratio stimulates respiration in combination with a drop in the ATP level, which is consumed mainly during biomass formation (Pérez-Torrado, 2004; Xu and Tsurugi, 2006; Pérez-Torrado *et al.*, 2009). In some industrial wine yeast production plants, fed-batch phases are initiated without consuming ethanol from the growth media, which considerably reduces the biomass yield.

Optimisation of biomass productivity requires an increase in both the specific growth rate and the biomass yield during the fed-batch phase to the highest values possible under sugar-limited cultivation. Generally, the growth rate profile during fed-batch cultivation is controlled primarily by the carbohydrate feedstock feed rate (Beudeker *et al.*, 1990). The control of optimum dissolved oxygen during the fed-batch phase is also essential to obtain a high biomass yield, and important studies have been done to optimise aeration control (Blanco *et al.*, 2008). Therefore sugar-limited cultivation in the presence of O_2 allows the full respiratory growth of *S. cerevisiae,* achieving much higher biomass yields than during the batch phase (Postma *et al.*, 1989). If the only objective is to maximise the biomass concentration starting with a sufficiently concentrated inoculum from the batch phase, it is necessary to grow cells at a rate as close to the critical growth rate as possible (μ_{crit}), which depends exclusively on the yeast strain (Valentinotti *et al.*, 2002), avoiding ethanol and acetate formation. Many of the parameters that have an impact on yeast's metabolic activities have to be controlled (Miskiewicz and Borowiak, 2005). The pH and temperature are important parameters to be controlled during this phase: maintaining pH constantly at around 4.5 by adjusting the pH automatically with acid/base solutions, and maintaining temperature at 30°C. Properly designed final fed-batch fermentations should also permit yeast cells maturation. This can be accomplished by stopping the feeding of nutrients at the end of fermentation, but allowing slight aeration to continue for an hour (Oura *et al.*, 1974). During this period, the substrate is completely assimilated and allows ripened cells to become more stable and avoids autolysis.

Many research efforts have focused on optimising fed-batch processes for baker´s yeast production with different aims (productivity, yeast quality, or energy saving) and most have been commonly done under laboratory conditions (Van Hoek *et al.*, 1998; Van Hoek *et al.*, 2000; Jansen *et al.*, 2005; Henes and Sonnleitner, 2007; Cheng *et al.*, 2008), but rarely under pilot plant conditions (Di Serio *et al.*, 2001; Lei *et al.*, 2001; Gibson *et al.*, 2007; Gibson *et al.*, 2008). They have all been designed to mainly analyse the fed-batch phase without considering the whole process. The first published study on the complete industrial process was the simulation of wine yeast biomass propagation by performing batch and fed-batch phases in only one bioreactor (Pérez-Torrado *et al.*, 2005). This simplification of the process enabled the study of yeast physiology from a molecular point of view with a bench-top design (Fig. 1), whose results display a good correlation with those obtained from pilot plants and this set of parameters for further investigation.

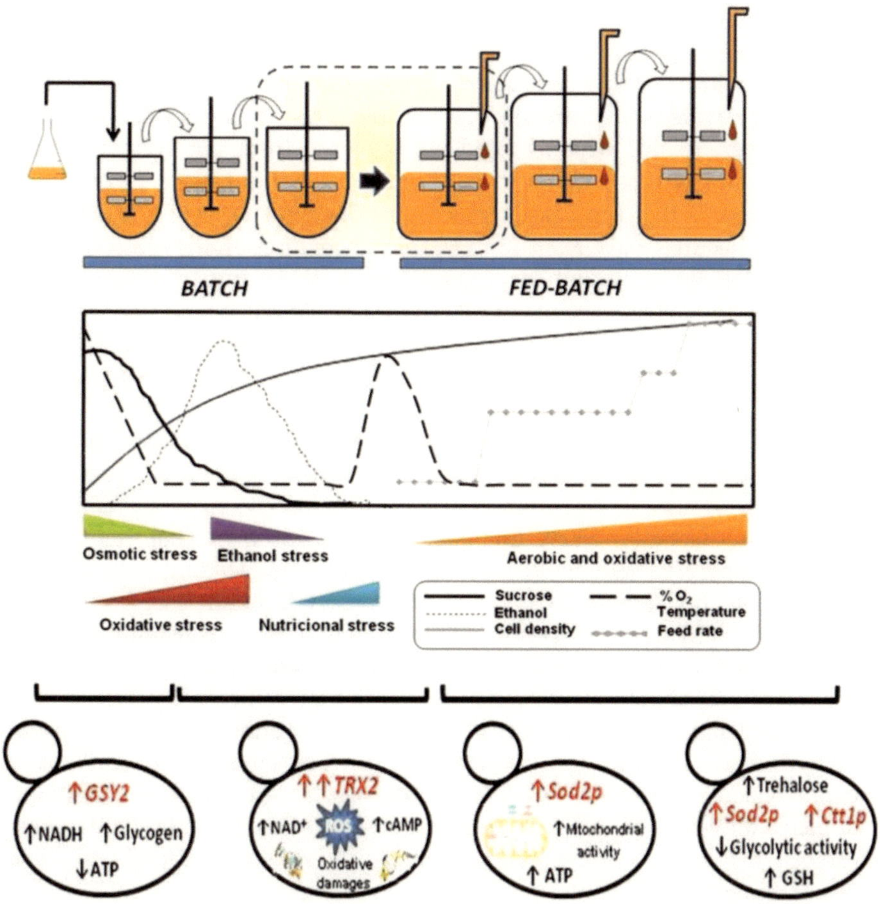

Fig. 1. Diagram of the different stages in the industrial yeast biomass propagation process.

The parameters employed throughout the process (sucrose and ethanol production / consumption, dissolved O_2, cell density and feed rate) have been adapted from Gómez-Pastor *et al.*, 2010b. The lower panel shows representative cellular states, along with the most relevant metabolites, proteins and gene expressions throughout biomass propagation.

4. Desiccation of wine yeasts

In contrast to baker's and brewer's yeast, seasonal wine production requires the development of highly stable dry yeast products. At the end of biomass propagation, wine yeast cells are recovered and dehydrated to obtain ADY (Chen and Chiger, 1985; Degre, 1993; Gonzalez *et al.*, 2005). After the maturation step, yeast cells are separated from fermented media by centrifugation, and are subjected to washing separations to reduce non-yeast solids, a necessary step because they affect the proper rehydration process of ADY for must fermentation. The separation process yields a slightly coloured yeast cream containing up to 22% yeast solids. After this step, the yeast cream can be stored at 4°C after adjusting the pH to 3.5 to avoid microbial contaminations. The cream yeast is further dehydrated to 30-35% solids by means of rotary vacuum filters or filter presses. The filtered yeast is usually mixed with emulsifiers prior to its extrusion into yeast strands. The yeast cake is extruded through a perforated plate, while particles are loaded into the dryer and dehydrated to obtain a product with very low residual moisture. Although several types of dryers exist (roto-louvre, belt dryers, spray dryers), the one most commonly used in industry is the fluidized-bed dryer. In this dryer, heated air is blown from the bottom through yeast particles at velocities which keep them in suspension. Air is treated to reduce its water content and to ensure that the yeast temperature does not exceed 35°C or 41°C during drying. Drying times may vary from 15 to 60 min depending on the mass volume and the used conditions. Finally, ADY with less than 8% residual moisture is vacuum-packaged or placed in an inert atmosphere, such as nitrogen and CO_2, to reduce oxidation. Depending on the strain, loss of viability is estimated at between 10% and 25% per year at 20°C. For this reason, manufacturers recommend storing ADY at 4°C in a dry atmosphere for a maximum 3-year period.

In order to produce an ADY product with acceptable fermentative activity and storage stability, several factors must be taken into account. The drying temperature and rate can be critical for yeast resistance to dehydration and rehydration (Beney *et al.*, 2000; Beney *et al.*, 2001; Laroche and Gervais, 2003). Some studies have shown that cell death during desiccation is strongly related to membrane integrity loss, leading to cell lysis during rehydration (Beney and Gervais, 2001; Laroche *et al.*, 2001; Simonin *et al.* 2007; Dupont *et al.*, 2010). A gradual dehydration kinetics, which allows a slow water efflux through the plasmatic membrane and homogenous desiccation, followed by a progressive rehydration during the starter preparation, have been related with high cell viability (Gervais *et al.*, 1992; Gervais and Marechal, 1994¸ Dupont *et al.*, 2010). The amount of cell constituents leaked during rehydration can also be reduced by adding emulsifiers, such as sorbitan monostearate (Chen and Chiger, 1985). Moreover, biomass propagation conditions have a major influence on yeast resistance to dehydration-rehydration. Several cultivation factors can affect cell resistance to desiccation, such as the substrate, growth phase and ion availability (Trofimova *et al.*, 2010).

5. Yeast stress along biomass production

Several classic studies have evaluated the energy, kinetic and yield parameters of the yeast biomass production process (Reed, 1982; Chen and Chiger, 1985; Reed and Nagodawithana, 1991; Degre, 1993). However, the biochemical and molecular aspects of yeast adaptation to adverse industrial growth conditions have been poorly characterised. In recent years, a substantial effort has been made to gain insight into yeast responses during the process. It was believed that industrial conditions were optimised to obtain the best performing yeast cells, but now we know that yeast cells endure several stressful situations that induce multiple intracellular changes and challenge their technological fitness (Attfield, 1997; Pretorius, 1997; Pérez-Torrado *et al.*, 2005). With wine yeast, moreover, the biomass is concentrated and dehydrated at the end of the process to obtain ADY yeasts that can be stored for long periods of time (Degre, 1993). Subsequently in a period of several hours during maturation and final drying processing, cells undergo nutrient limitation and a complex mixture of different stresses (thermic, osmotic, oxidative, etc.) (Garre *et al.*, 2010). As a result, these dynamic environmental injuries seriously affect biomass yield, fermentative capacity, vitality, and cell viability (Attfield, 1997; Pretorius, 1997; Pérez-Torrado *et al.*, 2005; Pérez-Torrado *et al.*, 2009).

Eukaryotic cells have developed molecular mechanisms to sense stressful situations, transfer information to the nucleus and adapt to new conditions (Hohmann and Mager, 1997; Estruch, 2000; Hohmann, 2002). Protective molecules are rapidly synthesised in stressful situations and transcriptional factors are activated, thus changing the transcriptional profile of cells. Many stress response genes are induced under several adverse conditions through sequence element STRE (stress-responsive element), which targets the main transcriptional factors Msn2p and Msn4p (Kobayashi and McEntee, 1993; Martinez-Pastor *et al.*, 1996). This pathway, also known as the "general stress response pathway", increases the expression of many different genes, including the well-studied *HSP12* and *GSY2* genes involved in protein folding and glycogen metabolism, respectively (Boy-Marcote *et al.*, 1998; Estruch, 2000). Furthermore, yeast cells have been seen to respond specifically to certain stresses. During thermal stress, transcriptional factor Hsf1p activates the transcription of genes, such as *STI1*, which code for those proteins that counteract protein denaturation and aggregation (Lindquist and Craig, 1988; Sorger, 1991). Aerobic growth during biomass propagation and pro-oxidants also generate reactive oxygen species (ROS), leading to several types of oxidative damage to cells (Gómez-Pastor *et al.*, 2010a). To neutralise the harmful effects of oxidative stress, proteins are generated, and they participate in two major functions: antioxidants (such as *GSH1, TRX2, CUP1*, and *CTT1*) to reduce proteins and eliminate ROS damage, and metabolic enzymes (such as *PMG1* and *TDH2*) that redirect metabolic fluxes to synthesise NADPH by slowing down catabolic pathways like glycolysis (Godon *et al.*, 1998). Another well-known specific stress response is the high-osmolarity glycerol response pathway (Brewster *et al.*, 1993), which induces the genes involved in glycerol synthesis (*GPD1, GPP2*) and methylglyoxal detoxification (*GLO1*). Intracellular accumulation of glycerol counteracts hyperosmotic pressure to avoid water loss (Hohmann, 2002). There are other stress response pathways that remain poorly understood, such as those involved in the adaptation to nutrient starvation. Large groups of well-known stress response genes and other genes with unknown functions, such as *YPG1*, are induced after exposure to one kind of stress, and are also involved in the protective mechanism against other different stresses, a phenomenon known as cross-protection (Coote *et al.*, 1991; Piper, 1995; Trollmo *et al.*, 1988; Varela *et al.*, 1992; Bauer and Pretorius, 2000). The molecular responses of laboratory *S. cerevisiae* strains to different stresses have been thoroughly studied, and a large body of knowledge is available (Gasch and Werner-Washburne, 2002; Hohmann and Mager, 2003). In addition, several approaches for the characterisation of stress responses under industrial conditions have been carried out for wine and lager yeasts (Pérez-Torrado *et al.*, 2005; Gibson *et al.*, 2007), and some correlations have been found between stress resistance of several yeast strains and their suitability for industrial processes (Beudeker *et al.*, 1990; Ivorra *et al.*, 1999; Aranda *et al.*, 2002; Pérez-Torrado *et al.*, 2002; Zuzuarregui *et al.*, 2005; Pérez-Torrado *et al.*, 2009; Gómez-Pastor *et al.*, 2010a). For these reasons, the study of stress responses under industrial conditions has become an important research field to improve our knowledge of not only complex industrial processes, but of yeast capabilities.

Given the antiquity of yeast fermentation processes, these microorganisms have evolved in natural stressing environments, which have favoured the selection of "domesticated" yeast that displays high stress resistance (Jamieson, 1998). Studies of brewing yeast under industrial fermentations have demonstrated the suitability of the marker gene expression as a tool to study yeast stress responses in industrial processes (Higgins *et al.*, 2003a). Monitoring stress-related marker genes, such as *HSP12, GPD1, STI1, GSY2* and *TRX2*, during bench-top growth trials of wine yeast biomass propagation have demonstrated that osmotic (*GPD1*) and oxidative stresses (*TRX2*) are the main adverse conditions that *S. cerevisiae* senses during this process (Pérez-Torrado *et al.*, 2005). Afterwards, a genome-wide expression analysis of the same process established stress-critical time points throughout the process based on the profiles of different oxidative stress response genes (Gómez-Pastor *et al.*, 2010b). Three relevant stressful points have been defined during biomass propagation: the first during the metabolic transition from fermentation to respiration in the batch phase; the second critical point is the end of the batch phase when previously produced ethanol is completely consumed; the third interesting point is the end of the fed-batch phase, after a long period under respiratory metabolism. Among these set points, metabolic transition during the batch phase is the most relevant as several genes relating to cell stress, especially those related to oxidative stress (*TRX2, GRX2* and *PRX1*), protein degradation, aerobic respiration and NADPH production, are induced while ribosomal proteins are dramatically repressed (Gómez-Pastor *et al.*, 2010b). Similar results have been observed in a genome-wide expression analysis during biomass propagation of brewer's yeasts , which also displays a strong induction of the genes involved in ergosterol biosynthesis and oxidative stress protection in initial industrial lager fermentation stages (Higgins *et al.*, 2003b; reviewed in Gibson *et al.*, 2007; Gibson *et al.*, 2008). However, while osmotic stress plays a role in initial biomass propagation stages as a result of the large amount of sugar in molasses, oxidative stress takes place throughout the process as a result of aeration (reviewed in Gibson *et al.*, 2007).

As mentioned earlier, an oxygen supply is necessary to generate yeast biomass and to ensure optimal physiological conditions for effective fermentation (Chen and Chiger, 1985; Reed and Nagodawithana, 1991; Hulse, 2008). Oxygen is required for lipid synthesis, which is necessary to maintain plasma membrane integrity and function, and consequently for both cell replication and the biosynthesis of sterols and unsaturated fatty acids. Despite its potential toxicity, eliminating oxygen in the first part of the batch phase diminishes biomass yield (Boulton *et al.*, 2000; Pérez-Torrado *et al.*, 2009) and avoids the expression of those genes related to oxidative stress response, such as *TRX2* and *GRE2,* which significantly increases oxidative cellular damage, such as lipid peroxidation, when the bioreactor is re-oxygenated to oxidise ethanol (Pérez-Torrado *et al.*, 2009). Clarkson *et al.* (1991) demonstrated that cellular antioxidant defences, such as Cu/Zn superoxide dismutase, Mn superoxide dismutase and catalase activities of brewing yeast strains, also change rapidly after adding or removing O_2 from fermentation.

During an industrial-scale propagation of wine and brewing yeasts, catalase and Mn superoxide dismutase activities increase as propagation proceeds (Martin *et al.*, 2003; Gómez-Pastor *et al.*, 2010a), indicating the importance of oxidative stress response throughout the process, whereas Sod1p (Cu/Zn superoxide dismutase) transiently accumulates at the end of the batch phase when ethanol is consumed (Gómez-Pastor *et al.*, 2010a). A study of different types of oxidative damage during wine yeast biomass propagation has revealed that lipid peroxidation considerably increases during the metabolic transition from fermentation to respiration, which decreases to basal levels during the fed-batch phase (Gómez-Pastor *et al.*, 2010a). Besides, the protein carbonylation analysis, one of the most important oxidative damages (Stadtman and Levine, 2000), has revealed different protein oxidation patterns during biomass propagation, which reach maximum global carbonylation levels at the end of the batch phase (Gómez-Pastor *et al.*, 2010a). As protein oxidation causes the loss of catalytic or structural integrity, further research into the specific oxidised proteins during biomass production should be done to correlate the detriment in fermentative capacity with specific damaged proteins. In addition, reduced glutathione, an important antioxidant molecule, varies during the process as is lowers during the metabolic transition, while oxidised glutathione increases. Then, reduced glutathione increases constantly in different stages of the process (Gibson *et al.*, 2006; Gómez-Pastor *et al.*, 2010a). Whether glutathione is directly affected by O_2 during biomass propagation remains unknown and requires further investigation.

The fed-bath phase is characterised by the accumulation of other important antioxidant molecules, such as trehalose and thioredoxin (Trx2p) (Pérez-Torrado, 2004; Gómez-Pastor, 2010), although the mRNA levels for the *TRX2* gene significantly increase during the batch phase metabolic transition (Pérez-Torrado *et al*, 2009). On the other hand, glycogen, a secondary long-term energy storage molecule which has been related to adaptation to the respiratory metabolism (Francois and Parrou, 2001), also accumulates at the end of the fed-batch phase (Pérez-Torrado, 2004). Studies using different dilution rates during the continuous cultivation of baker´s yeast have shown that the accumulation of trehalose and glycogen has a negatively effect as it increases dilution rates, which is also detrimental for fermentative capacity and cellular responses to heat stress during dehydration (Ertugay and Hamaci, 1997; Garre *et al.*, 2009). Despite a high biomass yield and the accumulation of several beneficial metabolites obtained during the fed-batch phase, *S. cerevisiae* dramatically diminished fermentative capacity after prolonged glucose-limited aerobic cultivation due to several glycolytic enzymes' diminished activity (Jansen *et al.*, 2005).

Proteomic studies have also been carried out to gain a better understanding of the fluctuations in the stress-related gene mRNA levels during biomass propagation and to correlate glycolytic enzyme activities with their corresponding protein levels. However, the proteomic data available from industrial processes are very limited and usually centre on bioethanol production (Cot *et al.*, 2007; Cheng *et al.*, 2008) or wine and beer fermentations (Trabalzini *et al.*, 2003; Zuzuarregui *et al.*, 2006; Salvadó *et al.*, 2008; Rossignol *et al.*, 2009). Recent proteomic studies performed by 2D-gel electrophoresis during wine yeast biomass propagation have revealed that several glycolytic enzyme isoforms increase during biomass production. This is probably due to the post-translational modifications after oxidative stress exposure (Gómez-Pastor *et al.*, 2010b; Costa *et al.*, 2002). Trabalzini *et al.* (2003) suggested that some specific isoforms of glycolytic/gluconeogenic pathway enzymes in wine strains of *S. cerevisiae* are involved in the physiological adaptation to different fermentation stresses. There have also been reports of the differential stress regulations of several proteins (Arg1p, Sti1p and Pdc1p) among different industrial strains possibly having important industrial implications for strain improvement and protection (Caesar *et al.*, 2007). It is interesting to note that biomass propagation experiments using a *trx2* deletion strain have shown a low number of several glycolytic enzyme isoforms and, consequently, an increase in oxidative cellular damage, such as lipid peroxidation and global protein carbonylation (Gómez-Pastor, 2010). During the metabolic transition in the batch phase, several proteins relating to oxidative stress are expressed (Prx1p, Ahp1p, Ilv5p, Pdi1p, Sod1p and Trr1p), which directly correlates with their mRNA levels observed for this growth stage (Gómez-Pastor *et al.*, 2010b). This scenario indicates adaptation to the new condition. In contrast, the genes coding for most of the heat shock proteins, chaperons (Mge1p, Hsp60p, Ssb1p and Ssc1p) and proteins related to ATP metabolism are specifically induced during the metabolic transition, but their protein levels decline throughout the process. The proteins with the highest expression levels at the end of the biomass propagation include Tdh1p, which codifies for glyceraldehyde-3-phosphate dehydrogenase, and Bmh1p and Bmh2p, homologues to the mammalian 14-3-3 proteins involved in global protein regulation at the post-translational level (Bruckmann *et al.*, 2007). The expression of these proteins at the end of biomass propagation is important as they control the translation of several glycolytic proteins (Fba1p, Eno1p, Tpi1p, Pck1p, Tdh1p, Tdh3p and Gpm1p), as well as the levels of those proteins involved in amino acid biosynthesis and heat shock proteins translation (Bruckmann *et al.*, 2007). This may explain the lack of correlation between the transcriptomic and the proteomic analyses for glycolytic enzymes during biomass propagation. Under oxidative stress, some glycolytic proteins (Tdh3p, Pdc1p, Ad1p and Eno1p) have been described to be specifically modified by oxidation (Le Moan *et al.*, 2006). This oxidation process could explain the loss of fermentative capacity observed in some commercial wine yeast industrial strains at the end of the biomass propagation process (Gómez-Pastor *et al.*, 2010a, b). Regarding this hypothesis, it is worth noting that the overexpression of the *TRX2* gene in industrial yeasts significantly increases the obtained biomass' fermentative capacity by improving the oxidative stress response during propagation, and by decreasing lipid and protein oxidation (Pérez-Torrado *et al.*, 2009; Gómez-Pastor *et al.*, 2010a, c). Figure 1 summarizes the different stresses affecting yeast cells during the biomass propagation process, especially those encountered during the batch phase, and shows the different cellular states with the most relevant metabolites, genes and proteins expressed in each propagation stage.

The industrial yeast biomass dehydration process also involves damaging environmental changes. As the biomass is being concentrated, water molecules are removed and temperature increases, all of which affect the viability and vitality of cells (Matthews and Webb, 1991). Dehydration is known to cause both cell growth arrest and severe damage to membranes and proteins (Potts, 2001; Singh *et al.*, 2005). Removal of water molecules causes protein denaturalisation, aggregation, and loss of activity in an irreversible manner (Prestrelski *et al.*, 1993). Additionally at the membrane level, desiccation is associated with an increased package of polar groups of phospholipids, and with the formation of endovesicles leading to cell lysis during rehydration (Crowe *et al.*, 1992; Simonin *et al.*, 2007). Yeasts have several strategies to maintain membrane fluidity (Beney and Gervais, 2001). One of them is to accumulate ergosterol, this being the predominant sterol in *S. cerevisiae*. Sterols have been proposed to maintain the lateral heterogeneity of the protein and lipid distribution in the plasma membrane because of the putative role they play in inducing microdomains, the so-called lipid rafts (Simons and Ikonen, 1997). Ergosterol synthesis has been related with yeast stress tolerance (Swan and Watson, 1998), and its beneficial role in the different processing steps of industrial yeast has been documented. Its synthesis during biomass production is critical to ensure suitable yeast ethanol tolerance in its later application in wine fermentation (Zuzuarregui *et al.*, 2005). Moreover, the addition of oleic acid and ergosterol during wine fermentation mitigates oxidative stress by reducing not only the intracellular content of reactive oxygen species, but oxidative damage to membranes and proteins, and enhancing cell viability (Landolfo *et al.*, 2010). Recently, experiments with a *erg6*Δ mutant strain, deficient in the ergosterol biosynthetic pathway and which accumulates mainly zymosterol and cholesta-5,7,24-trienol instead of ergosterol, have shown that the nature of sterols affects yeast survival during dehydration, and that

resistance to dehydration-rehydration cycles can be restored with ergosterol supplementation during the anaerobic growth of the *erg6*Δ mutant (Dupont *et al.*, 2010).
Recent phenomic and transcriptomic analyses during the desiccation of a laboratory strain have indicated that this process represents a complex stress involving changes in about 12% of the yeast genome (Ratnakumar *et al.*, 2011). Under these conditions, the induction of 71 genes grouped into the "environmental stress response" category was observed, suggesting a role of the general stress transcription factors Msn2p and Msn4p in the desiccation stress response. Furthermore, the phenomic screen looking for genes that are beneficial to desiccation tolerance has identified several of the transcriptional regulators or protein kinases involved in oxidative (*ATF1, SKN7*) and osmotic (*HAL9, MSN1, MSN2, MSN4, HOG1, PBS2, SSK2*) stress responses. Although studies with lab strains generate interesting information about the desiccation process, an analysis of stress marker genes during dehydration in ADY production has revealed that inductions of gene expressions in wine yeast T73 are generally moderate, although statistically significant, in some steps, such as hot air drying and final product (Garret *et al.*, 2010). One such example is the induction of osmotic stress marker *GPD1* due to water loss. However, despite the yeast biomass losing approximately 95% of water content during this dehydration process, *GPD1* induction is not as important as previously observed in lab yeast strains under osmotic stress (Pérez-Torrado *et al.*, 2002). These data are in agreement with the robustness of industrial yeasts strains compared to laboratory strains (Querol *et al*, 2003), and also with the well-known relevance of biomass propagation conditions to confer resistance to subsequent suboptimal conditions (Bisson *et al.*, 2007). One interesting aspect in the same study carried out by Garre and coworkers (2010) is that the highest induction is displayed by oxidative stress marker *GSH1* that codes for γ-glutamilcysteine synthetase activity. This observation is supported by: i) significant inductions of the other genes involved in oxidative stress response, such as *TRR1* and *GRX5*, ii) rise in the cellular lipid peroxidation level, iii) increased intracellular glutathione accumulation, and iv) a peak of its oxidized form GSSG during the first minutes of drying. In addition, a genomic analysis of an oenological-dried yeast strain has shown a strong induction of the other genes related with oxidative stress response, such as *CTT1, SOD1, SOD2, GTT1* and *GTT2* (Rossignol *et al.*, 2006). Currently, free radical damage is emerging as one of the most important injuries during dehydration. Several studies with laboratory yeast strains have shown considerable ROS accumulation during dehydration that results in protein denaturation, nucleic acid damage and lipid peroxidation (Espindola *et al.*, 2003; Pereira *et al.*, 2003; França *et al.*, 2005, 2007). Antioxidant systems appear to be interesting targets affecting yeast's desiccation tolerance. Several examples using lab strains have been shown. Overexpression of antioxidant enzymes genes, such as *SOD1* and *SOD2*, increases yeast survival after dehydration (Pereira *et al.*, 2003), whereas a mutant without cytosolic catalase activity is more sensitive to water loss (França *et al.*, 2005). Glutathione seems to play a significant role in the maintenance of intracellular redox balance because glutathione-deficient mutant strains are much more oxidised after dehydration than the wild-type strain, and they show high viability loss (Espindola *et al.*, 2003). Furthermore, addition of glutathione to *gsh1* cells restores survival rates to control strain levels. Remarkably, the overexpression of the *TRX2* gene in wine yeast has proved a successful strategy to improve fermentative capacity and to produce lower levels of oxidative cellular damage after dry biomass production than its parental strain (Pérez-Torrado *et al.*, 2009; Gómez-Pastor *et al.*, 2010a).

The accumulation of some metabolites has been related to yeasts' resistance to drying and subsequent rehydration. One of them is the amino acid proline. This amino acid exhibits multiple functions *in vitro*: it enhances the stability of proteins, DNA and membranes, inhibits protein aggregation, and acts as a ROS scavenger; but its functions *in vivo*, particularly as a stress protectant, are poorly understood. Although *S. cerevisiae* cells do not accumulate this amino acid in response to stresses, it has been recently shown with laboratory strains that proline-accumulating mutants are more tolerant than wild-type cells to freezing, desiccation, oxidative, or ethanol stress (reviewed in Takagi, 2008; Kaino and Takagi, 2009). Self-cloning has been used to construct the baker's yeasts that accumulate proline by carrying the disruption of the *PUT1* gene involved in the degradation pathway, and expressing a mutant *PRO1* gene that encodes a less sensitive γ-glutamate kinase to feedback inhibition in order to enhance biosynthetic activity. The engineered yeast strain shows enhanced freeze tolerance in doughs (Kaino *et al.*, 2008). A recent transcriptomic analysis of air-dried cells has suggested activated transport and metabolic processes to increase the intracellular concentration of proline during yeast desiccation (Ratnakumar *et al.*, 2011).

Interestingly, wine yeasts accumulate large amounts of disaccharide trehalose, usually in the 12-20% range of cell dry weight (Degre, 1993) although higher percentages have been detected in industrial stocks (Garre *et al.*, 2010). Trehalose content has been proposed as one of the most important factors to affect dehydration survival. Baker's yeasts with 5% of trehalose are 3 times more sensitive to desiccation than those cells accumulating 20% of trehalose (Cerrutti *et al.*, 2000). The main function of this metabolite is to act as a protective molecule in stress response. This effect can be achieved in two ways: by protecting membrane integrity through the union with phospholipids (reviewed in Crowe *et al.*, 1992); by preserving the native conformation of proteins and preventing the aggregation of partially denatured proteins (Singer and Lindquist, 1998a). The indispensability of this metabolite to survive dehydration is a controversial subject. Some studies have suggested that its presence is essential and needed in both sides of the membrane to confer suitable protection (Eleuterio *et al.*, 1993; Sales et al, 2000). However, these results are argued alongside the *tps1* mutant's dehydration resistance, which is unable to synthesise trehalose, as other authors have indicated (Ratnakumar and Tunnacliffe, 2006). On the other hand, dehydration tolerance conferred by trehalose seems to be also related to its ability to protect cellular components from oxidative injuries (Benaroudj *et al.*, 2001; Oku *et al.*, 2003; Herdeiro *et al.*, 2006; da Costa Morato *et al.*, 2008; Trevisol *et al.*, 2011). The addition of external trehalose during dehydration reduces intracellular oxidation and lipid peroxidationand increases the number of viable cells after dehydration (Pereira *et al.*, 2003). Moreover, the compensatory trehalose accumulation observed in *hsp12*Δ mutants confers a higher desiccation tolerance than the parent wild-type cells, which is the result of increased protection by mutant cells against reactive oxygen species (Shamrock and Lindsey, 2008). Some studies have proved the applicability of this metabolite to improve industrial yeast tolerance to dehydration. A clear and simple example is that of Elutherio and co-workers (1997), where the trehalose accumulation induced by osmotic stress in the species *Saccharomyces uvarum* var. *carlsbergensis* before dehydration is enough to achieve survivals of up to 60% after drying, whereas the stationary cells presenting low trehalose levels are unable to survive. The construction of trehalose-overaccumulating strains by removing degradative activities emerges as a useful strategy for industrial yeasts (Kim *et al.*, 1996). Studies done with laboratory strains have shown that the deletion of genes *ATH1* and *NTH1*, respectively encoding acid and neutral trehalase activity, improve yeast cells viability after dehydration, which is provoked by hyperosmotic stress (Garre *et al.*, 2009). Similar approaches using baker's yeast have also been successful, and defective mutants in neutral or acid trehalase activities exhibit higher tolerance levels to dry conditions than the parent strain, as well as increased gassing power of frozen dough (Shima *et al.*, 1999).

6. Conclusions

In the last few decades, the yeast biomass production industry has contributed with many advanced approaches to traditional technological tools with a view to studying the physiology, biochemistry and gene expression of yeast cells during biomass growth and processing. This has provided a picture of the determinant factors for the commercial product's high yield and fermentative fitness. Cell adaptation to adverse industrial conditions is a key element for good progress to be made in biomass propagation and desiccation, and towards the characterisation of specific stress responses during industrial processes to clearly indicate the main injuries affecting cell survival and growth. One major aspect of relevance in the complex pattern of molecular responses displayed by yeast cells is oxidative stress response, a network of mechanisms ensuring cellular redox balance by minimising structural damages under oxidant insults. Different components of this machinery have been identified as being involved in cellular adaptation to industrial growth and dehydration, including redox protein thioredoxin, redox buffer glutathione and several detoxifying enzymes such as catalase and superoxide dismutase, plus protective molecules like trehalose which play a relevant role in dehydration.

7. Future prospects

In spite of the sound knowledge available on molecular responses to exogenous oxidants, the endogenous origin of oxidative stress in yeast biomass production, given the metabolic transitions required for growth under the described multistage-based fermentation conditions and desiccation, makes it challenging to search for the specific targets undergoing oxidative damage during both biomass propagation and desiccation, and to correlate this damage with physiologically detrimental effects. Based on the currently global data available and the use of potent analytical and genetic manipulation tools, further research has to be conducted to (i) define specific oxidised proteins and to know how this oxidation affects fermentative efficiency, (ii) identify new key elements in stress response, which can be manipulated to improve it and can be also used as markers to select suitable strains for biomass production, (iii) analyse the effects of potential beneficial additives, such as antioxidants, on yeast cells' ability to adapt to stress, and then yeast biomass' yield and fermentative fitness in industrial production processes.

8. Acknowledgement

This work has been supported by grants AGL 2008-00060 from the Spanish Ministry of Education and Science (MEC). E.G. was a fellow of the FPI program of the Spanish Ministry of Education and Science, R.G-P was a predoctoral fellow of the I3P program from the CSIC (Spanish National Research Council).

9. References

Alexander, M.A. & Jeffries, T.W. (1990). Respiratory efficiency and metabolize partitioning as regulatory phenomena in yeasts. *Enzyme Microbe. Technol* Vol. 12, pp. 2-29.

Aranda, A.; Querol, A. & del Olmo, M. (2002). Correlation between acetaldehyde and ethanol resistance and expression of HSP genes in yeast strains isolated during the biological aging of sherry wines. *Arch.Microbiol.*, Vol. 177, No. 4, pp. 304-312.

Arshad, M., Khan, Z.M., Khalil-ur-Rehman, Shah, F.A. & Rajoka, M.I. (2008). Optimization of process variables for minimization of byproduct formation during fermentation of blackstrap molasses to ethanol at industrial scale. *Lett Appl Microbiol.* Vol. 47, No. 5, pp. 410-414.

Attfield, P. V. (1997). Stress tolerance: the key to effective strains of industrial baker's yeast. *Nat.Biotechnol.*, Vol. 15, No. 13, pp. 1351-1357.

Bauer, F.F. & Pretorius, I.S. (2000). Yeast stress response and fermentation efficiency: how to survive the making of wine - A review. *South African Journal of Enology and Viticulture* Vol. 21, pp. 27-51.

Beiroti, A. & Hosseini, S.N. (2007) Production of baker's yeast using date juice. *Sheng Wu Gong Cheng Xue Bao.* Vol. 23, No. 4, pp. 746-50.

Benaroudj, N., Lee, D.H. & Goldberg, A.L. (2001). Trehalose accumulation during cellular stress protects cells and cellular proteins from damage by oxygen radicals. *J.Biol.Chem.*, Vol. 276 No. 26, pp. 24261-24267.

Beney, L., Martinez, d.M., I, Marechal, P.A. & Gervais, P. (2000). Influence of thermal and osmotic stresses on the viability of the yeast *Saccharomyces cerevisiae. Int.J.Food Microbiol.*, Vol. 55, No. 1-3, pp. 275-279.

Beney, L. & Gervais, P. (2001). Influence of the fluidity of the membrane on the response of microorganisms to environmental stresses. *Appl Microbiol Biotechnol.* Vol. 57, No. 1-2, pp. 34-42.

Beney, L., Marechal, P.A. & Gervais, P. (2001). Coupling effects of osmotic pressure and temperature on the viability of *Saccharomyces cerevisiae. Appl.Microbiol.Biotechnol.*, Vol., 56, No. 3-4, pp. 513-516.

Beudeker, R.F.; van Dam H.W.; van der Plaat, J.B. & Vellenga, K. (1990). Developments in bakers' yeast production. In: *Yeast Biotechnology and Biocatalysis* (Verachtert, H. and De Mot, R., Eds.), 103-146, Marcel Dekker, New York, NY.

Bisson, L.F., Karpel,.J. E, Ramakrishnan, V. & Joseph, L. (2007). Functional genomics of wine yeast *Saccharomyces cerevisiae. Adv Food Nutr Res.*, Vol. 53, pp. 65-121.

Blanco, C. A.; Rayo, J. & Giralda, J. M. (2008). Improving industrial full-scale production of baker's yeast by optimizing aeration control. *J.AOAC Int.*, Vol. 91, No. 3, pp. 607-613.

Boulton, C. (2000). Trehalose, glycogen and sterol. In: *Brewing Yeast Fermentation Performance.* 1st ed (SmartKA, Ed), 10-19, Blackwell Science, Oxford, UK.

Boy-Marcote, E.; Perrot, M.; Bussereau, F.; Boucherie, H. & Jacquet, M. (1998). Msn2p and Msn4p control a large number of genes induced at the diauxic transition which are repressed by cyclic AMP in *Saccharomyces cerevisiae. J. Bacteriol* Vol. 180, No.5, pp. 1044-1052.

Brewster, J. L.; de, V. T.; Dwyer, N. D.; Winter, E. & Gustin, M. C. (1993). An osmosensing signal transduction pathway in yeast. *Science*, Vol. 259, No. 5102, pp. 1760-1763.

Bruckmann, A.; Hensbergen, P. J.; Balog, C. I.; Deelder, A. M.; de Steensma, H. Y. & van Heusden, G. P. (2007). Post-transcriptional control of the *Saccharomyces cerevisiae* proteome by 14-3-3 proteins. *J.Proteome.Res.*, Vol. 6, No. 5, pp. 1689-1699.

Caesar, R.; Palmfeldt, J.; Gustafsson, J. S.; Pettersson, E.; Hashemi, S. H. & Blomberg, A. (2007). Comparative proteomics of industrial lager yeast reveals differential expression of the cerevisiae and non-cerevisiae parts of their genomes. *Proteomics.*, Vol. 7, No. 22, pp. 4135-4147.

Cerrutti, P., Segovia de, H.M., Galvagno, M., Schebor, C., & del Pilar, B.M. (2000). Commercial baker's yeast stability as affected by intracellular content of trehalose, dehydration procedure and the physical properties of external matrices. *Appl.Microbiol.Biotechnol.*, Vol. 54, No. 4, pp. 575-580.

Chen, S.L. & Chiger, M. Production of baker's yeast. (1985). In *Comprehensive Biotechnology* ed. Moo-Young, M., Blarch, H.W., Drew, S. and Wang, D.I.C. pp. 429-462. New York: Pergamon Press.

Cheng, J. S.; Qiao, B. & Yuan, Y. J. (2008). Comparative proteome analysis of robust *Saccharomyces cerevisiae* insights into industrial continuous and batch fermentation. *Appl.Microbiol.Biotechnol.*, Vol. 81, No. 2, pp. 327-338.

Clarkson, S.P.; Large, P.J.; Boulton, C.A. & Bamforth, C.W. (1991). Synthesis of superoxide dismutase, catalase and other enzymes and oxygen and superoxide toxicity during changes in oxygen concentration in cultures of brewing yeast. *Yeast* Vol. 7, No. 2, pp. 91–103.

Coote, P. J.; Cole, M. B. & Jones, M. V. (1991). Induction of increased thermotolerance in *Saccharomyces cerevisiae* may be triggered by a mechanism involving intracellular pH. *J.Gen.Microbiol.*, Vol. 137, No. 7, pp. 1701-1708.

Costa, V. M.; Amorim, M. A.; Quintanilha, A. & Moradas-Ferreira, P. (2002). Hydrogen peroxide-induced carbonylation of key metabolic enzymes in *Saccharomyces cerevisiae*: the involvement of the oxidative stress response regulators Yap1 and Skn7. *Free Radic.Biol.Med.*, Vol. 33, No. 11, pp. 1507-1515.

Cot, M.; Loret, M. O.; Francois, J. & Benbadis, L. (2007). Physiological behaviour of *Saccharomyces cerevisiae* in aerated fed-batch fermentation for high level production of bioethanol. *FEMS Yeast Res.*, Vol. 7, No. 1, pp. 22-32.

Crowe, J.H., Hoekstra, F.A. & Crowe, L.M. (1992). Anhydrobiosis. *Annu.Rev.Physiol*, Vol. 54, pp. 579-599.

da Costa Morato, N.D., da Silva, C.G., Mariani, D., Fernandes, P.N., Pereira, M.D., Panek, A.D. & Eleutherio, E.C. (2008). The role of trehalose and its transporter in protection against reactive oxygen species. *Biochim.Biophys.Acta*, Vol. 1780, No. 12, pp. 1408-1411.

De Deken, R. H. (1966). The Crabtree effect: a regulatory system in yeast. *J.Gen.Microbiol.*, Vol. 44, No. 2, pp. 149-156.

Degre, R. (1993). Selection and commercial cultivation of wine yeast and bacteria. In: *Wine Microbiology and Biotechnology* (Fleet GH, Ed.) 421–448. Harwood Academic Publishers, Chur, Switzerland.

Di Serio, M.; Tesser, R. & Santacesaria, E. (2001). A kinetic and mass transfer model to simulate the growth of baker's yeast in industrial bioreactors. *Chem. Eng. J* Vol. 82, pp. 347-354.

Dombek, K. M. & Ingram, L. O. (1987). Ethanol production during batch fermentation with *Saccharomyces cerevisiae*: changes in glycolytic enzymes and internal pH. *Appl.Environ.Microbiol.*, Vol. 53, No. 6, pp. 1286-1291.

Dupont, S., Beney, L., Ritt, J.F., Lherminier, J. & Gervais, P. (2010). Lateral reorganization of plasma membrane is involved in the yeast resistance to severe dehydration. *Biochim.Biophys.Acta*, Vol. 1798, No. 5, pp. 975-985.

Dupont, S., Beney, L., Ferreira, T. & Gervais, P. (2010). Nature of sterols affects plasma membrane behavior and yeast survival during dehydration. *Biochim.Biophys.Acta*.

Eleutherio, E.C., Araujo, P.S. & Panek, A.D. (1993). Role of the trehalose carrier in dehydration resistance of Saccharomyces cerevisiae. *Biochim.Biophys.Acta*, Vol. 1156, No. 3, pp. 263-266.

Eleutherio, E.C., Maia, F.M., Pereira, M.D., Degre, R., Cameron, D. & Panek, A.D. (1997). Induction of desiccation tolerance by osmotic treatment in *Saccharomyces uvarum var. carlsbergensis*. *Can.J.Microbiol.*, Vol. 43, No. 5, pp. 495-498.

Ertugay, N. & Hamamci, H. (1997). Continuous cultivation of bakers' yeast: change in cell composition at different dilution rates and effect of heat stress on trehalose level. *Folia Microbiol.(Praha)*, Vol. 42, No. 5, pp. 463-467.

Espindola, A.S., Gomes, D.S., Panek, A.D. & Eleutherio, E.C. (2003). The role of glutathione in yeast dehydration tolerance. *Cryobiology*, Vol. 47, No. 3, pp. 236-241.

Estruch, F. (2000). Stress-controlled transcription factors, stress-induced genes and stress tolerance in budding yeast. *FEMS Microbiol.Rev.*, Vol. 24, No. 4, pp. 469-486.

França, M.B., Panek, A.D. & Eleutherio, E.C. (2005). The role of cytoplasmic catalase in dehydration tolerance of *Saccharomyces cerevisiae*. *Cell Stress.Chaperones.*, Vol. 10, No. 3, pp. 167-170.

França, M.B., Panek, A.D., & Eleutherio, E.C. (2007). Oxidative stress and its effects during dehydration. *Comp Biochem.Physiol A Mol.Integr.Physiol*, Vol. 146, No. 4, pp. 621-631.

François, J. & Parrou, J. L. (2001). Reserve carbohydrates metabolism in the yeast *Saccharomyces cerevisiae*. *FEMS Microbiol.Rev.*, Vol. 25, No. 1, pp. 125-145.

Garre, E.; Pérez-Torrado, R.; Gimeno- Alcañíz, J. V. & Matallana, E. (2009). Acid trehalase is involved in intracellular trehalose mobilization during postdiauxic growth and severe saline stress in *Saccharomyces cerevisiae*. *FEMS Yeast Res.*, Vol. 9, No. 1, pp. 52-62.

Garre, E.; Raginel, F.; Palacios, A.; Julien, A. & Matallana, E. (2010). Oxidative stress responses and lipid peroxidation damage are induced during dehydration in the production of dry active wine yeasts. *Int.J.Food Microbiol.*, Vol. 136, No. 3, pp. 295-303.

Gasch, A. P. & Werner-Washburne, M. (2002). The genomics of yeast responses to environmental stress and starvation. *Funct.Integr.Genomics*, Vol. 2, No. 4-5, pp. 181-192.

Gervais, P., Marechal, P.A. & Molin, P. (1992). Effects of the kinetics of osmotic pressure variation on yeast viability. *Biotechnol.Bioeng.*, Vol. 40, No. 11, pp. 1435-1439.

Gervais, P. & Marechal, P.A. (1994). Yeast resistance to high-levels of osmotic-pressure-influence of kinetics. *J. Food Engineer*, Vol. 22, pp. 399-407.

Gibson, B. R.; Lawrence, S. J.; Boulton, C. A.; Box, W. G.; Graham, N. S.; Linforth, R. S. & Smart, K. A. (2008). The oxidative stress response of a lager brewing yeast strain during industrial propagation and fermentation. *FEMS Yeast Res.*, Vol. 8, No. 4, pp. 574-585.

Gibson, B.R.; Smith, J.M.; Lawrence, S.J.; Shelton, N.; Smith, J.M. & Smart, K.A. (2006). Oxygen as toxin: oxidative stress and brewing yeast physiology. *Belgian J Brew Biotechnol.*, Vol. 31, No. 1, pp. 25–36.

Gibson, B.R.; Lawrence, S.J.; Leclaire, P.R.; Powell, C.D. & Smart, K.A. (2007). Yeast responses to stresses associated with industrial brewering handling. *FEMS Microbiol. Rev.*, Vol 31, No. 5, pp.535-569.

Godon, C.; Lagniel, G.; Lee, J.; Buhler, J. M.; Kieffer, S.; Perrot, M.: Boucherie, H.; Toledano, M. B. & Labarre, J. (1998). The H_2O_2 stimulon in *Saccharomyces cerevisiae*. *J.Biol.Chem.*, Vol. 273, No. 35, pp. 22480-22489.

Gómez-Pastor R. (2010). Estrés oxidativo en la producción de levaduras vínicas. Implicación del gel TRX2. Universitat de València. Valencia; PhD Thesis.

Gómez-Pastor, R.; Pérez-Torrado, R. & Matallana, E. (2010a). Improving yield of industrial biomass propagation by increasing the Trx2p dosage. *Bioeng.Bugs.*, Vol. 1, No. 5, pp. 352-353.

Gómez-Pastor, R.; Pérez-Torrado, R.; Cabiscol, E., & Matallana, E. (2010b). Transcriptomic and proteomic insights of the wine yeast biomass propagation process. *FEMS Yeast Res.*, Vol. 10, No. 7, pp. 870-884.

Gómez-Pastor, R.; Pérez-Torrado, R.; Cabiscol, E.; Ros, J. & Matallana, E. (2010c). Reduction of oxidative cellular damage by overexpression of the thioredoxin TRX2 gene improves yield and quality of wine yeast dry active biomass. *Microb.Cell Fact.*, Vol. 9, p. 9.

González, R., Muñoz, R. & Carrascosa, A.V. (2005). Producción de cultivos iniciadores para elaborar el vino. (Carrascosa, A.V, Muñoz, R. &. González, R, Ed.) 318–341. AMV Ediciones, España.

Haarasilta, S. & Oura, E. (1975). Effect of aeration on the activity of gluconeogenetic enzymes in *Saccharomyces cerevisiae* growing under glucose limitation. *Arch.Microbiol.*, Vol. 106, No. 3, pp. 271-273.

Henes, B. & Sonnleitner, B. (2007). Controlled fed-batch by tracking the maximal culture capacity. *J.Biotechnol.*, Vol. 132, No. 2, pp. 118-126.

Herdeiro, R.S., Pereira, M.D., Panek, A.D. & Eleutherio, E.C. (2006). Trehalose protects *Saccharomyces cerevisiae* from lipid peroxidation during oxidative stress. *Biochim.Biophys.Acta*, Vol. 1760, No. 3, pp. 340-346.

Higgins, V. J.; Beckhouse, A. G.; Oliver, A. D.; Rogers, P. J. & Dawes, I. W. (2003b). Yeast genome-wide expression analysis identifies a strong ergosterol and oxidative stress response during the initial stages of an industrial lager fermentation. *Appl.Environ.Microbiol.*, Vol. 69, No. 8, pp. 4777-4787.

Higgins, V. J.; Rogers, P. J. & Dawes, I. W. (2003a). Application of genome-wide expression analysis to identify molecular markers useful in monitoring industrial fermentations. *Appl.Environ.Microbiol.* , Vol. 69, No. 12, pp. 7535-7540.

Hohmann, S. & Mager, W.H. (1997). Yeast stress responses. MBIU R.G. Landes Company, USA.

Hohmann, S. & Mager, W.H. (2003). Yeast stress responses. Springer, New York, USA.

Hohmann, S. (2002). Osmotic stress signaling and osmoadaptation in yeasts. *Microbiol.Mol.Biol.Rev.*, Vol. 66, No. 2, pp. 300-372.

Hongisto, H.J. & Laakso, P. (1978). 20th General Meeting, American Society of Sugar Beet Technologists, San Diego, 26 February–2 March.

Hulse, G. (2008). Yeast Propagation, in Brewing Yeast Fermentation Performance, Second Edition (ed K. Smart), Blackwell Science, Oxford, UK, doi: 10.1002/9780470696040.ch23.

Ivorra, C.; Pérez-Ortin, J. E. & del Olmo, M. (1999). An inverse correlation between stress resistance and stuck fermentations in wine yeasts. A molecular study. *Biotechnol.Bioeng.*, Vol. 64, No. 6, pp. 698-708.

Jamieson, D. J. (1998) .Oxidative stress responses of the yeast *Saccharomyces cerevisiae. Yeast*, Vol. 14, No. 16, pp. 1511-1527.

Jansen, M. L.; Diderich, J. A.; Mashego, M.; Hassane, A.; de Winde, J. H.; ran-Lapujade, P. & Pronk, J. T. (2005). Prolonged selection in aerobic, glucose-limited chemostat cultures of *Saccharomyces cerevisiae* causes a partial loss of glycolytic capacity. *Microbiology*, Vol. 151, No. 5, pp. 1657-1669.

Kaino, T. & Takagi, H. (2009). Proline as a stress protectant in the yeast *Saccharomyces cerevisiae*: effects of trehalose and PRO1 gene expression on stress tolerance. *H.Biosci Biotechnol Biochem*, Vol. 73, No. 9, pp. 2131-2135.

Kaino, T., Tateiwa, T., Mizukami-Murata, S., Shima, J. & Takagi, H. (2008). Self-cloning baker's yeasts that accumulate proline enhance freeze tolerance in doughs. *Appl Environ Microbiol*, Vol. 74, No. 18, pp. 5845-5849.

Kim, J., Alizadeh, P., Harding, T., Hefner-Gravink, A. & Klionsky, D.J. (1996). Disruption of the yeast *ATH1* gene confers better survival after dehydration, freezing, and ethanol shock: potential commercial applications. *Appl.Environ.Microbiol.*, Vol. 62, No. 5, pp. 1563-1569.

Kobayashi, N. & McEntee, K. (1993). Identification of cis and trans components of a novel heat shock stress regulatory pathway in *Saccharomyces cerevisiae. Mol.Cell Biol.*, Vol. 13, No. 1, pp. 248-256.

Kopsahelis, N., Nisiotou, A., Kourkoutas, Y., Panas, P., Nychas, G.J. & Kanellaki, M. (2009). Molecular characterization and molasses fermentation performance of a wild yeast strain operating in an extremely wide temperature range. *Bioresour Technol.* Vol. 100, No.20, pp. 4854-62.

Landolfo, S., Zara, G., Zara, S., Budroni, M., Ciani, M. & Mannazzu, I. (2010). Oleic acid and ergosterol supplementation mitigates oxidative stress in wine strains of *Saccharomyces cerevisiae. Int J Food Microbiol,* Vol. 141, No. 3, pp. 229-235.

Laroche, C., Beney, L., Marechal, P.A. & Gervais, P. (2001). The effect of osmotic pressure on the membrane fluidity of *Saccharomyces cerevisiae* at different physiological temperatures. *Appl.Microbiol.Biotechnol.*, Vol. 56, No. 1-2, pp. 249-254.

Laroche, C. & Gervais, P. (2003). Achievement of rapid osmotic dehydration at specific temperatures could maintain high *Saccharomyces cerevisiae* viability. *Appl.Microbiol.Biotechnol.*, Vol. 60, No. 6, pp. 743-747.

Le Moan, N. Clement, G.; Le Moan, S.; Tacnet, F. & Toledano, M. B. (2006).The *Saccharomyces cerevisiae* proteome of oxidized protein thiols: contrasted functions for the thioredoxin and glutathione pathways. *J.Biol.Chem.*, Vol. 281, No. 15, pp. 10420-10430.

Lei, F.; Rotbøll, F. & Jørgensen, S.B. (2001). A biochemically structured model for *Saccharomyces cerevisiae. J. Biotechnol* Vol. 88, No. 3, pp. 205-221.

Lindquist, S. & Craig, E. A. (1988). The heat-shock proteins. *Annu.Rev.Genet.*, Vol. 22, pp. 631-677.

Maemura, H.; Morimura, S. & Kida K. (1998). Effects of aeration during the cultivation of pitching yeast on its characteristics during the subsequent fermentation of wort. *J Inst Brew.* Vol. 104, pp. 207–211.

Martin, V.; Quain, D.E. & Smart, K.A. (2003). Brewing yeast oxidative stress responses: impact of brewery handling. *Brewing Yeast Fermentation Performance.* 2nd edn (SmartKA, ed), 61–73. Blackwell Science, Oxford, UK.

Martinez-Pastor, M. T.; Marchler, G.; Schuller, C.; Marchler-Bauer, A.; Ruis, H. & Estruch, F. (1996) .The *Saccharomyces cerevisiae* zinc finger proteins Msn2p and Msn4p are required for transcriptional induction through the stress response element (STRE). *EMBO J.*, Vol. 15, No. 9, pp. 2227-2235.

Matthews, T.M. & Webb, C., (1991). Culture systems. In: *Biotechnolgy Handbooks 4: Saccharomyces* (Tuite, M.F., Oliver, S.G., Eds.) 249–282. Plenum Press.

Miskiewicz, T., & Borowiak D. A logistic feeding profile for a Fed-Batch baker's yeast process. (2005). *Electronic Journal of Polish Agricultural University* Vol. 8, No.4, pp. 1-13.

Oku, K., Watanabe, H., Kubota, M., Fukuda, S., Kurimoto, M., Tsujisaka, Y., Komori, M., Inoue, Y. & Sakurai, M. (2003). NMR and quantum chemical study on the OH...pi and CH...O interactions between trehalose and unsaturated fatty acids: implication for the mechanism of antioxidant function of trehalose. *J Am Chem Soc,* Vol. 125, No. 42, pp. 12739-12748.

Oura, E. (1974). Effect of aeration intensity on the biochemical composition of baker's yeast. I. Factors affecting the type of metabolism. *Biotechnol.Bioeng.*, Vol. 16, No. 9, pp. 1197-1212.

Pereira, E.J.; Panek, A.D. & Eleutherio, E.C. (2003). Protection against oxidation during dehydration of yeast. *Cell Stress.Chaperones.*, Vol. 8, No. 2, pp. 120-124.

Pérez-Torrado R. (2004). Estudio y mejora del proceso de producción industrial de levaduras vínicas. Universitat de València. Valencisa ; PhDThesis.

Pérez-Torrado, R.; Bruno-Barcena, J. M. & Matallana, E. (2005). Monitoring stress-related genes during the process of biomass propagation of *Saccharomyces cerevisiae* strains used for wine making. *Appl.Environ.Microbiol.*, Vol. 71, No. 11, pp. 6831-6837.

Pérez-Torrado, R.; Gimeno-Alcaniz, J. V. & Matallana, E. (2002). Wine yeast strains engineered for glycogen overproduction display enhanced viability under glucose deprivation conditions. *Appl.Environ.Microbiol.*, Vol. 68, No. 7, pp. 3339-3344.

Pérez-Torrado, R.; Gómez-Pastor, R.; Larsson, C. & Matallana, E. (2009). Fermentative capacity of dry active wine yeast requires a specific oxidative stress response during industrial biomass growth. *Appl.Microbiol.Biotechnol.*, Vol. 81, No. 5, pp. 951-960.

Piper, P. W. (1995). The heat shock and ethanol stress responses of yeast exhibit extensive similarity and functional overlap. *FEMS Microbiol.Lett.*, Vol. 134, No. 2-3, pp. 121-127.

Potts, M. (2001). Desiccation tolerance: a simple process? *Trends Microbiol.* Vol. 9, No. 11, pp. 553-559.

Postma, E.; Verduyn, C.; Scheffers, W. A. & Van Dijken, J. P. (1989). Enzymic analysis of the crabtree effect in glucose-limited chemostat cultures of *Saccharomyces cerevisiae*. *Appl.Environ.Microbiol.* , Vol. 55, No. 2, pp. 468-477.

Prestrelski, S.J., Tedeschi, N., Arakawa, T. & Carpenter, J.F. (1993). Dehydration-induced conformational transitions in proteins and their inhibition by stabilizers. *Biophys J*, Vol. 65, No. 2, pp. 661-671.

Pretorius, I.S. Utilization of polysaccharides by *Saccharomyces cerevisiae*. (1997). In: *Yeast Sugar Metabolism*. F.K. Zimmermann & K.-D. Entian (Eds.), 459-501. Technomic Publishing Co., Inc., Lancaster, Pennsylvania, USA.

Pronk, J. T.; Yde, S. H. & Van Dijken, J. P. (1996). Pyruvate metabolism in *Saccharomyces cerevisiae. Yeast*, Vol. 12, No. 16, pp. 1607-1633.

Querol, A., Fernández-Espinar, M.T., del Olmo, M. & Barrio, E. (2003). Adaptive evolution of wine yeast. *Int. J. Food Micr*, Vol. 86, pp. 3–10.

Ratnakumar, S. & Tunnacliffe, A. (2006). Intracellular trehalose is neither necessary nor sufficient for desiccation tolerance in yeast. *FEMS Yeast Res.*, Vol. 6, No. 6, pp. 902-913.

Ratnakumar, S., Hesketh, A., Gkargkas, K., Wilson, M., Rash, B.M., Hayes, A., Tunnacliffe, A. & Oliver, S.G. (2011). Phenomic and transcriptomic analyses reveal that autophagy plays a major role in desiccation tolerance in *Saccharomyces cerevisiae*. *Mol.Biosyst.*, Vol. 7, No. 1, pp. 139-149.

Reed, G. & Nagodawithana, T. W. (1988). Technology of Yeast Usage in Winemaking. Am. J. Enol. Vitic., Vol. 39, No. 1, pp. 83-90.

Reed, G. & Nagodawithana, T. W. (1991). Baker's Yeast Production. In: *Yeast Technology* (Reed, G. & Nagodawithana,T. W., Eds.), 261-314, Van Nostrand Reinhold, New York.

Reed, S. I. (1982). Yeast genetics. *Science*, Vol. 215, No. 4537, pp. 1233-1234.

Rizzi, M.; Baltes, M.; Theobald, U. & Reuss, M. (1997). In vivo analysis of metabolic dynamics in *Saccharomyces cerevisiae*: II. Mathematical model. *Biotechnol.Bioeng.*, Vol. 55, No. 4, pp. 592-608.

Rossignol, T.; Kobi, D.; Jacquet-Gutfreund, L. & Blondin, B. (2009). The proteome of a wine yeast strain during fermentation, correlation with the transcriptome. *J.Appl.Microbiol.*, Vol. 107, No. 1, pp. 47-55.

Rossignol, T., Postaire, O., Storaï, J. & Blondin, B. (2006). Analysis of the genomic response of a wine yeast to rehydration and inoculation. *Appl Microbiol Biotechnol*, Vol. 71, No. 5, pp. 699-712.

Sales, K., Brandt, W., Rumbak, E. & Lindsey, G. (2000). The LEA-like protein HSP 12 in *Saccharomyces cerevisiae* has a plasma membrane location and protects membranes against desiccation and ethanol-induced stress. *Biochim.Biophys.Acta*, Vol. 1463, No. 2, pp. 267-278

Salvadó, Z.; Chiva, R.; Rodríguez -Vargas, S.; Rández -Gil, F.; Mas, A. & Guillamón, J. M. (2008). Proteomic evolution of a wine yeast during the first hours of fermentation. *FEMS Yeast Res.*, Vol. 8, No. 7, pp. 1137-1146.

Shamrock, V.J. & Lindsey, G.G. (2008). A compensatory increase in trehalose synthesis in response to desiccation stress in Saccharomyces cerevisiae cells lacking the heat shock protein Hsp12p. *Can.J.Microbiol.*, Vol. 54, No. 7, pp. 559-568.

Shima, J., Hino, A., Yamada-Iyo, C., Suzuki, Y., Nakajima, R., Watanabe, H., Mori, K. & Takano, H. (1999). Stress tolerance in doughs of *Saccharomyces cerevisiae* trehalase mutants derived from commercial Baker's yeast. *Appl.Environ.Microbiol.*, Vol. 65, No.7, pp. 2841-2846.

Shima, J.; Kuwazaki, S.; Tanaka, F.; Watanabe, H.; Yamamoto, H.; Nakajima, R.; Tokashiki, T. & Tamura, H. (2005). Identification of genes whose expressions are enhanced or reduced in baker's yeast during fed-batch culture process using molasses medium by DNA microarray analysis. *Int.J.Food Microbiol.*, Vol. 102, No. 1, pp. 63-71.

Simonin, H., Beney, L. & Gervais, P. (2007). Sequence of occurring damages in yeast plasma membrane during dehydration and rehydration: mechanisms of cell death. *Biochim.Biophys.Acta*, Vol. 1768, No. 6, pp. 1600-1610.

Simons, K. & Ikonen, E. (1997). Functional rafts in cell membranes. *Nature*, Vol. 5, No. 6633, pp. 569-572.

Singer, M.A. & Lindquist, S. (1998). Thermotolerance in *Saccharomyces cerevisiae*: the Yin and Yang of trehalose. *Trends Biotechnol.* Vol. 16, No. 11, pp. 460-468.

Singh, J., Kumar, D., Ramakrishnan, N., Singhal, V., Jervis, J., Garst, J.F., Slaughter, S.M., DeSantis, A.M., Potts, M. & Helm, R.F. (2005). Transcriptional response of *Saccharomyces cerevisiae* to desiccation and rehydration. *Appl.Environ.Microbiol.*, Vol. 71, No.12, pp. 8752-8763.

Sorger, P. K. (1991). Heat shock factor and the heat shock response. *Cell*, Vol. 65, No. 3, pp. 363-366.

Stadtman, E. R. & Levine, R. L. (2000). Protein oxidation. *Ann.N.Y.Acad.Sci.*, Vol. 899, pp. 191-208.

Swan, T.M. & Watson K. (1998). Stress tolerance in a yeast sterol auxotroph: role of ergosterol, heat shock proteins and trehalose. *FEMS Microbiol Lett.* Vol. 1, No. 1, pp. 191-197.

Takagi, H. (2008). Proline as a stress protectant in yeast: physiological functions, metabolic regulations, and biotechnological applications. *Appl.Microbiol.Biotechnol.*, Vol. 81, No. 2, pp. 211-223.

Trabalzini, L.; Paffetti, A.; Ferro, E.; Scaloni, A.; Talamo, F.; Millucci, L.; Martelli, P. & Santucci, A. (2003). Proteomic characterization of a wild-type wine strain of *Saccharomyces cerevisiae. Ital.J.Biochem.*, Vol. 52, No. 4, pp. 145-153.

Trevisol, E.T., Panek, A.D., Mannarino, S.C. & Eleutherio, E.C. (2011). The effect of trehalose on the fermentation performance of aged cells of *Saccharomyces cerevisiae. Appl.Microbiol.Biotechnol.*

Trofimova, Y., Walker, G. & Rapoport, A. (2010). Anhydrobiosis in yeast: influence of calcium and magnesium ions on yeast resistance to dehydration-rehydration. *FEMS Microbiol.Lett.*, Vol. 308, No. 1, pp. 55-61.

Trollmo, C.; Andre, L.; Blomberg, A. & Adler, L. (1988). Physiological overlap between osmotolerance and thermotolerance in*Saccharomyces cerevisiae. FEMS Microbiol Lett* Vol. 56, No. 3, pp. 321–326.

Valentinotti, S.; Holmberg, U.; Srinivasan, B.; Cannizzaro, C.; Rhiel, M.; von Stockar, U. & Bonvin, D. (2002). Optimal operation of fed-batch fermentations via adaptive control of overflow metabolite. *Control Eng Pract.* Vol. 11, No. 6, pp. 665-674.

Van der Aar P.C.; Van Verseveld, H.W. & Stouthamer, A.H. (1990). Stimulated glycolytic flux increases the oxygen uptake rate and aerobic ethanol production, during oxido-reductive growth of *Saccharomyces cerevisiae*. *J. Biotechnol* Vol. 13, pp. 347–359.

Van, H. P.; de, H. E.; Van Dijken, J. P. & Pronk, J. T. (2000). Fermentative capacity in high-cell-density fed-batch cultures of baker's yeast. *Biotechnol.Bioeng.*, Vol. 68, No. 5, pp. 517-523.

Van, H. P.; Van Dijken, J. P. & Pronk, J. T. (1998). Effect of specific growth rate on fermentative capacity of baker's yeast. *Appl.Environ.Microbiol.*, Vol. 64, No. 11, pp. 4226-4233.

Varela, J. C.; van, B. C.; Planta, R. J. & Mager, W. H. (1992) .Osmostress-induced changes in yeast gene expression. *Mol.Microbiol.*, Vol. 6, No. 15, pp. 2183-2190.

Vemuri, G.N.; Eiteman, M.A.; McEwen, J.E.; Olsson, L. & Nielsen, J. (2007). Increasing NADH oxidation reduces overflow metabolism in *Saccharomyces cerevisiae. Proc. Natl. Acad. Sci.*, Vol. 104, No. 7, pp. 2402-2407.

Verduyn, C. (1991). Physiology of yeasts in relation to biomass yields. *Antonie Van Leeuwenhoek*, Vol. 60, No. 3-4, pp. 325-353.

Vu, V.H. & Kim K. (2009). High-cell-density fed-batch culture of Saccharomyces cerevisiae KV-25 using molasses and corn steep liquor. *J Microbiol Biotechnol.* Vol. 19, No. 12, pp. 1603-11.

Woehrer, W. & Roehr, M. (1981). Regulatory aspects of bakers' yeast metabolism in aerobic fed-batch cultures. *Biotechnol. Bioeng* Vol. 23, No. 3, pp. 567–581.

Xandé, X., Archimède, H., Gourdine, J.L., Anais, C. & Renaudeau, D. (2010). Effects of the level of sugarcane molasses on growth and carcass performance of Caribbean growing pigs reared under a ground sugarcane stalks feeding system. *Trop Anim Health Prod.* Vol. 42, No. 1, pp. 13-20.

Xu, Z. & Tsurugi, K. (2006). A potential mechanism of energy-metabolism oscillation in an aerobic chemostat culture of the yeast *Saccharomyces cerevisiae. FEBS J.*, Vol. 273, No. 8, pp. 1696-1709.

Zuzuarregui, A.; Carrasco, P.; Palacios, A.; Julien, A. & del Olmo, M. (2005). Analysis of the expression of some stress induced genes in several commercial wine yeast strains at the beginning of vinification. *J.Appl.Microbiol.*, Vol. 98, No. 2, pp. 299-307.

Zuzuarregui, A.; Monteoliva, L.; Gil, C. & del Olmo, M. (2006). Transcriptomic and proteomic approach for understanding the molecular basis of adaptation of *Saccharomyces cerevisiae* to wine fermentation. *Appl.Environ.Microbiol.*, Vol. 72, No. 1, pp. 836-847.

12

Biomass Alteration of Earthworm in the Organic Waste-Contaminated Soil

Young-Eun Na[1], Hea-Son Bang[1], Soon-Il Kim[2] and Young-Joon Ahn[2]
[1]*National Academy of Agricultural Science and Technology, Rural Development Administration, Suwon 441–707,*
[2]*WCU Biomodulation Major, Department of Agricultural Biotechnology, Seoul National University, Seoul 151–921, Republic of Korea*

1. Introduction

Earthworm populations show a considerable amount of variability in time and space, with mean densities and biomass ranging from less than 10 individuals and 1 g m^{-2} to more than 1,000 individuals and 200 g m^{-2} under favourable conditions. Earthworms have been considered to play a great role in soil-formation processes and in monitoring soil structure and fertility (Lavelle & Spain, 2001) because they may increase the mineralisation and humification of organic matter by food consumption, respiration and gut passage (Edwards & Fletcher, 1988; Lavelle & Spain, 2001) and may indirectly stimulate microbial mass and activity as well as the mobilisation of nutrients by increasing the surface area of organic compounds and by their casting activity (Emmerling & Paulsch, 2001). However, within particular climatic zones, earthworm assemblages, with fairly characteristic species richness, composition, abundance and biomass, can often be recognised in broadly different habitat types, such as coniferous forest, deciduous woodland, grassland and arable land (Curry, 1998).

Agriculture is facing a challenge to develop strategies for sustainability that can conserve non-renewable natural resources, such as soil, and enhance the use of renewable resources, such as organic wastes. It has been estimated that 357,861 tons of organic sludge daily were produced in South Korea in 2009 (Anon., 2009). The production and use of organic compounds have also risen rapidly over the last four decades. Organic compounds which are released either through direct discharge into the sewer system, or indirectly through run-off from roads and other surfaces are found in sewage sludge (Halsall et al., 1993). As a suitable bioindicator of chemical contamination in soil, earthworms are easy, fast and economical merits to handle. Especially, analysis of their tissues may also provide an excellent index of bioavailability of heavey metals in soils (Helmke et al., 1979; Pearson et al., 2000).

Although the acute earthworm toxicity test developed by Edwards (1984) has been widely used and an internationally accepted protocol was also used for assaying the chemical toxicity of contaminants in soils (Organisation for Economic Cooperation and Development [OECD], 1984), the chronic toxicity test to detect subtle effects of contaminants on them by long-term exposure has not been fully achieved (Venables et al., 1992). Based upon these tests, lots of information on heavy metal uptake, toxicity and accumulation by various earthworm species have been produced. Therefore, earthworms could fill the gap by being used as potential biomarkers of ecotoxicity to various chemicals, including organic contaminants.

This chapter is particularly focused on the hazardous effects on composition, numbers and biomass of Megascolecid and Moniligastrid earthworms, which are dominant groups in South Korea, of 8 consecutive yearly applications of three levels of four different organic sludges and pig manure compost as a positive reference using field lysimeters and microcosms.

2. Legal criteria of inorganic pollutants

Many countries have been trying to prepare a regulatory limit to the use of organic wastes, such as food wastes or sludge, into crop production system in the light of their rapid increase. The regulatory system for the agricultural use of organic waste in South Korea is defined as soil concentration limits for potentially toxic elements (PTEs) to safeguard human health and crop yields. Despite legal limits, the damage of crop in the agricultural soil frequently occurs with organic waste for long-term application and with sub-quality compost made from sewage sludge.

The control system in the application of sludge to farmland varies according to country (Table 1). In South Korea, the control system for the application of sludge to farmland primarily depends upon heavy-metal concentrations that are similar to those in developed countries. Legally allowed limit values for PTEs— such as copper (Cu), zinc (Zn), chromium (Cr), cadmium (Cd), lead (Pb) and nickel (Ni) —were 400, 1,000, 250, 5, 130 and 45 mg kg^{-1}, respectively, under the Fertilizer Management Act in South Korea (Anon., 2010a).

The control system for soil intoxication limit levels primarily depends upon heavy-metal concentration. The limit levels in South Korea are Cu 50, Zn 300, Cr 4, Cd 1.5, Pb 100 and Ni 40 mg kg^{-1} under the Soil Environmental Conservation Act (Anon., 2007). In Japan, Cu must be less than 125 mg kg^{-1}, Cr 0.05 mg l^{-1} or less, Cd 0.4 mg kg^{-1} or less and Pb 0.01 mg l^{-1} or less (Ministry of the Environment Government of Japan, 1994).

In many countries, current rules for controlling the use of organic wastes on agricultural land have been criticized because they apparently do not take into consideration of the potential adverse effects of inorganic heavy metals and organic compounds produced in organic waste-treated soils on soil organisms (McGrath, 1994). The regulatory limit to the application of industrial waste on farmland only depends upon the level of PTEs in South Korea. However, PTEs limit may not be an adequate regulation protocol since organic wastes contain lots of inorganic and organic contaminants (Ministry of Agriculture, Fisheries and Food [MAFF], 1991).

An overall assessment of the soil contamination caused by inorganic and organic compounds of organic waste has been, therefore, attempted by ascribing qualitative description of the apparent risk and developing the integrated hazard assessment system (Hembrock-Heger, 1992). Available options for dealing with sludge include application to agricultural land, incineration, land reclamation, landfill, forestry, sea disposal and biogas. Of these, the application to agricultural land is the principal way for deriving beneficial uses of organic sludge by recycling plant nutrients and organic matter to soil for crop production (Coker et al., 1987). Also, agricultural use provides a reliable cost-effective method for sludge disposal. Recycling (81.7%) is the largest means of waste disposal, with 11.1% land deposition, 5.2% incineration and 2.0% sea disposal in South Korea (Anon., 2009). As an alternative way of waste disposal, the Fertilizer Management Act was revised to make it possible to apply industrial and municipal wastes into farmland in December 1996 in South Korea (Anon., 2006).

Country	Parameter (mg kg of dry matter^{-1})							
	As	Hg	Pb	Cd	Cr	Cu	Zn	Ni
South Korea[a]	45	2	130	5	250	400	1000	45
USA[b]	75	57	840	85	3000	4300	7500	420
Canada[c]	13	0.8	150	3	210	400	700	62
EU[d]	-	1-1.5	50-300	1-3	-	50-140	150-300	30-75
Belgium[e]	-	1	120	1.5	70	90	300	20
Denmark[e]	25	0.8	120	0.8	100	1000	4000	30
France[e]	-	10	800	20	1000	1000	3000	200
Netherlands[e]	-	0.3	100	1	50	90	290	20
Sweden[e]	-	2.5	100	2	100	600	800	50
Germany[f]	-	8	900	10	900	800	2500	200
UK[g]	-	1	200	1.5	100	200	400	50
Switzerland[h]	-	1	120	1	100	100	400	30
Australia[i]	20	1	150-300	1	100-400	100-200	200-250	60
New Zealand[j]	20	2	300	3	600	300	600	60

a Anon. (2010a)
b USEPA (2000)
c Canadian Council of Ministers of the Environment [CCME] (2005)
d Anon. **(2010b)**
e Brinton (2000)
f Anon. (2010c)
g British Standards Institution [BSI] (2011)
h Anon. (2010d)
i Anon. (1997)
j New Zealand Water and Waste Association [NZWWA] (2003)

Table 1. Criteria of the inorganic pollutants in compost or sewage sludge for application to the arable land in 14 selected countries

3. Importance of earthworm

3.1 Role in soil

Earthworms have a critical influence on soil structure, forming aggregates and improving the physical conditions for plant growth and nutrient uptake. They also improve soil fertility by accelerating decomposition of plant littre and soil organic matter. Earthworms are the most important invertebrates in this initial stage of the recycling of organic matter in various types of soils. Curry & Byrne (1992) demonstrated that the decomposition rate of straw which was accessible to the earthworms was increased by 26–47% compared with straw from which they were excluded. Organic matter that passes through the earthworm gut and is digested in their casts is broken down into much finer particles, so that a greater surface area of the organic matter is exposed to microbial decomposition. Martin (1991) reported that casts of the tropical earthworms had much less coarse organic matter than the surrounding soil, indicating that the larger particles of organic matter were fragmented during passage through the earthworm gut. Earthworm species, such as *Lumbricus terrestris*, are responsible for a large proportion of the overall fragmentation and incorporation of littre in many woodlands and farmland of the temperate zone, which resulted in the formation of mulls. As a result, the surface littre and organic layers are mixed thoroughly with the mineral soil (Scheu & Wolters, 1991).

The numbers of earthworm burrows have been counted between 50 and 200 burrows m^{-2} on horizontal surfaces (Edwards et al., 1990). Earthworms not only improve soil aeration by their burrowing activity, but they also influence the porosity of soils. Earthworm burrows was found to increase the soil-air volume from 8% to 30% of the total soil volume (Wollny, 1890). In one soil, earthworm burrows comprise a total volume of 5 litres m^{-3} of soil, making a small but significant contribution to soil aeration (Kretzschmar, 1978). Water infiltration was from 4 to 10 times faster in soils with earthworms than in soils without earthworms (Carter et al., 1982). They bring large amounts of soil from deeper layers to the surface and deposit as casts on the surface. The amounts which turned over in this way greatly differ with habitats and geographical regions, ranging from 2 to 268 tons ha^{-1} (Beauge, 1912; Roy, 1957). The importance of this turnover, which was discussed first by Darwin (2009), can be seen by comparing the profile of a stratified mor soil (with few earthworms) with that of a well-mixed mull soil. Blanchart (1992) reported in a formation of aggregates that under natural conditions with or without earthworms, large aggregates (>2 mm) comprised only 12.9% of soil with no earthworms, whereas in soil with worms, large aggregates comprised 60.6% of soil after 30 months in the field. Devliegher & Verstraete (1997) introduced the concepts of nutrient enrichment process and gut associated process. They noted that earthworms are performing these two different functions that may have contrasting their effects on soil microbiology, chemistry and plant growth. Earthworms, such as *L. terrestris*, incorporate and mix surface organic matter with soil and increase biological activity and nutrient availability. However, they also assimilate nutrients from soil and organic matter as these materials pass through their gut.

3.2 Occurrence of earthworm in Korean soil ecosystem

The earthworm fauna of South Korea is dominated by the family Megascolecidae and identified 101 species, with 12 species in Lumbricidae, 9 species in Moniligasteridae and 80 species in Megascolecidae (Fig. 1) (Hong, 2000, 2005; Hong et al., 2001). In general, earthworms are classified into three types based upon life style and burrowing habit (Bouché, 1972). The epigeal forms (e.g., *Lumbricus rubellus* and *Eisenia fetida*) hardly burrow in soil at all, but inhabit decaying organic matters on the surface, including manure or compost heaps. The endogenous species (e.g., *Allolobophora chlorotica* and *Allolobophora caliginosa*) produce shallow branching burrows in the organo-mineral layers of the soil. Lastly, the anectic forms (e.g., *L. terrestris* and *Allolobophora longa*) are deep burrowing species, producing channels to a depth of one meter or more. Megascolecidae species identified in Korean ecosystem come under anectic forms. Occurrence of earthworms in agroecosystem appeared the most individuals of *Amynthas agrestis*, *Amynthas heteropodus* and *Amynthas koreanus* (Hong & Kim, 2007).

(A) (B) (C)

Fig. 1. Representative earthworms in Lumbricidae (A), Moniligasteridae (B) and Megascolecidae (C) in South Korea

3.3 Biomonitor for biological hazard assessment on soil contamination

Concerns about contamination of soil and detrimental effects of contaminants on the living environment have resulted in a strong and growing interest in soil organisms among environmental scientists and legislators. Legislation in many countries has recently focused on the need of sensitive organisms from the soil environment for environmental monitoring. Many toxic materials have been accumulated along with food webs. The decomposer levels are frequently the first to be affected since the organic matter and the soil are the ultimate sink for most contaminants. Ecologically, earthworms are near the bottom of the terrestrial tropic levels. The effects of contaminants on earthworms which were kept in soil in the laboratory have been studied (Edwards & Thompson, 1973). These tests tended to produce consistent and reproducible results because 10 individuals of *E. fetida* were used and these worms were an intimate contact with pesticides. van Hook (1974) demonstrated that earthworms could serve as useful biological indicators of contamination because of the fairly consistent relationships between the concentrations of various contaminants and mortality of earthworm. The basic requirements of finding a species easy to rear and genetically homogeneous could be fulfilled by using representatives of the species, although there have been arguments for the use of *Eisenia andrei* or a genetically controlled single strain of the *E. fetida* complex (Bouché, 1992). Callahan et al. (1994) have suggested that *E. fetida* may be a representative of the species, *Allolobophora tuberculata, Eudrilus eugeniae* and *Perionyx excavatus* based upon the concentration-response relationship for 62 chemicals when applying the Weibull function. Habitational earthworms, including *E. fetida,* are useful as biological indicator species in the ecological sense or a more useful biomonitor species. It has been proposed that *A. heteropodus* could be adopted as a bioindicator in agroecosystem because of dominant species in South Korea (Kim et al., 2009).

4. Effects of organic waste sludge application on earthworm biology

4.1 Composition and biomass of earthworms

Four different types of organic waste sludge used in this study were as follows: municipal sewage sludge (MSS) collected from sewage treatment plants on Gwacheon (Gyeonggi Province, South Korea); industrial sewage sludge (ISS) collected from industrial complex on Ansan (Gyeonggi Province); alcohol fermentation processing sludge (AFPS) collected from Ansan industrial complex; and leather processing sludge (LPS) collected from sewage treatment plant on Cheongju (Chungbuk Province, South Korea). Pig manure compost

(PMC) was purchased from Anjung Nong-hyup, Anjung (Gyeonggi Province). These materials were collected in early March 1994 and kept in deep freezers (-60°C) to be applied annually from 1994 to 2001.

Lysimeters which composed of 45 concrete plots (1.0 m length, 1.0 m width and 1.1 m depth) (Fig. 2) were made in the upland field of Suwon (Gyeonggi Province) in March 1993. Each plot was uniformly filled with the same sandy loam soil without earthworms up to the ground surface in mid-May 1993. Three levels (12.5, 25 and 50 tons of dry matter ha^{-1} $year^{-1}$) of test materials were applied to each plot twice annually for 8 consecutive years (mid-March 1994 to mid-March 2001) and mixed into the soil of a depth of 15 cm. PMC served as a standard for comparison in lysimeter tests. A randomized complete block design with three replicates was used. Two radish, *Raphanus sativus*, cultivars (jinmialtari and backkyoung) were cultivated in every spring and autumn, respectively. Planting densities were 12 × 15 cm in spring and 25 × 30 cm in autumn with one plant. Other practices followed standard *Raphanus* culture methods without application of any mineral fertilizer and pesticide. The lysimeters were covered with a nylon net to prevent any access by birds or animals.

Fig. 2. Field lysimeters

Earthworms were collected from each of the 45 lysimeter plots from an area of 1 m^2 up to 0.3 m depth by hand sorting in mid-October 1997 and mid-October 2001 as described previously (Callaham & Hendrix, 1997). They were immediately transported to the laboratory in plastic containers and separated into juveniles and adults with a clitellum. The earthworm numbers, composition and biomass were investigated before they were fixed in a 10% formalin solution. Earthworm species identification followed Hong & James (2001), Kobayashi (1941) and Song & Paik (1969).

Pollution index (PI) was determined according to the method of Jung et al. (2005), PI = [∑(heavy metal concentration in soil tolerable level–1) number of heavy $metal^{-1}$]. Tolerable level of Cu, Zn, Cr, Cd, Pb and Ni were 125, 700, 10, 4, 300 and 100 mg kg^{-1} in Korean soil, respectively (Anon., 2007). PI values are employed to assess metal pollution in soil and indicate the average on ratios of metal concentration over tolerable level. A soil sample is judged as contaminated by heavy metal when PI value is greater than 1. Total toxic unit of PTEs was calculated by threshold level described under the Soil Environmental Conservation Act (Anon., 2007) in South Korea as follows: ∑ (Cu 50 + Zn 300 + Cr 4 + Cd 115 + Pb 100 + Ni 40). Bonferroni multiple-comparison method was used to test for significant differences among treatments in the fresh biomass of earthworms and pollution indices (SAS Institute, 2004). Correlations between accumulated pollutant contents and observed earthworm numbers and biomass were estimated from the Pearson correlation coefficients using SAS. pH values, heavy-metal contents and pollution indices of 8 consecutive yearly applications of three levels of four different organic waste materials and PMC in field lysimeters were reported previously (Na et al., 2011).

Effects on earthworm composition of 8 consecutive yearly applications of four organic waste materials and PMC were investigated using field lysimeters (Table 2). Earthworm composition in all treatments varied according to waste material examined, treatment level and application duration. Of 390 adults collected from 45 plots, earthworms were classified into 2 families (Megascolecidae and Moniligastridae), 2 genera (*Amynthas* and *Drawida*) and 5 species (*Amynthas agrestis, Amynthas hupeiensis, Amynthas sangyeoli, Drawida koreana* and *Drawida japonica*). The number of earthworm species in MSS-, ISS-, LPS-, AFPS- and PMC-treated soils was 2, 2, 2, 3 and 5, respectively. The dominant species were *A. agrestis, A. hupeiensis, A. sangyeoli* and *D. japonica* in the sludge treatments 4 years after treatment but was replaced with *A. hupeiensis* in all the plots 8 years after treatment. This finding indicates that *A. hupeiensis* was more tolerant to toxic heavy metals than other earthworm species. In ISS- and LPS-treated soils, the proportion of juveniles appeared was 67–100% 4 years after treatment, but no juveniles was observed 8 years after treatment.

At 4 years after treatment, effect of test waste material (F = 16.91; df = 4,44; P < 0.0001) and treatment level (F = 4.09; df = 2,44; P = 0.0268) on the number of earthworms was significant (Table 2). The material by level interaction was also significant (F = 2.63; df = 8,44; P = 0.0258). At 8 years after treatment, effect of test waste material (F = 17.33; df = 4,15; P < 0.001) and treatment level (F = 11.00; df = 3,29; P < 0.001) on the number of earthworms was significant. The material by level interaction was also significant (F = 20.53; df = 8,44; P < 0.001). The number of earthworms was significantly reduced in 25 and 50 ton MSS treatments, 25 and 50 ton AFPS treatments and 12.5 and 25 ton PMC treatments 4 years after treatments than 8 years of treatments. The total number of earthworms collected 4 and 8 years after treatment was as follows: MSS-treated soil, 66/29; ISS-treated soil, 4/2; LPS-treated soil, 15/1; AFPS-treated soil, 30/11; and PMC-treated soil, 127/439.

Earthworm biomass collected from 45 plots during the 8-year-investigation period is given in Fig. 3. The biomass in all treatments was dependent upon waste material examined, treatment level and application duration. At 4 years after treatment, effect of test waste material (F = 49.45; df = 4,44; P < 0.0001) and treatment level (F = 5.80; df = 2,44; P = 0.0074) on the earthworm biomass was significant. The material by level interaction was also significant (F = 3.88; df = 8,44; P = 0.0031). At 8 years after treatment, effect of test waste material (F = 165.13; df = 4,44; P < 0.0001) and treatment level (F = 14.39; df = 2,44; P < 0.0001) on the earthworm biomass was significant. The material by level interaction was also significant (F = 19.77; df = 8,44; P < 0.0001). Significant increase in biomass of soil treated with 50 ton PMC ha^{-1} $year^{-1}$ was observed 8 years after treatment.

Material[a]	Rate[b]	Species	Individuals of species		Total number[d]		P-value[e]
			4 YAT[c]	8 YAT	4 YAT	8 YAT	
MSS	12.5	*A. sangyeoli*	3	1	10	16	0.0132
		A. hupeiensis	3	8			
		Juvenile	4	7			
	25	*A. sangyeoli*	4	0	22	8	0.0006
		A. hupeiensis	5	5			
		Juvenile	13	3			
	50	*A. sangyeoli*	4	0	34	5	0.0038
		A. hupeiensis	5	4			
		Juvenile	25	1			
ISS	12.5	*A. agrestis*	1	0	3	2	0.7247
		A. hupeiensis	0	2			
		Juvenile	2	0			
	25	Juvenile	1	0	1	0	0.3739
	50		0	0	0	0	
LPS	12.5	*D. japonica*	1	0	8	0	0.0907
		Juvenile	7	0			
	25	*A. hupeiensis*	0	1	5	1	0.2302
		Juvenile	5	0			
	50	Juvenile	2	0	2	0	0.1161
AFPS	12.5	*A. sangyeoli*	3	0	10	9	0.9019
		A. hupeiensis	3	4			
		D. japonica	0	2			
		Juvenile	4	3			
	25	*A. sangyeoli*	4	0	9	0	0.0065
		A. hupeiensis	1	0			
		Juvenile	4	0			
	50	*A. sangyeoli*	5	0	11	2	0.0031
		A. hupeiensis	2	1			
		Juvenile	4	1			
PMC	12.5	*A. agrestis*	2	1	24	63	0.0069
		A. hupeiensis	6	40			
		D. japonica	4	2			
		Juvenile	12	20			

Table 2 (Continued)

Material[a]	Rate[b]	Species	Individuals of species		Total number[d]		P-value[e]
			4 YAT[c]	8 YAT	4 YAT	8 YAT	
PMC	25	*A. agrestis*	0	1	34	117	0.0054
		A. sangyeoli	7	0			
		A. hupeiensis	14	84			
		D. japonica	3	2			
		D. koreana	0	2			
		Juvenile	10	28			
	50	*A. sangyeoli*	0	2	69	259	0.2066
		A. hupeiensis	24	70			
		D. japonica	10	19			
		D. koreana	7	18			
		Juvenile	28	150			

[a] Abbreviations are same as in the text
[b] Tons of dry matter ha^{-1} $year^{-1}$
[c] Years after treatment plots
[d] The combined number of earthworms in the three replicate plots
[e] *t*-test

Table 2. Earthworm numbers and composition of 4 and 8 consecutive yearly applications (twice annually) of three levels of four different organic waste materials and pig manure compost using field lysimeters

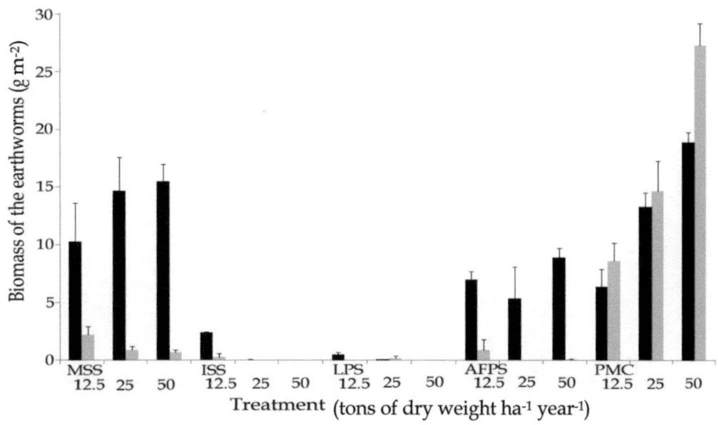

Fig. 3. Earthworm biomass of 4 (■) and 8 (■) consecutive yearly applications (twice annually) of three levels of four different organic waste materials and pig manure compost using field lysimeters.

To evaluate potential toxic effects of residual heavy metals, total toxic units of PTEs were determined (Fig. 4). The total toxic units in all treatments varied with waste material examined, treatment level and application duration. At 4 years after treatment, effect of test waste material (F = 34872.4; df = 4,44; $P < 0.0001$) and treatment level (F = 60.24; df = 2,44; $P < 0.0001$) on the the total toxic units of PTEs was significant. The material by level interaction was also significant (F = 2601.2; df = 8,44; $P < 0.0001$). At 8 years after treatment, effect of test waste material (F = 52439.5; df = 4,44; P < 0.0001) and treatment level (F = 28451.0; df = 2,44; P < 0.0001) on the the total toxic unit of PTEs was significant. The material by level interaction was also significant (F = 13057.2; df = 8,44; P < 0.0001).

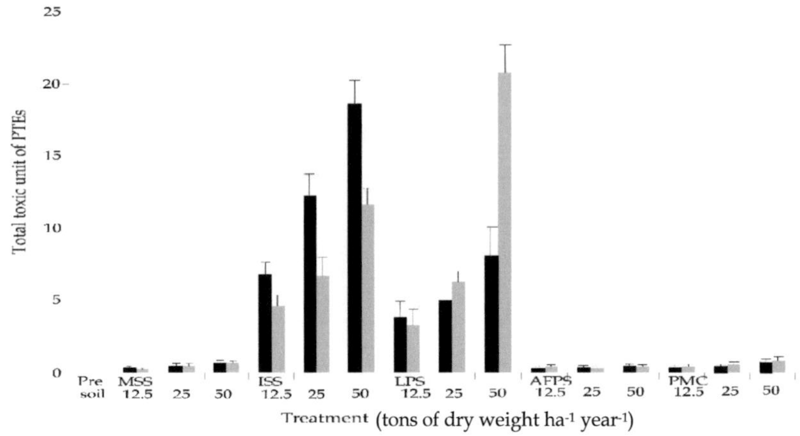

Fig. 4. Total toxic units of potentially toxic elements (PTEs) of 4 (■) and 8 (■) consecutive yearly applications (twice annually) of three levels of four different organic waste materials and pig manure compost using field lysimeters. Abbreviations are same as in the text

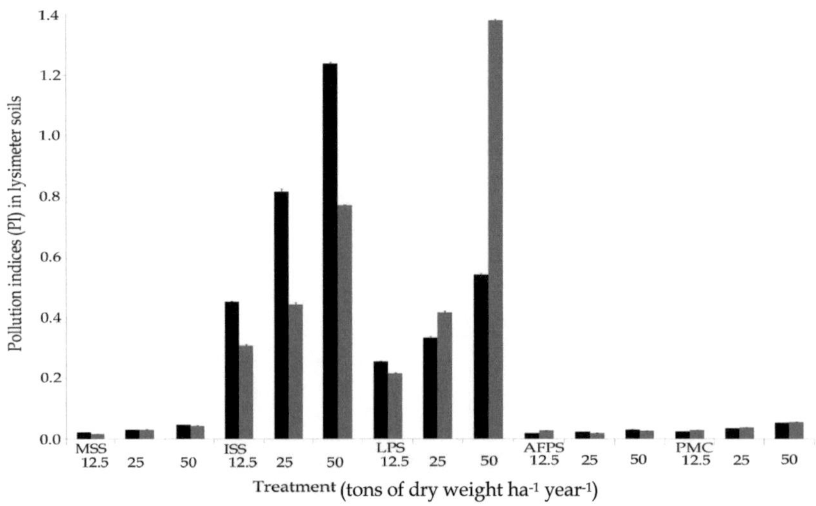

Fig. 5. Pollution indices of 4 (■) and 8 (■) consecutive yearly applications (twice annually) of three levels of four different organic waste materials and pig manure compost using field lysimeters. Abbreviations are same as in the text

PI values of lysimeter soils sampled during the 8-year-investigation period are reported in Fig. 5. At 4 years after treatment, effect of test waste material (F = 34047.6; df = 4,44; P < 0.0001) and treatment level (F = 5957.3; df = 2,44; P < 0.0001) on the the total toxic unit of PTEs was significant. The material by level interaction was also significant (F = 2505.3; df = 8,44; P < 0.0001). At 8 years after treatment, effect of test waste material (F = 48793.6; df =

4,44; $P < 0.0001$) and treatment level ($F = 26515.1$; df = 2,44; $P < 0.0001$) on the the total toxic unit of PTEs was significant. The material by level interaction was also significant ($F = 12190.9$; df = 8,44; $P < 0.0001$). There was significant difference in PI values between the treatment duration. Particularly, PI value of ISS-treated soil was higher 8 years after treatment than 4 years after treatment, while PI value of LPS-treated soil was higher 4 years after treatment than 8 years after treatment.
Correlation between total toxic unit of PTEs and PI and earthworm individuals and biomass was determined (Table 3). At 4 years after treatment, earthworm individuals were correlated negatively with the total toxic unit of PTEs ($r = -0.509$) and PI ($r = -0.508$). At 8 years after treatment, earthworm individuals were correlated negatively with the total toxic unit of PTEs ($r = -0.265$), but were not correlated negatively with PI.
At 4 years after treatment, earthworm biomass was correlated negatively with the total toxic unit of PTEs ($r = -0.673$) and PI ($r = -0.672$) (Table 3). At 8 years after treatment, earthworm biomass was correlated negatively with the total toxic unit of PTEs ($r = -0.308$), but were not correlated negatively with PI.

Parameter	Correlation coefficient (r)			
	Earthworm individuals		Earthworm biomass	
	4 YAT[a]	8 YAT	4 YAT	8 YAT
Total toxic unit of PTEs	-0.509	-0.265	-0.673*	-0.308
PI	-0.508*[b]	-0.265	-0.672*	-0.280

[a] Years after treatmen
[b] *0.001<P<0.05 treatment

Table 3. Correlation between total toxic unit of potentially toxic elements (PTEs) and pollution indicies (PI) and earthworm individuals and biomass 4 and 8 years after treatment

The impact of heavy metals and sludge on lumbricid earthworms, particularly *E. fetida* and *L. terrestris*, has been well noted. Heavy metals cause mortality and reduce fertility, cocoon production and viability, growth, composition and biomass, and bioaccumulation and bioavailability of earthworms. The toxic values of heavy metals to earthworms vary according to an earthworm acute toxicity test. Based upon an artificial soil test, Spurgeon et al. (1994) determined no observed-effect concentrations (NOECs) for *E. fetida* exposed to heavy metals. The estimated NOEC values were 39.2 mg Cd kg^{-1}, 32 mg Cu kg^{-1}, 1,810 mg Pb kg^{-1} and 199 mg Zn kg^{-1}. In soil contaminated by effluent containing Cr, the rate of 10 mg kg^{-1} was fatal to *Peretima posthuma* and other species (Abbasi & Soni, 1983). Copper caused higher mortality than Pb or Zn against *E. fetida* at the same rate and the LC_{50} and NOEC values for Cd could not be determined since no significant mortality was observed at the highest test rate (300 μg g^{-1}) (Spurgeon et al., 1994).
Although heavy metals did not show direct lethal effects to earthworms, they can sensitively cause their reproduction and sperm count reduction and low hatching success of cocoons. *Lumbricus terrestris* worms exposed in artificial soil to sublethal concentrations of technical chlordane (6.25, 12.5 and 25 ppm) and cadmium nitrate (100, 200 and 300 ppm) exhibited significant reduction in spermatozoa from testes and seminal vesicles (Cikutovic et al., 1993). *Eisenia fetida* worms grew well in the lead-contaminated environment and produced cocoons at the same rate as the control worms, but the hatchability of these cocoons was much lower, indicating that lead toxicity affects reproductive performance by major

spermatozoa damage (Reinecke & Reinecke, 1996). In addition, Zn, Mn and Cu produced slower growth, later maturation and fewer or no cocoons. Reinecke and Reinecke (1997) have shown the structural damage of spermatozoa, including breakage and loss of nuclear and flagellar membranes, thickening of membranes, malformed acrosomes and loss of nuclear material, and the results are associated with heavy metals, such as Pb and Mn. The toxicity order of metals on reproduction in earthworms is Cd, Cu, Zn and Pb. Similar results have been found in *E. fetida* exposed to a geometric series of concentrations of Cd, Cu, Pb and Zn in artificial soil and the effects of Cd and Cu on the reproductive rate were particularly acute (Spurgeon et al., 1994).

It has been well known that earthworms are able to inhabit soils contaminated with heavy metals (Becquer et al., 2005; Li et al., 2010; Maity et al., 2008) and can accumulate undesirably high concentration of heavy metals (Cu, Zn, Pb and Cd) that may give adverse effects on livestock (Hobbelen et al., 2006; Oste et al., 2001). Earthworms (*L. rubellus* and *Dendrodrilus rubidus*) sampled from one uncontaminated and 15 metal-contaminated sites showed significant positive correlations between earthworm and total (conc. nitric acid-extractable) soil Cd, Cu, Pb and Zn concentrations (Morgan & Morgan, 1988). The important factor in the accumulation of heavy metals in earthworms is bioavailability by uptake (Dai et al., 2004; Spurgeon & Hopkin, 1996) because there are significant correlations between the concentrations of heavy metal accumulated in earthworms and bioavailable metal concentrations of field soils (Hobbelen et al., 2006). Earthworm metal bioaccumulation and bioavailability have been well reviewed by Nahmani et al. (2007). There were positive relationships between earthworm tissue and soil metal concentrations and also earthworm tissue and soil solution metal concentrations with slightly more significant relationships between earthworm tissue and soil metal concentrations 42 days after treatment. Recently, Li et al. (2010) reported the positive logarithmic relationship between the bioaccumulation factors of *E. fetida* to heavy metals and the exchangeable metal concentration of pig manure. The differences in these accumulation and availability among earthworms may, in part, play a role in affecting their population density and genetic adaptation living in metal-contaminated soils.

However, Lee (1985) suggested that the differences in the relative toxicity of compounds may explain some of the conflicting data in the literature on the concentrations which have deleterious effects on earthworms. For instance, very high concentrations of lead that influence growth and reproduction of earthworms may be attributable more to the very low solubility of lead compounds that are found in soils and the ability of earthworms to sequester absorbed lead than to any lower toxicity of lead compared with other heavy metals. It has been suggested that *E. fetida* may regulate the concentration of zinc in their body tissue through allowing rapid elimination by binding zinc using metallothioneins in their chloragogenous tissue (Cotter-Howells et al., 2005; Morgan & Morris, 1982; Morgan & Winters, 1982; Prento, 1979). High tolerance of earthworms to cadmium poisoning may also result from detoxification by metallothionein proteins in the posterior alimentary canal (Morgan et al., 1989). In addition, heavy metals have high affinity for glutathione, metallothioneines and enzymes of intermediary metabolism and heme synthesis (Montgomery et al., 1980). The metals Zn, Pb, Bi and Cd which are not consistently prevailing toxicants were most accessible to earthworms and Cu, Zn and Cr were also accumulated in earthworm tissue and the contaminated soils imparied earthworm reproduction and reduced adult growth, while elevated superoxide dismutase activity suggested that earthworms experienced oxidative stress (Berthelot et al., 2008).

Lead, copper and zinc may inhibit *d*-aminolevulinic acid dehydratase (*d*-ALAD) which is a key enzyme in heme synthesis by lowering haemoglobin concentration in earthworm blood. Replacement of zinc, a protector of the active site of *d*-ALAD, by lead may result in its inhibition.

Soil pH has been comprehensively identified as the single most important soil factor controlling the availability of heavy metals in sludge-treated soils (Alloway & Jackson, 1991). Soil pH is also one of the most important factors that limit the species, numbers and distribution of earthworms (Dunger, 1989; Edwards & Bohlen, 1996; Satchell & Stone, 1972) because it may affect the survival of adults and thus production and avoidance behaviour of juveniles (Aorim et al., 1999, 2005). van Gestel et al. (2011) reported that soil pH and organic matter content determine molybdenum toxicity to enchytraeid worm, *Enchytraeus crypticus* A higher pH resulted in a decreased sorption of the molybdate anion, and it caused increased bioavailability and toxicity.

A lot of studies concerning the effects of heavy metals on earthworms in terms of mortality, loss of weight, fertility, cocoon production, cocoon viability and growth were carried out during short-term experiments (14 or 21 days) in artificial soils contaminated with metal solution containing a single metallic element. Recently, Na et al. (2011) studied the effects of long-term (8 years) application of four organic waste materials on earthworm numbers and biomass. They reported that earthworm individuals were correlated positively with pH (r = 0.37) and negatively with heavy metals (r = -0.36 to -0.55) with the exception of Zn 4 years after treatment, while earthworm individuals were correlated positively with pH (r = 0.46) and negatively with Pb (r = -0.41) but positively with Zn (r = 0.59) 8 years after treatment. Earthworm biomass was correlated negatively with heavy metals (r = -0.43 to -0.72) with the exception of Zn 4 years after treatment, while earthworm biomass was correlated positively with pH (r = 0.57) and negatively with Pb (r = -0.50) and Ni (r = -0.30) but positively with Zn (r = 0.68) 8 years after treatment.

4.2 Effects of hexane extractable material on composition and biomass of earthworm

United States Environmental Protection Agency [USEPA] 9071B method (1998) was used to extract relatively non-volatile hydrocarbons from 45 lysimeter soils treated twice annually with three levels of four different organic waste materials and pig manure compost tested for 8 consecutive years, as stated in section 4.1. The extracts were generally designated hexane extractable material (HEM) because the solvent used was hexane. Soils were acidified with 0.3 ml of concentrated HCl and dried over magnesium sulfate monohydrate. After drying in a fume hood, HEM was extracted for 4 hr using a Soxhlet apparatus which was attached a 125 ml boiling flask containing 90 ml of hexane. Solvent was then concentrated under vacuum for less than 30 min at 35°C. The extracts were cooled in a desiccator for 30 min, and HEM concentrations were calculated by the formula, HEM (mg kg of dry weight^{-1}) = (A × 1000)/BC, where A is gain in weight of flask (mg), B is weight of wet solid (g) and C is dry weight fraction (g of dry sample g of sample^{-1}).

HEM amounts varied with treatment level and organic waste examined (Fig. 6). At 8 years after treatment, effect of test waste material (F = 49.45; df = 4,14; P < 0.001) and treatment level (F = 4.09; df = 2,30; P = 0.028) on the HEM was significant. The material by level interaction was also significant (F = 2.63; df = 8,44; P = 0.0258). Particularly, the amount of HEM in PMC-treated soil was the lowest of any of test materils at all treatment levels.

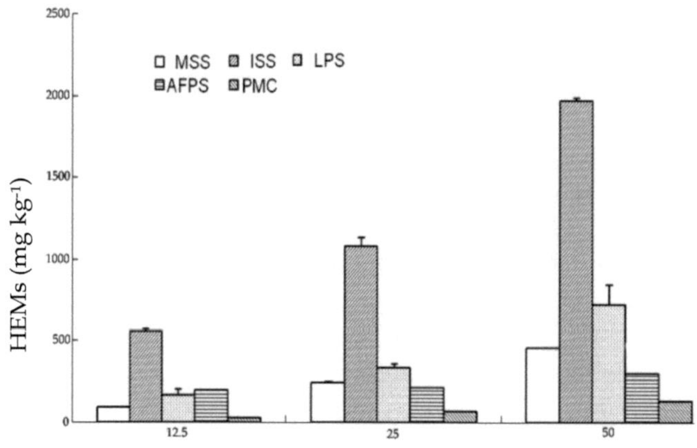

Fig. 6. Hexane extractable material (HEM) contents of 8 consecutive yearly applications (twice annually) of three levels of four different organic waste materials and pig manure compost using field lysimeters. Abbreviations are same as in the text

Correlation between HEM content (Fig. 6) and earthworm individuals and biomass (Table 2) was determined. At 8 years after treatment, earthworm individuals were negatively correlated with HEM (r = -0.313) and earthworm biomass (r = -0.335).

In general, organic compounds existed in sewage sludge have been potentially transferred to sludge-amended agricultural soils, and most organic compounds have been solved in hexane solvent. HEMs from sewage sludges contain a variety of contaminants, such as hydrocarbons, grease, plant or animal oils, wax, soap, polychlorinated biphenyls (PCBs) and polycyclic aromatic hydrocarbons (PAHs) (Hua et al., 2008; Stevens et al., 2003). Drescher-Kaden et al. (1992) reported that 332 organic contaminants (e.g., pyrene, benzo(*a*)pyrene, benzene and toluene) with potential to exert soil contamination were identified in German sewage sludges. Hembrock-Heger (1992) found that the concentrations of PAHs and PCBs appeared to be highest in soils treated with sewage sludge for 10 years. According to the United Kingdom Water Research Centre Report No. DoE 3625/1 on the occurrence, fate and behaviour of some of organic pollutants in sewage sludge (Sweetman et al., 1994), there was no evidence of any significant problems arising from organic contaminants in sludges applied to agricultural land.

Of some waste sludge and PMC applied into red pepper fields in South Korea from 2003 to 2004, the highest contents of HEM and PAHs were observed in cosmetic and pharmaceutical industy sludge, respectively, and the cosmetic industry sludge affected remarkedly growth of red pepper, which resulted in 25-60% of yield reduction (Lee, 2006). These results indicate that PMC may contain a lot of polar compounds with functional groups, such as COO^-, O^-, NR_2H, COOH or OH, to be more easily metabolized by various soil-born organisms, including earthworm. Water drained from processing of ISS, LPS, MSS and AFPS may contain more non-soluble compounds than that of PMC. Considering the hexane fraction obtained from PMC containing plentiful P or N atom (Na, 2004), it may be biodegradable by long-term exposure to a variety of soil organisms owing to biological uses. In general, most hydrophobic compounds are accumulative and difficult to biodegrade them introducing into environments because most aliphatic hydrocarbons retain unfavorable large ΔG (minus value) with increase in chain length.

4.3 Toxicity of soil contamination level to *E. fetida* in microcosms

Each microcosm was made of commercially available high-density stable polyethylene container (14 cm length, 14 m width and 7 m depth) with 36 pores (1 mm diameter) of lid. Soils sampled in microcosms treated with three levels (12.5, 25 and 50 tons of dry matter ha^{-1} year^{-1}) of MSS, ISS, LPS, AFPS and PMC for 4 consecutive years (twice annually) were sieved gently through a 2 mm mesh sieve. In a preliminary experiment, ~40% of water holding capacity was optimal for microcosm test. Amount of 300-g fresh soil was hydrated to ~40% of water holding capacity. Hydrated water required to achieve the desired hydration was calculated according to the method of Greene et al. (1988). Ten earthworms were placed into each microcosm. The microcosms were kept in the controlled chamber at 20°C and 60±5% relative humidity under a 16:8 h light:dark cycle. Mortalities were assessed by emptying the test soil onto a tray and sorting the worms from the soil. Earthworms were considered to be dead if their bodies and anterior did not move or respond when they prodded with fine wooden dowels. Live worms were placed back into their original microcosms. The numbers of live and dead worms in each microcosm were recorded every 2 weeks and the dead worms were discarded. A randomized complete block design with three replicates was used. Mortality percentages were transformed to arcsine square root values for analysis of variance. The Bonferroni multiple-comparison method was used to test for significant differences among the treatments (SAS Institute, 2004).

Toxic effects of MSS, ISS, LPS, AFPS and PMC treatments on *E. fetida* in microcosm tests were evaluated (Table 4). All treatments did not affect any adverse effects on the organisms 2 weeks after treatment. At 4 weeks after treatment, effect of test waste material (F = 3.73; df = 4,44; P = 0.0141) on the mortality was significant but that of treatment level (F = 1.83; df = 2,44; P = 0.1785) was not significant. The material by level interaction was also significant (F = 2.34; df = 8,44; P = 0.0436). At 8 weeks after treatment, effect of test waste material (F = 200.90; df = 4,44; P < 0.0001) and treatment level (F = 5.37; df = 2,44; P = 0.0101) on the the mortality was significant. The material by level interaction was also significant (F = 9.49; df = 8,44; P < 0.0001). After 16 weaks after treatment, effect of test waste material (F = 124.11; df = 4,44; P < 0.0001) and treatment level (F = 9.73; df = 2,44; P = 0.0006) on the mortality was significant. The material by level interaction was also significant (F = 63.42; df = 8,44; P < 0.0001).

Heimbach et al. (1992) demonstrated that there is a good correlation (r = 0.86) between LC_{50} values of pesticides from an artificial soil test and the number of earthworms collected from a standardized field test. Our present and previous studies indicate that microcosm soil test using earthworms can predict results from a field test for assessing side effects occurred by long-term exposure of soil contaminants. Burrows & Edwards (2002) have been tried to use integrated soil microcosm based upon earthworms to predict effects of pollutants on soil ecosystems.

5. Future perspectives

Due to the predicted impacts of climate change, many farmers are increasingly concerned about severe soil compaction and water stagnation on their fields. To prevent the deterioration of arable soils, appropriate soil management stratigies have to be developed. Earthworms are an important component of the soil biodiversity and their positive effects on soil structure are well-known. A variety of functional groups of earthworms can be restored through decrease in a soil disturbance and occurence of crop residues in the upper soil. Investigating earthworm biomass and population is a very complex process and therefore consists of various methods of sampling. However, it is difficult to conduct efficient investigations due to horizontally aggregated earthworm populations and their complex phenologies. Of the vaiorus methods, hand sorting which involves sorting through soil samples by hand is one of the most earliest popular sampling methods. The soil washing method is more effective in sorting out cocoons and smaller earthworms. This method consists of a combination of washing and sieving soil samples, with a possible flotation stage. Another method that is used for soil sampling is the electrical method which consists of inserting an electrode into the groud causing earthworms to surface due to the electrical pulse in the soil. These methods, however, usually result in disrupting earthworm biomass and population, killing and injuring them and affecting their habitats. Considering the relationships between heavy metals and earthworms inhabiting in contaminated soils, it needs to adjust study focus on the long-term effects of multiple elements, not one heavy metal, to earthworms. However, these methods make it difficult to consistantly study and investigate a selective biomass during long periods.

With future developements in terms of remote sensing used for detecting the small- or large-scale acquisition of information of an object or phenomenon, these issues will no longer serve as a problem because biomass can be studied without any need of disruption. Although underground remote sensing technologies are in use, they have not yet been applied to the investigation of living organisms, such as earthworms. For that reason, we believe that scientists and remote sensing developers should put their heads together to optimize remote sensing equipment to the investigation of underground living organisms. These advancements will significantly help researchers to consistantly study a select biomass and calculate the amount of toxic materials that are being inserted into the soil more accurately.

Treatment[a]	Rate[b]	% mortality (mean ± SE) at weeks after treatment		
		4	8	16
MSS	12.5	0	6 ± 3.3	53 ± 6.0
	25	3 ± 3.3	7 ± 6.7	93 ± 6.0
	50	10 ± 10.0	13 ± 3.3	70 ± 5.2
ISS	12.5	37 ± 18.6	60 ± 5.8	87 ± 6.0
	25	7 ± 6.7	97 ± 3.3	100
	50	0	97 ± 3.3	100
LPS	12.5	0	30 ± 5.8	97 ± 3.0
	25	0	3 ± 3.3	97 ± 3.0
	50	0	27 ± 8.8	100
AFPS	12.5	0	0	33 ± 6.0
	25	0	17 ± 3.3	20 ± 5.2
	50	0	0	90 ± 9.0
PMC	12.5	0	0	17 ± 7.9
	25	0	0	20 ± 0.0
	50	3 ± 3.3	3 ± 3.3	3 ± 3.0

[a,b] Tons of dry matter ha^{-1} $year^{-1}$

Table 4. Accumulative mortality of *Eisenia fetida* earthworms in microcosm soils treated twice annually with three levels of four different organic waste materials and pig manure compost tested for 4 consecutive years

6. Conclusion

The long-term applications of organic waste materials containing heavy metals and HEMs affected the establishment of Megascolecid and Moniligastrid earthworms in field. The biomass of earthworms in lysimeter and microcosm soil tests would provide valuable tools for establishing the integrated hazard assessment system for organic wastes. Future research is needed to establish additional soil physico-chemical characteristics, particularly those that might influence heavy-metal bioaccumulation and bioavailability and physical habitat such as compaction and soil water holding capacity across treatment through time course.

7. Acknowledgement

This work was supported by the Rural Development Administration and WCU (World Class University) programme (R31-10056) through the National Research Foundation of Korea funded by the Ministry of Education, Science and Technology.

8. References

Abbasi, S.A. & Soni, R. (1983). Stress-Induced Enhancement of Reproduction in the Earthworm *Octochaetus pattoni* Exposed to Chromium and Merandy: Implications in Environmental Management. *International Journal of Environmental Studies*, Vol. 22, No. 1, pp. 43-47, ISSN 0020-7233

Alloway, B.J. & Jackson, A.P. (1991). The Behaviour of Heavy Metals in Sewage Sludge-Amended Soil. *Science of the Total Environment*, Vol. 100, pp. 151-179, ISSN 0048-9697

Amorim, M.J.B.; Römbke, J. & Soares, A.M.V.M. (2005). Avoidance Behaviour of *Enchytraeus albidus*: Effects of Benomyl, Carbendazim, Phenmedipham and Different Soil Types. *Chemosphere*, Vol. 59, No. 4, pp. 501-510, ISSN 0045-6535

Amorim, M.J.; Sousa, J.P.; Nogueira, A.J.A. & Soares, A.M.V.M. (1999). Comparison of Chronic Toxicity of Lindane (γ-HCH) to Enchytraeus albidus in Two Soil Types: the Influence of Soil pH. *Pedobiologia*, Vol. 43, No. 6, pp. 635-640, ISSN 0031-4056

Anon. (1997) *Environmental Guidelines: Use and Disposal of Biosolids Products*, New South Wales Environment Protection Authority, ISBN 0731037928, Sydney, New South Wales, Australia

Anon. (2006). *Fertilizer Management Act*, Rural Development Administration, Republic of Korea

Anon. (May 2007). *Soil Environment Conservation Act*, In: *Environmental Laws*, 12.04.2011, Available from http://eng.me.go.kr/board.do?method=view&docSeq=8094&bbsCode=law_law_law

Anon. (2009). *Statistics on Production and Outlets of Sewage Sludge*, Ministry of Environment, Gwacheon, Gyeonggi, Republic of Korea

Anon. (2010a). *Official Standards of Fertilizer, Notification No. 2010-33*, Rural Development Administration, Suwon, Gyeonggi, Republic of Korea

Anon. (February 2010b) Richtlinie des Rates vom 12. Juni 1986 über den Schutz der Umwelt und insbesondere der Boden bei der Verwendung von Klärschlamm in der Landwirtschaft (86/278/EWG), In: *Verwendung von Klärschlamm in der*

Landwirtschaft, 12.04.2011, Available from http://europa.eu/legislation_summaries/agriculture/environment/128088_de.htm

Anon. (November 2010c) Klärschlammverordnung (AbfKläV), In *Klärschlammverordnung*, 12.04.2011, Available from http://de.wikipedia.org/wiki/K1%C3%A4rschlammverordnung

Anon. (2010d) Schweizerische Qualitätsrichtlinie 2010 der Branche für Kompost und Gärgut, In: *Qualitätsrichtlinie 2010 für Kompost und Gärgut*, 12.04.2011, Available from http://www.Kompostverband.ch/LinkClickaspx?fileticket=2sYYyf4Hb10%3D&tabid=982&language=fr-FR

Beauge, A. (1912). Les vers de terre et la fertilite du sol. *Journal of Agricultural Pratt*, Vol. 23, pp. 506-507

Becquer, T.; Dai, J.; Quantin, C. & Lavelle, P. (2005). Sources of Bioavailable Trace Metals for Earthworms from a Zn-, Pb- and Cd-Contaminated Soil. *Soil Biology & Biochemistry*, Vol. 37, No. 8, pp. 1564-1568, ISSN 0038-0717

Berthelot, Y.; Valton, E.; Auroy, A.; Trottier, B. & Robidoux, P.Y. (2008). Integration of Toxicological and Chemical Tools to Assess the Bioavailability of Metals and Energetic Compounds in Contaminated Soils. *Chemosphere*, Vol. 74, No. 1, pp. 166-177, ISSN 0045-6535

Blanchart, E. (1992). Restoration by Earthworms (Megascolecidae) of the Macroaggregate Structure of a Destructured Savanna Soil under Field Conditions. *Soil Biology & Biochemistry*, Vol. 24, No. 12, pp. 1587-1594, ISSN 0038-0717

Bouché, M.B. (1972). *Lombriciens de France: Ecologie et Systematique*, INRA Publications, Paris, France

Bouché, M.B. (1992). Earthworm Species and Ecotoxicological Studies, In: *Ecotoxicology of Earthworms*, P.W. Greig-Smith, H. Becjer, P.J. Edwards & F. Heimback, (Eds.), pp. 20-35, Intercept, ISBN 9780946707409, Andover, Hampshire, UK

Brinton, W.F. (2000). *Compost Quality Standards & Guidelines*, Woods End Research Labratory, New York, NY, USA

BSI. (2011) *PAS 100:2011 Specification for Composted Materials*, British Standards Institution, ISBN 978-0-580-65307-0, London, UK

Burrows, L.A. & Edwards, C.A. (2002). The Use of Integrated Soil Microcosms to Predict Effects of Pesticides on Soil Ecosystems. *European Journal of Soil Biology*, Vol. 38, No. 3-4, pp. 245-249, ISSN 1164-5563

Callaham, M.A. & Hendrix, P.F. (1997). Relative Abundance and Seasonal Activity of Earthworms (Lumbricidae and Megascolecidae) as Determined by Hand-sorting and Formalin Extraction in Forest Soils on the Southern Appalachian Piedmont. *Soil Biology & Biochemistry*, Vol. 29, No. 3-4, pp. 317-321, ISSN 0038-0717

Callahan, C.A.; Shirazi. M.A. & Neuhauser, E.F. (1994). Comparative Toxicity of Chemicals to Earthworms. *Environmental Toxicology and Chemistry*, Vol. 13, No. 2, pp. 291-298, ISSN 1552-8618

Carter, A.; Heinonen, J. & de Vries, J. (1982). Earthworms and Water Movement. *Pedobiologia*, Vol. 23, No. 6, pp. 395-397, ISSN 0031-4056

CCME. (2005). *Guidelines for Compost Quality*, Canadian Council of Ministers of the Environment, ISBN 1-896997-60-0, Winnipeg, Manitoba, Canada

Cikutovic, M.A.; Fitzpatrick, L.C.; Venables, B.J. & Goven, A.J. (1993). Sperm Count in Earthworms (*Lumbricus terrestris*) as a Biomarker for Environmental Toxicology: Effects of Cadmium and Chlordane. *Environmental Pollution*, Vol. 81, No. 2, pp. 123-125, ISSN 0269-7491

Coker, E.G.; Hall, J.E.; Carlton-Smith, C.H. & Davis, R.D. (1987). Field Investigations into the Manurial Value of Lagoon-Matured Digested Sewage Sludge. *Journal of Agricultural Science*, Vol. 109, No. 3, pp. 467-478, ISSN 0021-8596

Cotter-Howells, J.; Charnock, J.M.; Winters, C.; Kille, P.; Fry, J.C. & Morgan, A.J. (2005). Metal Compartmentation and Speciation in a Soil Sentinel: the Earthworm, *Dendrodrilus rubidus*. *Environmental Science and Technology*, Vol. 39, No. 19, pp. 7731-7740, ISSN 0013-936X

Curry, J.P. (1998). Factors Affecting Earthworm Abundance in Soil, In: *Earthworm Ecology*, C.A. Edwards, (Ed.), pp. 37-64, Taylor & Francis, ISBN 1-884015-74-3, Boca Raton, Florida, USA

Curry, J.P. & Byrne, D. (1992). The Role of Earthworms in Straw Decomposition and Nitrogen Turnover in Arable Land in Ireland. *Soil Biology & Biochemistry*, Vol. 24, No. 12, pp. 1409-1412, ISSN 0038-0717

Dai, J.; Becquer, T.; Rouiller, J.H.; Reversat, G.; Bernhard-Reversat, F.; Nahmani, J. & Lavelle, P. (2004). Heavy Metal Accumulation by Two Earthworm Species and Its Relationship to Total and DTPA-Extractable Metals in Soils. *Soil Biology & Biochemistry*, Vol. 36, No. 1, pp. 91-98, ISSN 0038-0717

Darwin, C. (2009). *The Formation of Vegetable Mould through the Action of Worms with Observations of Their Habits*, Cambrige University Press, ISBN 978-1-108-00512-8, New York, NY, USA

Drescher-Kaden, U.; Bruggeman, R.; Matthes, B. & Matthies, M. (1992). Contents of Organic Pollutants in German Sewage Sludges, In: *Effects of Organic Contaminants in Sewage Sludge on Soil Fertility, Plants and Animals*, J.E. Hall, D.R. Sauerbeck & P. L'Hermite, (Eds.), pp. 14-34, Office for Official Publications of the European Communities, ISBN 9282638782, Luxembourg, Luxembourg.

Devliegher, W. & Verstraete, W. (1997). The Effect of *Lumbricus terrestris* on Soil in Relation to Plant Growth: Effects of Nutrient-Enrichment Processes (NEP) and Gut-Associated Processes (GAP). *Soil Biology & Biochemistry*, Vol. 29, No. 3-4, pp. 341-346, ISSN 0038-0717

Dunger, W. (1989). The Return of Soil Fauna to Coal Mine Areas in the German Democratic Republic, In: *Animals in Primary Succession: The Role of Fauna in Reclaimed Land*, J.D. Majer, (Ed.), pp. 307-337, Cambridge University Press, ISBN 0521334004, Cambridge, UK

Edwards, C.A. (1984). Report on the Second Stage in Development of a Standardized Laboratory Method for Assessing the Toxicity of Chemical Substances to Earthworms, Doc EUR 9360 EN, In: *The Artificial Soil Test*, Revision 4, Commission of the European Communities, DG X1/AL/82/43, Brussels, Luxembourg

Edwards, C.A. & Thompson, A.R. (1973). Pesticides and the Soil Fauna. *Residue Review*, Vol. 45, pp. 1-79, ISSN 0080181

Edwards, C.A. & Fletcher, K.E. (1988). Interactions between Earthworms and Micro-organisms in Organic-Matter Breakdown. *Agriculture Ecosystems & Environment*, Vol. 24, No. 1-3, pp. 235-247, ISSN 0167-8809

Edwards, C.A. & Bohlen, P.J. (1996). *Biology and Ecology of Earthworms* (3rd edition), Chapman & Hall, ISBN 0412561603, New York, NY, USA

Edwards, W.M.; Shipitalo, M.J.; Owens, L.B. & Norton, L.D. (1990). Effect of *Lumbricus terrestris* L. Burrows on Hydrology of Continuous No-till Corn Fields. *Geoderma*, Vol. 46, No. 1-3, pp. 73-84, ISSN 0016-7061

Emmerling, C. & Paulsch, D. (2001). Improvement of Earthworm (Lumbricidae) Community and Activity in Mine Soils from Open-cast Coal Mining by the Application of Different Organic Waste Materials. *Pedobiologia*, Vol. 45, No. 5, pp. 396-407, ISSN 0031-4056

Greene, J.C.; Bartels, C.L.; Warren-Hicks, W.J.; Parkhurst, B.R.; Linder, G.L.; Peterson, S.A. & Miller, W.E. (1988). *Protocols for Short Term Toxicity Screening of Hazardous Waste Sites*, US EPA Report 600/3-88/029, pp. 1-112, US Environmental Protection Agency, Washington, DC, USA

Halsall, C.; Burnett, V.; Davis, B.; Jones, P.; Pettit, C. & Jones, K.C. (1993). PCBs and PAHs in U.K. Urban Air. *Chemosphere*, Vol. 26, No. 12, pp. 2185-2197, ISSN 0045-6535

Heimbach, F.; Greig-Smith, P.W.; Becker, H.; Edwards, P. J. & Heimbach F. (1992). Effects of Pesticides on Earthworm Populations: Comparison of Results from Laboratory and Field Tests, In: *Ecotoxicology of Earthworms*, P.W. Greig-Smith, H. Becjer, P.J. Edwards & F. Heimback, (Eds.), pp. 299-302, Intercept, ISBN 9780946707409, Andover, Hampshire, UK

Helmke, P.A.; Robarge, W.P.; Korotev, R.L. & Schomberg, P.J. (1979). Effects of Soil-Applied Sewage Sludge on Concentrations of Elements in Earthworms. *Journal of Environmental Quality*, Vol. 8, No. 3, pp. 322-327, ISSN 0047-2425

Hembrock-Heger, A. (1992). Persistent Organic Contaminants in Soils, Plants and Food, In: *Effects of Organic Contaminants in Sewage Sludge on Soil Fertility, Plants and Animals*, J.E. Hall, D.R. Sauerbeck & P. L'Hermite, (Eds.), pp. 78-89, Office for Official Publications of the European Communities, ISBN 9282638782, Luxembourg, Luxembourg

Hobbelen, P.H.F.; Koolhaas, J.E. & van Gestel, C.A.M. (2006). Bioaccumulation of Heavy Metals in the Earthworms *Lumbricus rubellus* and *Aporrectodea caliginosa* in Relation to Total and Available Metal Concentrations in Field Soils. *Environmental Pollution*, Vol. 144, No. 2, pp. 639-646, ISSN 0269-7491

Hong, Y. (2000). Taxonomic Review of the Family Lumbricidae(Oligochaeta) in Korea. *Korean Journal of Systematic Zoology*, Vol. 16, No. 1, pp. 1-13, ISSN 1018-192X

Hong, Y. (2005). The Earthworm Fauna from Ulleungdo Island, Korea with One New Species of the Genus *Amynthas* (Oligochaeta: Megascolecidae). *Korean Journal of Systematic Zoology*, Vol. 21, No. 2, pp. 129-136, ISSN 1018-192X

Hong, Y. & James, S.W. (2001). New Species of Korean *Amynthas* Kinberg, 1867 (Oligochaeta, Megascolecidae) with Two Pairs of Spermathecae. *Revue Suisse de Zoologei*, Vol. 108, No. 1, pp. 65-93, ISSN 0035-418X

Hong, Y. & Kim, T.H. (2007). Occurrence of Earthworm in Agro-Ecosystem. *Korean Journal of Environmental Biology*, Vol. 25, No. 2, pp. 88-93, ISSN 0254-8704

Hong, Y.; Kim, T.H. & Na, Y.E. (2001). Identity of Two Earthworms Used in Vermiculture and Vermicompositing in Korea: *Eisenia andrei* and *Perionyx excavates*. *Korean Journal of Systematic Zoology*, Vol. 17, No. 2, pp. 185-190, ISSN 1018-192X

Hua, L.; Wu, W.X.; Liu, Y.X.; Tientchen, C.M. & Chen, Y.X. (2008). Heavy Metals and PAHs in Sewage Sludge from Twelve Wastewater Treatment Plants in Zhejiang Province. *Biomedical & Environmental Sciences*, Vol. 21, No. 4, pp. 342-352, ISSN 0895-3988

Jung, G.B.; Kim, W.I.; Lee, J.S.; Lee, J.S.; Park, C.W. & Koh, M.H. (2005). Characteristics of Heavy Metal Contamination in Residual Mine Tailings near Abandoned Metalliferous Mines in Korea. *Korean Journal of Environmental Agriculture*, Vol. 24, No. 3, pp. 222-231, ISSN 1225-3537

Kim, T.H.; Hong, Y. & Choi, N.J. (2009). Selection of Earthworm for Bioindicators in Agroecosystem. *Korean Journal of Environmental Biology*, Vol. 27, No. 1, pp. 40-47, ISSN 0254-8704

Kobayashi, S. (1941). Earthworm of Korea (II). *Science Report of the Tohoku Imperial University 4th Series (Biology)*, Vol. 16, No. 2, pp. 147-156, ISSN 1937-240

Kretzschmar, A. (1978). Quantification écologique des galeries de lombriciens: techniques et premiéres estimations. *Pedobiologia*, Vol. 18, No. 1, pp. 31-38, ISSN 0031-4056

Lavelle, P. & Spain, A. (2001). *Soil Ecology*, Kluwer Scientific Publications, ISBN 0-7923-7123-2, Amsterdam, the Netherlands

Lee, K.E. (1985). *Earthworms: Their Ecology and Relationships with Soils and Land Use*, Academic Press, ISBN 978-0124408609, Sydney, Australia

Lee, S.H. (2006). *Chemical and Biological Evaluation of Organic Wastes for Agricultural Utilization*, PhD Dessertation, Chungbuk National University, Cheongju, Chungbuk, Republic of Korea

Li, L.; Xu, Z.; Wu, J. & Tian, G. (2010). Bioaccumulation of Heavy Metals in the Earthworm *Eisenia fetida* in Relation to Bioavailable Metal Concentrations in Pig Manure. *Bioresource Technology*, Vol. 101, No. 10, pp. 3430-3436, ISSN 09608524

MAFF. (1991). *Code of Good Agricultural Practice for the Protection of Water, Air and Soil* PB0587, MAFF Publications, London, UK

Maity, S.; Padhy, P.K. & Chaudhury, S. (2008). The Role of Earthworm *Lampito mauritii* (Kinberg) in Amending Lead and Zinc Treated Soil. *Bioresource Technology*, Vol. 99, No. 15, pp. 7291-7298, ISSN 09608524

Martin, A. (1991). Short- and Long-Term Effects of the Endogenic Earthworm *Millsonia anomala* (Omodeo) (Megascolecidae, Oligochaeta) of Tropical Savannas on Soil Organic Matter. *Biology and Fertility of Soils*, Vol. 11, No. 3, pp. 234-238, ISSN 0178-2762

McGrath, S.P. (1994) Effects of Heavy Metals from Sewage Sludge on Soil Microbes in Agricultural Ecosystems, in: *Toxic Metals in Soil-Plant Systems*, S.M. Ross, (Ed.), pp. 247–274, John Wiley & Sons, ISBN 978-0-471-94279-5, Chichester, West Sussex, UK

Ministry of the Environment Government of Japan. (February 1994). Environmental Quality Standards for Soil Pollution, In: *Water/Soil/Ground Environment*, 04.04.2011, Available from http://www.env.go.jp/ en/water/soil/sp.html

Montgomery, R.; Dryer, R.L.; Conway, T.W. & Spector, A.A. (1980). *Biochemistry, A Case-Oriented Approach* (3rd edition), YB Medical Publishers, ISBN 0-8016-3470-9, London, UK

Morgan, A.J. & Morris, B. (1982). The Accumulation and Intracellular Compartmentation of Cadmium, Lead, Zinc, and Calcium in Two Earthworm Species (*Dendrobaena rubida* and *Lumbricus rubellus*) Living in a Highly Contaminated Soil. *Histochemistry*, Vol. 75, No. 2, pp. 269-285, ISSN 0018-2222

Morgan, A.J. & Winters, C. (1982). The Elemental Composition of the Chloragosomes of Two Earthworm Species *Lumbricus terrestris* and *Allolobophora longa* Determined by Electron Probe X-Ray Microanalysis of Freeze-Dried Cryosections. *Histochemistry*, Vol. 73, No. 4, pp. 589-598, ISSN 0018-2222

Morgan, J.E. & Morgan, A.J. (1988). Earthworms as Biological Monitors of Cadmium, Copper, Lead and Zinc in Metalliferous Soils. *Environmental Pollution*, Vol. 54, No. 2, pp. 123-138, ISSN 0269-7491

Morgan, J.E.; Norey, C.G.; Morgan, A.J.; & Kay, J. (1989). A Comparison of the Cadmium-Binding Proteins Isolated from the Posterior Alimentary Canal of the Earthworm *Demlrodrilus rubidus* and *Lumbricus rubellus*. *Comparative Biochemistry and Physiology Part C*, Vol. 92, No. 1, pp. 15-21, ISSN 1532-0456

Na, Y.E. (2004). *Hazard Assessment of Organic Waste-Contaminated Soil Using Earthworm*, PhD Dessertation, Seoul National University, Seoul, Republic of Korea

Na, Y.E.; Bang, H.S.; Kwon, S.I.; Kim, M.H. & Ahn, Y.J. (2011). Harzadous Effects of Eight Years of Application of Four Organic Waste Materials on Earthworm Numbers and Biomass in Field Lysimeters. *Archives of Environmental Contamination and Toxicology*, Vol. 60, No. 1, pp. 99-106, ISSN 0090-4341

Nahmani, J.; Hodson, M.E. & Black, S. (2007). A Review of Studies Performed to Assess Metal Uptake by Earthworms. *Environmental Pollution*, Vol. 145, No. 2, pp. 402-424, ISSN 0269-7491

NZWWA. (2003) *Guidelines for the Safe Application of Biosolids to Land in New Zealand*, New Zealand Water and Waste Association, ISBN 1-877134-43-0, Wellington, New Zealand

OECD. (1984). Test No. 207: Earthworm Acute Toxicity Tests, In *OECD Guidelines for Testing of Chemicals, Section 2: Effects on Biotic Systems*, Organisation for Economic Cooperation and Development, ISBN 9789264070042, Paris, France

Oste, L.A.; Dolfing, J.; Ma, W.C. & Lexmond, T.M. (2001). Cadmium Uptake by Earthworms as Related to the Availability in the Soil and the Intestine. *Environmental Toxicology and Chemistry*, Vol. 20, No. 8, pp. 1785-1791, ISSN 0730-7268

Pearson, M.S.; Maenpaa, K.; Pierzynski, G.M. & Lydy, M.J. (2000). Effects of Soil Amendments on the Bioavailability of Lead, Zinc and Cadmium to Earthworms. *Journal of Environmental Quality*, Vol. 29, No. 5, pp. 1611-1617, ISSN 0047-2425

Prento, P. (1979). Metals and Phosphate in the Chloragosomes of *Lumbricus terrestris* and Their Possible Physiological Significance. *Cell and Tissue Research*, Vol. 196, No. 1, pp. 123-134, ISSN 0302-766X

Reinecke, A.J. & Reinecke, S.A. (1996). The Influence of Heavy Metals on the Growth and Reproduction of the Compost Worm *Eisenia fetida* (Oligochaeta). *Pedobiologia*, Vol. 40, No. 5, pp. 439-448, ISSN 0031-4056

Reinecke S.A. & Reinecke A.J. (1997). The Influence of Lead and Managenese on the Spermatozoa of *Eisenia fetida* (Oligochaeta). *Soil Biology & Biochemistry*, Vol. 29, No. 3-4, pp. 737-742, ISSN 0038-0717

Roy, S.K. (1957). Studies on the Activities of Earthworms. *Proceedings of the Zoological Society of Bengal*, Vol. 10, pp. 81-98, ISSN 0373-5893

SAS Institute. (2004). *SAS OnlineDoc*, Version 8.01, SAS Institute, Cary, North Carolina, USA

Satchell, J.E. & Stone, D.E. (1972). Colonization of Pulverished Fuel Ash Sites by Earthworms. *Publicazione del Centro Pirenacio de Biologica Experimentalis*, Vol. 9, pp. 59-74

Scheu, S. & Wolters, S. (1991). Influence of Fragmentation and Bioturbation on the Decomposition of Carbon-14-Labelled Beech Leaf Litter. *Soil Biology & Biochemistry*, Vol. 23, No. 11, pp. 1029-1034, ISSN 0038-0717

Song, M.J. & Paik, K.Y. (1969). Preliminary Survey of the Earthworms from Dagelet Isl., Korea. *Korean Journal of Zoology*, Vol. 12, No. 1, pp. 13-21, ISSN 0440-2510

Spurgeon, D.J. & Hopkin, S.P. (1996). Effects of Variations of the Organic Matter Content and pH of Soils on the Availability and Toxicity of Zinc to the Earthworm *Eisenia fetida*. *Pedobiologia*, Vol. 40, No. 1, pp. 80-96, ISSN 0031-4056

Spurgeon, D.J.; Hopkin, S.P. & Jones, D.T. (1994). Effects of Cadmium, Copper, Lead and Zinc on Growth, Reproduction and Survival of the Earthworm *Eisenia fetida* (Savigny): Assessing the Environmental Impact of Point-Source Metal Contamination in Terrestrial Ecosystems. *Environmental Pollution*, Vol. 84, No. 2, pp. 123-130, ISSN 0269-7491

Stevens, J.L.; Northcott, G.L.; Stern, G.A.; Tomy, G.T. & Jones, K.C. (2003). PAHs, PCBs, PCNs, Organochlorine Pesticides, Synthetic Musks, and Polychlorinated *n*-Alkanes in U.K. Sewage Sludge: Survey Results and Implications. *Environmental Science & Technology*, Vol. 37, No. 3, pp. 462–467, ISSN 0013-936X

Sweetman, A.; Rogers, H.R.; Watts, C.D.; Alco, R. & Jones, K.C. (1994). *Organic Contaminents in Sewage Sludge: Phase III (Env 9031)*, Final Report to the Department of the Environment, Water Research Centre (WRc) Report No. DoE 3625/1, WRc Medmenham, Wycombe, UK

USEPA. (1998). n-Hexane Extractable Material (HEM) for Sludge, Sediment, and Soil Sample, Method 9071B, In *Test Methods for Evaluating Solid Waste, Volume 1A: Laboratory Manual Physical/Chemical Methods*, United States Environmental Protection Agency, Washington, DC, USA

USEPA. (February 2000). *Maximum Allowable Metal Concentrations in Sludge Amended Soils USEPA 503 Regulations (USEPA 1997)*, 503 Regulations (USEPA 1997), Available from http://www.tpub.com/content/ArmyDOE/doercl2/doercl20003.htm

van Gestel, A.M.C.; Borgman, E.; Verweij, R.A. & Ortiz, M.D. (2011). The Influence of Soil Properties on the Toxicity of Molybdenum to Three Species of Soil Invertebrates. *Ecotoxicology & Environmental Safety*, Vol. 74, No. 1, pp. 1-9, ISSN 0147-6513 van Hook, R.I. (1974). Cadmium, Lead and Zinc Distributions between Earthworms and Soils: Potentials for Biological Accumulation. *Bulletin of Environmental Contamination and Toxicology*, Vol. 12, No. 4, pp. 509-512, ISSN 0007-4861

Venables, B.J.; Fitzpatrick, L.C. & Goven, A.J. (1992). Earthworms as Indicators of Toxicity, In: *Ecotoxicology of Earthworms*, P.W. Greig-Smith, H. Becjer, P.J. Edwards & F. Heimback, (Eds.), pp. 197-206, Intercept, ISBN 9780946707409, Andover, Hampshire, UK

Wollny, E. (1890). Untersuchungen über die Beeinflussung der Fruchtbarkeit der Ackerkrume durch die Tätigkeit der Regenwürmer. *Forschung Agricole Physik*, Vol. 13, pp. 381-395

13

Plant Biomass Productivity Under Abiotic Stresses in SAT Agriculture

L. Krishnamurthy, M. Zaman-Allah, R. Purushothaman,
M. Irshad Ahmed and V. Vadez
International Crops Research Institute for the Semi-Arid Tropics (ICRISAT),
Patancheru 502 324, Andhra Pradesh
India

1. Introduction

1.1 Prevalence of abiotic stresses in SAT agriculture

The semi-arid tropics (SAT) include parts of 48 countries in the developing world: in most of India, locations in south east Asia, a swathe across sub-Saharan Africa, much of southern and eastern Africa, and a few locations in Latin America (Fig 1). Semi-arid tropical regions are characterized by unpredictable weather, long dry seasons, inconsistent rainfall, and soils that are poor in nutrients. Sorghum, millet, cowpea, chickpea, pigeonpea and groundnut are the vital crops that feed the poor people living in the SAT.

Environmental stresses represent the most limiting factors for agricultural productivity. Apart from biotic stresses caused by plant pathogens, there are a number of abiotic stresses such as extremes temperatures, drought, salinity and radiation which all have detrimental effects on plant growth and yield, especially when several occur together (Mittler 2006).

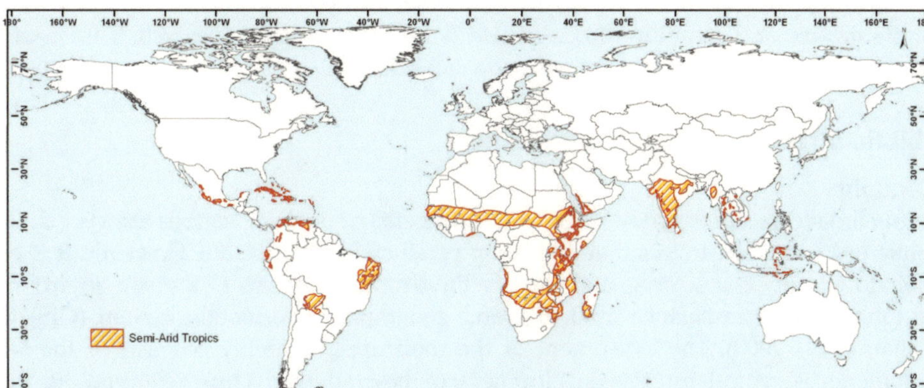

Fig. 1. Distribution of semi-arid tropical regions in the world (Source: http://www.fao.org/sd/EIdirect/climate/EIsp0002.htm)

Drought and soil salinity are the most prevailing abiotic stresses that curtail crop productivity in the SAT. Arable lands are lost every year due to desertification and salinization, as a result of sparse and seasonal rainfall and mismanagement of the natural resource base for agriculture (Evans, 1998). Expansion of irrigation does not seem feasible in many countries in Asia, the Middle East, and North Africa, where most of the available and easily accessible water resources have been already utilized. Furthermore, irrigated soils are affected by salinity with significant subsequent yield losses. Desertification may be aggravated by both extensive farming due to demographic pressure and the regional climatic changes. Hence, there is a need for the breeding programs to assign high priority for the development of crops with tolerance to both drought and salinity stress. The genetically complex control of these stresses in the plant genome may be facilitated through the manipulation of specific genes governing the component characteristics needed to achieve tolerance to salt or drought in plant crops.

1.2 Plant biomass productivity as affected by drought and salinity stress

Plant biomass is primarily a product of photosynthesis, a process needing carbon dioxide, water as bi-products and solar radiation as the energy source and mineral nutrients as basic blocks. In majority of the instances carbon dioxide and solar radiation never limit biomass production while abiotic stresses like water deficit and soil salinity very often do. Plant response to abiotic stress is one of the most active research topics in plant biology due to its practical implications in agriculture, since abiotic stresses (mainly drought and high soil salinity) are the major cause for the reduction in crop biomass and yield worldwide, especially in the SAT.

Plants are extremely sensitive to changes resulting from drought or salinity, and do not generally adapt quickly (Lane and Jarvis 2007). Plants also adapt very differently from one another, even from a plant living in the same area. When a group of different plant species was prompted by a variety of different stress signals, such as drought or cold, each plant responded uniquely. Hardly any of the responses were similar, even though the plants had become accustomed to exactly the same home environment (Mittler 2006). Abiotic stresses can come in many forms. The occurrence of many of these abiotic stresses is unpredictable, however, in agricultural management point of view, drought and soil salinity are relatively more predictable and common in occurrence demanding focused research. Therefore, the scope of this chapter is limited to drought and soil salinity.

2. Abiotic stresses and crop productivity

2.1 Drought

The agroclimatic and production-system environments of the SAT regions are very diverse. The inherent water constraints that limit crop production are variable. However, it is quite possible to broadly characterize and classify the drought patterns of a given environment using long-term water-balance modeling and geographic information system (GIS) tools (Chauhan et al., 2000). The assessment of the moisture-availability patterns of the target environments is critical for the development of best adapted crop genotypes to target environments and to identify iso-environments of drought patterns. As mentioned earlier, SAT environments are often characterized by a relatively short growing season in a generally dry semi-arid climate, with high average temperatures and potential evaporation rates. Soils are moderate to heavy, with low to moderate levels of available water content to the plants. In addition, the dry season at this location is generally rain-free, with a high mean air temperature and vapour-pressure deficits. This season provides an ideal screening environment to expose plants to controlled drought-stress treatments by regulating the timing and quantity of irrigation (Bidinger et al., 1987; Johansen et al., 1994).

Drought stress is a major limiting factor at the initial phase of plant growth and establishment. The usual effects of drought on the development of a plant are a lowered production of biomass and/or a change in the distribution of this biomass among the different organs. In addition, plant productivity under drought stress is strongly related to the processes of dry matter partitioning and temporal biomass distribution (Kage et al., 2004). Reduction of biomass due to water stress is common in both cereals and legumes, although genotypic variation does exist. In general, cereals biomass production is less affected by drought than legumes.

The types of drought occurrence is usually categorized as early, intermittent and terminal depending on the growth phase of the plant when the water deficit becomes acute. For example, long duration pigeonpea, a crop usually sown at the first onset of south Asian monsoon rains, experiences all the three types of drought.

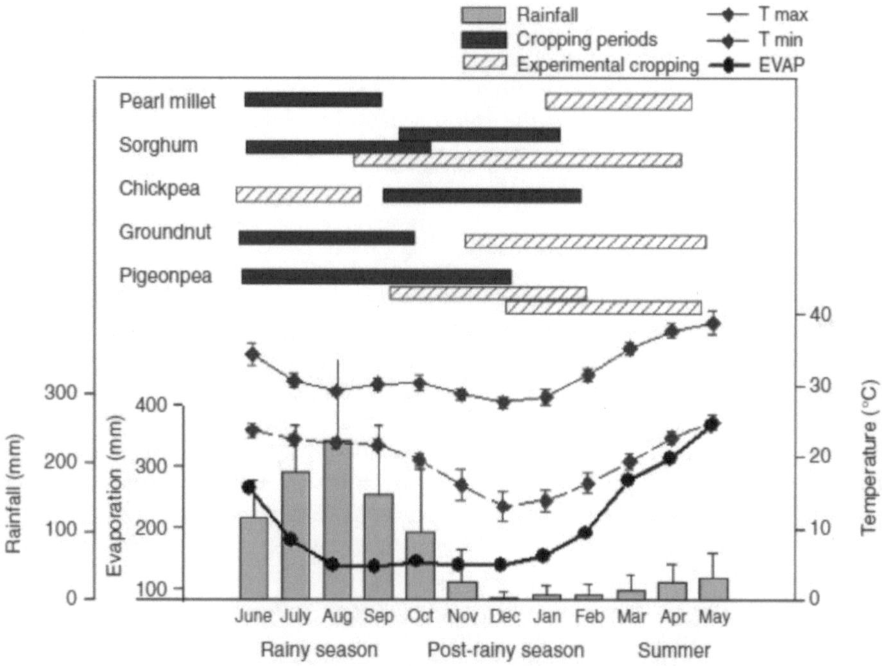

Fig. 2. Long-term average climate conditions (1974-2000) and cropping schedule at ICRISAT, Patancheru (17°N 78° E, msl 542M), India (Source: Serraj et al. 2003).

At the early seedling stages of the crop, lack of water can adversely affect seedling growth and occasionally kill seedlings and reduce the plant population. Similar lack of water for a period of time at the later stages can affect leaf area expansion and subsequently the root and shoot growth causing intermittent set backs and relief. However at later stages once the rains cease the plants during their reproductive growth phases tend to rely on the constantly receding soil moisture leading to increasing levels of terminal drought stress affecting largely the reproductive plant parts. This may reduce the number of pod/spikelet bearing sites or the number of seeds formed in a pod/spike or the size of the developing seeds. On the other hand, pearl millet, sorghum, groundnut and pigeonpea sown in the rainy season experience intermittent drought while chickpea that is primariely grown postrainy experiences the terminal drought (Fig 2).

2.1.1 Cereals

Pearl millet

Most of the pearl millet, either as grain or fodder crop, is grown in the arid and semi-arid zones of south Asia and West Africa where the soils are prone to drought stress or soil salinity problems. The main target environment for the pearl millet drought work of ICRISAT and its partners in India is the pearl millet growing area of the north-western states of Rajasthan, Gujarat and Haryana, where postflowering stress, either alone or in combination with preflowering stress, is a very common feature of the environment (van Oosterom et al., 1996). The focus of pearl-millet research has thus been on terminal drought as it is also the most damaging to grain yield (Bidinger et al., 1987). As an example of the magnitude of yield loss under drought, pearl millet yields were reduced by 0-16% with the intermittent drought across years that was imposed as a preflowering stress (stressed from 12 days after emergence till flowering) whereas it was reduced by 55 to 67% with a post-flowering terminal drought stress (Bidinger et al. 1987). Pearl millet yields were reduced during the dry season compared to the rainy season by about 14 %. However the shoot biomass was reduced by 12% under normal photoperiod while it was not affected under extended photoperiod (van Oosterom et al. 2002).

Sorghum

Sorghum, a major grain and forage crop, is one of the most extensively adapted crops to the semi-arid tropics. The rainfall during the crop season could vary from 300 to 2000 mm. Terminal-drought stress is the most serious constraint to sorghum production worldwide. In sub-Saharan Africa, drought at both seedling establishment and grain-filling stages is also very common. In India, sorghum is grown during the rainy and the post-rainy seasons. The variable moisture environment during the rainy season can have a severe impact on biomass and grain yield, affecting both preflowering and postflowering stages. Characterizing drought in post-rainy season sorghum is simpler, compared with the intermittent drought experienced by rainy season crops. This is because much of the rainfall is received before the planting of the crop, which is therefore grown almost entirely on stored soil moisture and exposed mostly to progressively increasing (terminal) water deficits. Therefore, the factors governing crop growth and water use in the post-rainy season, i.e. radiation, temperature, vapour pressure and potential evaporation, are relatively stable and predictable, so that simulation modeling of both crop growth and the effects of various crop traits is quite feasible. In a set of NILs (Near Isogenic Lines) of sorghum the overall mean yield reduction due to preflowering drought stress was only 4% while that of the post flowering drought was 37% (Ejeta et al. 1999).

2.1.2 Legumes

Chickpea

Chickpeas (*Cicer arietinum* L.) sown at the end of the rainy season, usually experience terminal drought stress as a consequence of growing on receding soil moisture conditions with a scanty or no rainfall condition during the crop growing season. When such drought stress was not allowed to occur with an optimum irrigation regime the shoot biomass productivity was near 5 t ha^{-1} with a seed yield of 2t ha^{-1}. However, under the normal receding soil moisture condition, the shoot biomass productivity ranged across years from 1.8 to 3.8 and the seed yield from 0.7 to 1.6 t ha-1 (Krishnamurthy et al. 2010).

Chickpea breeding program at ICRISAT has placed high emphasis on development of early and extra early maturing varieties so that these can escape terminal drought. The early maturing crop, however, cannot accumulate enough total plant biomass due to reduced total photosynthetic period compared to the relatively longer duration varieties.

Terminal drought reduces both shoot biomass and yield in chickpea. For example the average shoot biomass reduction of 40 cultivated chickpea genotypes due to terminal drought was 44 to 61 % across two years whereas the grain yield reductions were 35 to 66% (Krishnamurthy et al. 1999). Similarly the average shoot biomass reduction of 216 (mini core) chickpea germplasm accessions due to terminal drought was 31 to 63 % across 3 years whereas the grain yield reductions were only 26 to 61% (Krishnamurthy et al. 2010). The relatively less reduction in grain yield under drought was due to an increased partitioning under the progressively built terminal drought stress.

Groundnut

Groundnut (*Arachis hypogaea* L.) is an important rainy-season crop in most of the production systems in the semi-arid tropical regions of south Asia and sub-Saharan Africa, where it is grown under varying agroecologies, either as a sole crop or intercropped with sorghum and pigeonpea. Groundnut yields are generally low and unstable under rain-fed conditions, due to unreliable rainfall patterns. Severity of drought stress depends on the stages of crop development and the duration of stress period (Wright and Nageswara Rao, 1994).

Improvement of transpiration efficiency (TE) is seen as a promising strategy to improve shoot biomass and pod yield productivity under episodes of intermittent drought. Efforts were made to identify simple and easily measurable traits that are closely associated with TE such as SCMR (Nageswara Rao et al., 2001; Sheshshayee et al., 2006), SLA (Nageswara Rao and Wright, 1994; Wright et al., 1994) and carbon isotope discrimination (Hubick et al., 1986; Farquhar et al., 1988; Wright et al., 1994). Recent works have demonstrated that root dry weight and SLA were important traits related to WUE under long term drought and considered useful as selection criteria for high WUE under long term drought (Songsri et al., 2009).

Groundnut pod yield productivity is more adversely affected by various seasonal droughts than the shoot biomass production. For example, in a field trial where the drought intensity and the timing is managed by withholding irrigation and providing a part by line source irrigation it was established that the drought occurring between emergence to peg initiation was rather beneficial, producing greater yields than the control. However the drought occurrence between the phases of start of flowering to start of seed growth had lead to a reduction of 13 to 49% in shoot biomass and 18 to 78% in pod yield. The drought stress from the start of seed growth to maturity (terminal drought) had caused a reduction of 16 to 73% for the shoot biomass and 24 to 95 % for the seed yield (Nageswara Rao et al. 1985).

Pigeonpea

Pigeonpea (*Cajanus cajan* (L.) Millspaugh) is a deep-rooted and drought-tolerant leguminous food crop grown in several countries, particularly in India and India accounts for about 80% of the total world pigeonpea production. It is grown mainly by resource poor farmers in

India south east Africa and, to a varying extent, throughout the tropics, usually under rain-fed conditions.
Pigeonpea can be exposed to intermittent drought stress during dry periods of the rainy season and to terminal-drought stress in the post-rainy season. Over the last two decades, shorter-duration pigeonpea (SDP) genotypes have been developed, with some genotypes capable of reaching maturity within 90 days (Nam et al., 1993). However, the developed short-duration genotypes are usually sensitive to intermittent drought. Considerable variation in tolerance to intermittent drought has been observed in short-duration pigeonpea lines and variation in sensitivity in relation to timing of drought stress has been established (Lopez et al. 1996). As in other crops, responses to intermittent drought stress have been shown to depend on the growth stage at which the stress occurs (Nageswara Rao et al. 1985). For example Nam et al. 1993 has shown that the drought incidences at flowering cause a large reduction in productivity than drought at preflowering stage or at pod fill stage. The shoot biomass reduction was 26 to 33% across years whereas the yield reduction was 30 to 48% (Nam et al. 1993).

2.2 Salinity

In the semi-arid agricultural areas of the world, soil salinization is closely linked to the extensive use of artificial irrigation, which in combination with extended dry seasons, very quickly turns formerly productive areas practically into deserts. In the future, this effect will even increase due to the high demand of water from other non agriculture sectors (i.e. industry, overpopulated cities), whereas the possibilities to increase any crop's productivity through irrigation will necessarily decrease. Apart from irrigated areas, salinity is a major management problem in many unirrigated rainfed areas.
Dryland salinity ranges from a slightly saline soil condition which reduces crop growth to extensive areas where cultivation is almost impossible. This constraint has been a threat to the land and water resources in several parts of the world including the SAT, although the seriousness of the problem well realized in recent years. All the crops are affected by salinity while they vary in their degree of response as some of them being tolerant while others are sensitive.

2.2.1 Cereals

Pearl millet

Soil salinity is a major problem for pearl millet [Pennisetum glaucum (L.) R. Br.] production in the arid and semi-arid zones of south Asia and West-Africa (Blummel et al. 2003). Pearl millet also remains as a potential crop to grow in the rice fallows of saline areas in south Asia, where typical increases of salinity levels during post-rainy season prevent crop production. Compared to other crop species, Pearl millet and its wild relatives are rated to be fairly tolerant to salinity (Maas and Hoffman 1977; Shannon 1984; Krishnamurthy et al. 2007) and provide an option while selecting crops that can be more profitably grown in saline soils.
Lack of a single reproducible screening protocol and lack of knowledge on trait(s) that confer yield under salinity is a great limitation to breeding tolerant varieties. Field screening under salinity stress may not be effective because of the extent of variability in salinity experienced within a single field and among plots even at shorter distances (Richards and Dennet 1980). Pearl millet seems to be sensitive at germination stage in ECe of 16 dS m^{-1} and

beyond but this sensitivity is to some extent compensated by the tillering capability (Dua 1989). However, it seems that salinity response estimated at germination stage does not correlate well with plant performance at later stages (Munns and James 2003; Krishnamurthy et al. 2007).

Na+ exclusion and grain K/Na ratios were suggested to be reliable traits for selection. However, their usefulness as selection criteria (Munns and James 2003; Poustini and Siosemardeh 2004) could not be emphasized when five cultivars in pearl millet used for this association study (Ashraf and McNeilly 1987) where as leaf Na+ contents or the K+/Na+ and the Ca++/Na+ ratios assessed with 100 ICRISAT breeding lines were found to explain the biomass productivity at flowering time (Krishnamurthy 2007). Therefore this relationship of Na-based ratios needs to be evaluated with a wider range of genotypes and in association with the grain yield. Overall, it seems that although various aspects have been related to tolerance, the variation in whole plant reaction to salinity has been suggested to provide the best means of initial isolation of salinity tolerant genotypes (Shannon 1984; Ashraf and McNeilly 1987).

Large genotypic variation was reported to exist in pearl millet for salinity response in terms of whole plant response (Ashraf and McNeilly 1987; 1992; Dua 1989). Moreover, availability of high levels of tolerance in other species of Pennisetum (Ashraf and McNeilly 1987; 1992; Muscolo et al. 2003) and within the *P. glaucum* (Dua 1989) offers a scope for understanding the traits related to tolerance and to integrate these tolerant crop species/genotypes into appropriate management programs to improve the productivity of the saline soils. A total shoot biomass productivity ranging from 9 to 12 t ha^{-1} and a grain yield from 3.1 to 4.9 t ha^{-1} recorded in normal Alfisol fields at Patancheru, India (van Oostrom et al. 2002) got reduced to an average of 3.3 t shoot biomass and 1.1 t ha^{-1} grain yield of 15 germplasm accessions when grown in a 10 dS m^{-1} saline vertisols at Gangavathi, Karnataka, India (Kulkarni et al. 2006).

Sorghum

Sorghum is characterized to be moderately tolerant to salinity (Maas, 1985; Igartua et al., 1995) with a large genotypic variation reported. It is considered relatively more salt tolerant than maize, the cereal crop ranking first in productivity globally (Maas, 1985). Therefore, sorghum has a good potential for salt affected areas (Ayers & Westcott, 1985; Igartua et al., 1994).

There are limited successes in enhancing crop yields under salinity stress as available knowledge of the mechanisms of salt tolerance has not been converted into useful selection criteria to evaluate a wide range of genotypes within and across species. Attempts have been made to evaluate salt tolerance at germination and emergence stages in grain sorghum (Igartua et al., 1994; Krishnamurthy et al. 2007), and large genotypic differences were reported, but this early evaluation appears to have little relation with overall performance under saline conditions (Munns et al., 2002; Krishnamurthy et al. 2007). Though Na+ exclusion and grain K+/Na+ ratios have been suggested to be reliable traits for selecting salt tolerant crops (Munns & James, 2003; Munns et al., 2002; Poustini & Siosemardeh, 2004; Netondo et al., 2004; Krishnamurthy et al. 2007), the value of that trait has not been used in a large scale. Therefore, there is a need to identify traits associated with salinity tolerance, and simple, high throughput, repeatable screening methods to evaluate large number of genotypes. In fact, the variation in whole-plant biomass responses to salinity was considered to provide the best means of initial selection of salinity tolerant genotypes (Shannon, 1984; Ashraf & McNeilly, 1987), prior to the evaluation on the basis of specific traits.

Some of the known salt tolerant genotypes (n=29) of sorghum have been reported to yield in the range of 1.5 to 4.2 t ha^{-1} in naturally occurring saline soils with an average ECe of 10 dS m^{-1} at the Agricultural Research Station, Gangavathi, Karnataka, India (Reddy et al. 2010). However the grain yield range was much superior (4.7 to 6.0 t ha^{-1}) for the hybrids that were tested along the germplasm lines under similar saline field conditions.

2.2.2 Legumes

Chickpea

Chickpea (*Cicer arietinum* L.) is sensitive to salinity (Flowers et al. 2010). The decline in the area sown to chickpea in traditional chickpea-growing areas of northern India and the Indo-Gangetic Plain (Gowda et al. 2009) is partly due to increased soil salinity and increased use of brackish water for irrigation. If this decline is to be reversed, then resistance of existing chickpea varieties to salinity needs to be improved. Since management options are often too expensive for small-holder farmers to adopt, breeding and selection of salinity-resistant varieties remains a more practical and immediate option.

Until recently, little genetic variation for salinity resistance had been observed in chickpea (Saxena 1984; Dua 1992; Johansen et al. 1990). However, recently a large range of variation (Vadez et al. 2007; Krishnamurthy et al. 2011) was found to exist in seed yield of 265 chickpea genotypes grown in artificially-salinized soils watered to field capacity with 80 mM sodium chloride. Further, it was found that the seed yield under salinity in chickpea was closely associated with time to flowering and to the seed yield under non-saline conditions.

Several reports have shown that the resistance to salinity in chickpea is related to the resistance of reproduction (Mamo et al., 1996; Katerji et al., 2001). Salinity resistance indeed had been shown to be associated with the capacity to maintain a large number of filled pods, rather than to the capacity to grow under salt stress (Vadez et al., 2007), indicating that salt stress may have a deleterious effect on flower/pod production and retention. Yet, reproductive success may have been conditioned by the late-sown conditions in which the previous work was carried out (Vadez et al., 2007) and needs to be validated with sowing at the normal sowing time.

As salinity is likely to be an increasing problem in a warming and drying world, especially for relatively sensitive crops such as chickpea, it is important to make sources of resistance available to the breeding community by systematically screening a representative set of germplasm. To date, only the mini-core collection of chickpea germplasm has been evaluated for salinity resistance (Vadez et al., 2007). This mini-core collection is based on morphological and agronomic traits (Upadhyaya and Ortiz 2001) and not a systematic screening for diversity of molecular markers. More recently, a reference collection of chickpea has been assembled using marker data from 50 SSR markers screened in over 3,000 genotypes (Upadhyaya et al., 2006). Although the reference collection includes all the germplasm in the mini-core collection, 89 additional entries of cultivated chickpea with additional molecular variability have been identified (Upadhyaya et al. 2008).

Groundnut

Groundnut is a very important oilseed crop globally and particularly in many developing countries of the SAT where salinity is an ever-increasing crop production constraint. It is not only the grain yield is important but also the protein-rich crop residues as dry fodder. In spite of the importance of the constraint as well as the crop very little has been published with groundnut being affected by soil salinity. In a salinity tolerance screening saturating soil once with with 80 mM NaCl solution and testing 288 groundnut genotypes/ germplasm accessions it has been found that the shoot biomass productivity was the least affected (0-30%) while the pod yield was affected by 50 to 100%. However there were genotypes that could produce pod yields >half of the control but these were very few (Srivastava 2006).

Pigeonpea

Pigeonpea is one of the major legume crops grown in the semi arid tropics, particularly in India. Its high sensitivity to salinity coupled with the dry growing environment pose a major constraint to crop production in certain areas. Salinity affects plant growth, development and yield of pigeonpea. However the quantum of work that had been carried out with pigeonpea under salinity is scarce. A study involving a tolerant (ICPL227) and a sensitive (HY3C) cultivated pigeon pea genotypes and some tolerant (*Atylosia albicans, A. platycarpa* and *A. sericea*) and sensitive (*Rynchosia albiflora, Dunbaria ferruginea, A. goensis* and *A. acutifolia*) wild relatives tested over a range of salinity levels (0, 4, 6, 8 and 10 dS/m) have shown that transpiration rate decreased with increasing salinity in tolerant and sensitive pigeon pea genotypes alike, while key difference was the greater salinity tolerance of *A. albicans, A. platycarpa* and *A. sericea* was associated with efficient sodium and chloride regulation in the plant system (Subbarao et al. 1990).

Shoot sodium concentrations of the tolerant wild species were found to be 5 to 10 times less than those of the sensitive species, while root sodium concentrations in the tolerant species were 2 to 3 times higher than in the sensitive species. Thus the efficiency of regulation of ion transport to shoots seemed to explain the differences in salinity response among pigeon pea genotypes and related wild species. Srivastava et al. (2007) assessed the morphological and physiological variation in pigeonpea for salinity tolerance in 300 genotypes, including the mini core collection of ICRISAT, wild accession and landraces from putatively salinity-prone areas worldwide. A large range of variation in salinity susceptibility index and the percent relative reduction (RR %) in both cultivated and wild accessions were shown to exist. Also less Na+ accumulation in shoot was indicative tolerance and this relationship was limited to the cultivated material. Some of the wild species reported tolerant are *C. platycarpus, C. scarabaeoides* and *C. sericea* whereas *C. acutifolius, C. cajanifolius* and *C. lineata* were more sensitive. In another study, six pigeonpea genotypes were tested under five different NaCl concentrations (0, 50, 100, 125, 150 mM) under controlled conditions. Salt concentration of 75 mM was identified to be the critical one as it reduced the biomass production by an average 50%. For pigeonpea, as SCMR was positively associated with higher biomass under salinity, SCMR was suggested to be an early indicator for salinity tolerance. The Na+ accumulation did not help to be of any indication of tolerance in pigeonpea.

3. Technology that can assist in estimating crop growth and productivity under abiotic stresses

Plant biomass is an important factor in the study of functional plant biology and growth analysis, and it is the basis for the calculation of net primary production and growth rate. The conventional means of determining shoot dry weight (SDW) is the measurement of oven-dried samples. In this method, tissue is harvested and dried, and then shoot dry weight is measured at the end of the experiment. For the measurement of biomass of a large number of plants, this method is time consuming and labor intensive. Also, since this method is destructive, it is impossible to take several measurements on the same plant at different time points. With the establishment of advanced technology facilities for high throughput plant phenotyping, the problem of estimating plant biomass of individual plants is becoming increasingly important. There are several technologies that can help to assess the effect of abiotic stresses like drought and soil salinity on plant growth while assisting in predicting crop yield under various environmental conditions.

3.1 Near-infrared spectroscopy on agricultural harvesters and spectral reflectance of plant canopy

The use of near-infrared spectroscopy on agricultural harvesters has the advantage of not being time and resources consuming. In contrast to conventional sample-based methods, near-infrared spectroscopy on agricultural harvesters secures a good distribution of measurements within plots and covers substantially larger amounts of plot material (Welle et al., 2003). Thus, this method reduces the sampling error and therefore, provides more representative measurements of the plot material.

Spectral reflectance of plant canopy is a non-invasive phenotyping technique that enables the monitoring with high temporal resolution of several dynamic complex traits, such as biomass accumulation (Montes et al., 2007). Investigations at the individual plant level under well controlled environmental conditions showed that spectral reflectance could be used to monitor plant photosynthetic pigment composition, assess the water status and detect abiotic or biotic plant stresses (Penuelas, and Filella, 1998; Chaerle, and Van Der Straeten, 2000).

Current methods for measuring biomass production in cereal plots involves destructive sampling which is not suitable for routine use by plant breeders where large numbers of samples are to be screened. The measurement of spectral reflectance using ground-based remote sensing techniques has the potential to provide a nondestructive estimate of plant biomass production. Quick assessment of genetic variations for biomass production may become a useful tool for breeders. The potential of using canopy spectral reflectance indices (SRI) to assess genetic variation for biomass production is of tremendous importance. The potential of using water-based SRI as a breeding tool to estimate genetic variability and identify genotypes with higher biomass production would be helpful to achieve higher grain yield in crops.

3.2 Infrared thermography

The integrator of drought is the plant water status (Jones, 2007), as determined by plant water content or water potential. A direct measurement of these variables is difficult and currently not possible in a high-throughput phenotyping approach. Probably the most commonly used technique in this context is thermal infrared imaging, or infrared thermography (IRT) to measure the leaf or canopy temperature.

Plant canopy temperature is a widely measured variable because it provides insight into plant water status. Although thermal imaging does not directly measure stomatal conductance, in any given environment stomatal variation is the dominant cause of changes in canopy temperature (Jones and Mann 2004).

Thermal imaging is becoming a high-throughput tool for screening plants for differences in stomatal conductance (Merlot et al. 2002). Thermal infrared imaging for estimating conductance has potential value as it can be used at the whole plant or canopy level over time. Leaf temperature has been shown to vary when plants are subjected to water stress conditions. Recent advances in infrared thermography have increased the probability of recording drought tolerant responses more accurately.

3.3 Magnetic resonance imaging (MRI) and positron emission tomography (PET)

These two methods are being used at Julich Plant Phenotyping Centre (Germany) to investigate root/shoot systems growing in sand or soil, with respect to their structures, transport routes and the translocation dynamics of recently fixed photoassimilates labelled with the short lived radioactive carbon isotope 11C. Quantitative MRI and PET data will help not only to study the differences between species, but also in phenotyping of cultivars or plant lines in which growth pattern, water relations or translocation properties are important traits with respect to plant performance (Jahnke et al. 2009). Therefore, MRI-PET combination can provide new insights into structure-function relationships of intact plants. It also allows monitoring of dynamic changes in plant properties, which has not been possible to assess systematically until now to understand plant performance such as resource use efficiency or biomass production.

3.4 RGB imaging

Digital image analysis has been an important tool in biological research and also has been applied to satellite images, aerial photographs as well as macroscopic images (Nilsson, 1995). The imaging method has been proposed to infer plant biomass accurately as a non-destructive and fast alternative to the conventional means of determining shoot dry weight. The approach predominantly cited in literature is the estimation of plant biomass as a linear function of the projected shoot area of plants using RGB images.

A relevant application of image analysis which has been used for decades is in the area of remote sensing forestry and precision agriculture in which the area of plant species cover and the biomass of the above-ground canopy are estimated from satellite and airborne images (Montès et al, 2000; Lamb and Brown, 2001).

These techniques have found a recent application in estimating the biomass of individual plants in a controlled environment and also in the field. There have been only a few reports on the application of image analysis techniques to estimate above-ground biomass of an individual plant. In these reports, the projected shoot area of the plants captured on two dimensional images was used as a parameter to predict the plant biomass (Tackenberg, 2007; Sher-Kaul et al, 1995; Paruelo et al, 2000).

3.5 Crop models and geographic information systems (GIS)

Numerous dynamic crop models have been developed for simulating crop growth in function of environmental factors (soil characteristics, climate) and of agricultural practices. Some of these models can be used for predicting crop biomass and yields and crop quality before harvest. For example the Geographic Information System (GIS) was successfully used to predict water-limited biomass production potential of various agro climatic zones of the world (Fig 3). It is very clear that the biomass producing potential of

SAT is between 300 to 600 g dry matter M^{-2} Y^{-1} that corresponds well with the observed annual productivities.

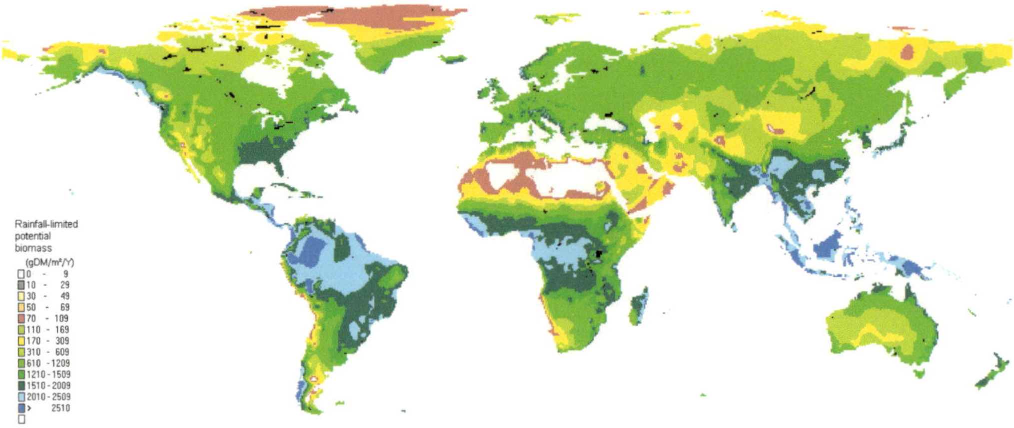

Fig. 3. Distribution of predicted rain-fall limited potential biomass production (Source: FAO-SDRN-Agrometeorology Group 1997.
http://www.fao.org/sd/EIdirect/climate/EIsp0061.htm)

The advent of remote sensing technology supported by Geographic Information System (GIS) has opened new vistas of improving agricultural statistics systems all over the world. The applications of Remote Sensing (RS) in the field of agriculture are wide and varied, ranging from crop discrimination, inventory, assessment and parameter retrieval, on one hand, to assessing long term changes and short-term characterization of the crop environment. The use of remote sensing for crop acreage and yield estimation has been well demonstrated through various studies all over the world, and has gained importance in recent years as a means of achieving these estimates possibly in a faster mode and at a cheaper cost (Murthy et al., 1996). An integrated methodology for providing area and yield estimation and yield forecasting models with small area estimates at the block level using satellite data has been developed (Singh and Goyal, 2000; Singh et al. 2002).

The remote sensing use for drought prediction can benefit from climate variability predictions. Recent research on crop-water relations has increasingly been directed towards the application of locally acquired knowledge to answering the questions raised on larger scales. However, the application of the local results to larger scales is often questionable. Crop simulation models, when run with input data from a specific field/ site, produce a point output. The scope of applicability of these simulation models can be extended to a broader scale by providing spatially varying inputs (soil, weather, crop management) and combining their capabilities with a Geographic Information System (GIS). The main purpose of interfacing models and GIS is to carry out spatial and temporal analysis simultaneously as region-scale crop behavior has a spatial dimension and simulation models produce a temporal output. The GIS can help in spatially visualizing the results as well as their interpretation by spatial analysis of model results.

4. Concluding remarks

4.1 Differential response of cereals and legumes to drought and salinity stress

Abiotic stresses (mainly drought and high soil salinity) are the major cause for the reduction in crop biomass and yield worldwide, especially in the SAT. Generally, Cereals are relatively better equipped to tolerate those stresses than the legumes, partly due to the carbon pathway differences between these two crop groups. Data collected using destructive measurements showed that under terminal drought the reduction of shoot biomass production in legumes can reach 50% especially in groundnut. In cereals, shoot biomass reduction is hardly above 40%.

Depending on the level of stress, both legumes and cereals may suffer from yield losses to a larger extent than shoot biomass reduction, however, in some cases, a better partitioning can help in a better yield. For example, reduction of chickpea seed yield due to terminal drought was recorded to be 26 to 61 % and the shoot biomass at maturity to be 31 to 63 % during three years of study using a large number of germplasm accessions. Whereas, the haulm yield of groundnut was reduced to 24 and 23% while the pod yield by 47 and 37% in the two years of field experimentation.

At a salinity level where the legumes would be completely dead, cereals like pearl millet and sorghum can thrive and be productive. However under salinity the larger adverse effect is on the reproductive growth than on the vegetative growth. Salinity affects plant growth and also equally the partitioning leading to a greater loss in seed yield. Reproductive biology is known to be more affected leading to greater yield damage. The partitioning to the root system plays a key role in tolerance to both drought and salinity.

4.2 Monitoring crop growth and productivity using remote sensing and GIS is key

The traditional approach of estimating the effect of a given abiotic stress on crop growth and productivity is becoming obsolete because of various reasons related to precision and up-scaling. Remote sensing data provide a complete and spatially dense observation of crop growth. This complements the information on daily weather parameters that influence crop growth. RS-crop simulation model linkage is a convenient vehicle to capture our understanding of crop management and weather with GIS providing a framework to process the diverse geographically linked data. Currently RS data can regularly provide information on regional crop distribution, crop phenology and leaf area index. This can be coupled to crop simulation models in a number of ways. CSM-RS linkage has a number of applications in regional crop forecasting, agro-ecological zonation, crop suitability and yield gap analysis and in precision agriculture.

In future the RS-CSM linkage will be broadened due to improvements in sensor capabilities (spatial resolution, hyper-spectral data) as well as retrieval of additional crop parameters like chlorophyll, leaf N and canopy water status. Thermal remote sensing can provide canopy temperatures and microwave data, the soil moisture. The improved characterization of crop and its growing environment would provide additional ways to modulate crop simulation towards capturing the spatial and temporal dimensions of crop growth variability.

5. Acknowledgement

The authors are thankful to the Bill & Melinda Gates Foundation for supporting this work through a grant (TL1) to the Generation Challenge Program.

6. References

Ashraf, M., McNeilly, T.M. (1987) Salinity effects on five cultivars/lines of pearl millet (*Pennisetum americanum* [L.] Leeke). *Plant and Soil* 103, 13- 19.

Ashraf, M., McNeilly, T.M. (1992) The potential for exploiting variation in salinity tolerance in pearl millet (*Pennisetum americanum* [L.] Leeke). *Plant Breeding* 104, 234- 240.

Ayers, R.S., Westcott, D.W. (1985) Water quality for agriculture. FAO Irrig. Drain. Paper No. 29 (1), Rome, Italy.

Bidinger, F.R., Mahalakshmi, V., Rao, G.D.P. (1987) Assessment of drought resistance in pearl millet [*Pennisetum americanum* (L.) Leeke]. I. Factors affecting yield under stress. *Australian Journal of Agricultural Research* 38, 37–48.

Blummel, M., Zerbini, E., Reddy, B.V.S., Hash, C.T., Bidinger, F.R., Khan, A.A. (2003) Improving the production and utilization of sorghum and pearl millet as livestock feed: progress towards dual-purpose genotypes. Field Crops Research 84, 143-158.

Chaerle, L. and van Der Straeten, D. (2000) Imaging techniques and the early detection of plant stress. *Trends in Plant Science* 5, 495–501.

Chauhan, Y.S., Bantilan, F.T., Padmaja, R., Ahmed, I., Moinuddin, M. (2000) Spatial and Temporal Distribution of Soil Water Availability in Bangladesh, Indo-Gangetic Plains of India, Pakistan and Sri Lanka. International Crops Research Institute for the Semi-Arid Tropics, Patancheru,India.

Dua, R.P. (1989) Salinity tolerance in pearl millet. *Indian Journal of Agricultural Research* 23, 9–14.

Dua, R.P. (1992) Differential response of chickpea (*Cicer arietinum*) genotypes to salinity. *Journal of Agricultural Sciences, Camb.* 119, 367-371.

Ejeta, G., Mohammed, A., Rich, P., Melakeberhan, A., Housley, T. L., Hess, D. E. (1999) Selection of specific mechanisms of resistance to Striga in Sorghum. In: Haussmann et al., (Eds.), Breeding for striga resistance in cereals. Proceedings of a workshop held at IITA, Ibadan, Nigeria, from 18-20 August. pp. (29)

Evans, L.T. (1998). Feeding the Ten Billion: Plants and Population Growth. Cambridge, UK: Cambridge University Press.

Farquhar, G.D., Hubick, K.T., Condon, A.G., Richards, R.A. (1988) Carbon isotope fractionation and plant water-use efficiency. In: Rundel P.W. et al., (Eds.), *Applications of stable isotope ratios to ecological research.* Springer-Verlag, New York. pp. 21-40.

Flowers, T.J., Gaur, P.M., Gowda, C.L.L., Krishnamurthy, L., Srinivasan, S., Siddique, K.H.M., Turner, N.C., Vadez, V., Varshney, R.K., Colmer, T.D. (2010) Salt sensitivity in chickpea. *Plant, Cell and Environment.* 33, 490-509.

Gowda, C.L.L., Parthasarathy Rao, P., Tripathy, S., Gaur, P.M., Deshmukh, R.B. (2009) Regional shift in chickpea production in India. In: Milestones in Food Legumes Research. Eds Masood Ali, Shiv Kumar. pp. 21-35. Indian Institute of Pulses Research, Kanpur 208 024, India.

Hubick, K.T., Farquhar, G.D., Shorter, R. (1986) Correlation between water-use efficiency and carbon isotope discrimination in diverse peanut (*Arachis*) germplasm. *Australian Journal of Plant Physiology* 13, 803-816.

Igartua, E., Gracia, M.P., Lasa, J.M. (1994) Characterization and genetic control of germination-emergence responses of grain sorghum to salinity. *Euphytica* 75, 185-193.

Igartua, E., Gracia, M.P., Lasa, J.M. (1995) Field responses of grain sorghum to a salinity gradient. *Field Crops Research* 42, 15-25.

Jahnke, S., Menzel, M.I., van Dusschoten, D., Roeb, G.W., Buhler, J., Minwuyelet, S., Blumler, P., Temperton, V.M., Hombach, T., Streun, M., Beer, S., Khodaverdi, M., Ziemons, K., Coenen, H.H., Schurr, U. (2009) Combined MRI-PET dissects dynamic changes in plant structures and functions. *Plant Journal* 59, 634-644.

Johansen, C., Saxena, N.P., Chauhan, Y.S., Subba Rao, G.V., Pundir, R.P.S., Kumar Rao, J.V.D.K., Jana, M.K. (1990) Genotypic variation in salinity response of chickpea and pigeonpea. *In*: Proceedings of the International Congress of Plant Physiology, Vol. 1, Eds Sinha, S.K., Sane, P.V., Bhargava, S.C., Agrawal, P.K. pp 977-983. Indian Society for Plant Physiology and Biochemistry, Indian Agriculture Research Institute, New Delhi, India.

Johansen, C., Krishnamurthy, L., Saxena, N.P. and Sethi, S.C. (1994) Genotypic variation in moisture response of chickpea grown under line-source sprinklers in a semi-arid tropical environment. *Field Crops Research* 37, 103–112.

Jones, P. D., Mann, M. E. (2004). Climate over past millennia. *Reviews of Geophysics* 42, 1-42.

Jones, H.G., (2007) Monitoring plant and soil water status: established and novel methods revisited and their relevance to studies of drought tolerance. *Journal of Experimental Botany* 58, 119-130.

Kage, H., Kochler, M., Stützel, H. (2004) Root growth and dry matter partitioning of cauliflower under drought stress conditions: measurement and simulation. *European Journal of Agronomy* 20, 379–394.

Katerji, N., van Hoorn, J.W., Hamdy, A., Mastrorilli, M., Oweis, T., Malhotra, R.S. (2001) Response to soil salinity of two chickpea varieties differing in drought tolerance. *Agricultural Water Management* 50, 83-96.

Krishnamurthy, L., Serraj, R., Hash, C.T., Dakheel, A.J., Reddy, B.V.S. (2007) Screening sorghum genotypes for salinity tolerance biomass production. *Euphytica* 156, 15-24.

Krishnamurthy, L., Serraj, R., Rai, K.N., Hash, C.T. , Dakheel, A.J. (2007) Identification of pearl millet (*Pennisetum glaucum* (L.) R. Br.]) lines tolerant to soil salinity. *Euphytica* 158, 179-188.

Krishnamurthy, L., Johansen, C., Sethi, S.C. (1999) Investigation of factors determining genotypic differences in seed yield of non-irrigated and irrigated chickpeas using a physiological model of yield determination. *Journal of Agronomy and Crop Science* 183, 9-17.

Krishnamurthy, L., Kashiwagi, J., Gaur, P. M., Upadhyaya, H. D., Vadez, V. (2010). Sources of tolerance to terminal drought in the chickpea (Cicer arietinum L.) minicore germplasm. *Field Crops Research* 119, 322-330.

Krishnamurthy, L., Gaur, P.M., Basu, P.S., Chaturvedi, S.K., Vadez, V., Rathore, A., Varshney, R., Gowda, C.L.L. (2011) Genetic variation for heat tolerance in the reference collection of chickpea (*Cicer arietinum* L.) germplasm. *Plant Genetic Resources* 9, 59-69.

Kulkarni, V.N., Rai, K.N., Dakheel, A.J., Mohammed Ibrahim, Hebbara, M. Vadez, V. (2006) Pearl millet germplasm adapted to saline conditions. *Journal of SAT Agricultural Research* 2.

Lamb, D.W., Brown, R.B. (2001) Precision Agriculture: Remote-Sensing and Mapping of Weeds in Crops. *Journal of Agricultural Engineering Research* 78, 117-125.

Lane, A., Jarvis, A. (2007) Changes in Climate will modify the Geography of Crop Suitability: Agricultural Biodiversity can help with Adaptation. eJournal 4(1), 1-12.

Lopez, F.B., Johansen, C., Chauhan, Y.S. (1996) Effects of timing of drought stress on phenology, yield and yield components of short duration pigeonpea. *Journal of Agronomy and Crop science* 177, 311-320.

Maas, E.V. (1985) Crop tolerance to saline sprinkling water. Plant and soil 89, 273-284.

Maas, E.V., Hoffman. G.J. (1977) Crop salt tolerance. Current assessment. *Journal of Irrigation and Drainage* 103, 115–134.

Mamo, T., Richter, C., Heiligtag, B. (1996) Salinity effects on the growth and ion contents of some chickpea (*Cicer arietinum* L) and lentil (*Lens culinaris medic*) varieties. *Journal of Agronomy and Crop Science* 176, 235-247.

Merlot, S., Mustilli, A. C., Genty, B., North, H., Lefebvre, V., Sotta, B., Vavasseur, A., Giraudat, J. (2002) Use of infrared thermal imaging to isolate Arabidopsis mutants defective in stomatal regulation. *The Plant Journal* 30, 601–609.

Mittler, Ron. (2006) Abiotic stress, the field environment and stress combination. *Trends in Plant Science* 11(1), 15–19.

Montès, N., Gauquelin, T., Badri, W., Bertaudière, V., Zaoui, E.H. (2000) A non-destructive method for estimating above-ground forest biomass in threatened woodlands. *Forest Ecology and Management* 130, 37-46.

Montes, J. M., Melchinger, A. E., Reif, J. C. (2007) Novel throughput phenotyping platforms in plant genetic studies. *Trends in Plant Science* 12, 433-436.

Munns, R., James, R.A. (2003) Screening methods for salinity tolerance: a case study with tetraploid wheat. *Plant and Soil* 253, 201-218.

Munns, R., Husain, S., Rivelli, A.R., James, R.A., Condon, A.G., Lindsay, M.P., Lagudah, E.S., Schachtman, D.P., Hare, R.A. (2002) Avenues for increasing salt tolerance of crops, and the role of physiologically based selection traits. *Plant and Soil* 247, 93-105.

Murthy, C.S., Thiruvengadachari, S., Raju, P.V., Lonna, S. (1996) Improved ground sampling and crop yield estimation using satellite data. *International Journal of Remote sensing* 17(5), 945-956.

Muscolo, A., Panuccio, M, R., Sidari, M. (2003) Effect of salinity on growth, carbohydrate metabolism and nutritive properties of kikuyu grass (*pennisetum clandestinum* Hochst). *Plant Science* 104, 1103-1110.

Nageswara Rao, R.C., Singh, S., Sivakumar, M.V.K., Srivastava, K.L., William, J.H. (1985) Effect of water deficit at different growth phase of peanut. II yield response. *Agronomy Journal* 80, 431-438.

Nageswara Rao, R.C., Talwar, H.S., Wright, G.C. (2001) Rapid assessment of specific leaf area and leaf nitrogen in peanut (*Arachis hypogaea* L.) using a chlorophyll meter. *Journal of Agronomy and Crop Science* 186, 175-182.

Nageswara Rao, R.C., Wright, G.C. (1994) Stability of the relationship between specific leaf area and carbon isotope discrimination across environments in peanut. *Crop Science* 34, 98-103.

Netondo, G.W., Onyango, J.C., Beck, E. (2004) Sorghum and salinity I. Response of growth, water relations, and ion accumulation to NaCl salinity. *Crop Science* 44, 707-710.

Nam, N. H., Chauhan, Y. S., Johansen, C. (1993) Comparison of extra-short-duration pigeonpea with short-season legumes under rainfed conditions on Alfsol. *Experimental Agriculture* 29, 307-316.

Nilsson, H. (1995) Remote Sensing and Image Analysis in Plant Pathology. *Annual Review of Phytopathology* 33, 489-528.

Paruelo, J.M., Lauenroth, W.K., Roset, P.A. (2000) Estimating Aboveground Plant Biomass Using a Photographic Technique. *Journal of Range Management* 53, 190-193.

Penuelas, J., Filella, I. (1998) Visible and near-infrared reflectance techniques for diagnosing plant physiological status. *Trends in Plant Science* 3, 151–156.

Poustini, K., Siosemardeh, A. (2004) Ion distribution in wheat cultivars in response to salinity stress. *Field Crops Research* 85, 125-133.

Reddy, B.V.S., Ashok Kumar, A., Sanjana Reddy, P., Ibrahim, M., Ramaiah, B., Dakheel, A.J., Ramesh, S., Krishnamurthy, L. (2010) Cultivar options for salinity tolerance in sorghum. *Journal of SAT Agricultural Research* 8.

Richards, R.A., Dennett, C.W. (1980) Variation in salt concentration in a wheat field. *Soil and Water Management* 44, pp. 8–9. University of California. Cooperative Extension.

Saxena, N.P. (1984) Chickpea. *In*: The Physiology of Tropical Field Crops, Eds Goldworthy and Fisher. pp. 419-452, John Wiley & Sons Ltd. New York

Serraj, R., Bidinger, F.R., Chauhan, Y.S., Seetharama, N., Nigam, S.N., Saxena, N.P. (2003) Water Productivity in Agriculture: Limits and Opportunities for Improvement (eds J.W. Kijne, R. Barker and D. Molden) CAB International 2003

Shannon, M.C. (1984) Breeding, selection and the genetics of salt tolerance. In: Staples, R.C., Toeniessen, G.H. (eds) Salinity tolerance in plants. John Wiley and Sons, New York, pp 231–254.

Sher-Kaul, S., Oertli, B., Castella, E., Lachavanne, J.B. (1995) Relationship between biomass and surface area of six submerged aquatic plant species. *Aquatic Botany* 51, 147-154.

Sheshshayee, M.S., Bindumadhava, H., Rachaputi, N.R., Prasad, T.G., Udayakumar, M., Wright, G.C., Nigam, S.N. (2006) Leaf chlorophyll concentration relates to transpiration efficiency in peanut. *Annals of Applied Biology* 148, 7-15.

Singh R., Goyal R.C., Pandey L.M., Shah S.K. (2000) Use of remote sensing technology in crop yield estimation survey-ii. Project report IASRI, New Delhi.

Singh, R., Semwal, D.P., Rai, A., Chhikara, R.S. (2002) Small area estimation of crop yield using remote sensing satellite data. *International Journal of Remote Sensing* 23(1), 49-56.

Songsri, P., Jogloy, S., Holbrook, C.C., Kesmala, T., Vorasoot, N., Akkasaeng, C., Patanothai, A. (2009.) Association of root, specific leaf area and SPAD chlorophyll meter reading to water use efficiency of peanut under different available soil water. *Agricultural Water Management* 96, 790-798.

Srivastava, N. (2006) Identification and field-testing of salinity tolerant groundnut in saline areas of India. Progress report for IFAR 2006. ICRISAT, Patancheru. http://www.ifar4dev.org/pdf/2006_COMP_ICRISAT_NAMITA_SRIVASTAVA.pdf

Srivastava, N., Vadez, V., Aruna, R., Nigam, S.N., (2007) Genetic Variability for salinity tolerance in groundnut (Arachis hypogaea). Presentation abstract 2002, pp 176,

"International Conference on Sustainable Agriculture for Food, Bio-energy and Livelihood Security".

Subbarao, G. V., Johansen, C., Jana, M. K., Rao, J. V. D. K. K. (1990) Physiological basis of differences in salinity tolerance of pigeonpea and its related wild species. *Journal of Plant Physiology* 137 (1), 64-71.

Tackenberg, O. (2007) A new method for non-destructive measurement of biomass, growth rates, vertical biomass distribution and dry matter content based on digital image analysis. *Annals of Botany* 99, 777-783.

Upadhyaya, H.D., Ortiz, R. (2001) A mini-core subset for capturing diversity and promoting utilization of chickpea genetic resources in crop improvement. *Theoratical and Applied Genetics* 102, 1292-1298.

Upadhyaya, H.D., Furman, B.J., Dwivedi, S.L., Udupa, S.M., Gowda, C.L.L., Baum, M., Crouch, J.H., Buhariwalla, H.K., Singh, S., (2006) Development of a composite collection for mining germplasm possessing allelic variation for beneficial traits in chickpea. *Plant Genetic Resources* 4, 13-19.

Upadhyaya, H.D., Dwivedi, S.L., Baum, M., Varshney, R.K., Udupa, S.M., Gowda, C.L.L., Hoisington, D., Singh, S. (2008) Genetic structure, diversity, and allelic richness in composite collection and reference set in chickpea (*Cicer arietinum* L.). *BMC Plant Biology* 8, 106.

Vadez, V., Krishnamurthy, L., Serraj, R., Gaur, P.M., Upadhyaya, H.D., Hoisington, D.A., Varshney, R.K., Turner, N.C., Siddique, K.H.M. (2007) Large variation in salinity tolerance in chickpea is explained by differences in sensitivity at the reproductive stage. *Field Crops Research* 104, 123-129.

Van Oostrom, E.J., Whitaker, M.L., Weltzien, R.E. (1996) Integrating genotype x environment analysis, characterization of drought patterns, and farmer preferences to identify adaptive plant traits for pearl millet. In: Cooper, M., Hammer, G.L. (eds.). *Plant Adaptation and Crop Improvement* CAB International/IRRI/ICRISAT, pp. 383-402.

Van Oostrom, E.J., O' Leary, G.J., Carberry, P.S., Craufurd, P.Q. (2002) Simulating growth, development, and yield of tillering pearl millet. III. Biomass accumulation and partitioning. *Field Crops Research* 79, 85-106.

Welle, R., Greten, W., Rietmann, B., Alley, S., Sinnaeve, G., Dardenne, P. (2003) Near-infrared spectroscopy on chopper to measure maize forage quality parameters online. *Crop Science* 43, 1407-1413.

Wright, G. C., Nageswara Rao, R.C. (1994) Groundnut water relations. In: The Groundnut Crop. Smartt, J., (Ed). Chapman and Hall, London, pp: 281-235.

Wright, G.C., Nageswara Rao, R.C., Farquhar, G.D. (1994) Water-use efficiency and carbon isotope discrimination in peanut under water deficit conditions. *Crop Science* 34, 92-97.

14

Aerobic Membrane Bioreactor for Wastewater Treatment – Performance Under Substrate-Limited Conditions

Sebastián Delgado, Rafael Villarroel,
Enrique González and Miriam Morales
Department of Chemical Engineering, Faculty of Chemistry, University of La Laguna, Spain

1. Introduction

It is widely known that many regions in the world have scarce water resources. In these areas the groundwater aquifers are also found to be in a critical condition as a result of over-exploitation. That is why, in such regions, the reuse of wastewater is a common practice and the competent authorities undertake multiple courses of action to encourage its reuse. Legislation implementing the reclaimed wastewater reuse is likewise very demanding in terms of quality and health and safety, which has resulted in the application of new technologies for water treatment and purification. Among the new emerging technologies appears the use of micro and ultrafiltration membranes as highly efficient systems, which are economically feasible for obtaining high quality recycled water.

Over the last two decades the technology of membrane bioreactors (MBRs) has reached a significant market share in wastewater treatment and it is expected to grow at a compound annual growth rate (CAGR) of 13.2%, higher than that of other advanced technologies and other membrane processes, increasing its market value from $ 337 million in 2010 to 627 million in 2015 (BCC, 2011). Aerobic MBRs represent an important technical option for wastewater reuse, being very compact and efficient systems for separating suspended and colloidal matter, which are able to achieve the highest effluent quality standards for disinfection and clarification. The main limitation for their widespread application is their high energy demand - between 0.45 and 0.65 kWh m^{-3} for the highest optimum operation from a demonstration plant, according to recent studies (Garcés et al., 2007; Tao et al., 2009).

The advantages of this process over the conventional activated sludge process are widely known (Judd, 2010), among these one of the most cited is the reduction in sludge production which results from operation at high solid retention time (SRT). However, its consequences for the structure and metabolism of the microbial suspensions need to be studied in detail.

Generally, we would expect that microorganisms subjected to severe substrate limitation should preferentially meet their maintenance energy requirements instead of producing additional biomass (Wei et al., 2003). This substrate limitation imposed on an MBR, by operating at low food-to-microorganism ratios (F/M), should modify the activity and characteristics of the sludge and could be the key factor for determining the process performance, particularly the membrane filtration (Trussell et al., 2006).

Biokinetic models are widely used to design activated sludge process. Knowledge of biokinetics parameters allows modelling of the process including the substrate biodegradation rate and biomass growth. At low growth conditions, as is demanded in MBRs, other processes apart from microbial growth have to be taken into consideration. These have been recognized as the maintenance energy requirement, endogenous respiration and subsequent cryptic growth (Van Loosdrecht & Hence, 1999). Macroscopically they cannot be perceived, but, from a practical point of view, the global process can be described by Pirt´s equation (Pirt, 1965).

Although there are several experiences with membrane bioreactors working without biomass purge (Rosenberger et al., 2002a; Pollice et al., 2004; Laera et al., 2005), none of these authors apply any kinetics models to describe process performance. Furthermore, these results were obtained in similar conditions, by treating raw municipal wastewater with a high substrate concentration, and it is interesting to compare this behaviour with an MBR treating wastewater with a low organic load. Additionally, not enough is known about the morphology and extracellular polymeric substance (EPS) production for total sludge retention and low F/M ratios.

The aim of this chapter is to summarize the current status of membrane bioreactor technology for wastewater treatment (Section 2.1). The advantages against the conventional activated sludge process and technological challenges are assessed (Section 2.2). Some design and operation trends, based on full-scale experience, are reviewed (Section 2.3). To discuss both fundamental aspects, biotreatment and filtration, some experimental results are presented. Special attention was given to the microbial growth modelling (Section 4.1.1), biomass characterisation (Sections 4.1.2 to 4.1.5) and membrane fouling mechanisms (Section 4.2). Some of these results have at the same time been compared with biomass from a conventional activated sludge process (CAS) operated in parallel.

2. Membrane bioreactor (MBR) technology

2.1 Current status and process description

The current penetration in the wastewater treatment market of the membrane bioreactors gives an idea of the degree of maturity reached by this technology. The most cited market analysis report indicates an annual growth rate of 13.2 % and predicts a global market value of $ 627 million in 2015 (BCC, 2011). Actually MBRs have been implemented in more than 200 countries (Icon, 2008). Particularly striking is the case of China or some European countries with an implementation rate of over 50% and 20%, respectively.

This technological maturity in urban wastewater market is also reflected in two main issues: the diversity of technology suppliers and the upward trend in plant size. Since 1990, the number of MBR membrane module products has grown exponentially until reaching over 50 different providers by the end of 2009 (Judd, 2010). However, globally, the market is dominated by three suppliers: Kubota, Mitsubishi Rayon and GE Zenon, which held about 85-90 % of the urban wastewater market (Pearce, 2008). In regard to the largest MBRs, there are 8 plants with a peak design capacity greater than 50 MLD (Table 1), all of them constructed before 2007 (Judd, 2010).

MBR technology is based on the combination of conventional activated sludge treatment together with a process filtration through a membrane with a pore size between 10 nm and 0.4 microns (micro/ultrafiltration), which allows sludge separation. The membrane is a barrier that retains all particles, colloids, bacteria and viruses, providing a complete disinfection of treated water. Furthermore, it can operate at higher concentrations of sludge (up to 12 g/l instead of the usual 4 g/l in conventional systems), which significantly reduces the volume of the reactors and sludge production.

Project	Technology	Date	DMDF (MLD)
Shending River, China	Beijing Origin Water	2010	120
Wenyu River, China	Asahi K/ Beijing Origin Water	2007	100
Johns Creek, GA	GE Zenon	2009	94
Beixiaohe, China	Siemens	2008	78
Al Ansah, Muscat, Oman	Kubota	2010	78
Peoria, AZ	GE Zenon	2008	76
Cleveland Bay, Australia	GE Zenon	2007	75
Sabadell, Spain	Kubota	2009	55
DMDF: Design maximum daily flow; MLD: Megalitres per day.			

Table 1. The largest 8 MBR plants (adapted from Judd, 2010).

Although there are two main process configurations of biomass rejection MBRs, submerged or immersed (iMBR) and sidestream (sMBR), the immersed configuration is the most widely used in municipal wastewater treatment due to lower associated costs of operation (e.g., Le-Clech et al., 2005a). In this configuration, the module is placed directly into the process tank and is thus less energy-intensive. As a result, it is only necessary to create a slight vacuum inside the membrane module, measured as transmembrane pressure (TMP), for filtration. For the immersed configuration, there are basically two types of commercial membrane modules available: flat sheet (FS), which is exemplified by the Kubota technology, and hollow fiber (HF) such as those supplied by GE Zenon or Mitsubishi Rayon. HF allows a higher packing density since it has a thinner space between membranes compared to FS. However, this makes it more susceptible to membrane clogging and/or sludging, and it can also make cleaning more difficult. Regarding the membrane material used for an iMBR, fluorinated and sulphonated polymers (polyvinylidene difluoride, polyethersulfone, in particular) dominate in commercial membrane MBR products (Santos & Judd, 2010).

For another approach to the analysis of technology maturity we might take a review of the research conducted on the MBR during the last decades. It is worth noting that considerable scientific interest has been aroused in recent years in this field. Santos et al. (2010) identified 1450 scientific papers published between 1990 and 2009, with a year-by-year increase of 20% from 1994 onwards. If we analyze this literature, the most cited research topic is membrane fouling (about 30%). In fact, scientific reviews have been published periodically that have analyzed in depth recent advances in the study of the mechanisms and factors that contribute to membrane fouling in MBR (Chang et al., 2002, Le-Clech et al, 2006, Meng et al ., 2009, Drews, 2010). Generally, these factors have been classified in four distinct groups: nature of the sludge, operating parameters, membrane/module characteristics and feed wastewater composition. However, although membrane fouling is an important issue in MBR operation, recent surveys of full-scale practitioners (Le-Clech et al., 2005b; Santos et al. 2010) show that pre-treatment and screening, membrane and aerator clogging, loss of membrane integrity, production of biosolids and other issues related to hydraulic overloading or system design, are of concern for MBR users.

2.2 Advantages and challenges

As already stated, MBRs represent an important technical option for wastewater treatment and reuse, being very compact and efficient systems for separation of suspended and colloidal matter and enabling high quality, disinfected effluents to be achieved. A key advantage of these MBR systems is complete biomass retention in the aerobic reactor, which decouples the sludge retention time (SRT) from the hydraulic retention time (HRT), allowing biomass concentrations to increase in the reaction basin, thus facilitating relatively smaller reactors or/and higher organic loading rates (ORL). In addition, the process is more compact than a conventional activated sludge process (CAS), removing 3 individual processes of the conventional scheme and the feed wastewater only needs to be screened (1-3 mm) just prior to removal of larger solids that could damage the membranes (Figure 1).

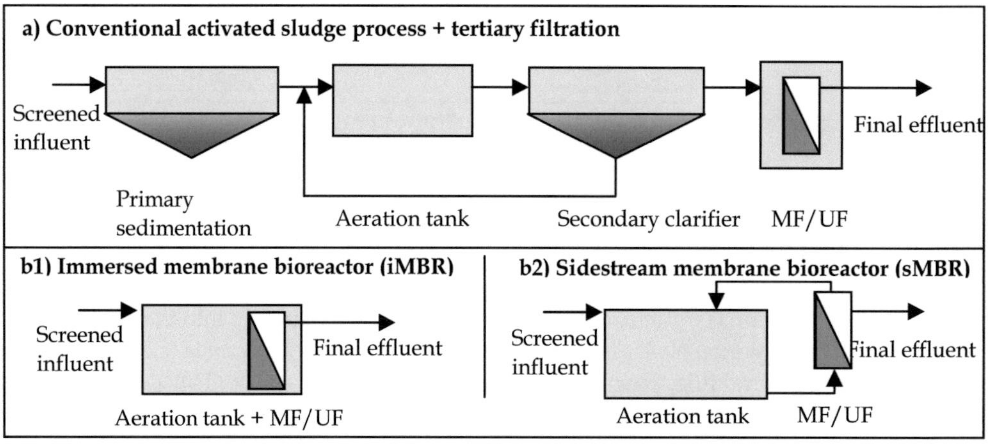

Fig. 1. Conventional activated sludge process (a) and MBR in both configurations: immersed (b1) and sidestream (b2)

Notwithstanding the advantages of MBRs, the widespread implantation is limited by its high costs, both capital and operating expenditure (CAPEX and OPEX), mainly due to membrane installation and replacement and high energy demand. This high energy demand in comparison with a CAS, is closely associated with strategies for avoiding/mitigating membrane fouling (70% of the total energy demand for iMBR) (Verrech et al., 2008; Verrech et al., 2010). Fouling is the restriction, occlusion or blocking of membrane pores or cake building by solids accumulation on the membrane surface during operation which leads to membrane permeability loss. The complexity of this phenomenon is linked to the presence of particles and macromolecules with very different sizes and the biological nature of the microbial suspensions, which results in a very heterogenic system. Meanwhile, the dynamic behaviour of the filtration process adds a particular complication to the fouling mechanisms (Le-Clech et al., 2006). Furthermore, permeability loss can also be caused by channel clogging, which is the formation of solid deposit in the voids of the membrane modules due to local breakdown of crossflow conditions (Figure 2). In addition, there are other operational problems, such as the complexity of the membrane processes (including specific procedures for cleaning), the tendency to form foam (partly due to excessive aeration), the smaller sludge dewatering capacity and the high sensitivity shock loads.

Fig. 2. a/b/c. Membrane module clogged. Debris can be observed located between the top headers modules forming a bridge between them (Morro Jable wastewater treatment plant, Canary Island, Spain; courtesy of CANARAGUA, S.A.)

For the immersed configuration, the operating strategy to control membrane fouling, (impacting directly or indirectly on CAPEX and OPEX) includes the following:

i. selecting an appropriate permeate flux,
ii. scouring of membrane surface by aeration,
iii. applying physical cleaning techniques, like backflushing (when permeate is used to flush the membrane backwards) and relaxation (when no filtration takes place), and
iv. applying chemical cleanings protocols, with different frequency and intensity (maintenance cleaning and recovery cleaning).

The fist concern, selecting an appropriate permeate flux, is determined by the classical trade-off problem: at higher fluxes CAPEX decreases while OPEX increases. High fluxes are desirable to reduce the membrane required (i.e. reduce CAPEX), however, membrane fouling increases with flux, which results in a higher membrane scouring demand and more frequent cleaning to control membrane fouling (i.e. increase OPEX). Furthermore, the correlation between membrane fouling and flux is not only influenced by hydrodynamics and cleaning protocols but also by feedwater characteristics and biological conditions. As a result, deciding a flux value depends on the analysis of empirical data obtained from pilot and full-scale experiments or available in the recent literature .

The second concern is membrane scouring. Ever since the iMBR appeared, air sparging has been widely used to mitigate fouling by constant scouring of the membrane surface (Cui et al., 2003) or by causing lateral fibre movement in HF configuration (Wicaksana et al., 2006). While the membrane fouling has been studied and mathematically modelled in classic filtration regimes (crossflow and dead-end) (e.g. Foley, 2006), the effect of turbulence induced by gas sparging in iMBR systems is still being assessed (Drews, 2010). As is well known, it has a clear contribution to minimizing the fouling problem, and therefore, a deeper understanding is extremely important in order to optimise aeration mode and rate, which has been proved to be one of its major operational costs.

The third concern is related to methods of physical cleaning (relaxation and backflushing) that have been incorporated as standard operation mode in MBRs. These techniques have successfully been proved to remove reversible fouling caused by pore blocking or sludge cake. For backflushing, the key parameters in the design of physical cleaning have been identified as frequency, duration, the ratio between these two parameters and its intensity (Le-Clech et al., 2006), and the same key parameters are expected for relaxation (with the exception of intensity). However, there is a knowledge gap in the inter-relationships between those parameters and the imposed permeate flux, especially when comparing both methods to obtain the same water productivity (Wu et al., 2008).

Finally, the fourth concern is chemical cleaning. Chemical cleaning is required when fouling cannot be removed by membrane surface scouring or physical cleaning methods. Although there are several types of chemical reagents used in membrane cleaning, in most full-scale facilities, two types of chemical reagents are commonly used: oxidants (e.g. NaOCl) for removing organic foulants (e.g. humic substances, proteins, carbohydrates), and organic acids (e.g. citric) for removing inorganic scalants. Basically, two objectives are pursued in the addition of chemical reagents: maintaining membrane permeability and permeability recovery. Maintenance cleaning is applied routinely via a chemically enhanced backflush where the reagent, at moderate concentration, is introduced with the permeate. In contrast, recovery cleaning is applied when the membrane permeability decreases until reaching non-operative values. The procedure consists of taking off the modules or draining off the membrane tanks to allow the membranes to be soaked in high concentrated reagents. Each MBR supplier has his own protocols which differ in concentrations and methods. Given its impacts on membrane lifetime and therefore on OPEX, there has recently been a growing interest in studying the influence of chemical cleaning procedures on membrane permeability maintenance and recovery (Brepols et al., 2008; Ayala et al., 2011). However, at the moment, the optimization of chemical cleaning protocols is far from being fully resolved.

2.3 Design and operation considerations

As was previously mentioned, the iMBR represents the most widely used configuration in large scale applications. This section gives some design and operation considerations including:

i. Pre-treatment,
ii. Design flux, hybrid systems and equalization tanks,
iii. Membrane fouling control and cleaning,
iv. Sludge retention time and biomass concentration, and
v. Membrane life

2.3.1 Pre-treatment

Membranes are very sensitive to damage with coarse solids such as plastics, leaves, rags and fine particles like hair from wastewater. In fact, a lack of good pre-treatment/screening has been recognised as a key technical problem of MBR operation (Santos and Judd, 2010a). For this reason fine screening is always required for protecting the membranes. Typically, screens with openings range between 1 mm (HF modules) to 3 mm (FS modules) are common in most facilities. However, data reported by Frechen et al. (2007) for 19 MBR European plants show a more conservative plant design by reducing the screen openings to 0.5-1.0 mm for both HF and FS. Regarding primary sedimentation, it was not economically viable for small-medium sized MBR plants (< 50.000 m^3/d), except for cases of retrofitting or upgrading of an existing CAS. However, for larger plants, given its advantages (smaller bioreactor volumes, reduced inert solids in the bioreactor, increased energy recovery, etc.), primary clarification can be considered. Its selection should be a compromise between energy and land cost.

2.3.2 Design flux, hybrid systems and equalization tanks

Membrane permeate flux is an important design and operational parameter that impacts significantly in CAPEX and OPEX. Typical operation flux rates for various full-scale iMBRs

applied to treat municipal wastewater treatment are over 19-20 l/h m^2 (Judd, 2010) with a peak flux (< 6 h) in the range 37-73 l/h m^2 (Asano et al., 2006).
A recent analysis of design and operation trends of the larger MBR plants in Europe (Lesjean et al., 2009), shows a broad difference between the design and operation flux. For Kubota systems, the designed maximum daily net fluxes are 14-48 l/h m^2 (mean at 32 l /h m^2) while for the GE Zenon modules they are 20-37 l/h m^2 (mean at 29 l/h m^2). However, it is interesting to note that for both systems the operation net flux is over 18 l/h m^2. Further differences are the same regardless of whether this is a new plant or a retrofit, or more or less conservative designs of a specific plant. In fact, the authors indicate that the averaged trend of the design maximum net flux and operation mean flux have moderately increased by only 3 l/h m^2 during the last 6 years. Given the impact of this discrepancy over CAPEX (i.e. higher membrane surface demand) and OPEX (i.e. higher membrane replacement costs) different solutions have been proposed: a plant has been designed in parallel to conventional activated sludge systems (hybrid systems), which can absorb the peak flows, or by addition of a buffer tank for flow equalisation.
In a comprehensive cost analysis of a large HF MBR plant, Verrecht et al. (2010) show the impact of both solutions on plant costs over the cycle life of the plant. While comparing a hybrid system with an MBR designed to manage maximum flow conditions, results indicate that the average energy demand for the full-flow MBR is 57% higher, as a result of under-utilization of the membrane available area and excess of membrane aeration. With regard to the adding of a buffering tank, the authors pointed out that the cost of buffering would be covered by reducing the required membrane surface area. However, this solution should increase the scale size of the plant by 10% compared to CAS treating the same flow. Therefore, the authors conclude that hybrid MBR plant is the most desirable option. Examples of some full-scale facilities with this hybrid system would be the Brescia plant with GE/Zenon in Italy, or the Sabadell plant with Kubota in Spain.

2.3.3 Membrane fouling control and cleaning

It is generally accepted that the optimal operation of an MBR depends on understanding membrane fouling (Judd, 2007). Abatement of fouling leads to elevated energy demands and has become the main contribution to OPEX (Verrech et al., 2008). In addition, uncertainty associated with this phenomenon has led to conservative plant designs where the supplied energy is so far to be optimised.
Traditional strategies for fouling mitigation such as air sparging, physical cleaning techniques (i.e backflushing and relaxation) and chemical maintenance cleaning have been incorporated in most MBR designs as a standard operating strategy to limit fouling. Air sparging, expressed as specific aeration demand SAD_m, takes a typical value for full-scale facilities between 0.30 Nm^3/h m^2 (FS configuration) to 0.57 Nm^3/h m^2 (HF configuration). Relaxation and backflushing (only for HF) are commonly applied for 30–130 seconds every 10–25 min of filtration (Judd, 2010). Frequent maintenance cleanings (every 2–7 d) are also applied to maintain membrane permeability. However, these pre-set fixed values of key parameters, based on general background or the recommendations of membrane suppliers, lead to under-optimised systems and results in loss of permeate and high energy demand.
Recently, several authors have proposed a feedback control system for finding optimal operating conditions. For example, Smith et al. (2006) have successfully validated a control system for backflush initiation by permeability monitoring. This system automatically adjusts the backflushing frequency as a function of the membrane fouling, which results in

a reduction of up to 40% in the backflushing water required. Ferrero et al. (2011) have used a control system at semi-industrial pilot scale trials based on monitoring membrane permeability, which achieved a energy saving between 7 to 21% with respect to minimun aeration recommended by membrane suppliers.

2.3.4 Sludge retention time (SRT) and biomass concentration

SRT contributes to a distinct treatment performance and membrane filtration, and therefore, to system economics. Specifically, these parameters act on biomass concentration (MLSS), generation of soluble microbial products (SMP) and oxygen transfer efficiency.
Increasing the SRT increases the sludge solids concentration and therefore, reduces bioreactor volume required. Furthermore, because of the low growth rates of some microorganisms (specifically nitrifying bacteria), a longer SRT will achieve a better treatment performance, as well as generating less sludge. In addition, it has been reported that high values of SRT can increase membrane permeability by decreasing SMP production (Trussel et al., 2006). Conversely, high solids concentration results in a higher viscosity of the microbial suspension (Rosenberger et al., 2002b), as a consequence, higher concentrations decrease air sparging efficiency and oxygen transfer rate to the microorganisms, resulting in a higher energy demand as well as increasing membrane fouling and the risk of membrane clogging. Given all of these factors, for economical reasons, most full-scale facilities are designed for MLSS range of 8-12 g/l and SRT range of 10-20 d (Asano et al., 2006; Judd, 2010).

2.3.5 Membrane life

As a consequence of being a relatively new technology, limited information on the life of membranes is available. However, analysis of the oldest plants evidence that membrane life can reach, or even exceed, 10 years (Verrech et al., 2010).
Recently, Ayala et al. (2011) has reported the effect of operating parameters on the permeability and integrity of cartridges taken from full-scale MBRs. Regarding permeability, a correlation of permeability loss and operation time was found, indicating that the membrane permeability reaches non-operative value after seven years of operation. The authors also suggested a significant effect of inorganic scaling on permeability loss. The correct functioning during membrane cartridge life, determined by the strength of the welding at its perimeter, appears to be related to the total volume of water permeated and the total mass of oxidant (NaOCl) used during chemical cleanings.

3. Experimental methodology

3.1 Experimental setup

The experimental unit consisted of a cylindrical 220 l submerged membrane bioreactor (MBR) equipped with a submerged hollow-fibre membrane of 0.03 μm rated pore diameter and 0.93 m^2 filtering surface area (ZeeWeed ZW10) supplied by GE Water & Process Technologies (Figure 3). The effluent (permeate) was extracted from the top header of the module under slight vacuum (transmembrane pressure lower than 0.12 bar). Fouling was controlled by coarse bubbling of air flow and by intermittent filtration of the permeate. The pilot plant (ZW10) was located in the wastewater treatment plant (WWTP) in Santa Cruz de Tenerife (Canary Islands, Spain).

3.2 Feedwater characteristics

The reactor was fed with screened (2.5 mm) municipal wastewater. The average feed concentrations are given in Table 2. The feedwater was characterized by a high biodegradable organic fraction (BOD_5/COD = 0.52-0.67). Also, suspended solids in the water had a high organic fraction (VSS/TSS = 0.85-0.95).

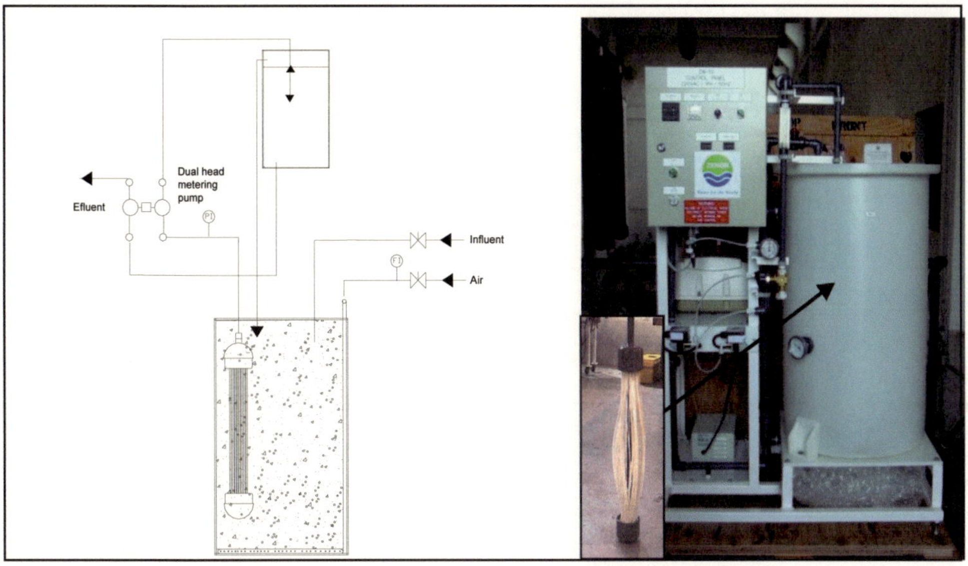

Fig. 3. Configuration and photograph of the pilot-MBR system, ZW10.

	COD mg/l	COD_s[a] mg/l	$N\text{-}NH_4^+$ mg/l	$N\text{-}NO_2^-$ mg/l	$N\text{-}NO_3^-$ mg/l	pH	TSS mg/l
Mean	879	262	70	0.07	2.0	8.1	830
Max.	1316	717	125	0.35	8.0	8.3	2200
Min.	270	137	33	0.03	1.0	7.7	150
[a] Samples were filtered through filter paper with a nominal pore size of 0.45 μm.							

Table 2. Mean concentrations of the feedwater

3.3 Operating conditions

Table 3 lists operating conditions. Permeate flux was incremented from 20 to 35 l/(h·m^2) in successive experimental runs. In order to maintain a constant HRT independent from the imposed permeated flux in each run, a peristaltic pump extracted from the permeate tank the flow rate necessary to maintain the required HRT and the excess of permeate was returned to the bioreactor (see Figure 1). Chemical cleaning of the membrane with sodium hypochloride (250 mg/l) was performed at the end of each experimental run.
Air was supplied through the bottom providing oxygen and stirring. The dissolved oxygen concentration was always above 1.5 mg/l in the reactor operated at 23 ± 2 °C.

3.4 Analytical methods

Dissolved oxygen (DO) was measured using a WTW 340i. Chemical oxygen demand (COD), ammonium-nitrogen ($N\text{-}NH_4^+$), total suspended solids (TSS), mixed liquor suspended solids

(MLSS), mixed liquor volatile suspended solids (MLVSS) were determined in conformity with the Standard Methods (American Public Health Association, 1992). Nitrite-nitrogen (N-NO_2^-) and Nitrate-nitrogen (N-NO_3^-) were measured by spectrophotometric methods with a HACH DR 2000. Microbial floc size was measured by Coulter LS100 (Coulter, UK). Proteins were determined as bovine albumin equivalent using the protein kit assay TP0300 supplied by Sigma, following the Lowry method (Lowry et al., 1951). Polysaccharides were measured as glucose equivalent by the Dubois` method (Dubois et al., 1956).

Parameters	Units	Value
Sludge retention time (SRT)	days	Infinite (without purge)
Hydraulic retention time (HRT)	hours	24.6
Filtration time	seconds	450
Duration of relax phase	seconds	30
Aeration rate per membrane area (SAD_m)	$Nm^3/h\ m^2$	1.9
Permeate flux	$l/h\ m^2$	20-35

Table 3. Operating conditions of the pilot-scale MBR

The oxygen uptake rate was measured by following the dissolved concentration with a membrane oxygen electrode in a medium without substrate (SOURe, endogenous). The sludge rheological properties were determined by using the concentric cylinder rotational viscosimeter Visco Star plus (FungiLab, Spain). The width of the annular gap was 1.0 mm. Measurements were done at 25 °C.

4. Experimental results

4.1 Biological process

4.1.1 Maintenance kinetics

Biomass concentration in the bioreactor is one of the most critical parameters in capital and operational costs of the process. It is known that increasing the biomass concentration reduces the bioreactor size and therefore, capital costs. However, high sludge concentration impacts on aeration efficiency (because of high viscosity) increasing membrane fouling propensity and, probably, membrane clogging (filling of the channels between the membranes with sludge solids). Therefore, a more frequent cleaning and higher aeration rate is necessary to maintain membrane permeability, which increments the operational costs. Therefore, fundamental knowledge of biomass development processes involved in the biological treatment of a MBR is required.

Figure 4 shows the typical trend of biomass evolution, expressed as total (MLSS) and volatile suspended solids (MLVSS), during the start-up and steady-state of an MBR operated without biomass purge. Biomass is developed from the microorganisms coming with the feed wastewater as the bioreactor had not been inoculated. During the initial period, biomass increased rapidly and then slower with increasing biomass concentration in the mixed liquor.

The first concern is the MLVSS/MLSS ratio, which remained within the range between 71 and 78%. It is important to note that, despite operating in conditions of total sludge retention, this ratio remains constant throughout the experiment, indicating no significant accumulation of inorganic matter in the sludge. This may be due to the fact that a small fraction of inorganic suspended solids in the feed (5-15%) is dissolved during the process and, therefore, does not accumulate in the sludge and leaves the system with the permeate.

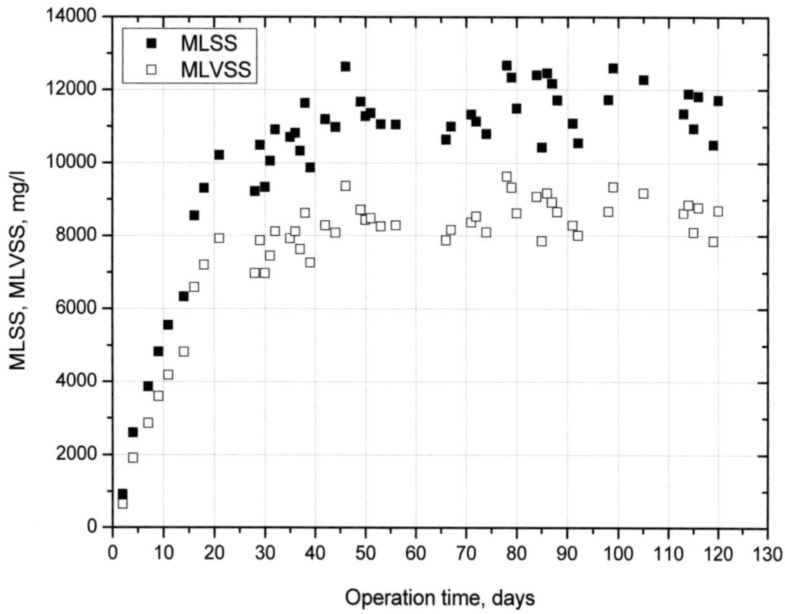

Fig. 4. Evolution of biomass concentration (MLSS and MLVSS) in the mixed liquor with operation time.

The second concern is the stabilisation value of the biomass concentration (MLSS and MLVSS), which is expected to depend on the hydraulic retention time (HRT) and COD removal, resulted in a stationary value of utilisation rate (U). Figure 5 shows the evolution of U with operation time where it can be observed that the system evolved until reaching a nearly constant value (0.083 ± 0.004 kg COD/kg MLVSS d). A symmetrical trend can also be observed for data obtained in a previously reported research (Delgado et al., 2010) in an MBR treating biological effluent from a WWTP. In that case, the MBR was inoculated and the initial biomass evolution was characterised by a lysis process. Afterwards, a stationary vale for U was reached (0.067 ± 0.004 kg COD/kg MLVSS d) independently of the fixed HRT value.

It is thought that the maintenance concept introduced by Pirt (1965) could be the reason for the equilibrium reached in the MBRs operated without biomass purge. Then, the utilisation rate can be described by the Pirt equation (1).

$$U = \frac{r_x}{Y \cdot X} + k_{m,S} \tag{1}$$

where r_s is the substrate removal rate, r_x is the biomass growth rate, Y is the true sludge yield, $k_{m,S}$ is the maintenance coefficient and X is the biomass concentration.

At very low growth rates (i.e. steady-state conditions), r_x can be neglected:

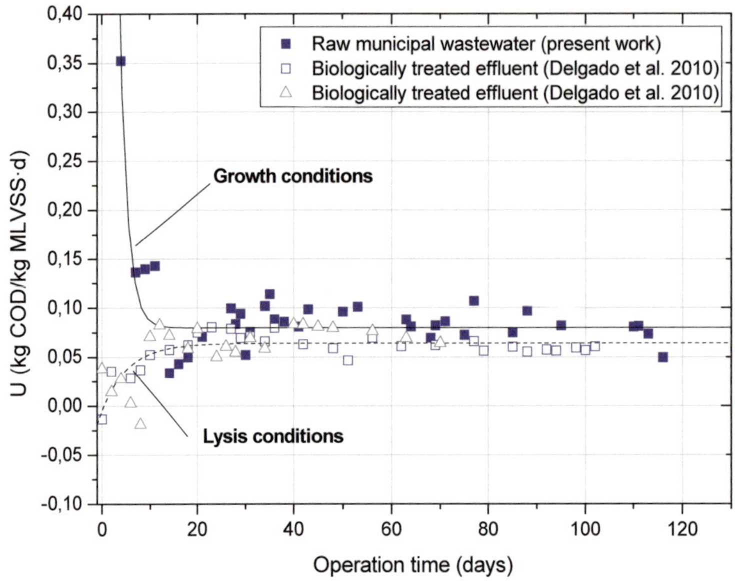

Fig. 5. Evolution of utilisation rate with operation time for MBRs treating different types of feed wastewaters.

$$U \approx k_{m,S} \tag{2}$$

Therefore, the stationary value of the utilisation rate is identical to the maintenance coefficient, which suggests that, in these substrate-limited conditions, microorganisms tend to minimize their energy requirements using the available substrate to satisfy their maintenance functions. For the presented data the best fitting parameter was $k_{m,S}$ = 0.0035 kg COD/kg MLVSS h.

4.1.2 Microbial activity: Specific endogenous oxygen uptake rate

The measurement of the oxygen demanded by the microorganisms is a parameter frequently used for assessing aerobic activity of microbial suspensions (Vanrolleghen et al., 1995). In this sense, Pollice et al. (2004) reported that the specific endogenous respiration rates are closely related to the organic loading rates (F/M). Table 4 shows specific endogenous oxygen uptake rates ($SOUR_e$) of sludge samples at steady-state conditions and other values reported in the literature. The $SOUR_e$ is considerably lower than the typical values, which confirms the maintenance energy requirement reached.

F/M, kg COD/ kg MLVSS d	$SOUR_e$, kg O_2/kg MLVSS d	Reference
-	0.118	Coello Oviedo et al., 2003
0.15	0.05	Pollice et al., 2004
0.08	0.01-0.05	Rodde-Pellegrin et al., 2002
0.09	0.0084 ± 0.03	This work

Table 4. Specific endogenous oxygen uptake rate of sludge samples

4.1.3 Sludge morphology

According to the literature, flocculant ability tends to be reduced when organic substrate is lacking (e.g. Wilen et a., 2000). In an MBR operated under substrate-limited conditions these conditions of stress are imposed and therefore a floc distribution characterised by a greater number of small flocs is expected. In addition, particle size distribution plays an important role in the formation of the cake on the membrane surface. A cake made with small particles has higher specific resistance and, therefore, is less permeable than the cake formed by larger particles (Defrance et al., 2000). As a consequence, it is crucial to analyze the effect of the several substrate-limited conditions imposed over the particle size of the flocs and the presence of small non-flocculating microorganisms in mixed liquor.

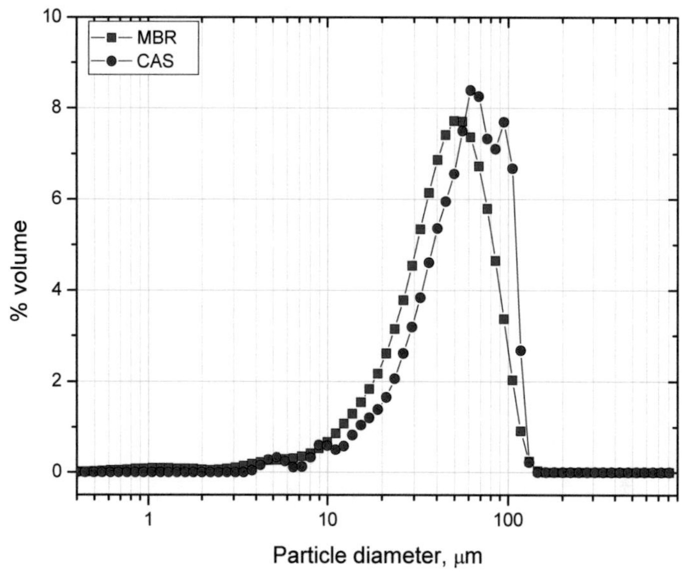

Fig. 6. Particle size distribution of MBR and CAS sludge samples.

Sludge morphology was analysed by optical microscope observations and by particle distribution measurements. In Figure 6 particle size distribution of a sludge sample at steady-state conditions is shown. Also, samples from a conventional activated sludge process (CAS) which treated the same influent were investigated and compared with the MBR sample. Figure 6 shows aggregates with bimodal distribution in CAS biomass, where 50 % of the particles have a size higher than 70 μm. In contrast, uniform and medium-sized flocs were observed in the MBR sludge, where 40 % of the particles were within the 15 to 50 μm range. Granulometric differences, which are a result of biomass separation by the membrane, are well documented in the literature (e.g. Cicek et al., 1999) and are attributable to effective particle retention by the membrane and high shear stress conditions due to air sparging for membrane fouling mitigation. Also, the low quantity of small non-floculating flocs (< 10 μm) could be due to the presence of higher organisms, which have traditionally been considered as predators that consume dispersed bacteria.

Alternatively, microscopic analysis of mixed liquor samples from the MBR is shown in Figure 7. The observations can be summarized into two main issues: firstly the absence of filamentous microorganisms, which can be linked to the process conditions, including high

dissolved oxygen and low readily biodegradable substrate concentrations (Martins et al., 2004). Secondly, as a result of the low organic loading conditions, higher organisms were also expected. In this sense, a significant quantity of worms (type *Aeolosoma hemprichi)* developed. Similar results were reported by Zhang (2000) where a high worm density resulted in a low sludge yield (0.10-0.15 kg MLSS/ kg COD). Worms are considered as predators with a great potential on sludge reduction and more attention has been paid to their effectiveness in wastewater treatment recently (Wei et al., 2003).
As already stated, to operate an MBR under substrate-limited conditions enhances the presence of worms that may lead to a substantial sludge reduction and improve biomass characteristics by removing small non-floculating flocs.

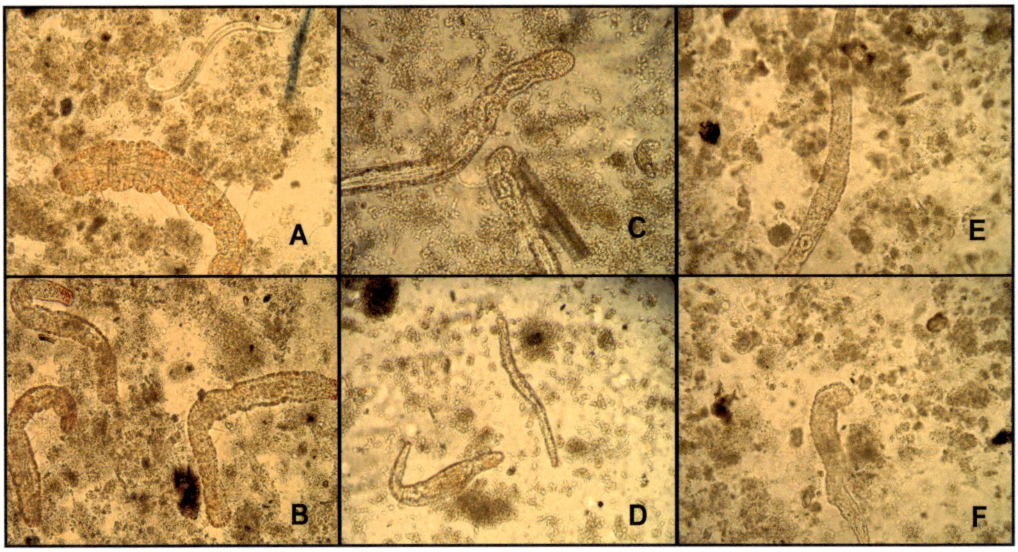

Fig. 7. Higher microorganisms found in MBR (A, B, D, F x20; C, E x40).

4.1.4 Rheological properties

Rheological properties are of crucial importance due to their effect on hydrodynamic conditions near the membrane. The rheological behavior of microbial suspensions has been described in the literature as non-Newtonian pseudoplastic fluids (Rosenberger et al., 2002b). When air is dispersed in a solid-liquid suspension a change can be seen in its rheological behavior due to the change in suspension structure: with increasing shear, the structure opens and biological aggregates are reorganized resulting in a decrease in viscosity. In addition, it is accepted that the microbial suspensions have a thixotropic nature, which means that the viscosity decreases with shear rate when samples are subject to shear stress. Rheology can be described by the Bingham model, the Ostwald model and the Herschel-Bulkley model represented by Eq. (3)-(5):

$$\mu_a = \frac{\tau_0}{dv / dr} + m \tag{3}$$

$$\mu_a = m \cdot \left(\frac{dv}{dr} \right)^{n-1} \tag{4}$$

$$\mu_a = \frac{\tau_0}{dv / dr} + m \cdot \left(\frac{dv}{dr} \right)^{n-1} \tag{5}$$

In these models μ_a is the apparent viscosity, *dv/dr* is the shear rate and τ_0, *m* and *n* are the model parameters. From the models we may deduce that the apparent viscosity can be described as a shear rate function.

Figure 8 shows one example of apparent viscosity reduction with the shear intensity. It decreases down to 75% when the shear varied from 13 to 130 s^{-1}. Additionally, plotting is shown according to the Bingham, Ostwald and Herschel-Bulkley models. In general, both the Ostwald model as well as the Herschel-Bulkley model fits quite well into the experimental data, while the Ostwald was selected because of its simplicity. From the equation of the curve (Figure 8) the parameter values for Ostwald model can be obtained:

$$n = 0.41$$

$$m = 122 \text{ mPa s}$$

where *n* is the flow behavior index and *m* is the consistency index.

Furthermore, as shown in Figure 8, apparent viscosity $(\mu_a)_{limit}$ can be perceived for higher values (> 130 s^{-1}). It does not decrease substantially with an increasing velocity gradient. Therefore, the effect of particle concentration on the viscosity can be evaluated by fitting the $(\mu_a)_{limit}$ to the sludge concentration, measured as MLSS concentration (Figure 9). As expected, microbial suspension viscosity also increased with the MLSS concentration. This behaviour is commonly accepted in the literature (e.g. Pollice et al., 2007).

Therefore, the following equation (Eq. (6)) can estimate the limit apparent viscosity as a function of the MLSS concentration.

$$\mu_{a_{limit}} = 1.1 \cdot 10^{-6} \cdot SSLM^{1.7} \tag{6}$$

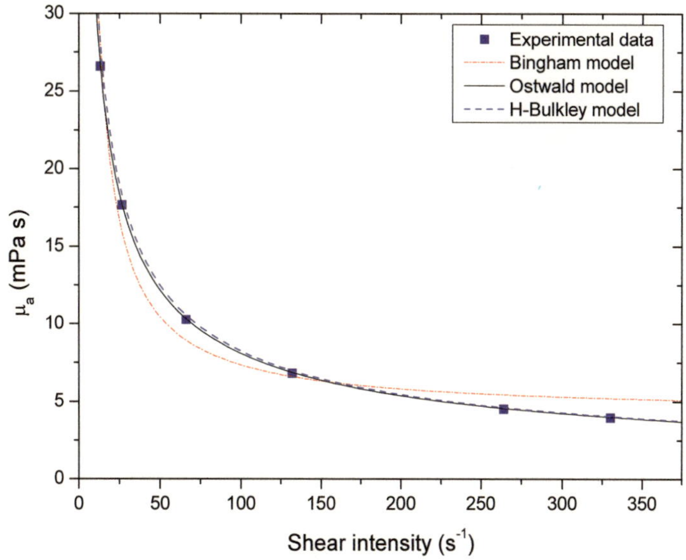

Fig. 8. Apparent viscosity against the shear intensity.

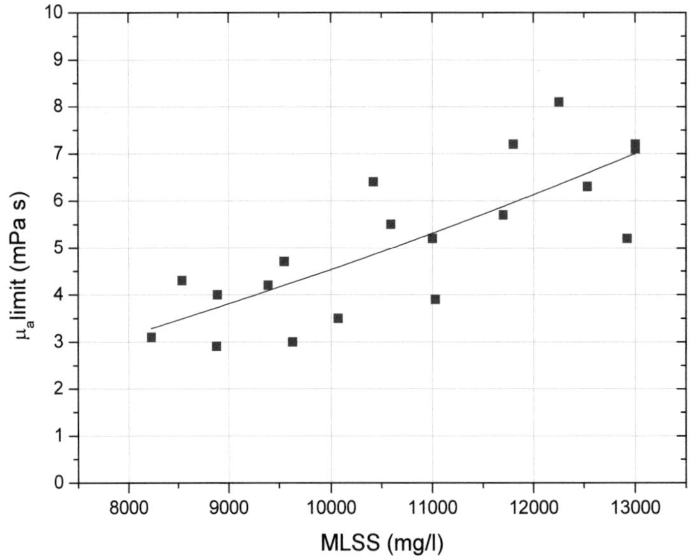

Fig. 9. Apparent viscosity limit (dv/dr = 264 s-1) against the MLSS

4.1.5 Analysis of the liquid phase. Extracellular polymeric substances

Extracellular polymeric substances (EPS) can be differentiated into two main types: bound EPS, which form the structure of the floc, and soluble EPS (often named soluble microbial products), which are soluble or colloidal form in the liquid medium. Recent studies have shown that the soluble and colloidal fraction plays an important role in membrane fouling (Drews, 2010). Their principle components are also generally recognised as proteins and polysaccharides (Sponza, 2002).

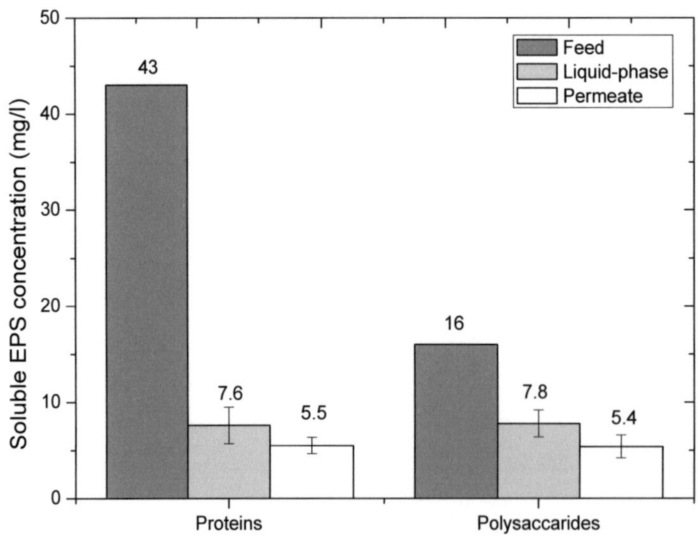

Fig. 10. Average soluble EPS concentration of feedwater, liquid-phase and permeate.

Figure 10 compares the average concentrations of proteins and polysaccharides in the feed wastewater, in the liquid-phase and in the permeate. A significant reduction in EPS can be observed in the liquid-phase in relation to feed (82% for proteins and 51% for polysaccharides), as a result of biological metabolism. On the other hand, the separation through the membrane of the polysaccharides is 31% and for the protein it is 28%, both remaining constant throughout the experimental test. These membrane retention values are similar to those found in the literature (Rosenberger et al., 2006).
A low concentration was unexpected in the liquid-phase, as the common trend is to suppose EPS accumulation resulting from polymer retention by the membrane (Masse et al., 2006). As a consequence specific microorganisms may be assumed to develop, which can degrade polysaccharides and proteins with a slow degradation rate.

4.2 Membrane performance

4.2.1 Membrane fouling characterisation: TMP profiles

As noted in the experimental procedure, all stages were performed using the same sequence of filtration and relaxation (450 s and 30 s, respectively). The experimental period was divided into five phases, each one operated at constant permeate flux. Membrane fouling was followed by measuring transmembrane pressure (TMP) evolution with operation time (Figure 11). Each phase finished when a pre-established TMP was reached.

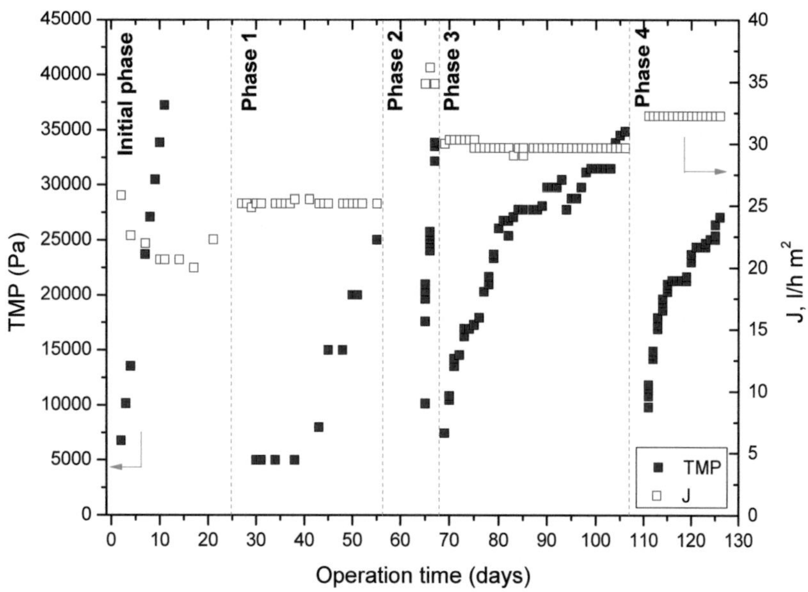

Fig. 11. Transmembrane pressure TMP and permeate flux J evolution with operation time

The initial period (Figure 11) showed a high rate of fouling (0.011 Pa/s) despite working with relatively low permeate flux (20-23 l/h m^2) and without reaching a high concentration of MLSS. This could be attributed to the initial biomass development until it obtained a high level of biological degradation. During this period, it was expected that microcolloidal and soluble species would have caused irreversible pore blocking, as a result of their small size (Di Bella et al., 2006). Afterwards, we assume that the developed biomass reaches steady-state conditions and degrades most of the colloidal and soluble matter. Therefore, feedwater characteristics and the level of physiological biomass seem to have a significant effect of fouling propensity.

4.2.2 Determination of sustainable flux

The fouling rate, measured as the slope of transmembrane pressure against filtration time, has been used in many works as a fouling quantification parameter in systems operated under constant permeate flux. Experimentally, it has been found that r_f depends exponentially on permeate flux (Figure 12). Therefore, a threshold flux value may be identified (32 l h^{-1} m^{-2}) above which the fouling increases at an unacceptable rate.

4.3 Physico-chemical and microbiological quality of the permeate

The physical and chemical quality of the permeate was assessed by the analysis of turbidity, COD and nitrogen compounds.

The permeate had an average turbidity value of 0.59 NTU, indicating a total retention of suspended solids and macro-colloidal matter. In addition, the low turbidity of the permeate registered during the whole experimental period showed that the membrane maintained its integrity.

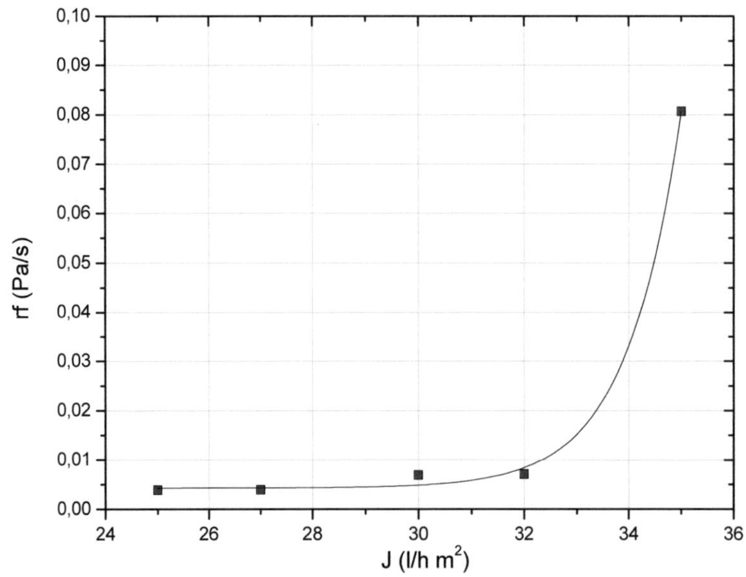

Fig. 12. Fouling rate against permeate flux.

The organic matter content was determined by measuring the COD in feed wastewater, in the permeate and in the liquid phase of the suspension. Soluble COD (COD_S) was obtained by filtering through a filter paper of 0.45 µm pore diameter. Figure 13 shows the COD of feedwater (COD feed), the soluble COD of feedwater (COD_s feed), the COD of the permeate (COD_P) and soluble COD of the liquid phase (COD_s reactor) versus operating time. Typical fluctuations of feed wastewater can be seem in a real treatment plant. These oscillations lessened considerably in the permeate and in the liquid phase.

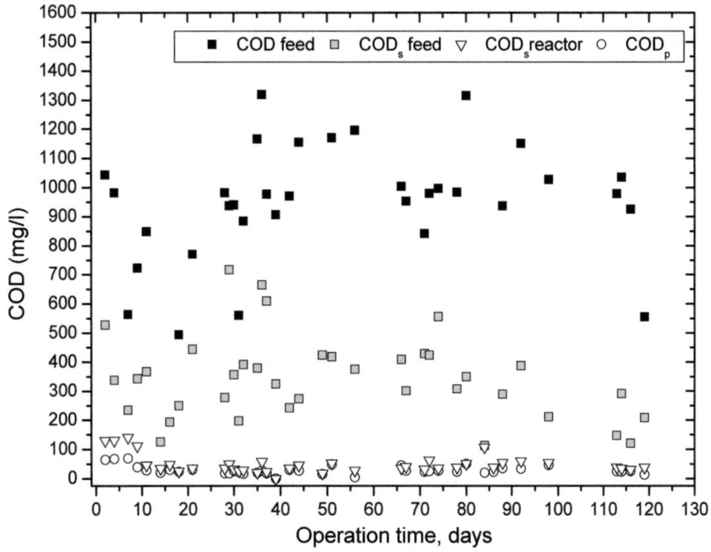

Fig. 13. COD evolution with operation time.

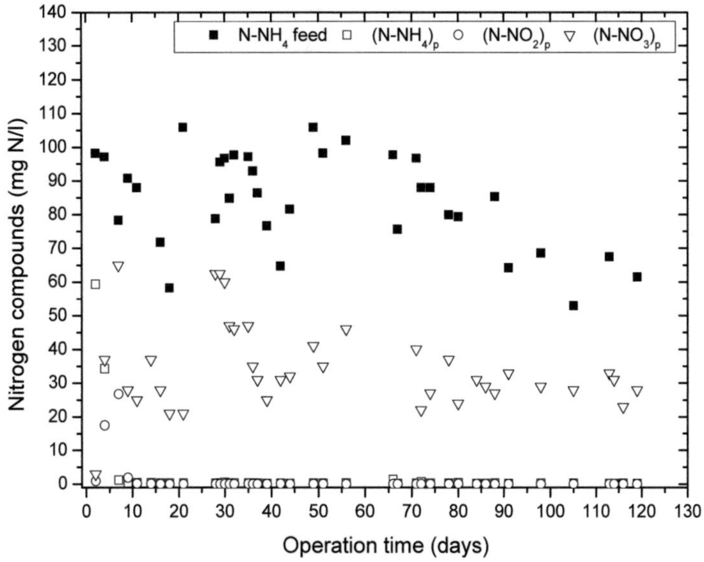

Fig. 14. Evolution of the nitrogen compounds with operation time.

As it is shown in Figure 13, there is a significant difference between the total and soluble COD of feed due to the presence of suspended solids. It was estimated that approximately 68% of the COD of the feed is in a particulate form. If the soluble COD of feed is compared with the soluble COD of the CODs liquid phase (CODs reactor) a removal efficiency close to 86% can be obtained, mainly due to biological degradation and only 6% is due to the membrane separation process. It should be noted that the BOD_5 was not analyzed because, through frequent and trustworthy analysis of the same water, the BOD_5/COD ratio was

confirmed to be approximately constant and equal to 0.75, so the COD analysis may be considered sufficient to determine the biodegradation produced.
Also, the evolution of the ammonium nitrogen concentration in feed wastewater ($N-NH_4$ feed) and the nitrogen compounds of the permeate ($(N-NH_4^+)_p$, $(N-NO_2^-)_p$, $(N-NO_3^-)_p$) were measured during the experimental period (Figure 14). As can be seen, the concentrations of nitrogen-nitrate in the permeate $(N-NO_3^-)_p$ were in the range of 15-45 mg/l, while nitrite and ammonia were completely removed. This is interpreted as a total oxidation of ammonium to nitrate.
As shown in Table 5, no bacterial contamination indicators, bacterial pathogens or parasites were detected in the permeate. This is attributed to the ultrafiltration membrane which has a pore diameter smaller than the size of bacteria and parasitic microorganisms, so that the membrane is an effective barrier. However, Table 5 shows the presence of viral indicators. Here, results indicate a great degree of removal (99.8% and 95.3% for somatic coliphages and F-RNA bacteriophages, respectively).

	Feed wastewater	Permeate (N = 3)
Bacteriological indicators		
Fecal coliform[1]	$7.7 \cdot 10^6$	absence
Escherichia Coli[1]	$7.3 \cdot 10^6$	absence
Enterococci[1]	$3.6 \cdot 10^6$	absence
Clostridium perfringens[1]	$1.1 \cdot 10^6$	absence
Indicators of pathogenic contamination		
Pseudomonas aeruginosa[1]	absence	absence
Salmonella sp. [1]	absence	absence
Viral indicators		
Somatic coliphages[2]	$3.2 \cdot 10^6$	$4.3 \cdot 10^3 \pm 1.6 \cdot 10^3$
F-RNA bacteriophages[2]	$2.3 \cdot 10^5$	$1.1 \cdot 10^4 \pm 1.6 \cdot 10^4$
Parasites		
Giardia lamblia [3]	absence	absence
Cryptosporidium sp. [3]	absence	absence
[1] CFU/100ml; [2] PFU/100ml; [3] No/100 ml. N= Number of samples		

Table 5. Feed wastewater and permeate microbial results.

Permeate microbial results proved that MBR systems are able to produce permeate of high microbial quality to be used in several applications such as land irrigation, agricultural activities etc., in accordance with local standards.

5. Conclusions

MBRs have been proven as efficient and versatile systems for wastewater treatment over a wide spectrum of operating conditions. The treatment performance of the MBR is better than in conventional activated sludge process. A high conversion of ammonium to nitrate

(>95%) and constant COD removal efficiency (80-98%) was achieved, regardless of the influent fluctuations. Microbial analysis of permeate showed the absence of bacterial indicators of contamination and parasitical microorganisms. At the same time, the membrane presented over 98% efficiency in the elimination of viral indicators.

Particularly interesting is the possibility of operating at maintenance energy level of the biomass, which significantly reduces sludge production. At these maintenance conditions, a minimal value for the carbon substrate utilization rate (0.07-0.1 kg COD kg^{-1} MLVSS d^{-1}) was found and the system was operated successfully at permeate flux between 30 and 32 l $h^{-1}m^{-2}$ and low physical cleaning frequency. As a result of carbon substrate limited conditions, EPSs were minimized and higher organisms appeared.

Biomass development at maintenance conditions can be well described by the kinetic model based on Pirt´s equation.

Although there are many practical experiences for MBR design and operation, there are still some aspects that are not completely understood. Without any doubt, the most cited is membrane fouling. The complexity of this phenomenon is linked to the presence of particles and macromolecules with very different sizes and the biological nature of the microbial suspensions which results in a very heterogenic system. Meanwhile, the dynamic behaviour of the filtration process adds a particular complication to fouling mechanisms. Therefore, further investigation is required so as to ascertain which component in the suspension is the primary cause of membrane fouling.

6. Acknowledgements

This work has been funded by the N.R.C. (MEC project CTM2006-12226). The authors also want to express their gratitude to the MEC for a doctoral scholarship, to GE ZENON, to CANARAGUA and to BALTEN for their support and finally to the Water Analysis Laboratory of the ULL Chemical Engineering Department for analytical advice.

7. Nomenclature

CAS	Conventional activated sludge process
COD	Chemical oxygen demand, mg O_2 /l
EPS	Extracellular polymeric substance
F/M	Feed to microorganisms ratio, kg COD/kg MLSS d
HRT	Hydraulic retention time, h
iMBR	Immersed membrane bioreactor
J	Permeate flux, l/h m^2
MLSS	Mixed liquor total suspended solids, mg/l
MLVSS	Mixed liquor volatile suspended solids, mg/l
NH_4-N	Ammonium nitrogen concentration, mg/l
NO_2-N	Nitrite nitrogen concentration, mg/l
NO_3-N	Nitrate nitrogen concentration, mg/l
SAD_m	Specific membrane aeration demand, Nm^3/h m^2
$SOUR_e$	Specific oxygen uptake rate in endogenous conditions, kg O_2/kg MLVSS d
SRT	Sludge retention time, days
TMP	Transmembrane pressure
U	Utilisation rate, kg COD/kg MLVSS d

8. References

APHA (1992). *Standard Methods for the examination of Water and Wastewater*, 18th ed. American Public Health Association/Water Environment Federation, Washington, DC, USA.

Asano, T.; Burton, F., Leverenz, H.; Tsuchinashi, R. & Tchobanoglous, G. (2006). *Water Reuse: Issues, Technologies and Applications.* Metcalf & Eddy/AECOM. ISBN: 978-0-07-145927-3. 1st ed.

Ayala, D.F.; Ferre, V. & Judd. S.J. (2011). Membrane life estimation in full-scale immersed membrane bioreactors. *Journal of Membrane Science* (in press), doi: 10.1016/j.memsci. 2011.03.013.

BCC. (2011). *Membrane bioreactors: global markets.* BCC Report MST047C. March 2011.

Brepols, C.; Dorgeloh, E.; Frechen, F.-B.; Fuchs, W. ; Haider, S.; Joss, A.; de Korte, K. ; Ruiken, C.; Schier, W.; van der Roest, H.; Wett, M. & Wozniak, T. (2008). Upgrading and retrofitting of municipal wastewater treatment plants by means of membrane bioreactor (MBR) technology. *Desalination,* Vol. 231, No. 1-3, pp. 20-26.

Chang, I.S.; Le-Clech, P.; Jefferson, B. & Judd, S. (2002). Membrane fouling in membrane bioreactors for wastewater treatment. *Journal of Environmental Engineering,* Vol. 128, No. 11, pp. 1018–1029.

Cicek, N., Franco, J.P., Suidan, M.T., Urbain, V., Manem, J. (1999). Characterization and comparison of a membrane bioreactor and a conventional activated-sludge system in the treatment of wastewater containing high-molecular-weight compounds. *Water Environ. Res.*, Vol. 71, No. 1, pp. 64-70.

Coello Oviedo, M.D., López-Ramírez, J.A., Sales Márquez, D. & Quiroga Alonso, J.M. (2003). Evolution of an activated sludge system under starvation conditions. *Chem. Eng. J.*, Vol. 94, pp. 139-146.

Cui, Z.F., Chang, S. & Fane, A.G. (2003). The use of gas bubbling to enhance membrane processes, *Journal of Membrane Science,* Vol. 221, pp. 1–35.

Defrance L., Jaffrin, M.Y.; Gupta, B.; Paullier, P. & Geaugey, V. (2000). Contribution of various constituents on activated sludge to membrane bioreactor fouling. *Bioresource Technology, Vol.* 73, pp. 105-112.

Delgado, S.; Villarroel, R.; González, E. (2010). Submerged Membrane Bioreactor at Substrate-Limited Conditions: Activity and Biomass Characteristics. ***Water Environment Research***, Vol. 82, No. 3, pp. 202-208.

Di Bella G., Durante, F., Torregrossa, M. & Viviani, G. (2006). The role of fouling mechanisms in submerged membrane bioreactor during the start-up. *Desalination,* Vol. 200, pp. 722-724.

Drews, A. (2010). Membrane fouling in membrane bioreactors—Characterisation, contradictions, cause and cures. *Journal of Membrane Science*, Vol. 363, No. 1-2, pp. 1-28.

Dubois, M.; Gilles, K.A.; Hamilton, J.K.; Rebers, P.A. & Smith, F. (1956). Calorimetric method for determination of sugars and related substances. *Anal Chem.*, Vol. 28, No. 3, pp. 350-356.

Ferrero, G.; Monclús, H.; Buttiglieri, G.; Comas, J. & Rodriguez-Roda, I. (2011). Automatic control system for energy optimization in membrane bioreactors. *Desalination*, Vol. 268, No. 1-3, pp. 276-280.

Foley, G. (2006) A review of factors affecting filter cake properties in dead-end microfiltration of microbial suspensions. *Journal of Membrane Science*, Vol. 274, pp. 38–46.

Frechen, F.B.; Schier, W.; & Linden. C. (2008). Pre-treatment of municipal MBR applications. *Desalination*, Vol. 231, No. 1-3, pp. 108-114.

Garcés, W.; De Wilde, C.; Thoeye & De Gueldre, G. (2007). Operational cost optimisation of MBR Schilde. *Proceedings of the 4th IWA International Membranes Conference, Membranes for Water and Wastewater Treatment*, Harrogate, UK, May 15-17.

Icon. (2008). *The 2009-2014 world outlook for membrane bioreactor (MBR) systems for wastewater treatment*. Icon Group Publications.

Judd, S. (2008). The status of membrane bioreactor technology. *Trends in Biotechnology*, Vol. 26, No. 2, pp. 109-116.

Judd, S. (2010). The MBR Book. Principles and Applications of Membrane Bioreactors in Water and Wastewater Treatment, Elvesier, ISBN: 978-0-08-096682-3, 2nd Ed, London.

Laera, G.; Pollice, A.; Saturno, D.; Giordano, C.; Lopez, A. (2005). Zero net growth in a membrane bioreactor with complete sludge retention. *Water Research*, Vol. 30 No. 20, pp. 5241-5249.

Le-Clech, P.; Chen, V. & Fane, T.A.G. (2006). Fouling in membrane bioreactors used in wastewater treatment. *Journal of Membrane Science*, Vol. 284, pp. 17–53.

Le-Clech, P.; Fane, A.; Leslie, G. & Childress, A. (2005b). The operator's perspective. *Filtration & Separation*, Vol. 42, pp. 20-23.

Le-Clech, P.; Jefferson, B. & Judd, S. J. (2005a). A comparison of submerged and sidestream tubular membrane bioreactor configurations. *Desalination*, Vol. 173, pp. 113-122.

Lesjean, B.; Ferre, V. ; Vonghia, E. & Moeslang, H. (2009). Market and design considerations of the 37 larger MBR plants in Europe. *Desalination Water Treat.*, Vol. 6, pp. 227-233.

Lowry, O.H.; Rosebrough, N.H.; Farr, A.L. & Randall, R.J. (1951). Protein measurement with the Folin phenol reagent. *J. Biol. Chem.*, Vol. 193, pp. 265-275.

Masse, A., Sperandio, M. & Cabassud, C. (2006). Comparison of sludge characteristics and performance of submerged membrane bioreactor and an activated sludge process at high solids retention time. *Water Research*, Vol. 40, pp. 2405-2415.

Meng, F.; Chae. S-R.; Drews, A.; Kraume, M.; Shin, H-S.; Yang, F. (2009). Recent advances in membrane bioreactors (MBRs): Membrane fouling and membrane material. *Water Research*, Vol. 43, pp.1489-1512.

Pearce, G. (2008). Introduction to membranes - MBRs: Manufacturers' comparison: part 1, *Filtration & Separation*, Volume 45, pp. 28-31

Pirt, S.J. (1965). The maintenance energy of bacteria in growing cultures. *Proc. R. Soc. London*, Vol. 163B, pp. 224-231.

Pollice A., Giordano, C., Laera, G., Saturno, D. & Mininni, G. (2007). Physical characteristics of the sludge in a complete retention membrane bioreactor. *Water Research*, Vol. 41, pp. 1832-1840.

Pollice A., Laera, G. & Blonda, M. (2004). Biomass growth and activity in a membrane bioreactor with complete sludge retention. *Water Research*, Vol. 38, pp. 1799-1808.

Rodde-Pellegrin M.L.; Winieswski, C.; Gramick, A.; Tazi, A. & Buisson, H. (2002). Respirometric needs of heterotrophic populations developed in an immersed membrane bioreactor working in sequenced aeration, *Biochemical Engineering Journal*, Vol. 11, pp. 2-12.

Rosenberger, S.; Krüger, U.; Witzig, R.; Manz, W.; Szewzyk, U.; Kraume, M. (2002a) Performance of a bioreactor with submerged membranes for aerobic treatment of municipal waste water. *Water Research*, Vol. 36, pp. 413-420.

Rosenberger, S.; Kubin, K. & Kraume, M. (2002b). Rheology of activated sludge in membrane bioreactors, *Engineering in Life Sciences*. Vol. 2, No. 9, pp. 269–275.

Rosenberger, S.; Laabs, C.; Lesjean, B.; Gnirss, R.; Amy, G.; Jekel, M. & Schorotter, J.C. (2006). Impact of colloidal and soluble organic material on membrane performance in membrane bioreactors for municipal wastewater treatment, *Water Research*, Vol. 40, pp. 710-720.

Santos, A. & Judd, S. (2010). The commercial status of membrane bioreactor for municipal wastewater. *Separation Science and Technology*, Vol. 45, No. 7, pp-850-857.

Santos, A; Ma, W. & Judd, S. (2010). Membrane bioreactors: Two decades of research and implementation. *Desalination* (in press), doi:10.1016/j.desal.2010.07.063.

Smith, P.; Vigneswaran, S.; Ngo, H.; Nguyen & H. Ben-Aim, R. (2006). Application of an automation system and a supervisory control and data acquisition (SCADA) system for the optimal operation of a membrane adsorption hybrid system. *Water Science and Technology*, Vol. 53, No. 179.

Sponza, D.T. (2002). Extracellular polymer substances and physicochemical properties of flocs in steady- and unsteady-state activated sludge systems. *Process Biochemistry*, Vol. 37, pp. 983-98.

Tao, G.; Kekre, K.; Oo, M-H.; Viswanath, B.; Lew, C-H.; Kan, L-M. & Seah, H. (2009). Large scale membrane bioreactor plant design (retrofit) and optimisation. *Proceedings of the 4th IWA Membrane Technology Conference*, Beijing, China, Sept 1-3.

Trusell, R.; Merlo, R.; Hermanowicz, S. & Jenkins, D. (2006). The effect of organic loading on process performance and membrane fouling in a submerged membrane bioreactor treating municipal wastewater. *Water Research*, Vol. 40, pp. 2675-2683.

Van Loosdrecht, M.C.M. & Hence, M. (1999). Maintenance, endogenous respiration, lysis, decay and predation. *Water Science and Technology*. Vol. 39, No.1, pp. 107-117.

Vanrolleghem P.A., van Daele, M. & Dochain, D. (1995). Practical identifiability of a biokinetic model of activated sludge respiration. *Water Research*, Vol. 29, pp. 2561-2570.

Verrecht, B.; Judd, S.; Guglielmi, G.; Mulder, J. W. & Brepols, C. (2008). An aeration energy model for an immersed membrane bioreactor. *Water Research*, Vol. 42, pp. 4761-4770.

Verrecht, B.; Maere, T.; Nopens, I.; Brepols, C. & Judd, S. (2010). The cost of a large-scale hollow fibre MBR. *Water Research*, Vol. 44, No. 18, pp. 5274-5283

Wei, Y.; van Houten, R.T.; Borger, A.R.; Eikelboom, D.H. & Fan, Y. (2003). Minimization of excess sludge production for biological wastewater treatment. *Water Research*, Vol. 37, pp. 4453-4467.

Wicaksana, F.; Fane, A.G. & Chen, V. (2006). Fibre movement induced by bubbling using submerged hollow fibre membranes, *Journal of Membrane Science*, Vol. 271, pp. 186–195.

Wilén, B-M.; Nielsen, J.; Keiding, K. & Nielsen, P. (2000). Influence of microbial activity on the stability of activated sludge flocs. *Colloids and Surfaces B: Biointerfaces*, Vol. 18, No. 2, pp. 145-156.

Wu, J., Le-Clech P., Stuetz, R., Fane, A., Chen, V. (2008). Effects of relaxation and backwashing conditions on fouling in membrane bioreactor. *Journal of Membrane Science*, Vol. 324, pp. 26–32.

Zhang S. (2000). *Polluted water treatment by the combining processes of membrane separation and biodegradation*. PhD thesis, Research Centre for Eco-Environmental Sciences, Chinese Academy of Sciences, China.

15

Rangeland Productivity and Improvement Potential in Highlands of Balochistan, Pakistan

Sarfraz Ahmad and Muhammad Islam
Arid Zone Research Centre, Quetta, Pakistan

1. Introduction

Pakistan has total land area of 88 million hectare (ha) and about 65% of this is rangelands. Five different types of range ecological zones (Sub-alpine and temperate, Sub-tropical humid, Sub-tropical sub-humid, Tropical arid and semi-arid deserts plains, and Mediterranean) have been described in Pakistan (Khan & Mohammad, 1987). These rangelands are the major feed source of about 97 million heads of livestock. Precipitation varies from 125 mm to over 1500 mm per annum. About 60 to 70% of monsoon rains received during the months of July to September while the winter rains occur from December to February (Khan, 1987).

Balochistan has a total area of 34 million ha of which only 4% (1.47 m ha) is under cultivation while 60% of the cultivated area is rainfed (Khan, 1987). Approximately, 93 % of this province (Fig. 1) is characterized as rangelands (FAO, 1983) Arid and semi-arid areas are falling within the rainfall zones of 50-200 mm and 250-400 mm, respectively (Kidd et al., 1988). Rainfall patterns are unpredictable with great variations. Like other arid and semiarid rangelands of the world, Balochistan ranges also provide a diversity of uses, including forage for livestock, wildlife habitat, medicinal plants, water storage and distribution, energy, minerals, fuel wood, recreational activity, wilderness and natural beauty.

Livestock rearing is the main activity of the inhabitants of Balochistan. Sheep and goats are the main livestock of the province. About 87% of the people in Balochistan directly or indirectly drive their livelihood from livestock rearing (Heymell, 1989). About 20 million sheep and goats population have been reported in Balochistan (GOB, 1996). Rangelands are the major feed source of these animals and approximately 90% of total feed requirements of sheep and goats were being met from rangelands (FAO, 1983). Overgrazing, drought, erosion, and human induced stresses caused severe degradation of rangelands in Balochistan (Islam et al., 2008; Hussain & Durrani, 2007). The degradation of rangelands includes changes in composition of desirable plant species, a decrease in rangeland diversity and productivity, reduction of perennial plant cover, and soil erosion (Milton et al., 1994).

In Balochistan, the mixed grass-shrub steppe is more common than single plant communities. The range vegetation types in Balochistan changes from south to north along the rainfall distribution. In South, shrub species *Haloxylon* species and *Artemisia* species while in north perennial grass species *Cymbopogon jwarancusa* and *Chrysopogon aucheri* are dominant. The fragile ranges of Balochistan are degrading very rapidly due to heavy grazing pressure, aridity, and human disturbances. However, still many of these ranges have potential for improvement by using grazing management practices, natural recovery of vegetation and artificial re-vegetation at suitable sites coupled with better water harvesting and conservation practices.

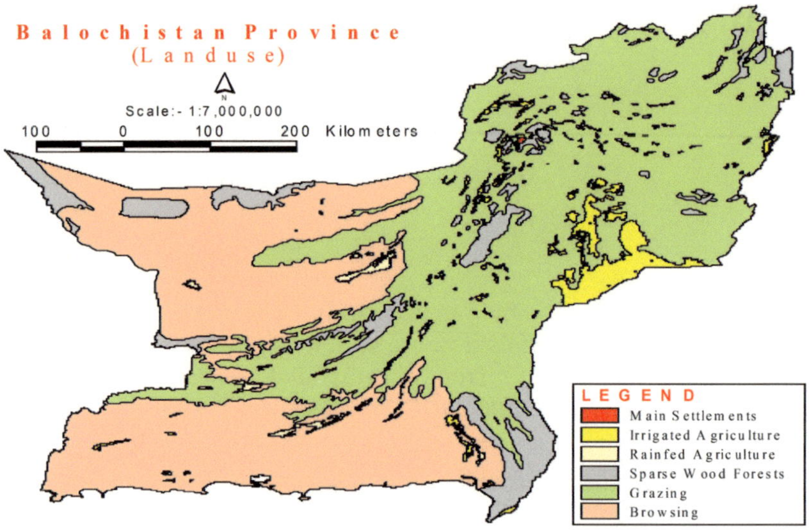

Fig. 1. Land use Patterns of Balochistan.

Natural re-vegetation practices particularly grazing management may restore vigor and accelerate the spread of desirable species (Vallentine, 1980). Grazing management alone may not accelerate the succession towards desirable species in arid and semiarid rangelands due to limited precipitation where artificial re-vegetation would involve the establishment of adapted species either by seed or transplanting seedlings (Roundy & Call, 1988). Restoration and rehabilitation are the two main procedures for regeneration of a depleted rangeland. Restoration or biological recovery means to bring the ecosystem to their pristine situation and rehabilitation or artificial recovery is the artificial establishment of a new type of vegetation different from the pristine native vegetation (Le Houerou, 2000). Biological or artificial recovery may include increase in biomass, plant cover, organic matter, soil micro and macro-organisms, better water intake and turnover, lower evaporation and runoff. Biological recovery may be obtained by protecting the target area from human and livestock intrusion. The purpose of rehabilitation of rangelands may be diverse like forage production, timber production, landscaping, wind breaks, sand dune fixation, and erosion control (Le Houerou, 2000).

A major concern of arid and semiarid ranges is the progressive reduction of secondary productivity and diversity (West, 1993) and how to manage these changes (Walker, 1993). The management and improvement of arid and semi-arid ranges is always a challenging job. Different theoretical models of rangelands have been developed and few are also being tested in different rangeland ecosystems of the world. However, the arid rangeland ecosystem of Balochistan is very dynamic where major climatic and agricultural changes are occurring. Hence many range management projects were carried out with little success.

Therefore, there is a need to re-look into research, policy and management issues for better productivity of rangelands and livestock.

1.1 Rangeland types

Balochistan can be divided into two zones regarding precipitation and grazing quality of the rangelands. The northern zone comprises the best ranges of the province located in the districts of Zhob, Loralai, Sibi, Nasirababd, Kohlu, Pishin, Quetta, Kalat, and the northern 18% of Khuzdar area. This zone, equivalent to only 38% of the total province area, carried 76.5% of the provincial livestock. The southern zone comprises the poorest ranges located in the rest of Khuzdar, Chagai, Khanar, Panjgur, Turbat, Gwadar and Lasbela district, which covers 62% of the province and carries only 23.5% of the livestock population (FAO, 1983). The high stocking rate and lack of grazing management in the Northern zone is rapidly depleting these ranges. Geomorphologically, the rangelands in Balochistan can be distributed into six types of landscapes, including mountains, uplands, piedmont, desert, flood plains and coastal plains. Muhammad (1989) divided rangelands of Balochistan into three main categories: Central Balochistan ranges, Western Balochistan Ranges, Eastern Balochistan Ranges. The biomass productivity varies from 30 to 380 kg/ha (Fig. 2.).

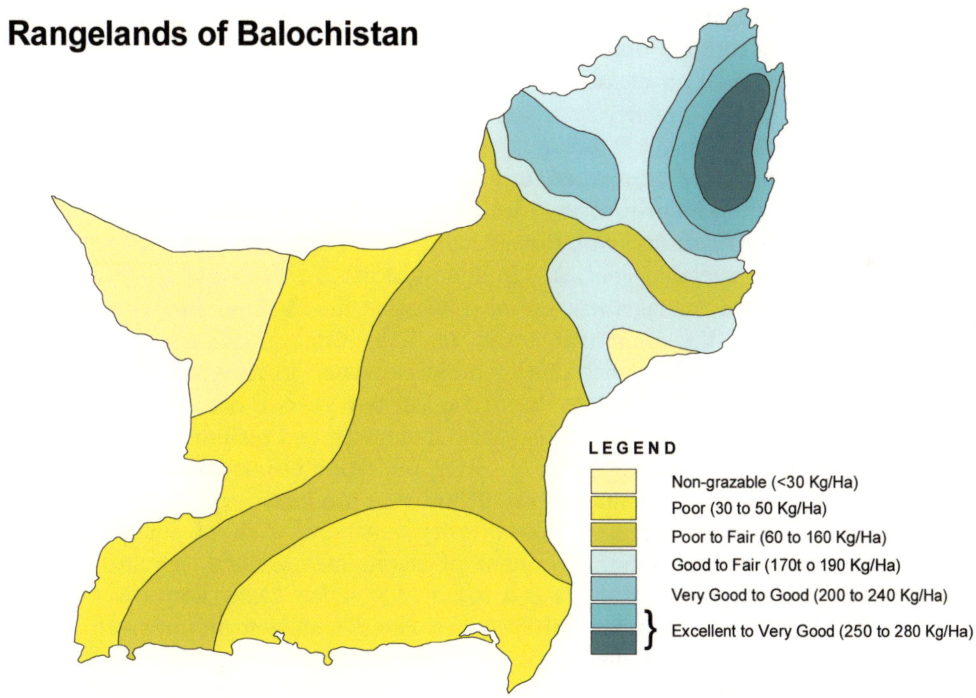

Fig. 2. Rangeland condition of Balochistan

1.2 Animal production and pastoral system

Generally three animal production systems (nomadic, transhumant, sedentary) are common in Balochistan. Most of the rangelands are used by nomadic and transhumant pastoral. According to an estimate only 30% sheep and goats are nomadic, 65% are transhumant and

5% sedentary (FAO, 1983). Nomadic flocks move continuously in search of forage. They have no agricultural land and migrate from uplands to lowlands in winter and come back again in spring to uplands. In lowlands they purchase generally sorghum crop for animal grazing. The size of a nomadic flock may vary from 200 to 700 sheep and goats. Transhumant flock owners have agricultural land and dryland agricultural activities. In winter some of them also migrate along with the families to lowlands. Sedentary flock owner raise few animals (5-20) on orchards, crop stubbles and also stale feeding. However, these systems are under transformation due to many factors like increase in livestock and human population, water mining for agriculture and orchards, changes in traditional migratory routes due to Afghan war. In a recent study, two new nomads groups (Commercial nomads and Nomad Transhumant) have been identified in Balochistan (MINFAL, 2000).

1.3 Range management issues

Range management problems in Balochistan are diverse and complex. The ranges of Balochistan are open and no one is responsible for management. Rangeland ownership is not clear or very poorly defined ownership. There are four major land ownership systems (Individual ownership, Tribal claims, Community ownership, State Ownership). Approximately 4% rangelands are under the Forest Department and the rest belongs to different groups. As a result of open grazing system the ranges are degrading very rapidly. The major range degradation factors are forage shortage, elimination of desirable range species, dominance of less preferred species, desertification, soil erosion, increased runoff and reduced infiltration (Fig. 3). Perennial grasses like *Chrysopogon aucheri* and *Cymbopogon jwarancusa* have completely eliminated in many ranges and are only found in some protected range areas. Similarly, many desirable shrub species like *Caragana ambigua, Stocksia brahvica, Berberis Balochistanica, Prunus eburnea* etc. have been replaced by *Haloxylon grifithii* and other unpalatable species. Limited information is available on rangeland resources, potential, and management options. Most of the Pastoral communities are in isolation especially in the mountain areas of Balochistan. Moreover, there is a transformation of these communities due to rapid extension in irrigated agriculture and changes in traditional migratory routes. From the last few years it has also been observed that to crop production on marginal lands is also increasing and resulting in conversion of rangelands into agricultural activities. Early spring migration of nomads from lowlands to highlands did not allow range plants for growth and seed production.

Generally, range management is a low priority area and lack of integrated range management approach and non-involvement of range management activities in other Natural Resource Management Projects is a common practice. Many Range Management Projects in Balochistan have adapted only technical range management approach ignoring the traditional customs, rights and local arrangements. Generally, most of the range management programs last two to three years. This duration is not sufficient to show any positive impact to communities on range management/improvement and livestock production. Removal of range vegetation for fuel wood is a major concern all over the Province and no alternate energy sources like solar cookers and other efficient cooking and heating devices are available. Recurrence of drought is a common phenomenon in Balochistan. However, no sound viable options are available to reduce the livestock mortality and rangeland degradation under drought conditions. Some productive ranges at present are under utilization due to non-availability of stock water. Community participation is one of the main factors for any successful range management Program. However, in Balochistan, very weak community participation in range management activities has been observed. Moreover, communities are in view that they there are no incentives for range management and they alone cannot bear the range management cost. Some other issues like limited research activities on all aspects of range management, lack of awareness, education and dissemination of knowledge, lack of trained manpower and reform in existing range management policies are also important for effective range management.

2. Materials and methods

2.1 Study sites

The experiments were conducted in three districts (Mastung: Siddiqabad, Loralai: Aghbar, Ziarat: Tomagh) of highlands of Balochistan. The research was conducted on degraded community rangelands. The selection of research sites were based on the availability of community rangelands, small ruminants and willingness of the communities for active participation in different range management activities. The Mastung district lies between 29° 03′ and 30° 13′ north and 66° 25′ and 7° 29′ east. The general topography of the district is mountainous, consisting of a series of parallel ranges running in a north-south direction. The district is severely cold during winter and hot during summer. Mean maximum and mean minimum temperatures of 36 °C and -3°C have been reported. Rains mostly occur during winter months. Loralai district lies between 29° 54′ to 30° 39′ north and 67° 44′ to 69° 40′ east. Topography of the district is mountains and hilly. Mean maximum and mean minimum temperatures of 38°C and 4°C. Rain occurs both during winter and summer months. Ziarat, Tomagh site located 15 km southwest of Sanjawi in Ziarat district. The mean annual precipitation at Tomagh is recorded 300 mm, which is distributed approximately 60% and 40% between winter and summer periods, respectively.

2.2 Traditional range management and knowledge

Information was collected in three districts. Data collection procedures include interviews, focused group discussion, and transect walk in the range areas. Fifty to sixty key informants from each site were involved on broad issues like traditional knowledge of range management. Dialogues with the communities were made to assess the existing range-livestock system, grazing patterns, and related information. Main focused areas were pastoralist knowledge on plants, grazing patterns, and migration patterns, collection of plants for winter season, communal grazing, and livestock management. Range productivity was also measured on the community lands.

2.3 Recovery of vegetation

Twelve parallel transects of 35 meter each were established at each site at a distance of 15 m apart each other. Three transects were used at each site for determination of forage production. Biomass estimates were made during the months of May/June (at optimal vegetation growth) to document the range productivity. At each transect four 1 x 5 m^2 subplots were established on alternate site of the transect line. The vegetation inside the 1 x 5 m^2 subplot was clipped at ground level, separated into leaves and wood, and oven dried. The dry matter forage production was converted into kg/ha. Descriptive analysis was used for calculation of dry forage production. Monthly, rainfall data were recorded from a rain gauge installed at Ziarat:Tomagh site while the rainfall data from Quetta site is used because due to non-availability of meteorological data of Mastung site.

2.4 Fodder shrubs plantations

Seedlings of *Atriplex canescens* and *Salsola vermiculata* were planted on degraded community rangeland during 2007. Initially, the seedlings of these species were raised in polythene bags at Arid Zone Research Centre, Quetta. Six to nine months old seedlings were transplanted on the community rangelands during late winter or early spring months. Micro-catchment water harvesting (MCWH) structures were developed on sloping lands. Contour-bunds were made by a tractor-mounted plough. Spacing between ridged was maintained at 15 m and two shrubs (Atriplex canescens and Salsola vermiculata) were planted in each micro-catchment basin with 2 m spacing. The number of shrubs in each strip ranged from 40-80. Shrub survival rate and shrub biomass was monitored. Shrub biomass production data were recorded during June 2010. Fifty shrubs from each species were randomly picked for recoding forage production. Harvested shrubs were separated into leaves and wood and oven dried for calculation of dry matter forage production. Descriptive analysis was used for calculation of average forage production of both species

3. Results and discussion

3.1 Traditional range management and knowledge

The communities were not observing any range management practices like resting the range area or rotational grazing. The rangelands are open and can also be used by the migratory nomads. In Tomagh, Ziarat, the livestock depends on grazing from April to November and the main vegetation is *Cymbopogon jwarancusa, Chrysopogon aucheri,* and *Saccharum grifthii.* From December to mid March the livestock owners also used dry Saccharum grass, dry maize, dry orchard leaves, green barley, and dry Alfalfa for livestock feeding. Pregnant herds and weak animals are also provided barley grains for two months. In case of severe drought or non-availability of forage the communities migrat the livestock to the nearby rangelands. Grazing is mostly carried out by young boys and girls and no shepherd hiring on monthly cost basis is common. Rangeland productivity in open areas is very low and ranges from 40-60 kg/ha. At Mastung, the common range vegetation is *Artemisia* species, *Haloxylon grifthii* and forbs. Generally, both annual and perennial grasses are missing in this range ecosystem. The communities utilize the range areas throughout the years both for grazing and fuel wood collection. The other feed resources include residuals of wheat, barley, and vegetables, dry orchard leaves, and dry sorghum or Alfalfa. Farmers also collect *Alhagi Camelorum* (dwarf shrub) either from fallow agricultural fields or range areas during summer months and store as a winter feed. Majority of the farmers stay throughout the year in same villages. However, some of them also migrate along with livestock towards lowlands of Balochistan during winter months. Rangeland productivity is very low and ranged from 40-70 kg/ha at various grazing areas. Shepherd hiring is common and mostly grazing is carried out by this method. The grazing price per animal ranged from Rs. 30-35 per month.

At Loralai, the range vegetation is dominated by perennial grasses like *Cymbopogon jwarancusa, Chrysopogon aucheri, Tetrapogon villosa,* and many annual grasses and forbs. The communities utilize the ranges throughout the year. This site has better range potential due to occurrence of monsoon rains. In case of monsoon rains, the grazing opportunities extended up to end of November. The nomads coming from Afghanistan are also passing through this site without any restriction on grazing. The other feed resources include residuals of wheat, barley, vegetables, and orchards. Rangeland productivity ranges from 70-100 kg/ha. Grazing is carried out on Shepherd labor sharing basis. The owner and shepherd make a contract for four years. The initial number of animals provided to the shepherd will remain the property of the owner. The agreement is made verbally and has binding on both the parties. The agreement generally consists of: shepherd will graze the animals for four years around village surroundings and/or long distances considering availability of forage and rangeland condition. The shepherd will get half of the male off-springs and 1/3 of the female young stock. The owner will provide 100 kg of wheat bag per month to the shepherd. The owner will provide two pairs of clothes and one pair of shoes per annum. After the expiry of the contract, the owner has the right to get initial number of animals from the herd and the remaining flock will be divided as per agreement i.e., male half and female 2/3 share. The expenditure made on medication of livestock rest with the owner.

Many pastoralists are willing to shift from pastoralists to crop cultivation and urban wages. Traditional knowledge is being gradually declining due to more attraction of the new generation in urban areas. Pastoralists at Tomagh try to maintain a diverse herd like both sheep and goats. Large animals (cattle) are very rare and one to two with few families for milk purposes. Large sheep and goat herds are considered as a prestige irrespective the quality of the herd. Sheep and goats are considered as a deposit in Bank account and can be cashed when required to meet the family requirements. The use of other animal products like hairs/wools are used to some extent at home for carpet making but the trend is decreasing due to easy availability of synthetic carpets at lower prices. Herd splitting, the practice of dividing the sheep/goats into separate herds depending on age is common at all three sites. Young sheep/goats after weaning separated and commonly grazed by young boys or girls.

Pastoralists at all sites pointed out that availability of experienced skilled person for grazing is also a major problem. They believe that herding is a specialized job and not everyone has the same aptitude and skills in herding. Mostly, the old men are involved on payment for this job but they cannot graze more distant pastures. The art of herding is disappearing very fast as more and more young people leave the remote range areas and prefer urban wages. Herding practices include night grazing, watering at morning and evening, camping at suitable sites to avoid predator danger, quick migration for opportunistic grazing, specialized sounds and cries needed to talk with sheep and goats.

Young boys and girls are responsible for herding sheep and goats while women are responsible for milking and making milk by-products. Pastoralists during migration consider quality, quantity of forage, water availability, household labor availability, cultural gatherings, tribal boundaries, disputes, and safe camp sites. Mobility is the best adapted and effective means of obtaining what livestock needed in an ever variable environment. In traditional content, the mobility is linked with traditional routes, tribal and social interactions and alliances with neighbors. Ethno-veterinary knowledge including management strategies to reduce reproductive wastage, use of medicinal plants in animal diseases are common at all the three sites. Pastoralists use local plants like the roots of *Berberis* species are boiled in water and given to sheep and goats for internal injuries.

Knowledge of local plants is more refined at all the sites. Pastoralist knew the local names of nearly all the plants of their areas. Communities were able to identify the preferred forage species and season of use. They distinguish between those that fatten livestock and improve their health. *Chrysopogon aucheri* is more preferred grass than *Cymbopogon jwarancusa*, wild olive leaves/fruits and *Alhagi Camelorum* are good for fattening of sheep and goats. The pastoralists were also able to identify the poisonous plants of their areas. The communities were also agreed that there is a shift in species composition like from preferred/palatable grasses to less preferred/un-palatable grasses and shrubs. The majority of the pastoralist was also in agreement that the changes in species composition is due to over grazing, removal of vegetation for fuel wood and Afghan nomadic flux during war. The animal health and productivity is an indirect method of rangeland assessment by the pastoralists. Pastoralists evaluate the range condition on the basis of animal performance like rumen fill, milk production and animal health.

3.2 Recovery of natural vegetation

Monthly rainfall from 2006 to 2010 of Quetta and Tomagh is presented in Table 1. Total annual rainfall at Quetta ranged from 105.8 to 247 mm while at Tomagh the total annual rainfall ranged from 214 to 462.6 mm. The dry matter forage production of different sites and years is presented in Table 2. The initial dry matter forage production during 2007 was 80, 60 and 184 kg/ha, respectively at Mastung, Ziarat and Loralai. Each year there were increasing trend of dry forage production and during 2010 the dry matter forage production was recorded 230, 485 and 864 kg/ha at Mastung, Zirata and Loralai, respectively (Table 2). Rainfall and its distribution during winter and spring, 2007 was comparatively better than 2006. The community degraded rangelands showed recovery potential at all sites. At Mastung the dominated range vegetation is Artemisia and *Haloxylon* species while at Loralai and Tomagh site perennial grasses (*Cymbopogon jwarancusa, Chrysopogon aucheri*) are dominated. The range recovery depends on the distribution of rainfall and management practices. The Loralai and Tomagh sites have better recovery potential of range vegetation due to occurrence of both winter and monsoon rains (Fig. 4).

	2006		2007		2008		2009		2010	
Months	Quetta	Ziarat (Tomagh)	Quetta	Ziarat (Tomagh)	Quetta	Ziarat (Tomagh)	Quetta	Ziarat (Tomagh)	Quetta	Ziarat (Tomagh)
January	22.2	0.0	13.0	6.4	117.6	0.0	59.2	56.0	29.8	2.0
February	7.8	34.4	105.2	148.0	10.2	0.0	45.4	49.2	45.2	0.0
March	32.4	68.6	28.3	86.8	0.0	3.2	31.4	118.2	9.6	25.6
April	7.4	20.2	14.8	0.0	9.6	30.4	30.7	50.8	9.0	1.6
May	5.9	0.0	0.0	0.0	0.0	20.0	11.4	0.0	10.2	43.4
June	0.0	4.96	42.5	116.3	5.6	0.0	0.0	94.0	2.0	36.6
July	3.6	22.4	12.2	0.0	0.0	20.8	0.0	44.0	0.0	104.0
August	69.1	88.8	0.0	0.0	14.0	95.2	0.0	0.0	0.0	162.4
September	0.0	37.2	0.0	0.0	0.0	0.0	0.0	2.4	0.0	14.4
October	0.0	0.0	0.0	0.0	0.0	0.0	0.0	0.0	6.4	0.0
November	44.8	114.8	5.8	0.0	0.0	0.0	0.0	0.0	0.0	0.0
December	54.6	71.2	17.0	0.0	12.1	45.2	45.4	15.7	0.0	0.0
Total	247.8	462.6	238.8	357.3	169.1	214.8	223.5	430.3	103.2	390.0

Table 1. Monthly Rainfall (mm) at Quetta and Tomagh

Districts	Dry Forage production (kg/ha)			
	2007	2008	2009	2010
Mastung	80 ± 5.10	171.91 ± 14.29	188.46 ± 11.07	230.0 ± 15.06
Ziarat	60 ± 11.76	133.48 ± 8.84	255.8 ± 12.57	484.8 ± 20.37
Loralai	184 ± 13.90	205.0 ± 22.36	630.0 ± 30.71	864.50 ± 47.71

Table 2. Improvement of Natural Vegetation and Increase in Forage Production as aresult of protection.

Arid rangelands of Balochistan characterized by highly unpredictable and variable rainfall events, behave as non-equilibrium system. This means that both climatic and grazing factors are important in any range management and improvement interventions. There are no universally accepted grazing strategies due to specific conditions of rangelands. However, resting, restricted grazing has proved for the recovery of natural range vegetation and forage improvement in many arid and semi-arid regions. The range vegetation of Balochistan has low reproductive potential due to the adaptive strategies of the plants for survival under extreme climatic conditions. The recovery potential is also very site specific like in case of Loralai, the grasses were heavily grazed but have shown good recovery potential under favourable conditions. The optimal growth time of grasses in Balochistan is from March to June, may be extended up to October in case of monsoon rains. Therefore, resting of vegetation during this time period is very essential for recovery and forage improvement. However, if the objectives were for seed production and re-generation than at least two to three years rest period must be provided (Ahmad et al., 2010; Ahmad et al., 2007; Ahmad et al., 2000 a,b,c). Accumulated dead material of perennial grasses can decline both productivity and nutritive value (Ahmad et al., 2009; Bano et al., 2009) therefore, a rotation grazing may yield better results than long term protection. Enhanced growth rate of grasses in response to grazing, fire and disturbance under favourable environments have been observed (Chapin & McNaughton, 1989).

Fig. 3. A degraded Rangeland

Fig. 4. Recovery potential of perennial grasses

Many rangeland areas in Balochistan still have potential of natural recovery if properly grazed. As a result of protection from grazing, it is evident from the results that the community rangelands are resilient and have potential of biological recovery subject to rainfall distribution and management practices. Range productivity is greatly influenced by fluctuations in rainfall, grazing pressure and nutrients (Olson & Richard, 1989; Scoones, 1995). Above ground net primary production can be used as an integrative attribute of eco-system function (McNaughton et al., 1989). Above ground net primary production is an important variable in natural resource management because it determines forage availability for both wild and domestic herbivores. Oesterheld et al., (1992) found a strong connection between stocking density and above ground primary production for South American Rangelands. The rate of biological recovery might be slow as expected in the arid and semiarid climatic zones. The rate of vegetation recovery is also related with the rainfall distribution during the optimal growing period rather than total rainfall distribution. Strong vegetation recovery response has been reported even under desert conditions with mean annual rainfall as 60-80 mm under deep and permeable soils (Le Houerou, 1992a). From Morocco to Iran the perennial ground cover and primary productivity were enhanced by a factor of 2-5 and in most cases, 3-4 within a few years either by total or partial protection (Le Houerou, 1992a). In West Asia and North Africa range exclosures from 11 countries showed that productivity in exclosures enhanced averaged by 2.8 times than the adjacent grazed areas (Le Houerou, 1998). However, very long-term protection may not yield better results due to accumulation of dead old material that may reduce the new fresh growth. Controlled grazing may produce similar or better results than exclosures in some cases (Le Houerou, 2000). The recruitment rate of grasses may not be achieved within two to three-year protection. The changes in species composition are very slow processes in arid and semiarid areas (West et al., 1984). Limited spring season rainfall (the optimal time of seedling recruitment) in Balochistan is the main factor for low seedling recruitment even under complete protection from grazing. According to long-term meteorological data analysis in Balochistan, it is observed that above-normal rainfall amounts that promoted spring seedling emergence occur with about 10% and less than 10% probability (Keatinge and Rees, 1988).

3.3 Fodder shrub plantation

Survival percentage of shrubs ranged from to 80 to 89% (Table 3). Average dry forage production of Atriplex canescens ranged from 349 to 670 kg/plant. Salsola vermiculata average dry forage production ranged from 112 to 225 kg/plant (Table 3).

Districts	Seedlings Planted		Survival %		Average Dry Forage Production/Plant	
	Atriplex canescens	Salsola vermiculta	Atriplex canescens	Salsola vermiculta	Atriplex canescens	Salsoal vermiculta
Mastung	6000	4000	80	85	350.20 ± 34.33	112.0 ± 15.37
Loralai	11000	5000	80	87	670.0 ± 63.13	225 ± 38.78
Ziarat	8000	5000	85	89	348.50 ± 22.09	205 ± 23.64

Table 3. Plantation of Fodder Shrubs on Community Rangelands, Survival % and dry matter forage production.

Atriplex canescens (Fourwing slatbush) has potential in highland areas of Balochistan due to cold and drought tolerant characteristics (Fig. 5). The biomass and productivity of Atriplex canescens is highly variable, depending upon the ecological condition of the soil and climate as well as the management applied. Artificial plantation of fourwing slat bush under rainfed conditions can yield up to 2000-4000 kg dry matter/ha/year in areas with mean annual rainfall of 200-400 mm under proper management (Le Houreou, 1992b). Average dry mass production of Atriplex canescens planataion in highlands of Balochistan after three years has been reported 1600 kg/ha (Afzal et al., 1992). The ratio between forage and wood in Atriplex species is about 50% which can be improved by appropriate management like pruning (Le Houreou, 1986). Young leaves and twigs show a much better forage quality, with higher nitrogen content and a lower amount of ashes and salts. The crude protein content in leaves of Atriplex canescens ranged 12-15% during mid winter (Thomson et al., 1987). One acre of fourwing slatbush might provide the supplemental protein requirements for 0.5 to 1 animal unit during a 90-day period (Ueckert, 1985). Like other halophytes, Atriplex canescens have low energy values because of high ash contents. Grazing of Atriplex canescens with wheat/Barley straw could lead to a well balance ration and fulfill the nutritional requirements of small ruminants (Mirza et al., 2004; Thomson et.al., 1997). *Salsoal vermiculata* commonly called saltwort is an exotic Mediterranean arid zone fodder species. This species belongs to the Chenopodiaceae family. S. vermiculata has the potential of self-regeneration and establishment under good rainfall years (Murad, 2000). *S. vermiculata* initiate new growth in late winter or early spring (depends on rainfall distribution) and provides a considerable amount of palatable forage for small ruminants. It is not an ever-green species, however, if sufficient rains occur during winter months it retains new vegetative growth. Maximum growth has been observed from April to May. Its height ranges between 35 and 110 cm. Crown cover ranges from 45 to 57 cm^2. Forage production ranged from 250-650 kg/ha with an equal amount of wood production (Ahmad et al., 2006). Crude protein content ranged from 15-18% (Ahmad and Islam , 2005).

Fig. 5. Atriplex canescens plantation

Fig. 6. Winter grazing of Atriplex Plantation

Artificial plantation is very costly interventions and must be carried out by considering the water availability for initial shrub survival, water harvesting techniques and availability of suitable seedlings. Plantation should be carried on highly degraded sites where no recovery chances of natural vegetation and non-availability of soil seed bank. Management of plantation is the critical aspect of success and failure (Fig. 6). Generally, any range plantation needs two to three years before grazing. The grazing period, intensity of grazing, and rest-period for recovery of biomass should also be given considerations for successful range improvement interventions. In Balochistan, the best utilization time of planted shrubs is the winter months along with dry ranges or wheat, barley stubbles.

4. Conclusions

The rangelands of Balochistan need an urgent and well-planned program in management and utilization to halt the degradation process leading towards desertification. Range management should also be based on knowledge of Pastoral communities, traditions, and local arrangements. Communities should be involved in range management planning and implementation processes. Formation of Pastoral communities or associations in major range areas may help in taking care of herd mobility, marketing of livestock, and maintenance of rangelands. Forage reserve block establishment on marginal lands with some Government incentives may ensure forage supply in winter or drought years. Supply of high production drought and cold tolerant fodder shrubs on minimum price should be introduced to complement native rangelands. These pastures may be used during the critical forage deficit period (winter months) and at the same time may allow some rest to the rangelands. Communities alone cannot bear the range management and improvement expenses. Therefore, some incentives may be provided for sustainable range management program.

5. Acknowledgment

This research study was carried with the financial support of Agricultural Linkages Programme (ALP-PARC-USDA). I am highly indebted the financial and technical support for carrying out range management and improvement intervention on the community rangelands. The assistance and cooperation of the ALP-Secretariat, PARC is highly appreciated. I am highly thankful the cooperation of all the communities involved in the range management activities.

6. References

Afzal, J.; Sultani, M.I. & Asghar, M. (1992). Growth of fourwing slatbush (*Atriplex canescens*) on the rangelands of upland Balochistan. *Pakistan Journal of Agricultural Research,* 13 (2): 180-183.

Ahmad, S.; Gul, S. Achakzai, AJK & Islam, M. (2010). Seedling growth response of *Seriphidium quettense* to water stress and non-water stress conditions. *ΦYTON International Journal of Experimental Botany* 79: 19-23.

Ahmad, S.; Islam, M. Bano, G. Aslam, S & Koukab, S. (2009). Seasonal variation in current season and dead biomass of *Chrysopogon aucheri* (Boiss) Stapf. and *Cymbopogon jwarancusa* (Jones) Schult. in highland Balochistan, *Pakistan. Pakistan Journal of Botany* 41 (2): 519-527.

Ahmad, S.; Gul, G, Islam, M & Athar, M. (2007). Seed dispersal and soil seed bank of *Seriphidium quettense* (Asteraceae) in highland Balochistan, Pakistan. *Botany Research Institute Texas* 1 (1): 569-575.

Ahmad, S.; Mirza, S.N & Islam, M. (2006). Evaluation of *Salsola vermiculata L.* for range improvement in highland Balochistan, Pakistan. *Pakistan Journal of Forestry* 56(2): 148-154.

Ahmad, S.; & Islam, M. (2005). Assessment of nutritional potential and performance of range species in Balochistan. ALP Project Report. Arid Zone Research Centre, Quetta.

Ahmad, S.; Call, C.A & Schupp, E.W. (2000a). Regeneration ecology of *Chrysopogon aucheri* and *Cymbopogon jwarancusa* in upland Balochistan. I. Morphology, viability and movement of seeds (spikelets). *Pakistan Journal of Biological Science* 3 (10): 1583-1587.

Ahmad, S.; Call, C.A & Schupp, E.W. (2000b). Regeneration ecology of *Chrysopogon aucheri* and *Cymbopogon jwarancusa* in upland Balochistan. II. Dispersal, predation and soil reserves of seeds (spikelets). *Pakistan Journal of Biological Science* 3 (11): 1880-1883.

Ahmad, S.; Call, C.A. Schupp, E.W & Mirza, S.N. (2000c). Regeneration ecology of *Chrysopogon aucheri* and *Cymbopogon jwarancusa* in upland Balochistan. III. Effect of precipitation and seedbed microhabitat on seedling recruitment. *Pakistan Journal of Biological Science* 3 (12): 2041-2047.

Chapin, F.S., & McNaughton, S.J. (1989). Lack of compensatory growth under phosphorus deficiencies in grazing adapted grasses from the Seregeti plains. *Oecologia* 79: 551-557.

FAO. 1983. Report of the assistance to rangeland and livestock development survey in Balochistan. *TCP/PAK 0107, FAO Technical cooperation program, Food and Agricultural Organization of the United Nations,* Pakistan.

Bano, G.; Islam, M. Ahmad, S. Aslam, S & Koukab, S. (2009). Seasonal variation in nutritive value of *Chrysopogon aucheri* (Boiss) Stapf. and *Cymbopogon jwarancusa* (Jones) Schult. in highland Balochistan, Pakistan. *Pakistan Journal of Botany* 41 (2): 511-517.

GOB. (1996). Livestock Census Report. *Government of Balochistan*. Quetta.

Heymell, V. (1989). Evaluation of pasture conditions in Nichara Union Council in relation to conservation of water stirage tanks. *Studies 09/89. Pak-German self help project for rural development. Balochistan,* Quetta, Pakistan.

Hussain, F.; & Durrani, M.J. (2007). Forage productivity of arid temperate Harboi rangeland, Kalat, Pakistan. *Pakistan Journal of Botany* 39 (5): 1455-1470.

Islam, M.; Ahmad, S. Aslam, S. & Athar, M. (2008). Mineral composition and anti-nutritional components of shrubs: Rangeland species from the upland Balochistan, Pakistan. *Agriculturae Conspectus Scientificus* 73 (1): 27-35.

Keatinge, J.D.H.; & Rees, D.J. (1988). An analysis of precipitation and air temperature records in the Quetta valley, Pakistan. The implication for potential improvement in agriculture productivity. *ICARDA Research Report No. 9, AZRI.*

Khan, C.M.A., & Mohammad, N. (1987). *Rangelands in Pakistan. United States-Pakistan Workshop on Arid lands Development and Desertification Control.* PARC, Islamabad.

Khan, S.R. A. (1987). *Rangelands in Pakistan. United States-Pakistan Workshop on Arid lands Development and Desertification Control.* PARC, Islamabad.

Kidd, C.H.R.; Rees, D.J. Keatinge, D.J.H. Rehman, F. Samiullah, A. & Raza, S.H. (1988). *Meteorological data analysis of Balochistan. Research Report No. 19.* ICARDA, Quetta, Pakistan.

Le Houerou, H.N. (1986). Salt tolerant plants of economic value in the Mediterranean basin (ecology, productivity, management and development potential). *Journal of Reclamation and Revegetation Research* 5: 319-341.

Le Houerou, H.N. (1998). Biodiversity: *Native plant resource conservation in Marmarica (N-W Egypt)*, 57 pp. ICARDA Facilitation Unit, Cairo, Egypt.

Le Houerou, H.N. (1992a). *An overview of vegetation and land degradation in world arid lands*, pp. 127-163, in H.E. Dregne, ed., Degradation and restoration of arid lands. Texas Tech University, Lubbock, Texas.

Le Houerou, H.N. (1992b). The role of saltbushes (Atriplex species) in arid land rehabilitation in the Mediterranean basis: a review. *Agroforestry Systems* 18:107-148.

Le Houerou. H.N. (2000). Restoration and rehabilitation of arid and semiarid Mediterranean ecosystems in north Africa and west Asia: A review. *Arid Soil Research and Rehabilitation* 14: 3-14.

Milton., S.J.; Richard, W. Dean, J. M. A. du. Plessis & Siegfried, W.R. (1994). A conceptual model of arid rangeland degradation. The escalating cost of declining productivity. *BioScience* 44 (2): 70-76.

Mirza, S.N.; Ahmad, S. Afzal, J. Islam, M. , Samad, N. & Khan, A.R. (2000). *Production and utilization of multipurpose shrub (Atriplex spp.) in highland Balochistan.* A review. Arid Zone Research Centre, Quetta. PP 1-28.

Mirza, S.N.; Islam, M & Ahmad, S. (2004). Nutritional quality of *Atriplex canescens*: A potential fodder shrub for the Mediterranean highland Balochistan. *Pakistan Journal of Arid Agriculture* 7 (2): 17-21.

MINFAL (2000). Socio-economic systems of Pastoralist communities of highland Balochistan, Pakistan, National Aridland development and Research Institute, Ministry of Food, Agriculture and Livestock, Islamabad. PP1-47.

Mohammad, N. (1989). *Rangeland management in Pakistan. Published by The International Center for Integrated Mountain Development.*

Murad, N. (2000). A study on Syrian steppe and forage shrubs. Fodder shrub development in arid and semiarid zones. *Proceeding of the workshop on native and exotic fodder shrubs in arid and semiarid zones (eds. Gintzburger G., M. Bounejmate & A. Nefzaoui)*, 27 Oct- 2 Nov. 1996, Hammamet, Tunisia, ICARDA, Aleppo, Syria.

Olson, Bret E.; & Richard, J.H. (1989). Crested wheatgrass growth and replacement following fertilization, thinning and neighbour plant removal. *Journal of Range Management* 42 (2): 93-97.

Roundy, B.; & Call, C.A. (1988). Re-vegetation of arid and semiarid rangelands. *Vegetation science applications for rangeland analysis and management.* Kluwer Academic Publishers, Dordrecht, London.

Thomson, E.F.; Mirza, S.N. Rafique, S. Afzal, J. Rasool, I. Atiq-ur-Rehman, Ershad, M. Hussain, A. Akbar, G & Alvi, A.S. (1997). *Utilization of Fourwing saltbush for the Arid Rangelands of Highland Balochistan, Pakistan. In: N. Haddad, R. Tutwiler and E. Thomson (eds.). Proceedings of Regional Symposium on Integrated Crop-Livestock Systems in the Dry Areas of West Asia and North Africa* held at Amman, Jordan from 6-10 November, 1995. ICARDA, Aleppo, Syria. 572 p.

Ueckert, D.N. (1985). Use of shrubs for rangeland revegetation. *In: proceedings of the International Rancher Roundup. Laredo, Texas. Editors: L.D. White, D.E. Guyn and T.R Troxel.* Texas Agricultural Extension Service, pages 190-196.

Vallentine, J.F. (1980). *Range development and improvement.* Second Edition. Brigham Young University Press, Provo, Utah.

Walker, B.H. (1993). Rangeland ecology: Understanding and managing change. *Ambio* 22(2-3): 80-87.

West, N.E. (1993). Biodiversity of rangelands. *Journal of Range Management* 46:2-13.

16

Effects of Protected Environments on Plant Biometrics Parameters

Edilson Costa[1], Paulo Ademar Martins Leal[2] and Carolina de Arruda Queiróz[3]
[1]*Professor Ph.D., Agricultural Engineer, State University of Mato Grosso do Sul-UEMS, Unit of Aquidauana*
[2]*Professor Ph.D., Agricultural Engineer, University of Campinas, College of Agricultural Engineering*
[3]*MSc in progress, Agronomist, Graduate Program in Agronomy, Crop Area, UEMS / Aquidauana-MS*
Brasil

1. Introduction

There is a high correlation between the type of greenhouse used for crop production with the system used for its production, especially with the type of container and substrate used. The same protected environment may present different responses in plant biometric parameters depending on the container volume and also the chemical and physical characteristics of a particular substrate. This relationship is expressed in greater or lesser accumulation of plant biomass.

Besides of the substrate and container type, other studies seek to improve the crop yield potentials and cropping systems associated with environmental control techniques, such as cooling and/or heating systems, use of CO2 for atmospheric enrichment, color screens systems and automated control of the atmospheric parameters.

Protected environments for crop production are generally constructed of low density polyethylene film (greenhouses), and shading screens, such as monofilament screens and aluminized thermal reflective screens (are widely used. In these types of environments growing in containers is preferred because it allows for better management of both water and nutrients (Grassi Filho & Santos 2004).

Changes in the microclimate inside the greenhouses caused by the use of polyethylene result in modification of the influence of air temperature, relative humidity and solar radiation on plant growth and development, and these are dependent on the intensity, duration and quality of solar radiation (Beckmann et al., 2006; Scaranari et al., 2008). These changes affect the plants physiology (Chavarria et al., 2009), and minimize the incidence of fungal diseases and therefore application of pesticides (Chavarria et al., 2007). In vineyards, where only the rows were covered with polyethylene film, Cardoso et al. (2008) found a reduction in evaporative demand.

According to Sganzerla (1987), the advantages that the greenhouses can provide to the protected plants are numerous, as long as these facilities are correctly used. Among these advantages some can be highlighted including harvesting crops of the season, higher product quality, early crop maturity, seedling production, better control of diseases and pests, conservation of raw materials and water, planting of selected varieties and considerable increase in production.

Despite the numerous advantages, greenhouses present poor thermal behavior since during the day elevated temperatures are observed and are difficultly avoided by natural ventilation, and at night temperatures often fall below the critical temperatures for the crops (Da Silva et al., 2000). For circumvent problems with high temperatures in greenhouses many producers use evaporative cooling systems, forcing air through a porous medium with a fan (pad-fan) or intermittent misting systems. These applications improve the thermal conditions and relative humidity during the hottest periods of the day.

Important aspects should be taken into consideration in the use of protected environments, such as knowing the different protection structures and their configurations and orientations, knowing the physiological responses of the crop to be cultivated within of the environment and knowing the energy and mass balance for the crop and its environment. This set of knowledge can aid in proper crop and environment management and obtain answers of the appropriate technology to be applied to the cropping system (Costa, 2004).

The parameters of leaf growth, area and mass characterize the plant biomass, so that it can be used to determine changes in carbohydrate assimilation by the plant during a season of the year (Butler et al., 2002), where the leaf area measures the plant biomass accumulation potential and leaf dry mass allows for determination of the capacity of the plant to increase its dry weight through photosynthesis.

Microclimate environmental modifications of the greenhouse and screen, i.e., the plastic covers for vegetative production, has promoted a positive impact on crops, increasing fruit yield, leaf area and quality of products produced (Buriol et al. 1997, Segovia et al., 1997).

The microclimatic effects of the protected environment influence the emergence, initial growth and development of fruit trees, vegetables, ornamental plants and forests. The objective of this study was to perform a literature review of authors who have researched comparisons between different environmental conditions and their correlation with plant performance.

2. Effects of environment on vegetables

Costa & Leal (2009) observed that in hydroponic production of lettuce, variety Vera, in three greenhouses, one without evaporative cooling and CO_2 injection, another with injection of CO_2 and without evaporative cooling and a third, with CO_2 injection and evaporative cooling (acclimatized), the environment with evaporative cooling and CO_2 injection promoted the best development of plants with larger leaves.

In acclimatized environment with evaporative cooling, Costa & Leal (2008) found greater accumulation of leaf biomass and greater leaf area of strawberry plants than in non-acclimatized environments, regardless of the season (Table 1).

For five cultivars of lettuce (Veronica, Vera, Cinderella, Isabela, Veneranda) under four different environmental conditions (Black screens with 30%, 40%, 50% shading and without the screen) in the region of Cáceres-MT/Brazil, Queiroz et al. (2009) found that the Veronica cultivar was the most productive during the winter of 2008 and shading of 40% was best for most cultivars.

Environment	ASO	NDJFM
LEAF AREA (LA) (mm^2)		
With cooling and carbon dioxide	66.78 A *	51.81 A
Without cooling and carbon dioxide	50.14 B	37.94 B
Without cooling and without carbon dioxide	53.72 B	35.51 B
LEAF FRESH MASS (LFM) (g)		
With cooling and carbon dioxide	1.71 A	1.16 A
Without cooling and carbon dioxide	1.19 B	0.83 B
Without cooling and without carbon dioxide	1.21 B	0.76 B
LEAF DRY MASS (LDM) (g)		
With cooling and carbon dioxide	0.41 A	0.30 A
Without cooling and carbon dioxide	0.29 B	0.22 B
Without cooling and without carbon dioxide	0.29 B	0.20 B

* Means in the same column followed by same letter do not differ by the Tukey test ($P < 0.05$).
Adapted from Costa & Leal (2008)

Table 1. Leaf area (LA), leaf fresh mass (LFM) and leaf dry mass (LDM) for the strawberry cultivar Tudla, during August-October (ASO) and November to March (NDJFM).

Cultivars of chicory (*Cichorium endivia* L.), AF-254 and Marina, produced under a natural environment and within a low tunnel constructed of white polypropylene in the region of Ponta Grossa-PR/Brazil, presented greater head mass in the low tunnel and a greater number of leaves in the natural environment. The AF-254 cultivar was more productive but more susceptible to tipburn in the protected environment (SA & Reghin, 2008).

Cunha et al. (2005) evaluated the radiation balance and yield of sweet pepper, hybrid Elisa, in a protected environment (a non-acclimatized greenhouse oriented in the NNW-SSE direction, covered with low density polyethylene film) and in a field located in Botucatu-SP/Brazil. The authors observed that plants in the protected environment present not only greater plant height and total dry matter during of total cycle, but also a greater leaf area index. However this environment showed less net energy for growth and development of the crop.

Interactions between greenhouse environments, substrates types and different cucumber hybrids were evaluated by Costa et al. (2010) and verified different behavior of the substrates in the different environments studied, noting that the seedling growth was affected by the environments and the substrates. Response of cucumber hybrids in terms of seedlings dry biomass depended on the substrate and the growing environment. The substrate "soil and coconut fiber" increased biomass accumulation in the greenhouse and nursery with black the monofilament screen. The substrate "soil and organic compost" showed greater aerial biomass in the nursery with the aluminized screen. Hybrid 'Safira' accumulated more root biomass in the substrate "soil and coconut fiber" and when using the screens. The hybrid 'Nikkei' accumulated higher root biomass in the nursery with the aluminized screen and in the substrate "soil and coconut fiber" and did not differ from the substrate "soil and saw-dust". Hybrids 'Aladdin F1' and 'Nobre F1' accumulated similar root biomass in the environments, where the 'Aladdin F1' had a higher accumulation of biomass in the substrates "soil and organic compound" and "soil and coconut fiber", while the hybrid 'Noble F1' showed greater accumulation in "soil and coconut fiber", showing no difference from "soil and saw-dust" (Tables 2 and 3).

	ADM (g)			RDM (g)		
**	A1	A2	A3	A1	A2	A3
S1	0.077 Aa *	0.089 Aa	0.073 Ba	0.027 Aa	0.030 Aa	0.030 Aa
S2	0.050 Ba	0.059 Ba	0.056 Ca	0.017 Bc	0.029 Aa	0.021 Bb
S3	0.041 Bc	0.067 Bb	0.090 Aa	0.019 Bc	0.023 Bb	0.031 Aa

* Means followed by same uppercase letters in the columns and same lowercase letters in the rows do not differ by the Tukey test at 5%;
** S1 = "soil + ground coconut fiber", S2 = "soil + saw-dust", S3 = "soil + organic compound"; A1 = greenhouse; A2 = nursery with black monofilament screen, A3 = nursery with aluminized screen.
Adapted from Costa et al. (2010)

Table 2. Aerial dry mass (ADM) and root dry mass (RDM) of cucumber seedlings at 23 days after sowing for the various substrates (S) and environments (A) studied.

	RDM (g)					
**	A1	A2	A3	S1	S2	S3
H1	0.022 Aa *	0.027ABa	0.025 Ba	0.027 ABa	0.018 Bb	0.029 Aa
H2	0.023 Ab	0.025 Bb	0.032 Aa	0.030 ABa	0.026 Aab	0.025 ABb
H3	0.018 Ab	0.032 Aa	0.028 ABa	0.032 Aa	0.022 ABb	0.024 ABb
H4	0.020 Aa	0.024 Ba	0.024 Ba	0.026 Ba	0.023 ABab	0.020 Bb

* Means followed by same uppercase letters in the columns and same lowercase letters in the rows do not differ by the Tukey test at 5%;
** H1 = Aladdin F1; H2 = Nikkei; H3 = Safira; H4 = Nobre F1; S1 = "soil + ground coconut fiber", S2 = "soil + saw-dust", S3 = "soil + organic compound"; A1 = greenhouse; A2 = nursery with black monofilament screen, A3 = nursery with aluminized screen.
Adapted from Costa et al. (2010)

Table 3. Root dry mass (RDM) of cucumber seedlings at 23 days after sowing for the various hybrids (H) in environments (A) and substrate (S) studied.

In tomato production in greenhouses with and without aluminized screen, Gent (2007) verified that the use of the screen with 50% shading increased commercial fruit production by 9% compared to the environment without the screen, verifying the beneficial use of this screen type in protected environments. Comparisons between the mobile aluminized screens with 40, 50 and 60% shading and the environment with polyethylene plastic film painted with lime, were evaluated by Fernandez-Rodriguez et al. (2001) in tomato production and it was found that the screens minimize energy consumption during periods of low temperatures.

With the objective of evaluating cucumber seedlings in function of environmental conditions, polystyrene trays with 72 and 128 cells and substrates with percentages of organic compound in Aquidauana-MS/Brazil, Costa et al. (2009c) conducted an experiment in six environmental conditions: plastic greenhouse with a height of 2.5 m; nursery with a black monofilament screen with 50% of shading and height of 2.5 m; nursery with an aluminized screen with 50% of shading and height of 2.5 m; nursery covered with native coconut palms with height of 1.8 m; plastic greenhouse with height of 4.0 m, zenithal opening and thermo-reflective screen over the black monofilament screen with 50% of shading and height of 3.5 m. The authors concluded that the greenhouses promoted better results for cucumber seedlings.

3. Effects of environments for fruit

In coffee conilon seedlings (*Coffea canephora*) with shading levels of 30%, 50%, 75% and full light, in the region of Alegre-ES/Brazil, it was found that the stem diameter was not influenced by the environment, but the height, the fresh and dry weight, volume and leaf area were greater where shading was 70% (Braun et al., 2007). But in coffee seedlings (*Coffea arabica* L.), Paiva et al. (2003) reported that of the with shading levels of 30%, 50% and 90%, 50% was most favorable, resulting in greater height, number of leaves and leaf area, consequently, greater vegetative growth.

Mezalira et al. (2009) when evaluating the effect of substrate, harvest period and environment of fig (*Ficus carica* L.) rooting in plots without cover, plots under low tunnel cover with plastic film (150 μ) and plots under a low tunnel with monofilament screen (50% shading) in Dois Vizinhos-PR/Brazil, observed the greatest root production in plots with the use of low tunnel with monofilament screen and the lowest in full sun.

	Fresh mass of the aerial portion (g)			
	Greenhouse	Monofilament screen	Aluminized screen	coconut palm
Soil + organic compost + vermiculite	0.52 Ac *	0.75 Ab	0.86 Aa	0.52 Ac
Soil + organic compost + sawdust	0.17 Bc	0.27 Cb	0.38 Ca	0.09 Bc
Soil + organic compost + vermiculite + sawdust	0.56 Ab	0.62 Bb	0.73 Ba	0.55 Ab
	Fresh mass of the root portion (g)			
Soil + organic compost + vermiculite	1.88 Ac	3.00 Ab	4.01 Aa	1.35 Ac
Soil + organic compost + sawdust	0.57 Bb	0.75 Bb	0.91 Ca	0.25 Bb
Soil + organic compost + vermiculite + sawdust	2.37 Aa	2.61 Aa	2.74 Ba	1.40 Ab
	Dry mass of root portion (g)			
Soil + organic compost + vermiculite	0.18 Ab	0.27 Aa	0.26 Aa	0.11 Ac
Soil + organic compost + sawdust	0.05 Bb	0.07 Ca	0.07 Ca	0.02 Bb
Soil + organic compost + vermiculite + sawdust	0.21 Aa	0.20 Ba	0.19 Ba	0.11 Ab

* Means followed by same uppercase letters in the columns and same lowercase letters in the rows do not differ by the Tukey test at 5%;
Adapted from Costa et al. (2009a).

Table 4. Interactions between environments and substrates for production of fresh mass of the aerial portion (FMAP), fresh mass of the root portion (FMRP) and dry mass of root portion (DMRP) for papaya seedlings, "Sunrise solo".

In Alegre-ES/Brazil, studies of germination and seedling production of guava (*Psidium guajava* L.) in full sun, environments covered with one, two and three screens showed that full sun and one screen promoted higher germination, rate of emergence, number of leafs, plant height and stem diameters, revealing that seedlings tend to develop less with increased levels of shading (Lopes & Freitas, 2009).

Araújo et al. (2006) evaluated the effects of three pots and three environmental conditions (greenhouse tunnel, nursery with a monofilament screen with 50% shading and natural environment) on the development of papaya (*Carica papaya* L.) cv. Sunrise Solo and concluded that the natural environment was most adequate for development of the seedlings at 45 days after sowing.

	Greenhouse	Monofilament screen	Aluminized screen	coconut palm
	Fresh mass of the aerial portion (g)			
polyethylene bag	5.50 Ac *	7.88 Ab	10.77 Aa	5.63 Ac
polystyrene trays	0.39 Ba	0.46 Ba	0.48 Ba	0.65 Ba
	Dry mass of the aerial portion (g)			
polyethylene bag	0.77 Ac	1.01 Ab	1.23 Aa	0.68 Ac
polystyrene trays	0.07 Ba	0.08 Ba	0.09 Ba	0.10 Ba
	Fresh mass of the root portion (g)			
polyethylene bag	2.67 Ac	3.71 Ab	4.57 Aa	1.57 Ad
polystyrene trays	0.55 Ba	0.54 Ba	0.53 Ba	0.43 Ba
	Dry mass of root portion (g)			
polyethylene bag	0.25 Ab	0.32 Aa	0.30 Aa	0.12 Ac
polystyrene trays	0.05 Ba	0.05 Ba	0.05 Ba	0.04 Ba

* Means followed by same uppercase letters in the columns and same lowercase letters in the rows do not differ by the Tukey test at 5%;
Adapted from Costa et al. (2009a).

Table 5. Interactions between environments and pots for production of fresh mass of the aerial portion (FMAP), dry mass of the aerial portion (DMAP), fresh mass of the root portion (FMRP) and dry mass of root portion (DMRP) for papaya seedlings, "Sunrise solo".

		Greenhouse	Monofilament screen	Aluminized screen	coconut palm
AFM	polyethylene bag	4.499 Ab *	7.703 Aa	7.159 Aa	3.937 Ab
	polystyrene trays	0.449 Ba	0.775 Ba	0.699 Ba	0.644 Ba
ADM	polyethylene bag	0.697 Ab	1.248 Aa	1.149 Aa	0.618 Ab
	polystyrene trays	0.087 Ba	0.161 Ba	0.140 Ba	0.186 Ba
RFM	polyethylene bag	1.063 Ab	1.539 Aa	1.435 Aa	0.589 Ac
	polystyrene trays	0.288 Ba	0.493 Ba	0.385 Ba	0.439 Aa
RDM	polyethylene bag	0.163 Ab	0.212 Aa	0.221 Aa	0.099 Ac
	polystyrene trays	0.054 Ba	0.064 Ba	0.057 Ba	0.067 Ba

* Means followed by same uppercase letters in the columns and same lowercase letters in the rows do not differ by the Tukey test at 5%;
Adapted from Costa et al. (2009b).

Table 6. Review of the analyses of mean aerial fresh mass (AFM), aerial dry mass (ADM), fresh root (RFM) and dry mass of root (RDM) in grams for the container (R) within environments (A); environments (A) inside the container (R) for the yellow passion fruit.

Costa et al. (2009a) when evaluating the production of papaya seedlings (*Carica papaya* L., cv 'Sunrise Solo') in a greenhouse with low density polyethylene film, nursery with black monofilament screen, nursery with aluminized screen and nursery with native coconut palm, using different substrates and containers in Aquidauana-MS/Brazil, observed that the best growth environment was the nursery with aluminized screen for leaf fresh weight, dry weight and fresh weight of the root system (Tables 4 and 5). The same treatments in the same region were applied on the development of passion fruit seedlings (*Passiflora edulis* Sims. *f. flavicarpa* Deg.) by Costa et al. (2009b), who found that the black monofilament screen environment provided good conditions for seedlings development. The environment with the aluminized screen also favored seedling growth (Tables 6 and 7).

		Greenhouse	Monofilament screen	Aluminized screen	coconut palm
ADM	Soil + organic compost + vermiculite	0.534 Ac *	0.955 Aa	0.788 Ab	0.545 Ac
	Soil + organic compost + sawdust	0.205 Bb	0.378 Ca	0.379 Ba	0.135 Bb
	Soil + organic compost + vermiculite + sawdust	0.437 Ab	0.781 Ba	0.767 Aa	0.526 Ab
RFM	Soil + organic compost + vermiculite	1.063 Aa	1.284 Aa	1.187 Aa	0.785 Ab
	Soil + organic compost + sawdust	0.292 Cab	0.411 Bab	0.435 Ba	0.176 Cb
	Soil + organic compost + vermiculite + sawdust	0.673 Bb	1.353 Aa	1.107 Aa	0.582 Bb

* Means followed by same uppercase letters in the columns and same lowercase letters in the rows do not differ by the Tukey test at 5%;
Adapted from Costa et al. (2009b).

Table 7. Review of the analyses of mean aerial dry mass (ADM) and the fresh root (RFM) in grams of substrate (S) within environments (A); environments (A) within the substrate (S) for passion fruit.

Initial growth of licuri seedling (*Syagrus coronata* (Mart.) Becc.), at luminosity levels of 30% (monofilament screen) and 100% (full sun) in the municipality of Feira de Santana-BA/Brazil showed greatest plant growth when subjected to 30% light intensity (Chapman et al., 2006).

Martelleto et al. (2008) studied the effect of the plastic covered greenhouse, shaded greenhouse with an additional monofilament screen (30%, over the plastic), shading with only the monofilament screen (30%) and the natural environment in development of papaya cv. Baixinho de Santa Amália ('Solo'), and concluded that growth is favored, both in terms of plant height and trunk diameter, foliage (number of leafs/plant) and leaf area inside the greenhouse without the additional monofilament screen (Tables 8 and 9).

Environment of cultivation	Plant height (cm)	Diameter of the trunk (cm)	Leaves number per plant	Leaf area (cm^2)
Greenhouse	183.8 A *	13.0 A	35.3 A	2077.7 A
Shaded greenhouse	174.8 B	10.0 B	35.4 A	1702.6 B
Screen	156.4 C	8.5 C	29.5 B	1376.3 D
Natural environment	144.2 D	10.0 B	29.4 B	1529.5 C
Coefficient of variation (%)	5.8	6.7	4.6	12.2

* Means followed by same uppercase letters in the columns and same lowercase letters in the rows do not differ by the Tukey test at 5%;

Table 8. Vegetative growth of the 'Baixinho de Santa Amália' papaya subjected to organic management in different cultivation environments, where the values of height and trunk diameter are relative to 12 months after transplanting the seedlings and the values of the leafs number per plant and leaf area correspond to monthly averages during one year of cultivation (Seropédica-RJ, 2004/2005).

Environment of cultivation	Number of fruits per plant	Fruit weight (kg per plant)	Average fruit weight (g)
Greenhouse	9.7 A *	3.53 A	364.7 A
Shaded greenhouse	7.3 B	2.01 B	276.1 D
Screen	4.6 C	1.39 C	302.8 C
Natural environment	6.5 B	2.12 B	326.1 B
Coefficient of variation (%)	20.9	22.2	9.8

* Means followed by same uppercase letters in the columns and same lowercase letters in the rows do not differ by the Tukey test at 5%;

Table 9. Commercial production of 'Baixinho de Santa Amália' papaya subjected to organic management in different cultivation environments where the values represent monthly averages during the first 12 months of harvest (Seropédica-RJ, 2004/2005).

Seedlings of tamarind (*Tamarindus indica*), in Lavras-MG/Brazil, were more vigorous when cultivated in the natural environment when compared to those produced in the greenhouse and nursery with black monofilament screen providing 50% shading (Mendonça et al., 2008).

In Flores da Cunha-RS/Brazil, grape yields (cv. Moscato Giallo), with and without plastic cover over the crop rows, was higher in the covered environment, with greater stability of production, but did not affect the relationship between shell and pulp mass of the berries. The film increased the daily temperature at the plant canopy, not affecting relative humidity, but decreasing the photosynthetic active radiation and wind speed (Chavarria et al., 2009).

Medina et al. (2002) found a better photosynthetic performance of citrus seedlings of the orange 'Pera' (*Citrus sinensis* Osbeck) and Rangpur lime (*Citrus limonia* Osbeck) in the greenhouse with the use of the termorrefletora screen applying 50% of shading (aluminized screen) below the polyethylene film, in comparison with the greenhouse without the screen. According to these authors, as well as increasing photosynthesis, the screen reduced the photosynthetically active radiation and leaf temperature. These effects were not only beneficial for the maintenance of proper stomatal aperture for gas exchange, but also for better functioning of the photochemical system under adverse conditions.

With the objective of evaluating biomass of passion fruit seedlings in function of environmental conditions and substrates with percentages of organic compound in Aquidauana-MS/Brazil, Sassaqui et al. (2008) conducted an experiment in six environmental conditions: greenhouse with a height of 2.5 m; nursery with black monofilament screen with 50% shading and height of 2.5 m; nursery with aluminized screen with 50% shading and height of 2.5 m; nursery covered with native coconut palm with height of 1.8 m; plastic greenhouse with height of 4.0, zenithal opening and mobile aluminized screen beneath the film at a height of 3.5 m. The authors concluded that the polyethylene film and aluminized screen together promoted better environmental conditions for the accumulation of biomass.

4. Effects of environments on forest species

Rubber rootstocks (Hevea spp.) in greenhouses covered with transparent low density polyethylene (LDPE), in the field protected by 50% mesh plastic screen as windbreaks and in the unprotected field (control) in Campinas-SP/Brazil, showed no differences in growth in the field with and without protection (Table 10). However, the greenhouse, compared to the control showed increased diameter (60%), height (108%), leaf area (266%) and dry weight (286%), and was the only environment that showed 60% of rootstock with a minimum diameter of 8.0 mm, suitable for grafting (Pezzopane et al., 1995).

	Control	Windbreaks	Greenhouse
Diameter (mm)	5.3 A *	5.5 A	8.4 B
Height (cm)	35 A	39 A	73 B
Leaf area (cm^2)	624 A	621 A	2283 B
Dry weight (g)			
- row system	1.4 A	2.2 A	5.4 B
- aerial portion	5.5 A	6.4	24.9 B
- total dry weight	6.9 A	8.6 A	30.3 B

* Means followed by same uppercase letters in the rows do not differ by the Tukey test at 5%; Adapted from Pezzopane et al. (1995)

Table 10. Mean values and results of the statistical analysis for growth measured in diameter, height, leaf area, average distance between shoots and average weight of dry matter.

With the objective to obtain information on an angelim seedling production system (*Andira fraxinifolia* Benth) in São Cristóvão-SE/Brazil, Carvalho Filho et al. (2004) studied two growth environment (50% shading and full sun), substrates and containers and concluded that the seedlings should be maintained in 50% shading and then be transferred to full sun.

Effects of greenhouse and full sun were studied using the parameters of emergence, mortality, stem diameter, plant height, leaf area and dry weight of araticum seedlings (*Annona crassiflora* Mart.) and it was verified that the stem diameter, plant height and leaf area were greater in the greenhouse and the other variables in full sun (Cavalcante et al., 2008).

The germination of the assacuzeiro (*Hura crepitans* L.) under 50% shading, greenhouse constructed of polypropylene and environment in full sun was studied by Effgen et al. (2005) in Alegre-ES/Brazil, who concluded that both 50% shade and the environment in full sun provided good conditions for germination.

In canafístula seedlings (*Cassia grandis* L.), subjected to full sun and 50% shading under the monofilament screen in São Cristóvão-SE/Brazil, it was observed that plant height, leaf number, stem diameter and dry weight leaf were greater under 50% shading with fast initial growth (Carvalho Filho et al., 2002).

Effects of shading levels of 0%, 30% and 50% in Lavras-MG/Brazil on growth, biomass allocation and total chlorophyll content of young plants of *Maclura tinctoria* (L.) D. Don ex Steud. (moreira), *Senna macranthera* (Collad.) Irwin et Barn. (fedegoso), *Hymenaea courbaril* L. var. stilbocarpa (Hayne) Lee et Lang. (jatobá) and *Acacia mangium* Willd. (acácia) revealed that the highest chlorophyll levels were observed in shaded conditions for all species; the chlorophyll a/b ratio in full sun and 50% shading showed no difference between species; in full sun, the fedegoso and moreira species showed greater growth; the diameter of the stem of moreira was smaller in full sun than 50% shading; the dry matter produced by moreira was greater than that of fedegoso, except in the shading level of 30% (Almeida et al., 2005).

Carvalho Filho et al. (2003) evaluated the effect of full sun and 50% shading environments on the production of jatoba seedlings (*Hymenaea courbaril* L.) in em São Cristóvão-SE/Brazil, and found that the emergence percentage was higher in full sun, recommending the production of seedlings in this environment. They also observed that for the other features there was interaction between environments, containers and substrates.

5. Effects of environments for flowers

In a greenhouse covered with transparent low density polyethylene, in Piracicaba-SP/Brazil, utilizing red, blue, black thermo-reflective screens (aluminized screen) all with 70% shading at 1.0 m above the cultivation bench, Holcman & Sentelhas (2006) evaluated the growth and development of the bromeliad (*Aechmea fasciata*) and concluded that the red screen resulted in the highest biometric values, however, the thermo-reflective screen was more favorable for the cultivation showing the best microclimate.

Seedlings of jasmine-oranges (*Murraya exotica* L) in full sun, under a white screen (30% shading) and black screen (50% shading), in São Cristóvão-SE/Brazil, presented higher emergence in full sun and under the white screen; higher rate of emergence and number of leaves were observed in full sun, and greater dry matter of aerial part was found under both screens (Arrigoni-Blank et al., 2003). It is recommended to produce seedlings of jasmine-orange first in full sun and after emerge under a white screen with 30% shading (Tables 11 and 12).

	Full sun	Clarite® 30%	Sombrite® 50%
Substrate	Germination rate		
Soil + sand 1:1	0.350 aA *	0.165 aB	0.144 aB
Soil + vermiculite + cattle manure 1:1:1	0.270 aA	0.317 aA	0.143 aB
Soil + sand + cattle manure 1:1:1	0.353 aA	0.181 bcB	0.128 aB
Sand + cattle manure 1:1	0.289 aA	0.286 abA	0.101 aB
	Plant height		
Soil + sand 1:1	2.77 aB	2.75 bB	4.00 aA
Soil + vermiculite + cattle manure 1:1:1	2.69 aB	3.46 aA	3.44 bA
Soil + sand + cattle manure 1:1:1	2.56 aB	2.80 bAB	3.16 bA
Sand + cattle manure 1:1	2.70 aB	3.19 abA	3.33 bA
	Number of leaves per plant		
Soil + sand 1:1	2.00 bB	2.54 aB	3.25 aA
Soil + vermiculite + cattle manure 1:1:1	2.52 abA	3.07 aA	3.05 abA
Soil + sand + cattle manure 1:1:1	2.21 abA	2.62 aA	2.75abA
Sand + cattle manure 1:1	2.75 aA	3.05 aA	2.50 bA

* * Means followed by same uppercase letters in the rows, and same lowercase letters in the columns do not differ by the Tukey test at 5%;
Adapted from Arrigoni-Blank et al. (2003)

Table 11. Mean values of the germination rate, plant height and number of leaves of jasmine orange (*Murraya exotica*) on different substrates and light conditions. São Cristóvão-SE, 2000.

	Full sun	Clarite® 30%	Sombrite® 50%
Substrate	Dry weight of leaves		
Soil + sand 1:1	0.067 bcC *	0.142 bB	0.216 aA
Soil + vermiculite + cattle manure 1:1:1	0.102 abC	0.232 aA	0.184 abB
Soil + sand + cattle manure 1:1:1	0.055 cB	0.150 bA	0.166 bA
Sand + cattle manure 1:1	0.131 aB	0.174 bA	0.182 abA
	Dry weight aerial part		
Soil + sand 1:1	0.091 bC	0.176 bB	0.271 aA
Soil + vermiculite + cattle manure 1:1:1	0.124 abC	0.284 aA	0.223 abC
Soil + sand + cattle manure 1:1:1	0.069 bB	0.184 bA	0.130 bA
Sand + cattle manure 1:1	0.155 aB	0.218 bA	0.222 abA
	Dry weight of roots		
Soil + sand 1:1	0.061 aB	0.090 bB	0.142 aA
Soil + vermiculite + cattle manure 1:1:1	0.087 aB	0.177 aA	0.106 aB
Soil + sand + cattle manure 1:1:1	0.062 aB	0.117 bA	0.110 aA
Sand + cattle manure 1:1	0.095 aB	0.129 bA	0.212 aAB

* * Means followed by same uppercase letters in the rows, and same lowercase letters in the columns do not differ by the Tukey test at 5%;
Adapted from Arrigoni-Blank et al. (2003)

Table 12. Mean values of dry weight of leaves, dry weight of the aerial part and dry weight of roots of jasmine orange (*Murraya exotica*) on different substrates and light conditions. São Cristóvão-SE, 2000.

6. Conclusions

There are diverse crops produced and evaluated with regards to different growing environments, where yields and qualities are influence by the type, size and shape of the environment, covering material, climate, location, seasonality, interactions with containers and substrates and other factors.

Polyethylene film and shading screens used either individually or together, minimize direct radiation to the plant, depending on the format of the environment and the time of day, preventing this radiation from causing damage to plant tissues.

Matrix planting of vegetables, fruit, flowers and forest species, as well as acclimation and production of seedlings often requires initial shading with screens that present different degrees of shading, therefore care must be taken to select the mesh so that it does not cause irregular plant growth.

The protected environment maximizes the productive potential of plants and to obtain successful yields correct management of the environment is necessary along with the use of trained labor.

7. Reference

Almeida, S. M. Z.; Soares, A. M.; Castro, E. M.; Vieira, C. V. & Gajego, E. B. (2005). Alterações morfológicas e alocação de biomassa em plantas jovens de espécies florestais sob diferentes condições de sombreamento. *Ciência rural*, Santa Maria, v. 35, n. 1, p. 62-68, ISSN 0103-8478.

Araujo, J. R. G.; Araújo Júnior, M. M.; Menezes, R. H. N.; Martins, M. R.; Lemos, R. N. S. & Cerqueira, M. C. M. (2006). Efeito do recipiente e ambiente de cultivo sobre o desenvolvimento de mudas de mamoeiro cv. Sunrise Solo. *Revista Brasileira de Fruticultura*, Jaboticabal, v. 28, n. 3, p. 526-529, ISSN 0100-2945.

Arrigoni-Blank, M. F.; Carvalho Filho, J. L. S.; Blank, A. F. & Santos Neto, A. L. (2003). Efeitos do substrato e luminosidade na emergência e desenvolvimento de mudas de jasmim-laranja (*Murraya exotica* l.). *Revista Ciência Agronômica*, Fortaleza, v. 34, n. 1, p. 5-12, ISSN 0045-6888..

Beckmann, M. Z.; Duarte, G. R. B.; Paula, V. A.; Mendez, M. E. G. & Peil, R. M. N. (2006). Radiação solar em ambiente protegido cultivado com tomateiro nas estações verão-outono do Rio Grande do Sul. *Ciência Rural*, Santa Maria-RS, v. 36, n. 1, p. 86-92, ISSN 0103-8478.

Braun, H.; Zonta, J. H.; Lima, J. S. S. & Reis, E. F. (2007). Produção de mudas de café 'conilon' propagadas vegetativamente em diferentes níveis de sombreamento. *Idesia*, Chile, v. 25, n. 3, p. 85-91, ISSN 0718-3429.

Buriol, G. A.; Luft, S. V. L.; Heldwein, A. B.; Streck, N. A. & Schneider, F. M. (1997). Efeito da ventilação sobre a temperatura e umidade do ar em túneis baixos de polietileno transparente e o crescimento da alface. *Revista Brasileira de Agrometeorologia*, Santa Maria, v.5, n.1, p.17-24, ISSN 0104-1347.

Butler, L. M.; Fernandez, G. E. & Louws, F. J. (January-March 2002). Strawberry plant growth parameters and yield among transplants of different types and from different geographic sources, grown in a plasticulture system. *Hortechnology*, v. 12, n°1, p.100-103, ISSN: 1063-0198.

Cardoso, L. S.; Bergamaschi, H; Comiran, F.; Chavarria, G.; Marodin, G. A. B.; Dalmago, G. A. & Santos, H. P. & Mandell, F. (2008). Alterações micrometeorológicas em vinhedos pelo uso de coberturas de plástico. *Pesquisa Agropecuária Brasileira*, Brasília-DF, v. 43, n. 4, p. 441-447, ISSN: 0100204X.

Carvalho, N. O. S.; Pelacani, C. R.; Rodrigues, M. O. S. & Crepaldi, I. C. (2006). Crescimento inicial de plantas de licuri (*Syagrus coronata* (Mart.) Becc.) em diferentes níveis de luminosidade. *Revista Árvore*, Viçosa, v. 30, n. 3, p. 351-357, ISSN (Versión impresa): 0100-6762. ISSN 1806-9088.

Carvalho Filho, J. L. S.; Arrigoni-Blank, M. F. & Blank, A. F. (janeiro/junho 2004). Produção de mudas de angelim (*Andira fraxinifolia* Benth.) em diferentes ambientes, recipientes e substratos. *Revista Ciência Agronômica*, Fortaleza, v. 35, n. 1, p. 61-67, ISSN 0045-6888.

Carvalho Filho, J. L. S.; Arrigoni-Blank, M. F.; Blank, A. F. & Rangel, M. S. A. (2003). Produção de mudas de jatobá (*Hymenaea courbaril* L.) em diferentes ambientes, recipientes e composições de substratos. *Revista Cerne*, Lavras, v. 9, n. 1, p. 109-118, ISSN 0104-7760.

Carvalho Filho, J. L. S.; Arrigoni-Blank, M. F.; Blank, A. F.; Santos Neto, A. L. & Amancio, V. F. (2002). Produção de mudas de *Cassia grandis* L. em diferentes ambientes, recipientes e misturas de substratos. *Revista Ceres*, Viçosa, v. 49, n. 284, p. 341-352, ISSN 0034-737X.

Cavalcante, T. R. M.; Naves, R. V.; Seraphin, J. C. & Carvalho, G. D. (2008). Diferentes ambientes e substratos na formação de mudas de araticum. *Revista Brasileira de Fruticultura*, Jaboticabal, v. 30, n. 1, p. 235-240, ISSN 0100-2945.

Chavarria, G.; Santos, H. P.; Sônego, O. R.; Marodin, G. A. B.; Bergamasch, H. & Cardoso, L. S. (2007). Incidência de doenças e necessidade de controle em cultivo protegido de videira. *Revista Brasileira de Fruticultura*, v. 29, n. 3, p. 477-482, ISSN 0100-2945.

Chavarria, G.; Santos, H. P.; Mandelli, F.; Marodin, G. A. B.; Bergamaschi, H. & Cardoso, L. S. (2009). Potencial produtivo de videiras cultivadas sob cobertura de plástico. *Pesquisa agropecuária brasileira*, Brasília, v. 44, n. 2, p. 141-147, ISSN 0100-204X.

Costa, E. (2004). *Avaliação da produção do morangueiro em sistemas hidropônicos, utilizando casas de vegetação com diferentes níveis tecnológicos*. 2004. 130 p. Tese (Doutorado em Construções Rurais e Ambiência) - Faculdade de Engenharia Agrícola, Universidade Estadual de Campinas, Campinas-SP.

Costa, E. & Leal, P. A. M. (2008). Avaliação da biomassa foliar de morangueiro hidropônico em diferentes ambientes protegidos. *Ciência e Agrotecnologia*, Lavras, v. 32, n. 6, 1941-1952, ISSN 1413-7054.

Costa, E. & Leal, P. A. M. (2009). Produção de alface hidropônica em três ambientes de cultivo. *Engenharia Agrícola*, Jaboticabal, v. 29, n. 3, p.358-369, ISSN 0100-6916.

Costa, E.; Rodrigues, E. T.; Alves, V. B.; Santos, L. C. R. & Vieira, L. C. R. (janeiro-março 2009b). Efeitos da ambiência, recipientes e substratos no desenvolvimento de mudas de maracujazeiro-amarelo em Aquidauana - MS. *Revista Brasileira de Fruticultura*, Jaboticabal, v. 31, n. 1, p. 236-244, ISSN 0100-2945.

Costa, E.; Santos, L. C. R. & Vieira, L. C. R. (outubro-dezembro 2009a). Produção de mudas de mamoeiro utilizando diferentes substratos, ambientes de cultivo e recipientes. *Engenharia Agrícola*, Jaboticabal, v. 29, n. 4, p. 528-537, ISSN 0100-6916.

Costa, E.; Vieira, L. C. R.; Rodrigues, E. T.; Machado, D.; Braga, A. B. P. & Gomes, V. A. (2009c). mbientes, recipientes e substratos na formação de mudas de pepino híbrido. *Agrarian*, Dourados, v. 2, n. 4, p. 95-116, ISSN 1984-252X ISSN da versão online: ISSN 1984-2538.

Costa, E.; Leal, P. A. M.; Gomes, V. A.; Machado, D. & Jara, M. C. (2010). Biomassa de mudas de pepinos híbridos conduzidos sob ambientes protegidos. *Bragantia*, Campinas, v.69, n.2, p.381-386, ISSN 0006-8705.

Cunha, A. R.; Escobedo, J. F.; Klosowski, E. S. & Galvani, E. (2001). Saldo de radiação e produtividade da cultura de pimentão em ambientes protegido e campo. In: Congresso Brasileiro de Biometeorologia, 3., 2001, Maringá-PR. *Resumos*... Maringá-PR: Sociedade Brasileira de Biometeorologia (SBBiomet).

Da Silva, E. T. & Schwonka, F. (2000) Comportamento da temperatura do ar sob condições de cultivo em ambiente protegido. Congresso Brasileiro De Engenharia Agrícola, 29., 2000, Fortaleza-CE. *Anais*... Jaboticabal-SP: Sociedade Brasileira de Engenharia Agrícola, SBEA..

Effgen, E. M.; Mendonça, A. R.; Bragança, H. B. N. & Martins Filho, S. (2005). Germinação de sementes de *Hura crepitans* L. em diferentes ambientes e diferentes substratos. In: Encontro Latino Americano de Iniciação Científica, 9., e Encontro Latino Americano de Pós-Graduação, 5., 2005, Vale do Paraíba-SP. *Resumos*... Vale do Paraíba-SP: Universidade do Vale do Paraíba..

Fernandez-Rodriguez, E. J.; Perez, D.; Camacho-Ferre, F.; Fernandez Vadillos, J. & Kenig, A. (2001). Effects of aluminized shading screens vs whitewash on tomato photochemical efficiency under a non heated greenhouse. *Acta Horticulturae*, v. 559, p. 279-284, ISSN 0567-7572.

Gent, M. P. N. (2007). Effect of Shade on Quality of Greenhouse Tomato. Acta Horticulturae, v. 747, p. 107-112, ISSN 0567-7572.

Grassi Filho, H. & Santos, C. H. (2004). Importância da relação entre os fatores hídricos e fisiológicos no desenvolvimento de plantas cultivadas em substratos. In: Barbosa, J. G.; Martinez, H. E. P.; Pedrosa, M. W. & Sediyama, M. A. N. (Eds.) *Nutrição e adubação de plantas cultivadas em substrato*. Viçosa-MG: UFV, p. 78-91.

Holcman, E. & Sentelhas, P. C. (2006). Crescimento e desenvolvimento de bromélias em ambiente protegido, cobertos com PEBD e diferentes malhas de sombreamento. In: Congresso Brasileiro de Meteorologia, 14., 2006, Florianópolis. Anais... Florianópolis-SC: Universidade Federal de Santa Catarina.

Lopes, J. C. & Freitas, A. R. (2009). Germinação de Sementes e Formação de Mudas de *Psidium guajava* L. (Goiabeira): Efeito de Sombreamento. *Revista Brasileira de Agroecologia*, Porto Alegre, v. 4, n. 2, p. 1939-1942, ISSN 1980-9735.

Martelleto, L. A. P.; Ribeiro, R. L. D.; Sudo-Martelleto, M.; Vasconcellos, M. A. S.; Marin, S. L. D. & Pereira, M. B. (2008). Cultivo orgânico do mamoeiro 'baixinho de Santa

Amália' em diferentes ambientes de proteção. *Revista Brasileira de Fruticultura*, Jaboticabal, v. 30, n. 3, p.662-666, ISSN 0100-2945.

Medina, C. L.; Machado, E. C.; Souza, R. P.; Ribeiro, R. V. & Silva, J. A. B. (2002). Photosynthesis response of citrus grown under reflective aluminized polypropylene shading nets. *Scientia Horticulturae*, Holanda, v. 96, n. 2, p. 115-125, ISSN: 0304-4238.

Mendonça, V.; Abreu, N. A. A.; Souza, H. A.; Teixeira, G. A.; Hafle, O. M. & Ramos, J. D. (2008). Diferentes ambientes e osmocote na produção de mudas de tamarindeiro (*Tamarindus indica*). *Ciência e agrotecnologia*, Lavras, v. 32, n. 2, p. 391-397, ISSN 1413-7054.

Mezalira, E. J.; Cassol, D. A.; Alegretti, A. L.; Nava, G. A. & Wagner Junior, A. (2009). Substrato, ambiência e época de coleta no enraizamento de estacas de figueira (*Ficus carica*). In: Seminário: Sistemas de Produção Agropecuária, 3., 2009, Dois Vizinhos-PR. Resumos... Dois Vizinhos-PR: Universidade Tecnológica Federal do Paraná.

Paiva, L. C.; Guimarães, R. J. & Souza, C. A. S. (janeiro-fevereiro 2003). Influência de diferentes níveis de sombreamento sobre o crescimento de mudas de cafeeiro (*Coffea arabica* L.). *Ciência e agrotecnologia*, Lavras, v. 27, n. 1, p. 134-140, ISSN 1413-7054.

Pezzopane, J. E. M.; Pedro Júnior, M. J. & Ortolani, A. A. (setembro-dezembro 1995). Uso de estufa com cobertura plástica e de quebra-ventos na produção de porta-enxertos de seringueira. *Scientia agrícola*, Piracicaba, v. 52, n. 3, p. 439-443, ISSN 0103-9016.

Queiroz, J. P. S.; Neves, L. G.; Seabra Junior, S. & Costa, A. J. M. Avaliação da produção " de genótipos de alface em diferentes ambientes, cultivadas no período de inverno em Cáceres/MT. In: Jornada científica da Unemat, 2., 2009, Barra dos Bugres-MT. *Resumos...* Barra dos Bugres-MT: Universidade Estadual de Mato Grosso, 2009.

Sá, G. D. & Reghin, M. Y. (2008). Desempenho de duas cultivares de chicória em três ambientes de cultivo. *Ciência e agrotecnologia*, Lavras, v. 32, n. 2, p. 378-384, ISSN 1413-7054.

Sassaqui, A. R.; Costa, E.; Gomes, V. A.; Machado, D.; Terena, T. F. S.; Cortelassi, J. A. S.; Albuquerque, V. Biomassa de mudas de maracujazeiro em substrato comercial na região de Aquidauana/MS. In: Congresso Brasileiro de fruticultura, 20., 2008, Vitória-ES. *Resumos...* Vitória-ES: Sociedade Brasileira de Fruticultura, 2008.

Scaranari, C.; Leal, P. A. M. & Pellegrino, G. Q. (2008). Estudo de simulações de microclimas em casas de vegetação visando à aclimatação de mudas micropropagadas de bananeira cv Grande Naine. *Revista Brasileira de Fruticultura*, Jaboticabal-SP, v. 30, n. 4, p. 1001-1008, ISSN 0100-2945.

Segovia, J. F. O.; Andriolo, J. L.; Buriol, G. A. & Schneider, F. M. (1997). Comparação do crescimento e desenvolvimento da alface (Lactuca sativa L.) no interior e no exterior de estufas de polietileno em Santa Maria, RS. *Ciência Rural*, Santa Maria-RS, v. 27, n. 1, p. 37-41, ISSN 0103-8478.

Sganzerla, E. (1987). *Nova agricultura*: a fascinante arte de cultivar com os plásticos. Porto Alegre-RS: Petroquímica Triunfo. 297 p.

Quality and Selected Metals Content of Spring Wheat (*Triticum aestivum* L.) Grain and Biomass After the Treatment with Brassinosteroids During Cultivation

Jaromír Lachman[1], Milan Kroutil[1] and Ladislav Kohout[2]
[1]*Department of Chemistry , Faculty of Agrobiology, Food and Natural Resources, University of Life Sciences in Prague,*
[2]*Department of Steroid Chemistry, Institute of Organic Chemistry and Biochemistry, Academy of Sciences of the Czech Republic, Prague*
Czech Republic

1. Introduction

Brassinosteroids (BRs) are plant natural polyhydroxysteroids supporting the plant growth; their structure resembles animal steroid hormones (Bajguz, 2010). In plants, steroid hormones serve as endogenous signaling molecules. Brassinosteroids act as positive growth regulators or as compounds responsible for plant stress tolerance. Phytoecdysteroids probably show an antifeedant activity (Kamlar et al., 2010). Brassinosteroids were classified as essential plant hormones nearly thirty years after the discovery of brassinolide (the first brassinosteroid) by Groove et al. (1979) in the rape (*Brassica napus* L.) pollen. Presence of brassinosteroids was demonstrated in many plant species including higher and lower plants and at the same time they were detected in parts of plants, e.g. pollen, seeds, leaves, stems, roots and flowers (Sakurai et al., 1999). Up to date it was characterized 70 compounds belonging to the class of brassinosteroids, among them 65 in free form and 5 conjugated (Zulo & Adam, 2002, Bajguz & Tretyn, 2003).

Brassinosteroids are phytohormones with pleiotropic effects. They influence growth, seed germination, cell elongation, photomorphogenesis and senescence (Upreti & Murti, 2004). In relation to the growth and growth regulators, the typical effect of brassinosteroids is coincidental elicitation of cell prolongation and division (Worley & Mitchell, 1971). Investigations confirm the ability of brassinosteroids quantitatively affect plant morphogenesis; this leads to the enhancement of number and growth productive lateral shoots and branches and thereby also to the enhancement of number of spikes, pods etc. (Sakurai et al., 1999). Brassinosteroids help to overcome stresses provoked by low and high temperature, drought, salt, infection, pesticides and heavy metals (Takematsu et al., 1986; Cutler, 1991; Kulaeva et al., 1991; Schilling et al., 1991; Hathout, 1996; Bajguz 2000; Anuradha & Rao, 2001; Krishna, 2003; Janeczko et al., 2005; Cao et al., 2005; Sharma & Bhardwaj, 2007 a, b; Kagale et al., 2007; Ali & Abdel-Fattah, 2006, Ali et al., 2007, 2008, Kroutil et al., 2010 a,b). Heavy metals give rise to antioxidant stress and brassinosteroids can it effectively reduce and induce enhancing of antioxidants under heavy metal stress (Hayat et al., 2007a). In term of the affecting of the uptake of minerals after treatment with brassinosteroids an increase of the content of minerals in aerial plant biomass was demonstrated (Nafie & El-Khallal, 2000) as well as the BRs ability to decrease uptake of heavy metals and accumulation of radioactive elements (Cs, Sr) by plants (Bajguz, 2000; Khripach et al., 1999).

In term of the affecting of the uptake of minerals after treatment with brassinosteroids an increase of the content of minerals in aerial plant biomass was demonstrated (Nafie & El-Khallal, 2000). Brassinosteroids can affect quality of plant products. Treatment with brassinosteroids at anthesis increased the starch content in rice kernels (Fujii & Saka, 2001); at tillering it increased the content of fatty acids in barley ectoplasts and the change of their rate (Khripach et al., 1999).

The aim of this work was to evaluate the ability of brassinosteroids to affect the quality parameters of spring wheat grain: change of the content of minerals in grain and the yield increase of spring wheat cultivated in rational-intensity conditions after brassinosteroids treatment. Another goal of this study was to evaluate the ability of brassinosteroids to lessen the uptake and accumulation of heavy metals (Cd, Pb, Zn, Cu) in spring wheat plants cultivated on contaminated soil of a polluted burdened region in the Czech Republic. Content of heavy metals was investigated in biomass, grains and straw of treated and control plants.

2. Material and methods

2.1 Plant material and conditions of cultivation

In the three-year period 2005-2007 spring wheat (*Triticum aestivum* L.) "Vánek" variety (maintenance of variety: Lochow-Petkus, GmbH, Germany, producer: Selekta, Inc., Czech Republic) was cultivated at 10 m^2 trial field plots outside environment (50°2'0"N, 14°36'54"E) on brown loamy soil. Every tested compound was applied in four replicates (10 m^2 field plots). There were sowed 217 kg of seeds per hectare. As foregoing crop broad bean was cultivated on the trial field before wheat plants every year and before wheat sowing the field was fertilized with the dose 60 kg N ha^{-1} with nitrogen-phosphorus-potassium fertilizer. Average content of minerals in trial field soil is described in Table 1.

In another experiment spring wheat, Vánek variety (maintenance of variety: Lochow-Petkus, GmbH, Germany, producer: Selekta, Inc., Czech Rep.) was cultivated for two years (2006, 2007) in pots in the outside environment. Plants were cultivated in the soil anthropogenic contaminated with heavy metals from the location Příbram, Central Bohemia, historically polluted from metal ores mining and smelting activities. Average content of minerals in contaminated soil are given in Table 2. Sowing was performed into the pots of 5 L volume filled with 5 kg of homogenized soil. Each pot was fertilized with the same dose of NPK (1.43 g N in the NH_4NO_3 form, 0.16 g P and 0.40 g K in the K_2HPO_4 form). The final number of plants in a pot was twenty. Plants were irrigated with demineralised water.

Weather conditions in cultivation period (from April to July) were similar in both years. Mean air temperature in both years was higher compared with the long-term normal. Mean precipitation in 2006 was higher in April, May and June, lower in July compared with the long-term normal. In April and May 2007 mean precipitation was by 25 per cent lower than normal and in June and July was higher compared with the long-term normal.

Depth of mould	N (NO_3^-)	N (NH_4^+)	N (total)	K	Mg	Ca	P	pH_{KCl}
	mg kg-1 DM							-
30 cm	21.1±2.1	0.4±0.04	21.5±2.2	264±13.2	132±6.6	3380±169	134±6.7	6.70±0.1
60 cm	4.8±0.5	0.4±0.04	5.2±0.5	185±9.3	141±7.1	2763±138	44±2.2	6.39±0.1

Table 1. Average content of minerals in the soil in field experiment

Soil „Příbram“	Cation H+ exchange capacity	pH_{KCl}	Cox	Zn	Cu	Cd	Pb
Unit	1 mmol kg-1	-	%	mg kg-1 DM			
Value	123	4.52±0.02	1.91±0.006	187±8.0	42.7±2.0	3.60±0.17	1321±71

Table 2. Content of selected metals and characteristics of used soil contaminated with heavy metals from the district of Příbram, Czech Republic in outside environment pot experiment

2.1.1 Brasssinosteroids and their treatment pattern

Plants were treated with eight different brassinosteroids (24-epibrassinolide; 24-epicastasterone; 4154 compound and five androstane and pregnane analogues of brassinosteroids marked KR1, KR2, KR3, KR4 and KR5) in 49-59 DC (growth phase referred to a *decimal code* for the growth stages of cereals - from visible awns to complete inflorescence emergence) (Zadoks et al., 1974). All brassinosteroids were applied in the form of 1 nmol L^{-1} of efficient compound in the water solution by spraying on all aerial biomass. Each of the tested brassinosterids was applied in four parallel replicates (4 x 10 m^2 field plots). Untreated plants were cultivated as well in tetraplicates as the control variant. Applied brassinosteroids (Fig. 1) were synthesized by the Institute of Organic Chemistry and Biochemistry of the Academy of Sciences of the Czech Republic. 24-epibrassinolide (24-epiBL) and 24-epicastasterone (24-epiCS) are naturally occurring plant phytohormones, compound 4154 is a synthetic brassinosteroid registered in the Czech Republic (Registration Nr. 294343, conferred on 4 Oct 2004) and the EU (Nr. 1401278, conferred on 28 Sep. 2005). Compounds KR1 - KR5 are synthetic permanently studied brassinolide analogues, which do not occur naturally in plants and will be published after finishing the synthesis of similar structures and protecting of these compounds by a patent (Vlašánková et al., 2009).

In pot experiment plants were treated with three brassinosteroids (24-epibrassinolide, 24-epicasterone, and 4154) in two different growth stages in four parallel replicates in each brassinosteroid. Plants or experimental pots were divided before the application of brassinosteroids into four groups that differed with growth stage in the date of treatment and number of brassinosteroids applications (Table 3).

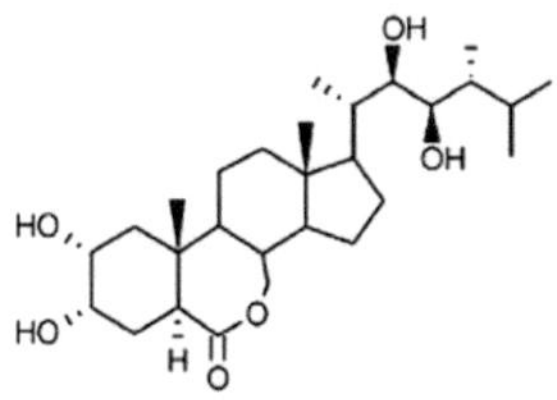

24-epibrassinolide (24-epiBL)

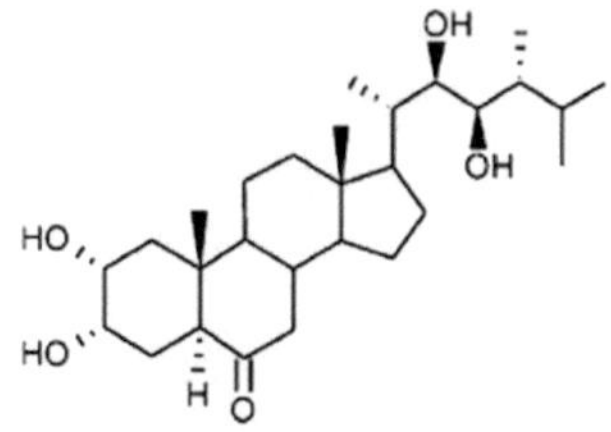

24-epicastasterone (24-epiCS)

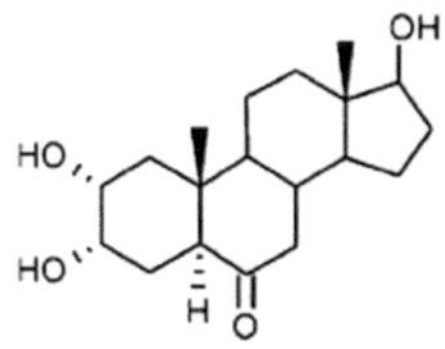

2α,3α,17β-trihydroxy-5α-androstan-6-one (4154)

Fig. 1. Chemical structure of brassinosteroids used for wheat treatment

Treatment*	Stage of brassinosteroids application DC	
	29-31	59-60
A-I	+	-
B-I	+	-
C-I	+	-
A-II	+	+
B-II	+	+
C-II	+	+
A-III	-	+
B-III	-	+
C-III	-	+
D-control	-	-

Table 3. Variants of analyzed spring wheat plants treated with brassinosteroids at different growth stages; *1st group of plants (pots A-I, B-I, C-I) was treated with brassinosteroids A (24-epibrassinolide), B (24-epicastasterone) and C (4154) once at the growth plant stage according to Zadoks growth scale 29-31 DC (off shooting); 2nd group (pots A-II, B-II, C-II) was treated with brassinosteroids two times, firstly in the plant growth stage 29-31 DC and again in the plant growth stage 59-60 DC (beginning of flowering); 3rd group (pots A-III, B-III, C-III) was treated once in the plant growth stage 59-60 DC (beginning of flowering); 4th group (D) consisted of untreated control plants

2.1.2 Harvest and sampling of plant material and grain

Plants of experimental field plots were harvested at physiological maturity (growth stage 90 DC) by a harvester thresher HEGE 140 (Hans-Ulrich Hege GmbH & Co, Germany). After the harvest, the grains were cleaned on the sieves by flow of air and then yield and thousand-grain weight were determined. Analytical samples were made according to methodology ISO 13690:1999 by quartering of cleaned grain. A common (average) analytical sample of four trial plots grain was prepared.

Sampling from the experimental pots was performed three weeks after the application of brassinosteroids in plant growth stages referred to a decimal code for the growth stages of cereals. The first sampling was performed in the plant growth phase 47-49 DC (visible awns), the second sampling in the growth stage 73-75 DC (30-50 % of final grain size). Grain and straw samples were taken in the growth phase Z90-92 (full ripeness). Green plants were taken from experimental pots cleaned up with distilled water and subsequently freeze-dried.

2.2 Methods of determination

2.2.1 Chemical and laboratory material and equipments

For dry decomposition there were used: nitric acid (65%, p.p., Lachema Neratovice CZ and Suprapur Merck, Germany), demineralised water (quality degree 1 according to EN ISO 3696 for the calibration of ICP-OES). Water calibration solutions with one element (Analytika, Ltd., CZ) were used for the calibration of F-AAS: Ca (1.000 ± 0.002 g L^{-1}) in 2% HCl, Cu (1.000 ± 0.002 g L^{-1}) in 2% HNO_3, Fe (1.000 ± 0.002 g L^{-1}) in 2% HNO_3, K (1.000 ± 0.002 g L^{-1}) in 2% HNO_3, Mg (1.000 ± 0.002 g L^{-1}) in 2% HCl, Mn (1.000 ± 0.002 g L^{-1}) in 2% HNO_3 and Zn (1.000 ± 0.002 g L^{-1}) in 2% HNO_3. For the testing of the dry decomposition method, a certified reference material NIST 8436 (Durum Wheat Flour) and internal reference material (IRM) from International Plant Analytical Exchange (IPE), RM Sample 3, Wheat 684, Quarterly Report 2000.3 were used. For the calibration of AAS and testing of the method of modified dry decomposition the following materials were used: calibration solutions with one element (Analytika, Ltd., CZ) 1.000 ± 0.002 g L^{-1} in 2% HNO_3 for elements Cu, Pb and Zn, while cadmium was dissolved in 2% HCl. Muffle oven (LM 112.10, MLW, Germany), heating plate ALTEC JRT 350 with temperature graduation per 10 °C and ultrasonic bath Elma Transonic T660/H were used for the dry decomposition of samples. Analyses of the metals were performed by atomic spectrometer VARIAN SpectrAA 110 (VARIAN A.G., Australia) with the possibility emission spectra and Varian SpectrAA 280Z atomic absorption spectrometer furnished with GTA 120 electrothermic atomizer. Laboratory hammer mills LM3100 and LM120, falling number bath FN 1500, Glutomatic 2200 and Gluten Index centrifuge 2015 made by Perten Instruments AB (Sweden) were used for the determination of Falling number, gluten content and gluten index. For protein determination there were used: nitric acid (HNO_3, 65%, p.a., Lachema Neratovice, CZ), automatic nitrogen analyzer Kjeltec system. A laboratory mill LM3100, lactic acid and bromphenol blue were used for the determination of sedimentation index (Zeleny sedimentation test).

2.2.2 Soil analysis

Soil pH was measured in suspension using 1:2.5 (w/v) ratio of soil and 0.2 M KCl at 20 ± 1°C by WTW pH 340i set. Available forms of nutrients (Ca, K, Mg and P) were determined using the Mehlich 3 soil extraction procedure (Mehlich, 1984; Zbíral, 2000) and organic nitrogen by the Kjeldahl method (Bremner, 1960).

2.2.3 Dry decomposition procedure

Samples of grain, straw and freeze-dried green plants were mineralized before analyses by dry thermal decomposition using SOP-3C (Standard Operation Procedure for Dry

Decomposition of Higher Plants and Green Algae) (Mader et al., 1998). Before dry decomposition, samples of freeze-dried plants, straw and grains were roughly ground in an IKA A11 Basic mill equipped with stainless steel working parts. Weight of the homogenized sample was about 1 g and each sample was analysed in two replicates. Initial temperature of the heater plate was 150 °C, final temperature was 350 °C. After cooling, the samples were combusted in a muffle oven at 480 °C; the ash was dissolved in 1.5 mL conc. HNO_3 (65% p.a.) and then repeatedly combusted at 480 °C. After the combustion, the samples (white ash) were dissolved in 5 mL of 1.5% HNO_3 after the addition of 1 mL conc. HNO_3 (65% p.a.).

2.2.4 FAAS determination (flame atomic absorption spectrometry) and ET-AAS (atomic absorption spectrometry with electrothermic atomization)

Determination of metals (Ca, Cu, Fe, K, Mg, Mn and Zn) content was performed with flame atomic absorption spectrometry (F-AAS) by calibration curve method. Atomization of samples proceeded in the flame acetylene/air; rate of injection of samples into the flame was 4.5 mL min^{-1}. Wavelengths used for the metals determination were 422.7, 285.2, 766.5, 213.9, 324.8, 279.5 and 248.3 nm for Ca, Mg, K, Zn, Cu, Mn and Fe, respectively. Determination of all metals content was performed with atomic spectrometer VARIAN SpectrAA 110. Limits of detection (LOD) and limits of quantification (LOQ) of the metals determination are given in Table 4. Determination of Cd, Zn and Cu was performed with flame atomic absorption spectrometry in samples prepared with dry decomposition. Atomization of samples was proceeded in the flame acetylene/air; rate of injection of samples into the flame was 4.5 mL min^{-1}. Wavelengths used for the metals determination were 228.8, 324.8 and 213.9 nm for Cd, Cu and Zn, respectively. Determination of Pb was performed with atomic absorption spectrometry with electrothermic atomization by Varian SpectrAA 280Z atomic absorption spectrometer furnished with GTA 120 electrothermic atomizer at wavelength 283.3 nm.

Parameter	Metal						
	Ca	Mg	K	Zn	Cu	Mn	Fe
LOD (mg kg^{-1})	1.0	0.03	0.08	0.09	0.01	0.15	0.18
LOQ (mg kg^{-1})	3.3	0.11	0.28	0.31	0.04	0.51	0.59

Table 4. Limits of detection (LOD) and limits of quantification (LOQ) of the metals determination

2.2.5 Determination of dry weight

Dry matter of straw samples was determined by drying at 105 ^{0}C in a laboratory oven and of that grain at 130 ^{0}C to constant weight (ISO 612).

2.2.6 Determination of wheat grain quality

Protein (N x 5.70) content was determined by Kjeldahl method by automatic nitrogen analyzer (methodology ISO 1871:1975). Falling number was determined according to ISO 3093:2004. Gluten content was determined by Glutomatic according to ISO 7495:1990. Sedimentation index of wheat flour (Zeleny sedimentation test) was determined according ISO 5529:1992. Determination of bulk density, called "mass per hectoliter" was performed according to ISO 7971-2:1995.

2.2.7 Replicates and statistical analysis

All variants were cultivated and treated in four replicates. Statistical evaluation was performed with ANOVA. Post-hoc analyses were performed by Tukey's HSD (Honestly Significant Difference) test ($p < 0.05$) for metals content and by Fisher's LSD (Least Significant Difference) test ($p < 0.05$) for grain quality parameters, thousand-grain weight and yield of grain.

3. Results

3.1 Content of Cd, Cu, Pb and Zn in aerial plant biomass (cultivation in experimental pots)

Experimental plants of the first group (A-I, B-I and C-I) and the second one (A-II, B-II and C-II) treated with BRs in the plant growth stage 29–31 DC did not differ in the growth stage 47 - 49 DC from untreated control and the plants of the third group (not treated in the stage 29 - 31 DC). The first differences in the content of investigated metals were shown in the plant growth stage 73 - 75 DC (Table 5). A distinct trend in copper content was not observed in the plant biomass. Content of lead decreased in all variants of treated plants. A decreased lead content was determined in the plants of the second group (as a whole treated two times) and the third (A-III, B-III and C-III) group (as a whole treated ones), in which the last BRs application was performed in growth stage 59–60 DC. In the first group that was treatedonly once in the stage 29–31 DC, lead content was higher than those in the other two groups. Similarly to lead content in the plants of the second group and the third group, lower cadmium and zinc contents were determined as related to the contents of the first group and in control plants (with the exception of plants treated with 4154 in the third group, where the lower Zn content was not determined). After the harvest of plants in the growth stage 90–92 DC (Table 5), a lower copper content in the first group and the third group (with the exception of plants treated with 4154 in the third group) was determined in plant straw. Likewise in the growth stage 73–75 DC, lower zinc content was determined in all plants of the second group and in the plants of the third group treated with 24-epiBL and 24-epiCS.

	Growth stage 73 - 75 DC				Growth stage 90 – 92 DC			
	Cu	Zn	Cd	Pb	Cu	Zn	Cd	Pb
D (control)	2.71	123	6.48	5.28	3.07	55.2	26.1	5.30
A-I	3.59*	130	5.91	3.59*	2.17*	47.4	29.3	4.70
B-I	3.35	129	5.42	2.37*	2.30*	46.8	30.6	4.80
C-I	1.60*	128	5.06*	3.39*	2.28*	46.0	31.7*	4.54
A-II	3.02	96.8*	4.01*	2.33*	2.48	33.7*	19.9*	3.56*
B-II	3.71*	100*	5.04*	1.70*	2.98	41.8*	26.5	3.65*
C-II	1.63*	108*	4.10*	2.95*	2.64	34.2*	25.6	4.45
A-III	2.52	92.0*	3.14*	1.36*	2.08*	26.3*	22.4	3.48*
B-III	1.94	93.5*	3.01*	1.88*	1.75*	26.2*	22.7	2.57*
C-III	3.56*	112	5.08*	2.68*	2.36	47.0	26.6	4.54

Table 5. Content of copper, zinc, cadmium and lead in plants (aerial biomass) treated with brassinosteroids and in untreated control in the growth stage 73 – 75 DC and 90 – 92 DC (mg kg^{-1} DM)

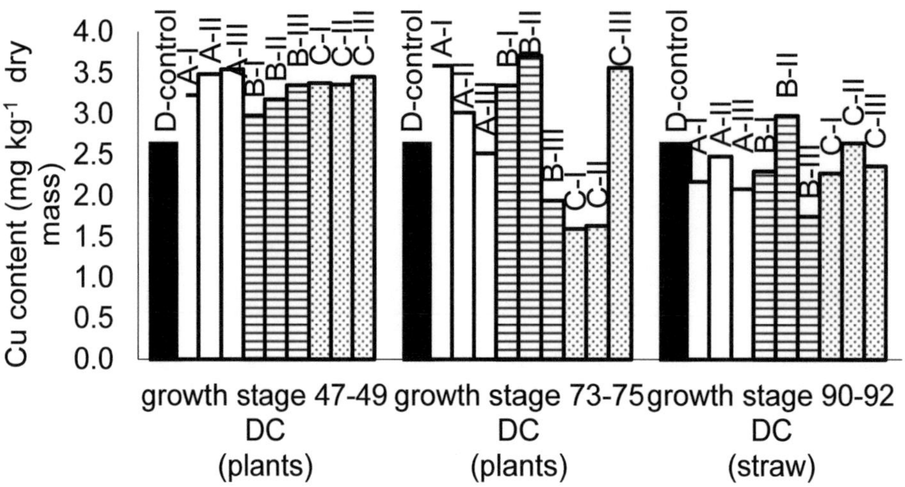

Fig. 2. Content of copper in aerial biomass of plants (mg kg-1 DM)

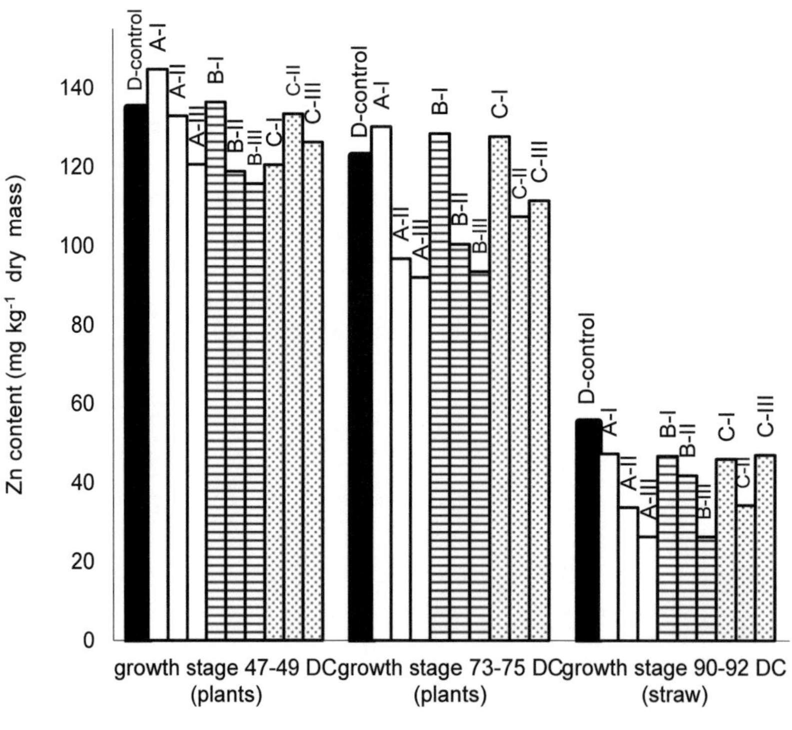

Fig. 3. Content of zinc in aerial biomass of plants (mg kg-1 DM)

No significant difference of the cadmium content in the growth stage 90–92 DC was found, with the exception of plants treated with 4154 in the first group and 24-epiBL in the second group. Lower lead content was determined in the plants of the second group and the third group treated with 24-epiBL and 24-epiCS. Copper content was affected more likely according to the actual and individual status of plants, however, in some cases these physiological processes could be affected by brassinosteroids treatment (Fig. 2). Zinc content in aerial biomass decreased during plant growth (Fig. 3). A significant decrease of cadmium content was determined after the applications of brassinosteroids in the growth stage 73–75 DC in the plants of the second group and the third group (Fig. 4). Lead was accumulated in the plant biomass of the control group during the all vegetation period (Fig. 5). Lower lead

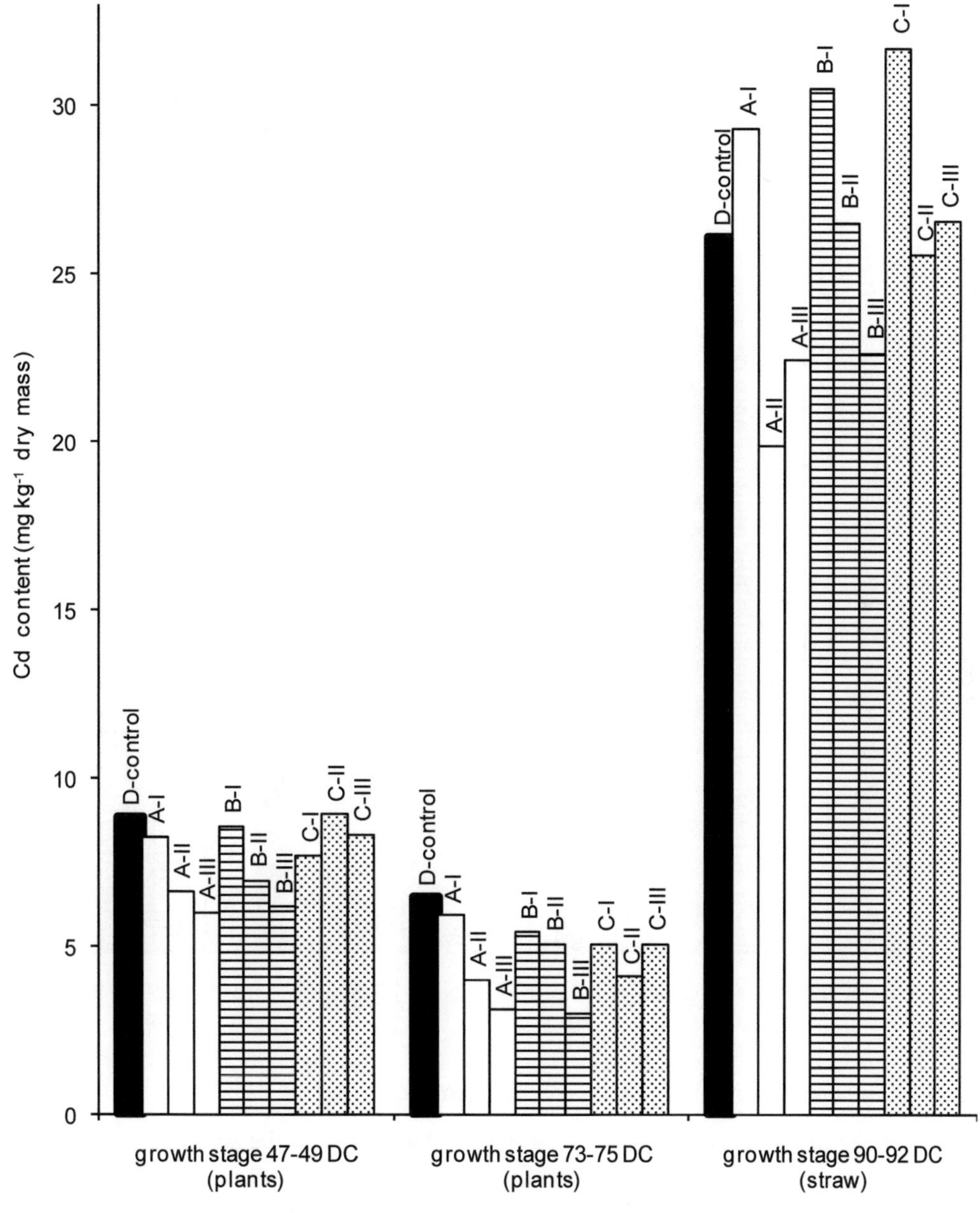

Fig. 4. Content of cadmium in aerial biomass of plants (mg kg^{-1} DM)

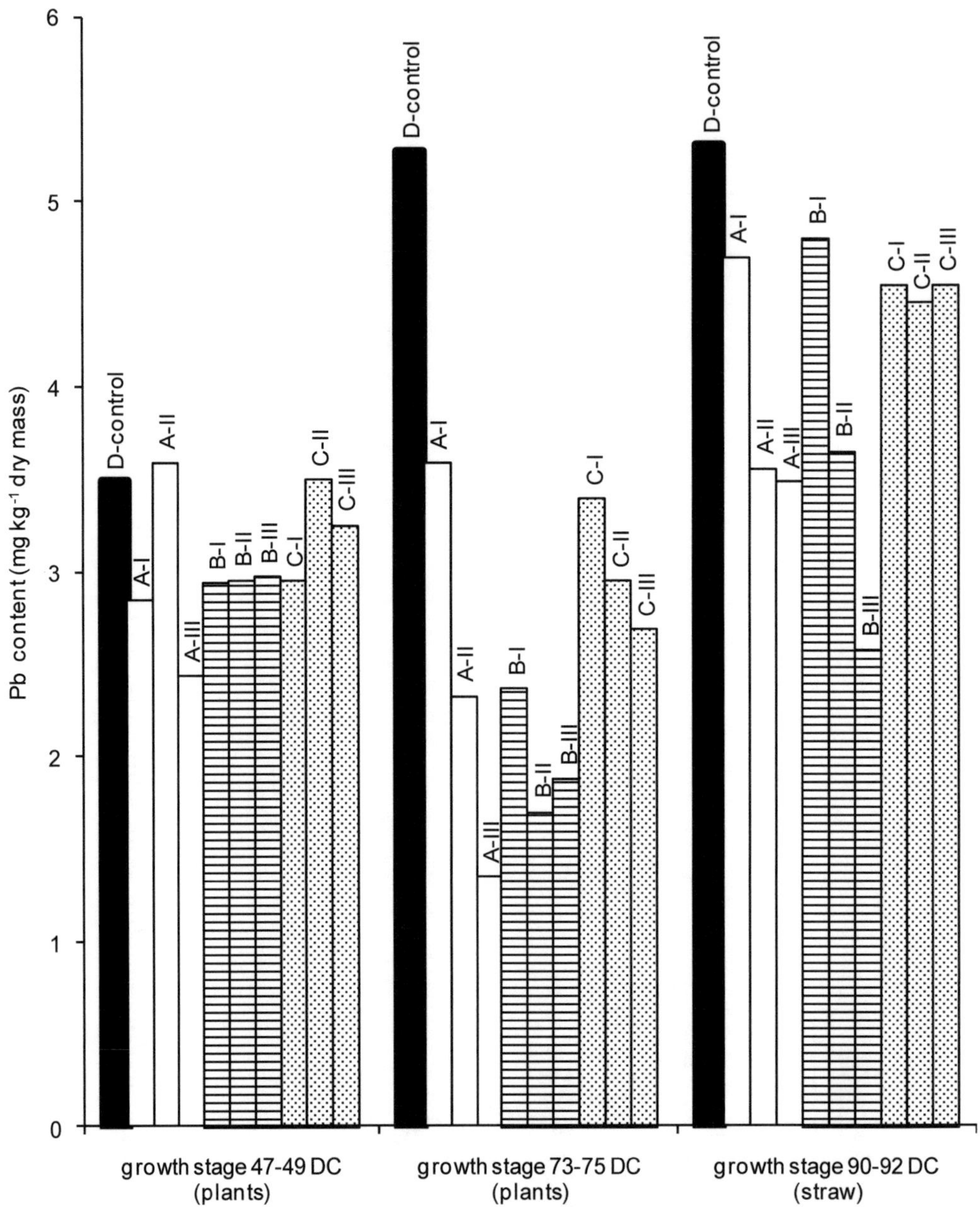

Fig. 5. Content of lead in aerial biomass of plants (mg kg^{-1} DM)

content in the stage 90–92 DC was found only in the plants of the second group and the third group that were treated with 24-epiBL and 24-epiCS. No significant difference was found in the plants of the first group that were treated in the growth stage 29–31 DC and in the treatment with brassinosteroid 4154.

3.2 Content of Cd, Cu, Pb and Zn in grains (cultivation in experimental pots)

After the application of BRs, lead content in grains decreased in the second and the third group (Table 5). While copper content significantly decreased in the plants of the third group following 24-epiCS treatment, the decrease of copper content was not statistically significant in other variants. Effect of brassinosteroids on the content of metals in grains of control plants is shown in Fig. 6.

	Copper	Zinc	Cadmium	Lead
D-control	4.88	15.6	11.7	1.87
A-I	4.15	15.9	12.9	1.33
B-I	3.90*	10.6	13.2	1.21
C-I	4.24	14.2	14.2*	1.76
A-II	4.49	14.7	14.2*	0.57*
B-II	3.95*	10.7	12.5	0.74*
C-II	4.72	14.9	11.5	0.49*
A-III	3.98*	12.3	12.1	0.97*
B-III	3.81*	16.3	12.0	0.57*
C-III	4.10*	17.0	12.6	0.83*

Table 5. Content of Cu, Zn, Cd and Pb in grain of plants treated with brassinosteroids and in untreated control (mg kg^{-1} DM); *statistically significant difference related to untreated control; for the used symbols of experimental variants see Table 2

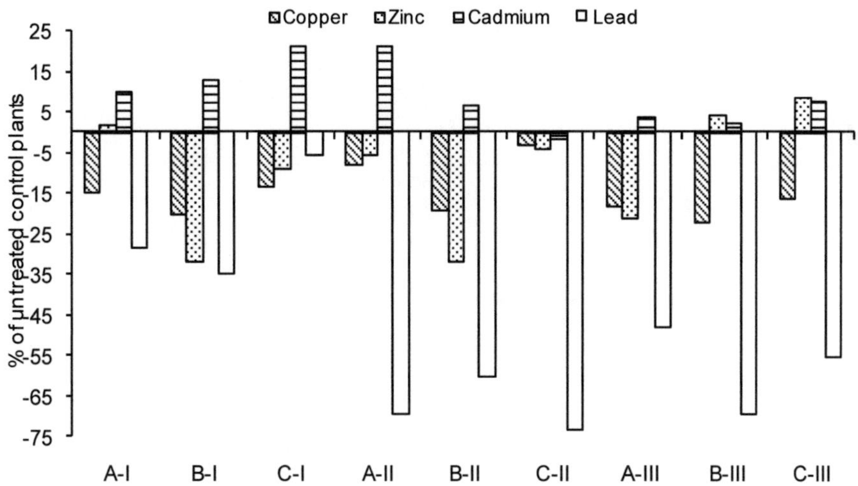

Fig. 6. Content of Cu, Zn, Cd and Pb in the grains of treated plants (in % of the content in untreated plants)

3.3 Content of Ca, Cu, Fe, K, Mg, Mn and Zn in wheat grain after BRs treatment (field experiment)

Values of metals content in wheat grain in individual years of cultivation are given in Table 6. The changes of the metals content in spring wheat grain were observed during the field experiments. Statistically significant differences in the total content of metals between years were found in the Ca, Mg, Mn and Fe contents. In potassium content, year 2007 differed from 2005 and 2006, while no difference was found between 2005 and 2006. Likewise, zinc content in 2005 differed from years 2006 and 2007. No statistically significant difference between 2005, 2006 and 2007 was found in copper content. In 2005, no differences in the selected metals content between untreated control plants and plants treated with BRs were determined. In 2006, potassium content increased in plants treated with 24-epiBL (by 22.2%), 4151 (by 31.2%) and KR1 (by 24.5%), while zinc content decreased in variants treated with 24-epiCS (by 14.5%) and KR1 (by 12.4%) as compared to the control variant. 2007, Mg, Mn and Fe contents decreased. In comparison with untreated control plants, there was lower magnesium content (by 11%) and manganese content (at least 7.5%) in variants treated with 24-epiCS, 4154 and with KR2–KR5. Different iron content was determined in variants treated with 24-epiBL, 24-epiCS and with KR1–KR5. Weather conditions were similar in all three years. Mean air temperature was higher as compared with the long-term normal value(Table 7) in whole three-year period. Mean precipitation in 2005 and 2006 was lower as

Variety	Year	Calcium	Magnesium	Potassium	Zinc	Copper	Manganese	Iron
untreated plants	2005	190.4	1337	3073	34.2	4.87	38.1	43.2
24-epiBL		187.3	1350	3089	36.6	4.93	37.9	44.0
24-epiCS		192.8	1353	3382	35.4	4.77	38.0	44.8
4154		190.8	1334	3330	37.8	4.77	36.9	44.6
KR1		191.5	1379	3097	37.4	4.82	37.5	45.8
KR2		188.4	1338	3062	34.9	4.80	38.2	46.0
KR3		186.7	1311	3146	35.2	4.77	36.8	44.3
KR4		189.6	1335	3445	35.7	4.90	38.4	46.4*
KR5		187.1	1351	3394	34.9	5.06	37.3	44.9
untreated plants	2006	304.5	1407	2591	37.3	4.76	44.6	46.1
24-epiBL		282.0	1439	3168*	33.4	4.77	46.4	44.8
24-epiCS		291.4	1428	3106	31.8*	4.91	46.2	47.2
4154		312.9	1456	3399*	33.5	4.92	45.2	47.2
KR1		287.7	1420	3226*	32.6*	4.75	43.5	46.1
KR2		301.3	1459	3174	35.2	4.82	44.2	47.8
KR3		294.3	1450	3111	35.2	4.70	42.4	48.4
KR4		315.1	1468	3117	33.2	4.84	46.7	47.7
KR5		306.7	1421	2898	34.4	4.78	45.7	48.1
untreated plants	2007	315.3	1262	3718	34.8	4.92	35.2	57.7
24-epiBL		313.9	1178	3533	34.4	4.83	34.0	49.4*
24-epiCS		320.5	1112*	3462	34.8	5.00	32.5*	49.6*
4154		327.7	1056*	3341	32.5*	4.77	32.3*	59.0
KR1		310.8	1145	3393	33.9	4.62	34.1	48.7*
KR2		314.7	976*	3314	33.1	4.80	29.3*	52.9*
KR3		318.2	1015*	3334	34.2	4.92	30.7*	65.7*
KR4		319.5	977*	3295	33.3	4.77	30.5*	48.6*
KR5		307.4	996*	3584	33.4	5.00	30.2*	48.5*

Table 6. Content of Ca, Cu, Fe, K, Mg, Mn and Zn (mg kg^{-1} DM) in spring wheat grain; *statistically significant difference (at the level of significance $p < 0.05$) between treated and untreated plants

compared with the long-term normal value, while close to long-term level in 2007. The results achieved in three-year period 2005–2007 indicate a possible effect of the year on metal content of grain affected probably by precipitation. In 2007, with usual precipitation level, contents of Fe, K, Mg and Mn were decreased. In 2005 and 2006 with below average precipitation, total content of metals were comparable or higher than content of metals in untreated control plants grain. Nevertheless, such hypothesis needs to be tested in further experiments.

	unit	Jan	Feb	Mar	Apr	May	Jun	Jul	Aug	Sep	Oct	Nov	Dec
Temperature	°C	-2.1	-0.8	3.4	8.2	13.4	16.3	18.2	17.5	14.0	8.6	3.2	-0.5
Precipitation	mm	28	27	31	46	65	74	74	72	49	41	34	34

Table 7. Long-term normal of mean air temperature and mean precipitation of cultivation area (50°2'0"N, 14°36'54"E)

Variety	Year	Bulk density (mass per hectolitre)	Falling number	Protein	Gluten	Sedimentation index (Zeleny test)
		kg hL^{-1}	sec	%	%	mL
untreated plants	2005	79.3	153.8	13.8	34.0	61.0
24-epiBL		79.4	138.8	13.9	34.4	58.3
24-epiCS		79.4	145.3	13.9	34.4	61.0
4154		79.6	146.7	13.8	34.2	58.8
KR1		79.3	156.3	13.9	34.4	61.8
KR2		79.4	151.3	14.0	34.6	58.3
KR3		79.5	160.8	13.7	33.9	57.3
KR4		79.4	153.5	13.5	33.2	58.0
KR5		79.4	152.5	13.6	33.5	55.5
untreated plants	2006	80.1	346.5	12.9	29.9	46.5
24-epiBL		79.9	335.8	12.2	28.1	42.8
24-epiCS		79.9	344.8	12.6	28.7	43.0
4154		80.1	346.8	12.9	29.4	45.5
KR1		80.2	350.3	12.6	28.7	42.5
KR2		80.2	348.8	12.8	29.1	45.8
KR3		79.9	353.8	12.8	29.5	46.3
KR4		79.9	344.0	12.5	28.7	43.8
KR5		80.0	332.5	12.9	30.6	46.3
untreated plants	2007	80.4	274.3	15.6	44.0	68.5
24-epiBL		80.6	290.5	15.5	44.1	70.5
24-epiCS		80.7	275.8	15.5	44.1	68.8
4154		80.5	273.0	15.4	43.4	68.8
KR1		80.6	282.8	15.3	44.4	69.0
KR2		80.3	274.3	15.3	43.7	68.8
KR3		80.5	286.0	15.3	43.5	68.8
KR4		80.6	269.0	15.4	43.6	69.3
KR5		80.6	269.0	15.3	43.8	68.3

Table 8. Quality parameters of grain after spring wheat treatment with brassinosteroids

Variety	Year	Yield (corrected on moisture 14 %)	1000 kernels weight
		t ha-1	g
untreated plants	2005	6.54	53.24
24-epiBL		6.09*	56.99*
24-epiCS		6.19	55.77*
4154		6.36	56.61*
KR1		6.20	54.99*
KR2		6.12	54.77*
KR3		6.41	54.91*
KR4		6.37	55.59*
KR5		6.33	55.38*
untreated plants	2006	7.10	45.04
24-epiBL		6.66	45.10
24-epiCS		7.12	45.10
4154		7.19	44.46
KR1		7.04	45.22
KR2		7.12	44.62
KR3		7.13	44.17
KR4		6.86	44.62
KR5		7.09	44.38
untreated plants	2007	4.57	45.60
24-epiBL		4.32	44.71
24-epiCS		4.60	45.74
4154		4.53	45.44
KR1		4.53	45.05
KR2		4.52	45.10
KR3		4.45	45.33
KR4		4.48	46.00
KR5		4.65	45.87

Table 9. Yield of grain and 1000 kernels weight after spring wheat treatment with brassinosteroids; *statistically significant difference (at the level of significance $p < 0.05$) between treated and untreated plants

3.4 Quality of wheat grain after BRs treatment (field experiment)

Values of determined qualitative parameters of food wheat grains (bulk density, falling number, protein content, gluten content and sedimentation index) were different according to the cultivation years; statistically significant difference has been proved between years. No statistically significant difference was observed between values of the grain qualitative parameters of plants treated with brassinosteroids or untreated. Average values of qualitative parameters of wheat grains in individual years are reported in Table 8.

3.5 Yield of wheat grain after BRs treatment (field experiment)

Yield of grain and values of thousand-grain weight (TGW) were different during the investigated period; a statistically significant difference between years was demonstrated. In 2005, an increase of TGW was determined in all treated variants. The yield per hectare decreased by 6.9% in the variant treated with 24-epiBL. In 2006 and 2007, no difference between grain yields of control and treated plants was observed. No difference was found also between TGW values. Average grain yields and TGW values are given in Table 9.

4. Discussion

Changes of metal composition of plants treated with brassinosteroids were reported in quite a few experiments. Most of these experiments are related to the ability of brassinosteroids to decrease the intake of heavy metals with plants. 24-epiBL at the concentration of 10^{-8} mol L^{-1} in combination with heavy metals blocked metal accumulation in algal cells (Bajguz, 2000) and treatment of *Brassica juncea* plants with 24-epiBL detoxified the stress generated by NaCl and/or $NiCl_2$ and significantly improved growth, the level of pigments and photosynthetic parameters (Ali et al., 2008b). After foliar application of brassinolide on tomato plants an increase in metals (P, K, Ca and Mg) in aerial parts of plants has been recorded (Nafie & El-Khallal, 2000). Our three-year results showed that after the brassinosteroids treatment of spring wheat some changes of the metals content were determined. However, these changes differed among the experimental years. Brassinosteroids application primarily affected content of K, Mg, Zn and Fe in grain. However, it did not affect Cu content. Brassinosteroids stimulate morphogenesis of plants which causes an increase in leaf area, number of leaves, dry and fresh mass of stems and roots and number of tillers and productive branches. Due to these effects on physiological processes in plants, an increase in the yield and quality of crops production has been observed (Sakurai et al., 1999). Yield increase depends on variety, climatic conditions, soil, application of fertilizers and also on frequency and dates of brassinosteroids application (Khripach et al., 2000, 2003; Janeczko et al., 2010). Different preparations (mixtures of natural 24-epibrassinolide and its synthetic isomers) especially used under unfavourable cultivation conditions cause an increase in yield of crops such as rice, maize, wheat, cotton, tobacco, vegetables and fruit. Exogenous brassinosteroids such as 24-epibrassinolide influences brassinosteroid balance in seedlings of wheat after soaking seeds, drenching or spraying plants and content of endogenous brassinosteroids brassinolide and castasterone varies with leaf insertion and plant age (Janeczko & Swaczynowá, 2010). The relative effects of brassinosteroids may be low, when the conditions under which plants are growing are generally favourable (Khripach et al., 2000). Treatment of barley cultivated in light-textured clay podzolic soil with brassinosteroids in a combination with nitrogen-phosphorus-potassium fertilizer (dose 60 kg N ha^{-1}) increased grain yield by 360 kg ha^{-1}; content of total protein in grain was not affected. However, in our experiments, where NPK fertilization at a dose of 60 kg N ha^{-1} was applied, no significant increase of grain yield per hectare has been proved. However, the application of brassinosteroids could reduce the negative effect of the stress factors on the yield and dry matter in wheat (Hnilička et al., 2007, Bajguz, 2009). In a greenhouse experiment with exogenously applied 24-epibrassinolide on two hexaploid wheat (*Triticum aestivum* L.) cultivars, S-24 (salt tolerant) and MH-97 (moderately salt sensitive), the application of 24-epibrassinolide increased plant biomass and leaf areal per plant of both cultivars under non-saline conditions. However, under saline conditions improvement in growth due to foliar application of 24-epibrassinolide was observed only in salt tolerant cultivar (Shahbaz et al., 2008). Drought stress and high temperature were found to have a negative effect on the amount of dry matter in the above-ground wheat biomass and the yield of grain and straw. Our results regarding the total protein content are in agreement with the results of experiments with wheat after exogenous plant treatment with 24-epibrassinolide, where as well no difference in soluble protein content between control and treated plants after brassinosteroid treatment was determined (Janeczko et al., 2010). In our experiments, no difference was recorded between treated plants and control plants in other qualitative parameters such as gluten content, sedimentation index and bulk density, which are affected more likely by varietals properties, or in Falling number, which is dependent on the harvest date and weather course during the harvest period.

Enhanced resistance of brassinosteroid-treated plants to extreme temperature, salt, pathogens and environmental stresses (heavy metals) was reported by Krishna (2003). The present study revealed the effect of brassinossteroids treatment on the accumulation of Cd, Cu, Pb and Zn contents in aerial wheat biomass or grains. The obtained results are in agreement with the results of Bajguz (2000), who observed that 24-epiBL at the concentration of 10^{-8} mol L^{-1} in combination with heavy metals blocked metal accumulation in algal cells. At metal concentrations of 10^{-6} – 10^{-4} mol L^{-1}, a combination with 24-epiBL appeared to have a stronger stimulatory effect on a number of cells than a single metal (a stronger inhibitory effect). The inhibitory effect on metal accumulation of 24-epiBL mixed with different heavy metals was arranged in the following order: zinc > cadmium > lead > copper. Our results obtained for spring wheat as an important crop confirm and are complementary to the results of Sharma & Bhardwaj (2007a, b), which describe the effects of 24-epiBL on plant growth, heavy metals uptake in the plants of *Brassica juncea* L. under heavy metal (Zn, Cu, Mn, Co and Ni) stress. 24-epiBL after the pre-germination treatment blocked copper metal uptake and accumulation in the plants. Likewise results of Anuradha & Rao (2007), obtained in a study on radish (*Raphanus sativus* L.) after the treatment with 24-epiBL and 28-homobrassinolide clearly indicated the inhibitory influence of brassinosteroids on the cadmium toxicity. Brassinosteroids supplementation alleviated the toxic effect of cadmium and increased the percentage of seed germination and seedling growth. Treatment with brassinosteroids regulates and enhances the activities of antioxidant enzymes ascorbate peroxidase, glutathione reductase, catalase, peroxidase and superoxide dismutase (Sharma, I. et al., 2010) and in drought stressed plants proline and protein content (Behnamnia et al., 2009). The application of brassinosteroids at low concentrations at a certain stage of development reduced significantly the metal absorption in barley, tomatoes and sugar beet. Our results indicate that for the decrease of heavy metals content in plants after the brassinosteroids application the growth stage of spring wheat is very important (Figs. 7 and 8).

The present study shows that the content of heavy metals in wheat plants is reduced variously in different growth stages. The plants of the second group and the third group contained in biomass at the growth stage 73–75 DC lower Pb content as compared to control

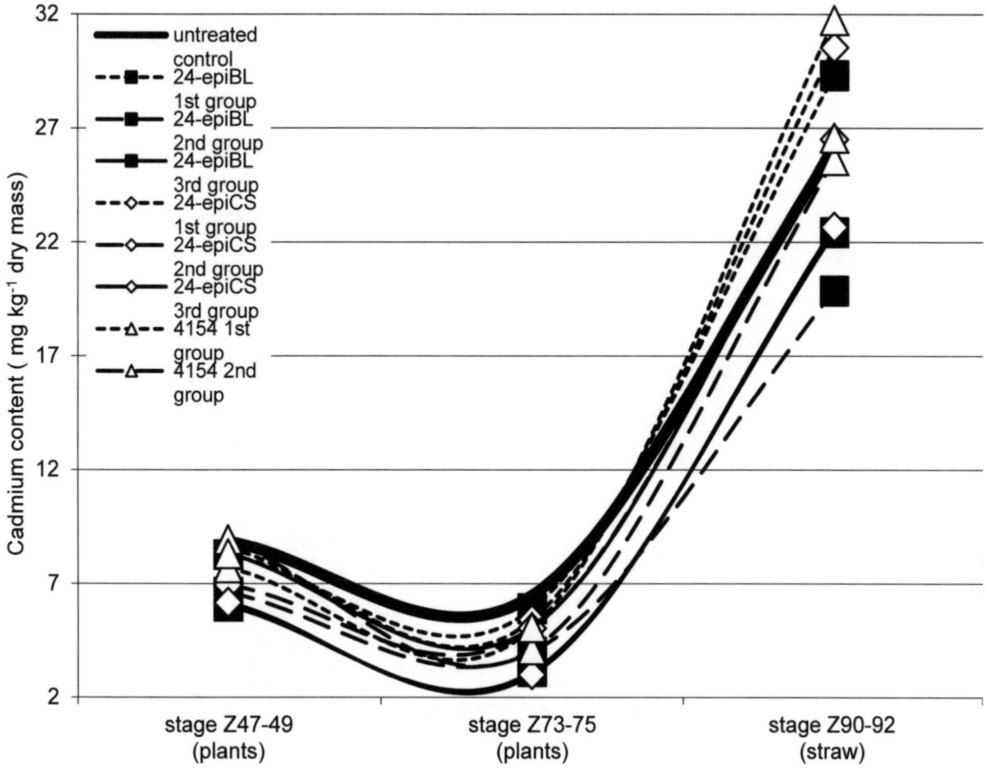

Fig. 7. Cd content in above ground biomass in untreated control and with BRs treated wheat variants; *1st group of plants (pots A-I, B-I, C-I) was treated with brassinosteroids A (24-epibrassinolide), B (24-epicastasterone) and C (4154) once in the growth plant stage according to Zadoks growth scale 29-31 DC (off shooting); 2nd group (pots A-II, B-II, C-II) was treated with brassinosteroids two times, firstly in the plant growth stage 29-31 DC and again in the plant growth stage 59-60 DC (beginning of flowering); 3rd group (pots A-III, B-III, C-III) was treated once in the plant growth stage 59-60 DC (beginning of flowering)

plants and the plants of the first group, which was treated with brassinosteroids last at the growth stage 29 - 31 DC. Also in the plants of the second group and the third group at the growth stage 73 - 75 DC lower Cd and Zn contents were determined (with the exception of brassinosteroid 4154 in the third group). The treatment of wheat plants with brassinosteroids 24-epiBL, 24-epiCS and 4154 at the plant growth stage 29–31 DC did not significantly influence content of the heavy metals in aerial plant biomass at the growth stage 47 – 49 DC. In the straw at the growth stage 90–92 DC, lower Pb and Zn contents were subsequently determined only in the plants treated with 24-epiBL and 24-epiCS (Zn also with the application of 4154 in the second group). Lower Cd content was determined only in the variant treated two times with 24-epiBL, which was considered as a highly active brassinosteroid. Lower Pb content was found in the grains of plants of the second group (treated two times in the stages 29-31 DC and 59-60 DC) and the third group (treated once in the stages 59-60 DC).

In terms of the content of heavy metals related to the number and growth stage of brassinosteroids applications, the most effective variants of treatment leading to decrease of metal content proved either double treatments in the growth stages 29 - 31 DC and 59 - 60 DC (plants of the second group) or one treatment only in the stage 59 - 60 DC (plants of the third group).

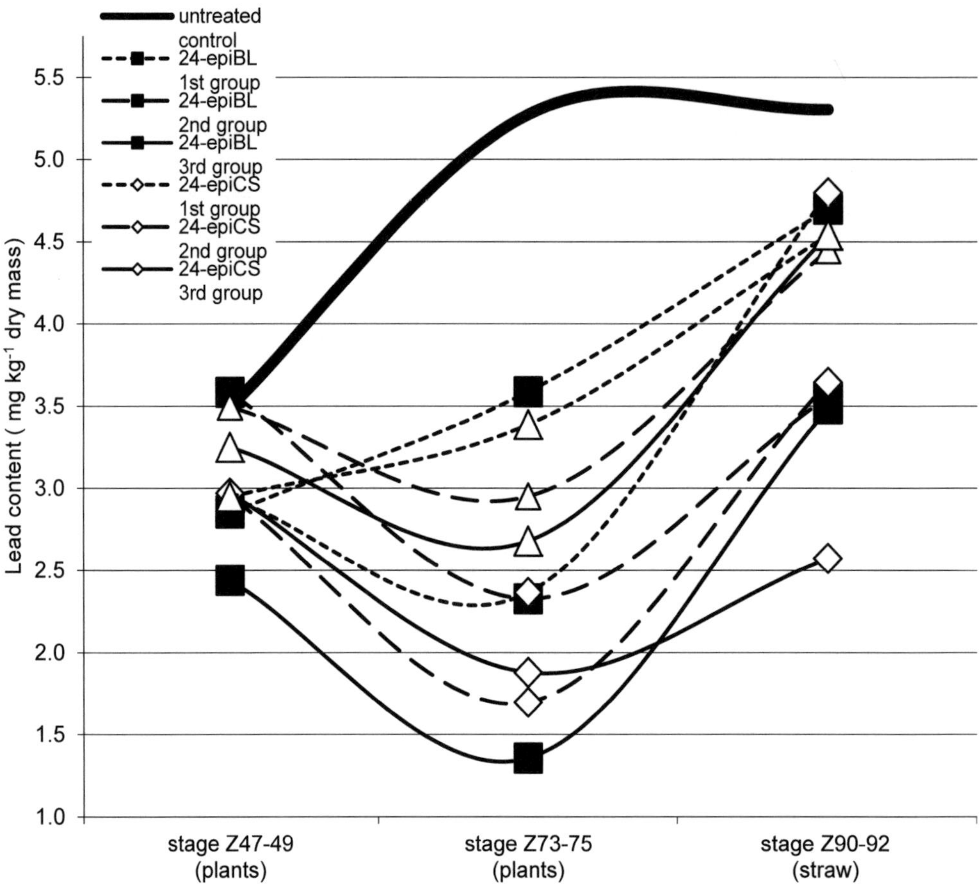

Fig. 8. Pb content in above ground biomass in untreated control and with BRs treated wheat variants (described in Fig. 7)

Brassinosteroids are able to manage plant water economy during a drought period by decreasing plant activity with a simultaneous conservation of the whole plant for more favourable conditions. Brassinosteroid-treated plants are then able to overcome the drought period in a much better condition than non-treated plants (Sasse, 1999). Their increase in net photosynthetic rate due to brassinosteroids application has already been observed in wheat, tomato and cucumber under normal condition and environment stresses (Ogweno et al., 2008; Shabaz et al., 2008; Xia et al., 2009; Yuan et al., 2010, Holá, 2010). Nowadays biological effects not only naturally occurring brassinosteroids, but also their androstane and pregnane analogues are widely synthesised and their biological effects studied (Hniličková et al., 2010) as well as their miscellaneous metabolic pathways in plants involving dehydrogenation, demethylation, epimerization, esterification, glycosylation, hydroxylation, side-chain cleavage and sulfonation (Bajguz, 2007). Because brassinosteroids control several important agronomic traits (Kang & Guo, 2010) such as flowering time, plant architecture, seed yield and stress tolerance, the genetic manipulation of brassinosteroids biosynthesis, conversion or perception offers a unique possibility of both changing plant metabolism and protecting plants from environmental stresses confirming the value of further research on brassinosteroids to improve productivity and quality of agricultural crops (Divi & Krishna, 2009) or their possible use for phytoremediation application (Barbafieri & Tassi, 2010).

5. Conclusion

From the perspective of minimal heavy metals content in biomass and grains related to the number of treatments and growth stage the most effective options of application of brassinolide treatment are those, which lead to a reduction in heavy metals in biomass: either dual treatment in growth stages DC 29-31 and DC 59-60 or single treatment only in the DC 59-60. Favourable is effective reduction of the content of heavy metals in the biomass of plants in grain milk stage (DC 73-75). After treatment of plants with brassinosteroids, when the plants are harvested for ensilage, the content of toxic metals was effectively reduced. Thus, treatment of plants with brassinosteroids can effectively reduce the content of heavy metals in plants (Cd and Pb) or harvested grain (Pb) of wheat and reduce the input of these contaminants into the food chain either cereal or meat products from the food industry. From the point of view of final effect on the content of the heavy metals in plant biomass and grains, the most suitable variant appears to be the single treatment in the growth stage 59–60 DC, which is economically preferable and its final effect does not differ remarkably from double treatments. Likewise lead content in grains decreased in the plants of the second group by 70–74% and of the third group by 48–70%. Thus, treatment of plants with brassinosteroids effectively decreased content of cadmium and lead in wheat plants (biomass) and content of lead in harvested grain and diminished in such way the input of these contaminants into the food chain.

Changes in the minerals content differed according to used brassinosteroid (variant) and investigated year; however unambiguous tendencies of changes or effects were not recorded. In comparison with control plants in the year 2005 the content of minerals in grain of treated plants did not differed significantly. In the year 2006 an increase of K after treatment with 24-epiBL, 4154 and KR1 compounds and a decrease of Zn content after treatment with 24-epiCS and KR1 compounds were recorded. In the year 2007 a decrease of Mg, Mn and Fe content was determined.

Similarly grain quality was not affected by the treatment with brassinosteroids in the investigated years. Content of proteins and gluten in the grains of treated and untreated plants was not significantly different. Similar results were obtained in the sedimentation index and bulk density. Falling number values differed depending on the date of harvest and year of cultivation; in comparison with control plants no difference was recorded. The hypothesis presented is that utilisation of brassinosteroids for plant treatment in the methods of agricultural management with a normal (rational) level of agricultural engineering is not effective. However, by contrast, their application could represent a high economic gain in all cases where the conditions for the cultivation of cereals are not quite ideal, e.g. under conditions of action of different environmental plant stressors, especially with cultivation on soils contaminated with heavy toxic metals or in different arrangements of agricultural engineering. The brassinosteroids-induced enhancement of photosynthetic capacity and regulation of antioxidant enzymes or growth could be under stress factors such saline conditions cultivar specific.

6. Acknowledgment

This study was supported by a grant project of the Ministry of Education, Youth and Sport *MSM 6046070901* of the Czech Republic and the Ministry of Agriculture of the Czech Republic *NAZV QH92111*.

7. References

Ali, A.A. & Abdel-Fattah, R.I. (2006). Osmolytes-antioxidant behaviour in *Phaseolus vulgaris* and *Hordeum vulgare* with brassinosteroid under salt stress. *Journal of Agronomy,* Vol. 5, No.1, (March 2006), pp. 167-174, ISSN 1812-5379

Ali, B., Hayat, S., Aiman-Hasan, S. & Ahmad, A. (2006). Effect of root applied 28-homobrassinolide on the performance of *Lycopersicon esculentum. Scientia Horticulturae,* Vol.110, No.3, (November 2006), pp. 267-273, ISSN 0304-4238

Ali, B., Hasan, S.A., Hayat, S., Hayat, Q., Yadav, S., Fariduddin, Q. & Ahmad, A. (2008). A role for brassinosteroids in the amelioration of aluminium stress through antioxidant system in mung bean (*Vigna radiata* L. Wilczek). *Environmental and Experimental Botany,* Vo.62, No.2, (March 2008), pp. 153-159, ISSN 0098-8472

Ali, B., Hayat, S., Ahmad, A., 2007. 28-Homobrassinolide ameliorates the saline stress in chickpea (*Cicer arietinum* L.). *Environmental and Experimental Botany,* Vol.59, No.2, (March 2007), pp. 217–223, ISSN 0098-8472

Anuradha, S. & Rao, S.S.R. (2001). Effect of brassinosteroids on salinity stress induced inhibition of seed germination and seedling growth of rice (*Oryza sativa* L.). *Plant Growth Regulation,* Vol. 33, No.2, (February 2001), pp. 151–153, ISSN 0167-6903

Anuradha, S. & Rao, S.S.R. (2007). The effect of brassinosteroids on radish (*Raphanus sativus* L.) seedlings growing under cadmium stress. *Plant, Soil and Environment,* Vol.53, No.11, (November 2007), pp. 465-472, ISSN 1214-1178

Bajguz, A. (2000). Blockage of heavy metals accumulation in *Chlorella vulgaris* cells by 24-epibrassinolide. *Plant Physiology and Biochemistry,* Vol.38, No.10, (October 2000), pp. 797-801, ISSN 0981-9428

Bajguz, A. (2007). Metabolism of brassinosteroids in plants. *Plant Physiology and Biochemistry,* Vol. 45, No.2, (February 2007), pp. 95-107, ISSN 0981-9428

Bajguz, A. (2009). Effect of brassinosteroids on the plant responses to environmental stresses. *Plant Physiology and Biochemistry,* Vol.47, No.1, (January 2009), pp. 1-8, ISSN 0981-9428

Bajguz, A. (2010). Brassinosteroids - occurrence and chemical structures in plants. In: *Brassinosteroids: A Class of Plant Hormone,* S. Hayat & A. Ahmad, (Eds.), 1-28, Springer Science + Business Media B.V., ISBN 978-007-0188-5, Heidelberg, Germany

Bajguz, A. & Godlewska-Zylkiewicz, B. (2004). Protective role of 20-hydroxyecdysone against lead stress in *Chlorella vulgaris* cultures. *Phytochemistry,* Vol.65, No.6, (March 2004), pp. 711-720, ISSN 0031-9422

Bajguz, A. & Tretyn A. (2003). The chemical structures and occurrence of brassinosteroids in plants. In: *Brassinosteroids. Bioactivity and Crop Productivity,* S. Hayat & A. Ahmad, (Eds.), 1-44, Kluwer Academic Publishers, ISBN 1-4020-1710-3, Dordrecht, The Netherlands

Barbafieri, M. & Tassi, E. (2010). Brassinosteroids for phytoremediation application. In: *Brassinosteroids: A Class of Plant Hormone,* S. Hayat & A. Ahmad, (Eds.), 403-438, Springer Science + Business Media B.V., ISBN 978-007-0188-5, Heidelberg, Germany

Behnamnia, M., Kalantari, K.M. & Ziaie, J. (2009). The effects of brassinosteroid on the induction of biochemical changes in *Lycopersicon esculentum* under drought stress. *Turkish Journal of Botany,* Vol.33, No.6, pp. 417-428, ISSN 1300-008X

Cao, S.Q., Xu, Q.T., Cao, Y.J., Qian, K., An, K., Zhu, Y., Hu, B.Z, Zhao, H.F. & Kuai, B.K. (2005). Loss-of- function mutation in DET2 gene lead to an enhanced resistance to oxidative stress in *Arabidopsis*. *Physiologia Plantarum,* Vol.123, No.1, (January 2005), pp. 57-66, ISSN 0031-9317

Cutler, G.C. (1991). Brassinosteroids through the looking glass: An appraisal. In: *Brassinosteroids: Chemistry, Bioactivity and Application,* H.G., Cutler, T., Yokota, G., Adam (Eds.), 155-163, ACS Symposium Ser., Vol. 474, American Chemical Society, ISBN 0-8412-2126-X, Washington, DC, USA

Divi, U.K. & Krishna, P. (2009). Brassinosteroid: a biotechnological target for enhancing crop yield and stress tolerance. *New Biotechnology,*Vol.26, No.3-4, Sp.Iss.2, (October 31, 2009), pp. 131-136, ISSN 1871-6784

Fujii, S. & Saka, H. (2001). Distribution of assimilates to each organ in rice plants exposed to low temperature at the ripening stage and the effect of brassinolide on the distribution. *Plant Production Science,* Vol.4, No.2 (June 2001), pp. 136-144, ISSN 1343-943X

Grove, M.D., Spencer, G.F., Rohwedder, W.K., Mandava, N., Worley, J.F., Warthen, J.D., Steffens, G.L., Flippen-Anderson, J.L. & Cook, J.C. (1979). Brassinolide, a plant growth-promoting steroid isolated from *Brassica napus* pollen. *Nature,* Vol.281, No.5728, pp. 216-217, ISSN 0028-0836

Hathout, T.A. (1996). Salinity stress and its counteraction by the growth regulator „brassinolide"in wheat plants (*Triticum aestivum* L. cultivar Giza 157). *Egyptian Journal of Physiological Sciences,* Vol.20, No.1-2, (December 1996), pp. 127-152, ISSN 0301-8660

Hayat, S., Ali, B. & Ahmad, A. (2006). Response of *Brassica juncea* to 28-homobrassinolide grown from the seeds exposed to salt stress. *Journal of Plant Biology,* Vol.33, No.1&2, (April 2006), pp.169–174, ISSN 1226-9239

Hayat, S., Ali, B., Hasan, S.A. & Ahmad, A. (2007a). Brassinosteroid enhanced the level of antioxidants under cadmium stress in *Brassica juncea*. *Environmental and Experimental Botany,* Vol.60, No.1, (May 2007), pp. 33-41, ISSN 0098-8472

Hayat, S., Ali, B., Hasan, S.A., Ahmad, A. (2007b). Effect of 28-homobrassinolide on salinity-induced changes in *Brassica juncea*. *Turkish Journal of Biology,* Vol.31, No.3, (August 2007), pp. 141-146, ISSN 1010-7576

Hniličk a, F., Hniličková, H., Martínková, J. & Bláha, L. (2007). The influence of drought and the application of 24-epibrassinolide on the formation of dry matter and yield in wheat. *Cereal Research Communications,* Vol.35, No.2, (April 30-May 05, 2007), pp. 457-460, ISSN 0133-3720

Hniličková, J., Kohout, L., Capdevila, E., Esteve, A., Vilaplana, M., Molist, M., Brosa, C., Swaczynowá-Oklestková, J. & Slavíková, B. (2010). The synthesis of androstane analogues with α-azido acid ester groups in position 17β. *Steroids,* Vol. 75, No.12, (December 2010), pp. 1005-1010, ISSN 0039-128X

Holá, D. (2010). Brassinosteroids and photosynthesis. In: *Brassinosteroids: A Class of Plant Hormone,* S. Hayat & A. Ahmad, (Eds.), 143-192, Springer Science + Business Media B.V., ISBN 978-007-0188-5, Heidelberg, Germany

Janeczko, A. & Swaczynowá, J. (2010). Endogenous brassinosteroids in wheat treated with 24-epibrassinolide. *Biologia Plantarum,* Vol.54, No.3, (September 2010), pp. 477-482 ISSN 0006-3134

Janeczko, A., Biesaga-Koscielniak, J., Oklesťkova, J., Filek, M., Dziurka, M., Szarek-Lukaszewska, G. & Koscielniak, J. (2010). Role of 24-epibrassinolide in wheat production: physiological effects and uptake. *Journal of Agronomy and Crop Science,* Vol. 196, No.4, (August 2010), pp. 311-321, ISSN 0931-2250

Janeczko, A., Koscielniak, J., Pilipowicz, M., Szarek-Lukaszewsa & G., Skoczowski, A. (2005). Protection of winter rape photosystem 2 by 24-epibrassinolide under cadmium stress. *Photosynthetica,* Vol.43, No.2, pp. 293-298, ISSN 0300-3604

Kagale, S., Divi, U.K., Krochko, J.E., Keller, W.A. & Krishna, P. (2007). Brassinosteroids confers tolerance in *Arabidopsis thaliana* and *Brassica napus* to a range of abiotic stresses. *Planta,* Vol.225, No.2, (January 2007), pp. 353-364, ISSN 0032-0935

Kamlar, M., Uhlík, O., Kohout, L., Harmatha, J & Macek, T. (2010). Steroid Phytohormones: Function, Mechanism of Action, Significance. *Chemické listy,* Vol.104, No.2, (February 2010), pp. 93-99, ISSN 0009-2770

Kang, Y.Y. & Guo, S.R. (2010). Role of brassinosteroids on horticultural crops. In: *Brassinosteroids: A Class of Plant Hormone,* S. Hayat & A. Ahmad, (Eds.), 269-288, Springer Science + Business Media B.V., ISBN 978-007-0188-5, Heidelberg, Germany

Khripach, V.A., Zhabinskii, V.N. & de Groot, A.E. (1999). *Brassinosteroids - A New Class of Plant Hormones,* Academic Press, ISBN 0-12-406360-8, San Diego, USA

Khripach, V.A., Zhabinskii, V.N. & de Groot, A.E. (2000). Twenty years of brassinosteroids: steroidal plant hormones warrant better crops for the XXI century. *Annals of Botany,* Vol.86, No.3, (June 2000), pp. 441-447, ISSN 0305-7364

Khripach, V.A., Zhabinskii, V.N & Khripach, N.B. (2003). New practical aspects of brassinosteroids and results of their ten-year agricultural use in Russia and Belarus. In: *Brassinosteroids, Bioactivity and Crop Productivity,* S., Hayat & A. Ahmad, (Eds.), 189-230, Kluwer Academic Publishers, ISBN 1-4020-1710-3, Dordrecht, The Netherlands

Krishna, P. (2003). Brassinosteroid-mediated stress responses. *Journal of Plant Growth Regulation,* Vol.22, No.4, (December 2003), pp. 289-297, ISSN 0721-7595

Kroutil, M., Hejtmánková, A. & Lachman, J. (2010a). Effect of spring wheat (*Triticum aestivum* L.) treatment with brassinosteroids on the content of cadmium and lead in plant aerial biomass and grain. *Plant, Soil and Environment,* Vol.56, No.1, (January 2010) pp. 43-50, ISSN 1214-1178

Kroutil, M., Hradecká, D., Kohout, L. & Lachman, J. (2010b). Quality and selected metals content of spring wheat (*Triticum aestivum* L.) grain after the treatment with brassinosteroids during cultivation. *Scientia Agriculturae Bohemica,* Vol.41, No.2, (June 2010), pp. 65-72, ISSN 0582-2343

Kulaeva, O.N., Burkhanova, E.A., Fedina, A.B., Khokhlova, V.A., Bokebayeva, G.A., Vorbrodt, H.M. & Adam, G. (1991). Effect of brassinosteroid on protein synthesis and plant-cell ultrastructure under stress condition. In: *Brassinosteroids: Chemistry, Bioactivity and Application,* H.G., Cutler, T., Yokota, G., Adam (Eds.), 155-163, ACS Symposium Ser., Vol. 474, American Chemical Society, ISBN 0-8412-2126-X, Washington, DC, USA

Mader, P., Száková, J. & Miholová, D. (1998). Classical dry ashing of biological and agricultural materials. Part II. Loses of analytes due to their retention in an insoluble residue. *Analusis,* Vol.26, No.3, (April 1998), pp. 121-129, ISSN 0365-4877

Nafie, E.M. & El-Khallal, S.M. (2000). Effect of brassinolide application on growth certain metabolic activities and yield of tomato. *Egyptian Journal of Physiological Sciences,* Vol.24, No.1, (January 2000), pp. 103-117, ISSN 0301-8660

Ogweno, J.O., Song, X.S., Shi, K., Hu, W.H., Mao, W.H., Zhou, Y.H., Yu, J.Q. & Nogues, S. (2008). Brassinosteroids alleviate heat-induced inhibition of photosynthesis by increasing carboxylation efficiency and enhancing antioxidant systems in *Lycopersicon esculentum, Journal of Plant Growth Regulation,* Vol.27, No.1, (March 2008), pp. 49–57, ISSN 0721-7595

Sakurai, A., Yokota, T. & Clouse, S.D. (1999). *Brassinosteroids - Steroidal Plant Hormones.* Springer-Verlag, Springer-Verlag, ISBN 4-431-70214-8, Tokyo, Japan

Sasse, J.M. (1999). Physiological actions of brassinosteroids. In: *Brassinosteroids – Steroidal Plant Hormones,* A., Sakurai, T., Yokota & S.D., Clouse (Eds.), 137-155, Springer-Verlag, ISBN 4-431-70214-8, Tokyo, Japan

Schilling, G., Schiller, C. & Otto, S. (1991). Influence of brassinosteroids on organ relations and enzyme activities of sugar beet plants, In: *Brassinosteroids. Chemistry, Bioactivity, and Application,* H.G., Cutler, T., Yokota, G. & Adam, (Eds.), 208-219, ACS Symposium Ser., Vol. 474, American Chemical Society, ISBN 0-8412-2126-X, Washington, DC, USA

Shahbaz, M. & Ashraf, M. (2007). Influence of exogenous application of brassinosteroid on growth and mineral nutrients of wheat (*Triticum aestivum* L.) under saline conditions. *Pakistan Journal of Botany,* Vol.39, No.2, (April 2007), pp. 513-522, ISSN 0556-3321

Shahbaz, M., Ashraf, M. & Athar, H.R. (2008). Does exogenous application of 24-epibrassinolide ameliorate salt induced growth inhibition in wheat (*Triticum aestivum* L.)? *Plant Growth Regulation,* Vol. 55, No.1, (January 2008), pp. 51-64 ISSN 0167-6903

Sharma, I., Pati, P.K. & Bhardwaj, R. (2010). Regulation of growth and antioxidant enzyme activities by 28-homobrassinolide in seedlings of Raphanus sativus L. under cadmium stress. *Indian Journal of Biochemistry & Biophysics,* Vol.47, No.3, (June 2010), pp. 172-177, ISSN 0301-1208

Sharma, P. & Bhardwaj, R. (2007a). Effects of 24-epibrassinolide on growth and metal uptake in *Brassica juncea* L. under copper metal stress. *Acta Physiologiae Plantarum,* Vol.29, No.3, (June 2007), pp. 259-263, ISSN 0137-5881

Sharma, P. & Bhardwaj, R. (2007b). Effect of 24-epibrassinolide on seed germination, seedling growth and heavy metal uptake in *Brassica juncea* L. *General and Applied Plant Physiology,* Vol.33, No.1-2, (June 2007), pp. 59-73, ISSN 1312-8183

Sharma, P., Bhardwaj, R., Arora, N. & Arora, H.K. (2007c). Effect of 28-homobrassinolide on growth, zinc metal uptake and antioxidative enzyme activities in *Brassica juncea* L. seedlings. *Brazilian Journal of Plant Physiology*, Vol.19, No.3, (July-September 2007), pp. 203-210, ISSN 1677-0420

Takematsu, T., Takeuchi, Y. & Choi, C.D. (1986). Effects of brassinosteroids on growth and yields of crops. In: *Brassinosteroids: Steroidal Plant Hormones*, A., Sakurai, T., Yokota & S.D., Clouse (Eds.), 137-161, Springer-Verlag, ISBN 4-431-70214-8, Tokyo, Japan

Upreti, K.K. & Murti, G.S.R. (2004). Effects of brassinosteroids on growth, nodulation, phytohormone content and nitrogenase activity in French bean under water stress. *Biologia Plantarum*, Vol.48, No.3, (September 2004), pp. 407-411, ISSN 0006-3134

Vardhini, B.V. & Rao, S.S.R. (2002). Acceleration of ripening of tomato pericarp discs by brassinosteroids. *Phytochemistry*, Vol.61, No.7, (December 2002), pp. 843-847, ISSN 0031-9422

Vlašánková, E., Kohout, L., Klemš, M., Eder, J., Reinöhl, V. & Hradilík, J. (2009). Evaluation of biological activity of new synthetic brassinolide analogs. *Acta Physiologiae Plantarum*, Vol. 31, No.5, (September 2009), pp. 987-993 ISSN 0137-5881

Worley, J.F. & Mitchell, J.W. (1971). Growth responses induced by brassins (fatty plant hormones) in bean plants. *Journal of the American Society for Horticultural Science*, Vol.96, No.3, pp. 270-273, ISSN 0003-1062

Xia, X.J., Huang, L.F., Zhou, Y.H., Mao, W.H., Shi, K., Wu, J.X., Asamim, T., Chen, Z.X. & Yu, J.Q. (2009). Brassinosteroids promote photosynthesis and growth by enhancing activation of Rubisco and expression of photosynthetic genes in *Cucumis sativus*, *Planta*, Vol.230, No.6, (November 2009), pp. 1185–1196, ISSN 0032-0935

Yuan, G,F., Jia, C.G., Li, Z., Sun, B., Zhang, L.P., Liu, N. & Wang, Q.M. (2010). Effect of brassinosteroids on drought resistance and abscisic acid concentration in tomato under water stress. *Scientia Horticulturae*, Vol.126, No.2, (September 2010), pp. 103-108, ISSN 0304-4238

Zadoks, J.C., Chang, T.T. & Konzak, C.F. (1974). A decimal code for the growth stages of cereals. *Weed Research*, Vol.14, No.6, pp. 415-421, ISSN 0043-1737

Zulo, M.A.T. & Adam, Q. (2002). Brassinosteroid phytohormones - structure, bioactivity and applications. *Brazilian Journal of Plant Physiology*, Vol.14, No.3, (September-December 2002), pp. 143-181, ISSN 1677-0420

18

Production of Enriched Biomass by Carotenogenic Yeasts - Application of Whole-Cell Yeast Biomass to Production of Pigments and Other Lipid Compounds

Ivana Marova[1], Milan Certik[2] and Emilia Breierova[3]
[1]*Brno University of Technology, Faculty of Chemistry, Centre for MaterialsResearch, Purkynova 118, 612 00 Brno,*
[2]*Slovak Technical University, Faculty of Chemical and Food Technology, Bratislava,*
[3]*Institute of Chemistry, Slovak Academy of Sciences, Bratislava,*
[1]*Czech Republic*
[2,3]*Slovak Republic*

1. Introduction

Yeasts are easily grown unicellular eukaryotes. They are ubiquitous microorganisms, occuring in soil, fresh and marine water, animals, on plants and also in foods. The environment presents for yeast a source of nutrients and forms space for their growth and metabolism. On the other hand, yeast cells are continuously exposed to a myriad of changes in environmental conditions. These conditions determine the metabolic activity, growth and survival of yeasts. Basic knowledge of the effect of environmental factors on yeast is important for understanding the ecology and biodiversity of yeasts as well as for control the yeast physiology in order to enhance the exploitation of yeasts or to inhibit or stop their harmful and deleterious activity.

The overproduction of some metabolites as part of cell stress response can be of interest to the biotechnology. For instance carotenogenic yeasts are well known producers of biotechnologically significant carotenoid pigments - astaxanthin, β-carotene, torulen, torularhodin and under stress conditions this carotenoid accumulation was reported to be increased. Knowledge of molecular mechanism of the carotenoid production stimulation can then lead to improvement of such biotechnological process. Red yeasts are able to accumulate not only carotenoids, but also ergosterol, unsaturated fatty acids, Coenzyme Q10 and other, which can contribute to the biomass enrichment. The use of this stressed biomass in feed industry could have positive effect not only in animal and fish feeds because of high content of physiologically active substances, but it could influence nutritional value and organoleptic properties of final products for human nutrition.

Yeast biomass, mainly in the form of *Saccharomyces cerevisiae*, represents the largest bulk production of any single-celled microorganism throughout the world. In addition to use of live yeast biomass for the leavening of bread dough, many other applications of yeast cells and yeast cell extracts have emerged. Most yeast biomass for industrial use is derived from *Saccharomyces cerevisiae*, but other yeasts have specific uses and may be grown on a range of substrates unavailable to *S.cerevisiae*. Some yeast strains are usable to industrial single-cell protein production from lignocellulose materials, methanol, n-alkanes, starch, oils and also other cheap carbon sources. Except compresses baker´s yeasts for baking, brewing, winemaking and distilling also other whole-cell yeast products are industrially used as animal feed, human and animal probiotics, as biosorbents for heavy metal sequestration and, also as nutritional trace element sources. Yeasts are rich sources of proteins, nucleic acids, vitamins and minerals but mostly with negligible levels of triglycerides.

Pigmented yeasts are used as feed and food colorants and, come of them, also as single cell oil producers. This chapter will be focused on controlled production of biomass and some interesting lipid metabolites of several non-traditional non-*Saccharomyces* yeast species. Growing interest in yeast applications in various fields coupled with significance of carotenoids, sterols and other provitamins in health and dietary requirements has encouraged "hunting" for more suitable sources of these compounds.

2. Production of enriched biomass by carotenoid-forming yeasts

2.1 Characterization of red (carotenogenic) yeasts

2.1.1 Taxonomy

Yeasts belong to the kingdom Fungi (Mycota) - a large group of eukaryotic organisms that includes microorganisms such as yeasts and moulds. Some species grow as single-celled yeasts that reproduce by budding or binary fission. Dimorphic fungi can switch between a yeast phase and a hyphal phase in response to environmental conditions. The fungal cell wall is composed of glucans and chitin. Another characteristic shared with plants includes a biosynthetic pathway for producing terpenes that uses mevalonic acid and pyrophosphate as chemical building blocks (Keller et al., 2005). Fungi produce several secondary metabolites that are similar or identical in structure to those made by plants. Fungi have a worldwide distribution, and grow in a wide range of habitats, including extreme environments such as deserts or areas with high salt concentrations or ionizing radiation, as well as in deep sea sediments. Some can survive the intense UV and cosmic radiation.

Around 100,000 species of fungi have been formally described by taxonomists, but the global biodiversity of the fungus kingdom is not fully understood. There is no unique generally accepted system at the higher taxonomic levels and there are frequent name changes at every level, from species upwards. Fungal species can also have multiple scientific names depending on their life cycle and mode (sexual or asexual) of reproduction. The 2007 classification of Kingdom Fungi is the result of a large-scale collaborative research. It recognizes seven phyla, two of which—the *Ascomycota* and the *Basidiomycota*—are contained within a branch representing subkingdom Dikarya (Hibbett, 2007).

The Ascomycota constitute the largest taxonomic group within the *Eumycota*. These fungi form meiotic spores called ascospores, which are enclosed in a special sac-like structure called an ascus. This phylum includes single-celled yeasts (e.g., of the genera *Saccharomyces, Kluyveromyces, Pichia,* and *Candida*), and many filamentous fungi living as saprotrophs, parasites, and mutualistic symbionts.

Some yeast species accumulate carotenoid pigments, such as β-carotene, torulene, and thorularodin which cause their yellow, orange and red colours and are therefore called red yeasts. Carotenogenic yeasts are a diverse group of unrelated organisms (mostly *Basidiomycota*) and the majority of the known species are distributed in four taxonomic groups: the *Sporidiobolales* and *Erythrobasidium* clade of the class *Urediniomycetes*, and *Cystofilobasidiales* and *Tremellales* of the class *Hymenomycetes* (Libkind et al., 2005). Along with the most known producer *Phaffia rhodozyma*, there is evidence of the capacity for carotene formation by other well-known pigmented yeasts of the genus *Rhodotorula* (order *Sporidiobolales*). The composition and amount of the carotenoid pigments in numerous natural isolates of the genera *Rhodotorula/ Rhodosporium* and *Sporobolomyces/Sporidiobolus* were studied in detail (Yurkov et al., 2008).

At this time the number of red yeasts species *Rhodotorula, Rhodosporidium, Sporidiobolus, Sporobolomyces, Cystofilobasidium, Kockovaella* and *Phaffia* are known as producers of carotene pigments. Many of these strains belong to oleaginous yeasts, some of them can effectively remove heavy metals from industrial effluents and detoxify certain pollutants. Studies with yeast mutants or carotenoid biosynthesis inhibitors have shown that carotenoid-deficient yeast strains are sensitive to free oxygen radicals or oxidizing environment, and that this sensitivity can be relieved by the addition of exogenous carotenoids (Davoli et al., 2004). The major yeast pigments are β-carotene, γ-carotene, torulene, torularhodin and astaxanthin (Dufosse, 2006).

2.1.2 Morphology and growth characteristics of main red yeast species

The genus *Rhodotorula* includes three active species; *Rhodotorula glutinis, Rhodotorula minuta* and *Rhodotorula mucilaginosa* (formerly known as *Rhodotorula rubra*) (Hoog et al., 2001). Colonies are rapid growing, smooth, glistening or dull, sometimes roughened, soft and mucoid (Figures 1 - 3). They are cream to pink, coral red, orange or yellow in color. Blastoconidia that are unicellular, and globose to elongate in shape are observed. These blastoconidia may be encapsulated. Pseudohyphae are absent or rudimentary. Hyphae are absent. *Rhodotorula glutinis* often called "pink yeast" is a free living, non-fermenting, unicellular yeast found commonly in nature. *Rhodotorula* is well known for its characteristic carotenoids "torulene, torularhodin and β-carotene. *Rhodotorula glutinis* is also reported to accumulate considerable amount of lipids (Perier et al., 1995).

The genus *Sporobolomyces* contains about 20 species. The most common one is *Sporobolomyces roseus* and *Sporobolomyces salmonicolor* (Hoog et al., 2001). *Sporobolomyces* colonies grow rapidly and mature in about 5 days. The optimal growth temperature is 25-30°C. The colonies are smooth, often wrinkled, and glistening to dull. The bright red to orange color of the colonies is typical and may resemble *Rhodotorula spp. Sporobolomyces* produces yeast-like cells, pseudohyphae, true hyphae, and ballistoconidia. The yeast-like cells (blastoconidia, 2-12 x 3-35 μm) are the most common type of conidia and are oval to elongate in shape. Pseudohyphae and true hyphae are often abundant and well-developed. Ballistoconidia are one-celled, usually reniform (kidney-shaped), and are forcibly discharged from denticles located on ovoid to elongate vegetative cells (Figures 4, 5) .

Among yeasts, *Rhodotorula* species is one of main carotenoid-forming microorganisms with predominant synthesis of β-carotene, torulene and torularhodin (Davoli et al., 2004; Libkind and van Broock, 2006; Maldonade et al., 2008). *Cystofilobasidium* (Figure 6) and *Dioszegia* were also found to synthesize these three pigments. Some of yeast carotenoids are modified with oxygen-containing functional groups. For example, astaxanthin is almost exclusively formed by *Phaffia rhodozyma* (*Xanthophyllomonas dendrorhous;* Frengova & Beshkova, 2009).

Nevertheless, although there are many strategies for stimulation of carotene biosynthetic machinery in yeasts, attention is still focused on unexplored yeast's habitats for selection of hyper-producing strains what is the important step towards the design and optimization of biotechnological process for pigment formation (Libkind & van Broock, 2006; Maldonade et al., 2008).

Studies on a number of fungi, including *Neurospora crassa, Blakeslea trispora, Mucor hiemalis, Mucor circinelloides* and *Phycomyces blakesleeanus* (oleaginous fungi with carotene-rich oil) have been published over the last twenty years (Dufosse, 2006). Fungal carotenoid content is relatively simple with dominat levels of β-carotene. Recent work with dimorphic fungal mutants *M. circinelloides* and *Blakeslea trispora* (Cerda-Olmedo, 2001) showed that these strains could be useful in a biotechnological production of carotenoids in usual fermentors.

In order to study yeast physiology under different conditions, it is important to know so called "reference parameters" which these yeasts possess under optimal condition. Red or carotenogenic yeasts are well known producers of valuable carotenoids. On agar plates they form characteristic yellow, orange and red coloured colonies. Red yeast can be of ellipsoidal or spherical shape (Figures 1 - 6). Under optimal conditions (28 °C, 100 rpm, permanent lightening) they are able to grow up in 5 to 7 days. The growth curve of *Rhodotorula glutinis* CCY 20-2-26 as well as other studied red yeast exhibited similarly typical two-phase character with prolonged stationary phase (Figures 7, 8) probably due to the ability of the yeast cells to utilize lipid storages formed during growth as additional energy source (Marova et al., 2010). The production of carotenoids during growth fluctuated and some local maxima and minima were observed. The maximum of beta-carotene production was obtained in all strains in stationary phase after about 80 hours of cultivation.

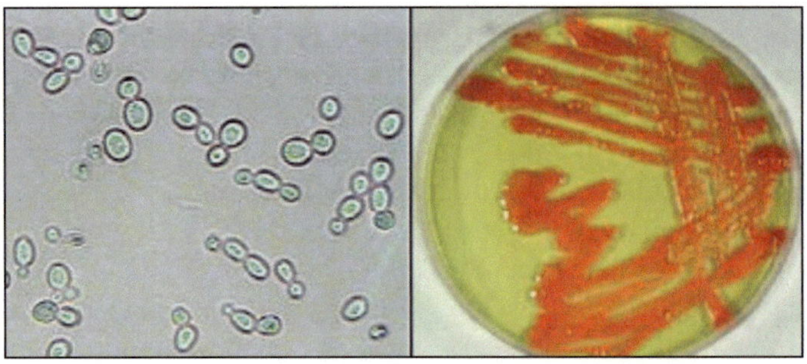

Fig. 1. Microscopic image and streak plate of *Rhodotorula glutinis*

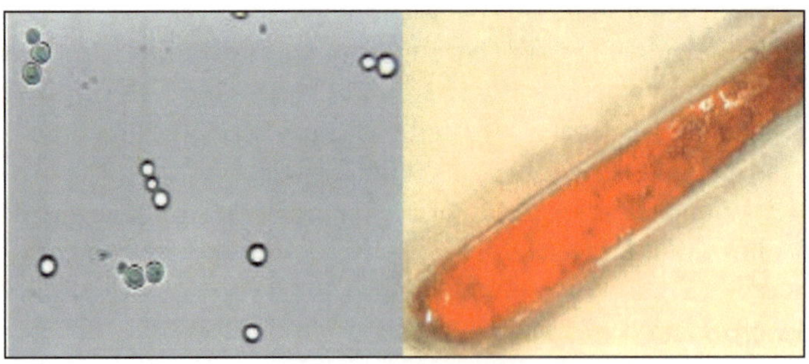

Fig. 2. Microscopic image and streak plate of *Rhodotorula rubra*

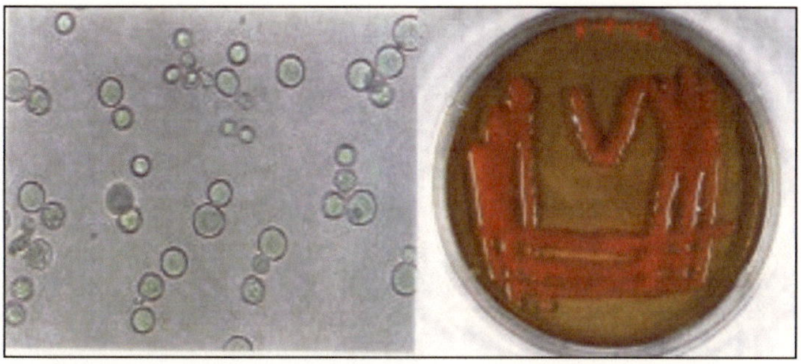

Fig. 3. Microscopic image of *Rhodotorula aurantiaca*

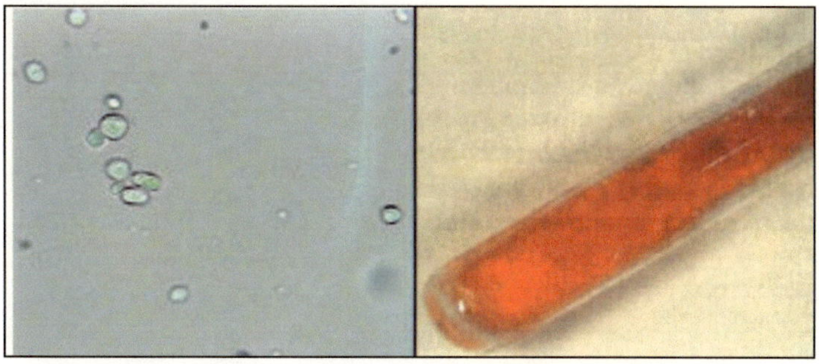

Fig. 4. Microscopic image and streak plate of *Sporobolomyces roseus*

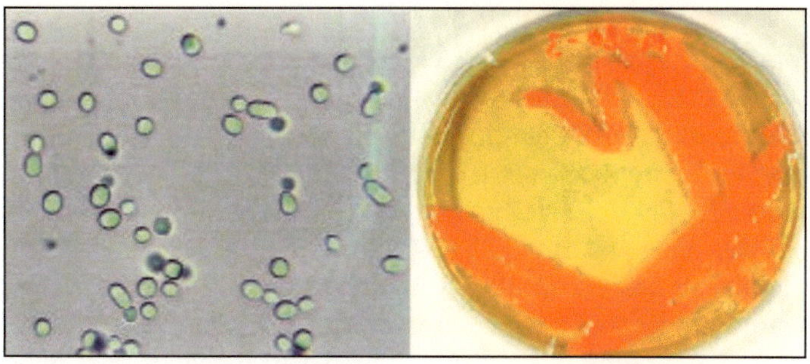

Fig. 5. Microscopic image and streak plate of *Sporobolomyces shibatanus*

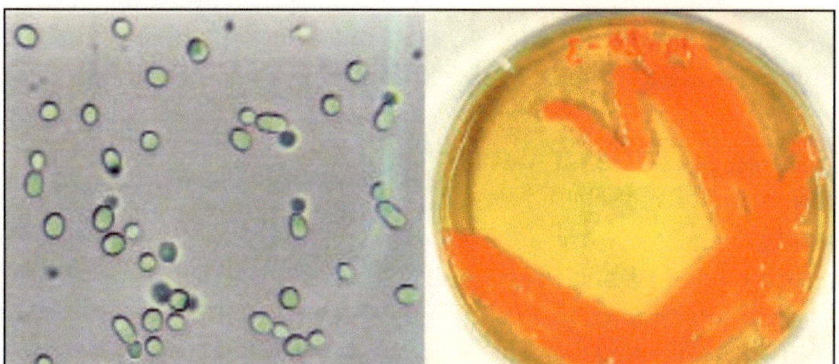

Fig. 6. Microscopic image and streak plate of *Cystofilobasidium capitatum*

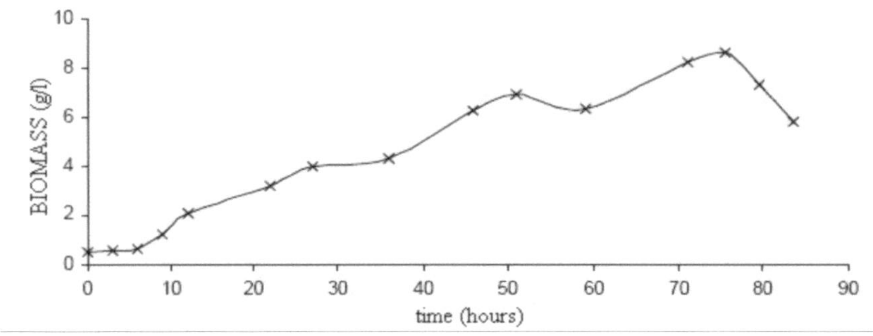

Fig. 7. Growth curve of *Rhodotorula glutinis*

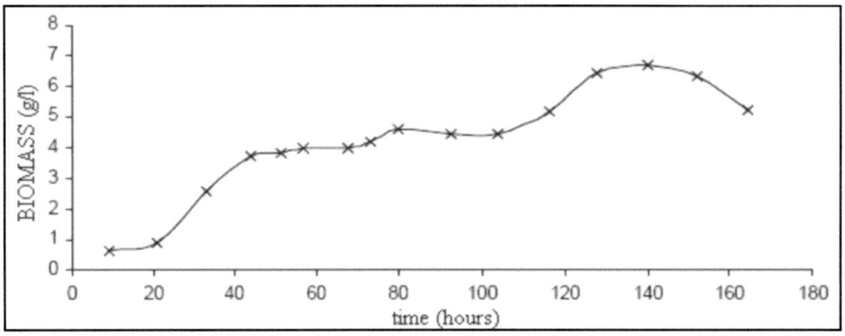

Fig. 8. Growth curve of *Sporobolomyces shibatanus*

Comparison of presented growth curves led to some partial conclusions about growth of red yeasts (Marova et al., 2010). All tested strains reached stationary phase after about 50 hours of cultivation. All strains also exhibited prolonged stationary phase with at minimum one, more often with several growth maxima. First growth maximum was observed in all strains after about 80 hours of growth. In strains followed for longer time than 100 hours additional growth maximum was observed after 105 - 140 hours. Carotenogenic yeasts probably utilize some endogenous substrates accumulated at the beginning of stationary phase. Growth maxima are mostly accompanied with carotenoid production maxima mainly in first 90 hours of cultivation. Cultivation in production media in presence of some stress factors or using waste substrates is recommended to carry out to first production maximum (about 80 - 90 hours) to eliminate potential growth inhibiton caused by nutrient starvation or toxic effect of stress. Longer cultivation can be also complicated by higher ratio of dead and living cells and in semi-large-scale and large-scale experiments also with higher production costs.

2.2 The main features of red yeast metabolism

Metabolism is the sum of cellular chemical and physical activities. It involves chemical changes to reactants and the release of products using well-established pathways regulated at many levels. Knowledge of such regulation in yeasts is crucial for exploitation of yeast cell physiology in biotechnology (Talaro & Talaro, 2001). At controlled cultivation conditions oleaginous red yeasts could be a good source (producer) of lipidic primary metabolites as neutral lipids, phospholipids and fatty acids and ergosterol, which is integrate part of yeast biomembranes.

Secondary metabolism is a term for pathways of metabolism that are not absolutely required for the survival of the organism. Examples of the products include antibiotics and pigments. The induction of secondary metabolism is linked to particular environmental conditions or developmental stages. When nutrients are depleted, microorganisms start producing an array of secondary metabolites in order to promote survival (Mann, 1990). Filamentous fungi and yeasts show a relatively low degree of cellular differentiation, but still they express a complex metabolism resulting in the production of a broad range of secondary metabolites and extracellular enzymes. This very high metabolic diversity has been actively exploited for many years. In terms of biotechnological application fungi and yeast have the advantage of being relatively easy to grow in fermenters and they are therefore well-suited for large-scale industrial production. Biomass enriched by suitable mixture of primary and secondary metabolites can be used too, mainly in feed and food applications (Mann, 1990, Walker 1998).

In general, biosynthesis of individual metabolites is governed by the levels and activities of enzymes employed to the total carbon flux through the metabolic system. Efficiency of that flow depends on the cooperation of individual pathways engaged in this process and which pathway is suppressed or activated varies with the growth medium composition, cultivation conditions, microbial species and their developmental stage. Because overall yield of metabolites is directly related to the total biomass yield, to keep both high growth rates and high flow carbon efficiency to carotenoids by optimal cultivation conditions is essential in order to achieve the maximal metabolite productivity (Certik et al., 2009).

2.2.1 The isoprenoid pathway

Isoprenoids occur in all eukaryotes. Despite the astonishing diversity of isprenoid molecules that are produced, there is a great deal of similarity in the mechanisms by which different species synthesize them. In fact, the initial phase of isoprenoid synthesis (the synthesis of isopentenyl pyrophosphate) appears to be identical in all of the species in which this process has been investigated. Thus, some early steps of isporenoid pathway could be used for genetic modification.

Starting with the simple compounds acetyl-CoA, glyceraldehyde-3-phopsphate, and pyruvate, which arise via the central pathawys of metabolism, the key intermediate isopentenyl diphosphate is formed by two independent routes. It is then converted by bacteria, fungi, plants and animals into thousands of different naturally occuring products. In fungi, carotenoids are derived by sequnce reactions via the mevalonate biosynthetic pathway. The main product 3-hydroxy-3-methylglutaryl-CoA (HMG-CoA) is finaly reduced to the mevalonic acid. This two-step reduction of HMG-CoA to mevalonate is highly controlled and is also a major control factor of sterol synthesis (Metzler, 2003). From prenyl diphosphates of different chain lengths, specific routes branch off into various terpenoid end products (Figure 9).

2.2.2 Carotenoid biosynthesis

Carotenoids are synthesized in nature by plants and many microorganisms. In addition to very few bacterial carotenoids with 30, 45, or 50 carbon atoms, C40-carotenoids represent the majority of the more than 600 known structures. Two groups have been singled out as the most important: the carotenes which are composed of only carbon and hydrogen; and the xanthophylls, which are oxygenated derivatives (Frengova & Beshkova, 2009). In the later, oxygen can be present as OH groups, or as oxy-groups or in a combination of both (as in astaxanthin). Hydroxy groups at the ionone ring may be glycosylated or carry a glycoside fatty acid ester moiety. Furthermore, carotenoids with aromatic rings or acyclic structures with different polyene chains and typically 1-methoxy groups can also be found. Typical fungal carotenoids possess 4-keto groups, may be monocyclic, or possess 13 conjugated double bonds (Britton et al., 1998).

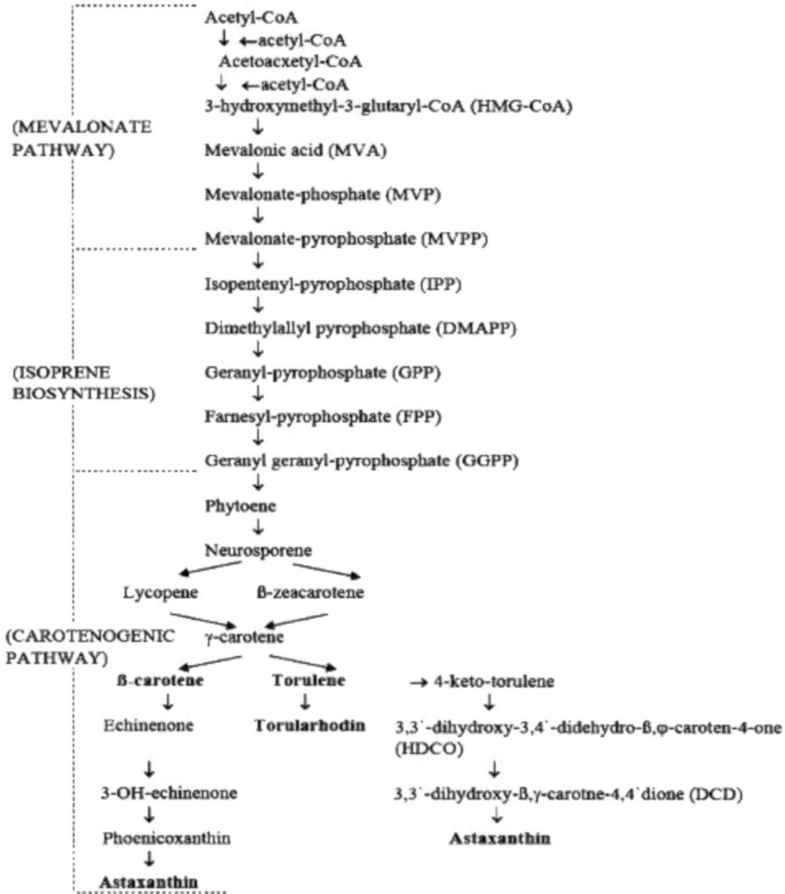

Fig. 9. Biosynthetic pathways from acetyl-CoA to β-carotene, torulene and torularhodin in *Rhodotorula* species and astaxanthin in *P. rhodozyma/X. dendrorhous* (Frengova & Beshkova, 2009)

All carotenoids are derived from the isoprenoid or terpenoid pathway. Carotenoids biosynthesis pathway commonly involves three steps: (i) formation of isopentenyl pyrophosphate (IPP), (ii) formation of phytoene and (iii) cyclization and other reactions of lycopene (Armstrong & Hearst, 1996). Before polyprenyl formation begins, one molecule of IPP must be isomerized to DMAPP. Condensation of one molecule of dimethylallyl diphosphate (DMADP) and three molecules of isopentenyl diphosphate (IDP) produces the diterpene geranylgeranyl diphosphate (GGDP) that forms one half of all C40 carotenoids. The head to head condensation of two GGDP molecules results in the first colorless carotenoid, phytoene. As Figure 9 shows, phytoene synthesis is the first committed step in C40-carotenoid biosynthesis (Britton et al., 1998; Sandmann, 2001). Subsequent desaturation reactions lengthen the conjugated double bond system to produce neurosporene or lycopene (Schmidt-Dannert, 2000).

Following desaturation, carotenoid biosynthesis branches into routes for acyclic and cyclic carotenoids. In phototrophic bacteria acyclic xanthophylls spheroidene or spheroidenone and spirilloxanthin, respectively are formed (Figure 9). Synthesis of cyclic carotenoids involves cyclization of one or both end groups of lycopene or neurosporene. Typically, β-rings are introduced, but formation of ε-rings is common in higher plants and carotenoids with γ-rings are found, for example, in certain fungi. Most cyclic carotenoids contain at least one oxygen function at one of the ring carbon atoms. Cyclic carotenoids with keto-groups at C4(C4´) and/or hydroxy groups at C3(C3´) (e.g. zeaxanthin, astaxanthin, echinenone and lutein) are widespread in microorganisms and plants (Schmidt-Dannert, 2000).

2.2.3 Ergosterol biosynthesis

Ergosterol, one of the most important components in fungal membranes, is involved in numerous biological functions, such as membrane fluidity regulation, activity and distribution of integral proteins and control of the cellular cycle. Ergosterol pathway is fungal-specific; plasma membranes of other organisms are composed predominantly of other types of sterol. However, the pathway is not universally present in fungi; for example, *Pneumocystis carinii* plasma membranes lack ergosterol. In *S. cerevisiae,* some steps in the pathway are dispensible while others are essential for viability (Tan et al., 2003).

Biosynthesis of ergosterol similarly to carotenoids and other isoprenoid compounds (e.g. ubiquinone), is derived from acetyl-CoA in a three-stage synthehtic process (Metzler, 2003). Stage one is the synthesis of isopenthenyl pyrophosphate (IPP), an activated isoprene unit that is the key building block of ergosterol. This step is identical with mevalonate pathway (Figure 9). Stage two is the condensation of six molecules of IPP to form squalene. In the stage three, squalene cyclizes in an astounding reaction and the tetracyclic product is subsequently converted into ergosterol. In the ergosterol pathway, steps prior to squalene formation are important for pathway regulation and early intermediates are metabolized to produce other essential cellular components (Tan et al, 2003). It should be noted that isoprenoid pathway is of great importance in secondary metabolism. Combination of C5 IPP units to squalene exemplifies a fundamental mechanism for the assembly of carbon skeletons in biomolecules. A remarkable array of compounds is formed from IPP, the basic C5 building block. Several molecules contain isporenoid side chains, for example Coenzyme Q10 has a side chain made ud of 10 isporene units.

2.2.4 Gene regulation of isoprenoid pathway branches

The isoprenoid pathway in yeasts is important not only for sterol biosynthesis but also for the production of non-sterol molecules, deriving from farnesyl diphosphate (FPP), implicated in N-glycosylation and biosynthesis of heme and ubiquinones. FPP formed from mevalonate in a reaction catalyzed by FPP synthase (Erg20p). In order to investigate the regulation of Erg20p in *Saccharomyces cerevisiae,* a two-hybrid screen was used for its searching and five interacting proteins were identified. Subsequently it was showed that Yta7p is a membrane-associated protein localized both to the nucleus and to the endoplasmic reticulum. Deletion of Yta7 affected the enzymatic activity of cisprenyltransferase (the enzyme that utilizes FPP for dolichol biosynthesis) and the cellular levels of isoprenoid compounds. Additionally, it rendered cells hypersensitive to lovastatin, an inhibitor of 3-hydroxy-3-methylglutaryl coenzyme A reductase (HMGR) that acts upstream of FPP synthase in the isoprenoid pathway. While HMGR is encoded by two genes, HMG1 and HMG2, only HMG2 overexpression was able to restore growth of the yta7- cells in the presence of lovastatin. Moreover, the expression level of the *S. cerevisiae* YTA7 gene was altered upon impairment of the isoprenoid pathway not only by lovastatin but also by zaragozic acid, an inhibitor of squalene synthase (Kuranda et al., 2009).

All enzymes involved in carotenoid biosynthesis are membrane-associated or integrated into membranes. Moreover, carotenoid biosynthesis requires the interaction of multiple gene products. At present more than 150 genes, encoding 24 different *crt* enzymes involved in carotenogenic branch of isoprenoid pathway, have been isolated from bacteria, plants, algae and fungi. The availability of a large number of carotenogenic genes makes it possible to modify and engineer the carotenoid biosynthetic pathways in microorganisms. A number of genetically modified microbes, *e.g. Candida utilis, Escherichia coli, Saccharomyces cerevisiae, Zymomonas mobilis*, etc. have been studied for carotenoid production (Wang et al. 2000; Schmidt-Dannert, 2000; Lee & Schmidt-Dannert, 2002; Sandmann 2001). However, lack of sufficient precursors (such as IDP, DMADP and GGDP) and limited carotenoid storage capability is the main task how to exploate these organisms as commercial carotenoid producers. Therefore, effort has been focuced on increasing the isoprenoid central flux and levels of carotenoid precursors. For example, overexpression of the IDP isomerase (*idi* - catalyzes the isomerization of IDP to DMAP) together with an archaebacterial multifunctional GGDP synthase (*gps* - converts IDP and DMADP directly to GGDP) resulted in a 50-fold increase of astaxanthin production in *E. coli* (Wang et al., 2000).

By combination of genes from different organisms with different carotenoid biosynthetic branches, novel carotenoids not found in any other pathway can be synthesized. Most *Mucor* species accumulate β-carotene as the main carotenoid. The crtW and crtZ astaxanthin biosynthesis genes from *Agrobacterium aurantiacum* were placed under the control of *Mucor circinelloides* expression signals. Transformants that exhibited altered carotene production were isolated and analyzed. Studies revealed the presence of new carotenoid compounds and intermediates among the transformants (Papp et al., 2006). *Fusarium sporotrichioides* was genetically modified for lycopen production by redirecting of the isoprenoid pathway toward the synthesis of carotenoids and introducing genes from the bacterium *Erwinia uredovora* (Leathers et al, 2004). Carotenoid biosynthetic pathway of astaxanthin producers of *Phaffia/Xanthophyllomyces* strains has also been engineered and several genes, such as phytoene desaturase, isopentenyl diphosphate isomerase and epoxide hydrolase were isolated and expressed in *E. coli* (Verdoes et al., 2003; Lukacz, 2006).

2.3 Some natural factors affecting growth and production of metabolites in red yeasts

2.3.1 Nutrition sources

Cellular organisms require specific internal conditions for optimal growth and function. The state of this internal milieu is strongly influenced by chemical, physical and biological factors in the growth environment. Understanding yeast requirements is important for successfull cultivation of yeast in the laboratory but also for optimalization of industrial fermentation process (Walker, 1998). Elemental composition of yeast cell gives a broad indication as to the nutritional reguirements of the yeast cell. Yeasts acquire essential elements from their growth environment from simple food sources which need to be available at the macronutrient level (approx. 10^{-3} M) in the case of C, H, O, N, P, K, Mg and S or at the micronutrient level (approx. 10^{-6} M) in the case of trace elements. Yeasts are chemoorganotrophs as they use organic compounds as a source of carbon and energy. Yeasts can use a wide variety of substances as nutrient sources. Decreasing availability of one substrate can, in many instances, be compensated by the utilisation of another (Xiao, 2005).

When a single essential nutrient becomes limiting and eventually absent, the cellular proliferative machinery is efficiently shut down and a survival program is launched. In the absence of any one of the essential nutrients, yeast cells enter a specific, non-proliferative state known as stationary phase, with the ultimate aim of surviving the starvation period. In the presence of a poor carbon source, starvation for nitrogen induces sporulation and in the presence of a good carbon source stimulates pseudohyphal growth (Gasch & Werner-Washburne, 2002). Starvation is a complex, albeit common, stress for microorganisms. The nutrients for which a cell can be starved include carbon and nitrogen, with other elements such as phosphate, sulphur, and metals being less commonly evaluated.

The environment presents for yeasts a source of nutrients and forms space for their growth and metabolism. On the other hand, yeast cells are continuously exposed to a myriad of changes in environmental conditions (referred to as environmental stress). These conditions determine the metabolic activity, growth and survival of yeasts. Basic knowledge of the effect of environmental factors on yeast is important for understanding the ecology and biodiversity of yeasts as well as to control the environmental factors in order to enhance the exploitation of yeasts or to inhibit or stop their harmful and deleterious activity (Rosa & Peter, 2005).

In order to improve the yield of carotenoid pigments and subsequently decrease the cost of this biotechnological process, diverse studies have been performed by optimizing the culture conditions including nutritional and physical factors. Factors such as nature and concentration of carbon and nitrogen sources, minerals, vitamins, pH, aeration, temperature, light and stress have a major influence on cell growth and yield of carotenoids. Because carotenoid biosynthesis is governed by the levels and activities of enzymes employed to the total carbon flux through the carotenoid synthesizing system, the efficient formation of carotenoids can also be achieved by construction of hyperproducing strains with mutagenesis and genetic/metabolic engineering (Frengova & Beshkova, 2009).

The efficiency of the carbon source conversion into biomass and metabolites, and the optimization of the growth medium with respect to its availability and price has been subject of intensive studies. Numerous sources including pentoses and hexoses, various disaccharides, glycerol, ethanol, methanol, oils, n-alkanes, or wide variety of wastes derived from agricultural have been considered as potential carbon sources for biotechnological production of carotenoids.Carotenoid pigment accumulation in most yeasts starts in the late logarithmic phase and continues in the stationary phase (typically for secondary metabolites), and the presence of a suitable carbon source is important for carotenoid biosynthesis during the nongrowth phase. Yeasts can synthesize carotenoids when cultivated in synthetic medium, containing various simple carbon sources, such as glucose, xylose, cellobiose, sucrose, glycerol and sorbitol. Studies on carotenogenesis have led to a growing interest in using natural substrates and waste products from agriculture and food industry: grape juice, grape must, peat extract and peat hydrolysate, date juice, hydrolyzed mustard waste isolates, hemicellulosic hydrolysates (Parajo et al., 1998), hydrolyzed mung bean waste flour, sugar cane juice, sugar cane and sugar-beet molasses, corn syrup, corn

hydrolysate, milk whey. In recent years, raw materials and by-products of agro-industrial origin have been proposed as low-cost alternative carbohydrate sources for microbial metabolite production, with the view of also minimizing environmental and energetic problems related to their disposal (Frengova & Beshkova, 2009).
The chemical composition and concentration of nitrogen source in medium might also be means of physiological control and regulation of pigment metabolism in microorganisms. Several inorganic and organic nitorgen sources as well as flour extracts and protein hydrolysates have been studied for improvement of carotenoid production. However, it seems that variation in carotene content in yeasts with regard to N-source used in a medium and the rate of pigment production is influenced by the products of catabolism of the nitrogen source rather than being the results of direct stimulation by the nitrogen compound itself (Certik et al., 2009, Somashekar & Joseph, 2000).

2.3.2 Environmental stress

Single-celled organisms living freely in nature, such as yeasts, face large variations in their natural environment. Environmental conditions that threaten the survival of a cell, or at least prevent it from performing optimally, are commonly referred to as cell stress. These environmental changes may be of a physical or chemical nature: temperature, radiation, concentrations of solutes and water, presence of certain ions, toxic chemical agents, pH and nutrient availability. In nature, yeast cells often have to cope with fluctuations in more than one such growth parameter simultaneously (Hohman & Mager, 2003). In industry, yeast stress has several very important practical implications. In brewing, for example, if yeast is nutrient-starved during extended periods of storage, certain cell surface properties such as flocculation capability are deleteriously affected (Walker, 1998).
Carotenogenic yeasts are considered to be ubiquitous due to its world-wide distribution in terrestrial, freshwater and marine habitats, and to its ability to colonize a large variety of substrates. They can assimilate various carbon sources, including waste materials as cheap substrates. The red yeast is able to grow under a wide range of initial pH conditions from 2.5 to 9.5 and over a wide range of temperatures from 5 to 26°C (Libkind et al., 2008; Latha et al., 2005). The most important consenquence of environmental stress in red yeast is stimulation of carotenoid and other secondary (as well as primary) metabolite production. Changes of ergosterol production, lipid content, glycerol and trehalose as well as membrane remodeling are described as a response to stress (Hohman & Mader, 2003). Carotenoid pigments accumulation in most yeasts starts in the late logarithmic phase and continues in the stationary phase and is highly variable. Carotenoid production depends on differences between strains of the same species and is strongly influenced by the cultivation conditions. Addition of stress factors into cultivation medium led to different changes of growth according to the yeast species, type of stress factor or growth phase, in which stress factors were added (Marova et al., 2004).
Carotenogenesis in many organisms is regulated by light. However, the intensity and protocol of illumination varies with the microorganism. Temperature is another important factor affecting the performance of cells and product formation. The effect of temperature depends on the species specificity of the microorganism and often manifests itself in quantity variations of synthesized carotenoids. It was reported that lower temperatures (25°C) seemed to favor synthesis of β-carotene and torulene, whereas higher temperatures (35°C) positively influenced torularhodin synthesis by *R. glutinis* (Frengova & Beshkova,

2009). The effect of aeration is dependent on the species of the microorganism. The aeration influenced not only the amount of carotenoids produced, but also the composition of individual pigments making up the total carotenoids (Simova et al., 2004). At higher aeration, the concentration of total carotenoids increased relative to the biomass and fatty acids in *R. glutinis*, but the composition of carotenoids (torulene > β-carotene > γ-carotene > torularhodin) remained unaltered. In contrast, *S. roseus* responds to enhanced aeration by a shift from the predominant β-carotene to torulene and torularhodin (Davoli, 2004). Also other inducers of oxidative stress such as irradiation and free radical generators have a significant effect on the carotenoid production. By UV mutagenesis of the pink yeast *R. glutinis* the yellow colored mutant 32 was obtained which produced 24-fold more total carotenoids (2.9 mg/g dry cells) and 120-fold more β-carotene than the wild-type in a much shorter time (Bhosale & Gadre, 2001). Production of carotenoids by *Rhodotorula glutinis* cells grown under oxidative stress was about 5–6 times higher than in wild-type (Marova et al., 2004; Marova et al., 2010).

Tolerance to deleterious factors (e.g., low pH) refers to a microorganism's ability to survive a stress. This phenomenon is described as adaptive response, induced tolerance, habituation, acclimatization or stress hardening. Once cells have been challenged with a mild stress, they become more resistant to severe stress. Also exposure to one type of stress has been demonstrated to lead to tolerance to other types of stress as well (cross-protection) (Hohman & Mager, 2003). When cells are shifted to stress environments, they respond with changes in the expression of hundreds or thousands of genes, revealing the plasticity of genomic expression. Some of the expression changes are specific to each new environment, while others represent a common response to environmental stress. Comparative analysis of the genomic expression responses to diverse environmental changes revealed that the expression of roughly 900 genes (around 14% of the total number of yeast genes) is stereotypically altered following stressful environmental transitions. The functions of these gene products may protect critical aspects of the internal milieu, such as energy reserves, the balance of the internal osmolarity and oxidation-reduction potential, and the integrity of cellular structures. The protection of these features by the stress gene products likely contributes to the cross-resistance of yeast cells to multiple stresses, in which cells exposed to a mild dose of one stress become tolerant of an otherwise-lethal dose of a second stressful condition (Hohman & Mager, 2003; Gasch & Werner-Washburne, 2002; Gasch et al., 2000).

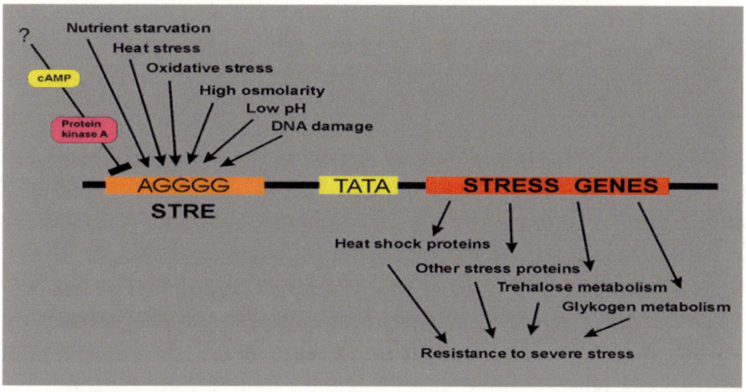

Fig. 14. Factors controlling stress response elements (STREs) and effects triggered by STRE activation in yeast (Walker, 1998)

A critical component of cell survival is maintaining a viable energy source. Glucose is the preferred carbon source in yeast, and upon stress, the cell induces a variety of genes that affect glucose metabolism. This includes genes encoding glucose transporters that serve to import external glucose into the cell and glucose kinases that activate the sugar for subsequent catabolism. In response to stressful environments, the fate of glucose is divided between trehalose synthesis, glycogen storage, ATP synthesis through glycolysis, and NADPH regeneration by the pentose phosphate shuttle (Hohman & Mager, 2003).

2.4 Strategies for improvement of carotenoid-synthesizing strains

2.4.1 Media compostion and cultivation mode

The production biotechnological process proceeds essentially in two stages: fermentation and product recovery. An important aspect of the fermentation process is the development of a suitable culture medium to obtain the maximum amount of desired product. In recent years, cheap raw materials and by-products of agro-industrial origin have been proposed as low-cost alternative carbohydrate sources for microbial metabolite production, with the view also of minimizing environmental and energetic problems related to residues and effluent disposal. For fermentation, seed cultures are produced from the original strain cultures and subsequently used in an aerobic submerged batch fermentation to produce a biomass rich in carotene pigment and other additional metabolites, e.g. ergosterol, metal ions etc. In the whole-cell strategy product isolation is not necessary and, moreover, complex biotechnological product in the form of slightly modified biomass could be obtained.

The traditional batch production system has the disadvantage of inducing the Crabtree effect (characterized by the synthesis of ethanol and organic acids as fermentation products), due to high concentrations of initial sugars, diminishing pigment and biomass yield. The strategy for solving this problem is the fed-batch culture. Maximum astaxanthin production (23.81 mg/l) by *P. rhodozyma* was achieved in fed-batch fermentation with constant pH = 6.0, 4.8 times greater that the one obtained in a batch culture and the biomass concentration (39.0 g/l) was 5.3 times higher than that in the batch culture (Ramirez et al., 2006). The maximum astaxanthin concentration by *X. dendrorhous* at fed-batch fermentation with pH-shift control strategy reached 39.47 mg/l, and was higher by 20.2 and 9.0% than that of the batch and fed-batch fermentation, respectively, with constant pH = 5.0. However, the maximal cell density at fed-batch fermentation with pH-shift control was 17.42 g dry cells/l, and was lower by 2.0% than that of fed-batch fermentation with constant pH = 5.0. As a result of the two stage fed-batch culture *P. rhodozyma*, cell and astaxanthin concentrations reached 33.6 g/l and 16.0 mg/l, respectively, which were higher when compared with batch culture. The final specific astaxanthin concentration (mg/g dry wt of cells) in the second stage was ca. threefold higher than that in the first stage and 1.5-fold higher than that in the dissolved oxygen controlled batch culture, indicating that the astaxanthin production was enhanced mush more in the second stage than in the first stage (Hu et al., 2007).

The astaxanthin production was enhanced by a high initial C/N ratio in the medium (second stage), whereas a lower C/N ratio was suitable for cell growth (first stage). A significant increase (54.9%) in astaxanthin production by *X. dendrorhous* was achieved in pulse fed-batch process when compared with batch process. The astaxanthin concentration was 33.91 mg/l in pulse fed-batch when compared with 30.21 mg/l in constant glucose fed-batch and 21.89 mg/l in batch fermentation. In contrast with this strain producing high yields of biomass and astaxanthin in pulse fed-batch process, another strain of *P. rhodozyma* demonstrated high astaxanthin-synthesizing activity during continuous fed-batch process (Hu et al., 2005). The utilization of continuous feeding showed to be the most efficient feeding method in fed-batch processes, as it did not lead to a reduction in the cellular astaxanthin concentration, as observed in the pulsed feeding. In the pulsed and continuous fed-batch processes, a cellular astaxanthin concentration of 0.303 mg/g biomass and 0.387 mg/g biomass, an astaxanthin concentration of 5.69 and 7.44 mg/l, a biomass concentration of 18.7 and 19.3 g/l were obtained, respectively.

Temperature was reported to control changes in enzyme activities that regulate metabolic activity in microorganisms. For example, *Rhodotorula glutinis* biosynthesized β-carotene more efficiently at lower temperature, whereas increased torulene formation was accompanied by higher temperature (Bhosale & Gadre, 2002). The reason might be found in γ-carotene that acts as the branch point of carotenoid synthesis. Subsequent dehydrogenation and decarboxylation leading to torulene synthesis is known to be temperature dependent since the respective enzymes are less active at lower temperature compared to the activity of β-carotene synthase. This is probable reason for an increase in the proportion of β-carotene at lower temperature in *Rhodotorula glutinis*. The moderately psychrophilic yeast *Xanthophyllomyces dendrorhous* also displayed a 50% increase in total carotenoids at low temperatures with elevated levels of astaxanthin (Ducrey Sanpietro & Kula, 1998).

Fed-batch co-cultures *R. glutinis–D. castellii* gave a volumetric production of 8.2 mg total carotenoid/l, about 150% of that observed in batch co-cultures and biomass concentration of 9.8 g/l which was about two times higher when compared with batch fermentation (Buzzini, 2001). The fedbatch technique maximized the specific growth rate of *R.glutinis*, resulted in higher biomass and minimized substrate inhibition of pigment formation. Molasses in the fed-batch mode led to increased biomass by 4.4- and 7-fold in double- and triple-strength feed, respectively when compared with 12.2 g/l biomass in batch fermentation. *R. glutinis* also produced a very high carotenoid concentration for double- and triple-strength feed supplement (71.0 and 185.0 mg/l, respectively), and was higher 2- and 3.7-fold of that observed in batch fermentation (Frengova & Beshkova, 2009).

2.4.2 Specific supplements and exogenous factors enhancing metabolic activity of red yeasts

There have been several reports on the enhancement of volumetric production (mg/l) as well as cellular accumulation (mg/g) of microbial carotenoid upon supplementation of metal ions (copper, zinc, ferrous, calcium, cobalt, alluminium) in yeasts and molds (Bhosale, 2004; Buzzini et al., 2005). Trace elements have been shown to exert a selective influence on the carotenoid profile in red yeasts. It may be explained by hypothesizing a possible activation or inhibition mechanism by selected metal ions on specific carotenogenic enzymes, in particular, on specific desaturases involved in carotenoid biosynthesis (Buzzini et al., 2005). The other explanation is based on observations that presence of heavy metals results in formation of various active oxygen radicals what, in a turn, induces generation of protective carotenoid metabolites that reduce negative behaviour of free radicals. Such strategy has been applied in several pigment-forming microorganisms to increase the yield of microbial pigments (Breierova et al., 2008; Rapta et al., 2005).

In order to achieve rapid carotenoid overproduction, various stimulants can be added to the culture broth. One group of such enhancers is based on intermediates of the tricarboxylic acid cycle which play an important role in metabolic reactions under aerobic conditions, forming a carbon skeleton for carotenoid and lipid biosynthesis in microbes. Because pigment increase is paralleled by decreased protein synthesis, restriction of protein synthesis is an important way how to shift carbon flow to carotenoid synthesis (Flores-Cotera & Sanchez, 2001). It was also proposed that high respiratory and tricarboxylic acid cycle activity is associated with production of large quantities of reactive species and these are known to enhance carotenoid production (An, 2001). It should be emphasized that the degree of stimulation was dependent on the time of addition of the citric acid cycle intermediate to the culture medium. Some fungi showed that addition of organic acids to media elevated β-carotene content and concomitantly decrease γ-carotene level with complete disappearance of lycopene (Bhosale, 2004).

Chemical substances capable of inhibiting biosynthetic pathways have been applied to characterize metabolic pathways and elucidate reaction mechanisms. In general, compounds that inhibit biosynthesis can act through various mechanisms, such as inhibiting the active site directly by an allosteric effect (reversible or otherwise), altering the regulation of gene expression and blocking essential biochemical pathways or the availability of cofactors, among other possibilities. From this view, number of chemical compounds including terpenes, ionones, amines, alkaloids, antibiotics, pyridine, imidazole and methylheptenone have been studied for their effect on carotene synthesis (Bhosale, 2004). In order to obtain commercially interesting carotenoid profiles, the effect of supplementation with diphenylamine (DPA) and nicotine in the culture media of *Rhodotorula rubra* and *Rhodotorula glutinis* was investigated. DPA blocks the sequence of desaturation reactions by inhibiting phytoene synthase, leading to an accumulation of phytoene together with other saturated carotenoids and nicotine inhibits lycopene cyclase, and consequently the cyclization reactions (Squina & Mercadante, 2005). Cultivation of *Xanthophyllomyces dendrorhous* in the presence of diphenylamine and nicotine at 4°C was reported to trigger interconversion of β-carotene to astaxanthin (Ducrey Sanpietro & Kula, 1998).

The addition of solvents such as ethanol, methanol, isopropanol, and ethylene glycol to the culture medium also stimulate microbial carotenogenesis. It should be noted that while ethanol supplementation (2%, v/v) stimulated β-carotene and torulene formation in *Rhodotorula glutinis*, torularhodin formation was suppressed (Bhosale, 2004). It was proposed that ethanol-mediated inhibition of torulene oxidation must be accompanied by an increase in β-carotene content suggesting a shift in the metabolic pathway to favor ring closure. Detailed studies revealed that ethanol activates oxidative metabolism with induction of HMG-CoA reductase, which in turn enhances carotenoid production. However, stimulation of carotenoid accumulation by ethanol or H_2O_2 was more effective if stress factors were employed to the medium in exponential growth phase than from the beginning of cultivation (Marova et al, 2004).

2.4.3 Mutagenesis

Mutagenesis is an alternative to classical strain improvement in the optimization of carotenoid production. Mutagenic treatment with *N*-methyl-*N*-nitro-*N*-nitrosoguanidine (NTG), UV light, antimycin, ethyl-methane sulfonate, *γ*-irradiation, high hydrostatic pressure have been used successfully to isolate various strains with enhanced carotenoid-producing activity. UV mutant *R.gracilis* has shown 1.8 times higher carotenoid synthesizing activity than that of the parent strain and the relative share of β-carotene in the total carotenoids was 60%. The yellow colored mutant 32 was also obtained by UV mutagenesis of the pink yeast *R. glutinis* and produced a large quantity of total carotenoids (2.9 mg/g dry cells), which was 24-fold higher accumulation of total carotenoids compared with the wild-type. Mutant 32 produced 120-fold more beta-carotene (2.05 mg/g dry cells) than the parent culture in a much shorter time (36 h), which was 82% (w/w) of the total carotenoid content. Later, after the treatments of five repeated cycles by high hydrostatic pressure of 300 MPa, the mutant *R. glutinis* RG6p was obtained, beta-carotene production of which reached 10.01 mg/l, increased by 57.89% compared with 6.34 mg/l from parent strain (Frengova & Beshkova, 2009).

A fivefold increase in beta-carotene accumulation was reported for yellow mutant *P. rhodozyma* 2-171-1 which was obtained after ethyl-methane sulfonate mutagenesis of dark red strain *P. rhodozyma*. This mutant is likely to be blocked in the oxidase step and therefore unable to perform the conversion of beta-carotene to echinenone and latter to astaxanthin. The UV-mutant *P. rhodozyma* PG 104 produced 46-fold more β-carotene (92% of total carotenoids) than the parent culture (2% of total carotenoids) and maximum beta-carotene yields were 1.08 mg/g dry cells and 9.95 mg/l. Using NTG mutagenesis two different strains of carotenoid accumulating *X. dendrourhous* mutants JH1 and JH2 were also isolated. Astaxanthin-overproducing mutant JH1 produced 4.03 mg astaxanthin/g dry cells, and this value was about 15-fold higher than that of wild-type. Mutant JH2 produced 0.27 mg beta-carotene/g dry cells, and this was fourfolds increase from that of wild-type and the mutant *X. dendrourhous* JH1 produced maximum astaxanthin concentration of 36.06 mg/l and 5.7 mg/g dry cells under optimized cultivation conditions (Kim et al., 2005).

To isolate a carotenoid-hyperproducing yeast, *P.rhodozyma* 2A2 N was treated by low-dose gamma irradiation below 10 kGy and mutant 3A4-8 was obtained. It produced 3.3 mg carotenoids/g dry cells, 50% higher carotenoid content than that of the unirradiated strain (antimycin NTG-induced mutant 2A2 N). Gamma irradiation produces oxygen radicals generated by radiolysis of water and could induce mutation of *P. rhodozyma* through a chromosomal rearrangement. A primary function of carotenoids in *P. rhodozyma* is to protect cells against singlet oxygen and these compounds have been demonstrated to quench singlet oxygen. Oxygen radicals have been known to cause changes in the molecular properties of proteins as well as enzyme activities. Thus, oxygen radicals generated by gamma irradiation might modify the pathway in astaxanthin biosynthesis of *P. rhodozyma* and cause an increase in carotenoid production of the mutant 3A4-8 isolated by gamma irradiation (Frengova & Beshkova, 2009).

2.4.4 Use of recombinant strains

One possibility for the improvement of the metabolic productivity of an organism is genetic modification. This strategy can be successful when an increase of the flux through a pathway is achieved by, e.g., the overproduction of the rate-limiting enzyme, an increase of precursors, or the modification of the regulatory properties of enzymes. In the carotenogenic yeasts, mevalonate synthesis, which is an early step in terpenoid biosynthesis, is a key point of regulation of the carotenoid biosynthetic pathway. In fact, addition of mevalonate to a culture of *X. dendrourhous* stimulated both astaxanthin and total carotenoid biosynthesis four times (from 0.18 to 0.76 mg/g and from 0.27 to 1.1 mg/g dry cells, respectively). This indicates that the conversion of HMG-CoA to mevalonate by HMG-CoA reductase is a potential bottleneck on the road to modified strains with higher astaxanthin content (Verdoes et al., 2003).

Like carotenoids, ergosterol is an isoprenoid and it is biosynthetically related to them by common prenyl lipid precursor, FPP. Astaxanthin production by *P. rhodozyma* strain was enhanced (1.3-fold) when sgualene synthase phenoxypropylamine-type inhibitor for sterol biosynthesis was added to the medium. The isolation and characteristic of the carotenogenic genes of yeasts facilitates the study of the effect of their overexpression on carotenoid biosynthesis. Use of recombinant DNA technology for metabolic engineering of the astaxanthin biosynthetic pathway in *X. dendrourhous* was described too. In several transformants containing multiple copies of the phytoene synthase-lycopene cyclase-encoding gene (*crtYB*), the total carotenoid content was higher (with 82%) than in the control strain. This increase was mainly due to an increase of the beta-carotene and echinenone content (with 270%), whereas the total content of astaxanthin was unaffected or even lower.

Alternatively, in recent years, several food-grade non-pigmented yeasts (*Saccharomyces cerevisiae, Candida utilis*) have been engineered in order to obtain strains possessing the ability to produce selected carotenoids (Verwaal et al., 2007). Identification of genes of enzymes from the astaxanthin biosynthetic pathway and their expression in a non-carotenogenic heterologous host have led to the overproduction of beta-carotene. The possibility of the use of *S. cerevisiae*as a host for efficient beta-carotene production by successive transformation with carotenogenic genes (*crtYB* which encodes a bifunctional phytoene synthase and lycopene cyclase; *crtI*, phytoene desaturase; *crtE*, heterologous GGPP synthase; *tHMGI*, HMG-CoA reductase) from *X. dendrorhous* was studied. Like *X. dendrorhous*, *S. cerevisiae* is able to produce FPP and converts it into GGPP, the basic building block of carotenoids. *S. cerevisiae*, the industrially important conventional yeast, cannot produce any carotenoid, while it synthesizes ergosterol from FPP by a sterol biosynthetic pathway. Conversion of FPP into GGPP is catalyzed by GGPP synthase encoded by *BTS1* gene in *S. cerevisiae*. Construction of a strain, producing a high level of beta-carotene (5.9 mg/g dry cells) was succesful. Oleaginous yeasts are also suitable host strains for the production of lipophilic compounds due to their high lipid storage capacity. Recently, the carotenoid-producing *Yarrowia lipolytica* has been generated by metabolic engineering. Acording to these results entire biosynthetic pathways can be introduced into new host cells through recombinant DNA technology and carotenoids can be produced in organisms that do not normally produce carotenoids.

2.5 Application of whole-cell yeast biomass to production of pigments and other lipid compounds

2.5.1 Carotenoid and ergosterol enriched biomass

Red yeasts are used predominantly as carotenoid producers and, thus, carotenoid-enriched biomass is the most frequently produced. The growing scientific evidence that carotenoid pigments may have potential benefits in human and animal health has increased commercial attention on the search for alternative natural sources. Comparative success in microbial pigment production has led to a flourishing interest in the development of fermentation processes and has enabled several processes to attain commercial production levels. An important aspect of the fermentation process is the development of a suitable culture medium to obtain the maximum amount of desired product. In recent years, cheap raw materials and by-products of agro-industrial origin have been proposed as low-cost alternative carbohydrate sources for microbial metabolite production, with the view also of minimizing environmental and energetic problems related to residues and effluent disposal.

During the produt recovery process, the biomass is isolated and transformed into a form suitable for isolating carotene, which can be further isolated from the biomass with appropriate solvent, suitably purified and concentrated. Using whole biomass as final product, isolation of metabolites is not necessary and other cell active components can be utilized. Nevertheless, cell disruption is recommended for better bioavailability of the most of lipid-soluble substance (Frengova & Beshkova, 2009). Several types of microbes have been reported to produce carotenoids and carotenoid-rich biomass; but only a few of them have been exploited commercially (Bhosale, 2004).

Among the few astaxanthin producing microorganisms, *Phaffia rhodozyma* (*Xanthophyllomyces dendrorhous*) is one of the best candidate for commercial production of pigment as well as enriched biomass. Therefore, many academic laboratories and several companies have developed processes which could reach an industrial level. *Phaffia/ Xanthophyllomyces* has some advantageous properties that make it attractive for commercial astaxanthin production: (i) it synthesizes natural form astaxanthin (3*S*,3'*S* configuration) as a principal carotenoid, (ii) it does not require light for its growth and pigmentation, and (iii) it can utilize many types of carbon and nitrogen sources (Lukacs et al, 2006; Dufosse, 2006). Studies on physiological regulation of astaxanthin in flasks cultivations was verified in bioreactors and the ataxanthin amount reached 8.1 mg/L (Dufosse, 2006). Enhanced production of the pigment was achieved during fed-batch fermentation with regulated additions of glucose and optimized fermentation condition finally yielded up to 20 mg astaxanthin/L (Certik et al., 2009). High carbon/nitrogen ratio induced amout of astaxanthin and C/N-regulated fed-batch fermentation of *P. rhodozyma* led to 16 mg astaxanthin/L. Thus, this strain can be considered as a potential producer of astaxanthin. In addition, to avoid isolation of astaxanthin from cells, two-stage batch fermentation technique was used (Fang & Wang, 2002), where *Bacillus circulans* with a high cell wall lytic activity was added to the fermentation tank after the accumulation of astaxanthin in *P. rhodozyma* was completed. Astaxanthin is the principal colorant in crustaceans, salmonids and flamingos. There is current interest in using *P.rhodozyma* biomass in aquaculture to impart desired red pigmentation in farmed salmon and shrimps.

Biotechnological production of β-carotene by several strains of the yeast *Rhodotorula* is currently used industrially. This yeast is convenient for large-scale fermentation because of its unicellular nature and high growth rate. Because *Rhodotorula glutinis* synthesizes β-carotene, torulene and torularhodin, the rate of production of the individual carotenoid depends upon the incubation conditions. Specially prepared mutants of *Rhodotorula* not only rapidly increased formation of torulene or thorularhodin, but amount of β-carotene reached the level of 70 mg/L (Sakaki *et al.*, 2000). Better strategy than isolation of individual pigments seems to be use of the whole enriched biomass to feed and food industry.

In our recent work exogenous stress factors were used to obtain higher production of carotenoids in *R. glutinis* CCY 20-2-26 strain. Physical and chemical stress factors were applied as single and in combination. Adaptation to stress was used in inoculum II. Short-term UV irradiation of the production medium led to minimal changes in biomass production. The production of carotenoids in *R. glutinis* cells was stimulated in all samples of exponentially growing cells when compared with control cultivation. In stationary phase, the production of carotenoids was induced only by 35-min irradiation. Ergosterol production exhibited very similar changes as β-carotene production both under temperature and UV stress. Our results are in good agreement with recent findings of the effect of weak white light irradiation on carotenoid production by a mutant of *R. glutinis* (Sakaki et al., 2000).

Using chemical stress, the influence of osmotic (2-10 % NaCl) stress, oxidative (2-10 mM H2O2) stress and combined effects of these stress factors on the morphology, growth and production of biomass, carotenoids and ergosterol by *R. glutinis* CCY 20-2-26 cells were studied (Marova et al., 2010). First, *R. glutinis* cells were exposed to higher concentration of stress factors added into the production medium. Further, low concentrations of NaCl and H_2O_2 were added to the inoculum medium or to both inoculum and production media. Exposition of red yeast cells to all tested stress factors resulted in higher production of carotenoids as well as ergosterol, while biomass production was changed only slightly. Under high stress 2-3 times increase of β-carotene was observed. The addition of low salt or peroxide concentration into the inoculation media led to about 2-fold increase of carotenoid production. In Erlenmeyer flasks the best effect on the carotenoid and ergosterol production (3- to 4-fold increase) was exhibited by the combined stress: the addition of low amount of NaCl (2 mM) into the inoculum medium, followed by the addition of H_2O_2 (5 mM) into the production medium. The production of ergosterol in most cases increased simultaneously with the production of carotenoids.

Cultivation of *R. glutinis* carried out in a 2-litre laboratory fermentor was as follows: under optimal conditions about 37 g/L of yeast biomass were obtained containing approx. 26.30 mg/L of total carotenoids and 7.8 mg/L of ergosterol. After preincubation with a mild stress factor, the yield of biomass as well as the production of carotenoids and ergosterol substantially increased. The best production of enriched biomass was obtained in the presence of peroxide in the inoculation medium (52.7 g/L of biomass enriched with 34 mg/L of carotenoids) and also in combined salt/peroxide and salt/salt stress (about 30–50 g/L of biomass enriched with 15–54 mg/L of total carotenoids and about 13-70 mg/L of ergosterol). *Rhodotorula glutinis* CCY 20-2-26 strain could be a suitable candidate for biotechnological applications in the area of carotenoid rich biomass production. Preliminary cultivation in a 2-litre laboratory fermentor after preincubation with stress factors in well-ballanced experiments led to the yield of about 40-50 g per litre of biomass enriched by 20-40 mg of β-carotene+lycopene sum (approximately 30–50 mg of total carotenoids per litre) and about 70 mg of ergosterol per litre. Addition of simple cheap stress factor substantially increased metabolite production without biomass loss. Therefore, this strain takes advantage of the utilization of the whole biomass (complete nutrition source), which is efficiently enriched for carotenoids (provitamin A, antioxidants) and also ergosterol (provitamin D). Such a product could serve as an additional natural source of significant nutrition factors in feed and food industry (Marova et al, 2010).

Our further work was focused on possiblity to use carotenogenic yeasts cultivated on alternative nutrition sources combined with stress factors (Marova et al., 2011). Both physiological and nutrition stress can be used for enhanced pigment production. Three red yeast strains (*Sporobolomyces roseus, Rhodotorula glutinis, Rhodotorula mucilaginosa*) were studied in a comparative screening study. To increase the yield of these pigments at improved biomass production, combined effect of medium with modified carbon and nitrogen sources (waste materials - whey, potato extract) and peroxide and salt stress was tested. The production of carotene-enriched biomass was carried out in flasks as well as in laboratory fermentor. The best production of biomass was obtained in inorganic medium with yeast extract. In optimal conditions tested strains differ only slightly in biomass production. Nevertheless, all strains were able to use most of waste substrates. Biomass and pigment production was more different according to substrate type. It was observed that addition of non-processed or processed whey or potato extract to media can increase beta-carotene production, while biomass production changed relatively slightly (Marova et al, 2011).

In *Rhodotorula glutinis* addition of whey substrate into production medium led to 3.5x increased production of beta-carotene without substantial changes in biomass. Non-processed whey or potato extract added to production media led to about 3x increase of beta-carotene production accompanied by biomass loss. The highest yield was reached after addition of lyophillized non-processed whey to INO II as well as to production media. Also potato extract added into INO II led to increased beta-carotene production while biomass yield was lower. *Sporobolomyces roseus* exhibited significant changes in biomass:carotene ratio dependent on whey substrate addition. Substantial biomass decrease in presence of lyophilized whey in INO II (under 5 g/L) was accompanied by very high beta-carotene yield (2.54 - 2.75 mg/g d.w.). Potato extract addition into production medium led to about 11-times increase of β-carotene production, while production of biomass was lower than in control. Preincubation of *S.roseus* cells with potato extract and following cultivation in production medium with 5% hydrogen peroxide led to about 20-times higher β-carotene production as in control, in this cultivation conditions biomass decreased only slightly. In general, total production of biomass by *S.roseus* was about 2-x lower as in *R.glutinis.* So, this is the reason why *S.roseus* CCY 19-4-8 cells is less suitable to enriched biomass production. *Rhodotorula mucilaginosa* CCY 20-7-31 seems to be relatively poor producer of carotenoids when compared with the other two strains. Production of biomass in this strain was more similar to *R.glutinis* (about 8 g/L). However, addition of potato extract into INO II combined with salt stress in production medium enabled to reach the highest biomass as well as β-carotene production observed in this strain yet (1.56 mg/g d.w.). It seems that this strain needs for optimal pigment/biomass production some additional nutrition factors which are no present in simple (but cheap) inorganic medium, but can be obtained from different waste substrates (also cheap).

In laboratory fermentor better producers of enriched biomass were both *Rhodotorula* strains. In experiments with *Rhodotorula glutinis* the production of yeast biomass in a laboratory fermentor was in most types of cultivation more than 30 grams per litre (about 3-times higher yield than in Erlenmeyer flasks; Table 1). The balance of cultivation in a fermentor in optimum conditions is as follows: we obtained about 37.1 g/l of biomass containing 17.19 mg per litre of β-carotene (see Table 1). The production of β-carotene was induced in most types of media combinations. High total yield of β-carotene was obtained in whey production medium (44.56 g/L of biomass; 45.68 mg of β-carotene per litre of culture). The highest total yield of β-carotene was obtained using combined whey/whey medium (51.22 mg/L); this cultivation was accompanied also with relatively high biomass production (34.60 mg/L). In experiments with *Sporobolomyces roseus* CCY 19-4-8 substantially higher production of biomass was obtained in fermentor when compared with cultivation in flasks. Mainly in whey medium about 3-times biomass increase (about 12 g/L) was reached and production of beta-carotene was mostly higher than in *R.glutinis*. Because of low biomass production, total yields were in *S.roseus* mostly lower than in *R.glutinis* cells. Yeast strain *Rhodotorula mucilaginosa* CCY 20-7-31 exhibited in most cases similar biomass production characteristics as *R.glutninis,* while pigment production was substantially lower (see Table 4). As the only substrate suitable for β-carotene production was found potato extract in INO II combined with 5% salt in production medium. Under these conditions 55.91 mg/L of β-carotene was produced in 30.12 g of cells per litre of medium (Marova et al, 2011).

The aim of all preliminary experiments carried out in laboratory fermentor was to obtain basic information about potential biotechnological use of the tested strains to the industrial production of β-carotene/ergosterol enriched biomass. The results of both *Rhodotorula* strains are very promising. The yield of *R.glutinis* CCY 20-2-26 biomass (37 - 44.5 g/L) produced in minimal cultivation medium was similar to the maximal biomass yield obtained in fed-batch cultivation of *Phaffia rhodozyma* (36 g/L), which is widely used as an industrial producer of astaxanthin (Lukacs et al., 2006). The maximal production of total carotenoids by used *P. rhodozyma* mutant strain was 40 mg/L, which is also similar to the yields obtained in *R. glutinis* CCY 20-2-26 cells grown in whey medium. The highest yields of pigments were obtained in *Rhodotorula glutinis* CCY 20-2-26 cells cultivated on whey medium (cca 45 g per liter of biomass enriched by 46 mg/L of beta-carotene) and in *Rhodotorula mucilaginosa* CCY 20-7-31 grown on potato medium and 5% salt (cca 30 g per liter of biomass enriched by 56 mg/L of beta-carotene). Such dried carotenoid-enriched red yeast biomass could be directly used in feed industry as nutrition supplement (Marova et al., 2011).

	Biomass			**Production of β-carotene**		
Substrate/stress factor	**R.g.** (g/l)	**S.r.** (g/l)	**R.m.** (g/l)	**β -carotene** (mg/l)	**β -carotene** (mg/l)	**β -carotene** (mg/l)
Control 0/0	37.14	17.00	26.55	17.93	3.25	4.31
0/whey deprot.*	44.56	9.59	27.06	45.68	23.36	8.80
0/potato	28.12	10.80	38.50	25.45	17.50	26.18
Whey*/ salt	40.86	8.16	18.35	28.00	14.23	10.81
Whey*/ whey	34.60	10.15	29.82	51.22	29.40	11.33
potato/salt	26.10	7.14	30.12	22.23	7.55	55.91
Potato/potato	18.56	6.28	28.48	22.48	6.13	27.23

Table 1. Production of beta-carotene enriched biomass in 2 L laboratory fermentor (Marova et al., 2011)

An alternative for utilization of some natural substrates for production of carotenoids by *Rhodotorula* species is the method of cocultivation. A widespread natural substrate is milk whey containing lactose as a carbon source. Carotenoid synthesis by lactose-negative yeasts (*R. glutinis, R. rubra* strains) in whey ultrafiltrate can be accomplished: by enzymatic hydrolysis of lactose to assimilable carbon sources (glucose, galactose) thus providing the method of co-cultivation with lactose-positive yeasts (*Kluyveromyces lactis*), producers of galactosidase or by creating conditions under which lactose is transformed into carbon sources (glucose, galactose, lactic acid) easily assimilated by the yeast when they were grown in association with homofermentative lactic acid bacteria or yogurt starter culture (Frengova & Beshkova, 2009). The maximum carotenoid yields for the microbial associations [*R. rubra* + *K.lactis; R. glutinis* + *Lactobacillus helveticus; R. rubra* + *L.casei; R. rubra* + (*L. bulgaricus* + *Streptococcus thermophillus*)] were as follows: 10.20, 8.10, 12.12, 13.09 mg/l, respectively. These yields are about five times higher than that of a lactose-positive strain *R. lactosa* cultivated in whey reported in literature (Frengova et al., 2004). *R. glutinis*–*Debaryomyces castellii* co-cultures was produced (5.4 mg carotenoids/l) about three times the amount of total carotenoids formed by the red yeast cultured alone in low hydrolyzed corn syrup (Buzzini, 2001) The author concluded that oligosaccharides and dextrins of syrup could be utilized for pigment production by *R. glutinis* after hydrolysis to maltose and glucose by the extracellular amylolytic enzymes produced by *D. castellii* DBVPC 3503 in co-cultures.

Rhodotorula species	Carbon source	Cultivation process	Cell mass (g/l)	Carotenes (mg/g dry cells)	Carotenes (mg/l culture)	References
R. glutinis	WLA 2	batch	8.12	8.20	66.32	Marova et al., 2011
R. glutinis	pastes + enzymes	batch	11.68	3,60	40.10	Marova et al., 2010
R. glutinis ATCC 26085	glucose	batch				Davoli et al., 2004
R. glutinis 32	glucose	batch	23.90	5.40	129.00	Bhosale & Gadre, 2001
R. glutinis 32	sugar cane molasses	fed-batch	78.00	2.36	183.00	Bhosale & Gadre, 2001
R. glutinis DBVPG 3853 *D. castellii* DBVPG 3503	corn syrup	fed-batch	15.30	0.54	8.20	Buzzini, 2001
R. glutinis TISTR	hydrolyzed mung bean waste flour	batch	10.35	0.35	3.48	Tinoi et al., 2005
R. glutinis 22P *L. helveticus* 12A	whey ultrafiltrate	batch	30.20	0.27	8.10	Frengova & Beshkova, 2009
R. mucilaginosa NRRL-2502	sugar-beet molasses	batch	4.20	21.20	89.0	Aksu & Eren, 2005
R. mucilaginosa NRRR-2502	whey	batch	2.40	29.20	70.0	Aksu & Eren, 2005

Table 2. Comparison of carotenoid production by *Rhodotorula* species cultivated on different waste substrates

As mentioned above, waste substrates and alternative nutrition sources were used to production of astaxanthin-enriched biomas sof *Xanthophyllomonas dendrorhous* sources (Lukacs et al, 2006; Dufosse, 2006). Batch culture kinetics of this yeast revealed reduction in biomass with glucose and lower intracellular carotenoid content with fructose. Figures were different when compared to sucrose. In contrast, specific growth rate constant stayed between 0.094 - 0.098 h^{-1}, irrespective of the carbon sources employed. Although the uptake rate of glucose was found to be 2.9-fold faster than that of fructose, sucrose was found to be a more suitable carbon source for the production of carotenoids by the studied strain. When sugar cane molasses was used, both the specific growth rate constant and the intracellular carotenoid content decreased by 27 and 17%, respectively. Compared with the batch culture using 28 g/L sugar cane molasses, fed-batch culture with the same strain resulted in a 1.45-fold higher cell yield together with a similar level of carotenoid content in *X. dendrorhous* SKKU 0107 (Park et al, 2008).

Phaffia rhodozyma NRRL Y-17268 cells were proliferated in xylose-containing media made from Eucalyptus wood. Wood samples were subjected to acid hydrolysis under mild operational conditions, and hydrolysates were neutralized with lime. Neutralized hydrolysates were treated with charcoal for removing inhibitors and then supplemented with nutrients to obtain culture media useful for proliferation of the red yeast *P.rhodozyma.* Biomass was highly pigmented and volumetric carotenoid concentrations up to 5.8 mg carotenoids/L (with 4.6 mg astaxanthin/L) were reached. Further experiments in batch fermentors using concentrated hydrolysates (initial xylose concentrations within 16.6 and 40.8 g/L) led to good biomass concentrations (up to 23.2 g cells/L) with increased pigment concentration (up to 12.9 mg total carotenoids/L, with 10.4 mg astaxanthin/L) and high volumetric rates of carotenoid production (up to 0.079 mg/L/h (Parajo et al., 1998).

In the future, other types of waste materials (for instance from winemarket) are intended to be tested as carbon sources for carotenogenesis in red yeasts (Table 2). Moreover application of an environmental stress in combination with waste materials can lead to overproduction of carotenoids and lipids and decrease cost of their production. Such strategies could result into production of yeast biomass rich not only in carotenoids and other provitamins, but also in other nutrition components (proteins, PUFA, metal ions etc.) that originate both from yeast cells and from cultivation substrates. This is the way to production of complex food additives based on naturally enriched yeast biomass.

2.5.2 Single-oil cell processes and lipid production by red yeasts

A number of microorganisms belonging to the genera of algae, yeast, bacteria, and fungi have ability to accumulate neutral lipids under specific cultivation conditions. The microbial lipids contain high fractions of polyunsaturated fatty acids and have the potential to serve as a source of significant quantities of transportation fuels (Subramaniam et al., 2010). Microorganisms possess the ability to produce and accumulate a large fraction of their dry mass as lipids. Those with lipid content in excess of 20% are classified as 'oleaginous' (Ratlege and Wynn, 2002).

Oleaginous yeasts have a fast growth rate and high oil content, and their triacylglycerol (TAG) fraction is similar to that of plant oils. These organisms can grow on a multitude of carbon sources (see above). Most oleaginous yeasts can accumulate lipids at levels of more than 40% of their dry weight and as much as 70% under nutrient-limiting conditions (Beopoulos et al., 2009). However, the lipid content and fatty acid profile differ between species. Some of the yeasts with high oil content are *Rhodotorula glutinis, Cryptococcus albidus, Lipomyces starkeyi,* and *Candida curvata* (Subramaniam et al., 2010). Newly, lipid production by the oleaginous yeast strain *Trichosporon capitatum* was described too (Wu et al, 2011). The main requirement for high lipid production is a medium with an excess of carbon source and other limiting nutrients, mostly nitrogen. Hence, production of lipids is strongly influenced by the C/N ratio, aeration, inorganic salts, pH, and temperature.

Yeasts are able to utilize several different carbon sources for the production of cell mass and lipids. In all cases, accumulation of lipids takes place under conditions of limitations caused by a nutrient other than carbon. Recently, production of lipids by the yeast *R. glutinis* on different carbon sources (dextrose, xylose, glycerol, mixtures of dextrose and xylose, xylose and glycerol, and dextrose and glycerol) was explored (Easterling et al., 2009). The highest lipid production of 34% TAG on a dry weight basis was measured with a mixture of dextrose and glycerol as carbon source. The fraction of unsaturated fatty acids in the TAGs was dependent on carbon source, with the highest value of 53% on glycerol and lowest value of 25% on xylose. With whey permeate for production of lipids by different yeast strains, *L. starkeyi* ATCC 12659 was found to have the highest potential of accumulating lipids among *Apiotrichum curvatum* ATCC 10567, *Cryptococcus albidus* ATCC 56297, *L. starkeyi* ATCC 12659, and *Rhodosporidium toruloides* ATCC. The yeast *L. starkeyi* is unique in that it is known not to reutilize the lipids produced by it and it produces extracellular carbohydrolases. Effect of C/N ratio on production of lipids by *L. starkeyi* and conditions favoring accumulation of lipids result in reduced growth of cells were confirmed. The cells could consume liquefied starch in batch culture and produced cells containing 40% lipids at a cell yield of 0.41 g dry weight per g starch. The yield on starch was higher than when glucose was used as carbon source (Subramaniam et al., 2010).

Culture temperature and pH influence the total cell number and lipid content in yeast cells. In minimal medium with glucose as carbon source, the yeast *L. starkeyi* accumulates large fractions of dry weight as lipids with a high yield in the pH range of 5.0–6.5. At higher temperatures, the cellular lipid content, the glucose conversion efficiency, and the specific lipid production rates in *L. starkeyi* were high, but the degree of fatty acid unsaturation was low (Subramaniam et al., 2010). Fastest growth of *L. starkeyi* cells occurred at 28°C (specific growth rate 0.158 h-1), and the lipid fraction in cells under these conditions was 55%. However, the fraction of oleic acid in the lipids increased from 52 to 60% of lipids when the accumulation phase temperature was reduced from growth temperature of 28–15°C. High lipid accumulation in cells of oleaginous yeast is obtained under limiting nitrogen concentration conditions. The oleaginous yeast *L. starkeyi* delivered lipid content of 68% at a C/N ratio of 150 compared to 40% in the presence of a C/N ratio of 60 while growing on digested sewage sludge (Subramaniam et al., 2010). The key fatty acids produced were C16:0, C16:1, C18:0, and C18:1. Accumulation of lipids by *Cryptococcus curvatus* cells also required a high C/N ratio of 50 in batch and fed-batch cultures (Hassan et al., 1996); the fatty acids produced were mainly oleic (C18:1), palmitic (C16:0), and stearic (C18:0). The highest fraction of stearic acid (18:0) in batch cultures was 14 and 19% in fed-batch culture.

Under optimal fermentation conditions in a batch reactor (100 g/L glucose as carbon source, 8 g/L yeast extract, and 3 g/L peptone as nitrogen sources, initial pH of 5.0, inoculation volume of 5%, 28°C temperature, and 180 rpm agitation in a 5-l bioreactor), *Rhodotorula glutinis* can accumulate lipids up to 49% of cell dry weight and 14.7 g/L lipid. In continuous culture, the cell biomass, lipid content, and lipid yield increase with decreasing growth rate. The yield 60.7% lipids in cells and 23.4 g l-1 lipid production in a continuous mode of operation was obtained (Subramaniam et al., 2010). In *R. toruloides* cultivated in fed-batch mode, oleic, palmitic, stearic, and linoleic acids were the main fatty acids (Li et al., 2007). Also in *R. mucilaginosa* TJY15a, 85.8% long-chain fatty acids were composed of palmitic, palmitoleic, stearic, oleic, and linolenic acids (Li et al., 2010). Under continuous culture conditions, nitrogen-limited medium and a dilution rate of about one-third of the maximum is recommended to achieve the maximum content of lipids in a microorganism (Dai et al., 2007). Mix cultivation of microalgae (*Spirulina platensis*) and yeast (*Rhodotorula glutinis*) for lipid production was studied (Xue et al., 2010). Mixing cultivation of the two microorganisms significantly increased the accumulation of total biomass and total lipid yield.

Oils and fats are primarily composed of triacylglycerols (TAGs). TAGs serve as a primary storage form of carbon and energy in microorganisms; their fatty acid composition is also superior to that of other cellular lipids (phospholipids and glycolipids) for biodiesel production (Subramaniam et al., 2010). Although fatty acids in microbial lipids range from lauric acid (C12:0) to docosahexaenoic acid (C22:6), palmitic (C16:0), stearic (C18:0), oleic (C18:1), and linoleic (C18:2) acids constitute the largest fraction. Of these, palmitic and oleic acids are the most abundant. Considering the saturated and unsaturated acid components, approximately 25–45% are saturated fatty acids, and 50–55% are unsaturated. Thus, the ratio of unsaturated to saturated fatty acids in microbial oils ranges between 1 and 2, which is somewhat similar to that in plant oils (such as palm). When cultivated under appropriately optimized conditions, microorganisms are capable of producing significant quantities of γ-linoleic (C18:2) and arachidonic (C20:4) acids. These fatty acids have high nutraceutical value, and microbial oils are generally marketed as extracted oils as health food. Technologically, the production of these high value compounds is accompanied by production of significant quantities of other neutral lipids. Hence, separation of non-nutraceutical fatty acids from the PUFA needs to be explored (Subramaniam et al., 2010).

Production of microbial lipids to biofuel production is limited by cost; economically viable biofuels should be cost competitive with petroleum fuels. The single-cell oil production cost depends mainly upon the species chosen for cultivation, lipid concentration within cells, and the concentration of cells produced. The cost of feed stock or carbon source required for the production of microbial lipids accounts for 60 to 75% of the total costs of the biodiesel. Thus, the cost of lipid production was influenced strongly by the cost of medium nutrients (50%) needed for cultivation of cells and the cost of solvent (25%) for the extraction of lipids from biomass. Hence, the economics of single-cell oil production can be improved by using carbon in wastes such as wastewater, municipal, and other carbonaceous industrial wastes and CO2 in flue gases from boilers and power plants. Economic analyses have indicated the need to minimize costs of medium components and for further research dealing with microbial systems capable of producing lipids at relatively high productivities in minimal media (Subramaniam et al., 2010).

Lipid production in *Rhodotorula* cells occurs over a broad range of temperatures and it can be considered an interesting genus for the production of single cell oils. The extent of the carbon excess had positive effects on triacylglycerols production, that was maximum with 120 g/L glucose, in terms of lipid concentration (19 g/L), lipid/biomass (68%) and lipid/glucose yields (16%). Both glucose concentration and growth temperature influenced the composition of fatty acids, whose unsaturation degree decreased when the temperature or glucose excess increased. Fatty acid profiles were studied in six carotenoid-producing yeast species isolated from temperate aquatic environments in Patagonia. The proportion of each FA varied markedly depending on the taxonomic affiliation of the yeast species and on the culture media used. The high percentage of polyunsaturated fatty acids (PUFAs) found in Patagonian yeasts, in comparison to other yeasts, is indicative of their cold-adapted metabolism (Libkind et al., 2004). The hydrolysis of triacylglycerols to free FA and glycerol by lipases from oleaginous yeasts as *R.glutinis* or *Yarrowia lipolytica* can have many prospective industial applications e.g. digestive acids, flavour modifications, interesterification of oils etc.

Growth and lipid modifications of pigment-forming yeasts of genus *Rhodotorula* and *Sporobolomyces* growing under presence of selenium recently were studied (Breierova et al.,

2008). Because some of the red yeasts also produce enough quantity of lipids, such selenized red yeasts might be considered as a valuable source of both carotene pigments and useful lipids. However, until date there has not been any available data dealing with effect of selenium on fatty acid alternations in microorganisms. Therefore the aim of our further study was to describe modification in fatty acid profile in various lipid structures of red yeasts grown under selenium addition to the cultivation medium. Sensitivities of all cultures to selenium were similar and yeasts commonly accepted up to 0.12 mM selenium ions. It should also be noted that addition of selenium to the media prolonged lag-phase of yeasts significantly probably as a consequence of adaptation on selenium presence (Certik M., unpublished data).

Total lipids, neutral lipids and the main membrane lipids (phosphatidylcholine, phosphatidylethanolamine, phosphatidylserine and phosphatidylinositol) of investigated yeasts consisted of mainly palmitic (C16:0), palmitoleic (C16:1), stearic (C18:0), oleic (C 18:1), linoleic (C18:2) and linolenic (C18:3) acids. Oleic acid was the main fatty acid almost in all investigated lipid structures, palmitic and stearic acids were also abundant in PE and in PS+PI fractions. Neutral lipids did not show such intensive changes in fatty acid composition as their polar counterparts. On the other hand, phosphatidylcholine displayed remarkable high amounts of C18:2 and C18:3 fatty acids in all investigated yeasts. Because conversion of oleic acid to its C18 di- and three-unsaturated metabolites is catalyzed by membrane-bound Δ^{12} and Δ^{15} fatty acid desaturases (Certik et al., 1998), it is tempting to speculate that biosynthesis of C18 unsaturated fatty acids in *Rhodotorula* and *Sporobolomyces* species is associated with phosphatidylcholine moieties. Microsomal PC was also found as the predominant site for fatty acid desaturations in other yeasts and fungi (Jackson et al., 1998).

Selenium in the medium without any doubt triggers a set of various mechanisms affecting overall metabolisms of yeasts. It is known that phospholipids as the basic structural elements of the membranes are sensitive to the environment alterations. Since fatty acids are the major constituents of the membrane lipids, modulation of number and position of double bonds in acyl chains by individual fatty acid desaturases play crucial role in preserving of suitable dynamic state of the bilayer. Preliminary results in *R. glutinis* demonstrate that selenium stimulates biosynthesis of C18 fatty acids as well as it promotes distribution unsaturated C18 fatty acids in the membrane lipids. These findings might be very useful for preparation of selenized red yeasts containing carotenoid pigments with enhanced accumulation of linoleic and linolenic acids. (Breierova et al 2008, Certik et al., 2009).

2.5.3 Production of red yeast biomass with accumulated metals

Heavy metals are natural components of the Earth´s crust. As trace elements, some heavy metals (e.g., copper, selenium, zinc) are essential to maintain the metabolism of the human body. However, at higher concentrations they can lead to poisoning. A special case of antioxidant/prooxidant behavior of carotenoids emerge in the presence of metals (e.g. metal-induced lipid peroxidation). In this case metal ions (Fe^{2+} or Cu^{2+}) react with hydroperoxides, via a Fenton-type reaction, to initiate free radical chain processes. There are several studies which indicate that β-carotene offers protection against metal-induced lipid oxidation. Presence of carotenoid in the reaction system not only decreases the free radical concentration, but also the reduction of Fe^{3+} to Fe^{2+} by carotenoids may occur. Recently free radical scavenging and antioxidant activities of metabolites produced by carotenogenic yeasts of *Rhodotorula* sp. and *Sporobolomyces* sp. grown under heavy metal presence were studied using various EPR experiments (Rapta et al., 2005). Since carotenogenic yeast differ each to other in resistance against the heavy metals due to their individual protective system, quenching properties and antioxidant activities of carotenoids yeasts were modulated by metal ions variously. Thus, activated biosynthesis of carotenoides by yeasts exposed to heavy metal presence could be in part explained by their scavenger characters (Rapta et al., 2005) as a protection against the harmful effect of the environment.

Several divalent cations (Ba, Fe, Mg, Ca, Zn and Co) have been demonstrated to act as stimulants for growth of *R. glutinis.* Trace elements have been shown to exert a selective influence on the carotenoid profile in *R. graminis*—Al3+ and Zn2+ had a stimulatory effect on beta-carotene synthesis, while Zn2+ and Mn2+ had a inhibitory effect on torulene and torularhodin synthesis (Buzzini et al, 2005). The observed effect of trace elements on the biosynthesis of specific carotenoids in red yeasts may be explained by hypothesizing a possible activation or inhibition mechanism by selected metal ions on specific carotenogenic enzymes, in particular, on specific desaturases involved in carotenoid biosynthesis. In a recent study, calcium, zink and ferrous salts were shown to have a stimulatory effect on volumetric production as well as cellular accumulation of carotenoids from the yeast *R. glutinis (*Bhosale & Gadre, 2001*)*. Divalent cation salts increased the total carotenoid content (mg/L) about two times. It can be assumed that this positive response was due to a stimulatory effect of cations on carotenoid-synthesizing enzymes, or to the generation of active oxygen radicalcals in the culture broth. In contrast, the addition of manganese salt in the presence of generators of oxygen radicals had an inhibitory effect on carotenoid formation in *X. dendrorhous* since manganese acts as a scavenger; however, this effect could be concentration dependent as manganese is also known to act as a cofactor for enzymes involved in carotenoid biosynthesis and thus enhances carotenoid accumulation at certain concentrations (Frengova & Beshkova, 2009).

Astaxanthin content was decreased significantly at >1 mg/L FeCl3 and growth of *P.rhodozyma* was poor at an FeCl3 concentration of <0.1–1.0 mg/L (An et al., 2001). Carotenoid production decreased in yeast with increasing Mn2+ concentration (0–10 mg/l) when succinate was used as the sole C source, but not when growth took place in the presence of glucose. The week oxygen radical scavengers Zn2+ and Cu2+ had no effect on carotenoid production by *P. rhodozyma,* whereas Cu2+ below 3.2 μM increased the astaxanthin content of cells *P. rhodozyma* but at the expense of a slightly decreased growth. In yeast, there are at least two intracellular enzyme systems requiring copper: cytochrome-*c*-oxidase and superoxide dismutase. These enzymes are probably related to the increased astaxanthin production seen in concentrations of Cu2+ below 3.2 μM. Copper deficit decreases the activity of antioxidant enzyme Cu,Zn-superoxide dismutase, as reported previously and may induce oxidative stress and astaxanthin synthesis because of diminished antioxidant defences. In contrast, iron below 1 μM decreased both the growth and astaxanthin content of cells *P. rhodozyma* (Flores-Cotera & Sanchez, 2001).

Selenium (Se) is a key trace element required in small amounts in humans and animals for the function of a number of Se-dependent enzymes; however, this element can also be toxic in larger doses. Se is incorporated into proteins to provide selenoproteins, which are important antioxidant enzymes; other selenoproteins participate in the regulation of thyroid function and play a role in the immune systém (Wang & Xu, 2008). Organically bound Se is considered as more bioavailable and suitable for dietary application than sodium selenite or podium selenate, the two inorganic forms of Se commonly used in the feed industry. Yeasts naturally incorporate Se into the biomass where it is stored as selenomethionine. The organic form of Se produced in yeasts is of the similar type as that obtained from food. Recently preparation of antioxidant formula based on carotenoid forming yeasts *Rhodotorula glutinis* and *Sporobolomyces roseus* that also efficiently accumulated selenium from the growth medium was reported (Breierova et al., 2008).

In the presence of Se, carotenogenic yeast strains produced less carotene pigments. The results obtained indicate that the most dramatic change was observed in the significantly lowered levels of β-carotene, while torularhodin and torulene contents decreased to a lesser extent (Breierova et al., 2008). Previously, it has been shown that Cd, Ni, and Zn induce the opposite effect and stimulate production of β-carotene. It was found that direct incorporation of Se into yeast cells during cultivation in Se-rich medium can not be used for preparation Se-enriched yeast biomass. Instead, cultivation of the yeasts and a subsequent treatment with sodium selenite during 24h should be applied. A non-lethal and simultaneously maximum tolerated concentration of Se was determined based on the growth curves of the individual strains. A 60-ppm concentration was used with all strains, and the distribution of Se in the cells, on the surface of cells, and in the exopolymers was analyzed. The maximum Se sorption was observed with the cells of species *Rhodotorula glutinis* CCY 20-2-26 (17 mg/g dry weight), while its exopolymers accumulated only 7% of the total adsorbed Se. The remaining Se was sorbed onto the fibrillar part of the cell wall and into the cells. Similarly, two other studied strains, CCY 19-6-4 and CCY 20-2-33, sorbed Se primarily into cells (63–74%) and the fibrillar part of cell wall (2–22%), whereas exopolymers bound only 12–32% of the total sorbed amount. The yeasts with high content of the carotenoid pigments and selenium may be used for the preparation of a new type of antioxidant formula that could be directly applied for various human and animal diets. Such a formula can only be produced by separate processes of the cultivation of red yeasts and a subsequent sorption of selenium into the cells (Breierova et al., 2008).

In general, there have been several reports on the enhancement of volumetric production (mg/l) as well as cellular accumulation (mg/g) of microbial carotenoid upon supplementation of metal ions (copper, zinc, ferrous, calcium, cobalt, alluminium) in yeasts and molds (Bhosale, 2004; Buzzini et al., 2005). Trace elements have been shown to exert a selective influence on the carotenoid profile in red yeasts. It may be explained by hypothesizing a possible activation or inhibition mechanism by selected metal ions on specific carotenogenic enzymes, in particular, on specific desaturases involved in carotenoid biosynthesis, in agreement with previous studies reporting activation or inhibition by metal ions in microbial desaturases (Buzzini et al., 2005). The other explanation is based on observations that presence of heavy metals results in formation of various active oxygen radicals what, in a turn, induces generation of protective carotenoid metabolites that reduce negative behaviour of free radicals. Such strategy has been applied in several pigment-forming microorganisms to increase the yield of microbial pigments (Rapta et al., 2005; Breierova et al., 2008).

2.5.4 Enrichment of red yeast biomass by specific isoprenoid compounds – ergosterol and Coenzyme Q10

In previous text main groups of biotechnologically important metabolites used for enrichment of red yeast biomass were described. Mainly carotenoids, ergosterol, lipids and metal accumulation in red yeast cells makes them attractive for industrial applications.

Ergosterol is provitamin D, part of was followed partly as the additional parameter of biomass quality and also to monitor the competition of two specialized branches of isoprenoid pathway, which is used for the biosynthesis of both carotenoids and sterols. The production of ergosterol was very similar to the production of β-carotene, even if these metabolites were formed in competitive branches of isoprenoid metabolic pathway (Marova et al., 2010). Practically simultaneous oscillation in carotenoid and ergosterol production under optimal conditions could be caused by the role of both metabolites in *R. glutinis* stress response. Carotenoids act as antioxidants and may prevent cells or cell membranes against negative effects of increased oxidative stress. Ergosterol is an integral component of yeast cell membranes, which are very sensitive to external stress. Recently it has been found that the major changes in intact cells of red yeast *Rhodotorula minuta* irradiated by UV-B were interpreted as combination of changes observed in the cell wall and membrane, the changes observed in the membrane preparations were attributed to ergosterol (Tan et al., 2003).

Ergosterol is a precursor of Vitamin D2 and it is also used for the production of cortisone (Metzler 2003). Now ergosterol as single product is commercially produced by yeast fermentation using *Saccharomyces cerevisiae* strains. The popular means to improve the ergosterol fermentation are optimization of the culture medium, screening of the high ergosterol producing strains. Different carbon sources, nitrogen sources and other nutrient materials had different influences on cell growth and accumulation of ergosterol in yeast biomass. A new yeast strain, obtained by way of protoplast fusion, increased the biomass to 2.45 g/100 ml (dry cell weight) and the ergosterol content to 3.07% (Frengova a Beshkova, 2009). It was reported that the synthesis of ergosterol was not determined by cell growth but by the oxygen consumption rate. Ethanol was formed in yeast fermentation and it had an obvious influence on the growth of yeast. In yeast culture process, glucose is preferred and when the glucose concentration reaches a low value, the cell growth is confined. Then after a short period of adaption, cells continue to grow by consuming the ethanol produced in the first phase as the carbon source. The whole process appeared to be a two-phase process. The ergosterol content increased when the specific growth rate decreased. The environmental and physiological parameters such as the dissolved oxygen, oxygen uptake rate of yeast cells culture had direct or indirect influences on the accumulation of ergosterol and the growth of yeast cells. The interaction relation might help to optimize the ergosterol fermentation. But until now little work has been reported on this relation (Tan et al., 2003).

Carotenoids are important natural pigments that play an essential role as accessory light-harvesting pigments and, especially, in protection against damage by photosensitized oxidation. Several yeast genera—*Rhodotorula, Sporobolomyces, Rhodosporidium,* and *Cryptococcus*—produce also coenzyme Q10 (CoQ10; Dimitrova et al., 2010). CoQ10 has a similar isoprenoid chain in its structure. It is also an interesting product for biotechnology. CoQ10 is present in all cells and membranes, and in addition to being a member of the mitochondrial respiratory chain, it also has several other functions of great importance for the cellular metabolism, such as participation in the extra-mitochondrial electron transport (plasma membranes and lysosomes), regulation of the mitochondrial permeability of transition pores, and regulation of the physicochemical properties of membranes. CoQ10, especially, is widely used as an essential component of ATP generation in the oxidative phosphorylation process and as an antioxidant preventing lipid peroxidation and scavenging superoxide. It has been proved that yeast CoQ10 is much better absorbed by the skin than the synthetic CoQ10. Peroxide reduction in the stratus corneum is considerably more pronounced after yeast CoQ10 application. Therefore, research efforts on the production of CoQ10 by microorganisms focus on the development of potent strains by conventional mutagenesis and metabolic engineering, analysis and modification of the key metabolic pathways, and optimization of fermentation strategies. Various microorganisms, including bacteria (e.g., *Agrobacterium, Rhodobacter*) and yeasts (e.g., *Candida, Rhodotorula*, and *Saitoella*), are reported as CoQ10 producers in patented laid-open applications purposely applied in pharmaceutical and cosmetic industry (Dimitrova et al., 2010, Yurkov et al., 2008).

Strains of basidiomycetous yeasts isolated from different sources were studied in order to determine the content of carotenoid pigments and ubiquinone Q10 for subsequent selection work to obtain producers of these substances. The high specific productivity of carotenoids (600–700 mg/g) was revealed in the representatives of the following species: *Cystofilobasidium capitatum, Rhodosporidium diobovatum, R. sphaerocarpum, Rhodotorula glutinis, Rhodotorula minuta,* and *Sporobolomyces roseus.* The ratio of the major pigments (torulene, torularhodine and β-carotene) in the representatives of different species was studied. Certain specific features of pigment formation in relation to the taxonomic position of the yeasts were determined. Eurybiont species with substantial ecological lability are the most active producers of carotenoids and ubiquinone Q10 among the epiphytes. It is the first time a comparative analysis of the coenzyme Q10 content in different taxa has been performed using several strains of the same species. The maximal coenzyme Q10 production (1.84 mg/g of dry biomass) was found in the yeast species *R. sphaerocarpum* (Yurkov et al., 2008).

2.5.5 Carotenoid-synthesizing yeasts—directions for their use

Because of the biological role of the carotenoids as vitamin A precursors in humans and animals and owing to their antioxidant properties and suspected activity in preventing some forms of cancer as well, carotenoid pigments represent a group of most valuable molecules for industrial applications of red yeasts. The pharmaceutical, chemical, feed and food industries have shown increased interest in the use of carotenoids, mainly as provitamin A, but also as natural food and feed colorants. Accordingly, the red yeast *P. rhodozyma* is currently used for the production of astaxanthin, an important carotenoid pigment that can be exploite in aquaculture to give an appealing pink color to the fresh of farmed salmonid fish, and it also helps to impart a desirable golden color to the egg yolk and fresh of poultry. Salmon farming is an industry that is growing and gradually replacing the world's wild salmon fisheries. The most expensive ingredient in salmonid feeds is astaxanthin, and though the actual revenues are privately held, it has been estimated that the market for astaxanthin in >US $100 milion per year (Frengova & Beshkova, 2009).

Similarly to *Xanthophyllomonas*, also other red yeast strains could be used for industrial puropses to pruduction of carotenoids - beta-carotene, torulene, lycopene, as well as further lipid metabolites produced in cells. In many works mostly *Rhodotorula glutinis* sems to be perspective strain. Combined enrichment of *Rhodotorula* biomass by provitamin A (carotenes) and provitamin D (ergosterol) could be used in food and feed supplements (Marova et al., 2010), aditional enrichment by Coenzyme Q10 is suitable product for cosmetics and could be used also in food and feed (Dimitrova et al., 2010). Formulas based on selenium-enriched red yeast biomass with enhanced carotenoid content could be used as nutrition suplement too (Breierova et al, 2008). There is also posibility to use oleaginous red yeasts to single cell oil production; in this case production of other lipid metabolites could be reduced and the main flow of acetylCoA will be directed to fatty acid and lipid biosynthesis (Dai et al., 2007).

One limitation impacting the industrial utility of *P. rhodozyma/ X. dendrourhous* or *Rhodotorula* species has been hindered absorption of carotenoids, due to the yeast`s thick cell wall. Because of presence of other specific biologically active compounds as well as high level of nutritionally sigificant yeast cell components (proteins, unsaturated fat, vitamins…) the best strategy is to disrupt cells and to use the whole biomass without isolation of individual compounds. The biotechnology industry has developed different means of active compounds liberation by the yeast including optimization of drying conditions, mechanical breakage, microwave treatment and enzyme treatment, as described below (Frengova & Beshkova, 2009).

When disrupted cells *P. rhodozyma,* without cell walls are added to the diets of animals, astaxanthin is readily absorbed from the gut; it effectively colors the fresh of penreared salmonids, and also helps impart a desirable golden color to the egg yolk and fresh of poultry. Astaxanthin in yeast (*X. dendrorhous*) prepared by spray drying and Xat-roller milling was well absorbed by laying hens and was successfully used as a pigmentation agent in animals (An, 2005). Specifically, when spray-dried and milled yeast was supplied in the feed (40 mg astaxanthin/kg feed), astaxanthin was successfully absorbed (1,500 ng/ml blood and 1,100 ng/g skin) by laying hens. Extrusion temperature did not affect utilization of dietary astaxanthin or rainbow trout fresh color significantly, but cell wall disruption of red yeast cells was critical to optimize carotenoid utilization. Increasing the degree of enzymatic cell wall disruption increased fresh astaxanthin concentrations from 2.2 to 6.7 mg/kg, redness values from 5.5 to 10.7, yellowness values from 11.7 to 16.7 and astaxanthin retentions in the muscle from 3.7 to 17.4%. A formulation of *P. rhodozyma* cells blended with ethoxyquin, lecithin and oil prior to drying also increased astaxanthin deposition in salmonid fish fresh and rainbow trout fresh when supplied in feed as an additive. Absorption and accumulation of biological astaxanthin were higher thah those of chemical astaxanthin, probably because of the high contents of lipids in the yeast (17%). Lipid peroxide formation in skin was significantly decreased by astaxanthin. The peroxide production in chickens fed chemical astaxanthin was markedly lowered compared to biological astaxanthin (Frengova & Beshkova, 2009) .

The levels of serum transaminase activities and of lipid peroxides in fish fed oxidized oil were significantly higher that those of the control fish fed non-oxidized oil. However, the supply of freeze-dried red yeast preparation considerably decreased both enzyme activities and lipid peroxides level. Furthermore, the serum lipid (triglycerides, total cholesterol and phospholipids) concentrations were also significantly decreased. Especially, the serum triglyceride level of fish fed the red yeast was as low as that of the control. Recently was found that Zn2+ ions induced changes in yeasts (*R. glutinis* and *R. rubra*) leading to more efficient scavenging and antioxidant capacities compared with Ni2+ ions, and antioxidants (carotenoids) present in yeast's walls showed higher ability to scavenge free radicals than those from inside the cells (Rapta et al., 2005). Later, the in vivo antioxidant and protective effects of astaxanthin isolated from *X. dendrorhous* against ethanol-induced gastric mucosal injury were established in animal models, especially rats (Kim et al., 2005). Oral administration of astaxanthin showed significant protection against ethanol-induced gastric lesion and inhibited elevation of the lipid peroxide levels in gastric mucosa. A histologic examination clearly indicated that the acute gastric mucosal lesion induced by ethanol nearly disappeared after pretreatment with astaxanthin (Frengova & Beshkova, 2009).

Chemopreventive and anticarcinogenic effects of carotenoids by *Rhodotorula* on the development of preneoplastic lesions during *N*-nitrosodiethylamine (DEN)-induced hepatocarcinogenesis in female Wistar strain rats were also studied (Bhosale et al., 2002). Spray-dried yeast *R. glutinis* (containing carotenoid pigments torulene, torularhodin and beta-carotene in proportion 58:33:2) showed significant effect on the prevention of liver tumor development. However, *R. glutinis* effects were relatively more significant in groups where *R. glutinis* was administered after DEN treatment, suggesting that *R. glutinis* is quite effective in the prevention of liver tumor development especially when administered after DEN treatment, indicating possible protective effects at the promotional stages.

3. Conclusions

Yeast is, due to its physiological properties, widely used in the food, feed, chemical and pharmaceutical industries for production of various valuable compounds. Red yeast is well known producer of carotenoids which are significant because of their activity as vitamin A precursors, colorants, antioxidants and possible tumor-inhibiting agents. Biological sources of carotenoids receive major focus nowadays because of the stringent rules and regulations applied to chemically synthesized/purified pigments. Compared with the extraction from vegetables, the microbial production of carotenoids is of paramount interest, mainly because of the problems of seasonal and geographic variability in the production and marketing of several of the colorants of plant origin. Moreover, red yeast is a rich source of other specific compounds - ergosterol, Coenzyme Q10, as well as unsaturated fatty acids, fats, proteins and vitamins and can be incorporated in feeds to enhance the nutritional value of yeast biomass. One limitation impacting the industrial utility of carotenogenic yeast has been complicated liberation and bioavailability of carotenoids and other active compounds, due to the yeast's thick cell wall.The biotechnological industry has developed different means of pigment liberation by the yeast including optimization of drying conditions, mechanical breakage, microwave treatment and enzyme treatment.

The other very important limitation involved in the practical exploitation of yeasts is the high cost of microbial production. The production cost could be reduced by increasing yields of product, as well as using less expensive substrates. There is a need to improve fermentation strategies. Biomass and metabolites production by red yeast is highly variable and can be influenced by cultivation conditions (light, temperature, pH, aeration etc.). Different approaches for improving the production properties of the yeast strains, such as environmental stress, mutagenesis or genetic modification, have been studied and optimized. The other possibility for production cost reduction is using various low-cost materials as carbon or nitrogen source. The potential of several waste materials (whey, potato mass, apple mass and various cereals) as substrates for carotenoid and ergosterol production by some yeast strains belonging to the genus *Rhodotorula* and *Sporobolomyces* were succesfully examined. Mild nutrition stress cause by several waste substrates was found to be the suitable induction factor for higher carotenogenesis and ergosterol production in red yeasts.

Environmental stress was reported to induce carotenoid, ergosterol and lipid production as part of red yeast stress response. Under stress cells posses altered phenotype biotechnologically significant and/or undesirable in a dose-dependent manner. Phenotypic profiling of the environmental stress responses demonstrates genetic susceptibility of yeast to environmental stress. Low concentrations of oxidative and osmotic stress, which can under specific conditions induce carotenogenesis, have no significant effect on yeast growth. Red yeast cultivated under osmotic and oxidative stress or on various waste substrates

shows no significant differences in cell morphology when compared with yeast cultivated in conventional glucose medium under optimal conditions. Thus, low environmental stress can be used for induction of carotenogenesis and use of non-toxic stress factors (salt, metals) can enable utilization o whole cell biomass to industrial use. Simple and cheap stress factor in relatively low concentration can substantially enhance biotechnologically significant metabolite production.
Growing interest in pigment and other metabolite applications in various fields coupled with their significance in health and dietary requirements has encouraged "hunting" for more suitable sources of these compounds. Due to restrictions, there is no possibility to apply carotenoids prepared by chemical synthesis for food, pharmaceutical and medical purposes. However, the success of microbial pigments, metabolites and single cell oils depends upon their acceptability in the market, regulatory approval, and the size of the capital investment required to bring the product to market. Therefore, the focus of biotechnology on highly valuable yeast biomass requires knowledge how microorganisms control and regulate the biosynthetic machinery in order to obtain metabolites and enriched biomass in high yield and at low price. From this view, attempts have been directed at the development and improvement of biotechnological processes for the utilization of red yeasts on an industrial scale. Current successes using mutation methods and molecular engineering techniques carried out over recent years have not only answered some fundamental questions related to pigment formation but has also enabled the construction of new microbial varieties that can synthesize unusual carotene metabolites. Elucidation of these mechanisms represents a challenging and potentially rewarding subject for the further research and may finally allow us to move from empirical technology to predictable carotenoid and/or isoprenoid metabolite design. Thus, the manipulation and regulation of red yeast metabolism open a large number of possibilities for academic research, demonstrates the enormous potential in its application and creates new economic competitiveness and market of microbial lipid compounds.

4. Acknowledgement

This work was supported by project "Centre for Materials Research at FCH BUT" No. CZ.1.05/2.1.00/01.0012 from ERDF. Finantial support was provided also by grants VEGA 1/0747/08 and VEGA 2/0005/10 from the Grant Agency of the Ministry of Education, Slovak Republic and by grant VVCE-0064-07 from the Slovak Research and Development Agency, Slovak Republic.

5. References

Aksu, Z. & Eren, A. T. (2005). Carotenoids production by the yeast Rhodotorula mucilaginosa: Use of agriculture wastesas a carbon source. *Process Biochemistry* Vol. 40, No.9, pp. 2985-2991, ISSN 1359-5113

An, G.H. (2001). Improved growth of the red yeast, Phaffia rhodozyma (Xanthophyllomyces dendrorhous), in thepresence of tricarboxylic acid cycle intermediates. *Biotechnology Letters* Vol. 23, pp. 1005–1009, ISSN 0141-5492

Armstrong, G. A. & Hearst, J. E. (1996). Carotenoids 2: Genetics and molecular biology of carotenoid pigment biosynthesis. *The FASEB Journal*, Vol.10, No.2, pp. 228-237, ISSN 0892-6638

Beopoulos, A., Cescut, J., Haddouche, R., Uribelarrea, J.L., Jouve, C.M. & Nicaud, J.M. (2009) Yarrowia lipolytica As a model for biooil production. *Progress in Lipid Research* Vol.48, pp. 375–387, ISSN 0163-7827

Bhosale, P. (2004). Environmental and cultural stimulants in the production of carotenoids from microorganisms. *Applied Microbiology and Biotechnology*, Vol.63, 351–361, ISSN 0175-7598

Bhosale, P. & Gadre, R. V. (2001a). Production of ß-carotene by a mutant of *Rhodotorula glutinis*. *Applied Microbiology and Biotechnology*,Vol.55, No.4, pp. 423-427, ISSN 0175-7598

Bhosale, P. & Gadre, R. V. (2001b). β-Carotene production in sugarcane molasses by a *Rhodotorula glutinis* mutant. *Journal of Industrial Microbiology and Biotechnology*, Vol.26, No.6, pp. 327-332, ISSN 1367-5435

Breierova, E.; Gregor, T.; Marova, I.; Certik, M. & Kogan, G. (2008). Enhanced antioxidant formula based on a selenium-supplemented carotenoid-producing yeast biomass. *Chemistry and Biodiversity*, Vol.5, No.3, pp.440-446, ISSN 1612-1872

Britton, G.; Liaen-Jensen, S. & Pfander H. (1998): *Carotenoids*. Volume 3: Biosynthesis and Metabolism, pp. 14-191. Birkhäuser Verlag Basel-Boston-Berlin ISBN 3-7643-5829-7

Buzzini, P. (2001). Batch and fed-batch carotenoid production by *Rhodotorula glutinis–Debaryomyces castellii* co-cultures in corn syrup. *Journal of Applied Microbiology*, Vol.90, No.5, pp. 843-847, ISSN 1364-5072

Cerda-Olmedo, E. (2001). Phycomyces and the biology of light and color. *FEMS Microbiology Reviews* Vol.25, pp.503–512, ISSN 0168-6445

Certik, M.; Hanusova, V.; Breierova, E.; Marova, I. & Rapta P. (2009). Biotechnological production and propertiesof carotenoid pigments In: *Biocatalysis and Agricultural Biotechnology* (415 p.), Chapter 25, pp.358-373. Tailor and Francis Group LLt, ISBN 978-1-4200-7703-2

Dai, C.; Tao, J.; Xie, F.; Dai, Y. & Zhao, M. (2007). Biodiesel generation from oleaginous yeast *Rhodotorula glutinis* with xylose assimilating capacity. *African Journal on Biotechnology* Vol.6, No.18, pp.2130–2134, ISSN 1684-5315

Davoli, P.; Mierau, V. & Weber R. W. S. (2004). Carotenoids and Fatty Acids in Red Yeasts *Sporobolomyces roseus* and *Rhodotorula glutinis*. *Applied Biochemistry and Microbiology*, 2004. 40(4): p. 392-397, ISSN 0003-6838.

Dimitrova, S.; Pavlova, K.; Lukanov, L. & Zagorchev P. (2010). Synthesis of Coenzyme Q10 and β-carotene by yeastsisolated from Antarctic soil and lichen in response to UV and visible radiations. *Applied Biochemistry and Biotechnology* Vol.162, pp.795–804, ISSN 0273-2289

Ducrey Sanpietro, L.M. & Kula, M.R. (1998). Studies of astaxanthin biosynthesis in *Xanthophyllomyces dendrorhous* (Phaffia rhodozyma). Effect of inhibitors and low temperature. *Yeast* Vol.14, pp.1007–1016, ISSN 0749-503X

Dufosse, L. (2006). Microbial Production of Food Grade Pigments. *Food Technology and Biotechnology*, Vol.44,pp. 313–321, ISSN 1330-9862.

Easterling, E.R.; French, W.T.; Hernandez, R. & Licha, M. (2009) The effect of glycerol as a sole and secondary substrate on the growth and fatty acid composition of *Rhodotorula glutinis. Bioresource Technology* Vol. 100, pp.356–361, ISSN 0960-8524

Fang, TJ., Wang, J.M. (2002). Extractibility of astaxanthin in a mixed culture of a carotenoid over-producing mutant of *Xanthophyllomyces dendrorhous* and *Bacillus circulans* in two-stage batch fermentation, *Process Biochemistry* Vol. 37, pp.1235–1245, ISSN 1359-5113

Flores-Cotera, L.B. & Sanchez, S. (2001). Copper but not iron limitation increases astaxanthin production by *Phaffia rhodozyma* in a chemically defined medium. *Biotechnology Letters* Vol.23, pp.793–797, ISSN 0141-5492

Frengova, G. I. & Beshkova, D. M. (2009). Carotenoids from *Rhodotorula* and *Phaffia*: yeasts of biotechnological importance. *Journal of Industrial Microbiology & Biotechnology*, Vol. 36, No.2, pp. 163-180, ISSN 1367-5435

Frengova, G.; Simova, E. & Beshkova, D. (2004). Use of whey ultrafiltrate as a substrate for production of carotenoids by the yeast *Rhodotorula rubra. Applied Biochemistry and Biotechnology* Vol.112, pp.133–141, ISSN 1367-5435

Gasch, A. P. & Werner-Washburne, M. (2002). The genomics of yeast responses to environmental stress and starvation. *Functional & Integrative Genomics*, September, Vol. 2, No.4-5, pp. 181-192, ISSN 1438-793X

Gasch, A. P.; Spellmann P. T.; Kao C. M.; Carmel-Harel, O.; Eisen M. B., Storz, G.; Botstein, D. & Brown, P. O. (2000). Genomic Expression Programs in the Response of Yeast Cells to Environmental Changes. *Molecular Biology of the Cell*, December, Vol. 11, No.12, pp. 4241-4257, ISSN 1059-1524

Hassan, M.; Blanc, P.J.; Granger, L.M.; Pareilleux, A. & Goma, G. (1996). Influence of nitrogen and iron limitationson lipid production by *Crytococcus curvatus* grown in batch and fed-batch culture. *Process Biochemistry* Vol.31, No.4, pp.355–361, ISSN 1359-5113

Hibbett, D.S. (2007). A higher level phylogenetic classification of the *Fungi. Mycological Research* Vol.111, No.5, pp.509–547, ISSN 0953-7562

Hohmann, S. & Mager, W.H. (2003). *Yeast Stress Responses*. Berlin: Springer-Verlag. 389 p. ISBN 3-540-43926-9

Hoog, G. S.; De Guarro, J.; Gene, J. & Figueras, M. J.(2001). *Atlas of Clinical Fungi.* Second edition. December 1, Washington: ASM Pressp. ISBN 9070351439

Hu, Z.C., Zheng, Y.G., Wang, Z. & Shen, Y.C. (2005). Effect of sugarfeeding strategies on astaxanthin production by *Xanthophyllomyces dendrorhous. World Journal of Microbiology and Biotechnology* Vol.21, pp.771–775, ISSN 0959-3993

Hu, Z.C., Zheng, Y.G., Wang, Z. & Shen, Y.C. (2007) Production of astaxanthin by *Xanthophyllomyces dendrorhous*ZJUT46 with fedbatch fermentation in 2.0 m3 fermentor. *Food Technology and Biotechnology* Vol. 45, pp.209–212, ISSN 1330-9862

Kim, J.H.;, Kang, S.W.; Kim, S.W. & Chang, H.I. (2005). High-level production of astaxanthin by *Xanthophyllomyces dendrorhous* mutant JH1 using statistical designs. *Bioscience Biotechnology and Biochemistry* 69:1743–1748.

Kim, J.H.; Choi, S.K.; Choi, S.Y.; Kim, H.K. & Chang, H.I. (2005). Suppresive effect of astaxanthin isolated from the *Xanthophyllomyces dendrorhous* mutant on ethanol-induced gastric mucosal injury in rats. *Bioscience Biotechnology and Biochemistry* Vol.69, pp.1300–1305, ISSN 0916-8451

Kuranda K, Grabinska K, Berges T, Karst F, Leberre V, Sokol S, François J, Palamarczyk G. (2009). The YTA7 gene is involved in the regulation of the isoprenoid pathway in the yeast Saccharomyces cerevisiae. *FEMS Yeast Research*, Vol.9, No.3, pp. 381-390, ISSN 1567-1364

Latha, B.V.; Jeevaratnam, K.; Murali, H. S & Manja, K. S. (2005): Influence of growth factors on carotenoid pigmentation of *Rhodotorula glutinis* DFR-PDY from natural source. *Indian Journal of Biotechnology*, Vol.4, pp. 353-357, ISSN 0972-5849

Leathers, T.D., Jones, J.D., Hohn, T.M. (2004): System for the sequential, directional cloning of multiple DNA sequences. *US patent 6,696,282*

Li, M.; Liu, G.L.; Chi, Z. & Chi, Z.M. (2010). Single cell oil production from hydrolysate of cassava starch by marine- derived yeast *R. mucilaginosa* TJY15a. *Biomass Bioenergy* Vol.34, pp.101–107, ISSN 0961-9534

Libkind, D.,, Brizzio, S. & van Broock, M. (2004). *Rhodotorula mucilaginosa*, a carotenoid producing yeast strain from a Patagonian high-altitude lake. *Folia Microbiologica* Vol.49, No.1, pp.19–25, ISSN 0015-5632.

Libkind, D., Sommaruga, R.; Zagarese, H. & van Broock M. (2005). Mycosporines in carotenogenic yeasts. *Systematic and Applied Microbiology*, Vol.28, No.8, pp. 749-754, ISSN 0723-2020

Libkind, D. & van Broock, M. (2006). Biomass and carotenoid pigment production by patagonian native yeasts. *World Journal of Microbiology & Biotechnology*, 2006, 22, 687–692, ISSN 0959-3993

Libkind, D.; Gadanho, M.; van Broock M., & Sampaio J. P. (2008). Studies on the heterogeneity of the carotenogenic yeast *Rhodotorula mucilaginosa* from Patagonia, Argentina. *Journal of Basic Microbiology*, Vol. 48, No.2), pp. 93-98, ISSN 0233-111X

Lukacs, G., Linka, B., Nyilasi, I. (2006). *Phaffia rhodozyma* and *Xanthophyllomyces dendrorhous*: Astaxanthin-producing yeasts of biotechnological importance. *Acta Alimentaria*, Vol.5, pp.99–107, ISSN 0139-3006

Maldonade, I.R.; Rodriguez-Amaya, D.B. & Scamparini, A.R.P.(2008). Carotenoids of yeasts isolated from the Brazilian ecosystem. *Food Chemistry* Vol.107, pp.145–150, ISSN 0308-8146

Mann, J. (1990). *Secondary Metabolism*. 2nd edition. September 27, USA: Oxford University Press. 390 p. ISBN 0198555296

Mantzouridou, F.; Roukas, T. & Kotzekidou, P. (2002). Effect of the aeration rate and agitation speed on b-Carotene production and morphology of *Blakeslea trispora* in a stirred tank reactor: mathematical modeling. *Biochemical Engineering Journal*, Vol.10, pp.123–135, ISSN 1639-703X

Marova, I.; Breierova, E.; Koci, R.; Friedl, Z.; Slovak, B & Pokorna J (2004) Influence of exogenous stress factors on production of carotenoids by some strains of carotenogenic yeasts. *Annals of Microbiology*, Vol. 54, No.1, pp. 73-85, ISSN 1590-4261

Marova, I.; Carnecka, M.; Halienova; A.; Koci, R. & Breierova, E.(2010). Production of carotenoid/ergosterol-supplemented biomass by red yeast *Rhodotorula glutinis* grown under external stress. *Food Technology and Biotechnology, Vol.*48, No.1, *p* p.56-61. ISSN 1330-9862

Marova, I.; Carnecka, M.; Halienova; Dvorakova, T. & Haronikova A.(2011). Use of several waste substrates for carotenoid-rich yeast biomass production. *Journal of Environmental Management*. Accepted 8.6.2011. DOI: 10.1016/j.jenvman.2011.06.018 ISSN 0301-4797

Metzler D.E. (2003): Biochemistry. *The Chemical Reactions of Living Cells*. Vol.2, p.1226-1271. Academic Press, Elsevier Science, USA. ISBN 0-12-492543-X

Papp, T.; Velayos, A.; Bartók, T.; Eslava, A.P; Vágvölgyi, C. & Iturriaga, E.A. (2006). Heterologous expression of astaxanthin biosynthesis genes in *Mucor circinelloides*. *Applied Microbiology and Biotechnology* Vol.69, pp.526–531, ISSN 0175-7598

Parajo, J.C.; Santos, V. & Vazajez, M. (1998). Production of carotenoids by *Phaffia rhodozyma* growing on media made from hemicellulosic hydrolysates of *Eucalyptus globulus* wood. *Biotechnology and Bioengineering* Vol. 59, No. 4, August 20, ISSN 0006-3592

Park, K.M., Song, M.W. & Lee, J.H. (2008). Production of carotenoids by β-ionone-resistant mutant of *Xanthophyllomyces dendrorhous* using various carbon sources. *Biotechnology and Bioprocess Engineering* Vol.13, pp. 197-203, ISSN 1226-8372

Perrier, V.; Dubreucq, E. & Galzy, P.(1995). Fatty acid and carotenoid composition of *Rhodotorula* strains. *Archives of Microbiology*, Vol. 164, No3, pp.173-179, ISSN 0302-8933.

Rapta, P., Polovka, M., Zalibera, M., Breierova, E., Zitnanova, I., Marova, I. & Certik, M. (2005). Scavenging andantioxidant properties of compounds synthesized by carotenogenic yeasts stressed by heavy metals - EPR spin trapping study. *Biophysical Chemistry*, Vol.116, pp.1-9, ISSN 0301-4622

Ratledge, C. & Wynn, J.P.(2002). The biochemistry and molecular biology of lipid accumulation in oleaginous microorganisms. *Advances in Applied Microbiology* Vol. 51, pp.1–51, ISSN

Rosa, C. & Peter, G. (2005). *Biodiversity and Ecophysiology of Yeasts (The Yeast Handbook)*. December 7, Berlin, Heidelberg, New York: Springer. 580 p. ISBN 3540261001

Sakaki, H.; Nakanishi, T.; Tada, A., Miki, W. & Komemushi, S. (2001) Activation of torularhodin production by *Rhodotorula glutinis* using weak white light irradiation. *Journal of Bioscience and Bioengineering* Vol. 92, pp.294–297, ISSN 1389-1723

Sandmann, G. (2001). Carotenoid Biosynthesis and Biotechnological Application. Archives *Biochemistry and Biophysics* Vol. 385, pp. 4–12, ISSN 0003-9861

Schmidt, I.; Schewe, H.; Gassel, S.; Jin, C.; Buckingham, J.; Hümbelin, M.; Sandmann, G. & Schrader, J. (2011).Biotechnological production of astaxanthin with *Phaffia rhodozyma/Xanthophyllomyces dendrorhous. Applied Microbiology and Biotechnology* Vol.89, No.1, pp.555–571, ISSN 0175-7598

Schmidt-Dannert, C. (2000). Engineering novel carotenoids in microorganisms. *Current Opinion in Biotechnology* Vol.11,pp.255–261, ISSN 0958-1669

Simova, E.D., Frengova, G.I., Beshkova, D.M. (2004). Synthesis of carotenoids by *Rhodotorula rubra* GED8 co-cultured with yogurt starter cultures in whey ultrafiltrate. *Indian Journal of Microbiology and Biotechnology* Vol. 31, pp. 115–121, ISSN 0972-5849

Somashekar, D. & Joseph, R (2000). Inverse relationship between carotenoid and lipid formation in *Rhodotorula gracilis*according to the C/N ratio of the growth medium. *World Journal of Microbiology & Biotechnology*, Vol.16, pp.491–493, ISSN 0959-3993

Squina, F.M. & Mercadante, A.Z. (2005). Influence of nicotine and diphenylamine on the carotenoid composition of *Rhodotorula* strains. *Journal of Food Biochemistry* Vol.29 pp. 638–652, ISSN 0145-8884

Subramaniam, R.; Dufreche, S.; Zappi, M. & Bajpai R. (2010). Microbial lipids from renewable resources: production andcharacterization. *Indian Journal of Microbiology and Biotechnology, Vol.* 37, pp.1271–1287, ISSN 0972-5849

Talaro, K. P. & Talaro, A. (2001). *Foundations in Microbiology*. Pasadena: McGraw-Hill. 889 p. ISBN 0-07-248864-6

Tan, T.; Zhang, M. & Gao, H. (2003). Ergosterol production by fed-batch fermentation of *Saccharomyces cerevisiae. Enzyme and Microbial Technology* Vol.33 pp. 366–370. ISSN 0141-0229

Tinoi, J.; Rakariyatham, N. & Deming, R.L. (2005). Simplex optimization of carotenoid production by *Rhodotorula glutinis* using hydrolyzed mung bean waste flour as substrate. *Process Biochemistry*, 40(7): p. 2551-2557, ISSN 0032-9592

Verdoes, J.C.; Sandmann, G.; Visser, H.; Diaz, M.; Van Mossel, M. & Van Ooyen, A.J.J. (2003) Metabolic engineering ofthe carotenoid biosynthetic pathway in the yeast *Xanthophyllomyces dendrorhous. Applied and Environmental Microbiology* Vol.69, pp.3728–3738, ISSN 0099- 2240

Verwaal, R.; Wang, J.; Meijnen, J.P.; Visser, H.; Sandmann, G.; Van den Berg, J. & Van Ooyen, A.J.J. (2007). High-level production of beta-carotene in *Saccharomyces cerevisiae* by successive transformation with carotenogenic genes from *Xanthophyllomyces dendrorhous. Applied and Environmental Microbiology* 73:4342–4350, ISSN 0099- 2240

Walker, G. M. (1998). *Yeast physiology and biotechnology*. West Sussex, England: John Willey & Sons Ltd. 362 p. ISBN 0-471-964468

Wang, C.; Oh, M.K. & Liao, J.C. (2000). Directed evolution of metabolically engineered *Escherichia coli* for carotenoid production. *Biotechnology Progress* Vol.16, pp.922–926 ISSN 1520-6033

Xiao, W. (2005). *Yeast Protocols (Methods in Molecular Biology)*. Second edition. September 1, Totowa, New Jersey: Humana Press. 392 p. ISBN 1588294374

Xue, F.; Miao, J.; Zhang, X. & Tan T.(2010). A New Strategy for Lipid Production by Mix Cultivation of *Spirulina platensis* and *Rhodotorula glutinis*. *Applied Biochemistry and Biotechnology* Vol. 160, pp.498–503, ISSN 0273-2289

Yurkov, A.M.; Vustin, M.M.; Tyaglov, B.V.; Maksimova, I.A. & Sinekoiy S.P. (2008). Pigmented *Basidiomycetous* Yeasts are a promising source of carotenoids and ubiquinone Q10. *Microbiology,* Vol. 77, No. 1, pp. 1–6.ISSN 0026-2617

Part 3

Usage

19

Biomass Burning in South America: Transport Patterns and Impacts

Ana Graciela Ulke[1], Karla María Longo[2] and Saulo Ribeiro de Freitas[3]

[1]*Departamento de Ciencias de la Atmósfera y los Océanos, Facultad de Ciencias Exactas y Naturales, Universidad de Buenos Aires, Buenos Aires*

[2]*Divisão de Geofísica Espacial, Instituto Nacional de Pesquisas Espaciais, São José dos Campos, São Paulo*

[3]*Centro de Previsão de Tempo e Estudos Climáticos, Instituto Nacional de Pesquisas Espaciais, Cachoeira Paulista, São Paulo*

[1]*Argentina*

[2,3]*Brazil*

1. Introduction

The Andes Mountains barrier and the interaction with the easterly trade winds, and the flow associated to the South Atlantic Subtropical High (SASH) are responsible of a key feature of the low-level atmospheric circulation and climate: the so called South American Low Level Jet (SALLJ). The SALLJ is a wind maximum immersed in a pole-ward and moist current with a cross stream mean dimension in the mesoscale, which has been identified as an efficient dynamical mechanism to transport heat and humidity from tropical to subtropical latitudes. The SALLJEX (South American Low Level Jet Experiment) field campaign provided a unique data set for the study and better understanding of the SALLJ (Vera et al., 2006). The SALLJ feeds and controls the life cycle of the mesoscale convective systems over an area that includes the Del Plata basin, and accounts for an important fraction of the precipitation in southern South America, thus influencing the water balance in the region (Nicolini et al., 2002; Saulo et al., 2000). The SALLJ has also being pointed as an important agent to transport and mix other biogeochemical components (Paegle, 1998).

The orographic control of the Andes favouring the poleward flow causes the persistency of the SALLJ all year round, being only episodically interrupted by mid-latitude transient systems arriving in the subtropical South America (SA) (James & Anderson, 1984; Nogues-Paegle et al., 1998). While during the summer this flow has a net poleward component, in the winter it has an eastward tendency up in the mid-latitudes, with an outflow toward the South Atlantic Ocean broadly ranging from 20° S to 40° S, strongly depending on the position of the SASH. Nogues-Paegle & Mo (1997) found an intraseasonal meridional seesaw of dry and wet conditions over tropical and subtropical South America during austral summer in which the South Atlantic Convergence Zone (SACZ) and the low-level stream intensify alternatively. Over the central and north bands of SA during the winter, the climate is strongly influenced by the northward motion of the Inter-tropical Convergence Zone (ITCZ) and the westward displacement of SASH, composing a scenario of a low levels high pressure system over the continent, with light winds and most of the convection being shifted to the northern part of the Amazon and very little precipitation.

This is the climatological scenario of the SA dry season that eases the Tropical Forest and Cerrado biomes anthropogenic crackdown, followed by the biomass burning. In fact, the vegetation fire activity had been since remote times incorporated, as a supposedly acceptable practice, by the local culture to expand pasture and crop lands and even as a regular agricultural harvest tool for some types of produce, such as sugar cane. Every year during the dry season hundreds of thousands of fire spots and the produced thick regional smoke plume, which covers an area of about 4-5 millions of square kilometres, have been detected by satellite observation over SA.

As the SALLJ drives an important mass exchange from the tropical Amazon to the sub-tropics it is predictable that this low level flow could as well play an important role intercommunicating regional climate changes in the Amazonian basin to the southern South American basins. This paper examines the mass exchange between the Amazon basin and the subtropical SA patronized by the SALLJ during the dry/burning season, when the transport of heat and moist occurs associated with the transport of biomass burning smoke aerosol particles.

2. Methods

A diagnose of the occurrence of the SALLJ events for the 2002 was performed based on the modified Bonner's first criterion for the strength and vertical shear of the wind field (Bonner, 1968; Saulo et al., 2000), using the 6-hourly analysis of the Global Data Assimilation System (GDAS) of the National Centers for Environmental Prediction (NCEP). This data set has one-degree horizontal resolution and is available every synoptic time (0000, 0600, 1200 and 1800 UTC), at 26 vertical pressure levels.

The information about fire spots over South America is obtained with remote sensors and after processing, it is freely available at http://www.cptec.inpe.br. The observations of aerosols in Buenos Aires that could give information of the intrusion of the regional smoke plumes consist on columnar aerosol content and derived quantities obtained from measurements at the CEILAP-BA (34.5° S, 58° W) (Buenos Aires) site of the AErosol RObotic NETwork (AERONET) from National Atmospheric and Science Administration (NASA) (http://aeronet.gfsc.nasa.gov).

The on-line atmospheric transport model CATT-BRAMS (Coupled Aerosol and Tracer Transport model to the Brazilian developments on the Regional Atmospheric Modeling System) was used to simulate the atmospheric transport of biomass burning smoke during the dry season of 2002. A detailed description of the CATT-BRAMS system can be seen at Freitas et al., 2009; Longo et al., 2010). The system considers the emission, transport and transformations of particulate matter (PM2.5) and gases (CO) and it is run operatively at Centro de Previsão de Tempo e Estudos Climáticos (CPTEC) with 40 km resolution over South America. It provides 72-hour predictions of the above mentioned aerosols and gases as well as the meteorological fields. Two SALLJ events were selected to perform a more in depth analysis of the transport patterns and the aerosol dispersion. The synoptic environment in which they took place was studied and the resulting spatial and temporal distributions of aerosols obtained with the CATT-BRAMS modelling system for each case were analysed.

3. Results

3.1 SALLJ and biomass burning in 2002

The occurrence of SALLJ in the 2002 was then determined and the pattern found was in agreement with previous studies for other years. Figure 1 shows the percent relative frequencies of SALLJ obtained for each month.

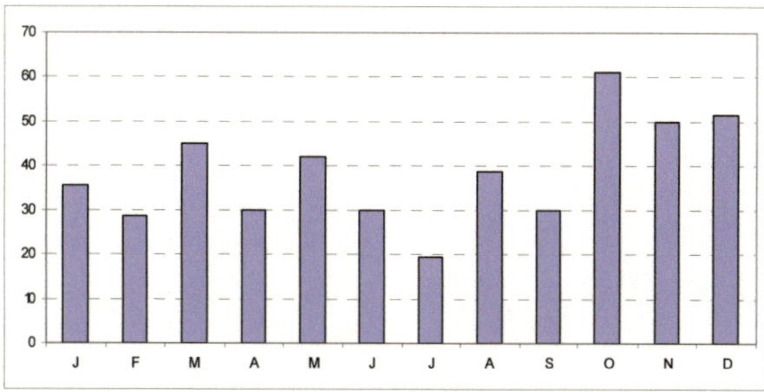

Fig. 1. Monthly relative frequencies of SALLJ (%) during 2002.

The low level flow was present all through the 2002 year, though presenting variability in its strength, frequency and location mainly related to the different synoptic conditions, and the greater scale climatological scenario. The higher frequencies of occurrence of SALLJ are observed in October and the lower in July. As previously mentioned, the aim of the present study is to relate the low-level jet east of the Andes with the dispersion of biomass burning products in South America. Figure 2 presents the number of fire spots in South America for each month in 2002. The important increase from August to October –namely the biomass burning season- is clearly evident. In consequence, we will restrict the further analysis to the events in those months.

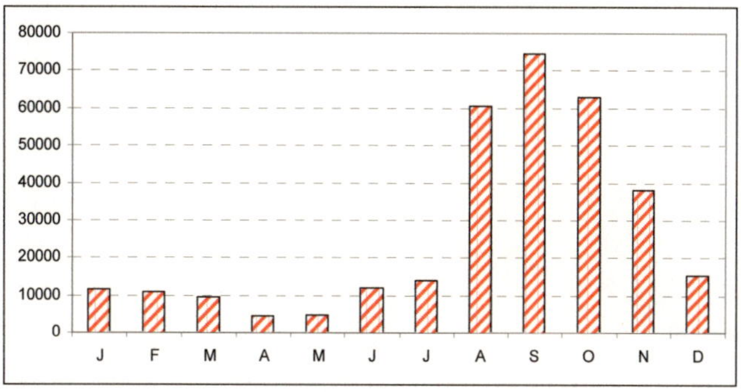

Fig. 2. Number of fires in South America per month during 2002.

The main characteristics of the mean low-level flow are depicted in the composite fields, obtained averaging the days that comprise the SALLJ events for the burning season months (Figure 3). August and October were characterized by a northerly oriented flow, when mainly the northeast of Argentina was under its influence. In September the mean pattern was more north-westerly oriented with an outflow towards the Atlantic Ocean, over passing the southern region of Brazil. August shows the southernmost penetration, greatest horizontal wind speed gradient and vertical wind speed shear. During this month, the events are less frequent but much stronger. In the opposite, in October, there is a higher recurrence of generally weaker events. The mean low level north-westerly flow organizes at about 15° S and extends southward reaching 30-35° S.

The associated circulation patterns in conjunction with the occurrence of biomass burning caused the transport of aerosols and gases towards different regions with diverse impacts. Figure 4 shows the composites of the modelled vertically integrated aerosol optical thickness at 500 nm (AOT500) and the flow pattern for the SALLJ events. The mean plume and flow are very well reproduced and the higher aerosol concentrations are directly related to the greater emissions during September and October.

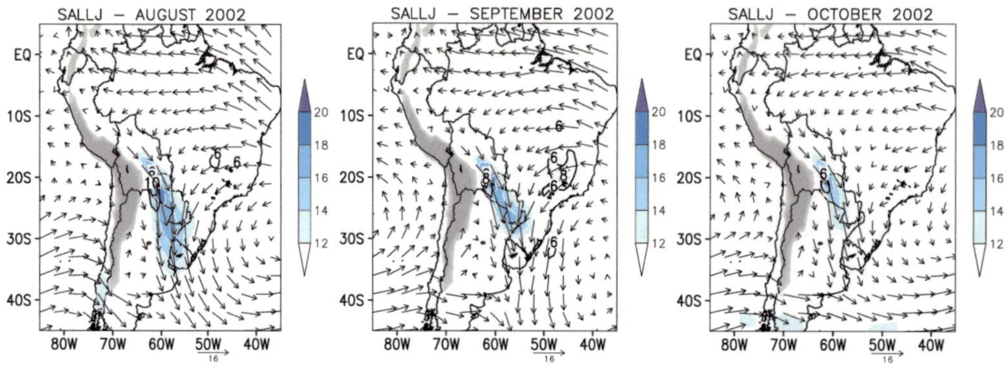

Fig. 3. Monthly composite fields for SALLJ during the biomass burning season: wind (*vector*); wind speed (*shaded*) at 850 hPa and wind shear between 850 hPa and 700 hPa (*black contours*). *Shaded*: wind intensity stronger than 12 m s^{-1}. *Black contours*: wind shear greater than 6 m s^{-1}. Terrain elevations higher than 1500 m are shown.

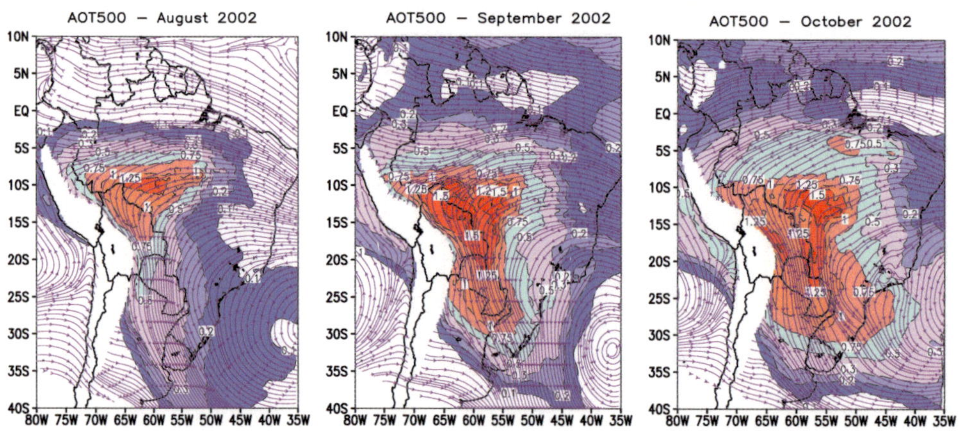

Fig. 4. Monthly composite fields for AOT500nm (*shaded*) and wind at 1400m (*streamlines*) for the SALLJ events during the biomass burning season. Fields are masked in terrain elevations higher than 1500 m.

The temporal behaviour of the AOT at the AERONET site in Buenos Aires is depicted in Figure 5 for the sub-samples SALLJ and NO-SALLJ along with the comparison with the CATT-BRAMS predicted values. The model is able to capture the evolution of the aerosol concentration. The underestimation of the values is linked to the comparison of point measurements and the model results resolution.

The relationship between the Ångström coefficient and the AOT is frequently used to get more information about the aerosol characteristics. The greater aerosol load observed during the SALLJ events is clearly associated to higher Ångström coefficients in agreement with the literature (Figure 6).

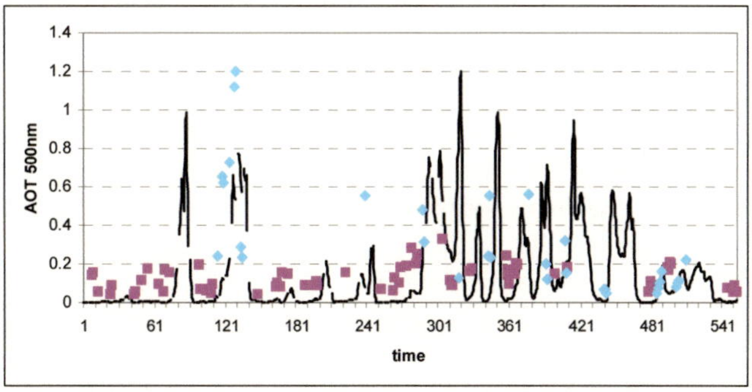

Fig. 5. Comparison of AOT 500nm obtained at the Buenos Aires AERONET site (*dots, diamond: SALLJ, square: NO-SALLJ*) and CATT-BRAMS model (*line*) from August to October 2002.

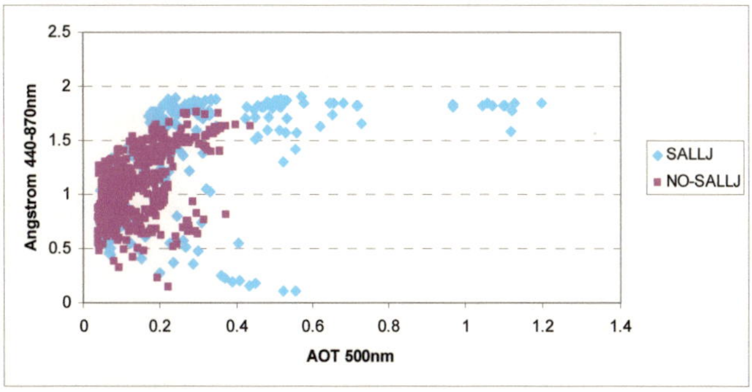

Fig. 6. Variation of the Ångström coefficient (440-870nm) with the AOT 500nm obtained from the measurements at the Buenos Aires AERONET site (*dots, diamond: SALLJ, square: NO-SALLJ*) from August to October 2002.

3.2 Case study: august 2002

A prolonged SALLJ event that occurred in conjunction with biomass burning took place from 23 to 28 August. The low-level jet had an important latitudinal extent and strength with a pattern that varied according to a baroclinic synoptic environment.

3.2.1 Meteorological environment and SALLJ features

Figure 7 depicts the 1000 hPa geopotential height and the 500/1000 hPa thickness fields for selected days during the event. On 23 August, the western branch of the SASH was over an important extension of SA and the low-level flow was from the N as far as 40° S. In the southernmost edge of SA, a baroclinic region -oriented NW to SE- was present and deep low-pressure systems were moving south-eastward. During the following day, a geopotential trough developed over central Argentina. The thickness field showed the associated maximum depth. There was a persistent N-NW flow over south-eastern SA. On 25 August, a further deepening of the trough over central Argentina occurred. The baroclinic region related to the cold front was located between 30° S and 40° S and moved towards the northeast. The low-pressure system behind the cold front weakened. There was a strong channelling of the low-level flow between the trough and the western region of the SASH. Twenty-four hours later, the baroclinic zone approached the southern region of Buenos Aires. A deep thickness trough was present over the eastern Pacific Ocean. Central

Argentina was still with the minimum geopotential. The flow from the north at low levels persisted. On 27 August, the situation was almost similar, with the new system strengthening and moving eastward and starting to surpass the Andes barrier. The northern low-level flow was still present over south-eastern SA and the low pressure further deepened over central Argentina. On 28 August, the system was able to reach eastern Argentina. The associated cold front presented a nearly north-south orientation and moved eastward towards the Atlantic Ocean. The low-pressure system over Argentina deepened and the low-level north-western flow persisted. During 29 August the cold sector of the front moved past Buenos Aires and Uruguay and the related surface cyclone, centred near 40° S and 55° W, deepened. The near-surface airflow over south-eastern SA was from the N-NE sector and from the S in Buenos Aires. On 30 August, the baroclinic region in the 500/1000 hPa thickness field was located at 30° S, with zonal orientation. The surface low-pressure system had its maximum depth at 0600 UTC and then started to fill while travelling to the east over the Atlantic Ocean. Central Argentina had relatively higher surface pressure. The near surface flow was from the S over northern Argentina. During the final day of the study period (31 August) the baroclinic region was in southern Brazil, co-located with a surface col region. Argentina had near surface southerly winds. The surface cyclone was in the occlusion stage at 1800 UTC.

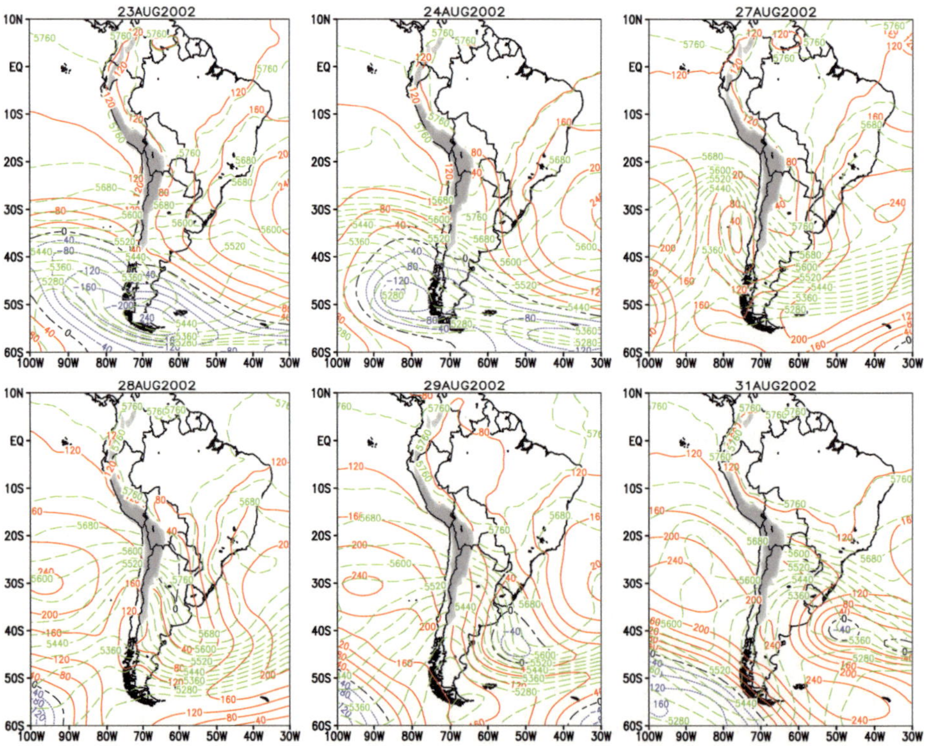

Fig. 7. Daily fields of 1000 hPa geopotential height (*red solid (positive), blue dot (negative) contours*) and 500/1000 hPa thickness (*green long dash contours*) (both every 40 mgp), from 23 to 31 August. Terrain elevations higher than 1500 m are shaded.

The wind field at 850 hPa and the regions that verified the modified Bonner criteria for some selected days are depicted in Figure 8. On 23 August, the affected region was from central

Bolivia to north-eastern Argentina, part of Uruguay and southern Brazil. The jet core was located over northern Paraguay. The related flow was from the N-NW sector. During the following day, the SALLJ had increased strength and vertical shear, as well as more spatial extension. The southernmost edge was near 40° S. On 25 August, the NW-SE region associated with the SALLJ had a greater latitudinal extension and the low-level flow was from the northwest and stronger due to the westward displacement of the Atlantic anticyclone. During the next day, the SALLJ had a smaller southward penetration and reached only 35° S. This was due to the advance of the cold front that was located past 40° S at the 850-hPa level over the ocean. The flow was more northerly oriented. The jet core was over western Paraguay and northern Argentina.

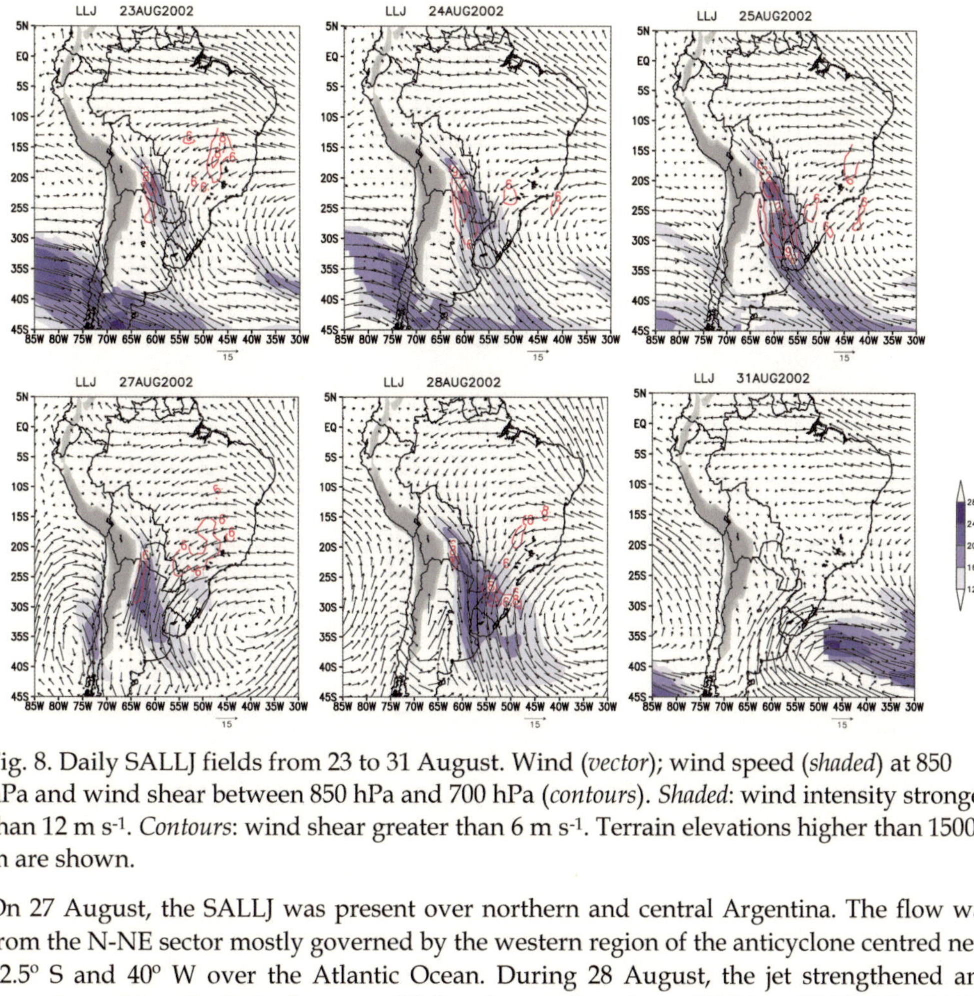

Fig. 8. Daily SALLJ fields from 23 to 31 August. Wind (*vector*); wind speed (*shaded*) at 850 hPa and wind shear between 850 hPa and 700 hPa (*contours*). *Shaded*: wind intensity stronger than 12 m s^{-1}. *Contours*: wind shear greater than 6 m s^{-1}. Terrain elevations higher than 1500 m are shown.

On 27 August, the SALLJ was present over northern and central Argentina. The flow was from the N-NE sector mostly governed by the western region of the anticyclone centred near 32.5° S and 40° W over the Atlantic Ocean. During 28 August, the jet strengthened and spread, reaching the latitudes near 45° S and extending from 65° W to 40° W. The SALLJ reinforced due to the new cold front that was located near 60° W at 1200 UTC with north-south orientation. On 29 August, the front reached Paraguay and south-eastern Brazil. The wind field at 850 hPa shows clearly the northwest wind ahead of the front whereas the winds behind were strong, from the southwest. The region spanned by the strongest winds has the typical shape of the frontal zone but the wind field did not verify the Bonner's criteria. During 30 August, the north-western edge of the frontal zone was over São Paulo, with the southerly winds blowing clear and dry air over South America up to 15° S. The convergence in the airflow is related to the surface cold front and the baroclinic region near São Paulo. The situation persisted on 31 August. SALLJ did not occur either. During this particular event, an important southward penetration of the low-level jet occurred and the associated moisture convergence at the exit region of the current favoured the development of convective systems south of 40° S, which strengthened mostly over the Atlantic Ocean. The interaction with the cold front further contributed to the convection.

3.2.2 Concentration behaviour

The evolution and spatial extent of the smoke plume is studied through the behaviour of the AOT500. Figure 9 shows the modelled AOT500 and the horizontal flow at 1400 m above the surface, at selected days during the analyzed period. On 23 August the smoke plume showed a relative maximum close to the emission sources, centred near 10° S and 60° W, with values higher than 2. The smoke plume had its greater longitudinal extension between

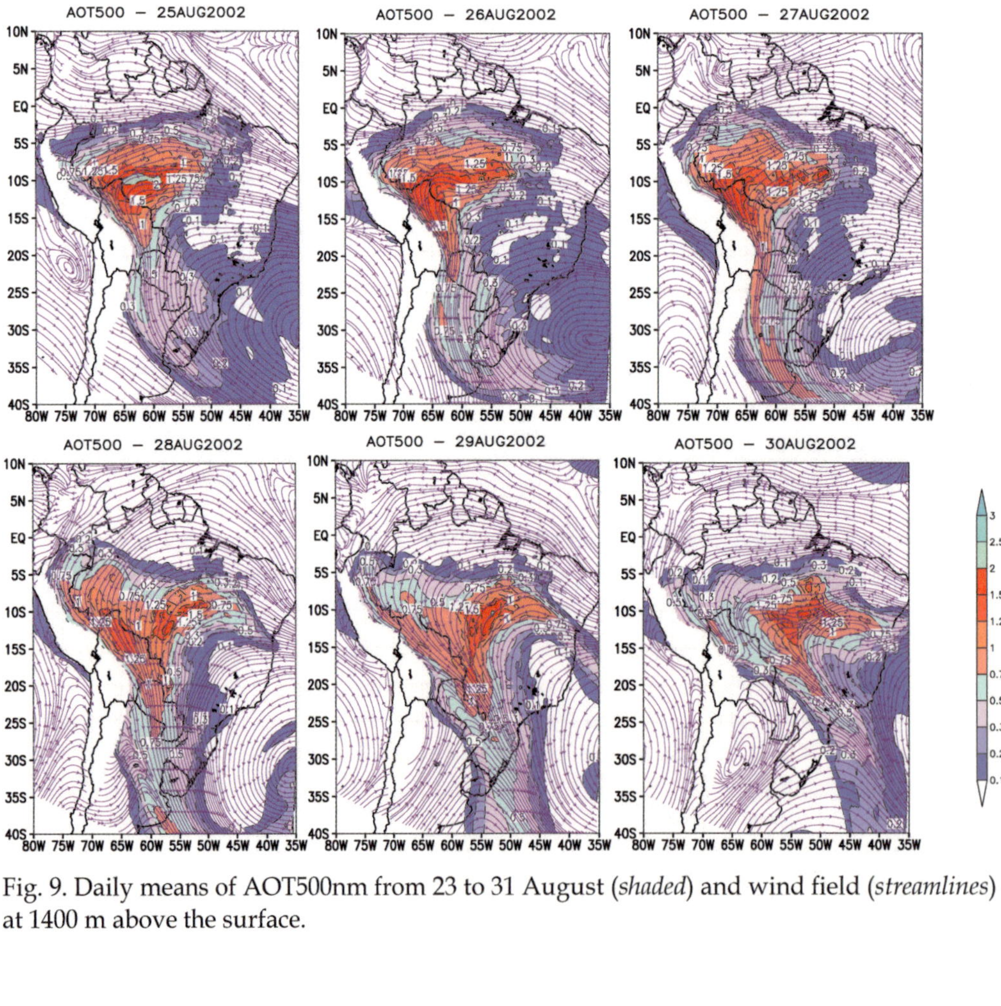

Fig. 9. Daily means of AOT500nm from 23 to 31 August (*shaded*) and wind field (*streamlines*) at 1400 m above the surface.

the Equator and 10° S. This feature is related with the dominance of the easterlies in that region. An outflow zone from South America towards the west is observed between 5° S and 10° S. To the south, at higher latitudes, the smoke plume had an important branch oriented from the NW to the SE reaching the latitude 20° S, with AOT values higher than 0.5. These features are due to the transport patterns at low and middle atmospheric levels, which were dominated by the flow at the western branch of the high-pressure system and the channelling effect of the Andes barrier.

On 24 August, the region with higher AOT values near the sources increased. The smoke plume had a greater latitudinal extent over Argentina, reaching 35° S. The core of the low-level jet was associated with a relative minimum. Contrarily, a relative maximum over northern Argentina appears west of the jet core. On 25 August the southern edge of the plume continues to travel towards higher latitudes and presents a shape associated with the anticyclonic circulation and the barrier effect of the Andes. Optical depths ranging from 0.3 to 0.5 cover NE of Argentina and Uruguay. An outflow region from South America towards the west is observed between 5° S and 15° S. On 26 August, Buenos Aires had AOT values between 0.5 and 0.75. The smoke plume has N-S orientation from latitudes near 15° S to 30° S. The greater values are observed near the sources, over central Brazil and Bolivia. On its southernmost extreme the plume shows a curvature associated with the high-pressure system centred over the Atlantic Ocean, near 32° S and 35° W. Córdoba is affected by aerosol optical thicknesses ranging from 0.75 to 1, which are higher than those at Buenos Aires.

On 27 August the smoke plume reached latitudes higher than 40° S. The AOT over Buenos Aires ranged from 0.75 to 1. On 28 August, the cold front succeeded in crossing the Andes and reached Argentina and afterwards, the plume started to be displaced towards the east but was still over Buenos Aires due to its pre-frontal location. During the next day, the smoke plume displaced towards the northeast, owing to the fast movement of the cold front, and reached southern Brazil. On 30 August the plume had clearly the shape of the frontal zone and reached São Paulo. During the next day the surface cold front was stationary over São Paulo. There is a region associated to the postfrontal anticyclone with a low-level recirculation of the aerosols towards the west of the plume centre. This occurs at the north-western edge of the frontal region, where the forced convection is weaker.

3.2.3 Meridional PM2.5 and water vapour transport

Vertical cross sections at latitudes 15° S, 25° S and 35° S, across the smoke plume contribute to depict the distribution of the meridional transport of PM2.5 (in $\mu gm^{-2}s^{-1}$) (Figure 10) and water vapour mixing ratio (in $gmkg^{-1}s^{-1}$) (Figure 11). The cross-sections clearly illustrate the role of the SALLJ as a transport mechanism. In general, the meridional transport of PM2.5 is limited to the layer between the surface and 4000 m and the higher values are near the emission sources. In the case of the water vapour the vertical extent is greater, reaching 8000m.

At 15° S (Figure 10a), on 23 August, the meridional flux of PM2.5 was mainly southward and on the layer between 1000 and 4000m, with the maximum located between 1500 and 2000m with values between -60 and -180 $\mu gm^{-2}s^{-1}$. The transport was on a narrow region east of the Andes range, centred at 65° W. During the following day, the level of maximum meridional transport was closer to the surface and the longitudinal extent increased. The values were similar than those on the previous day. On 25 August, two relative maxima were present, one close to the surface at 65° W and the other one between 1500 and 3500m above the ground at 62.5° W, ranging from -60 to -300 $\mu gm^{-2}s^{-1}$. During the following day, the conditions were almost similar, but the maximum close to the surface weakened. During 27 August, the southward transport east of the Andes was comparable, ranging from -180 to -240 $\mu gm^{-2}s^{-1}$ and a secondary maximum west of the mountain range near 3000m was present. On 28 and 29August, the southward transport was mainly in the layer between the surface and 3000m, with a longitudinal extent from 72.5° W to 55° W. The maximum values varied from -240 to -360 $\mu gm^{-2}s^{-1}$. A narrow elevated maximum occurred, at upper levels and 40° W. On 30 August, the southerlies reached this latitude, and a northward transport occurred near the surface from 65° W to 50° W with values between 80 and 120 $\mu gm^{-2}s^{-1}$. The southward transport persisted at upper levels close to the Andes, but gradually vanished according to the cold front movement. During 31 August, the northward flux near the surface prevailed, ranging from 80 to 140 $\mu gm^{-2}s^{-1}$.

At 25° S (Figure 10b), on 23 August, the meridional flux showed a maximum of -120 $\mu gm^{-2}s^{-1}$ centred at 60° W. The next day the maximum flux occurred westward, at 62.5° W and ranged from -120 to -240 $\mu gm^{-2}s^{-1}$. During 25 August, the location was similar and the values increased, varying from -180 to -360 $\mu gm^{-2}s^{-1}$. On the following two days, one region of maximum transport was located close to the Andes from surface up to 3500 m, with values that ranged from -300 to -660 $\mu gm^{-2}s^{-1}$ and the second one, was near the surface centred at 57.5° W, varied from -60 to -180 $\mu gm^{-2}s^{-1}$ and spanned ten degrees east of 65° W.

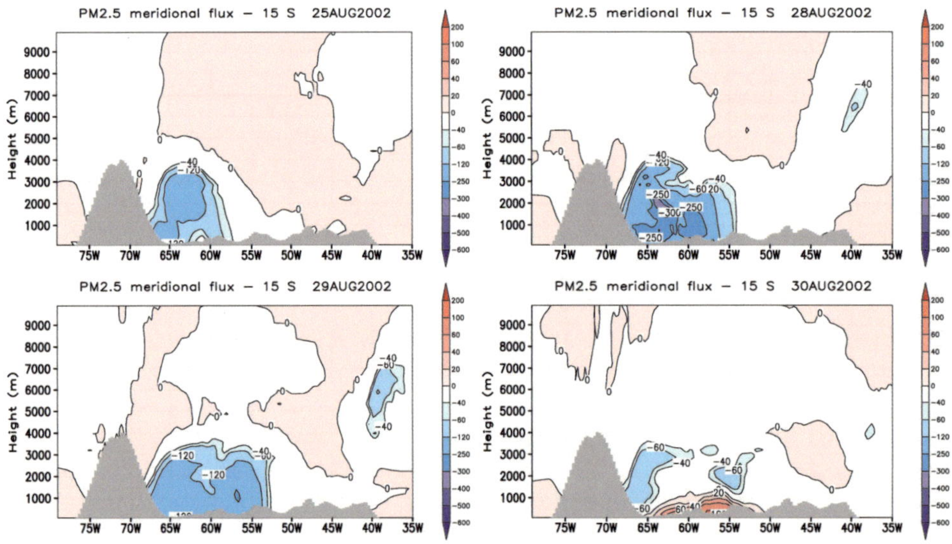

Fig. 10a. Vertical cross-sections at 15° S of PM2.5 meridional transport ($\mu gm^{-2}s^{-1}$) against the height above the surface. Terrain height profile is included.

During 27 August, there is also a transport towards the south between 3000 and 4000m west of the Andes. The next day, the transport had similar longitudinal and vertical span and values from -360 to -600 $\mu gm^{-2}s^{-1}$. By 29 August the southward flux was -180 to -600 $\mu gm^{-2}s^{-1}$ between 62.5° and 50° W and the northward transport was centred at 60° W, ranging from 60 to 260 $\mu gm^{-2}s^{-1}$. During 30 August the northward flux occurred between 63° and 50° W and values from 20 to 60 $\mu gm^{-2}s^{-1}$ and towards the south in upper levels at 47° W ranging from -180 to -80 $\mu gm^{-2}s^{-1}$. The last day of the studied period had very light northward transport smaller and equal than 20 $\mu gm^{-2}s^{-1}$, and southward flux in upper levels from 45° to 40° W with a maximum of -60 $\mu gm^{-2}s^{-1}$.

Figure 10c illustrates the PM2.5 meridional flux at 35° S. The southward flux started on 24 August, the plume was near the surface between 60° and 50° W, with values from -60 to -120 $\mu g m^{-2} s^{-1}$.

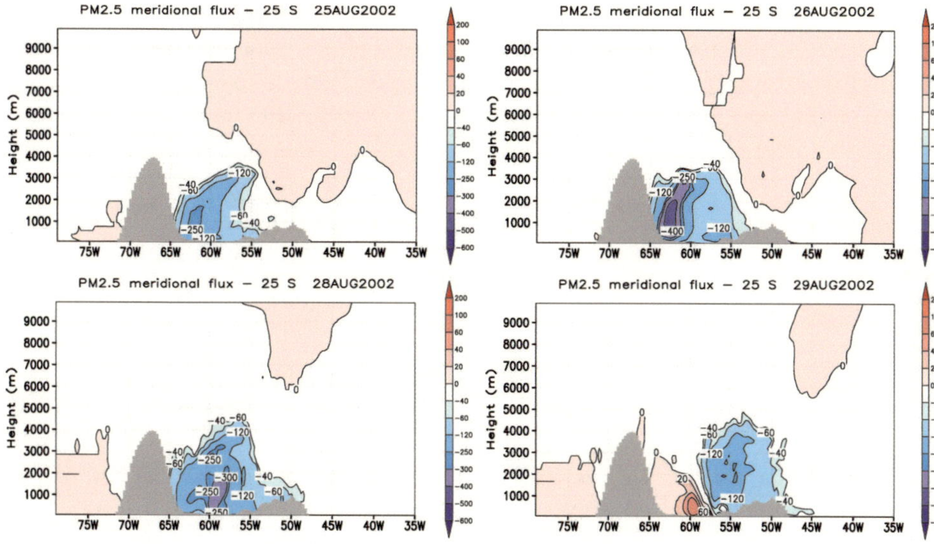

Fig. 10b. Vertical cross-sections at 25° S of PM2.5 meridional transport ($\mu g m^{-2} s^{-1}$) against the height above the surface. Terrain height profile is included.

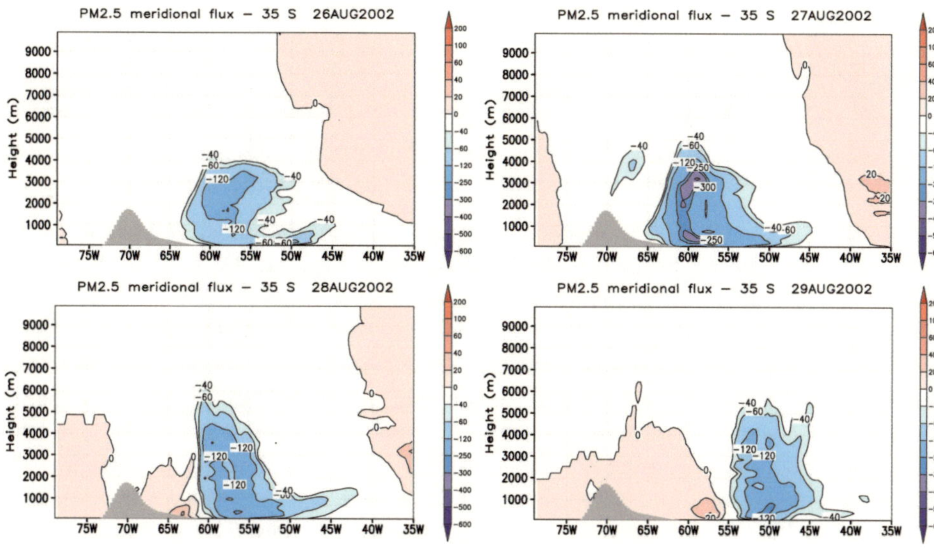

Fig. 10c. Vertical cross-sections at 35° S of PM2.5 meridional transport ($\mu g m^{-2} s^{-1}$) against the height above the surface. Terrain height profile is included.

During the next day, two maxima appeared, one located near the surface and the other one centred at 2000m and values ranging from -60 to -120 $\mu g m^{-2} s^{-1}$. During 26 August, the upper level maximum, centred at 2500m and east of 60° W strengthened, the values ranged from -60 to -360 μgm^{-2}s^{-1}. On 27 August the southward transport was widespread and ranged from -60 to -420 μgm^{-2}s^{-1}. On the following day, the smoke transport extended up to 5000m, remaining towards the south and east of 60° W and the maximum values ranged from -60 to -360 μgm^{-2}s^{-1}. Close to the mountains a northward transport occurred near the surface, with values between 20 and 100 μgm^{-2}s^{-1}. By 29 August the plume was over the Atlantic Ocean and the northward transport was west of 55° W, ranging from 40 to 60 μgm^{-2}s^{-1}. During the next two days the flux gradually disappeared at this latitude due to the fast movement of the cold front.

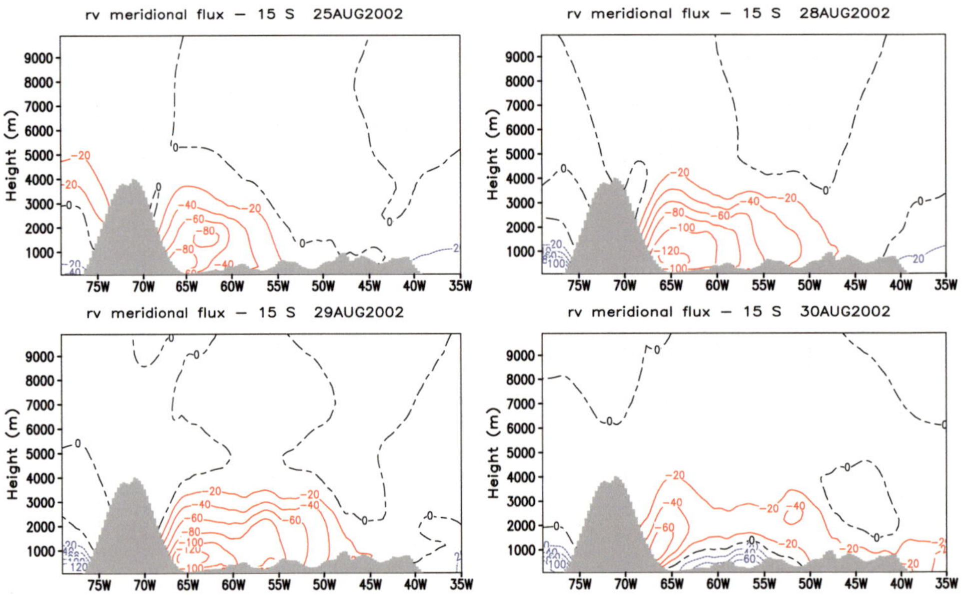

Fig. 11a. Vertical cross-sections at 15° S of water vapour mixing ratio meridional transport (g m kg^{-1} s^{-1}) against the height above the surface. Terrain height profile is included.

Figure 11 shows the vertical cross sections at similar latitudes, but illustrates in this case, the water vapour meridional transport. At 15° S (Figure 11a) on 23 August there was a prevalence of the southward transport of water vapour, spanning from 72.5^{o} W to 47° W, from the surface up to 3000m, and the maximum flux centred at 1500m with a mean daily value of -60 gmkg^{-1}s^{-1}. The northward transport took place over the oceans near the surface. The next day the pattern was similar and the value of the meridional flux increased. On the following three days the longitudinal extent of the zone with southward flux was narrower and the values -80 and -60 gmkg^{-1}s^{-1} respectively. West of the Andes, at upper levels the water vapour southward flux also occurred. The northward transport over the oceans was still present. On 28 and 29 August the longitudinal extent increased as well as the value of the maximum flux, the difference is the location near the surface. The northward water vapour transport increased over the Pacific Ocean. During 30 August, the incursion of the cold front caused a northward flux near the surface between 65° and 50° W. The flux from the north was restricted next to the Andes centred at 1000m. The following day the pattern was nearly similar, with a decrease in the southward transport.

At 25° S (Figure 11b) from 23 to 28 August, there was a southward flux at all longitudes east of the Andes from the surface up to middle levels in the troposphere.

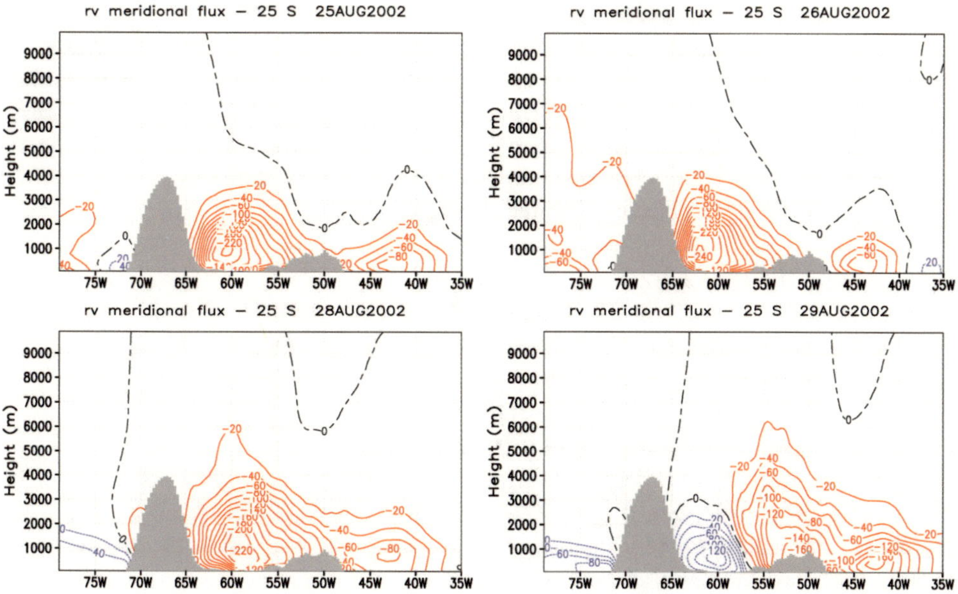

Fig. 11b. Vertical cross-sections at 25° S of water vapour mixing ratio meridional transport (g m kg^{-1} s^{-1}) against the height above the surface. Terrain height profile is included.

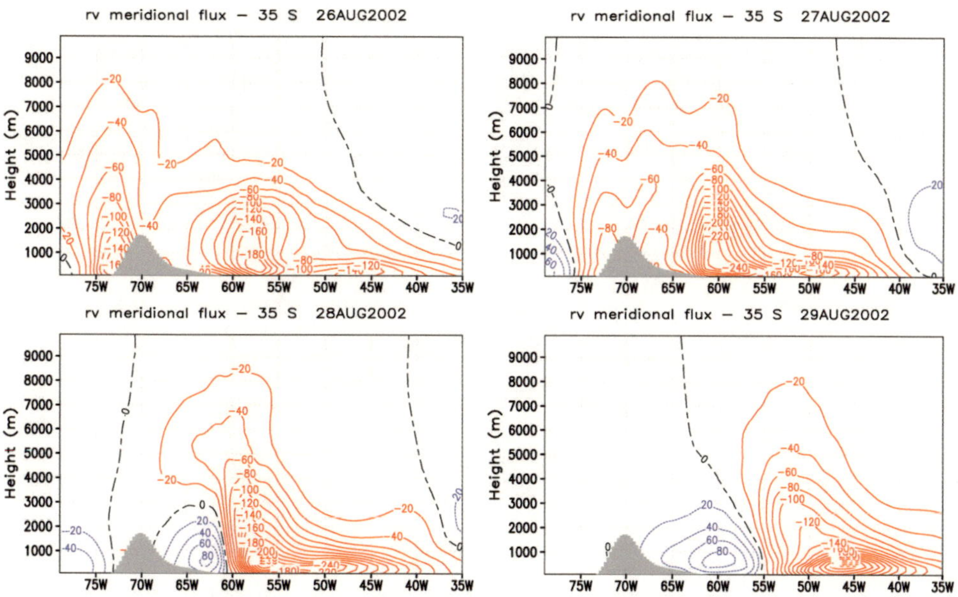

Fig. 11c. Vertical cross-sections at 35° S of water vapour mixing ratio meridional transport (g m kg^{-1} s^{-1}) against the height above the surface. Terrain height profile is included.

The values ranged from -140 to -260 $gmkg^{-1}s^{-1}$. West of the mountain range, the southward flux also occurred on 26 and 27 August reaching a daily maximum of -80 $gmkg^{-1}s^{-1}$. From 29 to 31 August the progression of the cold front caused a northward flow that varied between 20 and 120 $gmkg^{-1}s^{-1}$ with a longitudinal range that moved to the east.

Figure 11c depicts the water vapour meridional transport at 35° S. The southward water vapour transport was present from 23 to 27 August from the surface up to 8000m and 75° W and 35° W, the maximum values varied from -100 to -260 $gmkg^{-1}s^{-1}$. The opposite transport directions associated with the surface cold front is sharply marked in the cross-sections on 28 and 29 August, and the maximum values are located near the surface. The next days showed the contrast in the air masses water vapour as well.

3.3 Case study: October 2002

This event extended from 17 to 21 October and was characterised by a variable low level flow pattern, which had a short SALLJ episode and a changing meteorological scenario, with transient perturbations of short duration.

3.3.1 Meteorological environment and SALLJ features

On 17 October, the 1000 hPa height shows the dominance of a post-frontal high pressure system over central Argentina (Figure 12). The surface front is located over central South America. On the south-western region of Argentina, the 500/1000 hPa depths show a baroclinic zone associated with a new frontal system.

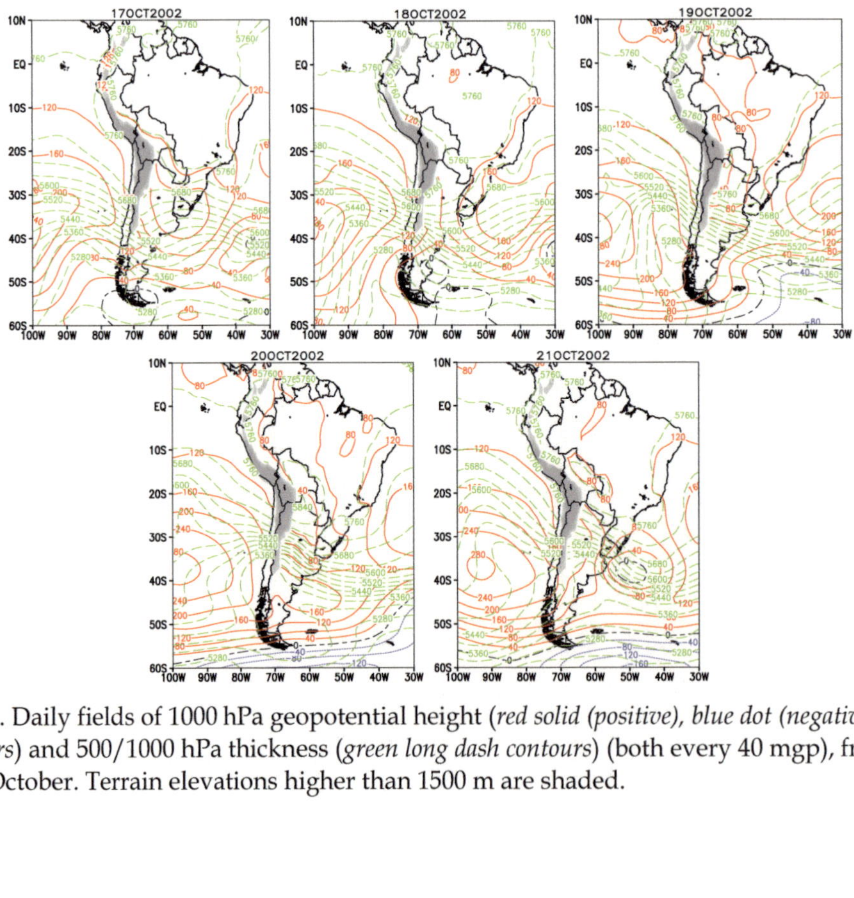

Fig. 12. Daily fields of 1000 hPa geopotential height (*red solid (positive), blue dot (negative) contours*) and 500/1000 hPa thickness (*green long dash contours*) (both every 40 mgp), from 17 to 21 October. Terrain elevations higher than 1500 m are shaded.

During the following day, the anticyclone moved to the Atlantic Ocean, centred about 40° W and 35° S. Behind the baroclinic zone, a low pressure system located near 65° W and 47° S, developed. A thickness through oriented from the NW to the SE, is observed over the Pacific Ocean associated to an upper air through. The low level flow over north-eastern Argentina was from the north. On 19 October, the surface low pressure region had a fast displacement towards the SE. On the other hand, an anticyclonic system moved eastward covering the southern region of Argentina. North of 30° S, central South America showed relatively lower pressures. By 20 October, the thickness through axis was over Los Andes Mountains and then moved eastward. The low pressure system on central-northern Argentina displaced to the east and accordingly, the flow near the surface turned and blew from the east over Buenos Aires. On 21 October, a low pressure system developed and evolved in agreement with the displacement of the pattern at upper levels. It is located around 40° S and 50° W. Argentina was under the influence of an extended anticyclone. The near surface flow was from the south.

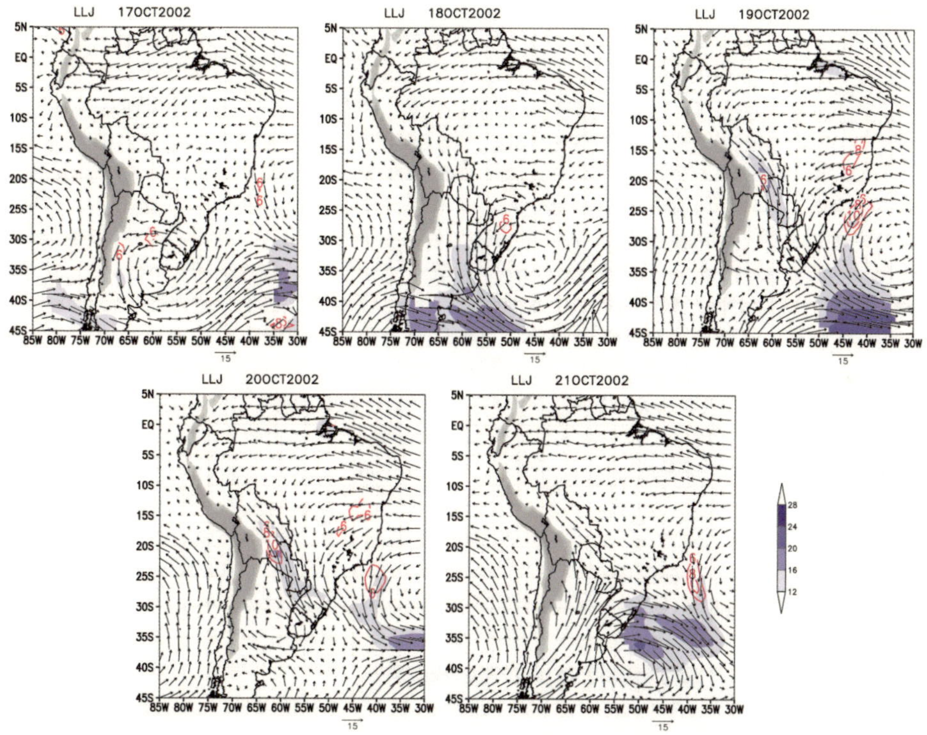

Fig. 13. Daily SALLJ fields from 17 to 21 October. Wind (*vector*); wind speed (*shaded*) at 850 hPa and wind shear between 850 hPa and 700 hPa (*contours*). *Shaded*: wind intensity stronger than 12 m s^{-1}. *Contours*: wind shear greater than 6 m s^{-1}. Terrain elevations higher than 1500 m are shown.

Figure 13 illustrates the 850 hPa flow and SALLJ features. On 17 October the low level flow associated to the post-frontal anticyclone centred over Buenos Aires is clearly shown. A very weak SALLJ is evident in the 850-700 layer, between Los Andes and the west of an anticyclone. The smaller wind intensities are observed over the biomass burning source regions. By 18 October, the low level flow strengthened and organized in a northerly current due to the approach from the southwest of the new cold front and the presence of the anticyclone now centred at 45° W and 35° S over the Atlantic Ocean. The 850 hPa winds did not satisfy the Bonner criteria. The north-western edge of the cold front is located near 35° S and 65° W. On 19 October the SALLJ spanned from central Bolivia to Paraguay and northern Argentina. The wind was from the north. Buenos Aires was behind the cold front. Another region with low level jet occurrence is over the Atlantic Ocean centred at 15° S. South of 30° S, the flow turned counter clockwise and acquired a north-western orientation ahead of the cold front. On 20 October, a SALLJ occurred, with its southern edge near 30° S. The front remained stationary over central Argentina. A low pressure system developed in the central region of Argentina whereas the exit region of the SALLJ was on southern Brazil. During the next day, there is a clear evidence of a strengthening and rapid displacement of the cold front that is oriented NW to SE. The low-level flow was from the south up to 20° S.

3.3.2 Concentration behaviour

On 17 October, the vertically integrated AOT clearly depicts the constraint on the southward displacement imposed by the cold front (Figure 14). The higher AOT are observed near the sources in close agreement with the regions in which the smaller wind speeds occurred. As the post-frontal anticyclone moves eastward, the southward transport of the smoke plume is favoured on its western region. In this particular case, the AOT values are low, indicative of relatively clean air, but the contrary might happen with greater emissions. Northern Argentina had AOT greater than 1. During the next day, with the displacement of the anticyclone towards the Atlantic Ocean and the further re-establishment of the north-western flow, AOT over 0.3 reached Buenos Aires. On 19 October, the smoke plume is narrower and the AOT greater than 1.25 reached southern Brazil. On the other hand, over Buenos Aires and Córdoba the AOT ranged from 0.2 to 0.5. During the next day, the greater AOT are observed near the source region. An interesting feature is that a relative minimum occurs in the same location than the SALLJ core over central Bolivia and northern Paraguay. On central Argentina, the development of the cyclonic circulation further helps the transport to the south on its eastern flank. AOT values ranging from 0.3 to 0.5 are predicted over Buenos Aires. On 21 October the strong south-westerly winds that blew over central Argentina caused the displacement of the smoke plume towards lower latitudes. The southern edge of the plume clearly shows the shape of the frontal region.

3.3.3 Meridional PM2.5 and water vapour transport

Figure 15 shows the PM2.5 meridional transport. At 20° S (Figure 15a), during 17 October, there was a northward transport in the layer ranging from near the surface to 1500m, between 65° W and 55° W. The values ranged from 20 to 220 $\mu gm^{-2}s^{-1}$. This agrees with the higher concentrations in the regional plume. Immediately above this maximum there was a southward flow reaching the upper troposphere. The maximum meridional transport towards the south was centred at about 2500m and 60° W, with values between -60 and -180 $\mu gm^{-2}s^{-1}$. This agrees with the flow pattern that was perturbed by the presence of the NW edge of the cold front. As the front moved north-eastward the northern meridional flow re-established co-located with the SALLJ. On the following day, the southward transport strengthened while the northward flow east of 60° W weakened, as well as its vertical extent. In this case the transport reached a value of 80 μgm^{-2}s^{-1}. The greater northern flow is observed in the longitudes between 65° W and 55° W centred at 2000m and reached a maximum of -240 μgm^{-2}s^{-1}. On 19 and 20 October the southward transport is dominant and the maximum values (-240 and -300 μgm^{-2}s^{-1}, respectively) appear closer to the surface with an eastward displacement. The pattern remained almost similar on 21 October, with a slight decrease in the southward transport.

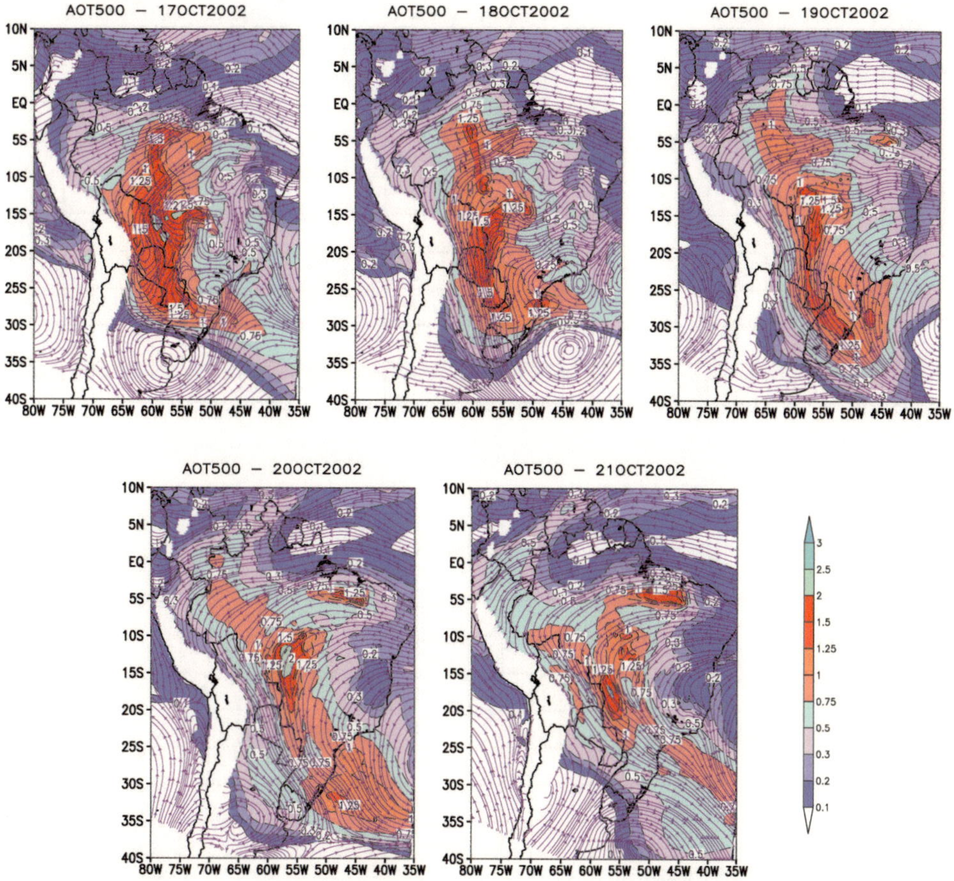

Fig. 14. Daily means of AOT500nm from 17 to 21 August (*shaded*) and wind field (*streamlines*) at 1400 m above the surface.

At the southernmost latitude considered in the vertical cross sections -30° S- (Figure 15b) during 17 October, the transport was from the south in the longitudes ranging from 60° W to 45° W from the surface up to 2000m, reaching a maximum value of -240 μgm^{-2}s^{-1}. On the next day, the flux was from the north in a layer from the surface up to middle troposphere, from 65° W and 50° W. The greatest value was -180 μgm^{-2}s^{-1} centred at 57° W and 1500m. The northward transport was smaller and over the Atlantic Ocean. During 19 October, the dominance of the southward transport was evident in the layer from the surface up to 3000m where had its greatest strength. The following day showed almost similar shape, with a slight decrease in the intensities.

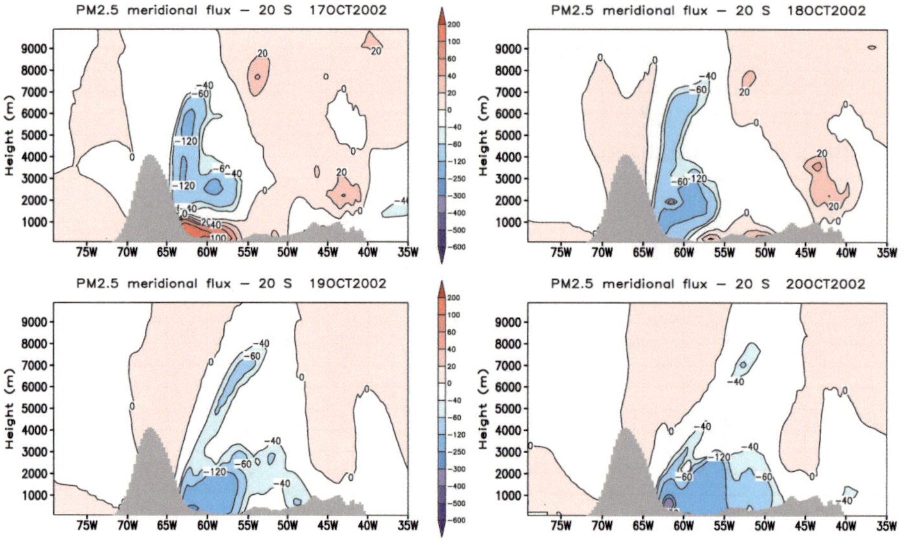

Fig. 15a. Vertical cross-sections at 20° S of PM2.5 meridional transport ($\mu g m^{-2} s^{-1}$) against the height above the surface. Terrain height profile is included.

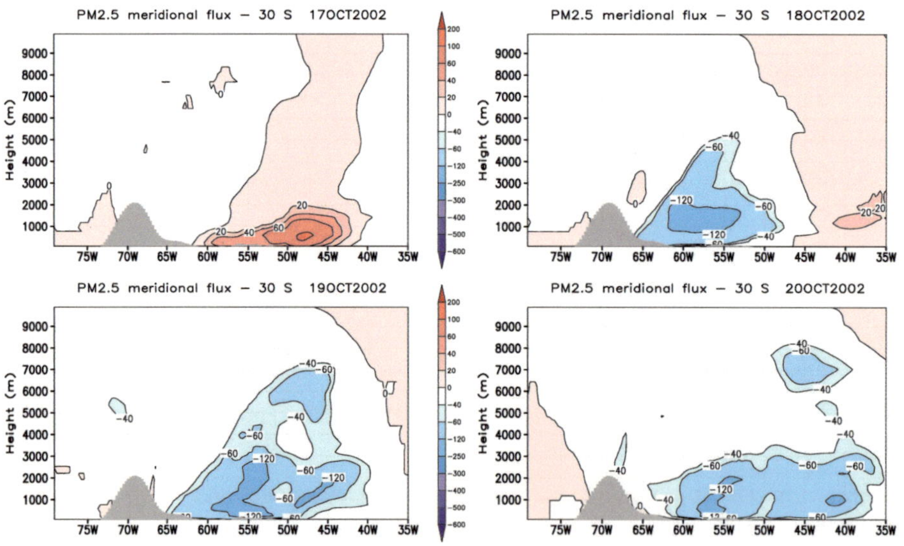

Fig. 15b. Vertical cross-sections at 30° S of PM2.5 meridional transport ($\mu g m^{-2} s^{-1}$) against the height above the surface. Terrain height profile is included.

On 21 October, the vertical cross section shows northward transport associated with the progression of the cold front, from 65° W to 55° W in the layer near the surface up to 1000m, and the opposite flux over the Atlantic Ocean, east of 45° W. The values reached 120 and -120 $\mu g m^{-2} s^{-1}$ respectively.

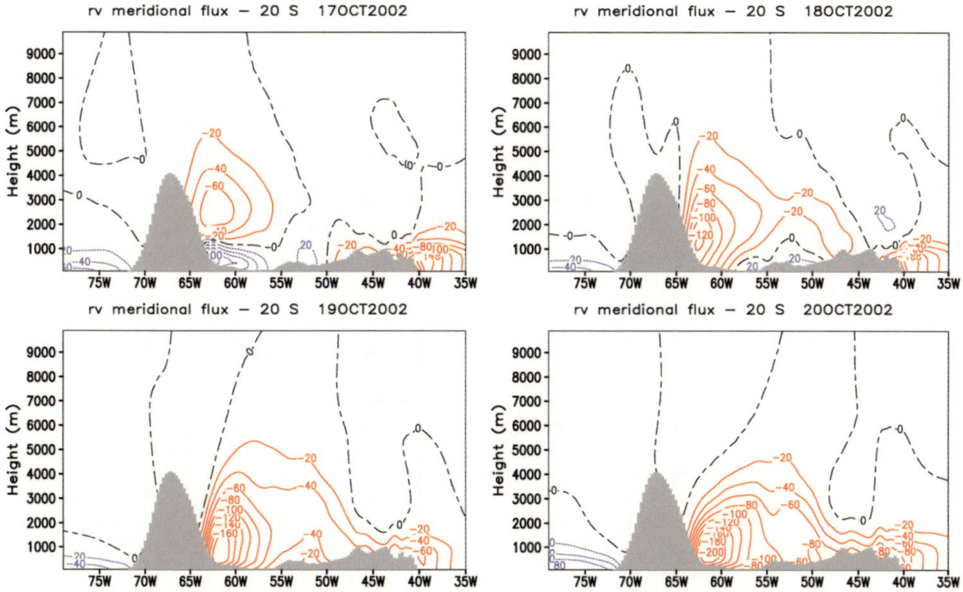

Fig. 16a. Vertical cross-sections at 20° S of water vapor mixing ratio meridional transport (g m kg^{-1} s^{-1}) against the height above the surface. Terrain height profile is included.

The water vapour meridional flow at 20° S (Figure 16a) on 17 October, showed opposite flows immediately east of the Andes, with northward water vapour flux near the surface up to 1000m and the contrary above this height. Contrarily to what happened with the PM2.5 transport, the southward transport east of 50° W was greater than that observed near the mountains, and this is related to the location of the water vapour and particulate sources. The southward transport reached values equal -80 gmkg^{-1}s^{-1} at 62.5° W and -140 gmkg^{-1}s^{-1} at -42.5° W during this day. On 18 October, the transport to the south was dominant with a strengthening of the maximum close to the Andes, with a mean daily value equal to -120 gmkg^{-1}s^{-1}. The next two days, in accordance with the occurrence of the SALLJ, the transport to the south was dominant at this latitude, with the highest value coincident with the jet core, reaching -220 gmkg^{-1}s^{-1}. On 21 October the region with southward flux moved slightly to the east, and the highest value was -160 gmkg^{-1}s^{-1}.

At 30° S (Figure 16b), on 17 October, there was northward transport near the surface from 60° W to 42° W, with a maximum value of 140 gmkg^{-1}s^{-1}. The flux to the south took place in a narrow region close to the Andes and reached -60 gmkg^{-1}s^{-1}. Another zone with southward transport was over the Atlantic. During the next day, the region with southward transport extended to 47° W, with the highest value below 1000m, centred at 55° W. An interesting feature is that the transport of water vapour and PM2.5 maximize in different altitudes and longitudes. This difference is also evident on 19 October, when the maximum water vapour transport reached -180 gmkg^{-1}s^{-1}. The following day, the southward flux had two maxima below 1000m, one centred at 57° W and the other one at 42° W. The values reached -180 gmkg^{-1}s^{-1}. On 21 October, 55° W marked the divide between the flux towards the north and the south in coincidence with the PM2.5 transport, but, once more, the layers of transport were different.

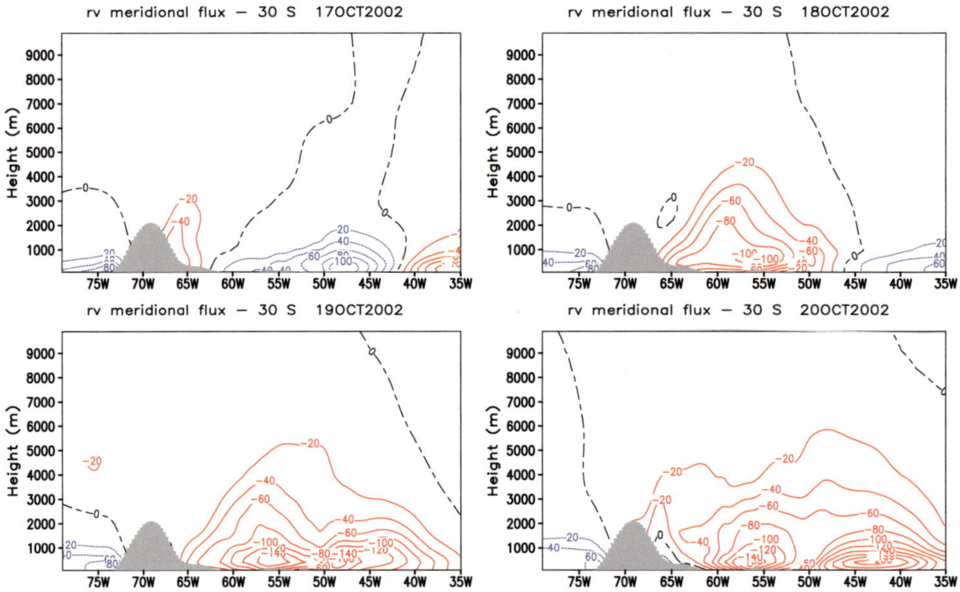

Fig. 16b. Vertical cross-sections at 30° S of water vapor mixing ratio meridional transport (g m kg^{-1} s^{-1}) against the height above the surface. Terrain height profile is included.

4. Discussion

The fire spots experience an important increase during the dry season in Tropical South America and the regional smoke plume is driven by the low level flow. The South American Low Level Jet is a frequent pattern that contributes and patronizes the dispersion and its importance was documented. The smoke plume can travel a long distance from the source region and cause several impacts on remote locations. Among these effects are the increase in the aerosol load and characteristics.

The pattern that emerges in the prolonged episode in August is that during the warm stage of the cold front incursion, the southward penetration of the smoke is favoured. The level of the transport is in close relationship with the maximum meridional wind that develops in the SALLJ event. Owing to the cold front displacement, there is a northward transport of the regional plume. Behind the cold front the air is clean. The horizontal transport mechanism is related to the tangential component of the wind, parallel to the frontal region. Therefore, ahead of the front, there is a preferred exit region from South America towards the Atlantic Ocean. Another interesting feature is that the material is forced to ascend at the frontal slope, and the level of maximum transport occurs at higher levels in the cold stage, so they are generally uncoupled from the surface and above the atmospheric boundary layer.

The regional transport of smoke is clearly shown. The smoke plume originated in the vegetation fires over tropical South America and was transported first westward, then deflected by the Andes barrier and finally southward, reaching mid-latitude regions farther south of 40° S. The cold front approach moved afterwards the polluted air mass towards southeastern Brazil and the Atlantic Ocean.

In the October episode, the short duration transient systems contributed to the dispersion and re-circulation of the smoke plume. The southward incursion of the smoke plume was

prevented by the fast displacement of a cold front, and in this case the exit to the Atlantic was observed over southern Brazil. The post-frontal anticyclonic circulations favoured the incursion of the plume over Argentina near the Andes.
It is worthy to point out that, in both cases, within the scenario of regional transport and interaction with the greater scale weather patterns, there is a mesoscale effect of the low level jet clearly evident in the region of the SALLJ core: a relative minimum in the AOT values. As regards the vertical distribution and preferred levels of dispersion, the importance of the SALLJ as a transport mechanism was demonstrated. The main difference between biomass burning products and water vapour is related to the longitudinal span of the transport, which arises from the spatial distribution of the sources. One distinctive feature is that the water vapour transport takes place at lower levels as compared with the particulate material transport.

5. Conclusion

A study of the relationship of the South American Low Level Jet east of the Andes and the regional transport of biomass burning products was carried out. The detailed three-dimensional structure and evolution of the meteorological and aerosols fields contributed to depict the preferred regions and levels in which the transport of the biomass burning products took place. The South American Low Level Jet is an agent to transport and mix biogeochemical substances and therefore, a possible impact on regional climate could occur in association with burning and destruction of the tropical rain forest. Biomass burning smoke effects must be included in climate models issuing to make any assessment of the regional climate change in the South American continent.

6. Acknowledgment

This research was partially funded by UBACyT X224 and ANPCyT PICT 08-1739 projects. NCEP is acknowledged for the meteorological analyses and Brent Holben for the AERONET data.

7. References

Bonner, W. D. (1968). Climatology of the low level jet. *Monthly Weather Review*, Vol.119, pp. 1575-1589, ISSN 0027-0644.

Freitas, S. R.; Longo, K. M.; Silva Dias, M. A. F.; Chatfield, R.; Silva Dias, P.; Artaxo, P.; Andreae, M. O.; Grell, G.; Rodrigues, L. F.; Fazenda, A. & Panetta, J. (2009). The Coupled Aerosol and Tracer Transport model to the Brazilian developments on the Regional Atmospheric Modeling System (CATT-BRAMS) Part 1: Model description and evaluation. *Atmospheric Chemistry and Physics*, Vol. 9, pp. 2843-2861, ISSN 1680-7316.

James, I. N. & Anderson, D. L. T. (1984). The seasonal mean flow and distribution of largescale weather systems in the southern hemisphere: the effects of moisture transport. *Quarterly Journal Royal Meteorological Society*, Vol. 110, pp. 943-966, ISSN 1477-870X.

Longo, K. M.; Freitas, S. R.; Andreae, M. O.; Setzer, A.; Prins, E. M. & Artaxo, P. (2010). The Coupled Aerosol and Tracer Transport model to the Brazilian developments on the

Regional Atmospheric Modeling System (CATT-BRAMS) Part 2: Model sensitivity to the biomass burning inventories. *Atmospheric Chemistry and Physics*, Vol. 10, pp. 5785-5795, ISSN 1680-7316.

Nicolini, M.; Saulo C.; Torres, J. C. & Salio, P. (2002). Enhanced precipitation over southeastern South America related to strong low-level jet events during austral warm season. *METEOROLOGICA*, Special Issue for the South American Monsoon System, Vol. 27, pp. 59-69, ISSN 0325-187X.

Nogues-Paegle, J. & Mo, K. C. (1997). Alternating wet and dry conditions over South America during summer. *Monthly Weather Review*, Vol. 125, pp. 279-291, ISSN 0027-0644.

Nogues-Paegle, J., K. C. Mo & Paegle J. (1998). Predictability of the NCEP-NCAR Reanalysis Model during Austral Summer. *Monthly Weather Review*, Vol. 126, pp. 3135-3152, ISSN 0027-0644.

Paegle, J. (1998). A comparative review of South American low level jets. *METEOROLOGICA*, Vol. 23, pp. 73-81, ISSN 0325-187X.

Saulo, C., Nicolini, M. & Chou, S. C. (2000). Model characterization of the South American low-level flow during 1997-1998 spring-summer season. *Climate Dynamics*, Vol. 16, pp. 867-881, ISSN 0930-7575.

Vera C. S. & collaborators (2006). The South American Low Level Jet Experiment, *Bulletin of the American Meteorological Society*, Vol. 87, pp. 63-77, ISSN 0003-0007.

20

The Chemistry Behind the Use of Agricultural Biomass as Sorbent for Toxic Metal Ions: pH Influence, Binding Groups, and Complexation Equilibria

Valeria M. Nurchi[1] and Isabel Villaescusa[2]
[1]*Department of Chemical Sciences, University of Cagliari,*
[2]*Department of Chemical and Agricultural Engineering, University of Girona,*
[1]*Italy*
[2]*Spain*

1. Introduction

Waters, because of human activities, are often characterized by different kinds of contamination. In this chapter we will deal with contamination due to toxic metal ions. To purify wastewaters from these pollutants different treatment processes are applied, which include chemical precipitation, chemical oxidation or reduction, electrochemical treatment, membrane filtration, ion exchange, carbon sorption, and coprecipitation/sorption. A number of these processes are extremely expensive and some of them are ineffective at low concentrations. Alternative cost effective technologies based on low cost sorbents are nowadays of great concern in the applied research. These low cost sorbents must be abundant in nature, easily available, and above all they have to fit the worldwide request of recycling. Certain waste products from agricultural operations may become inexpensive sorbents and the potential of some of these wastes for the removal of a number of metal ions has been extensively investigated.

The use of these wastes as sorbents fulfills two important scopes for the protection of environment: the reuse of waste materials and the detoxification of wastewaters.

The biomass source depends on the agricultural production prevailing in the geographical areas where pollution and subsequent decontamination process take place.

The real challenge in the field of biosorption is to identify the chemical mechanism that governs metal uptake by biosorbents. Vegetal biomaterials, constituted principally by lignin, cellulose and by a non-negligible portion of fatty acid as major constituents, can be regarded as natural ion-exchange materials. Furthermore, the functional groups on the biomaterial surface, such as hydroxyl, carbonyl, amino, sulphydryl and carboxylic groups, allow the sorption of metal ions by strong coordination. Therefore, identification of the functional groups can help in shedding light on the mechanism responsible for metal uptake. Also some factors affecting the sorption process such as particle size, pH, metal ion concentration, agitation time, and kinetics must be investigated. The results obtained contribute to the knowledge of the overall process that takes place.

No doubt that metal removal from waste water by biomass requires a multidisciplinary approach (as do environmental sciences in general). The efforts of analytical chemists and solution equilibrium experts can give an important contribution to the knowledge and optimization of these processes.
The study of the chemical characteristics (complex formation constants, hydrolysis,...) of binding groups present on the biomass is of paramount importance to identify the mechanisms of metal sequestration, and to predict the selectivity towards the different cations, the strength of binding and the influence of pH on the sorption processes.

2. An overview of environmental pollution

Many elements play a double role in the physiology of living organisms; some are indispensable, while most of them are toxic at elevated concentrations. The concern on the potential toxic effects of metal ions has been increasing in recent years. As a result of industrial activities and technological development, heavy metals released into the environment pose a significant threat to environment and public health because of their toxicity, accumulation in the food chain and persistence in nature.
In the sixties of last century the importance of controlling the concentration of toxic metal ions in waters for human use became apparent after the Four Big Pollution Diseases of Japan, a group of manmade diseases all caused by environmental pollution due to improper handling of industrial wastes by Japanese corporations.
Two of the Four Big Pollution Diseases of Japan, Minamata (1932-1968) and Niigata disease (1965), were due to mercury poisoning. The first one, first discovered in Minamata in 1956, is a neurological disease characterized by ataxia, numbness in the hands and feet, general muscle weakness, narrowing of the field of vision and damage to hearing and speech, and in extreme cases, insanity, paralysis, coma and death. This poisoning was caused by the release of methyl mercury in the industrial wastewater from the Chisso Corporation's chemical factory. The highly toxic mercury has been bio-accumulated in shellfish and fish in Minamata Bay and the Shiranui Sea, and human and animals deaths continued over more than 30 years. In March 2001, 2265 victims had been officially recognized (1784 of whom had died) and, in addition, individual payments of medical expenses and a medical allowance had been provided to 10072 people in Kumamoto, Kagoshima and Niigata for their mercury related diseases (http://www.nimd.go.jp/english/index.html).

2.1 Main anthropogenic sources of toxic element pollution and their health effects

Environmental pollution, strictly interconnected to industrial spread, started in the most advanced countries. It is now diffused all over the world with a significant predominance in the emerging industrialized states. Varying factors contribute to the location of a large number of "potential polluting" industries in these countries due to the quite recent industrialization: source of raw materials (mines, forests, ...), water availability, ready availability of manpower and its lower incidence on cost, laws not yet as restrictive as in advanced industrial countries. Actually, most raw matter is treated locally, not only for their natural resources, but also because of the lower cost of preliminary treatments. These treatments are the most hazardous, the heaviest and above all the most polluting.
In order to have a clear picture of the main anthropogenic sources of metal, or better said toxic element in general, pollution and their health effects, the sources, uses, correlated health disorders, and suggested concentration limits are reported in the following sections for each main polluting toxic element.

2.1.1 Aluminium

The element aluminium (atomic weight 26.98) is a silver white metal (density 2.7 g/mL). In its inorganic compound it presents only two oxidation states: 0, +3. Aluminium is the most abundant metal in the Earth's crust, and the third most abundant element, after oxygen and silicon. Because of its extremely low redox-potential potential in nature, it is found combined in over 270 different minerals as oxides or silicates. Aluminium is remarkable for low density and for its ability to resist corrosion due to the phenomenon of passivation.

Structural components made from aluminium and its alloys are vital to the aerospace industry and are very important in transportation and building. Aluminium compounds are widely used in the paper industry, in the dye production, in the textile industry, in processed food, and as a component of many cosmetic and pharmaceutical preparations.

Soluble aluminium salts have demonstrated toxic effects in elevated concentrations. Its toxicity can be traced to deposition in bone and the central nervous system. Because aluminium competes with calcium for absorption, increased amounts of dietary aluminium may contribute to osteopenia (reduced skeletal mineralization). In very high doses, aluminium can cause neurotoxicity. In a smaller amount it can give in susceptible people contact dermatitis, digestive disorders, vomiting or other symptoms upon contact or ingestion.

Owing to limitations in the animal data as a model for humans and the uncertainty surrounding the human data, a health-based WHO guideline value cannot be derived; however, practicable levels based on optimization of the coagulation process in drinking-water plants using aluminium-based coagulants are derived: 0.1 mg/L or less in large water treatment facilities, and 0.2 mg/L or less in small facilities (World Health Organization [WHO], 2008).

2.1.2 Arsenic

The element arsenic exists in three allotropes: grey arsenic, density 5.73 g/mL; yellow arsenic, density 1.93 g/mL; and non stable black amorphous arsenic, density 4.73 g/mL. Arsenic (atomic weight 74.92) shows metallic as well as non metallic properties. In its inorganic compound it presents different oxidation states: -3, 0, +3, +5. It is released into the air by volcanoes and is a natural contaminant of some deep-water wells. Arsenic is used to preserve wood, as a pesticide, to produce glass, in copper and other metal manufacturing, in the electronics industry and in medicine.

Occupational exposure to arsenic is common in the smelting industry (in which arsenic is a by-product) and in the microelectronics industry. Low-level arsenic exposure takes place in the general population through the use of inorganic arsenic compounds in common products such as wood preservatives, pesticides, herbicides, fungicides, and paints; through the consumption of foods treated with arsenic-containing pesticides; and through the burning of fossil fuels in which arsenic is a contaminant. The toxicity depends on its valence oxidation state and on its form inorganic or organic. In general, inorganic arsenic is more toxic than organic arsenic, and trivalent arsenite is more toxic than pentavalent and zero-valent arsenic. Arsenic, particularly in its trivalent form, inhibits critical sulphydryl-containing enzymes. In the pentavalent form, the competitive substitution of arsenic for phosphate can lead to rapid hydrolysis of the high-energy bonds in compounds such as ATP. The normal intake of arsenic by adults primarily occurs through ingestion and averages around 50 μg/d. After absorption, inorganic arsenic accumulates in the liver,

spleen, kidneys, lungs, and gastrointestinal tract. It is then rapidly cleared from these sites but leaves a residue in keratin-rich tissues such as skin, hair, and nails.
Guide line value for drinking water is 0.01 mg/L. It is a provisional value, as there is evidence of a hazard, but the available information on heath effects is limited (WHO, 2008).

2.1.3 Cadmium

Cadmium (atomic weight 112.41) is a silver white metal (density 8.65 g/mL). The oxidation states are 0, +2. The main uses of cadmium were steel production, non-ferrous metal production, refining, cement manufacture, cadmium plating, battery manufacture, waste and combustion, and phosphate fertilizers. Nowadays, because of concerns about its environmental toxicity, the use of cadmium has drastically decreased. About two thirds of the cadmium in use today come from nickel-cadmium batteries, the rest from pigments, metal plating and the plastic industry. It is a lot like lead and mercury, in that it accumulates both in the environment and in the body, causing long-term damage to life.
Cadmuim toxicity can manifest in a variety of syndromes, as hypertension, renal dysfunction, bone defects, hepatic injuries, lung damage, and reproductive effects. The maximum acceptable cadmium in drinking water is 0.003 mg/L (WHO, 2008).

2.1.4 Chromium

Chromium (atomic weight 51.99) is a lustrous, brittle, hard silver-gray metal (density 7.14 g/mL). It exists in different oxidation states: -2, 0, +2, +3, +6. Chromium is mainly used in steel production and in chrome plating. Its products are also used in leather tanning, printing, dye production, pigments, wood preservatives, and many others.
The respiratory and dermal toxicity of chromium are well-documented. Workers exposed to chromium have developed nasal irritation (at <0.01 mg/m^3, acute exposure), nasal ulcers, perforation of the nasal septum (at ~2 μg/m^3, subchronic or chronic exposure) and hypersensitivity reactions and "chrome holes" of the skin. Among the general population, contact dermatitis has been associated with the use of bleaches and detergents. Compounds of both Cr(VI) and Cr(III) have induced developmental effects in experimental animals that include neural tube defects, malformations, and fetal deaths. The speciation of chromium has become of relevant interest because of the association Cr(VI)-cancer. The different toxicity of the two forms Cr(VI) and Cr(III) are now under examination, even if at the moment the WHO Guidelines report the provisional value 0.05 mg/L referred to total chromium (WHO, 2008).

2.1.5 Copper

Copper (atomic weight 63.54) is ductile, lustrous, reddish metal (density 8.92 g/mL). The main application of copper is in electrical industry (transformers, generators, and transmission of electricity). Pollution derives from copper mining, brass manufacture, electroplating industries and from the use of its compounds in agriculture. Copper is known as one of the highest mammalian toxic compounds; inhalation of copper containing sprays is linked with an increase in lung cancer among exposed workers. Copper sulphate is widely used as an algaecide in water supply reservoirs affected by blooms of blue-green algae.
The maximum acceptable copper in drinking water is 2 mg/L (WHO, 2008).

2.1.6 Lead

Lead (atomic weight 207.19) is a bluish-grey, soft, dense metal (density 11.34 g/mL). The oxidation states are 0, +2, +4. Lead is extremely resistant to corrosion and is a poor conductor of electricity. Large quantities of lead, both as the metal and as the dioxide, are used in storage batteries. Lead is also used in cable covering, as ammunition, as electrodes, in solder and as roofing material. The metal is used as shielding from radiation, e.g. in x-ray rooms and nuclear reactors. Lead oxide is also used in the manufacture of fine crystal glass. Historically, lead was used in plumbing. Tetraethyl lead was used as an anti-knock agent in petrol, and as an additive in paints. These uses have been reduced recently because of environmental concerns about cumulative lead poisoning. Although lead is one of the most useful of all the metals, used since antiquity because of its wide distribution and its easiness to be extracted and to work with, it is also the metal that has the most damaging effects on human health.

Environmental contamination by lead probably dates back to Bronze Age. It can enter the human body through the uptake of food (65%), water (20%) and air (15%). Human activities, such as fuel combustion, industrial processes and solid waste combustion contribute to the rise of lead concentrations in the environment. Lead interferes with a variety of body processes and is toxic to many organs and tissues including heart, bones, intestines, kidneys, and reproductive and nervous systems. It interferes with the development of the nervous system and is therefore particularly toxic to children, causing potentially permanent learning and behavior disorders. Occupational exposure is a common cause of lead poisoning in adults. Lead can reach water through the corrosion of pipelines in water transportation systems. WHO Guidelines limit for lead in drinking water is 0.01 mg/L (WHO, 2008).

2.1.7 Mercury

Mercury (atomic weight 200.59) is a heavy, liquid at room temperature, silvery colored metal (density 13.53 g/mL). It presents the three oxidation states 0, +1, +2. The most modern uses are in batteries and cells. The Castner-Kellner process, that produces chlorine and sodium hydroxide, requires mercury in the entire process. It is furthermore used in thermometers, thermostats, switches, vacuum pumps, fluorescent and energy-saving lights, tooth fillings and electrical components. Many compounds of mercury have been used as medicines since many ages. However, in recent years, as awareness about the toxicity of mercury has increased amongst people, most of the medicines have become obsolete. Mercurochrome (used in cuts and wounds) and Thimerosal (as an dental amalgamation) are the compounds that are no more used in many countries. Mascara, an ingredient of cosmetics, contains some amounts of Thimerosal. During the past ten years mercury consumption has shown a strong upward trend. The major proportion can be accounted for by the chloro-alkali industry, from which mercury is released into the environment. Most of it finds its way to watercourses exposing aquatic ecosystems where mercury accumulates. The use of seed-dressings containing mercury is decreasing, although this use of mercurial's is still considerable, and in view of findings in other countries elevated mercury levels in seed-eating birds and their predators must be expected. Many states in the US are now very strict against the use of mercury in cosmetics and medicines. Mercury in the form of gaseous vapors is used in mercury vapor lamps, neon signs and fluorescent lamps.

Biological properties of mercury are very important and include these characteristics: inhaled mercury is more dangerous than ingested mercury; human workers and handlers of mercury may become contaminated and mercury-diseased; elemental and inorganic mercury can be transformed to the extremely toxic methyl-mercury (CH_3Hg^+) by some microbes; mercury accumulates in living organisms, cells, tissues, organs and organisms; mercury can damage immune cells and tissues, and organs such as brain, heart, kidneys, lungs; mercury can be concentrated in the environment and then magnified upwards along the food chain (bioaccumulation and bio-magnification); all compounds of mercury, except those not soluble in water, are to be considered poisonous regardless of the manner of inhalation or ingestion. Mercury limit in drinking water is 0.006 mg/L (WHO, 2008).

2.1.8 Nickel

Nickel (atomic weight 58.69) is a ductile, malleable, silver-white metal (density 8.91 g/mL). It presents the oxidation states -1, 0, +1, +2, +3, +4. More than 70% of nickel produced annually is devoted to the production of alloys; nickel is used in a variety of electrolytic procedures, in the manufacture of batteries and in welding procedures, as a catalyst in large scale processes, and in the glass and ceramics industry. In addition to 8.5 million tons per year of nickel in the atmosphere due to natural sources, 43 million tons are released by anthropogenic activities. Population exposed at soluble nickel concentration < 1 µg m^{-3} has no respiratory cancer risk, which is related to exposure to concentrations greater than 1 mg m^{-3} (workers in nickel industries). Dermal sensitivity to nickel is presented by 10-20 % of female and 1 % of male population. The nickel content in surface water ranges from 2 to 20 µg/L. The limit for nickel in drinking water is 0.07 mg/L (WHO, 2008).

2.1.9 Zinc

Zinc (atomic weight 65.41) is a soft, bluish-white metal (density 7.14 g/mL). It presents the oxidation states 0, +2. Zinc and its products are widely used in alloy production, as anticorrosion coatings of steel and iron, in electrical devices, in rubber and tire industries, in paints, in pesticides and as chemical reagents in a number of applications. Zinc is the second most abundant trace metal in the human body: it appears in the active site of a variety of enzymes and many of the metabolic consequences of its deficiency are related to a diminished activity of zinc metallo-enzymes. Zinc is relatively nontoxic, even if daily doses greater than 100 mg during several months may lead to different disorders. Zinc imparts an undesirable astringent taste to water. Water containing zinc at concentrations in the range 3–5 mg/L also tends to appear opalescent and develops a greasy film when boiled. This feature allows the high zinc limit 3 mg/L in drinking water (WHO, 2008).

3. Interaction between biomass and metal ions

The capacity of a given biomass to absorb toxic metal ions has been traditionally quantified using either Langmuir, Freundlich, Langmuir–Freundlich isotherms, or different alternative models. These isotherms were developed under chemical assumptions that are not generally met in biosorption processes.

The main reason for their extended use is that they describe satisfactorily experimental data. They can be used for predictions, although they do not take into account external parameters, such as the pH or ionic strength. Langmuir equation

$$q_{eq} = q_{max}\, b\, C_{eq} / (1 + b\, C_{eq}) \quad (1)$$

is the simplest and the one used by the most of authors. In this equation, q_{eq} is the amount of metal ion sorbed at equilibrium, C_{eq} the equilibrium concentration of metal ion in solution and *b* is the Langmuir constant related to the energy of sorption, which reflects quantitavely the affinity between the biomass and the metal ion. The parameter q_{max} represents the maximum capacity of the biomass to absorb a given metal ion and it is usually determined by fitting the isotherm experimental data to the equation model. The q_{max} values are quite almost expressed as milligrams of sorbed metal ion respect to the weight in grams of dry sorbent.

The q_{max} values reported in an our recent paper (Nurchi & Villaescusa, 2008), based on the survey of last ten years of literature, lie in the ranges 2.81-285.7 mg/g for Cd^{2+}, 11.7-32.00 mg/g for Cu^{2+}, 8.45-73.76 mg/g for Pb^{2+}, 1.78-35 mg/g for Zn^{2+}, 7.9-19.56 mg/g for Ni^{2+}, 17.2-126.9 mg/g for Cr(VI), and 3.08 mg/g for Cr^{3+}. These quantities look more similar when expressed in molar concentrations (0.025-2.5 mmol/g for Cd^{2+}, 0.185-0.50 mmol/g for Cu^{2+}, 0.04-0.36 mmol/g for Pb^{2+}, 0.027-0.53 mmol/g for Zn^{2+}, 0.13-0.34 mmol/g for Ni^{2+}, 0.33-2.44 mmol/g for Cr(VI), and 0.06 mmol/g for Cr^{3+}) and the maximum quantity of metal ion sorbed by a gram of sorbent is of the order of 0.5 mmoles (values five times higher are found for Cd^{2+} and Cr(VI), which could be considered a reasonable result if we consider the large variability in materials and experimental conditions (particle size, pH, temperature, etc.).

In order to better characterize the behavior of a given sorbent, the use of chemical (mmol/g) instead of technical (mg/g) units has to be recommended whenever comparisons have to be made. The results obtained in this way actually contain information on the number of coordinating sites, which can be of great utility to make provisional forecasts of the binding capacity of different metal ions, without restraints due to their atomic mass.

In literature different variables (particle size, temperature, pH, exchange and so on), and different kinetics and thermodynamic models (Langmuir, Freundlich, ...) are taken into account. In the following sections 5 and 6 we will discuss the effect of temperature and pH on the sorption process. In order to design sorption processes, it is important to predict the rate at which a pollutant is removed from an aqueous solution. The rate constant and reaction order must be determined experimentally. It is usually necessary to carry out experimental studies varying several parameters such as metal ion and sorbent concentration, agitation speed, particle size, and temperature. Fitting the experimental results allows determining the kinetic mechanism, e.g. film diffusion, kinetic sorption, diffusion sorption or a combination of these processes. The kinetic models most used in biosorption studies were widely discussed in an intersting review by Ho et al., 2000.

4. Identification of functional groups and their role in metal sorption

The sorption of metal ions by biomass occurs via functional groups on its surface by one or more mechanisms. All the sorbents derived from different by-products of agriculture share a common network of lignin and cellulose, and differ for the presence of functional groups which characterize each single biomass. As said before, identification of the functional groups is crucial for understanding the mechanism that governs the sorption process. Indeed, each functional group presents its own coordinating abilities toward the different metal ions. These coordinating abilities can be rationalized in term of the hard/soft character both of the binding group and of the metal ion. In order to highlight the importance of each different binding group in the mechanism of metal ion adsorption, the percent incidence drawn out from 1997 to nowadays literature is presented in Fig. 1.

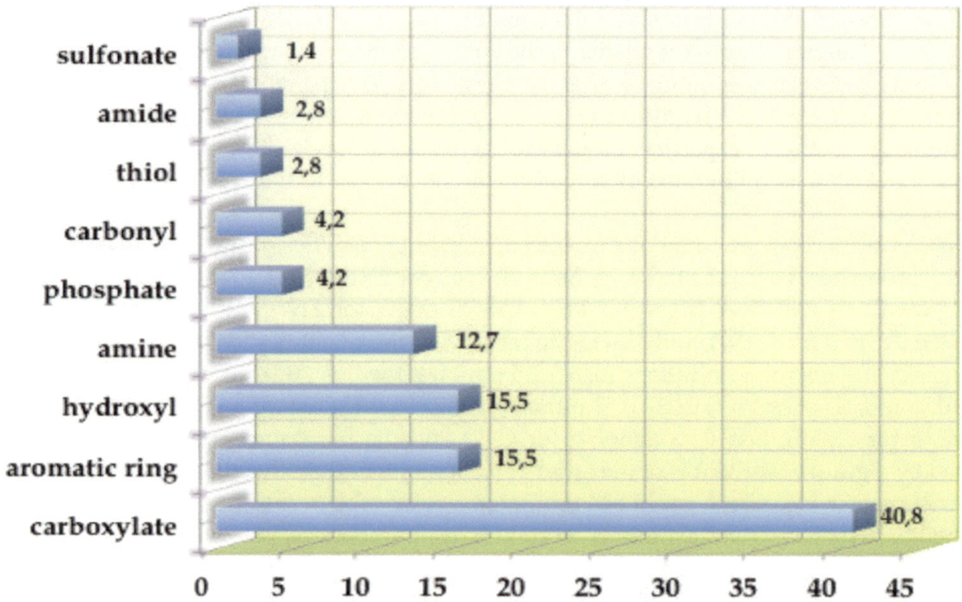

Fig. 1. Incidence of the different binding groups on biomass surface involved in metal ion complexation.

Potentiometric titrations, chemical treatments of the sorbent, alkaline and alkaline-earth metal ion release and spectroscopic techniques are the procedures widely followed to reveal the binding groups. A brief survey of these methods is presented in the next sections.

4.1 Potentiometric titrations

Potentiometric titrations measure the acid-base properties of the sorbent and the ionic exchange properties with regard to H^+ and OH^- ions. The presence of acid and basic sites determines the sorbent amphoteric properties and, depending on the pH, the functional groups can be either protonated or deprotonated. Active site concentrations are generally determined by acid-base potentiometric titration of the adsorbent and related modeling. Acidity constants found in the literature can be considered as mean values, which are representative of the class of the functional groups. Potentiometric titrations can also be used to determine the pH at the point zero charge (pH_{pzc}) of biomass. pH_{pzc} is the pH at which the sorbent surface charge takes a zero value as the charge of the positive surface sites is equal to that of the negative ones.

The knowledge of pH_{pzc} allows one to hypothesize on the ionization of functional groups and their interaction with metal species in solution; at solution pH's higher than pH_{pzc} the sorbent surface is negatively charged and could interact with metal positive species while at pHs lower than pH_{pzc} the solid surface is positively charged and could interact with negative species. Carboxylic groups were found to be the most involved, in the majority of cases, where potentiometric titration was used to elucidate the functional groups on biomass responsible for metal ions sorption. This fact is in part expected on the basis of their easiest deprotonation in the 2 - 6 pH range which is the most suitable for metal sorption.

4.2 Chemical treatment of sorbent surface

The contribution of each functional group can be evaluated by chemical treatment. It consists in carrying out chemical reactions that selectively block different functional groups on the sorbent surface. The most common chemical modifications are esterification of carboxylic and phosphate groups, methylation of amines, and modification of mercapto groups. Carboxylic groups can be alkylated by reaction with methanol or ethanol in acidic media, while amines by reaction with formaldehyde and formic acid. Alkylation of both functional groups prevents their participation in metal biosorption, thus reducing the biosorption efficiency.

Chemical treatments were also used to selectively extract different compounds, such as fats or polyphenols, in order to improve metal sorption. A report on the application of these methods can be found in a work of Nurchi at al., 2010.

4.3 Alkaline and alkaline-earth metal ion release

Vegetal biomaterial can be viewed as a natural ion-exchange material that primarily contains weak acidic and basic groups on its surface. One of the common procedures to investigate whether ion-exchange is the mechanism responsible for metal sorption is to determine the concentration of alkaline and alkaline-earth metal ions or protons (when the sorbent is pretreated with acid) released from the sorbent to the solution after metal uptake. The determination of the concentration of ions released into the solution (M: Na^+, K^+, Ca^{2+}, Mg^{2+}, H^+) allows the balance of the concentration of the absorbed toxic metal ion (M^*), through a charge balance, not explicitly reported in equation (2).

$$R-M + M^* \leftrightarrows R-M^* + M \tag{2}$$

On the solid material the appearance of the sorbed metals, associated with the disappearance of alkaline and alkaline-earth metal ions, can be followed by Scanning Electron Microscopy (SEM) coupled with energy dispersive X-ray analysis (EDAX). This technique greatly contributes to indicate that ion exchange takes place between alkaline and alkaline-earth metal ions on the sorbent and the toxic metal ions in the solution.

4.4 Spectroscopic analysis

Useful information on the role of functional groups on metal sorption can be reached by non-destructive spectroscopic methods, observing the modifications induced by the metal on the spectra of the pure adsorbent.

4.4.1 Fourier transform infrared spectroscopy (FTIR)

FTIR is one of the most used techniques. Infrared Spectroscopy belongs to the group of molecular vibrational spectroscopies which are molecule-specific, and give direct information about the functional groups, their kind, interactions and orientations. Its sampling requirements allow the gain of information from solids, and in particular from solid surfaces. Even if historically IR has been mostly used for qualitative analysis, to obtain structural information, nowadays instrumental evolution makes non-destructive and quantitative analysis possible, with significant accuracy and precision. The shift of the bands and the changes in signal intensity allow the identification of the functional groups involved in metal sorption. Using this technique, carbonyl, carboxylic, aromatic, amine, and hydroxyl groups has been found to be involved in metal uptake by different biosorbents.

4.4.2 Diffuse reflectance infrared fourier transform spectroscopy (DRIFTS)

DRIFTS occurs when light strikes on the surface of a material and is partially reflected and transmitted. The light that penetrates the material may be absorbed or reflected out again. The diffuse reflectance (radiation reflected from an absorbing material) is thus composed of surface-reflected and bulk re-emitted components, and contains information relative to the structure and composition of the sample. Even if DRIFTS has been not of large use, it has found interesting applications on verifying the enhancement of cadmium sorption capacity by juniper wood when carbonyl groups were substituted by sulfonic groups and on determining that Cr^{3+}, Cu^{2+} and Zn^{2+} were sorbed onto the organic polymeric fraction of olive mill wastewater by ion exchange between alkaline and alkaline-earth metal ions and protons bound to carboxylic groups.

4.4.3 X-ray absorption spectroscopy (XAS)

XAS specifically examines the local structure of elements in a sample. The structure of a material is deduced on theoretical basis, but usually the interpretation of XAS spectra is founded on databases of known structures. This technique is useful in the case of heterogeneous samples and a wide variety of solid materials can be examined directly and non-destructively. Also the structure of amorphous phases can be easily achieved, as the local structure does not depend on long-range crystalline order. The application of XAS varies from the trace element concentration up to that of major elements. So it is useful to speciate trace elements adsorbed on the surface of biomass. X-ray absorption spectroscopy consists in the absorption of high energy X-rays by an atom in a sample. This absorption takes place at the energy corresponding to the binding energy of the electron in the sample. The interaction of ejected electrons with the surrounding atoms produces the observed spectrum. (XAS) and extended X-ray absorption fine structure (EXAFS) were used to ascertain the ligands involved in metal binding and the coordination environment for Cr^{3+} bound to *alfalfa shoot* biomass by Tiemann et al., 1999, and by Gardea-Torresday et al., 2002.

4.4.4 X-ray photoelectron spectroscopy (XPS)

XPS, introduced by the Nobel Prize winner Siegbahn in 1949, is the main technique used for qualitative and quantitative elemental analysis of surfaces. It provides significant information on the chemical bonding of atoms. The absorption of high-energy electromagnetic radiation (X-ray or UV) by surfaces leads to the emission of photoelectrons; those generated in the outermost layers emerge from the surface into the vacuum and can be detected. The measure of the kinetic energy of the emitted photoelectrons allows the determination of the binding energies of electrons and the intensity function (number of photoelectrons vs. kinetic energy), and quantitative results are obtained from the knowledge of the number of atoms involved in the emission process.

Ashkenazy et al., 1997, using X-Ray photoelectron spectroscopy (XPS) pointed out the involvement of nitrogen in lead sorption and the lead-oxygen interaction at the carboxyl group on the basis of the decrease in nitrogen concentration and of the shift of oxygen peak. The same technique confirmed that chromium was sorbed onto grape stalks in both its trivalent and hexavalent forms, and allowed the ascertainment of the oxidation state of chromium bound on pine needles. Furthermore it was used to explain the increase of cadmium and lead sorption onto baker's yeast after modification of sorbent surface by cross linking cysteine.

4.4.5 Scanning-electron microscopy (SEM)

SEM is a useful technique in the study of both the natural sorbent morphology and its modification derived from sorbate interactions. SEM is an electron microscope, which provides images of the sample surface by scanning it with a high-energy beam of electrons. The electron interactions with the atoms of the sample produce signals that contain information about topography, morphology, and composition of the sample surface. The samples must be electrically conductive, at least on their surface, for conventional SEM imaging. Nonconductive samples are coated with an ultra-thin layer of electrically-conducting material; this coating prevents the accumulation of static electric charges on the sample surface during electron irradiation. Magnification of the imaging can be controlled over a range of up to 6 orders of magnitude from about x25 to 250,000 times. When coupled with energy dispersive X-ray analysis (EDAX), the atom concentrations on the sorbent surface can be determined. This enables the confirmation of a mechanism of ion exchange, generally investigated by determining the concentration of alkaline and alkaline-earth metal ions released from the sorbent after metal sorption.

5. Effect of temperature

In studies on heterogeneous material, requiring long equilibration times, it is hard to perform reliable calorimetric measurements. Thus, only carrying out experiments at variable temperature can give information on how this parameter affects the sorption of metal ions. From the limited extent of studies at variable temperature, only controversial conclusions can be reached. Most studies have been carried out at a fixed room temperature (20 or 25 °C). Some studies point out a low temperature influence or, at least, in a limited temperature range, giving evidence that ion exchange is the mechanisms responsible for the sorption process. Nevertheless, Kapoor and Viraraghavan, 1997, remarked that biosorption reactions are normally exothermic, which indicates that sorbent capacity increases with decreasing temperature. Conversely, Romero-González et al., 2005, found that the sorption capacity of Agave lechuguilla leaves for Cr(VI) sorption increased on increasing the temperature from 10 to 40 °C, justifying this endothermicity with Cr(VI) reduction to Cr(III). Malkoc and Nuhoglu, 2007, confirmed the endothermicity of Cr(VI) sorption on tea factory waste, metal uptake increasing as temperature increas from 25 °C to 60 °C. The favorable temperature effect was attributed to a swelling effect within the internal structure of the sorbent enabling the large metal ions Cr(VI) to penetrate further.

6. Effect of pH on sorption

As we have already discussed in section 4.3, one of the mechanisms involved in the sorption of positively charged metal species is ion-exchange. Vegetal biomaterials (constituted principally by lignin and cellulose as major constituents and by a non negligible portion of fatty acid, bearing functional groups such as alcohol, ketone and carboxylic groups that can be involved in complexation reactions with metallic cations) can be viewed as natural ion-exchange materials. These materials primarily contain weak acid and basic groups on the surface, whose ionization degree strongly depends on the pH of the solution. Several authors have performed potentiometric titrations to investigate acid-base properties on the surface of biosorbents and to determine the number of active sites for metal ion sorption.
The strong pH dependence of the sorption parameters can depend on several factors, which can be simplified as follows:

1. behaviour and speciation of metal ions;
2. dependence of the acid-base characteristics of the adsorbing material on the pH;
3. dependence of the interaction metal ion-sorbent on the pH.

As far as point 1 is concerned, we report a statement made by Baes and Mesmer, 1976, in their classical book on the hydrolysis of cations: "*soluble hydrolysis products are important when cation concentrations are very low and can profoundly affect the chemical behaviour of the metals; the formulas and charges of the hydrolysis products formed in such systems can control such important aspects of chemical behaviour as:*

a. *sorption of the dissolved metals in mineral and soil particles;*
b. *tendency of metal species to coagulate colloidal particles;*
c. *solubility of the hydroxide (or oxide) of the metals;*
d. *extent to which the metals can be complexed in solution or extracted from solution by natural agents;*
e. *oxydizability or reducibility of the metals to another valence state.*"

Based on these considerations, we demonstrate the influence of pH on sorption taking as an example the behaviour of one of the most important toxic metal ion, lead, in presence of different coordinating groups. Firstly we take into account the hydrolysis of this metal ion at two different concentrations, 100 mg/L and 0.05 mg/L, i.e. at concentration in strong polluted water and at concentration equal to EU recommended value for drinking water (Fig. 2). At 100 mg L^{-1}, the species $Pb(OH)^+$ (pH> 6) and the polynuclear species $Pb_3(OH)_4^{2+}$ and $Pb_6(OH)_8^{4+}$(pH >7) are formed before hydroxide precipitation occurs at pH~9.5; at 50 µg L^{-1}, Pb^{2+} do not form precipitates and only the mononuclear species are formed instead of the polynuclear ones observed at 100 mg L^{-1}. Metal ion hydrolysis equilibria, as well as hydroxide precipitation, can help explain the dependence of metal ion sorption on the pH. In most cases, the observed pH dependence lies in a range in which the metal ion is completely insensitive to the acidity of the medium. In metal ion sorption, pH effects are commonly accounted for by charge variations on the sorbent surface: protonation of basic sites or dissociation of acidic groups. According to the majority of authors a negative charge favours metal ion sorption by an ionic exchange mechanism or by electrostatic interactions, i.e. the sorption is completely determined by the acid-base behaviour of the functional groups on the surface of the adsorbing material.

The real behaviour is certainly far more complex and can be rationalised in terms of metal ion coordination by surface binding groups. The presence of phenolic, carboxylic, catecholic, amino, and mercapto groups on the surface is well known. As a working hypothesis we can imagine that the different binding groups on the solid particles, dispersed in the metal ion solution, behave as different ligands. With this simplifying assumption, we can consider our system as set of solution equilibria. In this assumption we can treat our system as solution equilibria between various ligands competing for a metal ion or for various metal ions. For example, a carboxylic group near a phenolic group on the surface can be assumed to behave as a salicylate ligand, limited to form only 1:1 chelates being anchored to a solid surface.

In the example showed in Fig. 3, we took into consideration three different coordinating groups as possible ligands for lead: COOH, hard, NH_2, intermediate, and SH, soft donors. Furthermore, we also considered all the possible combination of them to obtain bidentate ligands, COOH-COOH; COOH-NH_2, COOH-SH, NH_2-NH_2, NH_2-SH, and SH-SH.

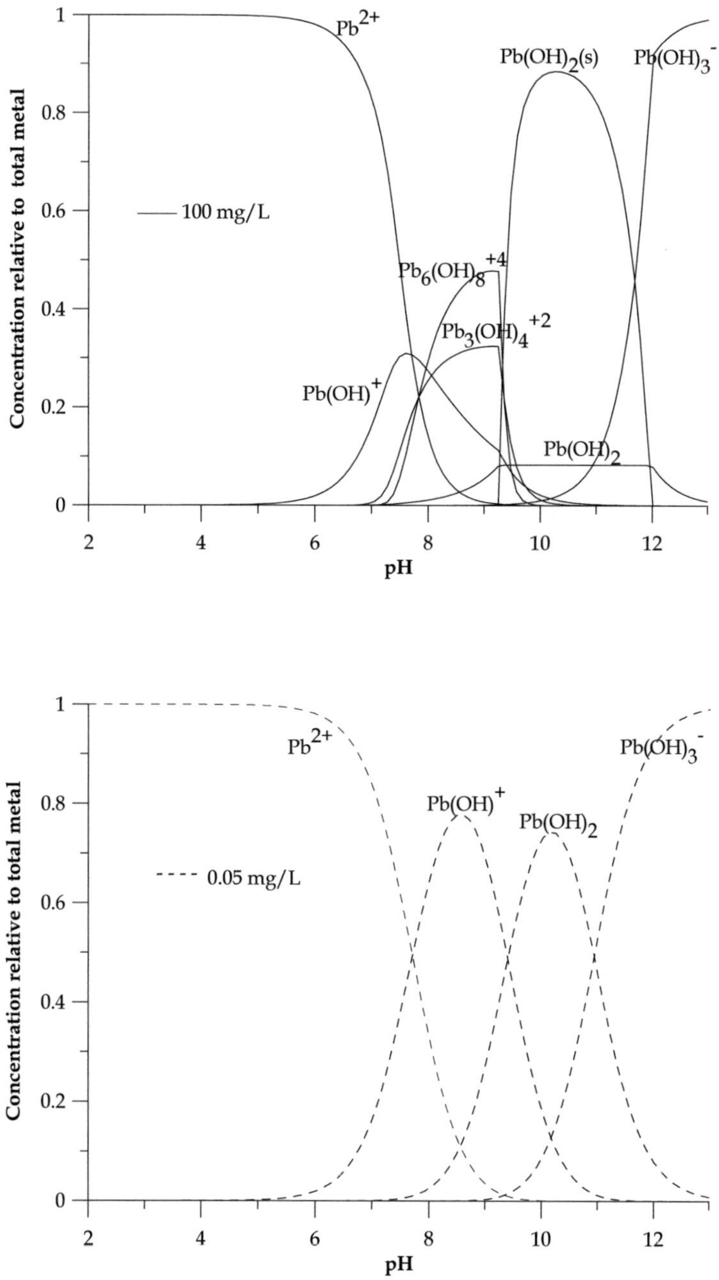

Fig. 2. Species distribution diagrams for Pb^{2+} hydrolysis at two different total concentration 100 mg/L (solid lines) and 0.05 mg/L (dashed lines).

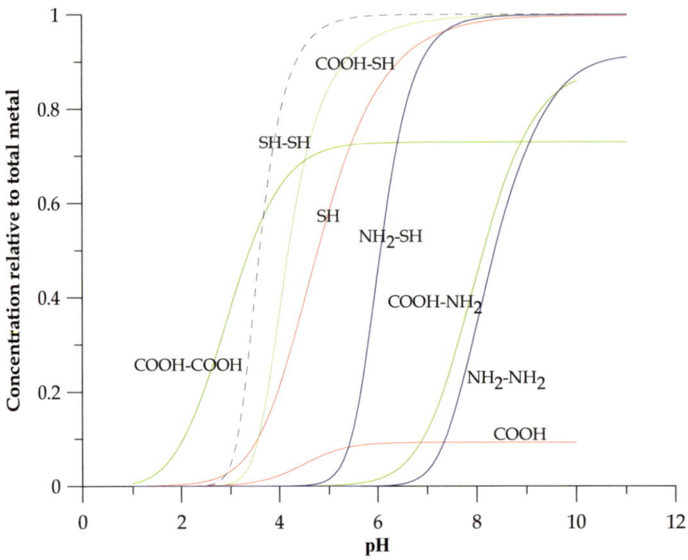

Fig. 3. Formation curves for complex formation between Pb^{2+} and various ligands, bearing the coordinating groups reported on the plots, calculated for 0.001 M solutions in both Pb^{2+} and ligand.

Starting from the distribution curves, obtained using the literature constants for lead complexes with different ligand bearing the above mentioned coordinating groups, some conclusions can be drawn. The soft metal Pb^{2+} ion prefers the soft SH group, which became completely coordinated in 4-6 pH range. No data is available in literature for a single NH_2-Pb interaction. The carboxylic group forms a weak complex in the pH range corresponding to its deprotonation. The addition of a second group (COOH or SH) to the starting SH favours lead coordination, while the addition of a NH_2 group has an adverse effect. Two vicinal COOH groups allow lead complexation at low pH values and act much better than a single COOH group, even if the per cent of complex formation is still much lower than that reached by SH groups. Regarding the coordinating properties related to the amino group, the complex formation, taking place at basic pH > 7, does not prevent the hydroxide formation.

7. Conclusion

The numerous studies on metal sorption by biomass are extremely spread: the investigation of the mechanism involved in metal ion sorption is performed by different techniques, methods and approaches that are related to the equipment availability in the researcher's laboratories and to the researcher education. The use of highly sophisticated and extremely expensive techniques, as mentioned in the above sections, enables one to obtain structural information on the sorbent morphology and indirect knowledge of the implied sorption mechanisms, by comparing some physical properties of the material before and after metal sorption. Even if little importance is given to the classical chemical methods, such as potentiometry and alkaline and alkaline-earth metal ion release, these on the contrary offer several advantages, such as the easy availability in all laboratories, the fact that they are fast, cheap, and friendly-used. The main benefit of these methods is the attainment of quantitative results, which allow the evaluation of the amount and the kind of functional groups involved and the amount of exchanged metal ions.

We hope that the achievements obtained from this enormous quantity of research works can lead in the coming years to a real outlet of practical applications, even if a lack of protocol or systematic approach in this kind of studies has to be remarked. Furthermore, the reached level of knowledge acquired should allow the classification of biomass on the basis of structural coordinating groups on its surface, essential to forecast their behavior toward the different toxic metal ions. Thank to this information, it will be possible to depict the strength of interaction and the pH range more useful for metal removal.

The application of biosorption for effluent detoxification will have a strong ecological impact, joining the advantage of recycling waste biomass and of purifying contaminated waters from toxic metal ions.

8. Acknowledgment

The authors express their gratitude to Professor Guido Crisponi for his help in writing this chapter, with encouraging discussions and useful suggestions.

9. References

Ashkenazy, R., Gottlieb, L., & Yannai, S. (1997). Characterization of acetone-washed yeast biomass functional groups involved in lead biosorption. *Biotechnology and Bioengineering,* Vol. 55, No. 1, (July 1997), pp. 1–10, ISSN 0006-3592.

Baes, C. F.Jr. & Mesmer, R.E. (1976). *The hydrolysis of cations,* J. Wiley & Sons, Inc., ISBN 0-471-03985-3, New York.

Ho, J Y.S., Ng, C.Y., & Mckay, G. (2000). Kinetics of pollutant sorption by sorbents: Review. *Separation and Purification Reviews,* Vol. 29, pp. 189-232, ISSN 1542-2119.

Kapoor, A., & Viraraghavan, T. (1997). Fungi as biosorbents, In: *Biorsorbents for metal ions,* Wasedaj & Foster, pp. 67-80, Taylor & Francis, ISBN 074840431, London.

Malkoc, E., & Nuhuglu, Y. (2007). Potential of tea factory waste for chromium(VI) removal from aqueous solutions: Thermodynamic and kinetic studies. *Separation and Purification Technology,* Vol. 54, No. 3, (May 2007), pp. 291-298, ISSN 1383-5866.

Nurchi, V.M., &Villaescusa, I. (2008). Agricultural biomasses as sorbents of some trace metals. *Coordination Chemistry Reviews,* Vol. 252, (May 2008), pp. 1178-1188, ISSN 00108545.

Nurchi, V.M., Crisponi, G., & Villaescusa, I. (2010). Chemical equilibria in wastewaters during toxic metal ion removal by agricultural biomass. *Coordination Chemistry Reviews,* Vol. 254, (September 2010), pp. 2181-2192, ISSN 00108545.

Romero-Gonzales, J., Peralta-Videa, J. R., Rodriguez, E., Ramirez, S. L.,& Gardea-Torresdey, J. L. (2005). Determination of thermodynamic parameters of Cr(VI) adsorption from aqueous solution onto Agave lechuguilla biomass. *The Journal of Chemical Thermodynamics,* Vol. 37, No. 4, (April 2005), pp. 343-347, ISSN 0021-9614.

Tiemann, K.J., Gardea-Torresdey, J.L., Gamez, G., Dokken, K., Sias, S., Renner, M.W., & Furenlid, L.D. (1999). Use of X-ray Absorption Spectroscopy and Esterification to

Investigate Cr(III) and Ni(II) Ligands in Alfalfa Biomass. *Environmental Science & Technology*, Vol.33, (December 1998), pp. 150-154, ISSN 0013-936X.
World Health Organization (Ed.). (2008). *Guidelines for drinking-water quality,* World Health Organization, ISBN 9241546743, Geneva.

21

Recycling of Phosphorus Resources in Agricultural Areas Using Woody Biomass and Biogenic Iron Oxides

Ikuo Takeda
Shimane University
Japan

1. Introduction

Phosphorus (P) is an essential element in plant nutrients, because many biochemical processes such as photosynthesis, respiration, and energy transfer depend on inorganic P or its organic derivatives. However, P is difficult for plants to obtain from the rhizosphere and P deficiency is one of the major limitations on crop production. This is because soluble P in soil, the primary P source for plants, is extremely low concentration (Condron et al., 2005) and significant portions of P in the soil are various organic complexes and unavailable (Raghothama, 2005). On a worldwide scale, land covering 5.7 billion hectares is estimated to be deficient in P for optimal crop production (Batjes, 1997). Since the soluble P in the soil is easily taken up by plants and microorganisms, continuous application of P fertilizer is necessary for crop production.

The global demand for P has increased 10-fold since the beginning of the 20th century (Cordell et al. 2009) and approximately 80% of the demand is for agricultural fertilizers (Steen, 1998). Thus, more P will be required as the world's population increases. However, there is concern that world P resources will be depleted in the next 50–100 years, because the reserves of high-grade phosphate rock are limited (Runge-Metzger, 1995; Steen, 1998; Smil, 2000; Stewart et al., 2005). Therefore, the recovery of P is essential for sustaining food production.

Figure 1 shows a conceptual illustration of the global P cycle, which is completed by P flux from the ocean to the land, and is intimately linked to global ocean circulation. The P derived from weathering or fertilizer application on the land is washed down in rivers and enters the ocean food chain. In deep ocean water (about 2,000–3,000 m in depth), the P concentration is considerably higher than that at the surface because dead fish and plankton fall on the ocean floor. However, the P-rich water is too deep for humans to exploit. In the deep ocean, the water flows from the Atlantic Ocean to the Pacific Ocean via the Antarctic and the Indian Ocean, while the surface water flows in the opposite direction. This movement is very slow; about 2000 years is required to complete this circulation. In some areas of the Pacific Ocean, the flow rises from the bottom to the surface, but this is rare phenomenon. Because the occurrence of this rising flow depends on a complex combination of sea currents, winds, and geographical features. Consequently, these selected areas are abundant in plankton and fish. However, the P flux from ocean to land occurs only via fishery and seabirds' droppings (guano), unlike nitrogen that can be released into the atmosphere via denitrification. In addition, the seabed is gradually transformed into land by the geological movement of the Earth's crust, but this occurs on a much longer time scale than do human activities. Therefore, the global P cycle is extremely limited.

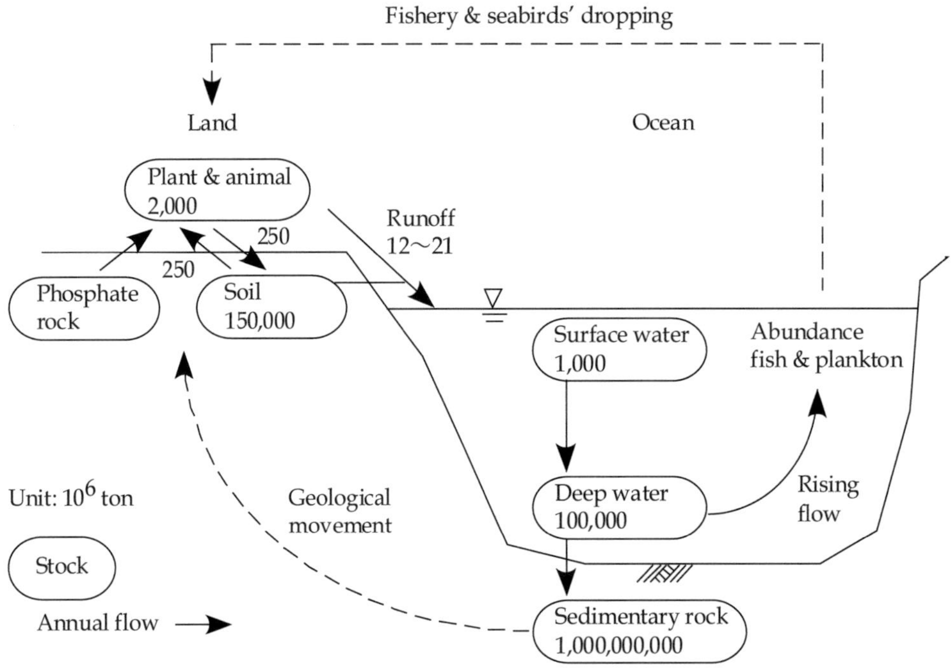

Fig. 1. Conceptual illustration of the global P cycle (data from Sumi, 1989)

Despite the limited nature of the P cycle, repeated applications of fertilizers and organic matter builds up nutrients in the soil. Strong relationships between the level of P monitored by soil tests and the amount of P lost in runoff have been reported (Pote et al., 1996; Sharpley, 1995). Thus, excessive application of P fertilizers contributes to eutrophication, which is sometimes responsible for the lack of clean water resources. From this viewpoint, the recovery of P is also essential.

The behavior of P in nature has been affected by iron (Fe) oxides since ancient time (Bjerrum & Canfield, 2002). In natural water bodies such as canals, swamps, and ponds with low oxygen groundwater seeps and circumneutral conditions, the accumulation of soft, reddish-brown sediment is often observed (Fig. 2). The essential compounds in this sediment are biogenic Fe oxides produced by microaerobic Fe-oxidizing bacteria (Emerson et al., 1999; Emerson & Weiss, 2004; James & Ferris, 2004) and this ferric substance in the sediment can adsorb P in a similar manner to abiotic P adsorbents of ferric compounds (Boujelben et al., 2008; Persson et al., 1996; Seida & Nakano, 2002; Zeng et al., 2004). Therefore, biogenic Fe oxides in nature are considered as one of the P resources. However, they have not yet been recognized as such, although they have been used for ferrous Fe removal in water treatment facilities (Pacini et al., 2005; Katsoyiannis & Zouboulis, 2004; Søgaard et al., 2001). This is because biogenic Fe oxides in natural water bodies are easily dispersed by water turbulence. In addition, it is difficult to collect only the Fe oxides as a P resource, because they usually accumulate only a few centimetres, and anaerobic and malodorous mud exists underneath (see Fig. 3). Moreover, the mud deposits that have existed for a long time may accumulate harmful substances such as heavy metals.

Fig. 2. Accumulation of reddish-brown soft sediment in an agricultural canal

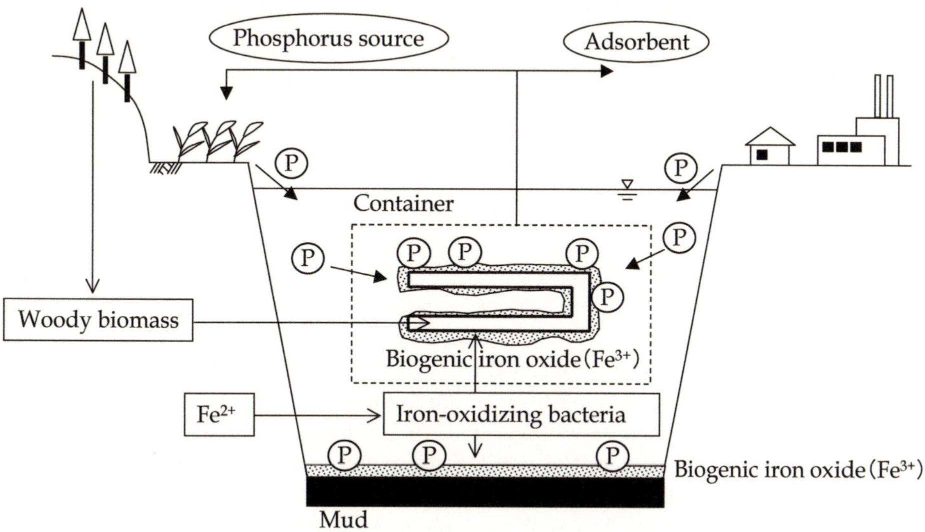

Fig. 3. Conceptual illustration of P recovery from natural water bodies using Fe-oxidizing bacteria and woody biomass.

A new method for the recovery of P from natural water bodies using Fe-oxidizing bacteria and woody biomass as a carrier has been proposed (Fig. 3). A woody carrier is immersed in water in which Fe-oxidizing bacteria are abundant and then removed several weeks later. In this chapter, this method was tested in an agricultural area, dominated by rice paddy fields, located in the eastern part of Shimane Prefecture, Japan. As the woody carrier, sawdust from the Japanese cedar and Japanese cypress were used. Since the accumulation of biogenic Fe oxides was observed throughout the year at several locations, the water quality at these points was monitored. In addition, heavy metals on the immersed carrier were also measured, because biogenic Fe oxides have the potential to also adsorb heavy metals such as arsenic (As), cadmium (Cd), chromium (Cr), mercury (Hg), lead (Pb), zinc (Zn), and nickel (Ni).

2. Material and methods

2.1 Water quality monitoring

Samples for water quality monitoring were collected at eight points on agricultural drainage canals (Fig. 4) on December 13, 2008. These points were located in the downstream area of the Hii River, Japan, at approximately 35° 24′ N and 132° 50′ E. The pH and oxidation-reduction potential (ORP) were monitored with a portable analyzer (Kasahara Chemical Instruments, KP-5Z). The Fe, P, and nitrogen (N) concentrations were analyzed in accordance with Japanese Industrial Standard (JIS) K 0102 (Namiki, 2003): total Fe (T–Fe) and dissolved Fe (D–Fe) were measured by the 1,10-phenanthroline method; total phosphorus (T–P) was measured by the ascorbic acid reduction molybdenum blue method after potassium peroxodisulfate decomposition; phosphate phosphorus (PO_4–P) was measured by the ascorbic acid reduction molybdenum blue method; total nitrogen (T–N) was measured by UV absorption spectroscopy after alkaline potassium peroxodisulfate decomposition; ammonium nitrogen (NH_4–N) was measured by the indophenol blue method; nitrate nitrogen (NO_3–N) was measured by ion chromatography (Shimadzu HIC-6A). The total organic carbon (TOC) concentration was measured by Shimadzu TOC-Vcsn system and suspended solids (SS) were measured by gravimetric analysis using glass-fiber filters (pore size = 0.45 μm; Advantec GS25).

2.2 Biomass carrier

Although the precise mechanism of Fe oxidation-deposition by Fe-oxidizing bacteria is not sufficiently understood (Pacini et al. 2005) and some of the species are characterized as autotrophic (Hallbeck & Pedersen, 1991; Imai, 1984), a substantial accumulation of biogenic Fe oxides was found on the surface of submerged aquatic plants in an agricultural drainage canal (Fig. 5). On the basis of this finding and some trial-and-error experiments, woody biomass (conifer heartwood) was used as the carrier for collecting biogenic Fe oxides. In particular, sawdust (particle size: 0.2–2 mm) of the Japanese cedar (*Cryptomeria japonica*) and the Japanese cypress (*Chamaecyparis obtusa*) were used, both of which are typical conifers found in Japan. The heartwood of the conifer contributes very little to secondary water pollution during the immersion test period, because it mainly consists of carbon, hydrogen, and oxygen and contains extremely small amounts of N and P (Jodai & Samejima, 1993). In addition, it contains a large amount of lignin, flavonoids, and phenols, which provide resistance to wood-decomposing fungi (Jodai & Samejima, 1993). Moreover, approximately 97% of the wood tissue of conifer heartwoods consists of tracheids, which are hollow elongated cells (Furuno & Watanabe, 1994). Thus, the sawdust is expected to have a large specific surface area.

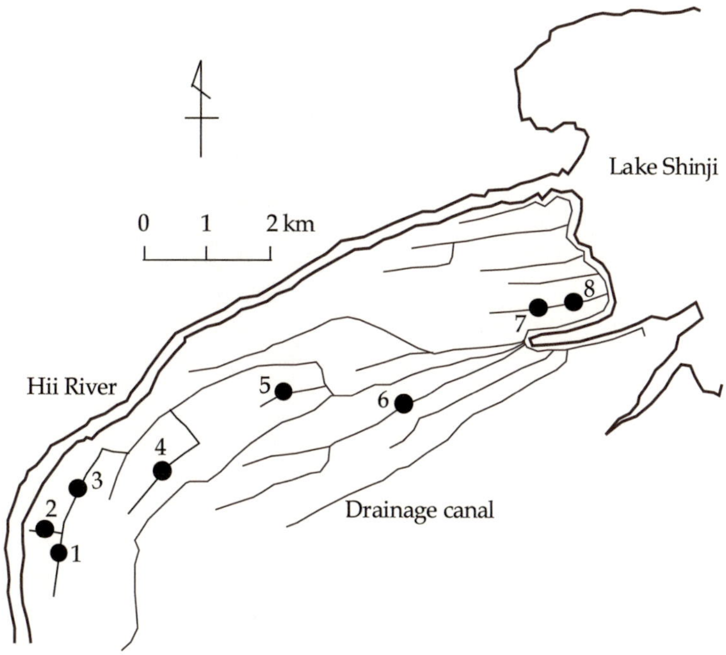

Fig. 4. Map of study site

Fig. 5. Accumulation of biogenic Fe oxides on the surface of submerged aquatic plant

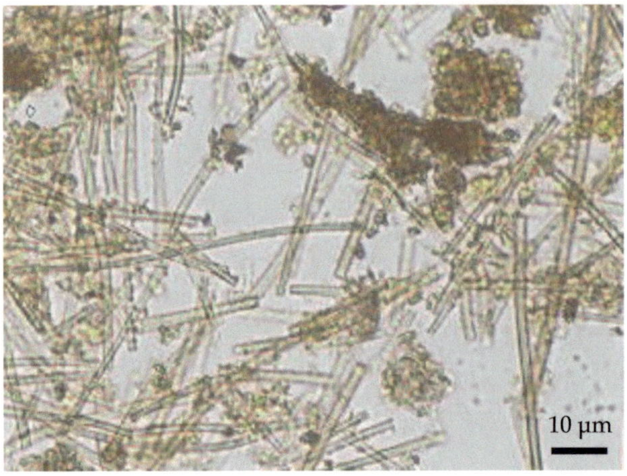

Fig. 6. Sheathed bacteria, *Leptothrix* spp.

2.3 Immersion test

The immersion test was conducted in an agricultural drainage canal (at point 1 in Fig. 4) where reddish-brown sediment accumulated, and sheathed bacteria (*Leptothrix* spp.) were found to be abundant (Fig. 6). The test was performed during the irrigation period for paddy fields (from May to September 2009) and the non-irrigation period (from October 2009 to April 2010), because the canal is mainly fed by drainage water from paddy fields via surface outlets and underdrains, and the water quality is affected by the paddy field irrigation.

In this test, the woody carrier was placed in a container of non-woven bag and lowered to the bottom of the canal. The carrier in the container was removed from the water after immersion for 4 weeks. The Fe collected on the immersed carrier was analyzed by the 1,10-phenanthroline method (Stucki & Anderson, 1981), and the P adsorbed on the Fe oxides was analyzed by the Bray-2 method (Byrnside & Sturgis, 1958). The Bray–2 P is a portion of the soil P and is one of the indexes of available P for plant uptake. In this study, the adsorbed P is expressed as g/kg instead of the conventional expression of Bray-2 P (mg P_2O_5/100 g dry material). In addition, the water samples were collected at weekly intervals and the water quality of D-Fe and PO_4-P was analyzed by the above-mentioned methods.

2.4 Elemental analysis

Elemental analysis of the immersed carrier was carried out by X-ray fluorescence spectrometry system (Shimadzu, EDX-720) at a voltage of 50 kV and a current of 1 mA.

3. Results and discussion

3.1 Water quality in the canals

Table 1 presents the water quality at eight points on the agricultural canals. At all points, the D–Fe concentration was much lower than the T–Fe concentration, and the same relationship was found between the PO_4–P concentration and T–P concentration. Therefore, most of the Fe and P in the water were associated with particulate matter. The average concentrations of

T-N and TOC were 2.864 and 2.180 mg/L, respectively, and the NO_2-N concentration was much lower than the T-N concentration.

Site	pH	ORP (V)	T-Fe (mg/L)	D-Fe (mg/L)	T-P (mg/L)	PO_4-P (mg/L)	T-N (mg/L)	NH_4-N (mg/L)	NO_2-N (mg/L)	NO_3-N (mg/L)	TOC (mg/L)
1	6.7	0.02	8.474	0.109	0.180	0.020	3.229	0.653	0.016	2.533	2.483
2	6.8	0.025	4.742	0.235	0.145	0.012	2.585	0.494	0.013	1.840	1.476
3	7.1	-0.026	13.222	0.600	0.319	0.022	2.627	1.293	0.003	1.080	1.833
4	6.9	0.049	12.612	0.085	0.226	0.022	1.972	1.568	0.002	0.169	2.257
5	6.7	-0.017	7.999	0.201	0.261	0.023	2.118	0.968	0.006	1.030	1.421
6	6.8	-0.017	14.783	0.061	0.529	0.033	2.330	1.324	0.013	0.630	2.077
7	6.8	-0.029	16.920	0.085	0.244	0.012	1.738	1.283	0.003	0.000	2.757
8	6.9	-0.015	12.171	0.071	0.180	0.002	6.309	1.114	0.036	4.470	3.138
Mean	6.8	-0.001	11.365	0.181	0.261	0.018	2.864	1.087	0.012	1.469	2.180

Table 1. Water quality at eight points on the agricultural canals

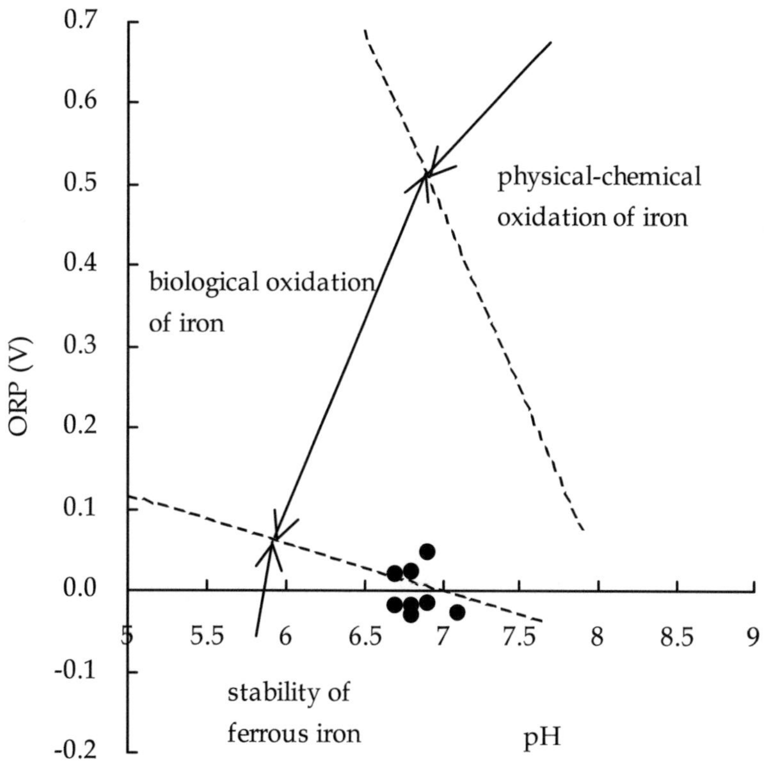

Fig. 7. Data plot on pH–OPR diagram (from Mouchet, 1992). Black dots represent the data monitored at the study sites of Fig. 4.

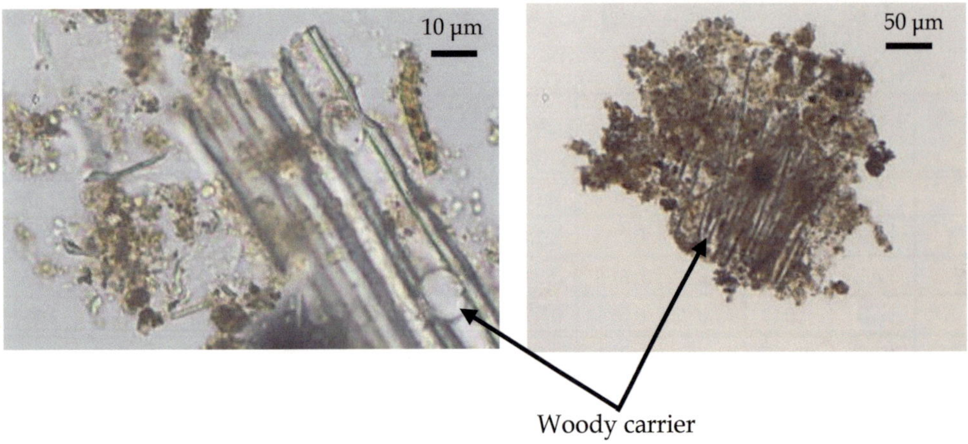

Fig. 8. Biogenic Fe oxides (brown mass) on woody carrier

Since the Fe oxidation was characterized with a pH–ORP diagram (Mouchet, 1992), the data from this study were plotted on it (Fig. 7). In this diagram, the pH–ORP area is divided into physical-chemical oxidation, biological oxidation, and stability of ferrous Fe. The data from this study were within the range of pH = 6.7 to 7.1 and ORP = −0.03 to 0.05 V, and were located near the boundary between the biological oxidation and the stable ferrous Fe area. Since the suitable aquatic conditions for the growth of Fe-oxidizing bacteria have been reported to be low concentration of oxygen and circumneutral pH (James & Ferris, 2004), the results of present study agree with this knowledge.

3.2 Fe and P on the carrier

The color on the woody carrier changed from light yellow to dark brown. Observation using a microscope revealed that biogenic Fe oxides produced by Fe-oxidizing bacteria had accumulated on the woody carrier (Fig. 8(a)). In many cases, the woody carriers were not easily visible because they had been completely covered by a mass of Fe oxides (Fig. 8(b)).
Figure 9 shows the Fe collected on the woody carrier and the D-Fe concentrations of the water at the site of the immersion test. The average accumulation of Fe on the Japanese cedar was 7.91 g/kg during the irrigation period and 6.74 g/kg during the non-irrigation period. The respective values for the Japanese cypress were 7.67 and 5.54 g/kg. There were no significant differences between the values during the irrigation and the non-irrigation period. The average D–Fe concentration during the irrigation period (0.952 mg/L) was much higher than that during the non-irrigation period (0.338 mg/L). There were no significant differences during the irrigation and non-irrigation period between the collected Fe for the Japanese cedar and the Japanese cypress (Fig. 10). When these values are expressed in parts per million (ppm), the Fe collected during the irrigation period was 7,910 ppm for the Japanese cedar and 7,670 ppm for the Japanese cypress, while the D–Fe concentration was 0.952 ppm. Therefore, the concentration of the Fe on the woody carrier was 8,000- to 8,300-fold greater than the Fe dissolved in the water. For the non-irrigation period, the degree of Fe concentration was 16,000- to 20,000-fold greater.
Figure 11 shows the P adsorbed on the woody carrier and the PO_4-P concentration. The average P adsorbed on the Japanese cedar carrier was 0.350 g/kg during the irrigation

period and 0.187 g/kg during the non-irrigation period. The respective values for the Japanese cypress were 0.332 and 0.172 g/kg. The differences between the values during the irrigation and non-irrigation periods were significant ($p < 0.05$). The average PO_4-P concentration of the water during the irrigation period (0.058 mg/L) was much higher than that during the non-irrigation period (0.022 mg/L). This is probably because the anaerobic conditions caused by flooded water on the paddy fields during the irrigation period lead to the reduction of ferric phosphate ($FePO_4$) compounds and the release of Fe^{2+} and phosphate (PO_4^{3-}) ions. There were no significant differences in the adsorbed P during the irrigation and the non-irrigation period between the Japanese cedar and the Japanese cypress (Fig. 12). When these values are expressed in ppm, the P adsorbed during the irrigation period was 350 ppm for the Japanese cedar and 332 ppm for the Japanese cypress, while the PO_4-P concentration was 0.058 ppm. Therefore, the concentration of the P on the woody carrier was 5,700- to 6,000-fold greater than the P dissolved in the water, and for the non-irrigation period, it was 7,800- to 8,500-fold greater.

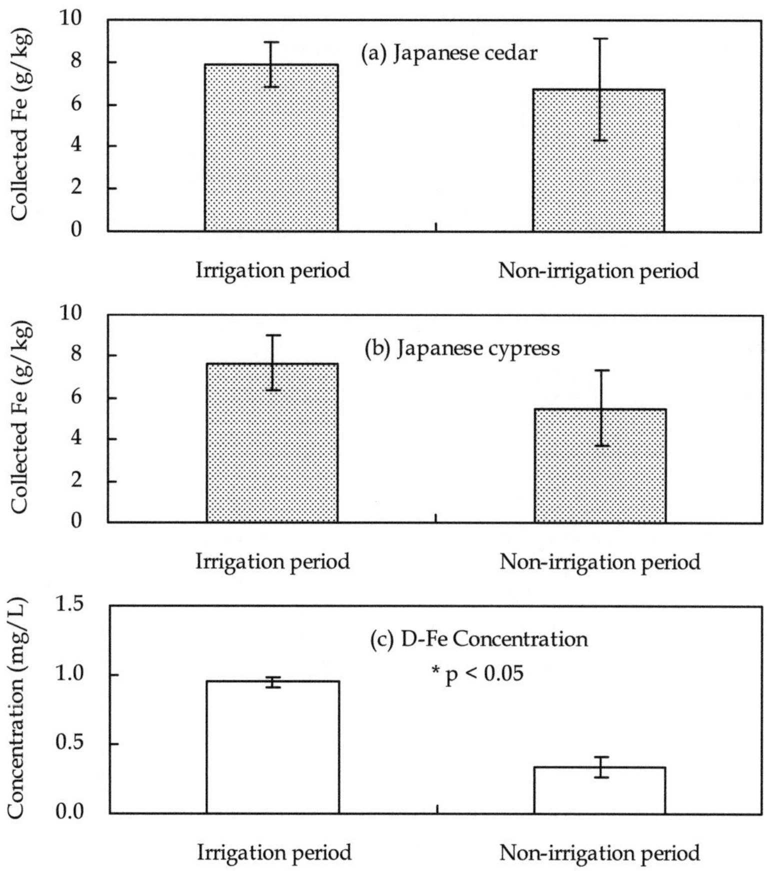

Fig. 9. Fe content after the immersion test. (a), (b): collected Fe after 4 weeks immersion; (c): D-Fe concentration of the water (means and standard errors, n=8)

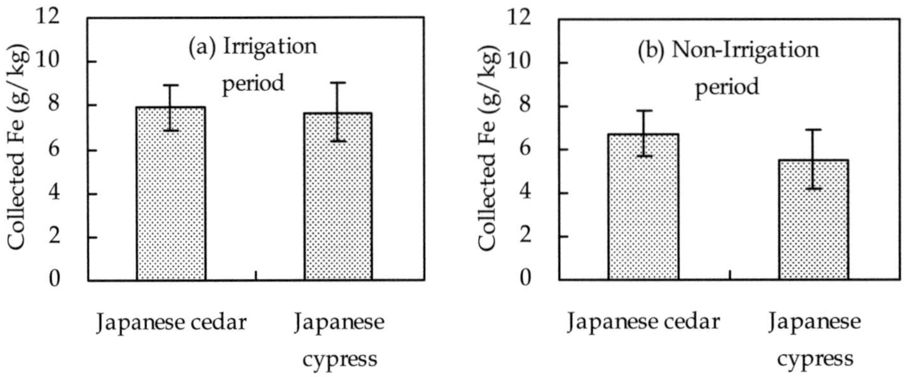

Fig. 10. Comparison of collected Fe between Japanese cedar and Japanese cypress (n=8)

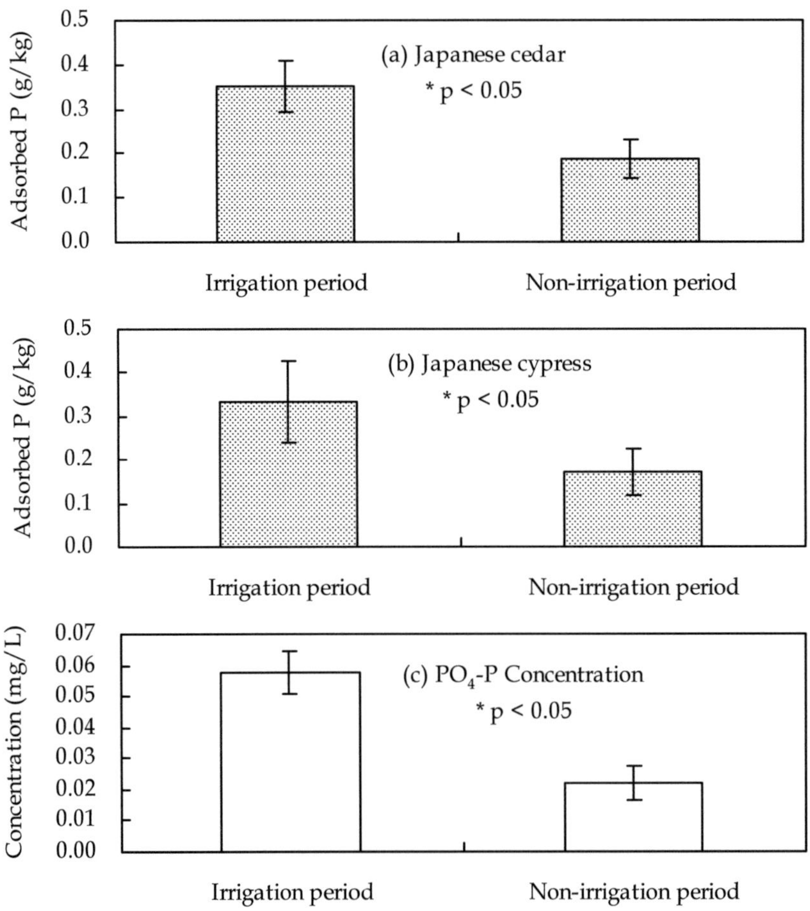

Fig. 11. P contents from the immersion test. (a), (b): adsorbed P after 4 weeks immersion; (c): PO_4-P concentration of the water (means and standard errors, n=8)

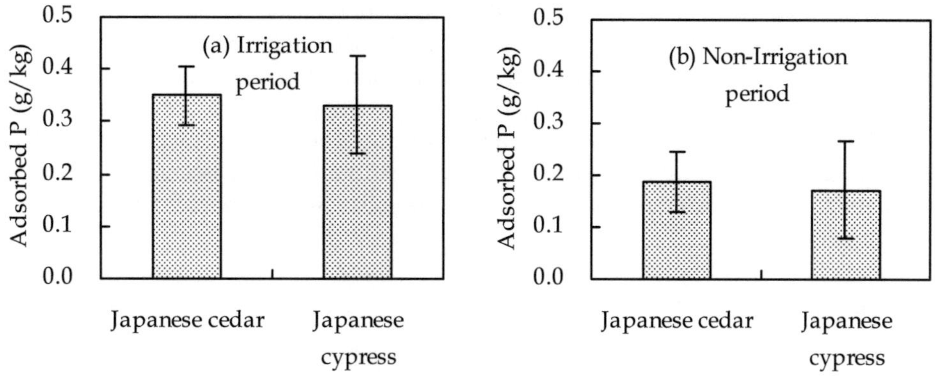

Fig. 12. Comparison of adsorbed P between Japanese cedar and Japanese cypress (n=8)

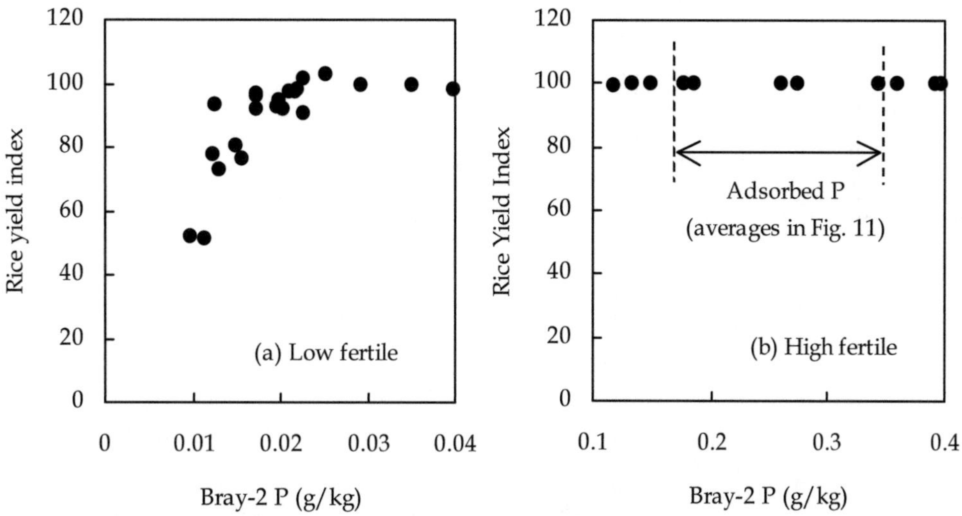

Fig. 13. P fertility of the immersed carrier in the relationship between the Bray-2 P in arable soils and the rice yield index (adapted from Komoto, 1984)

Figure 13 shows the P fertile position of the immersed carrier on the relationship between the Bray-2 P in arable soils and the rice yield index (adapted from Komoto, 1984). In low-fertility soil (Fig. 13(a)), the yield index increases with Bray-2 P, but does not increase over the fertile level of 0.025 g/kg of Bray-2 P. As shown in Fig. 13(b), soils containing greater than 0.1 g/kg are categorized as high-fertility soil. The P values from this study were between 8- and 17-fold higher than the required level (0.025 g/kg) and categorized in the range of high-fertility soil. Therefore, the immersed carrier had obtained sufficient P fertility.

3.3 Heavy metals on the carrier

Figure 14 shows an example of an X-ray fluorescence spectrum of the immersed carrier. Fe was the main species detected, although silicon (Si), calcium (Ca), aluminum (Al), P, sulfur (SO_4), potassium (K), chlorine (Cl) were also present. Heavy metals were not detected on most of the carriers, but traces of Pb and Zn were detected in some samples (Table 2). However, they were well below regulation levels set out in the Fertilizers Regulation Act (Ministry of Agriculture, Forestry and Fisheries, 2007) and the Guidelines against Heavy Metal Accumulation in Arable Soil (Environment Agency, 1984). This was probably because the study site was in a rural area that had not been contaminated by heavy metals and also because the immersion period was too short for these metals to accumulate.

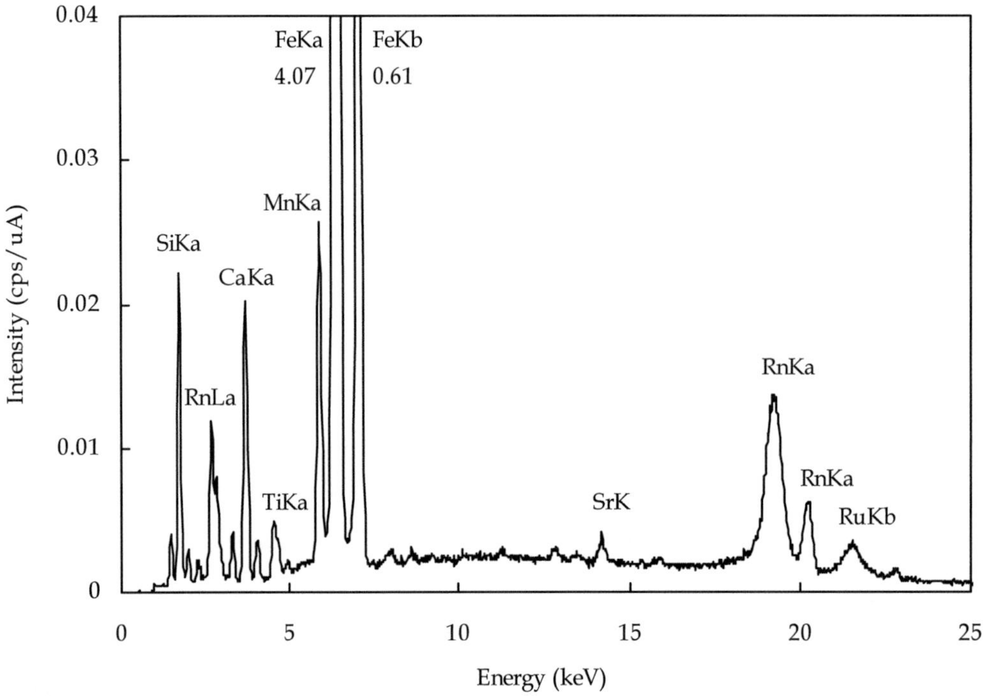

Fig. 14. X-ray fluorescence spectrum of immersed carrier

Element	Concentration (mg/kg)	Regulation value (mg/kg)
As	ND	50*
Cd	ND	5*
Cr	ND	500*
Hg	ND	2*
Ni	ND	300*
Pb	5.3	100*
Zn	4.0	120**
Cu	ND	125**

* Ministry of Agriculture, Forestry and Fisheries, 2007
** Environment Agency, 1984

Table 2. Heavy metal concentrations in immersed carrier (maximum for n=45)

3.4 Possible further applications

The findings reported in this chapter have been obtained from a specific region in Japan. However, Fe is the third most abundant metal found in the soil (Spark, 1995), and Fe-oxidizing bacteria are not rare (Emerson et al., 1999; Emerson & Weiss, 2004; James & Ferris, 2004). Thus, this method can be applicable in many places, provided suitable aquatic conditions supporting the growth of Fe-oxidizing bacteria (low concentration of oxygen and circumneutral pH) are available. In addition, the immersed woody carrier can be applied directly to agricultural land in the form of a fertilizer, without P extraction procedures, which are commonly required for P recovery methods. Therefore, this method is a low-cost technique that should contribute to P resource recycling and the improvement of the aquatic environment, if adopted on a large scale.

4. Conclusions

A new method of P recovery from natural water bodies using Fe-oxidizing bacteria and woody biomass (Japanese cedar and Japanese cypress) was applied in an agricultural canal during irrigation and non-irrigation periods. The amounts of P adsorbed on the carrier during these periods were 0.332–0.350 and 0.172–0.187 g/kg, respectively, while the PO_4-P concentrations of the water were 0.058 and 0.022 mg/L. Expressed these values in parts per million, the P adsorbed on the carrier was 5,700- to 8,500-fold more concentrated than the P dissolved in water. The P on the carrier was 8- to 17-fold higher than the required level for sufficient fertility to support rice production, and it was categorized in the range of high-fertility soil. Some traces of heavy metals adsorbed on the carrier were detected, but they were much lower than the regulation levels. In addition, the woody carrier can be applied directly to agricultural land without P extraction. Therefore, this method is a low-cost technique that should contribute to P resource recycling and the improvement of aquatic environment.

5. Acknowledgement

This study was partially supported by a grant from the Shimane University Priority Research Project and a Grant-in-Aid for Scientific Research from the Japan Society for the Promotion of Science (#20380179).

6. References

Batjes, N. H. (1997). A world data set of derived soil properties by FAO-UNESCO soil unit for global modeling. *Soil Use Manage*, 13, pp.9-16

Bjerrum, C. & Canfield, D. (2002). Ocean productivity before about 1.9 Gyr ago limited by phosphorus adsorption onto iron oxides. *nature*, 417, pp.159-162

Byrnside, D. S. & Sturgis. M. B. (1958). Soil phosphorus and its fractions as related to response of sugarcane to fertilizer phosphorus. *Louisiana Agricultural Expansion Station Bulletin*, 513, pp.56-66

Boujelben, N., Bouzid, J., Elouear, Z., Feki, M., Jamoussi, F., Montiel, A. (2008). Phosphorus removal from aqueous solution using iron coated natural and engineered sorbents. *Journal of Hazardous Materials*, 151, pp.103-110

Cordell, D., Drangert, J. & White, S. (2009). The story of phosphorus: Global food security and food for thought. Global Environmental Change, 19, pp.292-305

Condron, L. M., Turner, B. L., & Cane-Menun, B. J. (2005). Chemistry and dynamics of soil organic phosphorus, In: *Phosphorus: Agriculture and the Environment*, Sims, J. T. & Sharpley, A. N. (eds.), pp.87-121, American Society of Agronomy, Crop Science Society of America, Soil Science Society of America, ISBN 978-0891181576, Madison, USA

Emerson, D, Weiss, J. V. (2004). Bacterial iron oxidation in circumneutral freshwater habitats: Finding from the field and the laboratory. *Geomicrobiology Journal*, 21, pp.405-414

Emerson. D. Weiss, J. V. & Megonigal, J. P. (1999). Iron-oxidizing bacteria are associated with ferric hydroxide precipitantes (Fe-plaque) on the roots of wetland plants. *Applied Environmental. Microbiology*, 65, pp.2758-2761

Environment Agency (1984). *Guidelines against Heavy Metal Accumulation in Arable Soil*, Environment Agency, Tokyo, Japan (in Japanese)

Furuno, T., Watanabe M. (eds.) (1994). *Wood Science 2: Tissue and Material*, Kaiseisya, ISBN 9784906165537, Tokyo, Japan (in Japanese)

Hallbeck, L. & Pedersen, K. (1991). Autotrophic and mixotrophic growth of Gallionella ferruginea. *Journal of General Microbiology*, 137, pp.2657-2661

Imai, W. (1984). *Autotrophic Bacteria*. Kagakudojin, ISBN 978-4759803525, Tokyo, Japan (in Japanese)

James, R. E. & Ferris, F. G. (2004). Evidence for microbial-mediated iron oxidation at a neutrophilic groundwater spring. *Chemical Geology*, 212, pp.301-311

Jodai, S. & Samejima, K. (eds.) (1993). *Wood Science 4: Chemistry*, Kaiseisya, ISBN 978-4906165445, Tokyo, Japan (in Japanese)

Katsoyiannis, I. A. & Zouboulis, A. I. (2004). Biological treatment of Mn(II) and Fe(II) containing groundwater: kinetic considerations and product characterization. *Water Research*, 38, pp.1922-1932.

Komoto, Y. (1984) Phosphorus fertility and yield in paddy soils, In: *Paddy Soil and Phosphate*, Japanese Society of Soil Science (ed.), pp.87-126, Hakuyusya, ISBN 482-6800711, Tokyo, Japan (in Japanese)

Ministry of Agriculture, Forestry and Fisheries, (2007) *Fertilizers Regulation Act*, Ministry of Agriculture, Forestry and Fisheries, Tokyo, Japan (in Japanese)

Mouchet, P. (1992). From conventional to biological removal of iron and manganese in France. *Journal of the American Water Works Association*, 84, pp.158-166.

Namiki, H. (ed.) (2008). *Analytical Method of Water Quality for Industrial Wastewater (JIS: Japan Industrial Standard (K0102)*, Japan Industrial Standard Association, ISBN 978-4542304123, Tokyo, Japan (in Japanese)

Pacini, V. A., Ingallinella, A. M. & Sanguinetti, G. (2005). Removal of iron and manganese using biological roughing up flow filtration technology. *Water Research*, 39, pp.4463-4475

Persson, P, Nilsson N. & Sjoberg, S. (1996). Structure and bonding of orthophosphate ions at the iron oxide-aqueous interface. *Journal of Colloid and Interface Science*, 177, pp.263-275

Pote, D. H., Daniel, T. C., Sharpley, A. N., Moore, P. A., Edwards, D. R. & Nichols, D. J. (1996) Relating extractable soil phosphorus to phosphorus losses in runoff. *Soil Science Society of America Journal*, 60, pp.855-859

Raghothama, K. G. (2005) Phosphorus and plant nutrition: an overview. In: Phosphorus: Agriculture and the Environment, Sims, J. T. & Sharpley, A. N. (eds.), pp.355-378, American Society of Agronomy, Crop Science Society of America, Soil Science Society of America, ISBN 978-0891181576, Madison, USA

Runge-Metzger, A. (1995). Cycle: Obstacles to efficient P management for improved global food security. In: SCOPE54 Phosphorus in the Global Environment, Tiessen, H. (eds.), pp.27-42, John Wiley & Sons, ISBN 978-0471956914 , New York, USA

Seida, Y. & Nakano, Y. (2002). Removal of phosphate by layered double hydroxides containing iron. *Water Research*, 36, pp.1306-1312

Sharpley, A. N. (1995) Identifying sites vulnerable to phosphorus loss in agricultural runoff. *Journal of Environmental Quality*, 24, pp.947-951

Smil, V. (2000). Phosphorus in the environment: natural flows and human interferences. *Annual Review of Energy and the Environment*, 25, pp.53-88

Søgaard, E. G., Aruna, R., Abraham-Peskir, J. & Koch, C. B. (2001). Conditions for biological precipitation of iron by Gallionella ferruginea in a slightly polluted ground water. *Applied Geochemistry*, 16, pp.1129-1137

Spark, D. L. (1995). *Environmental Soil Chemistry*, Academic Press, ISBN 978-0126564457, San Diego, USA

Steen, I. (1998). Phosphorus availability in the 21 Century: management of a non-renewal resource. *Phosphorus and Potassium*, 217, pp.25-31.

Stewart, W. M., Hammond, L. L. & Kauwenbergh, S. J. (2005). Phosphorus as a natural resource, In: Phosphorus: Agriculture and the Environment, Sims, J. T., Sharpley, A. N. (eds.), pp.3-22, American Society of Agronomy, Crop Science Society of America, Soil Science Society of America, ISBN 978-0891181576, Madison, USA

Stucki. J. W. & Anderson W. L. (1981). The quantitative assay of minerals for Fe^{2+} and Fe^{3+} using 1-10 phenanthroline. *Soil Science Society of American Journal*, 45, pp.633-637.

Sumi., S. (1989). Material cycle and air environment, Kagaku, 59, pp.125-132 (in Japanese)

Zeng, L., Li, X. & Liu, J. (2004). Adsorptive removal of phosphate from aqueous solutions using iron oxide tailings. *Water Research*, 38, pp.1318-1326

22

Sweet Sorghum: Salt Tolerance and High Biomass Sugar Crop

A. Almodares[1], M. R. Hadi[2] and Z. Akhavan Kharazian[1]
[1]*Department of Biology, University of Isfahan,*
[2]*Department of Biology, Sciences and Research Branch of Fars,*
Islamic Azad University,
Iran

1. Introduction

Soil salinity is one of the main problems for plant growth in agriculture, especially in countries where crops should be irrigated (Ahloowalia et al., 2004). Soil salinity has been considered a limiting factor to crop production in arid and semi arid regions of the world (Munns, 2002). Saline soils are estimated about 5 - 10% of the world's arable land (Szabolcs, 1994), and the area affected by salinity is increasing steadily (Ghassemi et al., 1995). Salt-affected soils are distributed throughout the world and no continent is free from the problem (Brandy and Weil, 2002). Globally, a total land area of 831 million hectares is salt-affected (Kinfemichael & Melkamu, 2008; FAO, 2000). However, soil salt accumulation can change with time and place, as a function of soil management, water quality (Almodares & Sharif, 2005), irrigation method, and the weather conditions. Salt accumulation is mainly related to a dry climate, salt-rich parent materials of soil formation, insufficient drainage and saline groundwater or irrigation water (Almodares et al., 2008a). Salts in soils are chlorides and sulfates of sodium, calcium, magnesium, and potassium that among them sodium chloride has the highest negative effect on the plant growth and development. Salinity causes slow seed germination, sudden wilting, and reduce growth, marginal burn on leaves, leaf yellowing, leaf fall, restricted root development, and finally death of plants. The inhibitory effects of salinity on plant growth include: (1) ion toxicity (2) osmotic influence (3) nutritional imbalance leading to reduction in photosynthetic efficiency and other physiological disorders. Among agricultural crops, sorghum (*Sorghum bicolor* L. Moench) is naturally drought and salt-tolerant crop that can produce high biomass yields with low input. Also, it can thrive in places that do not support corn, sugarcane and other food crops. In addition, sweet sorghum has potential uses (six F) such as: food (grain), feed (grain and biomass), fuel (ethanol production), fiber (paper), fermentation (methane production) and fertilizer (utilization of organic byproducts), thus it is an important crop in semi-aired and aired regions of the world. Sorghum is grown on approximately 44 million hectares in 99 countries (ICRISAT, 2009). An estimation of the world-wide tonnage produced in 2007-2008 is shown in Table 1. The increasing cost of energy and deplete oil and gas reserves has created a need for alternative fuels from renewable sources. The consumption of biofule may reduce greenhouse gases. Also it can be replaced with lead tetraethyl or MTBE (Methyl tert-butyl ether) that are air and underground water pollutants, respectively (Almodares & Hadi, 2009). Plants are the best choice for biofule global demands. Currently, ethanol production is based on sugar or starch of crops such as sorghum, corn, sugarcane, wheat and etc. In comparison with other crops, carbohydrate content of sweet sorghum stalk and its grain starch is similar to sugarcane and corn, respectively but its water and fertilizer requirements are much lower than both sugarcane and corn. Thus, in many tropical and temperate countries where sugarcane and corn cannot be grown, a growing interest is being focused on the potential of sweet sorghum to produce bioethanol feed stock (Almodares et al., 2006, 2008d). Sweet sorghum biomass has rich fermentable sugars such as sucrose, glucose, and fructose so it is an excellent raw material for fermentative production (Almodares et al., 2008d). The total soluble sugars can be increase in sweet sorghum with increasing salinity level and sucrose content could be an indicator for its salt tolerance. (2008b). Salt-stressed sorghum plants additionally accumulate organic solutes, like proline, glycinabetaine, sugars, etc. (Lacerda et al., 2001). These organic solutes may contribute to osmotic adjustment, protecting cell structure and function, and/or may serve as metabolic or energetic reserve (Hasegawa et al., 2000). Inorganic and organic solutes concentrations maintained during salt stress, therefore, they may be important during the salt stress recovery period (Pardossi et al., 1998). Since sweet sorghum is more salt tolerant than sugarcane and corn which currently are the main sources of bioethanol production. Therefore, it is suggested to plant sweet sorghum for biofule production in hot and dry countries to solve problems such as increasing the octane of gasoline and to reduce greenhouse gases.

Country	**Production** (tonnes x 1000)	**% of Total**
United States	12,827	20
Nigeria	10,000	16
India	7,780	12
Mexico	6,100	10
Sudan	4,500	7
Ethiopia	3,230	5
Argentina	2,900	5
Australia	2,691	4
China	1,900	3
Burkina Faso	1,800	3
Brazil	1,700	3
Other countries	*6,880*	*12*
Total	**62,308**	100

Table 1. World Sorghum Production 2007-2008 (Quotation from U.S. Grain Council, 2008).

2. Salinity problem and ways to resolve it

About 7% of the world's total land area is affected by salt, as is a similar percentage of its arable land (Ghassemi et al., 1995). Salinity is often accompanied by other soil properties, such as sodicity and alkalinity, which exert their own specific effects on plant growth. There are three ways in which salinity stress of crops could be reduced; 1- Farm management practices; 2- Screening; 3- Breeding which will be discussed in the followings:

2.1 Farm management practices

All irrigation waters contain some dissolved salts. Thus, soil salinization may be expected by crop irrigation. Removal of salts from the root zone may be the most effective way to eliminate the effects of salinity. However, it is expensive and requires good drainage system. It is not always possible to carry out this operation; thereby a number of other different ways could be considered such as:

a. Soil Reclamation; in a case Na ions are the major cause of soil salinity, it may be replaced with Ca ions by adding of gypsum (calcium sulfate) to the soil.
b. Reduction of the salt from seed germination zone; Seed germination and seedling establishment are the most sensitive stages to salinity. A number of approaches have been used. 1) Removal of surface soil (Qureshi et al., 2003). 2) Pre-sowing irrigation with good quality water (Goyal et al., 1999). 3) Planting seed on the ridge shoulders rather than on the ridge top of the furrow. 4) Planting in a pre-flooded field with good quality water (Goyal et al., 1999).
c. Reducing soil salinity by adding mulch, organic matter or deep tillage to the soil.

2.2 Screening

Salinity and waterlogging co-exist in the lower reaches of several river basins throughout the world, affecting agricultural production and the livelihoods of the affected communities (Wichelns and Oster, 2006). Efforts being made to overcome salinity and waterlogging problems by consist of engineering solutions such as installation of a drainage system to manage the drainage effluent generated by irrigated agriculture. This is a long term strategy; however drainage installation is expensive. The areas under salt-affected and waterlogged soils are expanding because of inappropriate on-farm water and soil management. Selection and cultivation of high-yielding salt-tolerant varieties of different crops is a potential interim strategy to fulfill the needs of the communities relying on these soils for their livelihoods (Ayers and Westcot, 1989). Many crops show intraspecific variation in response to salinity. Sorghum is moderately salt-tolerant. Generally, substantial genotypic differences exist among sorghum cultivars in response to salinity stress (Sunseri et al., 2002; Netondo et al., 2004).

2.2.1 Screening methods based on growth or yield

Screening large numbers of genotypes for salinity tolerance in the field is difficult, due to spatial heterogeneity of soil chemical and physical properties, and to seasonal rainfall distribution. Frequently, short-term growth experiments have revealed little difference between genotypes that differ in long-term biomass production or yield. Many short-term growth experiments measuring whole shoot biomass revealed little difference between plant genotypes in their response to salinity, even between those known to differ in long-term biomass production or yield (Rivelli et al., 2002). Longer-term experiments are necessary to detect genotypic differences in the effects of salinity on growth: it is necessary to expose plants to salinity for at least two weeks, and sometimes several months (Munns et al., 1995). Even with rice, a fast growing and salt sensitive species, it is necessary to grow plants for several weeks to be confident of obtaining reproducible differences in salinity tolerance between genotypes (Zhu et al., 2001).

2.2.2 Screening methods based on damage or tolerance to very high salinity levels

Techniques that can handle large numbers of genotypes include: germination or plant survival in high salinity, leaf injury as measured by membrane damage (leakage of ions from leaf discs), premature loss of chlorophyll (using a hand-held meter), or damage to the photosynthetic apparatus (using chlorophyll fluorescence). These methods can identify genotypes able to germinate, or survive, in very high salinities (over 200 mM NaCl), but do not discriminate between genotypes in their ability to tolerate the low or moderate salinities typical of many saline fields (50–100 mM NaCl). A major limitation to the use of injury or survival to identify salt-tolerant germplasm arises when the cause of injury is not known.

2.2.2.1 Screening methods based on physiological mechanisms

Because of the complex nature of salinity tolerance, as well as the difficulties in maintaining long-term growth experiments, trait-based selection criteria are recommended for screening techniques (Noble and Rogers, 1992). Traits used for screening germplasm for salinity tolerance have included Na+ exclusion, K^+/Na^+ discrimination (Asch et al., 2000) and Cl^- exclusion (Rogers and Noble, 1992). The relationship between salinity tolerance and K^+/Na^+ discrimination was also considered, because K^+/Na^+ rather than Na^+ alone has been used as an index of salinity tolerance for cultivar comparisons in wheat (Chhipa and Lal, 1995) and rice (Zhu et al., 2001). One of the mechanism of salinity tolerance that could be considered was tissue tolerance of high internal Na^+ concentrations. Tissue tolerance cannot be measured directly, and is difficult to quantify. Yet it is clearly important; overexpression of vacuolar Na^+/H^+ antiporter that sequesters Na^+ in vacuoles improved the salinity tolerance of Arabidopsis, tomato and brassica (Aharon et al., 2003).

2.3 Breeding

Breeding programs for new varieties of sweet sorghum suited to semi arid tropics, temperate areas with rainy summer, Mediterranean areas with dry summer and soil salinity, are under development (Cosentino, 1996).

3. Why sweet sorghum?

3.1 Agricultural advantages

3.1.1 Salt tolerance

Sorghum is characterized as moderately tolerant to salinity (Almodares and Sharif, 2005; Almodares and Sharif, 2007). Salinity reduces sorghum growth and biomass production . Salinity greatly reduced sorghum growth and this effect was more pronounced at 250 mM than at 125 mM NaCI (Ibrahim, 2004). However it was reported that sorghum growth was significantly reduced at all salinity levels from 50 to 150 mM (El-Sayed et al., 1994). Imposition of salt stress resulted in decreases in the percentage of seeds germinated (Almodares et al., 2007), although the strongest decline in germination occurred at the highest salt concentration (Table 2). Nevertheless, the development of high-yielding salinity tolerant sorghums is the best option to increase the productivity in soils (Igartua et al. 1994). Similarly, Gill et al. (2003) observed a great reduction in germination rate due to salt stress, in sorghum seeds at 37 ◦C in NaCl (−1.86MPa).

Relative percent germination(%)in osmotic potential (Mpa)created by NaCl						
Cultivars	-0.4	-0.8	-1.2	-1.6	-2.0	-2
IS 9639	48d	4e	0f	0e	0b	0b
Sova	87.5abc	70abc	30de	12.5de	7.5b	7.5b
Vespa	80abc	51.5bcd	17ef	3de	0b	0b
S 35	83abc	74.5ab	54.5bcd	8.5de	3b	3b
M 81E	73bc	85.5a	36de	0e	0b	0b
IS 19273	81abc	46.5cd	29.5de	0e	0b	0b
IS 6936	87abc	77a	33.5de	5de	0b	0b
MN 1500	72.5bc	47.5cd	20ef	2.5de	0b	0b
Sumac	100a	62.5abcd	67.5abc	47.5ab	45a	45a
IS 686	63cd	40d	66abc	14de	0b	0b
SSV 108	87.5abc	85a	72.5ab	25bcde	5b	5b
Roce	87abc	74ab	89.5a	42abc	34.5a	34.5a
Sofrah	89.5ab	84a	53bcd	23.5bcde	5.5b	5.5b
Satiro	95ab	42d	32de	0e	5b	5b
IS 2325	89.5ab	77a	46cd	28bcd	0b	0b
E 36-1	62.5cd	42.5d	30de	2.5de	0b	0b
IS 6973	85.5 abc	74.5ab	71.5ab	20cde	23ab	23ab
SSV84	94.5ab	84.5a	64bc	64a	0b	0b

Values of letters (a, b,...) within each column followed by the same letter are not significantly different at 5% level, using Duncan multiple rang test.

Table 2. Effects of salinity on relative percent germination in 18 sweet sorghum cultivars (Quotation from Samadani et al., 1994).

According to Prado et al. (2000), the decrease in germination may be ascribed to an apparent osmotic 'dormancy' developed under saline stress conditions, which may represent an adaptive strategy to prevent germination under stressful environment. Germination time delayed with the increase in saline stress and root growth was more sensitive to salt stress than was germination (Gill et al., 2003). It seems that grain weight is related to salt tolerance in sweet sorghum. It showed that higher total seedling dry weight was obtained with larger seed size in 18 sweet sorghum cultivars under salt stress (Table 3 and Fig. 1). The presence of large genotypic variation for tolerance to salinity is reported in sorghum (Maiti et al, 1994). Sorghum seems to offer a good potential for selection, as intraspecific variation for germination under saline conditions (Table 2) or in the presence of other osmotic agents that has already been reported. Selection of salt tolerant cultivars is one of the most effective methods to increase the productivity of salinity in soils (Ali et al., 2004). By using these salt tolerant plants in breeding they produced progranuned an improved plant having higher chlorophyll concentration, more leaf area, early and better yield potential etc. The advancement of salinity tolerance during the early stages of sorghum growth been successfully accomplished through selection.

Cultivar	Thousand Grain Weight (g)	Total Seedling Fresh Weight (mg/20grain)
IS 9639	18.75	79
Sova	19.77	197
Vespa	15.35	180
S 35	30.63	349
M 81E	14.59	127
IS 19273	27.69	267
IS 6936	34.33	418
MN 1500	24.59	192
Sumac	12.63	81
IS 686	17.15	194
SSV 108	39.61	381
Roce	17.16	159
Sofrah	16.68	170
Satiro	15.21	246
IS 2325	31.35	335
E 36-1	33.33	434
IS 6973	38.52	344
SSV84	40.05	524

Table 3. Thousand Grain Weight (g) of 18 sweet sorghum cultivars and Total Seedlings Fresh weight (mg/20 grain) grown in osmotic potential (-0.4 Mpa) of NaCl after 12 day treatment (Quotation from Samadani et al., 1994).

Genotypes possessing salt tolerance characteristics will help in boosting up plants production in salt-affected soils (Ali et al., 2004). Azhar and McNeilly (1988) found that, for salinity tolerance of young sorghum seedlings, both additive and dominant effects were involved, the latter being of greater importance. Attempts have been made to evaluate salt tolerance at the germination and emergence stages in sorghum (Igartua et al., 1994). In fact, the variation in whole-plant biomass responses to salinity was considered to provide the best means of initial selection of salinity tolerant genotypes (Krishnamurthy et al, 2007). The presence of large genotypic variation for tolerance to salinity reported in sorghum (Krislmamurthy et al., 2007). There are large genotypic variations for tolerance to salinity in sorghum (Table 4). The other possible solution could be either using physical or biological practice (Gupta and Minhas, 1993). Sudhir and Murthy (2004) reviewed both multiple inhibitory effects of salt stress on photosynthesis and possible salt stress tolerance mechanisms in plants. Salinity reduced relative growth rates and increased soluble carbohydrates, especially in the leaves of salt sensitive genotype (Lacerda et al., 2005). In addition salt-stressed sorghum plants additionally accumulate organic solutes, like proline, glycinabetaine, sugars, etc. (Lacerda et al., 2001). The total soluble sugar increased in sorghum sap with increasing salinity level (Ibrahim, 2004; Almodares et al., 2008a). Sucrose content of plant parts is an indicator of salt tolerance (Juan et al., 2005). The imposition of strong water or salt stresses in sorghum has been demonstrated to be accompanied to an increase in the sugar levels of embryos, which may help in osmoregulation under stress conditions (Gill et al., 2003). The fructose level is always higher than glucose and sucrose levels in response to various salinity treatments (Gill et al., 2001; Almodares et al., 2008a).

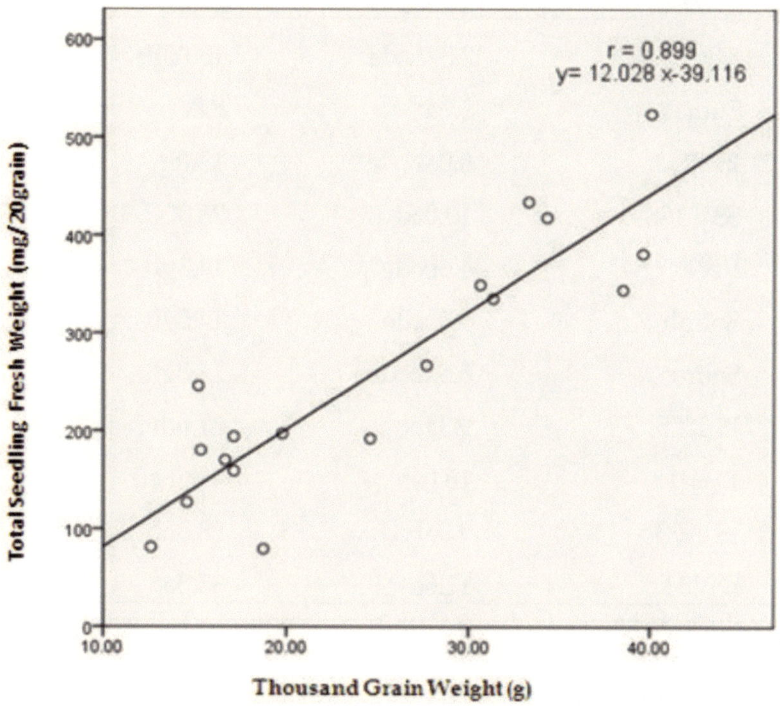

Fig. 1. Correlation between total seedling fresh weight and thousand grain weight in sweet sorghum (Quotation from Samadani et al., 1994).

3.1.1.1 Mechanisms of salt tolerance in crops

Sodium is the major cation that accumulated in roots and stems as salinity increased (Meneguzzo et al., 2000). It is evident that salt tolerance is associated with low uptake of Na^+(Santa-Maria and Epstein, 2001), partial exclusion (Colmer et al., 1995) and compartmentalization of salt in the cell and within the plant (Ashraf, 1994). The preferential accumulation in roots over shoots may be interpreted as a mechanism of tolerance in at least two ways.

	Dry Weight (mg)	
Cultivar	Root	shoot
IS 9639	3.5de	13.5h
Sova	5.3bcde	16.3gh
Vespa	4.0cde	14.5gh
S 35	7.3abcde	24.0cde
M 81E	6.1abcde	15.3gh
IS 19273	7.6abcde	19.6efg
IS 6936	10.3ab	29.6b
MN 1500	7.3abcde	18.0fgh
Sumac	3.0e	8.0i
IS 686	6.0abcde	13.0h
SSV 108	10.0ab	28.0bc
Roce	4.6bcde	14.3gh
Sofrah	5.5bcde	14.5gh
Satiro	6.5abcde	16.0gh
IS 2325	9.3abc	21.6def
E 36-1	10.0ab	26.6bcd
IS 6973	9.0abcd	28.3bc
SSV84	11.6a	35.3a

Values of letters (a, b,...) within each column followed by the same letter are not significantly different at 5% level, using Duncan multiple rang test.

Table 4. Root and shoot dry weight of 18 sweet sorghum cultivars that grown in osmotic potential (-0.4 Mpa) of NaCl through 12 day (Quotation from Samadani et al., 1994).

First, maintenance of a substantial potential for osmotic water uptake into the roots and second, restricting the spread of Na^+ to shoots (Renault et al., 2001). High Na^+ levels in the external medium greatly reduce the physicochemical activity of dissolved calcium and may thus displace Ca^{2+} from the plasma membrane of root cells. In turn, displacement of Ca^{2+} from root membranes by Na^+ affects Na/K uptake selectivity in favor of sodium. A low Ca^{2+} concentration under saline conditions may severely affect the functions of membranes as barriers to ion loss from cells (Boursier and Läuchli, 1990). Various organic and inorganic solutes such as K^+, Na^+, Cl^-, proline, and glycinebetaine have been reported to contribute to such osmotic adjustment (Saneoka et al., 2001). Salinity inhibits the accumulation of K^+ and Ca^{2+} in roots and stems. The negative effect of NaCl on the allocation of K^+, Ca^{2+}, and Mg^{2+} to the leaf tissues may contribute to their deficiency and the accompanying metabolic perturbations. The altered ion and water relations have a severe impact on the photosynthetic performance of the plant (Netondo et al., 2004). Many plants accumulate high levels of free proline in response to osmotic stress. This amino acid is widely believed to function as a protector or stabilizer of enzymes or membrane structures that are sensitive to dehydration or ionically induced damage. The salt stress caused increases in proline levels. Several investigations have shown that, besides other solutes, the level of free amino acids, especially proline, increases during adaptation to various environmental stresses. Plant salt tolerance has been generally studied in relation to regulatory mechanisms of ionic and osmotic homeostasis (Ashraf and Harris, 2004). In addition to ionic and osmotic components, salt stress, like other abiotic stress, also leads to oxidative stress through an increase in Reactive Oxygen Species (ROS), such as superoxide (02-), hydrogen peroxide (H_2O_2) and hydroxyl radicals (OH) (Mittler, 2002). It has been reported that most abiotic stress including NaCI salt stress impose injury in plants by osmotic stress, ionic stress and generating reactive oxygen species (Shalata and Tal, 1998). During oxidative stress, the excess production of Reactive Oxygen Species (ROS) causes membrane damage that eventually leads to cell death. For protection against ROS, plants contain antioxidant enzymes such as superoxide dismutase (SOD), catalase (CAT), ascorbate peroxidase (APX), guaiacol peroxidase (GPX) and Glutathione Reductase (GR) or as well as a wide array of non-enzymatic antioxidants (Blokhina et al., 2003). SOD is the major 02^- scavenger and its enzymatic action results in H202 and O2 formation. The H202 produced is then scavenged by CAT and several classes of peroxidases. CAT, which is found in peroxisomes, cytosol and mitochondria, dismutates H_20_2 to H_20 and O_2 (McKersie and Leshem, 1994). Sorghum is a salt tolerant plant therefore it seems that it uses some of the above mechanisms for its adaptation to salt and drought stress.

3.1.2 High yield in drought and salinity regions

Sorghum is the 5th grain crop grown based on tonnage, after maize, wheat, rice, and barley (CGIAR website, 2009) with a high yield of biomass (Almodares et al., 1994; Gardner et al., 1994). Sweet sorghum like grain sorghum produces grain 3-7 t/ha (Almodares et al., 2008e). But the essence of sweet sorghum is not from its seed, but from its stalk, which contains high sugar content (Almodares et al., 2008c). In general, it can produce stalk 54-69 t/ha (Almodares et al., 2008d).

Genotypes	Stem Yield (t ha-1)	Brix (%)	Sucrose (%)	Purity (%)
Cultivars:				
Roce	39.14	21.96	14.39	66.71
Vespa	84.53	20.99	13.05	74.59
Brandes	77.14	18.72	8.92	46.39
MN1500	83.71	20.71	12.00	57.59
E36-1	48.00	18.26	13.41	76.02
Soave	61.57	20.73	13.46	65.00
M81-E	103.57	16.01	10.26	65.10
Sumac	44.43	21.12	12.85	60.10
Sofrah	85.57	19.63	12.61	64.05
SSV-108	62.85	22.25	13.97	62.26
SSV-94	70.14	20.64	11.75	57.12
SSV-96	62.00	22.54	13.71	60.10
Theis	100.14	19.10	7.26	37.59
Foralco	97.71	20.40	12.64	60.83
Rio	95.00	22.36	16.06	71.31
S-35	58.43	19.78	11.58	58.75
Turno	39.86	11.16	6.00	35.86
Satiro	27.86	17.16	10.33	60.02
Wary	126.42	15.84	7.85	49.40
Lines:				
IS 686	61.43	16.54	9.00	54.39
IS 16054	51.85	21.07	11.73	55.83
IS 18154	42.14	19.04	12.71	66.71
IS 6962	43.00	23.01	13.61	58.85
IS 9639	54.00	21.77	14.31	65.23
IS 2325	59.57	20.70	14.28	60.18
IS 6973	33.43	22.85	14.21	61.88
IS 4546	56.43	22.03	13.05	60.12
IS 19273	46.28	20.29	15.04	73.69
IS 4354	33.86	17.66	9.80	55.28
Hybrids:				
A1 x IS 6973	83.28	16.46	9.53	57.17
A13 x IS 1273	97.00	21.18	14.26	66.78
A1 x IS 19261	88.13	18.69	11.82	63.04
A1 x IS 14446	87.13	16.51	10.51	62.89
A45 x IS 14446	124.13	17.95	13.36	74.06
A1 x IS 19273	128.85	17.82	13.00	73.51
A13 x IS 14446	113.56	14.32	10.73	74.40

Table 5. Mean comparisons among 36 sweet sorghum cultivars, lines and hybrids regarding stem yield, Brix , Sucrose and purity (Almodares and Sepahi, 1996).

Besides having rapid growth, high sugar accumulation (Almodares and Sepahi, 1996), and biomass production potential (Almodares et al., 1994), sweet sorghum has wider adaptability (Reddy et al., 2005). Many factors could increase biomass in sweet sorghum such as: fertilizer (Almodares et al., 2006, 2008d, 2009, 2010), irrigation regimes (Almodares & Sharif, 2007), cultivars (Table 5), plant population density (Solymani et al., 2010), planting dates (Almodares et al., 1997) (Table 6), harvesting stages (Almodares et al., 2010), climatic conditions, etc.

Almodares et al. (2006) reported that application of nitrogen-fertilizer siginficantly increased leaf area, leaf dry weight, stem dry weight, total dry weight, paincle dry weight and paincle dry length of sweet sorghum cultivars. Almodares et al., 2010 reported that among nitrogen treatments, application of 100 kg ha-1 urea at planting and 200 kg ha-1 urea at 4 leaf stage had the highest aconitic acid (0.26%) and invert sugar (3.44%).

Cultivars	Stalk yield t/ha		sucrose yield t/ha	
	May 5	june 5	May 5	June 5
Soave	46.1cd	33.0de	5.7cd	4.2b
Brandes	*54.4bc*	*49.8b*	*3.7ef*	*4.2b*
Theis	56.7bc	56.8ab	2.8fg	2.3cd
MN1500	66.6ab	49.7b	6.5bc	4.2b
Wray	71.8a	63.1a	7.2ab	5.3ab
Keller	68.7a	54.6ab	8.5a	5.7a
Dale	75.5a	47.7bc	4.8de	2.1cd
Vespa	51.7c	42.4cd	6.1bcd	5.2ab
NK326	35.5d	17.8d	3.0f	0.7c
NK300	23.8e	25.2e	1.5g	1.0d

* Mean comparisons were made using Student Newman Keuls' test. Means with the same letter (a, b, ...) within a column are not significantly difference at 5% level.

Table 6. Mean comparisons* between the ten sweet sorghum cultivars for the two planting dates and two characteristics of economical importance (Almodares et al., 1994).

3.1.3 Low water requirement

In the semiarid regions, water and salinity stresses are increasingly becoming primary limiting environmental conditions which restrict successful establishment of crops. Sorghum is tolerant of low input levels and essentially for areas that receive too little rainfall for most other grains (Table 7). Increased demand for limited fresh water supplies, increasing use of marginal farmland, and global climatic trends, all suggest that dry land crops such as sorghum will be of growing importance to feed the world's expanding populations. Generally lower water demands for sorghum than maize, versus their equal ethanol yields, suggests that sorghum will be of growing importance in meeting grain-based biofuels needs. In many tropical and temperate countries where sugarcane cannot be grown, a growing interest is being focused on the potential of sweet sorghum to produce bio-ethanol feed stocks (Avant, 2008) specially that salinity and drought tolerance are major features of sweet sorghum with low water requirements for high yields. One of the main reactions to drought stress is closing of stomata. The C4 plant such as sweet sorghum, in opposite to the C3, are able to utilize very low concentration of carbon dioxide which enables them to assimilate CO_2 even during considerable stomatal closure (El Bassaru, 1998). This might be one of the probable reasons for the difference in resistance to stress between both plant groups. Photosynthesis is a complex process; therefore, it is possible that a number of elements in the C3 and the C4 may differ in resistance to drought.

	Sugarcane	Sugar beet	Sweet sorghum
Crop duration	About 7 months	About 5 - 6 months	About 4 months.
Growing season	Only one season	Only one season	One season in temperate and two or three seasons in tropical area.
Soil requirement	Grows well in drain soil	Grows well in sandy loam; also tolerates alkalinity	All types of drained soil.
Water management	36000 m^3/h	18000 m^3/h	12000 m^3/h
Crop management	Requires good management	Greater fertilizer requirement; requires moderate management	Little fertilizer required; less pest and disease complex; easy management.
Yield per ha	70 - 80 tons	30 - 40 tons	54 - 69 tons.
Sugar content on weight basis	10 - 12%	15 - 18%	7 - 12%.
Sugar yield	7 - 8 tons/ha	5 - 6 tons/ha	6 - 8 tons/ha.
Ethanol production directly from juice	3000 - 5000 L/ha	5000 - 6000 L/ha	3000 L/ha.
Harvesting	Mechanical harvested	Very simple; normally manual	Very simple; both manual and through mechanical harvested.

Table 7. Comparison of Sugarcane, Sugar Beet, and Sweet Sorghum in Iran (Almodares & Hadi, 2009).

3.2 Biofuel advantages

3.2.1 Bioethanol production from sweet sorghum

Sweet sorghum is a crop for producing energy which not only produce food, but also energy, feed and fiber (Almodares & Hadi, 2009). The chief sugars present in sorghum are monosaccharides: glucose and fructose, and disaccharides: sucrose. Fermentable carbohydrates in sweet sorghum stalks comprise approximately 80% soluble sugars and 20% starch. To optimize production of ethanol from sweet sorghum grain requires both liquefying and saccharifying enzymes (Rooney and Waniska, 2000). Therefore, it seems that using carbohydrates in the stalk (sucrose and invert sugar) is suitable for ethanol production for biofuel production because these carbohydrates are easily converted to ethanol (Fig 2). Although, ethanol can be produced from sweet sorghum grain (Fig. 2) but it needs more process for converting it's starch to glucose that later will be converted to ethanol (Jacques et al., 1999). In addition, the produced baggase after juice extraction can be used for ethanol production (Jacques et al., 1999) or animal feed. However, presently it is not economically feasible to produce ethanol from sweet sorghum baggase (Drapcho et al., 2008).

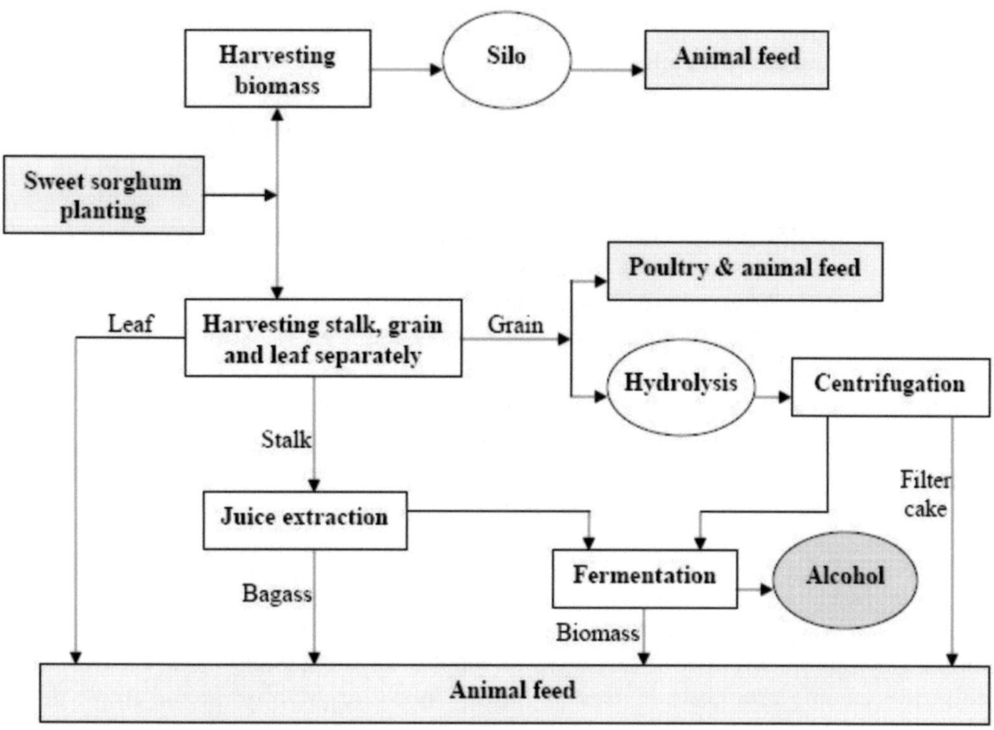

Fig. 2. Proposed layout for ethanol production and by-product from sweet sorghum (Almodares & Hadi, 2009).

3.2.2 The important of ethanol in biofuel

One method to reduce air pollution is to oxygenated fuel for vehicles. MTBE (Methyl tert-butyl ether) is a member of a group of chemicals commonly known as fuel oxygenates (Fischer et al., 2005). It is a fuel additive to raise the octane number. But it is very soluble in water and it is a possible human carcinogenic (Belpoggi et al., 1995). Thereby, it should be substituted for other oxygenated substances to increase the octane number of the fuel. Presently, ethanol as an oxygenated biomass fuel is considered as a predominant alternative to MTBE for its biodegradable, low toxicity, persistence and regenerative characteristic (Cassada et al., 2000). In most countries, gasoline supply is an ethanol blend, and the importance of ethanol use is expected to increase as more health issues are related to air quality. Ethanol may be produced from many high energy crops such as sweet sorghum, corn, wheat, barely, sugar cane, sugar beet, cassava, sweet potato and etc (Drapcho et al., 2008). Like most biofuel crops, sweet sorghum has the potential to reduce carbon emissions. Therefore, it seems that sweet sorghum is the most suitable plant for biofuel production than other crops under hot and dry climatic conditions. In addition, possible use of bagasse as a by-product of sweet sorghum include: burning to provide heat energy, paper or fiber board manufacturing, silage for animal feed or fiber for ethanol production. However, since

sweet sorghum is at a relatively early stage of its development, continued research was needed to obtain better genetic material and match local agro-economic conditions. The challenge is to harvest the crop, separate it into juice and fiber, and utilize each constituent for year-round production of ethanol.
Sweet sorghum juice is assumed to be converted to ethanol at 85% theoretical, or 54.4 liter ethanol per 100 kg fresh stalk yield. Potential ethanol yield from the fiber is more difficult to predict (Rains et al., 1993). The emerging enzymatic hydrolysis technology has not been proven on a commercial scale (Taherzadeh and Karimi, 2008). One ton of corn grain produces 387 L of 182 proof alcohol while the same amount of sorghum grain produces 372 L (Smith and Frederiksen, 2000). Sorghum is used extensively for alcohol production (Gnansounou et al., 2005), where it is significantly lower in price than corn or wheat (Smith and Frederiksen, 2000). The commercial technology required to ferment sweet sorghum biomass into alcohol has been reported in china (Gnansounou et al., 2005). One ton of sweet sorghum stalks has the potential to yield 74 L of 200- proof alcohol (Smith and Frederiksen, 2000). Therefore, it seems that because ethanol can be produced from both stalk and grain of sweet sorghum (Fig. 2), so it is the most suitable crop for ethanol production using for biofuel comparing to other crops such as corn or sugarcane.

4. Food and feed

Sorghum is an important food cereal in many parts of worldwide. According to the U.S. National Sorghum Producers Association (2006), approximately 50% of the world production of sorghum grain is used as human food. Sorghum grain is a staple diet in Africa, the Middle East, Asia and Central America where its processed grain may be consumed in many forms including porridge, steam-cooked product, tortillas, baked goods, or as a beverage (CGIAR, 2009). China and India account for almost all of the food use of sorghum in Asia, in other parts of the world, sorghum grain is used mainly as an animal feed. It has the distinct advantage (compared to other major cereals) of being drought-resistant and many subsistence farmers in these regions cultivate sorghum as a staple food crop for consumption at home (Murty and Kumar, 1995). Therefore sorghum acts as a principal source of energy, protein, vitamins and minerals for millions of the poorest people living in these regions (Klopfenstein and Hoseney, 1995). The improvement of sorghum nutrient availability is critical for food security. Cereal scientists and sorghum food processors are thus faced with the challenge of identifying the factors that adversely affect, and developing processing procedures that improve sorghum protein digestibility. Most parts of the sorghum plant are used as animal feed. Growing sorghum may be grazed, or the aerial parts of the plant may be ensiled or dried and fed as stover or silage for ruminant animals. Whole sorghum grain is cracked, ground, or steam flaked and fed to poultry, swine, dairy and beef cattle as a source of energy. Crop residues are a major animal feed resource in many crop–livestock farming systems. They are very useful in ameliorating the problem of inadequacy of feeds for ruminant livestock during the dry season. Although useful as dry season feeds, crop residues, particularly those of cereal origin, are low in protein and energy content (Agyemang et al., 1998). The stover of sorghum also is used as fodder for animals. The nutrient composition of sorghum grain is presented in Table 8.

Measurement	NRC[1]	*Range of mean values*
Moisture, %	10.0–12.5	*9.2–12.5*
	% DM[2]	
Crude Protein[3]	10.1–12.6	*10.1–15.6*
Total fat[4]	3.0–3.3	*0.8–4.3*
Ash	1.9–2.0	*1.5–3.3*
Nitrogen-free extract (NFE)[5]		*70.8–84.9*
Crude Fibre	2.6–3.0	*1.7–6.9*
NDF[6]	10.9–23.0	*10.9–23.0*
ADF[7]	5.0–9.3	*5.0–9.0*

[1] NRC- Nutrient Requirements for Poultry
[2] DM= Dry mater
[3] Crude protein = Nitrogen x 6.25
[4] Total fat as measured by ether extract
[5] NFE = 100 - (ash + ether extract + crude protein + crude fibre)
[6] Neutral detergent fibre
[7] Acid detergent fibre

Table 8. Proximate analysis of S. bicolor grain (dry matter basis) (Quotation from OECD, 2010)

4.1 By-products of sorghum processing

The by-product of sorghum ethanol production is distillers' grains. Table 9 presents the available nutritional information for wet and dry sorghum distillers' grains, and dry grains plus solubles. Distiller's dried grains with solubles contain all fermentation residues, including yeast, remaining after ethanol is removed by distillation (Shurson, 2009).

	Wet distillers'	Dry distillers'	Dry distillers' + solubles
DM, %[1]	23.5–35.3	91.4	91.4
	(% DM basis)		
Crude Protein	31.2–31.6	32.9	31.4
Ash	2.5		
Total fat	11.3–13.3	13.0	11.8
ADF [2]	28.5	28.4	
NDF [3]	41.3–45.4	45.8	51.1
NSC [4]	9.2	3.3	
Starch	10.2		7.4

[1] Dry matter; [2] Acid detergent fibre ; [3] Neutral detergent fibre ; [4]Non-structural carbohydrate

Table 9. Nutrient composition of sorghum distillers' grains (Quotation from OECD, 2010).

5. Conclusion

It is clear that biomass production for biofuel from sweet sorghum is the best choice to be implement under hot and dry climatic conditions regarding both economic and environmental considerations. Because, sweet sorghum has higher tolerance to drought

(Tesso et al., 2005), water logging , and salt (Almodares et al., 2008a, 2008b), alkali, and aluminum soils; It may be harvested 3-4 month after planting and planted 1-2 times a year (in tropical areas); Its energy output/fossil energy input is higher than sugarcane, sugar beet, corn, wheat and etc… specially in temperate areas; It is more water use efficient (1/3 of water used by sugarcane at equal sugar production); Its production can be completely mechanized and Its bagasse has higher nutritional value than the bagasse from sugarcane, when used for animal feeding. Also, by implementing agricultural practices such as adequate water and fertilizers, suitable cultivars or hybrids, crop rotation, pest management and etc… can increase productivity with focus on biofuel production from its biomass (Reddy et al., 2005). In addition, sweet sorghum has high amount of sucrose (Almodares and Sepahi, 1996) and invert sugar (Almodares et al., 2008c) which are easily converted to ethanol (Prasad et al., 2007). Therefore, it seems that sweet sorghum biomass is the most suitable raw material for biofuel production in arid regions of the world. This awareness should push government of the countries with such climatic conditions to promote the development of projects for fuel ethanol production from sweet sorghum biomass.

6. References

Aharon, GS., Apse ,MP., Duan, SL.,Hua, XJ., Blumwald, E. (2003)Characterization of a family of vacuolar Na+/H+ antiporters in Arabidopsis thaliana. Plant Soil;253:245-256.

Ahloowalia, BS.(2004) Global impact of mutation-derived varieties. Uphytica 135: 187–204.

Ali, Y., Aslam Z., Ashraf M. Y and Tahir G. R (2004). Effect of salinity on chlorophyll concentration, leaf area, yield and yield components of rice genotypes grown under saline environment. International Journal of Environmental Science and Technology, 1(3):221-225.

Almodares , A., Hadi , M. R. and Dosti B.(2008a). The effects of salt stress on growth parameters and carbohydrates contents in sweet sorghum. Research Journal of environmental science, 2 (4):298-304.

Almodares ,A,. Sepahi ,A .(1996) Comparison among sweet sorghum cultivars, lines and hybrids for sugar production. Ann. Plant Physiol. 10: 50-55.

Almodares ,A,. Sepahi, A., Dalilitajary ,H., Gavami ,R. (1994) Effect of phenological stages on biomass and carbohydrate contents of sweet sorghum cultivars Ann. Plant Physiol. 8: 42-48.

Almodares ,A., Sharif ,ME. (2007) Effects of irrigation water qualities on biomass and sugar contents of sugar beet and sweet sorghum cultivars. J. Environ. Biol. 28: 213-218.

Almodares ,A., Sharif, ME. (2005) Effect of water quality on yield of sugar beet and sweet sorghum. J. Environ. Biol. 26: 487-493.

Almodares, A and Hadi , M. R.(2009). Production of Bioethanol from Sweet Sorghum - A Review. African Journal of Agricultural Research Vol. 4 (9), pp. 772 - 780, September, 2009

Almodares, A., Taheri, R., Hadi, MR., Fathi , M . (2006). The Effect of Nitrogen and Potassium Fertilizers on the Growth Parameters and the Yield Components of Two Sweet Sorghum Cultivars. Pakistan. Biol. Sci. 9: 2350-2353

Almodares, A., Hadi, MR., Ahmadpour, H .(2008b) Sorghum stem yield and soluble cabohdrates under phonological stages and salinity levels Afircan J. Biotech. 7: 4051-4055.

Almodares, A., Hadi, MR., Dosti , B. (2007a) Effects of Salt Stress on Germination Percentage and Seedling Growth in Sweet Sorghum Cultivars. J. Biol. Sci. 7: 1492-1495.

Almodares, A., Jafarinia M., & Hadi, M. R. (2009) The Effects of Nitrogen Fertilizer on Chemical Compositions in Corn and Sweet Sorghum. American-Eurasian J. Agric. & Environ. Sci., 6 (4): 441-446.

Almodares, A., Ranjbar M. and Hadi, M. R (2010) The Effects of Nitrogen Treatments and Harvesting Stages on the Aconitic Acid, Invert Sugar and Fiber in Sweet Sorghum Cultivars. Journal of Environmental Biology 31(6) 1001-1005.

Almodares, A., Sepahi, A., Shirvani, M .(1997) The effects of planting date and genotype on carbohydrate production from sweet sorghum in the south of Iran. Ann. Plant Physiol. 11: 1-5.

Almodares, A., Taheri, R., Adeli, S .(2008c) Stalk yield and carbohydrate composition of sweet sorghum [Sorghum bicolor (L.) Moench] cultivars and lines at different growth stages. J. Malesian Appl. Biol. 37: 31-36.

Almodares, A., Taheri, R., Chung, M., Fathi, M .(2008d). The effect of nitrogen and potassium fertilizers on growth parameters and carbohydrate content of sweet sorghum cultivars. J. Environ. Biol. 29: 849-852

Almodares, A., Taheri, R., Safavi, V.(2008e) Sorghum, Botany, Agronomy and Biotechnology, Jahad Daneshgahi of University of Isfafan, Isfahan.

Asch, F., Dingkuhn, M., Dǫrffling, K. and Miezan, K.(2000). Leaf K/Na ratio predicts salinity induced yield loss in irrigated rice. Euphytica, 113: 109–118.

Ashraf, M. (1994). Breeding for salinity tolerance in plants. Critical Review in Plant Science, 13: 17-42.

Ashraf, M., Harris PJC (2004). Potential biochemical indicators of salinity tolerance in Plants. Plant Sci. 166(1): 3–16.

Avant,B (2008).Hightonnagededicatedenergycrops.In:ProceedingsWorld Biofuels markets Congress, Bruxelles (BE), 12–14.

Ayers, R.S., and D.W. Westcot (1989). Water quality for agriculture. Irrigation and Drainage Paper No. 29, Rev. 1, FAO, Rome.

Azhar, F.M & McNeilly, T. (1988).The genetic basis of variation for salt tolerance in *Sorghum bicolor* (L.) Moench, seedlings. Plant Breed 101: 114–121

Belpoggi ,F., Soffritti ,M., Maltoni, C. (1995) Methyl tertiary-butyl ether (MTBE) - A gasoline additive - causes testicular and lymphohaematopoetic cancers in rats. Toxicol. Ind. Health 11: 119-149.

Beta, T., Rooney, L., and Taylor, J (2000). Effect of chemical conditioning on the milling of high-tannin sorghum. Journal of the Science of Food and Agriculture. 80:2216-2222.

Blokhina, O., Virolainen ,E., Fagerstedt, KV. (2003) Antioxidants, oxidative damage and oxygen deprivation stress: a review. Ann Bot (Lond) 91: 179–194

Boursier, P., Lauchli,A (1990). Growth responses and mineral nutrient relations of salt stressed sorghum. Crop Sci. 30, 1226–1233.

Brady, N. C. & Weil R. R (2002). The Nature and Properties of Soils. Thirteenth Edition, Pearson Education Inc, USA, 823p.

Cassada, DA., Zhang, Y., Snow ,DD., Spalding ,RF. (2000) Trace analysis of ethanol, MTBE, and related oxygenate compounds in water using solid-phase microextraction and gas chromatography/mass spectrometry. Analy. Chem. 72: 4654-4658.

CGIAR (Consultative Group on International Agricultural Research) (2009) website http://www.cgiar.org/.

Chhipa, B.R and Lal P .(1995) Na/K ratios as the basis of salt tolerance in wheat. Aust. J. Agric. Res. 46, pp. 533–539

Colmer, T. D., Epstein, E. and Dvorak, J (1995). Differential solute regulation in leaf blades of various ages in salt sensitive wheat and salt tolerant wheat × *Lophopyrum elongatum* (Host) love amphiploid. Plant Physiology, 108: 1715–1724.

Cosentino, S.L (1996) .Crop physiology of sweet sorghum (Sorghum bicolor (L.)Moench).In: Proceedings of First European Seminaron Sorghum for Energy and Industry, 1-3 April, Toulose, France, pp. 30–41.

Cramer, G., Laughli, A and Epstein, E. (1986) Effects of NaCl and CaCl2 on Ion Activities in Complex Nutrient Solutions and Root Growth of Cotton'Plant .Physiol. 81, 792-797

Drapcho ,CM., Nhuan, NP., Walker, TH. (2008) Biofuels Engineering Process Technology, The McGraw-Hill companies, Inc, USA

Duodu, K.G., Taylor ,J.R.N., Belton ,P.S., Hamaker ,B.R.(2003) Factors affecting sorghum protein digestibility. Journal of Cereal Science 38 (2003) 117–131

El-Sayed, H., El-Haddad, M and. Oleary,J.W (1994).Effect of salinity and K/Na ratio of irrigation water on growth and solute content of *Atriplex amnicola* and Sorghum bicolor. Irrig. Sci., 14: 127-133.

FAO/AGL.; [Online] Available http://www.fao.orglag/agl/agll/spush/topic?.htm.; November,2000.

Fischer, A., Oehm , C., Selle, M., Werner, P. (2005) Biotic and abiotic transformations of methyl tertiary butyl ether (MTBE). Environ. Sci. Pollut. Res. Int. 12: 381-386.

Ghassemi, F., Jakeman, A.J., and Nix, H.A (1995), Salinisation of land and waterresources: Human causes, extent, management and case studies: Wallingford,UK, CAB International, 544

Gill, P.K., Sharma A. D.., Singh P., Bhullar S. S (2001). Effect of various abiotic stresses on the growth, soluble sugars and water relations of sorghum seedlings grown in light and darkness. Bulg. J. Plant Physiol., 27(1–2): 72–84.

Gill,P.K.,Sharma,A.D.,Singh,P.,Bhullar,S.S (2003).Changes in germination, growth and soluble sugar contents of Sorghum bicolor (L.) Moench seeds under various abiotic stresses. Plant Growth Regul. 40, 157–162.

Goyal , S., Chander ,K.. Mundra ,M. C., Kapoor, K. K. (1999) Influence of inorganic fertilizers and organic amendments on soil organic matter and soil microbial properties under tropical conditions Biol Fertil Soils 29 :196–200

Gupta, A. K., Sharma S. K., Sindhu S. S. and Yadav S (2003). Effect of soil salinity on mineral and proline contents of chrysanthemum. Haryana Journal of Horticultural Sciences, 32(3/4): 224-227

Hasegawa,P.M.,Bressan,R.A.,Zhu,J.-K.,Bohnert,H.(2000). PlantCellular and molecular responses to high salinity. Ann. Rev. Plant Physiol. PlantMol.Biol.51,463–499.

Ibrahim, A. H. (2004).Efficacy of exogenous glycine betaine application on sorghum plants grown under salinity stress. Acta Botanica Hungarica, 43(3-4): 307-318

ICRISAT (International Crops Research Institute for the Semi-Arid Tropics) (2009), website http://www.icrisat.org/.

Igartua, E., M.P. Gracia and J.M. Lasa, (1994). Characterization and genetic control of germination-emergence responses of grain sorghum to salinity. Euphytica, 76: 185-193.

Jacques,K., Lyons ,TP., Kelsall ,DR .(1999) The Alcohol Textbook.3rd Edition. 388pp.

Juan, M., Rivero R. M., Romero L and Ruiz J. M (2005). Evaluation of some nutritional and biochemical indicators in selecting salt-resistant tomato cultivars. Environmental and Experimental of Botany, 54(3): 193-201.

Kinfemichael, G and Melkamu,G.(2008). Response of some lowland growing sorghum (Sorghum bicolor L. Moench) accessions to salt stress during germination and seedling growth. African Journal of Agricultural Research Vol. 3 (1), pp. 044-048.

Klopfenstein,C.F.,Hoseney,R.C(1995).Nutritionalpropertiesofsorghum and the millets. In: Dendy, D.A.V., (Ed.), Sorghum and Millets: Chemistry and Technology,American Association of Cereal Chemists, St Paul, MN, pp. 125–168.

Krishnamurthy, L., Serraj R., Hash C. T., Dakheel A. J and Reddy B.V. S (2007). Screening sorghum genotypes for salinity tolerant biomass production. *Euphytica,* 156 (1-2):15-24.

Lacerda, C. F., Cambraia J., Oliva M. A., Ruiz H. A (2005). Changes in growth and in solute concentrations in sorghum leaves and roots during salt stress recovery. Environmental and Experimental Botany, 54: 69–76.

Lacerda, CF., Cambraia, J., Oliva,MA., Ruiz, HA. (2001) Plant growth and solute accumulation and distribution in two sorghum genotypes, under NaCl stress. Rev. Bras. Fisiol.Veg. 13:270-284.

Maiti,R.K.,Ibarra,M.R.,Sandoval,N (1994).Genotypic variability in glossy sorghum lines for resistance to drought, salinity and temperature stress at the seedling stage. J. Plant Physiol. 143,241–244.

McKersie, B.D. and Leshem, Y.Y. (1994). Stress and Stress Coping in Cultivated Plants. Kluwer Academic Publishers, Dordrecht, Netherlands

Meneguzzo, S., Navari-Izzo, F. and Izzo, R (2000). NaCl effects on water relations and accumulation of minerals in shoots, roots and cell sap of wheat seedling. Journal of Plant Physiology, 156: 711-716.

Mittler, R (2002). Oxidative stress, antioxidants and stress tolerance. Trends Plant Sci. 7, 405–410.

Munns, R., 2002. Comparative physiology of salt and water stress. Plant Cell and Environment, 25: 239- 250.

Munns,R., Schachtman ,D.P. & Condon ,A.G. (1995) The significance of a two-phase growth response to salinity in wheat and barley. Australian Journal of Plant Physiology 22, 561–569.

Murty, D.S., Kumar, K.A (1995).Traditional uses of sorghum and millets.In: Dendy, D.A.V., (Ed.), Sorghum and Millets: Chemistry and Technology, American Association of Cereal Chemists, St Paul, MN, pp. 185–221.

Netondo, G. W, Onyango J. C and Beck E(2004). Sorghum and Salinity: II. Gas Exchange and Chlorophyll Fluorescence of Sorghum under Salt Stress. Crop Sci. 44:806-811

OECD (Organisation de Coopération et de Développement Économiques) 2010. Consensus ocument on Compositional Considerations for New Varieties of GRAIN SORGHUM [Sorghum bicolor (L.) Moench]: Key Food and Feed Nutrients and Anti-nutrients (http://www.oecd.org/ehs/).

Pardossi, A. , Malorgio, F. , Oriolo, D. , Gucci, R. , Serra, G. and Tognoni, F(1988) Water relations and osmotic adjustment in Apium graveolens during long-term NaCl stress and subsequent relief. *Physiol. Plant.* 102 , pp. 369-376.

Prado,F.E., Boero,C., Gallardo,M., Gonzalez, J.A (2000). Effect of NaCl on germination, growth and soluble sugar content in Chenopodium quinoa willd seeds. Bot. Bull.Acad. Sin. 41, 27–34.

Qureshi, MS., Richards, B., Hay, A., Tsai, C., McBride, M., Baveye, P., Steenhuis,T. (2003). Effect of microbial activity on trace element release from sewage sludge, *Environmental Science & Technology*, Vol. 37, No. 15, pp. 3361–3366

Renault, S., Clare, C., Franklin, JA., Zwiazek ,JJ. (2001) Effects of NaCl and Na_2SO_4 on red-osier dogwood (*Cornus stolonifera* Michx) seedlings. Plant Soil 233*:* 261–268*DOI*

Richards, KD., snowden, KC., Gardner, RC. (1994) wali6 and wali7: genes induced by aluminum in wheat (Triticum aestium L.) roots. Plant Physiol 105:1455–1456.

Rivelli, A.R., Lovelli, S., Perniola, M (2002). Effects of salinity on gas exchange, water relations and growth of sun flower(Helianthusannuus). Funct. Plant Biol. 29, 1405–1415.

Rogers, M.E., Noble, C.L(1992) Variation in growth and ion accumulation between two selected populations of *Trifolium repens* L. differing in salt tolerance. - Plant Soil. 146: 131-136,

Samadani, B., B. Kholdbarin. and A. Almodares.(1994). Salt tolerance of sweet sorghum cultivars and their mechanisms. M. Sc. thesis in biology (plant physiology). Shiraz University - Iran, pages 45, 60, 82.

Saneoka, H., Ishiguro, S and Moghaieb , E.(2001)Effect of salinity and abscisic acid on accumulation of glycinebetaine and betaine aldehyde dehydrogenase mRNA in Sorghum leaves (*Sorghum bicolor*).Journal of Plant Physiology Volume 158, Issue 7, Pages 853-859.

Santa-Maria, G. E.; Epstein, E (2001) Potassium/sodium selectivity in wheat and amphiploid cross wheat × Lophopyrum elongatum. Plant Sci. 160: 523–534.

Shalata,A., Tal ,M.(1998).The effect of salt stress on lipid peroxidation and antioxidants in the leaf of the cultivated tomato and its wild salt-tolerant relative Lycopersicon pennellii. Physiologia Plantarum104, 169–174

Shurson, J. (2009), Distillers Grains By-products in Livestock and Poultry Feeds, Overview: Distiller's Dried Grains with Solubles, Dept. of Animal Science, College of Food, Agriculture and Natural Resource Science, University of Minnesota, St. Paul, MN, available online at http://www.ddgs.umn.edu/overview.htm.

Smith ,CW., Frederiksen ,RA (Eds.) (2000) Sorghum: Origin, history, technology, and production, John Wiley and Sons, New York Gnansounou E, Dauriat A, Wyman CE (2005) Refining sweet sorghum to ethanol and sugar: economic trade-offs in the context of North China. Bioresource Technol. 96: 985-1002.

Smith, CW., Frederiksen, RA (Eds.) (2000) Sorghum: Origin, history, technology, and production, John Wiley and Sons, New York.

Solymani, A., A.Almodares and L. Naranjani. (2010). Effects of plant populations on stalk yields and sucrose content of two sweet sorghum cultivars. Iranian j. Field Crop Res. 8 (3):455-464.

Sudhir, P., Murthy S. D (2004). Effects of salt stress on basic processes of photosynthesis. photosynthetica ,42(4):481-486.

Sunseri, F ., Palazzo ,D., Montemurro, N and Monte murro, F. (2002). Salinity tolerance in sweet sorghum. (Sorghum bicolor L. Manch). Field performance under salt stress. Ital. J. Agron., 2(2): 111-1

Szabolcs, I .(1994). Soil salinization. In: M. Pessarakli, Editor, Handbook of Plant Crop *Stress*, Marcel Dekker, New York, pp. 3–11.

Taherzadeh, MJ., Karimi, K .(2008) Bioethanol: Market and Production Processes. In: Biofuels Refining and Performance (Ed. A Nag). The Mc Graw- Hill, USA Rains GC, Cundiff JS, Welbaum GE (1993) Sweet sorghum for a piedmont ethanol industry. In: New crops (Eds. J Janick & JE Simon). Wiley New York.

Tesso, TT., Claflin, LE., Tuinstra ,MR.(2005) Analysis of Stalk Rot Resistance and Genetic Diversity among Drought Tolerant Sorghum Genotypes. Crop Sci. 45: 645-652.

U.S. Grains Council (2008), Sorghum, Washington DC., available online at http://www.grains.org/sorghum.

U.S. National Sorghum Producers Association (2006), website http://www.sorghumgrowers.com/.

Wichelns, D., and J.D. Oster (2006). Sustainable irrigation is necessary and achievable, but direct costs and environmental impacts can be substantial. Agricultural Water Management, 86: 114-27.

Zhu, J. (2001) Plant salt tolerance. Trends in Plant Science 6, 66–71

23

From a Pollutant Byproduct to a Feed Ingredient

Elisa Helena Giglio Ponsano[1], Leandro Kanamaru Franco de Lima[2] and Ane Pamela Capucci Torres[1]
[1]*Unesp Univ Estadual Paulista, Faculty of Veterinary Medicine, Araçatuba,*
[2]*Brazilian Agricultural Research Corporation,*
Embrapa Fisheries and Aquaculture, Palmas,
Brazil

1. Introduction

Industrial activities have always been associated to the economic development of nations and their population. Nevertheless, they are also associated to the generation of industrial byproducts, generally considered undesirable due to the environmental damage they impose to society (Pipatti et al., 2009). Industrial byproducts have variable characteristics and compositions, since they are directly dependent on crude matter essence, kind of processing, facilities characteristics and volume of output, among so many other factors. Nowadays, the broad range of industries spread all over the world in an effort to supply the necessity of global population makes evident the need for the adoption of strategies capable of equilibrating economic development and environmental preservation as a way of reaching a sustainable industrial production (Parente & Silva, 2002).

In that way, transformation industries are currently searching for productive technologies of low environmental impact, which include practices like minimization of byproducts generation and/or recuperation and recycling of these residues, so aiming at the optimization of industrial processes (Juskaitė-Norbutienė et al., 2007; Leite & Pawlowsky, 2005; Souza & Silva, 2009). The adoption of such technologies is a differential for the establishment and maintenance of industries in the current social and economic world scenery (Leite & Pawlowsky, 2005).

The management of industrial byproducts generally combines techniques as recuperation, treatment and safe disposal. Regarding to liquid waste, also called wastewater or effluent, treatments performed in the food industry generally consist of physical, chemical and biological operations. Physical treatments provide the removal of suspended solids and the separation of oils and fats by means of filtration, grading, sedimentation or floating techniques, while chemical treatments provide the removal of dissolved matter and even of microorganisms by using different chemicals (Giordano, 2006). The biological treatments, in turn, count on the ability of bacteria, fungi, micro algae and protozoa in transforming organic matter into new cells, called biomass, and gases (Arvanitoyannis & Tserkezou, 2009; Giordano, 2006). This kind of treatment simulates the natural remediation processes that occur in nature and brings as an advantage the production of compounds with particular applications, which may be appropriately separated and used for distinct purposes (Liu, 2007). Microbial biomass, for instance, has been considered as an alternative source of proteins for foods and feeds and may be produced in different substrates, including effluents from industries and farms (Nasseri et al., 2011).

Some organisms may be used for the removal of organic matter from agro industrial residues yielding a biomass with potential for use in animal feeding, such as the phototrophic bacteria (Azad et al., 2003; Izu et al., 2001; Ponsano et al., 2008). Purple Non Sulfur Bacteria (PNSB), for example, are phototrophic bacteria commonly found in rivers, ponds, lakes and wastewater treatment systems, that can grow both as photoautotroph and photoheterotroph under anaerobic-light or microaerobic-light conditions (Choorit et al., 2002; Kantachote et al., 2005). Some PNSB also can grow in the dark using fermentation when they are in anaerobic environments or respiration when in aerobiosis (Devi et al., 2008; Kantachote et al., 2005; Kim et al., 2004; Ponsano et al., 2002a). Due to the ability of phototrophic bacteria to utilize diverse metabolic activities in different substrates and growth conditions, they find a role in the depollution of wastewaters from food industries, still producing a biomass rich in proteins, vitamins and carotenoids that may be used in the supplementation of animal feed (Carlozzi & Sacchi, 2001; Izu et al., 2001; Kantachote et al., 2005; Ponsano et al., 2002a, 2003a, 2004a, b; Zheng et al., 2005 a, b).

Rubrivivax gelatinosus, formerly named *Rhodocyclus gelatinosus* is a PNSB commonly found in many wastewaters in which it grows as an autotrophic or a heterotrophic, depending on light and oxygen conditions (Ponsano et al., 2003a, 2008). As the bacterium produces oxycarotenoids as photosynthetic pigments, its biomass can find use as a pigmenting additive in animal production, as previously suggested and tested by Ponsano et al. (2002b, 2003b, 2004a, b) and Polonio et al. (2010).

The use of pigmenting additives in animal production is justified by the fact that animals are unable to synthesize their own carotenoids and therefore, rely on dietary supply to achieve their natural pigmentation (Gouveia et al., 2003). The effectiveness of oxycarotenoids or xanthophylls in providing pigmentation to animals is possible because these carotenoids have the ability to deposit on different parts in animal bodies, such as muscles, fat, skin, feather, legs, ovaries and eggs (Ponsano et al., 2002b, 2004b).

Primarily, pigmenting additives were added into food formulations in order to replace color lost during the industrialization processes but, when the remarkable acceptance of consumers for well colored products was identified, industries started coloring a broad range of food items, reaching consumers desire and so improving its sales (Calil & Aguiar, 1999). In case of poultry and fish production, for instance, either natural or synthetic additives are used when intensive rearing is adopted and/or when feed ingredients are poor in xanthophylls, so lacking in color in the final products. The most used synthetic additives for this purpose are apocarotenoic acid ethyl ester, canthaxanthin and astaxanthin, which show good stability and deposition rates on animal tissues. Nevertheless, more and more consumers around the world have been showing their preference for natural additives, what stimulates the search for natural sources of pigments, like those from biotechnological production. Among natural xanthophylls used in animal production, those from plants, algae, bacteria and yeasts have been previously described in literature (Akiba et al., 2000, 2001; Bosma et al., 2003; Gouveia et al., 1996; Liufa et al., 1997; Perez-Vendrell et al., 2001; Toyomizu et al. 2001).

The great acceptance that fish finds among consumers due to its nutritional and sensorial properties guarantees its market and yet claims for increases in production, which has been supplied by the aquaculture (Lem & Karunasagar, 2007). Nevertheless, fish is a perishable food and so requires the application of methods for its preservation, such as fermentation, refrigeration, freezing, canning, smoking, drying and others, that may be performed separately or in combinations. As it happens in any other food industry, fish processing generates great amounts of wastewaters with variable Chemical Oxygen Demand which depends on fish species, fish products and methods of processing, since water is involved in several stages of manufacturing, like butchering, evisceration, filleting, salting, cooking, canning, freezing, sterilization and cleaning operations (Arvanitoyannis & Kassaveti, 2008; Liu, 2007). The utilization of these effluents for the biomass production is an alternative for minimizing costs with treatment and environmental impacts. Moreover, in case the composition of the biomass finds an appropriate purpose, it can represent extra profits for the industry.

So, the hypothesis to be tested in this chapter is that an industrial byproduct may undergo a biological treatment yielding a product with application. The objective of this chapter was to describe a study on the transformation of a fish processing wastewater into a product with potential of use in animal rearing.

2. Study conduction

2.1 Wastewater characterization and treatment

Tilapia fish processing wastewater used in the experiment was donated by Tilapia do Brasil Inc. (Buritama City, SP, Brazil) and was made up of effluents from killing, scaling, gutting, cleaning, skinning, filleting and freezing operations, and also from cleaning operations, which were gathered and roughly filtered (grating), averaging 10,000 L h^{-1}.

Crude wastewater was analyzed for turbidity, total solids (TS), pH, total nitrogen (TN) and oils and greases (OG), according to standard methods (American Public Health Association, American Water Works Association, Water Pollution Control Federation [APHA, AWWA and WPCF], 2005). Chemical Oxygen Demand (COD) was determined by chemical digestion (HR digestion solution for COD 0-1500 ppm; DRB200; DR2800; Hach), based on the protocol developed by Jirka & Carter (1975).

Before being used as a substrate for the bacterial growth, the wastewater was filtered in a 50 μm mesh fast filter (Gardena 1731; 3,000 L h^{-1}) for the withdrawal of gross particles and heat treated (Incomar LTLT tank) at 65°C/30 min to eliminate pathogenic agents and repress the level of competing microorganisms. After that, wastewater was cooled to room temperature and so it was ready to receive the bacterial inoculum.

Microbiological analyses of crude and heat treated wastewater comprised mesophilic aerobic bacteria, total and fecal coliforms, molds and yeasts, *Aeromonas* spp and *Salmonella* spp, and were performed according to standard methodology (APHA, AWWA and WPCF 2005).

2.2 Bacterial inoculum preparation

Rubrivivax gelatinosus previously isolated from poultry slaughterhouse wastewater and characterized by morphological and biochemical tests was used in this experiment. The cells were maintained in Pfennig medium containing (per liter): 0.5 g KH_2PO_4; 0.4 g $MgSO_4.7\ H_2O$; 0.4 g NaCl; 0.4 g NH_4Cl; 0.05 g $CaCl_2.2H_2O$; 1.0 g sodium acetate, 0.2 g yeast extract; 0.005 g ferric citrate; 10.0 mL trace elements solution ($FeSO_4.7H_2O$ 200 mg; $ZnSO_4.7H_2O$ 10 mg; $MnCl_2.4H_2O$ 3 mg; H_3BO_3 30 mg; $CoCl_2.6H_2O$ 20 mg; $CuCl_2.2H_2O$ 1 mg; $NiCl_2.6H_2O$ 2 mg; Na_2MoO_4. $2H_2O$ 3 mg); 20.0 g bacteriological agar; 10.0 ml biotin sol. (0.0015%) and 10.0 ml thiamine-HCl sol. (0.005%). The pH was adjusted to 7.0 before autoclaving at 121°C for 15 min.

For the initial inoculum preparation, cells were grown in Pfennig liquid medium with the same pH and composition described above but bacteriological agar, under anaerobiosis (fully filled screw-crap tubes), 32 ± 2°C and 1,400 ± 200 lux for approximately 3 days, until a slight red color arose.
For the final inoculum, an aliquot from initial inoculum was transferred at 1% (v/v) to the same medium and incubation was carried out under the same conditions described before, until optical density at 600 nm reached 0.5 (Ponsano et al., 2003a).

2.3 Biomass preparation and recuperation

The bacterial inoculum was added, at 1% (v/v), to 100 L of treated wastewater. Cultivation was accomplished in anaerobiosis inside 100 L glass reactors at 32 ± 2°C and 2,000 ± 500 lux for seven days.
For the biomass recuperation, the culture was filtered at 0.2 µm, 1.5 m^3 h^{-1} and 4.5 bar (Frings), giving origin to a concentrate containing the cells and a permeate. The concentrate was centrifuged at 3,400 g for 30 min at 5°C (Incibras Spin VI) and the resulting slime was frozen at - 40°C and lyophilized (Liobras L 101) for 48 h. Hand grinding was performed to obtain the power biomass. Procedures were repeated six times.

2.4 Process analyses

Cell mass concentration was determined from 20 mL of concentrate, after successive centrifugation (900 g/15 min) and washing cycles followed by drying at 80°C until it gets constant weight.
For productivity determination, it was considered the mean production of dry biomass per liter per day.
TN, OG, COD and pH determinations in permeate were accomplished as previously described for crude wastewater (APHA, AWWA and WPCF, 2005).

2.5 Biomass analyses

For the microbiological characterization by biomass, total and fecal coliforms, molds and yeasts, coagulase-positive staphylococci, *Aeromonas* spp and *Salmonella* spp were investigated according to methodologies described by Vanderzant & Splittstoesser (1992).
For the proximate composition of biomass, the concentrations of moisture, lipids, proteins and ash were determined according to Association of Official Analytical Chemists (1995).
Amino acid determinations were carried out before and after acid hydrolysis (5 mg of extract) with a mixture containing 6 mol L^{-1} of HCl and 5% phenol/water (0.08 mL) for 72 h at 110°C. Samples were dried, diluted with citrate buffer pH 2.2 and filtered in a GV Millex Unity (Millipore). Amino acids analyses were performed by cation-exchange chromatography using a Shimadzu LC-10A/C-47A, sodium eluents and post-column derivatization with o-phthaldialdehyde. Identification and quantification were accomplished by the comparison of retention time and area of each amino acid with a standard containing 16 amino acids (100 nmol mL^{-1}), respectively (Fountoulakis & Lahm, 1998).
The biomass color attributes *L* (lightness), *C* (chroma) and *h* (hue) were obtained from the average of three consecutive pulses launched from the optical chamber of the MiniScan XE Plus (Hunter Lab) using illuminant D65 and 2° observer, after calibration with black and white standards (Commission Internationale de l'Éclairage, 1986).
For the determination of oxycarotenoids, an adaptation of Valduga (2005) methodology was used. Pigments were extracted from biomass with dimetilsulfoxide at 55°C/30 min and

alternated cycles of ultrasound at 40 kHz (Unique/USC 1800A) and shaking (Phoenix/P-56). Next, a mixture containing acetone: methanol (7:3, v/v) was added, tubes were centrifuged at 3.400 g and 5°C/10 min and the supernatant was transferred to a 50 mL volumetric flask. Successive extractions were performed until no color remained in cells or solvent. Final dilutions were made up with methanol and the quantification of oxycarotenoids was accomplished at 448 nm (Hitachi U-1000/U-1100). Total carotenoids were estimated according to Davies (1976) using the absorption coefficient of carotenoids suggested by Liaaen-Jensen & Jensen (1971).

3. Main findings of the study

The microbiological investigation on crude and treated wastewaters showed a sharp decrease in indicator organisms after heat treatment (Table 1).

Aeromonas spp are spread in aquatic environments, what may explain the presence of such organism in the crude effluent. Nevertheless, some species such as *A. hydropila* and *A. salmonicida* may be responsible for lethal infections in fish, bringing considerable economic losses to aquaculture (Maluping et al., 2005; Vieira, 2003) and some others have been described as emergent pathogens for humans (Vieira, 2003). So, the presence of this microorganism in the crude wastewater claims for periodic control in aquaculture, slaughter and processing of tilapia fish, as a way of avoiding financial injury to the fish industry and to consumers.

The presence of *Salmonella enterica* subsp. *enterica* serotype *Typhi* was detected in the wastewater, which represents a potential risk to public health and reveals deficient sanitary conditions during manipulation in the industry, since man is the natural reservoir of this serotype. This bacterium may be transmitted by water and foods contaminated with human feces, causing a serious infectious disease (Franco & Landgraf, 1996).

Microbiological analysis	Crude wastewater	Treated wastewater[2]
Mesophilic aerobic bacteria (CFU* mL^{-1})	8.5×10^5	7.0
Moulds and yeasts (CFU mL^{-1})	4.6×10^3	6.0
Total coliforms (MPN** mL^{-1})	1.0×10^5	<1.0
Fecal coliforms (MPN mL^{-1})	0.41	<1.0

[1]Mean values. [2]Filtration (50 μm)/heat treatment (65 °C/30 min). *Colony Forming Units. **Most Probable Number.

Table 1. Microbiological characteristics of tilapia fish industrial wastewater[1]

Heat treatment was able to eliminate contaminants and pathogenic microorganisms detected in the crude wastewater, so reducing competition for substrate during *Rubrivivax gelatinosus* cultivation.

The knowledge on the wastewater physicochemical properties reveals its suitability for discharge. Total solids, for instance, represent dissolved or suspended substances, both of organic or inorganic structures and, if too high, may cause damages to water bodies and aquatic organisms. Turbidity units indicate the transparency of the wastewater and the presence of colloids that, when excessive, may alter the aspect of streams and rivers and so prevent photosynthetic organisms' metabolism. The acidic or alkaline characteristic of the wastewater is defined by pH and, together with temperature, find an important role on the control of biotechnological processes. Nitrogen in wastewaters may derive from synthetic

detergents used during cleaning operations or from protein degradation. Although this element may be essential to most living organisms, in high concentrations it may cause the proliferation of aquatic plants in water bodies and effluents. Oils and greases in wastewaters may originate from industrial kitchens, mechanic repairs garages, boilers and other equipments, as well as from raw material. They can easily be oxidized and so exhale bad odors in the environment. COD is an indirect measure of organic compounds concentration in wastewaters and so, reflects its pollutant load (Giordano, 2004; Liu, 2007).
The physicochemical data found for crude tilapia fish processing wastewater (Table 2) indicate the need for previous treatments for a safe discharge, according to Brazilian legislation. On the other hand, the presence of such organic matter in the wastewater was important to ensure the growth of *R. gelatinosus* with the resulting production of cells and oxycarotenoids.

Physicochemical parameter	Quantity
Effluent volume (l day^{-1})	120,000
Effluent flow (l h^{-1})	11,000 to 15,000
Temperature (°C)	20.3 ± 0.23
Total solids (g L^{-1})	1.5 ± 0.32
Turbidity (TU)	35.7 ± 2.25
pH	9.4 ± 0.09
Total nitrogen (mg L^{-1})	813.3 ± 54.65
Oils and greases (mg L^{-1})	1,166.3 ± 68.52
COD (mg L^{-1})	1,127.5 ± 33.84

[1]Mean values and standard errors.

Table 2. Physicochemical characteristics of crude tilapia fish industrial wastewater[1]

The physicochemical characteristics of wastewaters presented herein differ from others previously reported. This happens because the particular characteristics of each industrial effluent derive from crude matter composition, season of the year, water supply, reuse procedures, factory installations and industrial processing techniques, among others (Liu, 2007). For settled and unsettled wastewater from sardine processing industry, for example, pH values from 6.2 to 6.3; 63,000 mg L^{-1} COD and 10.88 mg L^{-1} TN were described (Azad et al., 2001; 2003). For white fish filleting plants, Arvanitoyannis & Kassaveti (2008) reported the generation of wastewater with 50 kg COD and Prasertsan et al. (1993) reported 5.3 to 8.3 pH; 5,950 to 157,080 mg L^{-1} COD; 19.30 to 82.22 g L^{-1} TS and 666 to 32,182 mg L^{-1} OG for effluents from different seafood processing plants. Concentrations around 4,300 mg L^{-1} COD, 800 mg L^{-1} OG and 6.2 to 7.0 pH also were reported for wastewater from fish processing operations by Giordano (2004).
Changes in tilapia fish wastewater physicochemical parameters after biomass recuperation comprised removals of 82% in COD, 48% in OG and 22% in TN and a decrease in pH to 7.9, rendering it suitable for discharge in the environment, according to Brazilian laws. So, the biomass production process itself worked as a biological treatment for the reduction of pollution in tilapia fish industry wastewater.
Mean cell mass production and productivity achieved with the biological treatment were 0.18 g L^{-1} and 0.0634 g L^{-1} day^{-1}, respectively. Prasertsan et al. (1993) credit the low cell production to the anaerobiosis/light cultivation conditions, in which the synthesis of oxycarotenoids is intensified. Other authors found higher cell mass concentrations when growing phototrophic organisms in industry wastewaters but, in those cases, initial organic matter and inoculum levels were higher than the ones used in this study and/or nutritional supplementation was

adopted (Azad et al., 2001, 2003; Prasertsan et al., 1997). In this study, we opted to maintain the original wastewater composition and to use a low inoculum level in an attempt to minimize costs and render the biomass production process feasible for the industry.

The microbiological investigation on *Rubrivivax gelatinosus* biomass indicated low counts on total coliforms (20.27 NMP g^{-1}), fecal coliforms (< 1.0 NMP g^{-1}) and molds and yeasts (1.2 x 10^3 UFC g^{-1}) and the absence of pathogenic organisms. This way, the product showed to be in agreement with Brazilian microbiological standards required for feed ingredients, which ensures its safe utilization.

Mean proximate composition of biomass and amino acid profile in the product are presented in Tables 3 and 4, respectively. As a typical feature of single cell proteins, the values indicate the high level of proteins in the biomass, which denotes its use in animal diets as a nutritional ingredient. Moreover, it also contained considerable amounts of all amino acids considered essential for animals, what reinforces the suggestion of its use in the supplementation of animal feeds in order to supply deficiencies that may cause, for instance, delay in protein utilization and reduction of growth, weight gain, feed conversion and immunity (Cyrino et al., 2004). In view of these findings, the bacterial biomass presents a potential for use as a nutritional ingredient for feeds.

Component	%
Moisture	4.55 ± 0.84
Ash	4.05 ± 0.66
Protein	57.39 ± 2.81
Lipids	11.08 ± 1.41

[1]Mean values and standard errors.

Table 3. Proximate composition of *Rubrivivax gelatinosus* biomass produced in tilapia fish industrial wastewater[1]

Amino acid	**Quantity (g 100 g^{-1})**
Aspartic acid	5.70 ± 2.35
Threonine	3.82 ± 1.50
Serine	2.81 ± 0.96
Glutamic acid	6.40 ± 2.29
Proline	2.93 ± 1.02
Glycine	3.46 ± 1.51
Alanine	5.32 ± 2.28
Valine	4.39 ± 1.84
Methionine	0.66 ± 0.29
Isoleucine	3.33 ± 1.43
Leucine	7.08 ± 2.41
Tyrosine	2.56 ± 0.96
Phenylalanine	3.43 ± 1.31
Histidine	1.92 ± 0.74
Lysine	4.52 ± 1.76
Arginine	3.85 ± 1.29

[1]Mean values and standard errors

Table 4. Amino acid composition of *Rubrivivax gelatinosus* biomass produced in tilapia fish industrial wastewater[1]

Oxycarotenoids content in the biomass was found to be 3.03 mg g^{-1} dry biomass, which conferred a dark red color to the power product (L = 22.42; C = 14.22; h = 25.48). This is in agreement with Prasertsan et al. (1997), who found concentrations of 2.13 to 3.90 mg of carotenoids per gram of dry biomass of *Rhodocyclus gelatinosus* produced in tuna processing wastewater.

The main photosynthetic pigments produced by *Rubrivivax gelatinosus* are bacteriochlorophyll *a* and carotenoids from alternative spirilloxanthin series, which contains spheroidene, hydroxyspheroidene and spirilloxanthin as the major representants (Holt et al., 2000). The blend among these pigments gives the bacterial cultures a reddish color (Ponsano et al., 2002a, 2003a, 2008) that remains in the dry biomass, since sensorial and nutritional properties of lyophilized products remain intact after drying process (Pereda et al., 2005). Considering that these pigments are oxycarotenoids and so have the ability to deposit in animal tissues, this feature of the biomass suggests its application as a pigmenting ingredient for the rearing of different animals.

The use of natural or synthetic oxycarotenoids for the rearing of animals is reported by many authors. Salmonids, for instance, are noble fish natural from cold waters in North Hemisphere, but that are being commercially farmed in many parts of the world. According to Baker & Günther (2004), in wild salmon, the natural carotenoid astaxanthin provides a majority of the color expected from this flesh. Nevertheless, for farmed salmonids, the same effect may be achieved by the use of pigmenting additives in rations. They may also be used for the raising of ornamental fish to increase skin color and beauty. For the raising of red *Cyprinus carpio* (*Kawari*), for instance, Gouveia et al. (2003) relate the utilization of carotenoids produced by micro algae *Chlorella vulgaris*.

For poultry products, the pigmentation varies according to market demand. In Mexico, Belgium, Italy, Peru and some regions in Brazil, for instance, the use of pigmenting ingredients in poultry production is a common practice since people prefer strong colors for broilers carcasses and egg yolks (Gouveia et al., 1996; Toyomizu et al., 2001). People often associate strong colors of a food item to safety and health and so look for strongly pigmented products. Taking it into account, Ponsano et al. (2002b, 2004a, b) added *Rhodocyclus gelatinosus* biomass produced in poultry slaughterhouse wastewater in broilers rations and found an increase in the color of breast meat. Polonio et al. (2010) used different concentrations of the same product in hens rations and found an improvement in yolks color, with no deleterious effects on birds performance. In the sensorial test, these authors identified the concentration of the biomass that, when used together with corn xanthophylls, provides a desired golden orange color to the yolks. Yet, Garcia et al. (2002) found an increase in yolks color, with no influence in the performance and eggs characteristics, when canthaxantin was used in hens diets.

Besides the pigmenting feature of oxycarotenoids, they are also known to exert benefits on animal health and welfare due to antioxidant properties. According to Baker & Günther (2004), evidences suggest that the carry-over of these pigments into the human food chain could be beneficial to human health too. In humans, the consumption of oxycarotenoids is associated to aging prevention and to the decrease of the risk of diseases related to the accumulation of free radicals (Bhosale, 2004; Bhosale; Bernstein, 2005). So, for further studies on the properties of *Rubrivivax gelatinosus* biomass, the antioxidant ability of its carotenoids will be considered.

4. Conclusion

In this chapter we showed the feasibility of using an industrial byproduct for the production of a biomass with potential of use in animal rearing, not only for being a source of natural pigments but also for having an elevated nutritional value. Moreover, we showed that the biomass production process worked as a biological treatment for the reduction of pollution in the industrial wastewater, requiring simple and feasible methods that can be operated in the industry, so minimizing byproducts and still rendering profits from the biomass commercialization.

5. Acknowledgements

Authors thank Tilapia do Brasil S/A Inc. for donating the effluent and students involved in the study, Lorrayne Bernegossi Polonio, Gabriela de Oliveira and Edson Francisco do Espírito Santo. Authors also thank Fapesp for financial support.

6. References

Akiba, Y., Sato K. & Takahashi, K. (2001). Meat color modification in broiler chickens by feeding yeast *Phaffia rhodozyma* containing high concentrations of astaxanthin. *Journal of Applied Poultry Research*, Vol.10, pp.154-161, ISSN 15370437

Akiba, Y., Sato, K., Takahashi, K., Toyomizu, M., Takahashi, Y., Tsunekawa, H., Hayasaka, Y. & Nagao, H. (2000). Availability of cell wall fractured yeast, *Phaffia rhodozyma*, containing high concentration of astaxanthin for egg yolk pigmentation. *Journal of Animal Science*, Vol.71, pp.255-260, ISSN 1525-3163

AOAC (1995). *Official methods of analysis*, Association of official analytical chemists, ISBN 0935584803, Washington, USA

APHA, AWWA & WPCF (2005). Standard methods for the examination of water and wastewater. American Public Health Association, ISBN 0875532071, Washington, USA

Arvanitoyannis, I.S. & Kassaveti, A. (2008). Fish industry waste: treatments, environmental impacts, current and potential uses. *International Journal of Food Science and Technology*, Vol.43, No.4, pp. 726-745, ISSN 13652621

Arvanitoyannis, I.S. & Tserkezou, P. (March 2009). *Waste management in the food industry: potential uses*, Retrieved from <http://www.scitopics.com/Waste_Management_for_the_Food_Industries.html>

Azad, S. A., Vikineswary, S., Chong, V. C. & Ramachandran, K. B. (2003). *Rhodovulum sulfidophilum* in the treatment and utilization of sardine processing wastewater. *Letters in Applied Microbioloy*, Vol.38, No.1, pp. 13-18, ISSN 1472765X

Azad, S.A., Vikineswary, S., Ramachandran, K.B. & Chong, V.C. (2001). Growth and production of biomass of *Rhodovulum sulfidophilum* in sardine processing wastewater. *Letters in Applied Microbioloy*, Vol. 33, pp. 264-268, ISSN 1472765X

Baker, R. & Günther, C. (2004). The role of carotenoids in consumer choice and the likely benefits from their inclusion into products for human consumption. *Trends in Food Science and Technology*, Vol.15, No.10, pp. 484-488, ISSN 09242244

Bhosale, P. (2004). Environmental and cultural stimulants in the production of carotenoids from microrganisms. *Applied Microbiology and Biotechnology*, Vol.63, No.4, pp. 351-361, ISSN 01757598

Bhosale, P.; Bernstein, P. S. (2005). Microbial xanthophylls. *Applied Microbiology and Biotechnology*, Vol.68, No.4, pp. 445–455, ISSN 01757598

Bosma, T.L., Dole, J.M. & Maness, N.O. (2003). Optimizing marigold (*Tagetes erecta* L.) petal and pigment yield. *Crop Science*, Vol.43, pp.2118-2124, ISSN 14350653

Calil, R. & Aguiar, J. (1999). *Aditivos nos alimentos*, R. M. Calil, ISBN 8590091716, São Paulo, Brasil

Carlozzi, P. & Sacchi, A. (2001). Biomass production and studies on *Rhodopseudomonas palustris* grown in an outdoor, temperature controlled, underwater tubular photobioreactor. *Journal of Biotechnology*, Vol.88, No.3, pp. 239-249, ISSN 01681656

Choorit, W., Thanakoset, P., Thongpradistha, J., Sasaki k. & Noparatnaraporn, N. (2002). Identification and cultivation of photosynthetic bacteria in wastewater from a concentrated latex processing factory. *Biotechnology Letters*, Vol.24, No.13, pp. 1055-1058, ISSN 15736776

Commision Internationale de l'Éclairage. (1986). Technical Report, Colorimetry, *Publ. CIE 15.2-1986*, CIE Central Bureau, Vienna, Austria

Cyrino, J. E. P., Urbinati, E. C., Fracalossi, D. M. & Castagnolli, N. (2004). *Tópicos especiais em piscicultura de água doce tropical intensiva,* TecArt, ISBN 8590468917, São Paulo, Brasil

Davies, B.H. (1976). Carotenoid. In: *Chemistry and Biochemistry of Plant Pigments,* T.W. Goodwin, (ed.), pp. 38-165, Academic Press, ISBN 0122899024, New York, USA

Devi, M.U., Padmavathi, V., Rani, T.S. & Reddy, K.J. (2008). Characterization and identification of anoxygenic phototrofic purple non-sulfur bacteria isolated from industrial waste waters. *International Journal of Agriculture, Environment & Biotechnology*, Vol.4, No.3, pp. 56-59, ISBN A100004267

Fountoulakis, M. & Lahm, H.W. (1998). Hydrolysis and amino acid composition of proteins. *Journal of Chromatography A*, Vol.826, pp. 109-134, ISSN 0021-9673

Franco, B.D.G.M. & Landgraf, M. (1996). *Microbiologia dos alimentos*. Atheneu, ISBN 8573791217, São Paulo, Brasil

Garcia, E.A., Mendes, A.A., Pizzolante, C.C., Gonçalves, H.C., Oliveira, R.P. & Silva, M.A. (2002). Efeitos dos níveis de cantaxantina na dieta sobre o desempenho e qualidade dos ovos de poedeiras comerciais. *Revista Brasileira de Ciência Avícola*, Vol.4, No.1, ISSN 1516635X

Giordano, G. (August 2004). *Tratamento e controle de efluentes industriais*, Retrieved from <http://www.ufmt.br/esa/Modulo_II_Efluentes_Industriais/Apost_EI_2004 _1ABES_Mato_Grosso_UFMT2. pdf>

Gouveia, L., Rema, P., Pereira, O. & Empis, J. (2003). Colouring ornamental fish (Cyprinus carpio and Carassius auratus) with microalgal biomass. *Aquaculture Nutrition*, Vol.9, No.2, pp. 123-129, ISSN 13652095

Gouveia, L., Veloso, V., Reis, A., Fernandes, H., Novais, J. & Empis, J. (1996). *Chlorella vulgaris* used to colour egg yolk. *Journal of the Science of Food and Agriculture*, Vol.70, No.2, pp. 167-172, ISSN 10970010

Holt, J.G., Krieg, N.R., Sneath, P.H.A. & Staley, J.T. (2000). Bergey's manual of determinative bacteriology. Lippincott Williams & Wilkins, ISBN 0683006037 Philadelphia, USA

Izu, K., Nakajima, F., Yamamoto, K. & Kurisu, F. (2001). Aeration conditions affecting growth of purple nonsulfur bacteria in an organic wastewater treatment process. *Systematic and Applied Microbiology*, Vol.24, No.2, pp. 294-302, ISSN 07232020

Jirka, A.M. & Carter, M.J. (1975). Micro semi-automated analysis of surface and wastewaters for chemical demand oxygen. Analytical Chemistry, Vol.47, pp.1937, ISSN 15206882

Juskaitè-Norbutienè, R., Miliutè, J. & Cesnaitis, R. (2007). Bio-degradable waste and by-products from food industry management systems in Lithuania: analysis, problems and improvement possibilities. *Environmental Research, Engineering and Management*, Vol.4, No.42, pp. 60-69, ISSN 20292139

Kantachote, D., Torpee, S. & Umsakul, K. (2005). The potencial use of anoxygenic phototrophic bacteria for treating latex rubber sheet wastewater. *Electronic Journal of Biotechnology*, Vol.8, No.3, ISSN 07173458

Kim, M.K., Choi, K., Yin, C., Lee, K., Im, W., Lim, J.H. & Lee, S. (2004). Odorous swine wastewater treatment by purple non-sulfur bacteria, *Rhodopseudomonas palustris*, isolated from eutrophicated ponds. *Biotechnology Letters*, Vol.26, No.10, pp. 819–822, ISSN 15736776

Leite, B.Z. & Pawlowsky, U. (2005). Alternativas de minimização de resíduos em uma indústria de alimentos da região metropolitana de Curitiba. *Engenharia Sanitária e Ambiental*, Vol.10, No.2, pp. 96-105, ISSN 14134152

Lem, A. & Karunasagar, I. (2007). Trade and safety in aquaculture products. *The FAO Aquaculture Newsletter*, No.38, pp. 13-16, ISSN 10203443

Liaaen-Jensen, N. & Jensen, A. (1971). Quantitative determination of carotenoids in photosynthetic tissues: total carotenoid content. In: *Methods in Enzymology: photosynthesis and nitrogen part A*, A.S. Pietro, (ed.), Vol.23, pp. 586-602, Elsevier, ISBN 0121818861, New York, USA

Liu, S.X. (2007) *Food and agricultural utilization and treatment*. Blackwell, ISBN 9780813814230, Ames, USA

Liufa, W., Xufang, L. & Cheng, Z. (1997). Carotenoids from *Alocasia* leaf meal as xanthophylls sources for broiler pigmentation. *Journal of Tropical Agriculture and Food Science*, Vol.37, pp.116-122, ISSN 13949829

Maluping, R.P., Lavilla-Pitogo, C.R., Depaola, A., Janda, J.M., Krovacek, K. & Greko, C. (2005). Antimicrobial susceptibility of *Aeromonas* spp., *Vibrio* spp. and *Plesiomonas shigelloides* isolated in the Philippines and Thailand. *International Journal of Antimicrobial Agents*, Vol.5, No.4, pp. 348-350, ISSN 09248579

Nasseri, A.T., Rasoul-Amini, S., Morowvat, M. H. & Ghasemi, Y. (2011). Single cell protein: production and process. *American Journal of Food Technology*, Vol.6, No.2, pp. 103-116, ISSN 1557458X

Parente, A.H. & Silva, E.A.B. (2002). Redução de efluentes líquidos na indústria alimentícia. *Revista Química & Tecnologia*, No.1, pp. 58-67

Pereda, J.A.O, Rodríguez, M.I.C., Álvarez, L.F., Sanz, M.L.G., Minguillón, G.D.G.F., Perales, L.H. & Cortecero, M.D.S. (2005). *Tecnologia de alimentos: componentes dos alimentos e processos*, Artmed, ISBN 8536304367, Porto Alegre, Brasil

Perez-Vendrell, A.M., Hernandez. J.M., Llaurado, L., Schierle, J. & Brufau, J. (2001). Influence of source and ratio of xanthophylls pigments on broiler chicken pigmentation and performance. *Poultry Science*, Vol.80, pp.320-326, ISSN 15370437

Pipatti, R., Chhemendra, S., Masato, Y., Alves, J.W.S., Gao, Q., Guendehou, G.H.S., Koch, M., Cabrera, C.L., Mareckova, K., Oonk, H., Scheehle, E., Smith, A., Svardal, P. & Vieira, S.M.M. (March 2009). Waste generation, composition and management data. In: *PCC. Guidelines for national greenhouse gas inventories*, Retrieved from <http://www.ipcc.nggip.iges.or.jp/public/2006gl/pdf/5_volume5/v5_2_ch2_waste_data.pdf>

Polonio, L. B., Ponsano, E. H. G., Pinto, M. F. & Garcia-Neto, M. (2010). Utilisation of bacterial (*Rubrivivax gelatinosus*) biomass for egg yolk pigmentation. *Animal Production Science*, Vol. 50, pp. 1-5, ISSN 18360939

Ponsano, E.H.G., Lacava, P.M. & Pinto, M.F. (2002a). Isolation of *Rhodocyclus gelatinosus* from poultry slaughterhouse wastewater. *Brazilian Archives of Biology and Technology*, Vol.45, No.4, pp. 445-449, ISSN 15168913

Ponsano, E.H.G., Lacava, P.M. & Pinto, M.F. (2003a). Chemical composition of *Rhodocyclus gelatinosus* biomass produced in poultry slaughterhouse wastewater. *Brazilian Archives of Biology and Technology*, Vol.46, No.2, pp. 143-147, ISSN 15168913

Ponsano, E.H.G., Paulino, C.Z. & Pinto, M.F. (2008). Phototrophic growth of *Rubrivivax gelatinosus* in poultry slaughterhouse wastewater. *Bioresource Technology*, Vol.99, No.9, pp. 3836-3842, ISSN 09608524

Ponsano, E.H.G., Pinto, M.F. & Garcia-Neto, M. (2003b). Biomassa de *Rhodocyclus gelatinosus* para a promoção da pigmentação de gemas de ovos. *Revista Higiene Alimentar*, Vol. 17, No.104/105, pp. 155-156, ISSN 01019171

Ponsano, E.H.G., Pinto, M.F., Garcia-Neto, M. & Lacava, P.M. (2002b). Evaluation of *Rhodocyclus gelatinosus* biomass for broiler pigmentation. *Journal of Applied Poultry Research*, Vol.11, No.1, pp. 77-82, ISSN 15370437

Ponsano, E.H.G., Pinto, M.F., Garcia-Neto, M. & Lacava, P.M. (2004a). Performance and color of broilers fed diets containing *Rhodocyclus gelatinosus* biomass. *Brazilian Journal of Poultry Science*, Vol.6, No.4, pp. 237-242, ISSN 1516635X

Ponsano, E.H.G., Pinto, M.F., Garcia-Neto, M. & Lacava, P.M. (2004b). *Rhodocyclus gelatinosus* biomass for egg yolk pigmentation. *Journal of Applied Poultry Research*, Vol.13, No.3, pp. 421-425, ISSN 15370437

Prasertsan, P., Choorit, W. & Suwanno, S. (1993). Isolation, identification and growth condition of photosynthetic bacteria found in seafood processing wastewater. *World Journal of Microbiology and Biotechnology*, Vol.9, No.5, pp. 590-593, ISSN 15730972

Prasertsan, P., Jaturapornipipat, M. & Siripatana, C. (1997). Utilization and treatment of tuna condensate by photosyntetic bacteria. *Pure and Applied Chemistry*, Vol.69, No.11, pp. 2439-2445, ISSN 13653075

Souza, M.R. & Silva, R.J. (May 2009). *A geração de resíduos industriais e sua destinação final*, Retrieved from <http://www.abepro.org.br/biblioteca/ENEGEP1997 _T6501.PDF>

Toyomizu, M., Sato, K., Taroda H., Kato T. & Akiba, Y. (2001). Effects of dietary spirulina on meat colour in muscle of broiler chickens. *British Poultry Science*, VoL.42, pp. 197-202, ISSN 14661799

Valduga, E. (December 2005). *Bioprodução de compostos voláteis e carotenóides por Sporodiobolus salmonicolor CBS 2636*, Retrieved from <http://www2.enq.ufsc.br/teses/d027.pdf>

Vanderzant, C. & Splittstoesser, D.F. (1992). *Compendium of methods for the microbiological examination of foods*, American Public Health Association, ISBN 087553175X, Washington, USA

Vieira, R.H.S.F. (2003). *Microbiologia, higiene e qualidade do pescado: teórica e prática*, Varela, ISBN 858551972X, São Paulo, Brasil

Zheng, S., Yang, M. & Yang, Z. (2005a). Biomass production of yeast isolate from salad oil manufacturing wastewater. *Bioresource Technology*, Vol.96, No.10, pp. 1183-1187, ISSN 09608524

Zheng, S., Yang, M., Yang, Z. & Yang, Q. (2005b). Biomass production from glutamate fermentation wastewater by the co-culture of *Candida halophila* and *Rhodothorula glutinis*. *Bioresource Technology*, Vol.96, No.13, pp. 1522-1524, ISSN 09608524

24

The Influence of Intercrops Biomass and Barley Straw on Yield and Quality of Edible Potato Tubers

Anna Płaza, Feliks Ceglarek, Danuta Buraczyńska
and Milena Anna Królikowska
University of Natural Sciences and Humanities in Siedlce
Poland

1. Introduction

Potatoes destined for direct consumption should be distinguished by a high trade yield with the best qualities. (Leszyński, 2002; Boligłowa and Gleń, 2003; Płaza and Ceglarek, 2009). In most European countries schemes for the verifiability of the potato product are introduced. The aim is to obtain good quality of potatoes, ensuring the reduction of harmful substances to human health and the natural environment (Spiertz et al., 1996). The beneficial effects of organic fertilization is noted here (Leszczyński, 2002; Boligłowa and Gleń, 2003; Makaraviciute, 2003; Płaza et. al., 2009).

Farmyard manure is a basic manure applied in potato cultivation (Batalin et.al., 1968; Kalembasa and Symanowicz, 1985; Rozrtopowicz, 1989). For many years its amount covered the demand, but now the situation has negatively affected due to the decline in livestock, especially cattle. Decreasing amount of farmyard manure, low profitability and the rationale for a system of integrated agriculture, tend to seek alternative, energy-efficient sources of biomass. As a result, a significant role is being attributed to green manures (Grześkiewicz i Trawczyński, 1997; Zając, 1997; Ceglarek et. al., 1998; Karlsson-Strese et. al., 1998; Płaza i in., 2009).

Green fertilizers were mentioned many times in literature. Batalin et. al. (1968), Roztropowicz (1989), Gruczek (1994), Dzienia and Szarek (2000) emphasize that the advantage of using this type of fertilization is high labor and energy saving in relation to its amount spent on works related to the application of farmyard manure. Estler (1991), Stopes et. al. (1995), Spiertz et. al. (1996), Karlsson-Strese et. al. (1998) and Songin (1998) show that the intercrops introduction into the cultivation is not only the production of biomass. They are also a kind of absorbent material to prevent leaching of nutrients into the deeper layers of soil and groundwater, which is important in protecting the agricultural environment.

From manuscripts connected with green fertilizers it is clear that among catch crops, undersown crops seem to be the cheapest source of organic matter because it does not require any additional costs associated with the cultivation and preparation of the soil before sowing, which is particularly troublesome in the cultivation of stubble crops (Ceglarek et. al., 1998). Seed cost is also low. As undersown the legumes are recommended to cultivate. The Renaissance intercrops from legumes is linked to the multilateral noticing them, valuable, but not fully used advantages of agronomic and biological properties. The rediscovery of these plants is associated with current global trends in agricultural techniques, aiming towards the promotion of proecological and ecological agriculture (Stopes et.al., 1995; Spiertz et.al., 1996; Karlsson-Strese et. al., 1998; Duer, 1999). White clover is distinguished by a high capacity of fixing atmospheric nitrogen, and a wide range of crops to allow its existence in a very different soil conditions have long been interested for researchers across Europe (Frye et. al., 1988). In Poland, there is little experimental data determining the suitability of this species to cultivation as undersown, designed for plowing, as a green manure in integrated potato cultivation. Researches of many authors (Batalin et. al., 1968; Gromadziński and Sypniewski, 1971; Zając, 1997; Ceglarek et. al., 1998) show that undersown legumes are quite unreliable in yielding. More similar are legume mixtures with grasses (Gromadziński and Sypniewski, 1971; Bowley et. al., 1984; Ceglarek et al,. 1998; Witkowicz, 1998; Płaza et. al., 2009). Reliable in yielding also are grasses grown in pure sowing. As a fast-growing plants and easily shading the soil interact with the position by weed reduction (Szymona et. al., 1983/1984; Sadowski, 1992; Karlsson-Strese et. al., 1998; Majda and Pawłowski, 1998; Kuraszkiewicz and Pałys, 2002).

An alternative source of biomass can also be stubble crops, which were mentioned in literature many times (Sadowski, 1992; Roztropowicz, 1989; Boligłowa and Dzienia, 1996; Grzeskiewicz and Trawczyński, 1997; Dzienia and Szarek, 2000). Recently, there has been an interest of the possibility of entering non-legume plants with a short growing season. It is recommended to sow fast-growing species, with good ability of shading, and not able to produce too large, aboveground woody mass. The most common are: white mustard, oil radish and phacelia (Allson and Amstrong, 1991; Boligłowa and Dzienia, 1997; Grześkiewicz and Trawczyński, 1997; Gutmański et. al., 1998). Among non-legume plants cultivated in stubble crop phacelia is distinguished by rapid growth, it produces a soft aboveground mass, easily frozen in winter. Is a phytosanitary plant. In Poland, previously carried out researches on fertilizing position of phacelia only in sugar beet cultivation (Nowakowski et. al., 1997; Gutmański et.al., 1999), still there is no experimental data evaluating its usefulness in the fertilization of potatoes.

Intercrops can be plowed down in autumn or left till spring in the form of mulch. The beneficial effects of intercrops plants left till spring in the form of mulch is to: protect the soil against wind and water erosion, gathering water from rainfall, slowing the process of mineralization of organic matter and prevent from nutrients leaching into the soil, reducing the cost of cultivation by eliminating plowing (Hoyt et. al., 1986; Gutmański et. al., 1999). It should be noted that the green fertilizers left till spring in the form of mulch causes a slight decrease in yield, but the improvement of the quality characteristics of the fertilized plants compared to fertilization applied in the traditional form.

Another substitute source of biomass can also be the straw left on the field after harvest of cereals (Szymankiewicz, 1993; Gruczek, 1994; Śnieg and Piramowicz, 1995; Dzienia and Szarek, 2000), especially used in combination with green fertilizers. Its addition to the legume biomass, not only does not reduce nitrogen losses, but also extends the period of green fertilizers acting (Nowak, 1982). In the case of non-legume plants effect of combined application of these forms of fertilization is not always positive (Dzienia, 1989; Sadowski, 1992). In Poland, there is little on this experimental data. Thus emerges the need for research aimed at comparing the impact of intercrops biomass, stubble crops both plowed down in autumn and left till spring in the form of mulch in combinations with straw or without straw, farmyard manure fertilization on yielding and chemical composition of edible potato tubers.

2. Material and methods

A field experiment was carried out in the years 2004-2007 at the Zawady Experimental Farm whose owner is the University of Natural Sciences and Humanities in Siedlce. The experimental site was Stagnic luvisol characterised by an average availability of phosphorus, potassium and magnesium. The experimental design was a split-block design with three replicates. Two factors were examined: I - intercrop fertilization: control object (without intercrop fertilization), farmyard manure (30 t ha^{-1}), undersown crop - biomass plowed down in autumn (white clover 18 kg ha^{-1}, white clover + Italian ryegrass 9 + 15 kg ha^{-1}, Italian ryegrass 30 kg ha^{-1}), stubble crop - biomass plowed down in autumn (phacelia 12 kg ha^{-1}), stubble crop - biomass left in the form of mulch until spring (phacelia 12 kg ha^{-1}). II. Straw fertilization: subblock without straw, subblock with straw.
Undersown crops were sown after planting spring barley cultivated for grain whereas stubble catch crops were planted after barley harvest. During spring barley harvest, on each plot straw yield was determined, and then the average its tests were taken in order to determine the content of macroelements (N - by Kjeldahl method, P - vanadium-molybdenum method, K and Ca - by flame photometry and Mg - by atomic absorption spectrometry) (Kerłowska-Kułas, 1993). In sub-block with straw fragmented straw was left and on sub-block without straw, straw was collected and brought out from field. On every plots with straw, with the exception of white clover undersown, compensatory dose of nitrogen was applied in the amount of 7 kg per 1 tonne of straw. Phacelia cultivated in stubble crop was sown in mid-August.
In the autumn, in random locations from each intercrop plot, the average sample of hay weight collected hay and crop residues of plants including their root mass, with a 30 cm layer to determine the yield of fresh weight. In collected plant material the content of dry matter was analyzed (by drier-weight method), and macroelements (N, P, K, Ca and Mg). Then on designated plots the cattle manure was transported, earlier the average sample was taken to determine the chemical composition.
In the first year following organic manuring edible potatoes Syrena cultivar was cultivated. In early spring mineral fertilizers were distributed, at the rates of 90 kg N, 39 kg P and 100 kg K per 1 ha. In the plots which had been ploughed in the autumn, mineral fertilizers were mixed with the soil using a cultivator equipped with a harrow whereas in the mulched plots, an application of a disc harrow was followed by a cultivator. Potatoes were planted in the third decade of April. In the integrated production system, a combination of mechanical and chemical control was applied. Until emergence, potato rows were earthed up and harrowed every 7 days; then just before emergence the herbicide mixture Afalon 450 SC in amount of 2 dm^3 ha^{-1} was sprayed, but after emergence (in the phase of 15-20 cm), when the weed infestation was noted herbicide Fusilade Super 125 EC in amount of 2 dm^3 ha^{-1} was sprayed. The Colorado potato beetle was and potato blight were controlled using, respectively, Fastac 10 EC (0.1 dm^3 ha^1) and the fungicide Ridomil MZ 72WP (2 dm^3 ha^{-1}). Potatoes were harvested in the second decade of September. During potato harvest, total and marketable yields were recorded in each plot, assuming that the marketable yield includes only healthy tubers with a diameter of more than 40 mm. Then 5-to-7-kg samples were collected from each plot to carry out their chemical analysis. In fresh mass the

following contents were determined: dry mass by drier-weight method, starch by the Reiman (Zgórska and Czerko, 1981), vitamin C using the Pijanowski method, reducing sugars and total sugar by Luffa-Schoorl method, nitrates by using an ion selective nitrate electrode and silver-silver chloride reference electrode (Rutkowska, 1981) and the content of glycoalkaloids by using the method of Bergersa (Bergers, 1980). Consumption value of potato tubers, ie the darkening of the raw and cooked tubers flesh, was evaluated according to the color plates in an inverted 9-point Danish scale, number 9 - marked the flesh intact, and the number one - the flesh is black. Changes in raw tubers flesh was evaluated after 4 hours from the time of slice potatoes and boiled at 24 hours. Flavor ratings were made using a 9-point scale, with scores 9 assumed to be very good, and a 1 as very poor (Zgórska and Frydecka-Mazurczyk, 1985). Each of the characteristics was subjected to analysis of variance according to the split-block linear model. Means for significant sources of variation were compared by the Tuckey test (Trętowski and Wójcik, 1991).

3. Results

3.1 Dry matter yield of researched organic fertilizer and the accumulation of macroelements

Amount of dry matter introduced into the soil by researched organic fertilizers was significantly differentiated (table 1). The biggest amount of the dry matter applied farmyard manure using jointly with straw and undersown intercrops with straw. Phacelia in combination with straw, irrespectively of its application introduced into the soil similar amount of dry matter as farmyard manure. However, intercrops and straw supplied the soil significantly less dry matter than farmyard manure.

Statistic analysis showed significant influence of the type of organic fertilizer on the amount of macroelements introduced into the soil (table 1). Indeed, the biggest amount of nitrogen supplied farmyard manure in combination with straw, white clover with straw and the mixtures of white clover mixed with Italian ryegrass also with the addition of straw. The amount of nitrogen supplied by white clover and the mixture of white clover with Italian ryegrass did not differ significantly from the amount of nitrogen supplied by farmyard manure. Other organic fertilizers introduced significantly less nitrogen than farmyard manure. Analyzing the amount of phosphorus applied by researched organic fertilizers, showed that only farmyard manure with straw provided that macroelement tha most. Comparable amount of phosphorus, as farmyard manure supplied white clover with straw, mixture of white clover with Italian ryegrass in combination with straw, and phacelia with straw. Other organic fertilizers introduced into the soil significantly less phosphorus than farmyard manure. The greatest amount of potassium supplied farmyard manure with straw and all intercrops also in combination with straw. Intercrops without straw supplied to the soil significantly less potassium than farmyard manure. Among researched organic fertilizer provided the most calcium applied farmyard manure with straw, white clover with straw, a mixture of white clover with Italian ryegrass and straw and phacelia with straw. Italian ryegrass in combination with straw provided a comparable amount of calcium, as farmyard manure. However, intercrops provided significantly less calcium than farm yard manure. Significantly more magnesium than farmyard manure provided farmyard manure used in combination with straw. However, intercrops in combinations without straw and with straw introduced into the soil significantly less magnesium than farmyard manure. The largest number of macroelements straw introduced into the soil.

Organic fertilization	Dry mass	Macroelements				
		N	P	K	Ca	Mg
Farmyard manure	7.8	162.0	48.3	132.6	63.8	40.2
White clover	5.3	157.7	32.0	112.4	49.3	24.1
White clover + Italian ryegrass	5.9	158.0	30.8	115.6	47.7	18.4
Italian ryegrass	6.3	114.5	26.9	109.1	35.3	13.6
Phacelia	4.4	112.8	37.8	92.7	43.8	21.0
Phacelia-mulch	4.5	112.9	38.0	92.9	43.9	21.2
Straw	4.2	32.8	11.2	76.4	27.0	9.9
Farmyard manure + straw	12.0	194.8	59.5	209.0	90.8	50.1
White clover + straw	9.5	190.5	43.2	188.8	76.3	34.0
White clover + Italian ryegrass + straw	10.1	190.8	42.4	192.0	74.7	28.3
Italian ryegrass + straw	10.5	147.8	38.1	185.5	62.3	23.5
Phacelia + straw	8.6	145.6	49.0	169.1	70.8	30.9
Phacelia-mulch + straw	8.7	145.7	49.2	169.3	70.9	31.1
$LSD_{0.05}$	1.0	11.7	5.9	10.7	5.5	3.2

Table 1. The amount of dry mass (t ha^{-1}) and macroelements (kg ha^{-1}) introduced into the soil by researched organic fertilizers (means from years 2000-2006)

3.2 Potato tubers yield

3.2.1 Total yield

Total yield of potato tubers was significantly modified by the examined factors and their interaction (table2). The highest yields of potato tubers were harvested from the objects

Catch crop fertilization	Straw fertilization		Means
	Subblock without straw	Subblock with straw	
Control object	27.4	36.2	31.8
Farmyard manure	42.8	41.7	42.3
White clover	43.0	46.2	44.6
White clover + Italian ryegrass	47.3	44.8	46.1
Italian ryegrass	37.4	36.3	36.9
Phacelia	44.7	43.0	43.8
Phacelia-mulch	42.6	44.2	43.4
Means	40.7	41.8	-
$LSD_{0.05}$			
Catch crop ferilization			1.0
Straw fertilization			0.9
Interaction			1.2

Table 2. Total field of potato tubers, t ha^{-1} (means from yaers 2005-2007)

fertilized with a mixture of white clover with Italian ryegrass, white clover, and phacelia both plowed down in autumn, and left till spring in the form of mulch. Only after Italian ryegrass applying total yield of potato tubers was significantly lower than recorded on control object. Straw fertilization also significantly modified the yield of potato tubers. At the sub-block with straw, potato tuber yield was significantly lower than recorded at the sub-block without straw. An interaction has been noted, which shows that the highest yield of potato tubers were obtained from the object fertilized with a mixture of white clover with Italian ryegrass and white clover with straw, and the smallest from control object, without intercrop fertilization.

3.2.2 Marketable yield

Statistical analysis showed a significant influence of examined factors and their interaction on the commercial yield of potato tubers (table 3). The highest yields were obtained from objects fertilized white clover, a mixture of white clover and Italian ryegrass and phacelia both plowed in the autumn, and left till spring in the form of mulch. Only on object fertilized with Italian ryegrass and on control object marketable yield of potato tubers was significantly lower than that recorded in farmyard manure. Straw fertilization also significantly differentiate commercial yield of potato tubers. At the sub-block with straw marketable yield of potato tubers was significantly higher than obtained in the sub-block without straw. An interaction has been noted, which shows that indeed the highest marketable yield was obtained from the object fertilized with a mixture of white clover with Italian ryegrass and white clover with straw, and the smallest from the control object without organic fertilization.

Catch crop fertilization	Straw fertilization		Means
	Subblock without straw	Subblock with straw	
Control object	17.8	27.0	22.4
Farmyard manure	38.6	37.2	37.9
White clover	39.4	45.6	42.5
White clover + Italian ryegrass	46.8	43.5	45.2
Italian ryegrass	28.9	28.1	28.5
Phacelia	43.9	41.2	42.6
Phacelia-mulch	38.4	42.0	40.2
Means	36.3	37.8	-
$LSD_{0.05}$			
Catch crop ferilization			0.9
Straw fertilization			1.0
Interaction			1.3

Table 3. Marketable field of potato tubers t ha^{-1} (means from years 2005-2007)

3.3 The quality of potato tubers

3.3.1 The dry matter content in potato tubers

The dry matter content in potato tubers was significantly differentiated by the intercrop fertilization, straw fertilization and their interaction (table 4). The highest concentration of dry matter characterized potato tubers fertilized with white clover, a mixture of white clover with Italian ryegrass and phacelia both plowed down in the autumn, as left till spring in the form of mulch.

The dry matter content in potato tubers fertilized with Italian ryegrass was significantly lower than in potatoes fertilized with farmyard manure. On control object, without organic fertilization dry matter content in potato tubers was significantly lower. Straw fertilization also significantly modified dry matter content in potato tubers. At the sub-block with straw potatoes distinguished by a higher concentration of dry matter than the tubers at sub-block without straw. From the interaction of researched factors showed that the highest content of dry matter was noted in potato tubers fertilized with white clover with straw, a mixture of white clover with Italian ryegrass in combinations without straw and with straw, phacelia in combination with straw, and phacelia used in the form of mulch with a straw or without the straw, and the lowest in potato tubers harvested from control object without organic fertilization.

Catch crop fertilization	Straw fertilization		Means
	Subblock without straw	Subblock with straw	
Control object	19.5	21.1	20.3
Farmyard manure	21.4	21.6	21.5
White clover	21.7	22.0	21.9
White clover + Italian ryegrass	22.1	22.3	22.2
Italian ryegrass	21.0	21.1	21.1
Phacelia	21.7	22.2	22.0
Phacelia-mulch	22.2	22.4	22.3
Means	21.4	21.8	-
$LSD_{0.05}$			
Catch crop ferilization			0.3
Straw fertilization			0.2
Interaction			0.4

Table 4. Dry matter content in potato tubers, % (means from years 2005-2007)

3.3.2 Dry matter yield of potato tubers

Dry matter yield of potato tubers was significantly modified by the intercrop fertilization, straw fertilization and their interaction (table 5). The highest dry matter yield of potato tubers was collected from the object fertilized with a mixture of white clover with Italian ryegrass, white clover and phacelia used in the form of mulch. Dry matter yield of potato tubers fertilized with phacelia did not differ significantly from the yield recorded on the farmyard manure. Only after the application of Italian ryegrass dry matter yield of potato tubers was significantly lower than that recorded on the farmyard manure. However, in this case, dry matter yield was significantly higher than that obtained on control object, without intercrop fertilization. Straw fertilization also significantly differentiate dry matter yield of potato tubers. On objects with straw dry matter yield of potato tubers was greater than on the objects without straw. There has been an interaction, which shows that the highest dry matter yield of potato tubers were obtained from the object fertilized with a mixture of white clover with Italian ryegrass in combinations without straw and with straw, white clover in combination with straw, and phacelia used in the form of mulch and also in combination with straw, and the smallest on control object, without intercrop fertilization.

Catch crop fertilization	Straw fertilization		Means
	Subblock without straw	Subblock with straw	
Control object	5.34	7.64	6.49
Farmyard manure	9.16	9.01	9.09
White clover	9.33	10.16	9.75
White clover + Italian ryegrass	10.45	9.99	10.22
Italian ryegrass	7.85	7.66	7.76
Phacelia	9.70	9.55	9.63
Phacelia-mulch	9.46	9.90	9.68
Means	8.76	9.13	-
$LSD_{0.05}$			
Catch crop ferilization			0.56
Straw fertilization			0.27
Interaction			0.59

Table 5. Dry matter yield, t ha^{-1} (means from years 2005-2007)

3.3.3 Starch content in potato tubers

Statistical analysis showed a significant effect of examined factors and their interaction on starch content in potato tubers (table 6). Intercrops fertilization of potato, with the exception

Catch crop fertilization	Straw fertilization		Means
	Subblock without straw	Subblock with straw	
Control object	13.2	13.9	13.6
Farmyard manure	14.0	14.1	14.0
White clover	13.7	14.0	13.9
White clover + Italian ryegrass	14.2	14.3	14.2
Italian ryegrass	14.4	14.5	14.4
Phacelia	14.5	14.7	14.6
Phacelia-mulch	14.6	14.8	14.7
Means	14.1	14.3	-
$LSD_{0.05}$			
Catch crop ferilization			0.2
Straw fertilization			0.1
Interaction			0.3

Table 6. Starch content in potato tubers, % (means from years 2005-2007)

of white clover caused a significant increase of starch content in potato tubers in comparison with farmyard manure fertilization. The starch content in potato tubers fertilized with white clover did not differ significantly from that observed in potato tubers fertilized with farmyard manure. However, on control object starch concentration in potato tubers was significantly lower than in tubers fertilized with farmyard manure. An interaction has been noted, which shows that the highest concentration of starch was noted in potato tubers fertilized with phacelia both plowed down in the autumn, and left till spring in the form of mulch in combination without straw and with straw and Italian ryegrass in combination with straw, and the lowest in potato tubers cultivated on control object.

3.3.4 Starch yield

Statistical analysis showed a significant effect of examined factors in experience on the starch yield of potato tubers (table 7). Intercrop fertilization caused a significant increase of starch yield in comparison with starch yield of potato tubers from the control object. The highest starch yield was obtained from the object fertilized with a mixture of white clover with Italian ryegrass, white clover, phacelia plowed down in autumn and left till spring in the form of mulch. Only after the application of Italian ryegrass the starch yield of potato tubers fertilized with Italian ryegrass was significantly lower than that recorded in the farmyard manure. Straw fertilization also modified the starch yield. At the sub-block with straw starch yield of potato tubers was significantly higher than at the sub-block without straw. An interaction has been shown that intercrop fertilization with straw fertilization, which shows that the highest yield of starch was obtained from the object fertilized with a mixture of white clover with Italian ryegrass and phacelia used in the form of mulch in combination with straw, and the smallest from control object, without intercrop fertilization.

Catch crop fertilization	Straw fertilization		Means
	Subblock without straw	Subblock with straw	
Control object	3.62	5.03	4.33
Farmyard manure	5.99	5.88	5.94
White clover	5.89	6.47	6.18
White clover + Italian ryegrass	6.72	6.41	6.57
Italian ryegrass	5.39	5.26	5.33
Phacelia	6.48	6.32	6.40
Phacelia-mulch	6.22	6.54	6.38
Means	5.76	5.99	-
$LSD_{0.05}$			
Catch crop ferilization			0.20
Straw fertilization			0.14
Interaction			0.21

Table 7. Starch yield, t ha^{-1} (means from years 2005-2007

3.3.5 Reducing sugars content in potato tubers

Statistical analysis showed a significant effect of examined factors on reducing sugars content in potato tubers (table 8). The highest concentration of reducing sugars noted in potato tubers harvested from control object. The highest concentration of reducing sugars noted in potato tubers harvested from control object. Intercrop fertilization significantly decreased reducing sugars content in potato tubers in comparison with their concentrations recorded in potatoes tubers harvested from control object. Indeed, the lowest content of reducing sugars was noted in potato tubers fertilized with phacelia both plowed down in the autumn, and left till spring in the form of mulch. Straw fertilization also significantly differentiate the concentration of reducing sugars in potato tubers. Higher its content was noted in potato tubers in the sub-block without straw than on the sub-block with straw.

Catch crop fertilization	Straw fertilization		Means
	Subblock without straw	Subblock with straw	
Control object	0.34	0.26	0.30
Farmyard manure	0.24	0.21	0.23
White clover	0.23	0.20	0.22
White clover + Italian ryegrass	0.17	0.15	0.16
Italian ryegrass	0.21	0.19	0.20
Phacelia	0.17	0.16	0.17
Phacelia-mulch	0.16	0.14	0.15
Means	0.22	0.19	-
$LSD_{0.05}$			
Catch crop ferilization			0.03
Straw fertilization			0.02
Interaction			n.s.

Table 8. Reducing sugars content in potato tubers, % (means from years 2005-2007)

3.3.6 The total sugar content in potato tubers

The total sugar content in potato tubers was significantly modified by intercrop fertilization and straw fertilization (table 9). Intercrop fertilization significantly decreased the concentration of total sugars in potato tubers. The lowest its content was recorded in potato tubers fertilized with a mixture of white clover with Italian ryegrass and phacelia plowed down in the autumn and left till spring in the form of mulch. The content of reducing sugars in potato tubers fertilized with white clover and Italian ryegrass did not differ significantly from their concentrations observed in tubers fertilized with farmyard manure. However, on control object, the content of total sugars in potato tubers was significantly higher than in the potato fertilized with farmyard manure. Straw fertilization also significantly modified the content of total sugars in potato tubers. At the sub-block without straw content of total sugars in potato tubers was significantly lower than at the sub-block with straw.

3.3.7 Vitamin C content in potato tubers

The vitamin C content in potato tubers was significantly differentiated by the examined factors of experiment and their interaction (table 10). Intercrop fertilization in comparison with control object caused a significant increase of vitamin C content in potato tubers. Indeed, the highest concentration of vitamin C were characterized in potato tubers fertilized with phacelia in the form of mulch and white clover. The vitamin C content in potato tubers fertilized with a mixture of white clover with Italian ryegrass and phacelia developed at a similar level as in the potato fertilized with farmyard manure. Straw fertilization also significantly differentiate the concentrations of vitamin C in potato tubers. On objects with straw the content of vitamin C in tubers was significantly higher than on the objects without straw. From the interaction between studied factors shows that the highest concentration of vitamin C were characterized by potato tubers fertilized with phacelia both plowed down in the autumn, and left till spring in the form of mulch, in combination without straw and with straw, and white clover and white clover with straw, and the lowest in potato tubers from control object.

Catch crop fertilization	Straw fertilization		Means
	Subblock without straw	Subblock with straw	
Control object	0.63	0.56	0.60
Farmyard manure	0.54	0.52	0.53
White clover	0.53	0.51	0.52
White clover + Italian ryegrass	0.48	0.42	0.45
Italian ryegrass	0.50	0.51	0.51
Phacelia	0.47	0.46	0.47
Phacelia-mulch	0.46	0.44	0.47
Means	0.52	0.49	-
$LSD_{0.05}$			
Catch crop ferilization			0.04
Straw fertilization			0.02
Interaction			n.s.

Table 9. The total sugar content in potato tubers, % (means from years 2005-2007)

Catch crop fertilization	Straw fertilization		Means
	Subblock without straw	Subblock with straw	
Control object	203.4	217.6	210.5
Farmyard manure	218.6	217.3	218.0
White clover	222.5	224.2	223.4
White clover + Italian ryegrass	219.4	222.5	221.0
Italian ryegrass	217.7	218.4	218.1
Phacelia	220.6	221.7	221.2
Phacelia-mulch	223.4	224.8	224.1
Means	217.9	220.9	-
$SLD_{0.05}$			
Catch crop ferilization			3.2
Straw fertilization			1.8
Interaction			4.3

Table 10. Vitamin C content in potato tubers, g kg^{-1} dry matter (means from years 2005-2007)

3.3.8 The total protein content in potato tubers

Statistical analysis showed a significant effect of examined factors and their interaction on total protein content in potato tubers (table 11). Intercrop fertilization significantly increased the concentration of total protein in potato tubers in relation to its content recorded in potatoes harvested from control object. Indeed, the highest concentration of total protein were characterized by potato tubers fertilized with white clover and with phacelia both plowed down in the autumn and left till spring in the form of mulch. The content of total protein in potato tubers fertilized with a mixture of white clover with Italian ryegrass did not differ significantly from that observed in potato tubers fertilized with farmyard manure. However, fertilization of potato with Italian ryegrass caused a significant decrease in total protein content in potato tubers in comparison with farmyard manure fertilization. Straw fertilization also significantly modified the concentration of total protein in potato tubers. On objects with straw total protein content in potato tubers was significantly higher on objects without straw. An interaction has been noted, which shows that the highest concentration of total protein was characterized by a potato fertilized with white clover, white clover with straw, and phacelia both plowed down in the autumn, and left till spring in the form of mulch, in combination, without straw and with straw, whereas the lowest potato tubers collected from the control object without intercrop fertilization.

Catch crop fertilization	Straw fertilization		Means
	Subblock without straw	Subblock with straw	
Control object	8.16	9.23	8.69
Farmyard manure	9.42	9.48	9.45
White clover	10.46	10.53	10.50
White clover + Italian ryegrass	9.45	9.56	9.51
Italian ryegrass	8.89	9.00	8.95
Phacelia	10.33	10.45	10.39
Phacelia-mulch	10.08	10.15	10.12
Means	9.54	9.77	-
$LSD_{0.05}$			
Catch crop ferilization			0.27
Straw fertilization			0.14
Interaction			0.43

Table 11. The content of total protein in potato tubers, % dry mass (means from years 2005-2007)

3.3.9 The content of true protein in potato tubers

The content of true protein in potato tubers was significantly differentiated by the intercrop fertilization, fertilization with straw and their interaction (table 12). The highest concentration of true protein in potato tubers was noted in potato tubers fertilized with phacelia and white clover both plowed down in the autumn, and left till spring in the form of mulch. The concentration of true protein in potato tubers fertilized with a mixture of white clover with Italian ryegrass remained at a similar level, such as on farmyard manure.

However, true protein content in potato tubers fertilized with Italian ryegrass was significantly lower than in tubers fertilized with farmyard manure. Straw fertilization also significantly differentiate true protein content in potato tubers. At the sub-block with straw concentration of true protein in potato tubers was significantly higher than on sub-block without straw. Investigated the interaction of factors we can see that the highest true protein content had potato tubers fertilized white clover, white clover with straw, and phacelia both plowed down in the autumn, and left till spring in the form of mulch in combination without straw and with straw, and the lowest potato tubers harvested from control object, without intercrop fertilization.

Catch crop fertilization	Straw fertilization		Means
	Subblock without straw	Subblock with straw	
Control object	3.67	4.72	4.20
Farmyard manure	4.92	5.06	4.99
White clover	5.74	5.83	5.79
White clover + Italian ryegrass	5.03	5.18	5.10
Italian ryegrass	4.38	4.45	4.42
Phacelia	5.54	5.66	5.60
Phacelia-mulch	5.43	5.54	5.48
Means	4.96	5.21	-
$LSD_{0.05}$			
Catch crop ferilization			0.26
Straw fertilization			0.14
Interaction			0.43

Table 12. The content of true protein in potato tubers, % dry mass (means from years 2005-2007)

3.3.10 Nitrate content in potato tubers

Statistical analysis showed significant effects of intercrop fertilization and interaction between intercrop fertilization and straw fertilization on the nitrate content in potato tubers (table 13). The highest concentration of nitrates was recorded in tubers harvested from control object. Intercrop fertilization caused a significant decrease of nitrate content in potato tubers. The lowest their concentration was noted in potato tubers fertilized with white clover, a mixture of white clover and Italian ryegrass and phacelia both plowed down in the autumn, and left till spring in the form of mulch. The nitrates content in potato tubers fertilized with Italian ryegrass, did not differ significantly from the concentrations observed in potato tubers fertilized with farmyard manure. An interaction has been noted which shows that the lowest content of nitrates was recorded in tubers fertilized with white clover and phacelia both plowed down in the autumn, and left till spring in the form of mulch, and the lowest on control object.

3.3.11 Glycoalkaloids content in potato tubers

The content of glycoalkaloids in potato tubers was significantly modified for examined factors and their interaction (table 14). Intercrop fertilization caused a significant decrease of glycoalkaloids in potato tubers in comparison with its concentrations observed in the potato from control object. The lowest content of glycoalkaloids was noted in potato tubers fertilized with white clover, a mixture of white clover with Italian ryegrass, phacelia both plowed down in the autumn, and left till spring in the form of mulch. The concentration of glycoalkaloids in potato tubers fertilized with Italian ryegrass did not differ significantly from that recorded in the potatoes fertilized with farmyard manure. Straw fertilization also significantly modified the content of glycoalkaloids in potato tubers. At the sub-block with

Catch crop fertilization	Straw fertilization		Means
	Subblock without straw	Subblock with straw	
Control object	147.0	141.2	144.1
Farmyard manure	109.2	122.2	115.7
White clover	92.3	84.9	88.6
White clover + Italian ryegrass	99.7	102.3	101.0
Italian ryegrass	108.3	118.6	113.5
Phacelia	88.2	107.4	97.8
Phacelia-mulch	95.4	88.6	92.0
Means	105.7	109.3	
$LSD_{0.05}$			
Catch crop ferilization			7.2
Straw fertilization			n.s.
Interaction			7.5

Table 13. Nitrate content in potato tubers, mg kg^{-1} of dry mass (means from years 2005-2007)

Catch crop fertilization	Straw fertilization		Means
	Subblock without straw	Subblock with straw	
Control object	63.9	58.2	61.1
Farmyard manure	54.4	54.2	54.3
White clover	46.3	44.1	45.2
White clover + Italian ryegrass	52.1	40.8	46.5
Italian ryegrass	55.2	54.6	54.9
Phacelia	47.5	46.6	47.1
Phacelia-mulch	47.2	45.7	46.5
Means	52.4	49.2	-
$LSD_{0.05}$			
Catch crop ferilization			3.1
Straw fertilization			0.4
Interaction			3.9

Table 14. Glycoalkaloids content in potato tubers, mg kg^{-1} of dry mass (means from years 2005-2007)

straw the concentration of glycoalkaloids in potato tubers was significantly lower than that recorded in the tubers of the sub-block without straw. Investigated the interaction of factors that were characterized it shows that the lowest content of glycoalkaloids in potatoes fertilized with white clover, white clover with straw, and phacelia both plowed down in the autumn, and left till spring in the form of mulch in combination without straw and with straw, and the highest potato tubers collected from the control object.

3.4 Consumption value of potato tubers

3.4.1 The darkening of raw potato tubers flesh

Statistical analysis revealed significant effects of intercrop fertilization and interaction of intercrop fertilization with straw fertilization on the darkening of raw potato tubers flesh (table15). Potatoes cultivated after intercrops showed less tendency to darkening of raw potato tubers flesh than tubers cultivated on control object. On control object fertilized with white clover, and with phacelia left till spring in the form of mulch noted significantly the lowest degree of darkening of raw potato tubers flesh. The darkening of tubers flesh fertilized with a mixture of white clover and Italian ryegrass, Italian ryegrass and phacelia plowed down in autumn remained at a similar level as the darkening of tubers flesh fertilized with farmyard manure. Differences between particular objects are within the limits of experimental error. There was an interaction, which shows the lowest degree of darkening of raw potato flesh was recorded in the object fertilized with phacelia in the form of mulch and white clover with straw, and the highest on control object.

Catch crop fertilization	Straw fertilization		Means
	Subblock without straw	Subblock with straw	
Control object	6.0	6.2	6.1
Farmyard manure	6.9	7.1	7.0
White clover	7.4	7.5	7.5
White clover + Italian ryegrass	7.0	7.1	7.1
Italian ryegrass	6.6	6.8	6.7
Phacelia	7.0	7.1	7.1
Phacelia-mulch	7.6	7.7	7.7
Means	6.9	7.1	-
$LSD_{0.05}$			
Catch crop ferilization			0.3
Straw fertilization			n.s.
Interaction			0.4

Table 15. The darkening of raw potato tubers flesh after 4 hours (means from years 2005-2007)

3.4.2 The darkening of cooked potato tubers flesh

The darkening of cooked potato tubers flesh was significantly modified by intercrop fertilization and the interaction of intercrop fertilization with straw fertilization (table16). The degree of darkening of cooked potato tubers fertilized with white clover, and phacelia left till spring in the form of mulch was the lowest. Darkening of cooked potato tubers flesh fertilized with a mixture of white clover with Italian ryegrass, Italian ryegrass and phacelia plowed down in the autumn did not differ significantly from the darkening of potato tubers flesh fertilized with farmyard manure. Only on control object the level of darkening of cooked potato tubers flesh was significantly lower than that recorded on the farmyard manure. An interaction of researched factors was noted, which shows that the lowest degree of darkening of cooked potato tubers flesh were recorded on the object fertilized with white clover, white clover with straw, and phacelia left till spring in the form of mulch with straw, and the highest on control object.

Catch crop fertilization	Straw fertilization		Means
	Subblock without straw	Subblock with straw	
Control object	7.0	7.1	7.1
Farmyard manure	7.8	7.9	7.9
White clover	8.1	8.3	8.2
White clover + Italian ryegrass	7.9	8.0	8.0
Italian ryegrass	7.6	7.7	7.7
Phacelia	7.7	7.9	7.8
Phacelia-mulch	8.1	8.2	8.2
Means	7.7	7.9	-
$LSD_{0.05}$			
Catch crop ferilization			0.2
Straw fertilization			n.s.
Interaction			0.4

Table 16. The darkening of cooked potato tubers flesh after 24 hours (means from years 2005-2007)

Catch crop fertilization	Straw fertilization		Means
	Subblock without straw	Subblock with straw	
Control object	5.4	5.5	5.5
Farmyard manure	6.3	6.4	6.4
White clover	8.0	8.2	8.1
White clover + Italian ryegrass	7.0	7.2	7.1
Italian ryegrass	6.5	6.6	6.6
Phacelia	7.1	7.2	7.2
Phacelia-mulch	7.5	7.7	7.6
Means	6.8	7.0	-
$LSD_{0.05}$			
Catch crop ferilization			0.2
Straw fertilization			n.s.
Interaction			0.3

Table 17. Savoriness of potato tubers, points (means from years 2005-2007)

3.4.3 Savoriness of potato tubers

Statistical analysis revealed significant effects of intercrop fertilization and interaction between intercrop fertilization and straw fertilization on savoriness of potato tubers (table17). Intercrop fertilization improved the savoriness of potato tubers in comparison with savoriness of potato tubers harvested from control object. The best savoriness had potato tubers fertilized with white clover, the mixture of white clover with Italian ryegrass, and phacelia both plowed down in the autumn, and left till spring in the form of mulch. Savoriness of potato tubers fertilized with Italian ryegrass did not significantly differ from savoriness of potato tubers fertilized with farmyard manure. There has been an interaction which shows that the best savoriness had potato tubers fertilized with white clover in combination without straw and with straw, and the worst potato tubers from control object.

4. Discussion

Shortage of farmyard manure due to the decline in farm animal stocks, low profitability and the rationale for integrated production tend to look for alternative and efficient ways of potato fertilization. The most important here are green fertilizers from undersown crops and stubble crops and straw left on field after cereal harvest. Selection of underplant crops as alternative sources of biomass, dictated the results of Batalina et al. (1968) and Ceglarka (1982). Batalin et al. (1968) initiated studies to evaluate the fertilizer value of underplant crops legumes, and Ceglarek (1982) conducted a thorough research on the determination of yield and chemical composition of crop residues of underplant crops. However Gutmański et al. (1998) have evaluated the value of fertilizer of oil radish, white mustard and phacelia used in sugar beet cultivation, which became the motivation for taking this type of research in potato cultivation. Under the conditions of this experiment, from the group of underplant crops yielding on the highest level was Italian ryegrass and a mixture of white clover with Italian ryegrass. The high biomass production of grasses also show results of Gromadziński and Sypniewski (1971), Zając and Witkowicz (1996), Ceglarka et al. (1998), Witkowicz (1998) and Kuraszewicza and Palys (2002). In own researches, phacelia grown in stubble intercrop yielded at a similar level as white clover cultivated as an intercrop. This is consistent with the results of Gromadziński and Sypniewski (1971), Witkowicz (1998), Trawczyński and Grześkiewicz (1997) and Nowakowski et. al. (1997), Ceglarek and Płaza, 2000). In the experiment the addition of straw to the intercrops caused a significant increase of the amount of dry matter and macronutrients.

Nowak (1982) indicates a predominance of green manure on the farmyard manure. This follows from the fact that the nutrients contained in green manure are generally more easily absorbed than the components of farmyard manure, due to rapid decomposition of organic matter. In this experiment, among intercrops the highest value of fertilizing showed undersown: a mixture of white clover with Italian ryegrass and white clover. Batalin et. al. (1968) the highest yields of potato tubers received after plowing the undersown of red clover and serradella, and Ceglarek et al. (1998) after plowing the mixtures of legume with Italian ryegrass. These differences are due to different rates of mineralization used forms of fertilization and the fact that the introduction into the soil with a mixture of larger amounts of biomass and macronutrients. According to Nowak (1982), during the decomposition of legumes may occur high losses of nitrogen. Depending on the temperature, humidity and time of decomposition, nitrogen losses could amount up to 50%. To prevent it, to the decomposing mass of legumes material rich in carbon should be added, such as grasses, in order to increase the C:N. In this experiment yields of potato tubers fertilized with Italian ryegrass were significantly smaller than in farmyard manure. However, in this case, tuber yields were significantly higher than those obtained on control object, without intercrop fertilization. The increase of tuber yield after plowing down the grass also found Sadowski (1992), Spiertz et al. (1996), Duer and Jończyk (1998) and Reust et al. (1999), but yields were lower than on the farmyard manure. This is because the introduction into the soil a large amount of biomass, with a low content of macronutrients (Sadowski 1992; Duer and Jończyk 1998). In addition, grasses have a wide ratio C:N. In this case, the less nitrogen mineralization, which is used primarily by soil microorganisms. In own research, the value of stubble crop fertilizer from phacelia plowed down in autumn and left till spring in the form of mulch equal the fertilizer value of farmyard manure. This is understandable because of non-legume stubble crops biomass of this plant was notable for its high content of macronutrients. This is confirmed by results of Dzienia (1989), Trawczyński and Grześkiewicz (1997) and Nowakowski, et al. (1997) and Różyło (2002).

In potato fertilization of stubble crops can also be used in the form of mulch. However, thus fertilizing the position, with the exception of phacelia, in terms of fertilizer could not match with farmyard manure. This is confirmed by research of Boligłowa and Dzienia (1996) and Dzieni and Szarka (2000) on potato fertilization by mulch from white mustard. In the system of integrated agriculture can recommend this method of fertilization, especially with phacelia mulch, while significantly reducing of costs. The beneficial effects of intercrops plants left on the field in the form of mulch slows the mineralization of organic matter, does not allow for leaching of nitrogen, stored water from the autumn-winter rainfalls, improves soil structure and enriches it in organic matter (Hoyt et al. 1986; Frye et al., 1988; Dzienia and Boligowa, 1993; Gutmański et al., 1999).

In that experiment fertilization with spring barley straw gave a lower effect than farmyard manure fertilization. This is consistent with the results of Sadowski (1992), Szymankiewicz (1993), Śnieg and Piramowicza (1995) and Ceglarek et al. (1998). However, its use combined with intercrop undersown of white clover and stubble crop left till spring in the form of mulch clearly strengthened its fertilising value. Potato tubers yields of after fertilization of these forms were comparable, in the case of white clover yields higher than those recorded on farmyard manure. Also Ceglarek et al. (1998) recommend the combined use of legumes as undersown.

Intercrops fertilization with straw affects not only for the amount of received yieldss, but also on quality, so reciprocal arrangement of the components involved in potato tubers (Roztropowicz, 1989; Grześkiewicz and Trawczyński, 1997; Boligłowa and Gleń 2003). The dry matter content and starch in potato tubers depends on the genetic factor, the distribution of rainfall and temperatures during the growing season and on agronomic factors, mainly from fertilizer (Rostropowicz, 1989; Grześkiewicz and Trawczyński 1997; Ceglarek et al., 1998; Dzienia and Szarek, 2000; Leszczyński 2002; Płaza and Ceglarek 2009; Makaraviciute 2003). In own studies, intercrop fertilization stimulated the content and dry matter yield of potato tubers and starch content and yield. The highest concentration of dry matter were characterized potatoes fertilized with mixture of white clover with Italian ryegrass and with phacelia plowed down in the autumn and left till spring in the form of mulch, and starch - potatoes fertilized with Italian ryegrass and phacelia plowed down and left till spring in the form of mulch. Research of Ceglarek et al. (1998) showed that potatoes fertilized with legume mixtures with Italian ryegrass include the most dry matter and Italian ryegrass fertilized include the most starch. Boligłowa and Gleń (2003) have not indicated significant differences between the starch content in potatoes fertilized with farmyard manure, and white mustard both plowed down in the autumn, as left till spring in the form of mulch. a Different view present Mazur and Jułkowski (1982) claiming that potato fertilization with legumes works better on the percentage starch content than with farmyard manure fertilization. In own studies, potato fertilization with stubble intercrop in the form of mulch increased the concentration of dry matter and starch in potato tubers as compared to that of intercrops plowed down in autumn. A similar relationship, but in sugar beet cultivation proved Gutmański et al. (1998). However Dzienia and Szarek (2000) and Boligłowa and Gleń (2003) found no significant differences between the starch content in potato tubers fertilized with farmyard manure, and white mustard both plowed down in the autumn, and left till spring in the form of mulch. Under the conditions of this experiment straw fertilization increased starch content in potato tubers, and in studies Gleń et al. (2002) did not decrease significantly the concentration of this component. Consumption potato tubers should contain about 0.3% reducing sugar, and 1% of total sugars. With increased content of total sugars, potatoes taste sweet (Głuska 2000; Leszczynski, 2000, 2002). In own studies, fertilization of potato with intercrop and straw caused a significant decrease in reducing sugars and total sugars in potato tubers as compared to the control object, without intercrop fertilization. Also, according to Leszczyński (2002) and Makaraviciute (2003) organic fertilizers reduce the concentration of sugars in potato tubers. However, the studies of Mondy and Munshi (1990) showed that enrichment of soil in substance abounds in nitrogen reduces the starch content and increases the sugar content in potato tubers. In own studies, potato fertilization with white clover did not result in significant differences in the amount of reducing sugars and total sugars as compared to farmyard manure fertilization.

In light of these studies used forms of organic fertilization stimulated the concentration of vitamin C in potato tubers. The highest concentrations of vitamin C were characterized in potatoes fertilized with white clover and phacelia both plowed down in the autumn, and left till spring in the form of mulch in combination without the straw and with straw. Also, the findings of other authors (Garwood et al. 1991; Weber and Putz 1999; Leszczyński 2002; Sawicka and Kuś 2002; Hamouz et al. 2005, 2007; Płaza and Ceglarek 2009) indicate a positive correlation between organic fertilization and vitamin C content in potato tubers.

In own researches, intercrop fertilization preferably affected on protein content in potato tubers. Also in the researches of Mazur and Jułkowskiego (1982), Sawicka (1991), Leszczyński (2002) and Sawicka and Kuś (2002) saw an increase in concentration of true protein in potato tubers cultivated in organic fertilizers. Most preferably, the discussed feature influenced white clover fertilization, also phacelia both plowed down in the autumn, and left till spring in the form of mulch in combination, without straw and with straw. A similar relationship has proved Wiater (2002). Potatoes cultivation in the position fertilized with legume plants and phacelia plants take larger amounts of nitrogen from soil than potatoes cultivated in position fertilized with green fertilizers. Nitrogen contained in the biomass of white clover and phacelia, is gradually mineralization is evenly shared to the potato crop, leading to total conversion of protein nitrogen. In own stuies, the lowest nitrate content was reported in potato tubers fertilized with white clover and phacelia both plowed down in the autumn and left till spring in the form of mulch. Only after Italian ryegrass applying nitrate content in potato tubers did not differ significantly from that recorded in potatoes fertilized with farmyard manure. The above relationship is explained by the fact that the biomass of white clover, or phacelia outside the higher content of nitrogen contained a few fibers which ensured its rapid degradation. Thanks to this all nutrients, including nitrogen available to potatoes plant are evenly distributed, allowing the total conversion of mineral nitrogen in protein nitrogen. This is consistent with the results of Dzienia et al. (2004) and Boligłowy and Gleń (2003), who showed that potato tubers fertilized with white mustard and rye straw contained significantly less nitrates than potatoes fertilized with farmyard manure. According to Leszczyński (2002) use of farmyard manure, whose chemical composition is not controlled, may increase for example nitrogen and other components content in the plant. However Boligłowa and Gleń (2003) showed that the nitrate content in potato tubers fertilized with white mustard developed at a similar level as in the potatoes fertilized with farmyard manure. In own studies the highest concentration of nitrates reported in potato tubers from the control object, only with mineral fertilization. This is due to the fact that mineral fertilizers, especially nitrogen increased the content of nitrogen compounds, mainly non-protein, including free amino acids, amines, ammonium nitrogen and nitrate nitrogen and reduces the share of protein in general (Wiater, 2002).

In this experiment the lowest concentration of glycoalkaloids in potatoes fertilized with white clover, a mixture of white clover and Italian ryegrass and phacelia both plowed down in the autumn, and left till spring in the form of mulch. According to Rudella et al. (2005) intercrop cultivation with a favorable ratio of carbon to nitrogen regenerates the soil environment, increases the humus content, the number of microorganisms, enzymes and other biologically active compounds in the soil, which inhibits the accumulation of harmful substances in potato tubers. In the experiment only after the applying of Italian ryegrass the concentration of glycoalkaloids in potato tubers was at the similar level as in the potato fertilized with farmyard manure. However, in this case the content of glycoalkaloids in tubers was significantly lower than that in potatoes cultivated without intercrop fertilization. Leszczyński (2002) shows that organic fertilizers reduce the harmful substances content in potato tubers by enriching the soil with organic substance which inhibits the synthesis process of glycoalkaloids. In own studies, straw fertilization also significantly differentiate the content of glycoalkaloids in potato tubers. On objects with straw the content of glycoalkaloids in potato tubers was significantly lower than on objects without straw. This is consistent with the results of research of Płaza et al. (2010). In this experiment the highest concentration of glycoalkaloids in potato tubers has been harvested from the control object, only with mineral fertilization. Also, the studies of Mondy and Munshi (1990), Hamouz et al. (2007), Kołodziejczyk et al. (2007) and Rytel et al. (2008) mineral fertilization increased the content of glycoalkaloids (solanine and chakoniny) in potato tubers. However, it should be noted that the potato in comparison with other crops have little ability to accumulate harmful substances for human. Moreover, the use of green manure and straw greatly reduces their concentration in comparison to traditional farmyard manure.

5. Conclusion

1. Among researched organic fertilizers the highest amount of dry matter and macroelements introduced into the soil farmyard manure with straw, white clover and straw and the mixture of white clover with Italian ryegrass and straw.
2. The largest potato yields were obtained from a combinations fertilized with a mixture of white clover with Italian ryegrass and white clover with straw.

3. Fertilization with straw from white clover undersown, and with phacelia left till spring in the form of mulch significantly increased potato tuber yield compared to the intercrop fertilization.
4. Intercrop and straw fertilization increased in potato tubers dry matter content, starch, total protein, true protein and vitamin C, and decreased the content of reducing sugars, total sugars, nitrates and glycoalkaloids.
5. Farmyard manure can be fully replaced in potato fertilization with substitutes, such as a mixture of white clover with Italian ryegrass, white clover and phacelia both plowed down in the autumn, and left till spring in the form of mulch in combinations without straw and with straw.

6. References

Allison, M. & Amstrong, M. (1991). The nitrate leaching problem - are catch crop the solution. *Br. Sugar Beet Rev.*, No. 3, pp. 8-11.

Batalin, M.; Szałajda, R. & Urbanowski, S. (1968). Wartość zielonego nawozu z poplonowych wsiewek roślin motylkowatych. *Pam. Puł.*, Nr 35, ss. 37-51.

Bergers, W. 91980). A rapid quantitative assai for solanidine glycoalkaloids In potatoes and industrial potato protein. *Potato Research*, No. 23, pp. 105-110.

Boligłowa, E. & Dzienia, S. (1996). Wpływ nawożenia organicznego i sposobu uprawy roli na plonowanie i jakość bulw ziemniaka. *Zesz. Nauk. AR Szczecin, Roln. LXII*, Nr 172, ss. 37-42.

Boligłowa, E. & Dzienia , S. (1997). Tendencje zmian w agrotechnice ziemniaka. Mat. konf. nauk. nt. „Nawozy roślinne w integrowanym systemie produkcji rolniczej". *AR Kraków*, ss. 51-56.

Boligłowa, E. & Gleń, K. (2003). Yielding and quality of potato tubers depending on the kind of organic fertilization and tillage methods. *Elec. Jour. Pol. Agroc. Univ. Ser. Agron.*, Vol. 1, No. 6, www.ejapau

Bowley, S. R..; Taylor, N.L. & Dougherty C.T. (1984). Physiology and morphology of red clover. *Adv. Agron.*, No. 37, pp. 317-347.

Ceglarek, F. (1982). Uprawa wsiewek międzyplonowych w zbożach. Cz. III. Ocena resztek pożniwnych wsiewek plonowych i ich wpływ na plon pszenicy jarej. *Zesz. Nauk. WSRP Siedlce, Ser. Rol.*, Nr 1, ss. 101-114.

Ceglarek, F.; Płaza, A.; Buraczyńska D. & Jabłońska-Ceglarek R. (1998). Alternatywne nawożenie organiczne ziemniaka jadalnego w makroregionie środkowo-wschodnim. Cz. II. Wartość odżywcza i konsumpcyjna ziemniaka. *Rocz. Nauk Rol.*, Ser. A. T. 113, Z. 3-4, ss. 189-201.

Ceglarek, F. & Płaza, A. (2000). Wartość konsumpcyjna bulw ziemniaka w zależności od rodzaju nawożenia organicznego. *Biul. IHAR*, Nr 213, ss. 117-123.

Dzienia, S. (1989). Wpływ masy organicznej na plonowanie roślin i chemiczne właściwości gleby lekkiej. *Zesz. Probl. Post. Nauk Rol.*, Nr 377, ss. 155-160.

Dzienia, S. & Boligłowa, E. (1993). Rola mulczowania w podnoszeniu żyzności i urodzajności gleby. *Post. Nauk Rol.*, Nr 1, ss. 107-111.

Dzienia, S. & Szarek, P. (2000). Efektywność uprawy bezpłużnej oraz międzyplonów i słomy w produkcji ziemniaka. *Zesz. Probl. Post. Nauk Rol.*, Nr 470, ss. 145-152.

Dzienia, S.; Szarek, P & Pużyński , S. (2004). Yielding and quality of potato tubers depending on the soil tillage systems and organic fertilization. *Zesz. Probl. Post. Nauk Rol.*, No. 500, pp. 235-242.

Duer, I. (1999). Plon suchej masy kilku odmian koniczyny uprawianej w ekologicznym i integrowanym systemie produkcji oraz akumulacja azotu w glebie. *Zesz. Nauk. AR Kraków,* Nr 347, ss. 69-77.

Duer, I & Jończyk, K. (1998). Potato fertilization cultivated in ecological farms. *Fragm. Agron.*, No. 1, pp. 85-95.

Estler, M. (1991). Conservation of soil and water by using a new tillage system for row crop. In: Cover crops for clean water. The proceedings of an international conference. *Jacson, Terenssee*, pp. 34-36.

Frye, W.W.; Bleviens, R.L.; Smith, M.S.; Corak, S.J. & Varco, J.J. (1988). Role of annual legume cover crops in efficient use water and nitrogen. No. 51, *Am. Soc. Agronom.* Medison.

Garwood, T.; Davles, D. & Hartley, A. (1999). The effect of winter cover crops on yielding of the following spring crops and nitrogen balance in a calcareous loam. *J. Agric. Sci.* Cambridge, No 132, pp. 1-11.

Gleń, K.; Boligłowa, E. & Pisulewski, P. (2002). Wpływ różnego rodzaju nawożenia organicznego na jakość bulw ziemniaka. Mat. konf. nauk. nt. „Ziemniak spożywczy i przemysłowy oraz jego przetwarzanie. Perspektywy ekologicznej produkcji ziemniaka w Polsce". *AR Wrocław*, ss. 100-101.

Głuska, A. (2000). Wpływ agrotechniki na kształtowanie jakości plonu ziemniaka. *Biul IHAR*, Nr 213, ss. 173-178.

Gruczek, T. (1994). Gospodarka bezobornikowa na glebie lekkiej. *Fragm. Agron.*, Nr 2, ss. 72-82.

Gromadziński, A. & Sypniewski, J. (1077). Przydatność różnych roślin do uprawy jako wsiewka poplonowa w żyto na ziarno i po życie na zielonkę. *Pam. Puł.*, Nr 68, ss. 93-103.

Grześkiewicz, H. & Trawczyński, C. (1997). Poplony ścierniskowe jako nawóz organiczny w uprawie ziemniaka. *Biul. Inst. Ziemn.*, Nr 48, ss. 73-82.

Gutmański, I.; Kostka-Gościniak, D.; Nowakowski, M.; Szymczak-Nowak, J. & Banaszak, K. (1998). Plonowanie i jakość przemysłowa buraka cukrowego w warunkach stosowania siewu w mulcz z międzyplonu. *Rocz. AR Poznań, CCCVII, Rol.*, Nr 52, ss. 255-262.

Gutmański, I.; Kostka-Gościniak, D.; Szymczak-Nowak, J.; Nowakowski, M. & Sitarski, A. (1999). Stan zachwaszczenia plantacji buraka cukrowego uprawianego z siewu w mulcz. *Folia Univ. Agric. Stetin. 195, Agric.*, Nr 74, ss. 105-114.

Hamouz, K; Lachman, J.; Dvořǎk, P. & Pivec, V. (2005). The effect of ecological growing on the potatoes yielding and quality. *Plant Soil Environ.*, No. 51, pp. 397-402.

Hamouz, K; Lachman, J.; Dvořǎk, P.; Duskova, O. & Cizek, M. (2007). Effect of conditions of locality, variety and fertilization on the content of ascorbic acid in potato tubers. *Plant Soil Environ.*, No. 53, pp. 252-257.

Hoyt, G.D. & Hargrove W.L. (1986). Legume cover crop for improving crop and soil management in the Southern United States. *Hortic. Sci.*, No. 21, pp. 34-39.

Jansen, G.; Flamme, W.; Shűller, K. & Vandrey, M. (2001). Tuber and starch quality of wild and cultivated potato species and cultivars. *Potato Res.*, No. 44, pp. 137-146.

Kalembasa, S. & Symanowicz, B. (1985). Wpływ nawożenia organicznego i mineralnego na plon i skład chemiczny wybranych odmian ziemniaków. *Zesz. Nauk. WSRP Siedlce, Ser. Rol.*, Nr 3, ss. 67-76.

Karlsson-Strese, E.M.; Rydberg, I.; Becker H.C. & Umaerus M. (1998). Strategy for Catch Crop Development. II. Screening of Species Undersown in Spring barley (*Hordeum vulgare* L.) with respect to catch crop Growth and Grain Yield. *Acta Agric. Scand., Sec. B. Soli and Plnat Sci.*, No. 48, pp 26-33.

Kerłowska-Kułas, M. (1993). Researches of the quality of food productions. *PWE, Warszawa.*

Kołodziejczyk, M.; Szmigiel, A.& Kiełbasa, S. (2007). Yielding and chemical composition of potato tubers in chageable conditions of fertilization. *Fragm. Agron.*, Nr 2, ss. 142-150.

Kuraszkiewicz, R. & Pałys, E. (2002). Wpływ roślin ochronnych na plon masy nadziemnej wsiewek międzyplonowych. *Annaless UMCS, Sec. E,* Nr 57. Ss. 105-112.

Leszczyński, W. (2000). Kryteria oceny jakości ziemniaka konsumpcyjnego i skrobiowego. Mat. konf. nauk. nt. „Ziemniaka spożywczy i przemysłowy oraz jego przetwarzanie". *AR Wrocław*, ss. 41-49.

Leszczyński, W. (2002). Zależność jakości ziemniaka od stosowania w uprawie nawozów i pestycydów.. *Zesz. Probl. Post. Nauk Rol.*, Nr 489, ss 47-64.

Majda, J. & Pawłowski, F. (1988). Plonowanie wsiewek życicy wielokwiatowej w zależności od ilości wysiewu i dawek azotu. Cz. I. Wsiewki w żyto zbierane na ziarno. *Rocz. Nauk Rol., Ser. A, T. 107,* Nr. 3, ss. 109-119.

Makaraviciute, A. (2003). Effect of organic and mineral fertilizers on the yield and quality of different potato varieties. *Agron. Res.*, No. 1, pp 197-209.

Mazur, T. & Jułkowski, M. (1982). Wpływ nawożenia organicznego i mineralnego na plonowanie, cechy jakościowe dwóch odmian ziemniaka uprawianego na glebie lekkiej. *Zesz. Nauk. ART Olsztyn, Ser. Rol.*, Nr 34, ss. 187-194.

Mondy, N.I. & Munshi, C.B. (1990). Effect of nitrogen fertilization on glicoalcaloid and nitra te kontent of potatoes. *Jour. Agric. Foood Chem.*, No. 38, pp. 565-567.

Nowak, G. (1982). Przemiany roślinnej materii organicznej znakowanej izotopem C_{14} w glebach intensywnie nawożonych. *Zesz. Nauk. AR Olsztyn, Ser. Rol.*, Nr 35, ss. 3-57.

Nowakowaski, M.; Gutmański, J.; Kostka-Gościniak, D. & Kaczorowski, G. (1997). Międzyplony ścierniskowe odmian gorczycy białej, rzodkwi oleistej i facelii błękitnej jako nawozy zielone i czynniki ograniczające mątwika burakowego w glebie. *Biul. IHAR*, nr 202, ss. 201-211.

Płaza, A. & Ceglarek, F. (2009). Tuber quality of edible potato fertilized with catch crops and barley straw. *Annales UMCS, Sec. E,* No. 3, pp. 79-90.

Płaza, A.; Ceglarek F.; Królikowska M.A. & Próchnicka M. (2009). Nawożenie ziemniaka jadalnego biomasą międzyplonów w warunkach środkowo-wschodniej Polski. *Biul IHAR,* Nr 254, ss. 137-144.

Płaza, A.; Ceglarek F. & Królikowska M.A. (2010). The influence of intercrops and farmyard manure fertilization in changeable weather conditions on consumption value of potato tubers. *Jour. Centr. Europ. Agric.*, No. 1, pp. 47-54.

Roztropowicz, S. (1989). Środowiskowe, odmianowe i nawozowe źródła zmienności składu chemicznego bulw ziemniaka. *Fragm. Agron.*, Nr 1, ss. 33-75.

Różyło, K. (2002). Wstępna ocena walorów konsumpcyjnych odmiany ziemniaka Irga różnie nawożonej na glebie lekkiej i ciężkiej. Mat. konf. nauk. nt. „Ziemniaka spożywczy i przemysłowy oraz jego przetwarzanie. Perspektywy ekologicznej produkcji ziemniaka w Polsce". *AR Wrocław*, ss. 97-98.

Rudella, C.A.; Davenport, J.R.; Evans, R.G., Hattendorf, M.J.; Alva, A.K. & Boydston, R.A. (2005). Relating potato yield and quality to yield scale variability in soil characteristics. *Amer. Jour. Potato Res.*, No. 79, pp. 317-323.

Rutkowska, U. (1981). Seleccted methods of study of composition and nutritional value of food. PZWL, Warszawa.

Rytel, E.; Lisińska, G. & Kozicka-Pytlarz, M. (2008). Wpływ sposobu uprawy na jakość konsumpcyjną ziemniaka. *Zesz. Probl. Post. Nauk Rol.*, Nr 530, ss. 259-269.

Sadowski, W. (1992). Porównanie efektywności obornika, słomy, nawozów zielonych i biohumusu w uprawie ziemniaka. Mat. konf. nauk nt. „Produkcyjne skutki zmniejszenia nakładów na agrotechnikę roślin uprawnych". *ART Olsztyn*, ss. 216-222.

Sawicka, B. (1991). Próba ustalenia niektórych czynników środowiska i zabiegów agrotechnicznych na ciemnienie miąższu bulw ziemniaka. *Biul IHAR*, Nr 179, ss. 67-74.

Sawiska, B. &Kuś, J. (2002). Zmienność składu chemicznego bulw ziemniaka w warunkach ekologicznego i integrowanego systemu produkcji. *Zesz. Probl. Post. Nauk Rol.*, Nr 489, ss. 273-282.

Smith, O. (2007). Potato quality. *Amer. Jour. Potato Res.*, No. 28, pp. 732-737.

Songin, W. (1998). Międzyplony w rolnictwie proekologicznym. *Post. Nauk Rol.*, Nr 2, ss. 43-51.

Spiertz, J.H.J.; Haverkort, A.J. & Vereijken P.H. (1996). Enviromentally safe and consumerfriendly potato production in The Netherlands. 1. Development of ecologically sound production systems. *Potato res.*, No. 39, pp. 371-378.

Stopes, C.; Milington, S. & Woolward, L. (1996). Dry matter and nitrogen accumulation by three leguminous green manure species and the yield of a following wheat crop in an organic production system. *Agric. Ecos. and Envir.*, No. 57, pp. 189-196.

Śnieg, L. & Piramowicz, W. (1995). Wpływ sposobu nawożenia na plonowanie ziemniaka i zboża jarego w ogniwie zmianowania. *Rocz. Nauk. Rol. , Ser. A, T. 111*, Nr 1-2, ss. 127-134.

Szymankiewicz, K. (1993). Wpływ stosowania pod ziemniaka słomy na plon i jego strukturę ze szczególnym uwzględnieniem bulw dużych. *Biul. Inst. Ziemn.*, Nr 43, ss. 93-103.

Szymona, J.; Krupiński, A. & Malicki L. (1983/1984). Resztki pożniwne niektórych roślin uprawianych na rędzinie. *Annales UMCS, Sec E*, Nr 7, ss. 77-87.

Trętowski, J & Wójcik, A.R. (1991). Methodology of agrocultural experiments. *Siedlce, WSRP*.

Weber, L. & Putz, B. (1999). Witamin C content In potato. Proceding 14 th triennial Conference of the European Association for Potato research. *Sorento, Italy*, pp. 230-231.

Wiater, J. (2002). Wpływ współdziałania niektórych odpadów z roślinami motylkowatymi na ilość i jakość białka ziemniaka. *Zesz. Probl. Post. Nauk Rol.*, Nr 484, ss. 743-752.

Witkowicz, R. (1998). Porównanie plonowania oraz wartości przedplonowej wsiewek roślin motylkowatych i traw na glebie lekkiej. *Rocz. AR Poznań, CCCVII, Ser. Rol.*, Nr 52, ss. 65-70.

Zając, T. (1997). Dobór roślin na nawozy zielone wraz z produkcyjna weryfikacją. Mat. konf. nauk. nt. „Nawozy roślinne w integrowanym systemie produkcji rolniczej". *AR Kraków*, ss. 5-12.

Zając, T. & Witkowicz, R. (1996). Skład chemiczny biomasy wsiewki koniczyny czerwonej na cele nawozowe. *Zesz. Nauk. AR Szczecin, Ser. Rol.*, Nr 172, ss. 553-559.

Zgórska, K. & Czerko, Z. (1981). Accurate and Simple metod of determining starch. Instrukcja wdrożeniowa. *Instytut Ziemniaka w Boninie.*

Zgórska, K. & Frydecka-Mazurczyk, A. (1985). Warunki agrotechniczne i przechowalnicze a cechy użytkowe bulw ziemniaka. *Biul. Inst. Ziemn.*, Nr 33, ss. 109-120.